D1806745

Lecture Notes in Computational Vision and Biomechanics

Volume 34

Series Editors

João Manuel R. S. Tavares ⓘ, Departamento de Engenharia Mecânica, Faculdade de Engenharia, Universidade do Porto, Porto, Portugal
Renato Natal Jorge, Faculdade de Engenharia, Universidade do Porto, Porto, Portugal

Research related to the analysis of living structures (Biomechanics) has been carried out extensively in several distinct areas of science, such as, for example, mathematics, mechanical, physics, informatics, medicine and sports. However, for its successful achievement, numerous research topics should be considered, such as image processing and analysis, geometric and numerical modelling, biomechanics, experimental analysis, mechanobiology and Enhanced visualization, and their application on real cases must be developed and more investigation is needed. Additionally, enhanced hardware solutions and less invasive devices are demanded. On the other hand, Image Analysis (Computational Vision) aims to extract a high level of information from static images or dynamical image sequences. An example of applications involving Image Analysis can be found in the study of the motion of structures from image sequences, shape reconstruction from images and medical diagnosis. As a multidisciplinary area, Computational Vision considers techniques and methods from other disciplines, like from Artificial Intelligence, Signal Processing, mathematics, physics and informatics. Despite the work that has been done in this area, more robust and efficient methods of Computational Imaging are still demanded in many application domains, such as in medicine, and their validation in real scenarios needs to be examined urgently. Recently, these two branches of science have been increasingly seen as being strongly connected and related, but no book series or journal has contemplated this increasingly strong association. Hence, the main goal of this book series in Computational Vision and Biomechanics (LNCV&B) consists in the provision of a comprehensive forum for discussion on the current state-of-the-art in these fields by emphasizing their connection. The book series covers (but is not limited to):

- Applications of Computational Vision and Biomechanics
- Biometrics and Biomedical Pattern Analysis
- Cellular Imaging and Cellular Mechanics
- Clinical Biomechanics
- Computational Bioimaging and Visualization
- Computational Biology in Biomedical Imaging
- Development of Biomechanical Devices
- Device and Technique Development for Biomedical Imaging
- Experimental Biomechanics
- Gait & Posture Mechanics
- Grid and High Performance Computing on Computational Vision and Biomechanics
- Image Processing and Analysis
- Image processing and visualization in Biofluids
- Image Understanding
- Material Models
- Mechanobiology
- Medical Image Analysis
- Molecular Mechanics
- Multi-modal Image Systems
- Multiscale Biosensors in Biomedical Imaging
- Multiscale Devices and BioMEMS for Biomedical Imaging
- Musculoskeletal Biomechanics
- Multiscale Analysis in Biomechanics
- Neuromuscular Biomechanics
- Numerical Methods for Living Tissues
- Numerical Simulation
- Software Development on Computational Vision and Biomechanics
- Sport Biomechanics
- Virtual Reality in Biomechanics
- Vision Systems
- Image-based Geometric Modeling and Mesh Generation
- Digital Geometry Algorithms for Computational Vision and Visualization

In order to match the scope of the Book Series, each book has to include contents relating, or combining both Image Analysis and mechanics.

Indexed in SCOPUS, Google Scholar and SpringerLink.

More information about this series at http://www.springer.com/series/8910

João Manuel R. S. Tavares ·
Renato Manuel Natal Jorge
Editors

VipIMAGE 2019

Proceedings of the VII ECCOMAS Thematic
Conference on Computational Vision
and Medical Image Processing,
October 16–18, 2019, Porto, Portugal

 Springer

Editors
João Manuel R. S. Tavares (iD)
Faculdade de Engenharia
Universidade do Porto
Porto, Portugal

Renato Manuel Natal Jorge
Faculdade de Engenharia
Universidade do Porto
Porto, Portugal

ISSN 2212-9391   ISSN 2212-9413   (electronic)
Lecture Notes in Computational Vision and Biomechanics
ISBN 978-3-030-32039-3   ISBN 978-3-030-32040-9   (eBook)
https://doi.org/10.1007/978-3-030-32040-9

© Springer Nature Switzerland AG 2019
This work is subject to copyright. All rights are reserved by the Publisher, whether the whole or part of the material is concerned, specifically the rights of translation, reprinting, reuse of illustrations, recitation, broadcasting, reproduction on microfilms or in any other physical way, and transmission or information storage and retrieval, electronic adaptation, computer software, or by similar or dissimilar methodology now known or hereafter developed.
The use of general descriptive names, registered names, trademarks, service marks, etc. in this publication does not imply, even in the absence of a specific statement, that such names are exempt from the relevant protective laws and regulations and therefore free for general use.
The publisher, the authors and the editors are safe to assume that the advice and information in this book are believed to be true and accurate at the date of publication. Neither the publisher nor the authors or the editors give a warranty, expressed or implied, with respect to the material contained herein or for any errors or omissions that may have been made. The publisher remains neutral with regard to jurisdictional claims in published maps and institutional affiliations.

This Springer imprint is published by the registered company Springer Nature Switzerland AG
The registered company address is: Gewerbestrasse 11, 6330 Cham, Switzerland

Preface

This book contains the full papers presented at VipIMAGE 2019—VII ECCOMAS Thematic Conference on Computational Vision and Medical Image Processing, which was held in Porto, Portugal, during the period 16–18 October 2019.

The event had 6 Invited Lectures and 70 contributed presentations originated from fifteen countries: Belgium, Brazil, Colombia, Czech Republic, France, Germany, Indonesia, Italy, Poland, Portugal, Spain, Tunisia, Turkey, UK and USA.

Techniques of Computational Vision have become predominant in our society. For instances, fully automatic or semi-automatic image-based systems have been developed to address tasks of surveillance, traffic analysis, recognition, inspection, motion analysis, human–machine interface, virtual reality and aided medical diagnosis and intervention.

One of the notable aspects of the Computational Vision domain is the inter- and multi-disciplinary. Actually, principles and methodologies of other sciences, such as informatics, mathematics, statistics, psychology, mechanics, medicine and physics, can be found embraced into this domain. Additionally, one of the major influences for the continual effort done in this field of the human knowledge is the high number of applications that can be found in medicine. For instance, computational algorithms can be applied on medical images for shape reconstruction, behaviour analysis, tissue characterization, 3D printing or computer-assisted intervention and therapy.

The main objective of these ECCOMAS Thematic Conferences on Computational Vision and Medical Image Processing, initiated in 2007, is to promote a comprehensive forum for discussion on the recent advances in the related fields in order to identify potential collaboration between researchers of different sciences. Henceforth, VipIMAGE 2019 brought together researchers representing fields related to biomechanics, biomedical engineering, Computational Vision, computer graphics, computer sciences, computational mechanics, electrical engineering, mathematics, statistics, medical imaging, medicine, sports and rehabilitation.

The included contributions covered a broad range of techniques for signal processing and analysis, image segmentation and matching, tracking and analysis of movement, machine learning and big data, medical imaging, computational bioimaging and visualization, computer-aided diagnosis, surgery, therapy and treatment, software development for image processing and analysis, Computational Vision, industrial applications, virtual reality, biomechanical modelling and simulation, cardiovascular, cerebrovascular and orthopaedic disease diagnosis, ontologies for medical image analysis and computer-assisted interventions, micro- and nanofluidics analysis, dental rehabilitation, tissue engineering and direct digital fabrication.

The conference co-chairs would like to take this opportunity to express gratitude for the support given by The International European Community on Computational Methods in Applied Sciences and The Portuguese Association of Theoretical, Applied and Computational Mechanics, and thank to all sponsors, to all members of the Scientific Committee, to all organizers of Thematic Sessions, to all Invited Lecturers, to all Session-Chairs and to all authors for submitting and sharing their knowledge.

João Manuel R. S. Tavares
Renato M. Natal Jorge
Conference Co-chairs

Invited Lecturers

During VipIMAGE 2019, six Invited Lectures were presented by six experts from different countries:

Aurélio Campilho	Universidade do Porto, Portugal
Daniela Iacoviello	Sapienza University of Rome, Italy
Francisco Vasconcelos	University College London, UK
João Paulo Papa	Universidade Estadual de São Paulo, Brazil
Jos Vander Sloten	KU Leven, Belgium
Wafa Skalli	Arts et Métiers ParisTech, France

Thematic Sessions

Under the auspicious of VipIMAGE 2019, nine Thematic Sessions were organized:

1. Cardiovascular, Cerebrovascular and Orthopaedic diseases: Imaging and Modelling
   Organizers:

Sónia I.S. Pinto	Universidade do Porto, Portugal
Luísa C. Sousa	Universidade do Porto, Portugal
Catarina F. Castro	Universidade do Porto, Portugal

2. Advances and Imaging Challenges in Micro and Nanofluidics
   Organizers:

Susana Catarino	Universidade do Minho, Portugal
Rui A. Lima	Universidade do Minho, Portugal
Graça Minas	Universidade do Minho, Portugal

3. Intersection between Image Processing and Machine Learning in Biomedical Applications
   Organizers:

   | Inês Domingues | Instituto Superior de Engenharia de Coimbra, Portugal |
   | Rui Nóbrega | INESC TEC/FEUP, Portugal |

4. Direct Digital Fabrication in Medicine: from digital data to physical models
   Organizers:

Henrique Almeida	Instituto Politécnico de Leiria, Portugal
Mário Correia	Instituto Politécnico de Leiria, Portugal
Joel Vasco	Instituto Politécnico de Leiria, Portugal
Rui Ruben	Instituto Politécnico de Leiria, Portugal

5. Computer Simulations and Visualization Applied to Tissue Engineering
   Organizers:

João Eduardo Ribeiro	Instituto Politécnico de Bragança, Portugal
Rui A. Lima	Universidade do Minho, Portugal
Pedro A. Martins	Universidade do Porto, Portugal

6. Parameterization of Reconstructed Organ Models
   Organizers:

   | Fethi Okyar | Yeditepe University, Turkey |
   | Ogulcan Guldeniz | Yeditepe University, Turkey |

7. Computational Vision and Image Processing applied to Dentistry
   Organizers:

André Correia	Universidade Católica Portuguesa, Portugal
Nuno Rosa	Universidade Católica Portuguesa, Portugal
Tiago Marques	Universidade Católica Portuguesa, Portugal

8. Network Neuroscience
   Organizers:

Sabina Tangaro	Istituto Nazionale di Fisica Nucleare - Sezione di Bari, Italy
Bellotti Roberto	Università degli Studi di Bari, Italy
Angela Lombardi	Istituto Nazionale di Fisica Nucleare - Section of Bari, Italy

9. Applications of Ontologies for Medical Image Analysis and Computer-Assisted Interventions

Michel Audette   Old Dominion University, USA

Scientific Committee

All works submitted to VipIMAGE 2019 were evaluated by an International Scientific Committee composed by 88 expert researchers from recognized institutions of 18 countries:

Abdelwahed Barkaoui	Université Internationale de Rabat, Morocco
Ahmed El-Rafei	Ain Shams University, Egypt
Alberto Gambaruto	Bristol University, UK
Alejandro F Frangi	University of Leeds, UK
Alexandre Cunha	California Institute of Technology, USA
Ali Fethi Okyar	Yeditepe University, Turkey
André R. S. Marçal	Universidade do Porto, Portugal
Angela Lombardi	Istituto Nazionale di Fisica Nucleare - Section of Bari, Italy
António Luís Pereira do Amaral	Instituto Superior de Engenharia de Coimbra, Portugal
Armando Sousa	Universidade do Porto, Portugal
Auzuir Ripado de Alexandria	Instituto Federal de Educação, Ciência e Tecnologia do Ceará, Brasil
Bellotti Roberto	Università degli Studi di Bari, Italy
Bhargab B. Bhattacharya	Indian Statistical Institute, India
Catarina F. Castro	Universidade do Porto, Portugal
Cristian A. Linte	Rochester Institute of Technology, USA
Daniela Iacoviello	Sapienza Università di Roma, Italy
Elisabetta Binaghi	University of Insubria, Italy
Evgin Goceri	Akdeniz University, Turkey
Fátima L. S. Nunes	Universidade de São Paulo, Brazil
Fethi Okyar	Yeditepe University, Turkey
Fiorella Sgallari	University of Bologna, Italy
Gerhard A. Holzapfel	Graz University of Technology, Austria
Giuseppe Placidi	University of L'Aquila, Italy
Graça Minas	Universidade do Minho, Portugal
Henrique Almeida	Instituto Politécnico de Leiria, Portugal
Henryk Palus	Silesian University of Technology, Poland
Hermerson Pistori	Universidade Católica Dom Bosco, Brazil
Inês Domingues	Universidade de Coimbra, Portugal
Isabel N. Figueiredo	Universidade de Coimbra, Portugal
J. Paulo Vilas-Boas	Universidade do Porto, Portugal
Jimmy T. Efird	East Carolina Heart Institute, USA

João Eduardo Ribeiro	Instituto Politécnico de Bragança, Portugal
João L. Vilaça	Instituto Politécnico do Cávado and Ave, Portugal
João Paulo Papa	Universidade Estadual de São Paulo, Brazil
Joel Vasco	Instituto Politécnico de Leiria, Portugal
John C. Brigham	Durham University, UK
Jolita Bernatavičienė	Vilnius University, Lithuania
Jorge A. Silva	Universidade do Porto, Portugal
Jorge S. Marques	Universidade de Lisboa, Portugal
Juan José Jiménez-Delgado	University of Jaén, Spain
Juan Roberto Jiménez Pérez	University of Jaen, Spain
Khan M. Iftekharuddin	Old Dominion University, USA
Koen Vermeer	Rotterdam Ophthalmic Institute, The Netherlands
Laurent Cohen	University Paris Dauphine, France
Leo Joskowicz	The Hebrew University of Jerusalem, Israel
Luís Paulo Reis	Universidade do Porto, Portugal
Luís Teixeira	Universidade do Porto, Portugal
Luísa C. Sousa	Universidade do Porto, Portugal
Lyuba Alboul	Sheffield Hallam University, UK
Manuel González-Hidalgo	University of the Balearic Islands, Spain
Manuele Bicego	Università degli Studi di Verona, Italy
Mário Correia	Instituto Politécnico de Leiria, Portugal
Mário Forjaz Secca	Universidade Nova de Lisboa, Portugal
Michel Audette	Old Dominion University, USA
Michela Cigola	University of Cassino & Southern Latium, Italy
Miguel Tavares Coimbra	Universidade do Porto, Portugal
Miguel Tavares da Silva	Universidade de Lisboa, Portugal
Mohamed Bakhouya	International University of Rabat, Morocco
Nilanjan Dey	Techno India College of Technology, India
Nuno Machado	Instituto Politécnico de Lisboa, Portugal
Ogulcan Guldeniz	Yeditepe University, Turkey
Paola Lecca	University of Trento, Italy
Paulo Eduardo Ambrósio	Universidade Estadual de Santa Cruz, Brazil
Pedro Bibiloni	Universitat de les Illes Balears, Spain
Pedro Martins	Universidade do Porto, Portugal
Pedro Moreira	Instituto Politécnico de Viana do Castelo, Portugal
Povilas Treigys	Vilnius University, Lithuania
Radim Kolar	Brno University of Technology, Czech Republic
Reneta Barneva	SUNY Fredonia, USA
Ricardo Vardasca	Universidade do Porto, Portugal
Roberto Bellotti	Università degli Studi di Bari, Italy
Rui A. Lima	Universidade do Minho, Portugal
Rui Nóbrega	Universidade do Porto, Portugal
Rui Rubem	Instituto Politécnico de Leiria, Portugal

Sabina Tangaro	Istituto Nazionale di Fisica Nucleare, Italy
Sanderson L. Gonzaga de Oliveira	Universidade Federal de Lavras, Brazil
Sandra Rua Ventura	Instituto Politécnico do Porto, Portugal
Sónia I. S. Pinto	Universidade do Porto, Portugal
Suresh Chandra Satapathy	Kalinga Institute of Industrial Technology, India
Susana Catarino	Universidade do Minho, Portugal
Susana Oliveira Branco	Instituto Politécnico de Lisboa, Portugal
Suzanne Shontz	University of Kansas, USA
Tinashe Mutsvangwa	University of Cape Town, South Africa
Yongjie Zhang	Carnegie Mellon University, USA
Yuri Bazilevs	Brown University, USA
Zbisław Tabor	Cracow University of Technology, Poland
Zeike Taylor	University of Leeds, UK
Zeyun Yu	University of Wisconsin at Milwaukee, USA

Acknowledgements

The editors and the conference co-chairs acknowledge the support towards the publication of the Book of Proceedings and the organization of the VII ECCOMAS Thematic Conference VipIMAGE to the following organizations:

- Universidade do Porto (UP)
- Faculdade de Engenharia da Universidade do Porto (FEUP)
- Instituto de Ciência e Inovação em Engenharia Mecânica e Engenharia Industrial (INEGI)
- European Community on Computational Methods in Applied Sciences (ECCOMAS)
- International Association for Computational Mechanics (IACM)
- Fundação para a Ciência e a Tecnologia (FCT)
- Associação Portuguesa de Mecânica Teórica Aplicada e Computacional (APMTAC)

Contents

Simulation and Modeling

Cardiovascular, Cerebrovascular and Orthopaedic diseases Imaging and Modelling

Signal and Image Processing

Colonic Polyp Identification Using Pareto Depth Anomaly Detection Algorithm

Isabel N. Figueiredo[1(✉)], Mahdi Dodangeh[1], Luís Pinto[1],
Pedro N. Figueiredo[2], and Richard Tsai[3,4]

[1] CMUC, Department of Mathematics, Faculty of Sciences and Technology,
University of Coimbra, Coimbra, Portugal
{isabelf,dodangeh,luisp}@mat.uc.pt
[2] Faculty of Medicine, CHUC (Centro Hospitalar e Universitário de Coimbra -
Department of Gastroenterology), Centro Cirúrgico de Coimbra,
University of Coimbra, Coimbra, Portugal
pnf11@sapo.pt
[3] Department of Mathematics and the Institute for Computational Engineering and
Sciences, The University of Texas at Austin, Austin, USA
ytsai@math.utexas.edu
[4] KTH Royal Institute of Technology, Stockholm, Sweden

Abstract. Colon cancer prevention, diagnosis, and prognosis are directly related to the identification of colonic polyps, in colonoscopy video sequences. In addition, diagnosing colon cancer in the early stages improves significantly the chance of surviving and effective treatment. Due to the large number of images that come from colonoscopy, the identification of polyps needs to be automated for effciency. In this paper, we propose a strategy for automatic polyp recognition, based on a recent multi-objective anomaly detection concept, which itself is based on Pareto Depth Analysis (PDA). Clinically, in medical images, polyps are diagnosed based on a few criteria, such as texture, shape and color. Few works use multi-criteria classification in a systematic way for polyp detection. In the present paper we use a PDA approach, to act as a binary classifier for the identification of colonic polyps. The results obtained in a medical dataset, of conventional colonoscopy images, consisting of short videos from 34 different patients, and 34 different polyps, with a total of 1360 different polyp frames, confirm that the proposed method clearly outperforms the single performance of each criterion.

Keywords: Image processing · Multicriteria optimization ·
Colonic polyp

1 Introduction

Diagnosing colon cancer in the early stages would improve significantly the relative survival rate of patients. It is also well known that conventional colonoscopy is a gold standard procedure for several colon pathologies. Detection of polyps or

© Springer Nature Switzerland AG 2019
J. M. R. S. Tavares and R. M. Natal Jorge (Eds.): VipIMAGE 2019, LNCVB 34, pp. 3–11, 2019.
https://doi.org/10.1007/978-3-030-32040-9_1

abnormal tissue in the colon is a sign of (possible) disease for a doctor. However, the size of the large bowel (1.5 to 2 m, in average, in adults) forces the imaging procedure to last for 30 to 60 min. Consequently, a common drawback of such a procedure, that produces a long video to be carefully analysed, has always the possible risk of human-error and therefore failures in polyp detection.

In the last decade, polyp detection is of much interest in computer vision and artificial intelligence. The main challenge is to create computer-aided diagnosis (CAD) systems, autonomous and reliable, for the identification of polyps or abnormal tissues in the colon, and ideally, in real-time. There are plenty of different CADs that rely either on hand-crafted descriptors or, more recently, on machine/deep learning techniques (for which the lack of labeled and extensive data is a serious drawback) - without being exhaustive we refer for instance to [2,4,9–11] and the references therein.

In real-life problems, multiple measures of similarity/dissimilarity are in general necessary to detect anomalies, because multiple criteria (as for instance, in medical image processing, texture, color, and shape) might lead to the identification of several, different and relevant abnormality patterns, undetectable with only one single criterion. The multi-objective anomaly detection concept [7] exploits this particular issue: it relies on the Pareto Depth Analysis (PDA) and declares anomalous those objects that are far from their nearest neighbor data sample objects, in terms of some appropriate predefined measures of similarity/dissimilarity involving the multiple criteria.

Concerning colonic polyps, these are objects that exhibit, in medical images, different texture and shape (and sometimes also color) from the surrounding normal mucosa. In this paper, we propose a strategy for automatic polyp recognition by using two different shape and texture criteria (that also embody, indirectly, a color criteria) and by applying the multi-objective Pareto depth anomaly (PDA) detection concept [7], for polyp identification.

The preliminary results obtained on a proprietary medical dataset confirm that the PDA strategy clearly outperforms the single performance of each criterion.

After this introduction, the PDA method is briefly described in Sect. 2, together with the chosen dissimilarity measures and the medical dataset. In Sect. 3 we report the results and the performance of the method. Finally, in Sect. 4 we draw some conclusions and indicate the future work.

2 Materials and Methods

2.1 Pareto Depth Anomaly Detection Algorithm

The Pareto depth anomaly detection algorithm (PDA), proposed by [7], is a multi-objective optimization methodology for (binary) classification. It relies on multiple similarity measures, nondominated sorting and a simple threshold method.

Nondominated sorting (sometimes called Pareto depth analysis) is a procedure for arranging a finite set $\mathcal{D}$ of points in $\mathbb{R}^N$ into successive and different layers, consisting of minimal (or nondominated) points, for the coordinatewise partial order $\leq$ on $\mathbb{R}^N$. A point $D = (D_i)_{i=1}^N$ in $\mathcal{D}$ is minimal (or nondominated) if no other point $\bar{D} = (\bar{D}_i)_{i=1}^N$ in $\mathcal{D}$ has smaller components than the components of D. It means $D \leq \bar{D}$ if and only if $D_i \leq \bar{D}_i$ for $i = 1, \ldots, N$. The layers are called Pareto fronts, or nondominated layers. The first one, denoted by $\mathcal{F}_1$, is the set of all nondominated points from $\mathcal{D}$, and the subsequent Pareto fronts, denoted by $\mathcal{F}_k$, for $k = 2, 3, \ldots$, consist of the minimal points from the set $\mathcal{D} \setminus (\mathcal{F}_1 \cup \mathcal{F}_2 \cup \ldots \mathcal{F}_{k-1})$ (that is, the set $\mathcal{D}$ excluding all the previous $k - 1$ nondominated layers). Moreover the index k of a front $\mathcal{F}_k$, to which a point D belongs, is called the Pareto depth of D.

In the anomalous detection algorithm, hereafter considered, the finite set $\mathcal{D}$ of points $\mathbb{R}^N$ is created from N different measures of dissimilarity (that represent, in the context of multi-objective optimization, N different objectives) and the successive Pareto fronts are used to generate a criterion for anomaly detection. Points lying in deeper Pareto fronts (these are fronts $\mathcal{F}_k$ with k large) are more prone to be anomalous, because these points are far from minimal points in $\mathcal{D}$.

We briefly outline the PDA algorithm that includes a training phase followed by a testing phase and finally the classification.

1. *Data* - Let d_i, $i = 1, \ldots, N$, be N dissimilarity measures, $S_{train} = \{X_1, \ldots, X_t\}$ a training set of objects, with size t, Y a testing object, and ρ a predefined positive number representing a threshold.
2. *Training phase*
   - Compute the set $\mathcal{D}$ of dyads D_{ij} defined by

$$D_{ij} = (d_1(X_i, X_j), \ldots, d_N(X_i, X_j)), \qquad \forall i, j \in \{1, \ldots, t\}.$$

   - Construct the different Pareto fronts by nondominated sorting, until each dyad is in a front.
   We recall that a dyad D_{ij} strictly dominates dyad D_{i*j*}, if $d_k(X_i, X_j) \leq d_k(X_i*, X_j*)$ for all $k = 1, \ldots, N$ and $d_l(X_i, X_j) < d_l(X_i*, X_j*)$ for some l. The set of optimal Pareto dyads, called the first Pareto front and denoted by $\mathcal{F}_1$, is the set of dyads in $\mathcal{D}$ that are not strictly dominated by an other dyads from $\mathcal{D}$. The next Pareto fronts are defined recursively by

$$\mathcal{F}_k = \text{Pareto front of } \mathcal{D} \setminus (\mathcal{F}_1 \cup \mathcal{F}_2 \cup \ldots \mathcal{F}_{k-1}), \qquad k = 2, \ldots, M.$$

   So, $\mathcal{F}_k$ is the set of dyads from $\mathcal{D} \setminus (\mathcal{F}_1 \cup \mathcal{F}_2 \cup \ldots \mathcal{F}_{k-1})$ that are not strictly dominated by any other dyads from $\mathcal{D} \setminus (\mathcal{F}_1 \cup \mathcal{F}_2 \cup \ldots \mathcal{F}_{k-1})$. M is the total number of Pareto fronts on dyads between training objects.
   A Pareto front $\mathcal{F}_k$ is said deeper than $\mathcal{F}_l$ if $k > l$.
3. *Testing phase* - Let Y be the testing object.
   - For each criterion d_i, define the number n_i, of nearest neighbor objects to Y with respect to at least one dissimilarity measure, in the training set S_{train}. Then create a total of $s = \sum_{i=1}^N n_i$ new dyads $D_1, D_2, \ldots, D_s$, corresponding to the N-tuples $(d_1(Y, X_k), \ldots, d_N(Y, X_k))$, with $k = 1, \ldots, s$, between Y and the union of the n_i nearest neighbors objects of S_{train} for each i.
   [Note that the value of n_i could be different in each criterium d_i].

- For each $k = 1, \ldots, s$, define the Pareto depth of dyad D_k as

$$\text{depth } (D_k) = \min\{j : \ D_k \text{ is below } \mathcal{F}_j\},$$

where "D_k is below $\mathcal{F}_j$" means that D_k strictly dominates at least a single dyad in $\mathcal{F}_j$.

Remark that if depth (D_k) is large (respectively, small) then D_k is in deep (respectively, shallow) Pareto fronts, and, consequently, Y is distant from (respectively, close to) the nearest neighbor objects X_k, with $k = 1, \ldots, s$, of the training set S_{train}, in terms of the dissimilarity measures d_i.

4. *Classification (simple threshold method)* - For the testing object Y compute the average depth of the s dyads associated to Y,

$$\text{average depth}(Y) = \frac{1}{s} \sum_{k=1}^{s} \text{depth } (D_k).$$

Y is an anomalous object if averagedepth$(Y) > \rho$, otherwise it is a normal object.

In this paper we use two criteria (or objectives), implemented by two functions representing the dissimilarity measures (d_1 and d_2 defined in the next Sect. 2.3). Given an image, the larger the value is for a criterion, the more likely we think that the image contains a polyp. Now if the value corresponding to one of the criteria is high but the other is very low, it is less likely that the image has a polyp. So in this logic, an image is more likely to contain a polyp if the values for both criteria are comparatively larger than the rest of the images - this image should lie on a deeper Pareto front.

2.2   Data Set

To perform this study a dataset from different conventional colonoscopy short videos of 34 different patients was collected by medical endoscopists, in Centro Cirúrgico de Coimbra, Coimbra, Portugal. For each video, a total of 40 frames were extracted by sampling the video every 10 frames, aiming to exclude very similar images. Thus, there is a total of 1360 polyp instances from the recorded 34 different polyps. From these videos we also extracted 1360 frames displaying only normal colonic mucosa.

2.3   Objects and Dissimilarity Measures

In our context the objects (defined in the algorithm) are the frames of the above mentioned dataset, hereafter denoted by F_i, for $i = 1, \ldots, 2 \times 1360$. These frames are firstly subjected to a pre-processing stage, that consists in the detection and removal of specular highlights followed by an inpainting technique (see [4] where an identical pre-processing procedure was adopted).

Colonic polyps can be distinguished from normal mucosa, in medical images, by two to three main features: shape (in general they exhibit roundish and protruded shape), texture (different patterns from normal mucosa) and occasionaly color (in general reddish). We decide to use two dissimilarity measures: one d_1 related to the shape descriptor, herein denoted by PF (meaning polyp function) and defined in [4,5], and a second one, d_2, related to the texture descriptor relying on the completed local binary pattern (CLBP) defined in [6]. Both measures involve implicitly the color feature, since we use the a-channel of the CIE-Lab color space of each frame F_i, herein denoted by F_i^a, as input for computing d_1 and d_2.

Thus, for any two frames F_i and F_j to be compared, the dissimilarity measure d_1 is just the absolute difference between the maxima of PF of the two frames, that is

$$d_1(F_i, F_j) = |\max_\Omega PF(F_i^a) - \max_\Omega PF(F_j^a)|,$$

where Ω is the pixel domain, and the dissimilarity measure d_2, which involves the comparison of two histograms corresponding to the CLBP of F_i^a and F_j^a, is the Kullback-Leiber divergence [8] of the related probability distributions $Q_{F_i^a}$ and $Q_{F_j^a}$ (obtained by normalizing the histograms), that is

$$d_2(F_i, F_j) = \sum_{y \in [0,L]} Q_{F_i^a}(y) \log \left(\frac{Q_{F_j^a}(y)}{Q_{F_i^a}(y)} \right),$$

with $y \in [0, L]$ denoting a generic point of the CLBP histogram domain.

We point out that we decided to discard a third dissimilarity measure explicitly related to a color feature, namely the absolute difference between the maximum of F_i^a and the maximum of F_j^a, because in our medical dataset it was not bringing any additional improvement to the performance of the PDA algorithm involving the two above defined measures d_1 and d_2.

3   Results

For conducting the experiments the medical dataset was randomly split into three subsets - training, validating and testing set - and such that the frames of a short video are included in only one subset and excluded from the other two subsets. Moreover, in each subset the number of frames with polyps is equal to the number of frames with normal mucosa.

To cross-validate, we have also permuted, randomly, the frames between the training and validating sets and repeated the experiments 10 times. At each experiment, the training set comprises 560 polyp frames from 14 short videos ($14 \times 40 = 560$) and 560 normal mucosa frames which in total corresponds to 1120 frames. The validating set consists of 400 polyp frames from 10 short videos ($10 \times 40 = 400$) and 400 normal mucosa frames, therefore a total of 800 frames. Finally the testing set contains 400 polyp frames from 10 short videos ($10 \times 40 = 400$) and 400 normal mucosa frames, thus a total of 800 frames.

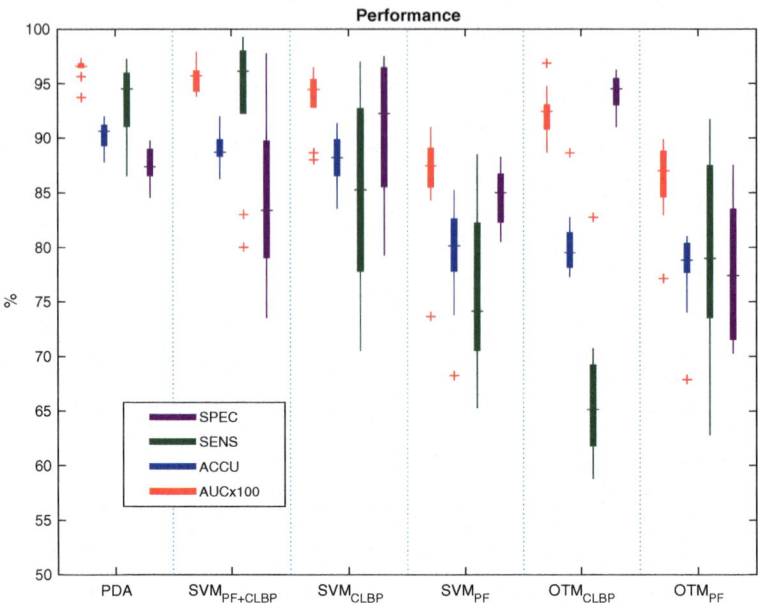

Fig. 1. Results of the cross-validating process (10 experiments resulting from the permutation between training and validating datasets): box plots for the validating dataset.

We have also compared the performance of the multiobjective algorithm PDA, with the approaches that use SVM (support vector machine classifier [3]) and TM (simple threshold method) with only one criterion - one that uses only the shape descriptor PF, and the other that uses only the texture descriptor CLBP.

The different performances are measured in terms of the standard metrics: sensitivity (SENS), specificity (SPEC), accuracy (ACCU) and area under the receiver operating characteristic (ROC) curve (AUC). Moreover, hereafter we denoted by optimal threshold method OTM, the threshold method TM with an optimal threshold, corresponding to the optimal operating point (OOP) in the ROC curve, that is the point that maximizes the accuracy.

We have used the results of the cross-validating 10 experiments to obtain the different parameters: the optimal operating point (OOP) in the different ROC curves, which are those that maximize the accuracy, the best number of nearest neighbors n_1 and n_2, whose values are 0 and 1, respectively. In particular, we found that the optimal threshold ρ defined in the PDA algorithm, for the current medical dataset, is 76.5.

Figure 1 displays the box plots corresponding to the results of the cross-validating process. There are six blocks of box plots, displaying the sensitivity, specificity, accuracy and area under the ROC curve. From left to right, the first block corresponds to PDA. The second, third, and fourth blocks depict box

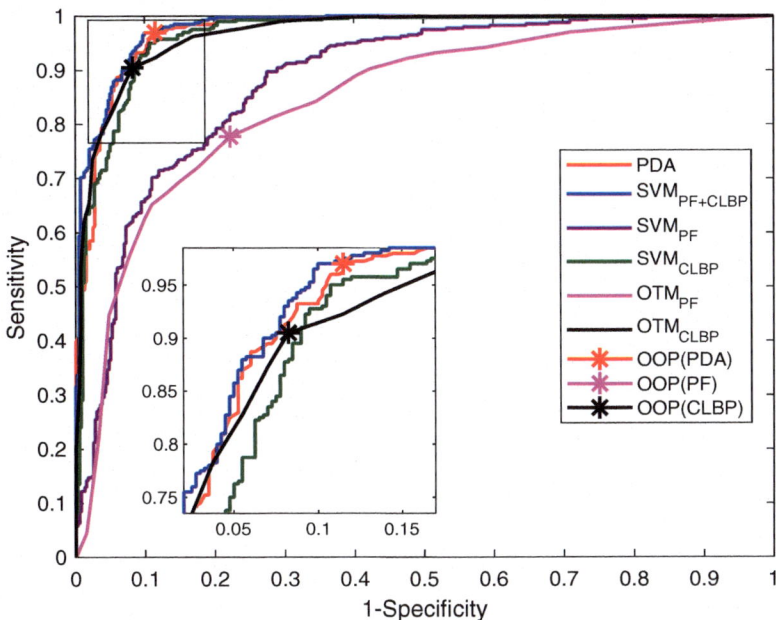

Fig. 2. Best ROC curves for each method in the cross-validating phase.

plots of SVM method where we used PF+CLBP, CLBP, and PF as descriptors, respectively. The fifth and sixth blocks are for OTM method, where CLBP and PF are the individual approaches, respectively.

The ROC curves for the cross-validating phase are depicted in Fig. 2 as well as some optimal operating points (OOP). For each method the ROC curve depicted in this Fig. 2 corresponds to the one with the largest AUC, among the 10 experiments.

Finally, Table 1 shows the results for the cross-validation experiments in the rows indicated by VL, where the values correspond to the maximum of each metric (AUC, ACCU, SENS and SPEC) as exhibited in Fig. 1, as well as, the results for the testing set, in the rows indicated by TS.

We remark that for each image we apply a pre-processing step that involves the removal of specular highlights and an inpainting technique (see begining of Sect. 2.3). Thus, the two last rows in this Table 1 show the execution time, measured in milliseconds (ms), per image, for each method. In row with the acronym "pp", for pre-processing, the exhibited time includes the pre-processing step, but in row with the acronym "d-pp-t", meaning discarding pre-processing time, it does not.

According to these results, one could say that the PDA method outperforms the individual approaches. In addition, the performance of PDA is similar to the performance of the SVM approach that includes the two criteria (PF and CLBP) regarding the cross-validating phase, as shown in Fig. 2 where the best

ROC curves in the cross-validation procedure are displayed. However, the PDA method performs better in the testing set as can be seen in Table 1. An analysis of this Table 1 indicates that the performance of SVM and individual approaches drop significantly from cross-validating to testing. In contrast, PDA shows a more stable and robust performance.

For all the experiments, it was used a workstation with an Intel(R) Core(TM) i7-6700 CPU@3.40GHz. The methods were coded with MATLAB® 2018b (The MathWorks, Inc., Natick, Massachusetts, United States).

Table 1. Results for the maxima in the box plots of Fig. 1 and for the testing set

		PF+CLBP		CLBP		PF	
		PDA	SVM	OTM	SVM	OTM	SVM
AUCx100	VL	97.33	97.90	96.86	96.49	89.88	91.03
ACCU	VL	92.00	92.00	88.63	91.38	81.00	85.25
	TS	91.38	88.13	75.75	87.00	66.13	72.13
SENS	VL	97.25	99.25	82.75	97.00	91.75	88.50
	TS	93.25	90.00	52.75	85.75	61.5	58.75
SPEC	VL	89.75	97.75	96.25	97.50	87.50	88.25
	TS	89.50	86.25	98.75	88.25	70.75	85.50
Time (ms)	pp	4308.42	4686.19	891.78	3729.45	850.25	1080.15
per image	d-pp-t	2568.22	2945.99	0.91	2838.58	0.92	230.82

4   Conclusion

In this paper, we propose an approach to detect polyps with the aid of multi-objective optimization. In particular, the method exploits a similarity-based anomaly detection algorithm. The obtained results on a proprietary medical dataset confirm that it clearly outperforms the single performance of each criterion.

Although it is an undeniable fact that it is extremely useful to process video images, taken from patients, with an automated procedure, as the one provided by the method described in this paper, the ideal would be a real-time method. In its present form, the proposed method has an execution time that is not compatible with real-time detection. However, it is possible to speed it up, using an appropriate reformulation, based on partial differential equations, compatible with real-time anomaly detection, as described in the very recent paper [1]. This will be the topic of our future research.

Acknowledgements. This work was supported by the FCT (Fundação para a Ciência e a Tecnologia, Portugal) research project PTDC/EMD-EMD/28960/2017, and also partially by the FCT grant UID/MAT/00324/2019.

References

1. Abbasi, B., Calder, J., Oberman, A.M.: Anomaly detection and classification for streaming data using PDEs. SIAM J. Appl. Math. **78**(2), 921–941 (2018)
2. Bernal, J., Sánchez, F.J., Fernández-Esparrach, G., Gil, D., Rodríguez, C., Vilariño, F.: WM-DOVA maps for accurate polyp highlighting in colonoscopy: validation vs. saliency maps from physicians. Comput. Med. Imaging Graph. **43**, 99–111 (2015)
3. Chang, C.C., Lin, C.J.: LIBSVM: a library for support vector machines. ACM Trans. Intell. Syst. Technol. **2**, 27:1–27:27 (2011). Software available at http://www.csie.ntu.edu.tw/~cjlin/libsvm
4. Figueiredo, P.N., Figueiredo, I.N., Pinto, L., Kumar, S., Tsai, Y.H.R., Mamonov, A.V.: Polyp detection with computer-aided diagnosis in white light colonoscopy: comparison of three different methods. Endosc. Int. Open **7**(02), E209–E215 (2019)
5. Figueiredo, P.N., Figueiredo, I.N., Prasath, S., Tsai, R.: Automatic polyp detection in pillcam colon 2 capsule images and videos: preliminary feasibility report. Diagn. Ther. Endosc. **2011**, Article ID 182435 (2011). https://doi.org/10.1155/2011/182435
6. Guo, Z., Zhang, L., Zhang, D.: A completed modeling of local binary pattern operator for texture classification. IEEE Trans. Image Process. **19**(6), 1657–1663 (2010)
7. Hsiao, K.J., Xu, K.S., Calder, J., Hero, A.O.: Multicriteria similarity-based anomaly detection using Pareto depth analysis. IEEE Trans. Neural Netw. Learn. Syst. **27**(6), 1307–1321 (2016)
8. Kullback, S., Leibler, R.A.: On information and sufficiency. Ann. Math. Stat. **22**(1), 79–86 (1951)
9. Mamonov, A.V., Figueiredo, I.N., Figueiredo, P.N., Tsai, Y.H.R.: Automated polyp detection in colon capsule endoscopy. IEEE Trans. Med. Imaging **33**(7), 1488–1502 (2014)
10. Tajbakhsh, N., Gurudu, S.R., Liang, J.: Automated polyp detection in colonoscopy videos using shape and context information. IEEE Trans. Med. Imaging **35**(2), 630–644 (2016)
11. Urban, G., Tripathi, P., Alkayali, T., Mittal, M., Jalali, F., Karnes, W., Baldi, P.: Deep learning localizes and identifies polyps in real time with 96% accuracy in screening colonoscopy. Gastroenterology **155**(4), 1069–1078 (2018)

Comparative Study of Dermoscopic Hair Removal Methods

Lidia Talavera-Martínez[1,2][✉], Pedro Bibiloni[1,2],
and Manuel González-Hidalgo[1,2]

[1] SCOPIA Research Group, Universitat de les Illes Balears, 07122 Palma, Spain
{l.talavera,p.bibiloni,manuel.gonzalez}@uib.es
[2] Balearic Islands Health Research Institute (IdISBa), 07010 Palma, Spain

Abstract. When analyzing dermoscopic images, the hairs and their shadows on the skin may occlude relevant information about the lesion at the time of diagnosis. As far as we know, there is no method that quantitatively evaluates the performance of hair removal algorithms. In this work, we present a hair removal benchmark of six state-of-the-art algorithms, each with a different approach to segment and inpaint the hair pixels. To evaluate the algorithms, 13 dermoscopic images without hair were selected from the PH2 database. Next, two different hair simulators, providing hairs with a wide range of characteristics, are applied to these images. The results obtained with the hair removal algorithms on the simulated hair samples can be contrasted with the reference hairless images. To quantitatively assess their efficacy, we use a series of performance measures that evaluate the similarity between the original hairless image and the one obtained by each of the algorithms. Also, a statistical test is used to check the superiority of a method with respect to the others.

Keywords: Benchmark · Image processing · Hair removal ·
Dermoscopy · Skin lesion · Inpainting

1 Introduction

Over the past decades, there has been a rapidly increasing incidence rate of both melanoma and non-melanoma skin cancers [22]. Although melanoma only represents 1.7% of the total number of cancer (excluding non-melanoma skin cancer) with nearly 300,000 new cases around the world in 2018 [3], it causes the majority of deaths (75%) related to skin cancer [2], being the most aggressive, metastatic and deadliest type of skin cancer. Melanoma, which is the 15th most commonly occurring cancer in men and women [3] is a disease caused by the abnormal and uncontrolled proliferation of melanocytes—cells that pigment the skin—, present in the basal layer of the epidermis, which have undergone a genetic mutation. In Europe, the estimated mortality caused by melanoma, in 2018, is approximately 3.8 per 100.000 men and women per year [1].

© Springer Nature Switzerland AG 2019
J. M. R. S. Tavares and R. M. Natal Jorge (Eds.): VipIMAGE 2019, LNCVB 34, pp. 12–21, 2019.
https://doi.org/10.1007/978-3-030-32040-9_2

Towards completing the clinical analysis and the diagnosis of the skin lesion at its earliest stage, physicians employ an examination technique known as dermoscopy. This approach is one of the most commonly used in-vivo, non-invasive skin imaging techniques. The process is based on examining the surface of the skin lesion using a specific optical system to magnify the skin, which is covered with mineral oil, alcohol or water to avoid the reflection of light on it. In some cases, dermoscopy can capture digital images of the cutaneous lesions. It improves the diagnostic accuracy up to 10–30% [11] compared to simple clinical observation, since it allows examining the subsurface structures.

To obtain an early, objective and reproducible diagnosis of the skin lesion, sophisticated computer-aided diagnosis (CAD) software have been developed, which focuses mainly on the acquisition of images, the elimination of artifacts (hairs, bubbles, etc.), the segmentation of lesion, the extraction and the selection of features and classification.

Hair detection and restoration are the major stages of the hair removal process. The former step detects and removes the hair pixels in the image, then the latter estimates the color and texture of the pixels of the skin under the detected hairs and replaces it. A variety of hair thickness, density and color can appear within the process. These hair's pixels inside the skin lesion usually occlude the underlying information of the boundary and texture structures, hindering the features of the lesions to be analyzed more effectively.

Hair removal stands out as one of the most studied topics within the pre-processing stage. Several algorithms have been presented in the literature addressing this problem. In [4], Abbas *et al.* compare hair removal methods without a reference image, relying on measures of how much the repaired pixels disturb the overall texture of the skin lesion. To our knowledge, there is no work that compares the performance of different hair removal approaches on dermoscopic images in terms of their similarity with a reference image.

Hence, our objective is to analyze how diverse hair removal strategies have been applied, using the same database, which would provide an objective comparison of the strengths and weaknesses, and thus of the quality of the methods. It should be noted that the efficacy of the method has a relevant role when it comes to a posterior pattern analysis, since it can lead to an alteration of the original texture of the pattern of the lesion.

The rest of the document is structured as follows. First, in Sect. 2 we introduce the description of the algorithms used along this work. Then, in Sect. 3 we establish the experimental framework in which we describe the database used, the results obtained and the statistical study that is carried out. Finally, in Sect. 4 we analyze on the results obtained.

2   Hair Removal Methods

This section focuses on the brief description of the algorithms that have been applied to perform hair removal in dermoscopic images. The six algorithms used are based on different hair detection and extraction methods, as well as inpainting processes. The choice of these methods is based first on their availability and second on the fact that they are widely used in the literature. They are summarized in Table 1.

Dullrazor© is a free hair removal software implemented by Lee et al. [10] in 1997. The algorithm consists of first identifying the hair regions through the morphological closing operator on each of the three RGB color channels separately, with three structuring elements having different directions. The binary hair mask is obtained by thresholding the absolute difference between the original color channel and the image generated by the closing. Once the hairs are removed, the authors replace the hair pixels by bilinear interpolation between two nearby hairless pixels and smooth the result using an adaptive median filter.

A decade later, in 2009, Xie et al. [23] proposed a method that is based first on improving the hair area by means of a morphological closing top-hat operator. Subsequently, the binary image is obtained through a statistical threshold. Next, to extract the hairs, the authors use the elongate feature function of connected regions. Finally, to restore the information occluded by the hair, they apply the image inpainting method based on PDE (partial differential equation), which is effective in preserving the linear geometry feature, since it realizes the diffusion of information through the difference between pixels.

Abbas et al. [4] in 2011, proposed an approach using the CIELab uniform color space where the hairs are detected by a Derivative of Gaussian (DoG). Then they are enhanced with morphological techniques to link broken hair segments, eliminate small and circular objects and fill the gaps between lines. Regarding the inpainting step of the selected pixels, the authors utilize a modified version of an inpainting method based on coherence transport.

After 2 years, in 2013, Huang et al. [9] implemented an approach that uses multiscale curvilinear matched filters for hair detection in grayscale images. Then, the hairs of the image are extracted by hysterisis thresholding. To recover the missing information left by the hairs removed, the authors apply region growing algorithms and the linear discriminant analysis (LDA) technique, based on the pixel color information in the CIELab color space.

Also, Toossi et al. [17] introduced a method that detects hairs by using an adaptive canny edge detector proposed in [15]. Previously, the image has been converted to a grayscale image via a PCA (Principal Component Analysis) and the noise is filtered with a Wiener filter. Once the image is preprocessed and the hairs detected, the authors apply a refining process with morphological operators to eliminate unwanted objects and obtain a smooth hair mask. The deleted pixels are repaired by applying a multi-resolution coherence transport inpainting method based on wavelets.

Finally, in 2017, Bibiloni et al. [7] introduced an approach for hair removal using soft color morphology operators in the CIELab color space. First, the contrast of the luminance of the image is improved with the CLAHE algorithm. Then, the hairs are detected by a combination of soft color closing top-hat operators, of which the size and orientation of the structuring elements can be adjusted according to the appearance of the hairs. Finally, the authors present an inpainting algorithm based on the arithmetic mean of the modified opening and closing morphological transformations to recover the missing pixels. For this algorithm, two versions, Bibiloni09 and Bibiloni11, have been used, in which the size of the kernel used to perform the inpainting is different, of 9×9 and 11×11, respectively.

Table 1. Detection and inpainting techniques employed in the literature to remove hair from dermoscopic images.

Year	Method	Hair segmentation	Inpainting method	Color space
1997	Lee *et al.* [10]	Grayscale closing	Bilinear interpolation	RGB
2009	Xie *et al.* [23]	Grayscale top-hat	Non-linear PDE	RGB
2011	Abbas *et al.* [4]	Derivative of Gaussians	Coherence transport	CIELab
2013	Huang *et al.* [9]	Conventional matched filters	Region growing algorithms and linear discriminant analysis	RGB
2013	Toossi *et al.* [17]	Canny edge detector	Coherence transport	RGB
2017	Bibiloni *et al.* [7]	Color top-hat	Morph. inpainting	CIELab

3 Experiments

In this section, we carry out a comparative study of the performance of the algorithms explained in Sect. 2. To do this, we first establish the experimental framework that describes the database used. Then, we analyze the results obtained by the algorithms from a qualitative point of view. Finally, we determine, from a quantitative point of view, which method or methods stand out from the rest by means of several performance measures and a statistical test.

3.1 Experimental Frame: Database

To evaluate the algorithms, we used 13 dermoscopic images extracted from the PH2 public database [12]. As reflected in Sect. 1, our goal is to determine which method provides the best hair removal results. To make the comparison effective, in this work we take as starting point the images of the PH2 database that do not have hair, these are used as reference images to compare the degree of adjustment of the results obtained by the automatic algorithms. Next, the following step is the hair simulation in those images. To this end, two different hair simulators have been used. First, the method based on generative adversarial networks by Attia *et al.* is applied [5] due to the realism of its results. We call this dataset DB GAN. To further validate the methods with different looking hair, another hair simulation algorithm has been used. This last approach is the one implemented by Mirzaalian *et al.* [14], whose software "HairSim" is available in [13], that creates less realistic hairs than the previous one, but which provides us hairs with very different characteristics. This dataset is denoted by DB SIM. In Fig. 1, we show an example of an original hairless image from our dataset, together with the simulations obtained using each of the hair simulators.

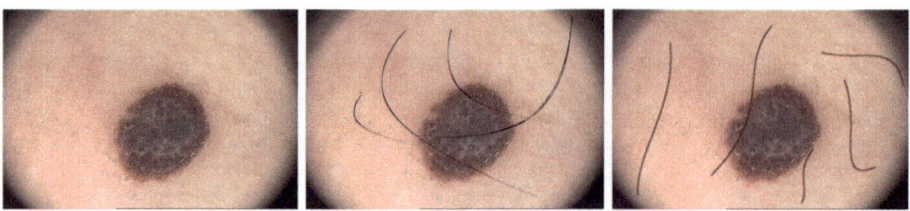

Fig. 1. Example of a hairless image of the PH2 database (left), hair simulation created by the deep neural network [5] (middle), and hair simulation created with the "HairSim" software [13] (right).

3.2   Graphical Results

As can be seen in Fig. 2, most algorithms seem successful in both the hair removal and the subsequent process of inpainting. However, Abbas *et al.*'s and Toossi *et al.*'s methods are not capable of detecting a large part of the hairs and, therefore, do not perform their segmentation properly. On the other hand, although the results of the rest of the methods are adjusted, at first sight, to the reference image, it is notable that all of them introduce some alterations, to a greater or lesser extent, with respect to the original texture present in the lesion. As a possible exception to this drawback, we could highlight the methods of Lee *et al.* and Xie *et al.*, which do not tend to blur features of the lesions, such as the streaks, in a noticeable way.

3.3   Quantitative Evaluation of the Results

In order to evaluate the quality of the results obtained by several algorithms, a visual comparison is not enough. Nowadays, the performance of the restoration is also measured quantitatively by several widely used objective performance measures. The automated evaluation of the results of the image restoration provides an objective and comparable evaluation.

A set of nine objective performance measures is applied to evaluate the similarity between the result obtained by each algorithm with respect to the original hairless image. The first subset is composed of the Mean Squared Error (MSE) [19], the Peak Signal-to-Noise Ratio (PSNR) [20], the Root Mean Squared Error (RMSE) [6], and the Structural Similarity Index (SSIM) [20]. These are pixel to pixel measurements, widely used in the literature. The second subset of performance measures consists of the Multi-Scale Structural Similarity Index (MSS-SIM) [21] and the Universal Quality Image Index (UQI) [18], which measure

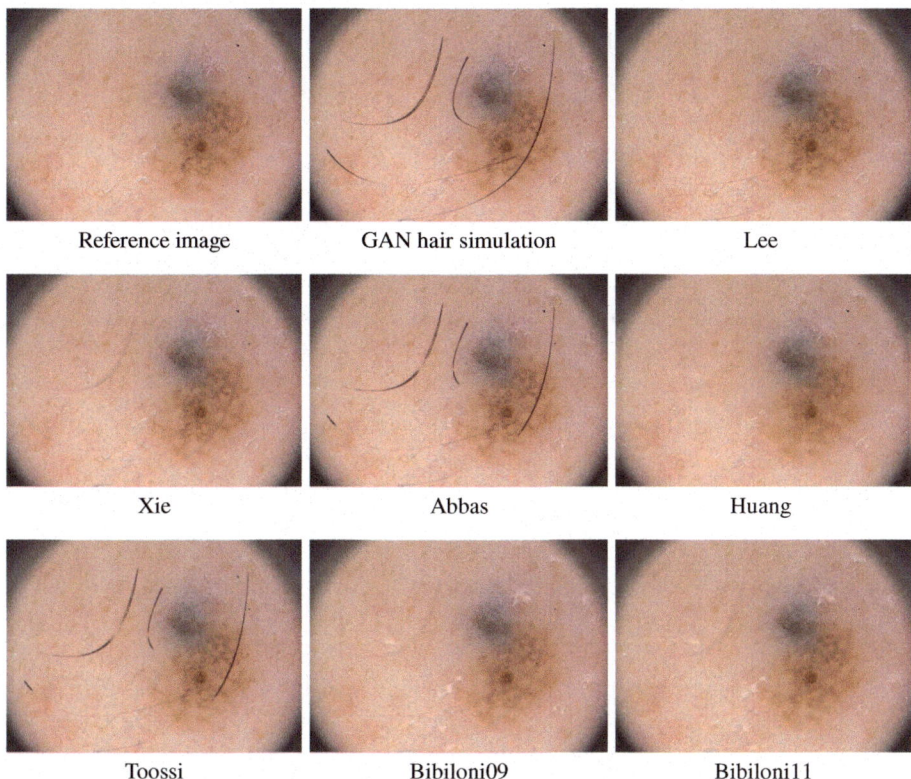

Reference image   GAN hair simulation   Lee

Xie   Abbas   Huang

Toossi   Bibiloni09   Bibiloni11

Fig. 2. Example of the hair removal results obtained by the different algorithms on DB GAN.

statistical features locally and then combine them. Finally, the measures Visual Information Fidelity (VIF) [16], the PSNR-HVS-M [8] and PSNR-HVS [8] have been designed to obtain more similar results to those perceived by the Human Visual System (HVS). This set of performance measures constitutes a representative selection of the state-of-the-art for the objective assessment of image restoration quality. In Table 2, we show the mean and standard deviations of the nine performance measures applied to both databases by the seven hair removal algorithms.

Once the performance objective measures have been obtained for all the images and for each hair removal algorithm, the next step is to study if one algorithm outperforms another one significantly. Fixing a similarity measure, we compare all the pairs of algorithms using a statistical test to contrast their means. Specifically, we have used the t-test if the samples pass the Shapiro-Wilk normality test, or if not, the Wilcoxon signed-rank test, both based on a 95% confidence level. So, we can determine which ones obtain the best results in the sense that no other algorithm outperforms them according to the statistical test.

Table 2. Mean and standard deviation of the similarity measures obtained for the different hair removal algorithms, and each of our datasets.

			Abbas	Lee	Huang	Bibiloni09	Bibiloni11	Toossi	Xie
MSE	DB GAN	μ	93.703	65.314	69.860	65.632	64.156	92.879	30.925
		σ	32.083	47.981	33.281	35.746	33.610	30.754	11.180
	DB SIM	μ	131.377	51.118	44.494	63.291	65.389	134.877	16.792
		σ	84.898	48.764	28.751	34.070	34.656	80.169	13.477
SSIM	DB GAN	μ	0.805	0.837	0.792	0.824	0.824	0.798	0.848
		σ	0.028	0.017	0.034	0.027	0.030	0.031	0.012
	DB SIM	μ	0.908	0.956	0.951	0.908	0.909	0.894	0.975
		σ	0.039	0.023	0.028	0.069	0.068	0.040	0.010
PSNR	DB GAN	μ	28.619	30.752	30.163	30.516	30.603	28.645	33.445
		σ	1.353	2.566	2.149	2.284	2.285	1.318	1.367
	DB SIM	μ	27.740	32.486	32.594	30.738	30.572	27.512	36.968
		σ	2.768	3.793	3.108	2.515	2.455	2.565	3.058
RMSE	DB GAN	μ	9.563	7.714	8.138	7.847	7.766	9.528	5.489
		σ	1.561	2.507	1.985	2.097	2.042	1.507	0.928
	DB SIM	μ	10.957	6.594	6.335	7.688	7.824	11.175	3.842
		σ	3.514	2.877	2.175	2.128	2.126	3.291	1.484
VIF	DB GAN	μ	0.316	0.343	0.265	0.316	0.315	0.303	0.363
		σ	0.041	0.031	0.042	0.032	0.034	0.041	0.038
	DB SIM	μ	0.758	0.838	0.838	0.584	0.589	0.713	0.920
		σ	0.077	0.072	0.079	0.198	0.197	0.079	0.024
UQI	DB GAN	μ	0.996	0.998	0.997	0.998	0.998	0.996	0.998
		σ	0.002	0.001	0.002	0.002	0.002	0.002	0.001
	DB SIM	μ	0.995	0.999	0.999	0.998	0.998	0.995	0.999
		σ	0.002	0.001	0.001	0.003	0.003	0.002	0.000
MSSSIM	DB GAN	μ	0.805	0.836	0.790	0.823	0.823	0.797	0.849
		σ	0.028	0.019	0.034	0.028	0.030	0.031	0.012
	DB SIM	μ	0.907	0.954	0.948	0.906	0.907	0.893	0.975
		σ	0.039	0.023	0.028	0.068	0.067	0.039	0.010
PSNR-HVS-M	DB GAN	μ	28.501	32.743	31.339	32.043	32.177	28.565	36.041
		σ	1.110	4.088	2.933	3.503	3.481	1.016	2.377
	DB SIM	μ	27.233	33.335	33.400	31.695	31.403	27.012	36.447
		σ	4.511	4.817	4.130	3.539	3.393	4.235	2.973
PSNR-HVS	DB GAN	μ	27.893	31.661	30.503	31.071	31.184	27.943	34.748
		σ	1.154	3.745	2.816	3.270	3.243	1.079	2.156
	DB SIM	μ	26.837	32.555	32.817	30.938	30.690	26.606	36.027
		σ	4.399	4.691	4.118	3.398	3.286	4.112	2.905

When there is a statistical difference between a pair of methods, the best method has a higher mean or a lower mean if the studied index is a performance or an error measure, respectively.

Table 3 summarizes the results obtained when applying the statistic test to all pairs of algorithms for each of the performance measures. As an example, the second row of Table 3, compares the algorithm Abbas with the algorithm Huang. For the measures of MSE, PSNR, RMSE, UQI, PSNR-HVS-M and PSNR-HVS Abbas' algorithm is statistically worse in mean than Huang's algorithm. In contrast, for the SSIM and MSSIM measures, we can not rule out the equality of their means, but the mean of Abbas' algorithm is statistically better than that of Huang's algorithm. Finally, for the VIF measure, the mean of Abbas' algorithm is better than that of Huang's algorithm.

Table 3. Classification of algorithms according to objective similarity measures. The results are as follows: ✓✓ if the population mean of the first algorithm is better than that of the second algorithm; ✓ if the equality of population means can not be ruled out, but the sample mean of the first algorithm is better than that of the second algorithm; ✗ if the equality of population means can not be ruled out, but the sample mean of the first algorithm is worse than that of the second algorithm; ✗✗ if the population mean of the first algorithm is worse than that of the second algorithm.

		MSE	SSIM	PSNR	RMSE	VIF	UQI	MSSSIM	PSNR-HVS-M	PSNR-HVS
Abbas vs. Lee	p-value	9.19e-03	2.78e-05	0.004649	1.07e-02	1.64e-03	1.16e-07	3.77e-05	5.77e-03	5.77e-03
	Statistical test	✗✗	✗✗	✗✗	✗✗	✗✗	✗✗	✗✗	✗✗	✗✗
Abbas vs. Huang	p-value	2.10e-02	0.382352	0.007884	1.21e-02	8.77e-03	6.21e-03	2.21e-01	5.88e-03	6.56e-03
	Statistical test	✗✗	✓	✗✗	✗✗	✓✓	✗✗	✓	✗✗	✗✗
Abbas vs. Bibiloni09	p-value	3.36e-03	0.02313	0.007132	7.13e-03	7.47e-02	2.37e-03	2.31e-02	5.77e-03	5.77e-03
	Statistical test	✗✗	✗✗	✗✗	✗✗	✗	✗✗	✗✗	✗✗	✗✗
Abbas vs. Bibiloni11	p-value	2.71e-03	2.31e-02	7.13e-03	4.65e-03	8.69e-02	2.98e-03	2.31e-02	5.77e-03	5.77e-03
	Statistical test	✗✗	✗✗	✗✗	✗✗	✓	✗✗	✗✗	✗✗	✗✗
Abbas vs. Toossi	p-value	3.45e-01	1.86e-05	3.82e-01	3.11e-01	1.47e-03	4.64e-02	6.78e-06	1.73e-01	4.63e-01
	Statistical test	✗	✓✓	✗	✗	✓✓	✗	✓✓	✗	✗
Abbas vs. Xie	p-value	1.47e-03	1.47e-03	1.47e-03	1.47e-03	1.47e-03	4.14e-06	1.47e-03	1.47e-03	1.47e-03
	Statistical test	✗✗	✗✗	✗✗	✗✗	✗✗	✗✗	✗✗	✗✗	✗✗
Lee vs. Huang	p-value	1.33e-01	1.87e-03	1.16e-01	1.33e-01	1.87e-03	1.92e-02	1.87e-03	3.92e-02	4.64e-02
	Statistical test	✓	✓✓	✓	✓	✓✓	✓✓	✓✓	✓✓	✓✓
Lee vs. Bibiloni09	p-value	6.00e-01	1.92e-02	4.22e-01	4.22e-01	7.13e-03	8.69e-02	3.92e-02	1.73e-01	2.21e-01
	Statistical test	✓	✓✓	✓	✓	✓✓	✓	✓✓	✓	✓
Lee vs. Bibiloni11	p-value	7.53e-01	3.30e-02	6.00e-01	6.00e-01	7.13e-03	1.33e-01	4.64e-02	3.11e-01	3.45e-01
	Statistical test	✓	✓✓	✓	✓	✓✓	✓	✓✓	✓	✓
Lee vs. Toossi	p-value	1.01e-02	1.35e-05	2.02e-03	1.07e-02	1.13e-04	1.47e-03	1.47e-05	5.77e-03	5.77e-03
	Statistical test	✓✓	✓✓	✓✓	✓✓	✓✓	✓✓	✓✓	✓✓	✓✓
Lee vs. Xie	p-value	5.77e-03	1.87e-03	3.73e-03	4.65e-03	1.47e-03	1.01e-01	1.47e-03	1.92e-02	1.92e-02
	Statistical test	✗✗	✗✗	✗✗	✗✗	✗✗	✗	✗✗	✗✗	✗✗
Huang vs. Bibiloni09	p-value	3.92e-02	1.87e-03	4.64e-02	7.63e-02	1.87e-03	1.92e-02	1.87e-03	2.77e-02	3.30e-02
	Statistical test	✗✗	✗✗	✗✗	✗	✗✗	✗✗	✗✗	✗✗	✗✗
Huang vs. Bibiloni11	p-value	3.30e-02	1.68e-05	1.07e-02	6.35e-02	1.87e-03	1.59e-02	1.04e-05	1.92e-02	2.31e-02
	Statistical test	✗✗	✗✗	✗✗	✗✗	✗✗	✗✗	✗✗	✗✗	✗✗
Huang vs. Toossi	p-value	2.22e-02	7.53e-01	8.51e-03	1.30e-02	5.46e-02	8.03e-03	7.01e-01	5.79e-03	6.28e-03
	Statistical test	✓✓	✗	✓✓	✓✓	✗	✓✓	✗	✓✓	✓✓
Huang vs. Xie	p-value	2.37e-03	1.47e-03	2.37e-03	2.37e-03	1.47e-03	1.87e-03	1.47e-03	1.87e-03	1.87e-03
	Statistical test	✗✗	✗✗	✗✗	✗✗	✗✗	✗✗	✗✗	✗✗	✗✗
Bibiloni09 vs. Bibiloni11	p-value	1.92e-02	5.07e-01	1.59e-02	1.92e-02	8.69e-02	3.92e-02	5.07e-01	2.31e-02	2.31e-02
	Statistical test	✗✗	✓	✗✗	✗✗	✓	✗✗	✓	✗✗	✗✗
Bibiloni09 vs. Toossi	p-value	3.49e-03	2.31e-02	8.77e-03	8.77e-03	2.77e-02	1.87e-03	2.31e-02	7.13e-03	7.13e-03
	Statistical test	✓✓	✓✓	✓✓	✓✓	✓✓	✓✓	✓✓	✓✓	✓✓
Bibiloni09 vs. Xie	p-value	5.77e-03	1.87e-03	3.73e-03	3.73e-03	1.47e-03	1.59e-02	1.47e-03	5.77e-03	4.65e-03
	Statistical test	✗✗	✗✗	✗✗	✗✗	✗✗	✗✗	✗✗	✗✗	✗✗
Bibiloni11 vs. Toossi	p-value	2.78e-02	2.31e-02	7.13e-03	4.65e-03	3.30e-02	4.65e-03	2.31e-02	5.77e-03	5.77e-03
	Statistical test	✓✓	✓✓	✓✓	✓✓	✓✓	✓✓	✓✓	✓✓	✓✓
Bibiloni11 vs. Xie	p-value	5.77e-03	1.87e-03	3.73e-03	5.77e-03	1.47e-03	2.77e-02	1.47e-03	5.77e-03	4.65e-03
	Statistical test	✗✗	✗✗	✗✗	✗✗	✗✗	✗✗	✗✗	✗✗	✗✗
Toossi vs. Xie	p-value	1.47e-03	1.47e-03	1.47e-03	1.47e-03	1.47e-03	4.30e-06	1.47e-03	1.47e-03	1.47e-03
	Statistical test	✗✗	✗✗	✗✗	✗✗	✗✗	✗✗	✗✗	✗✗	✗✗

From the results shown in Table 3 we can conclude that all measures point to a single best algorithm, which is the algorithm presented by Xie. In the same way, we can see that Lee's algorithm significantly outperforms Abbas' and Toossi's algorithms in all measurements. However, in comparison with Bibiloni's algorithm, it is statistically better with regard to the SSIM, VIF and MSSIM measurements, but it can not be rejected that the two methods have a statistically equivalent performance for the rest of the measurements. In the comparison between the algorithms of Huang and Bibiloni, it is the latter that, in its two versions, surpasses the first statistically. Finally, it is reflected that the Abbas and Toossi algorithms provide statistically worse results that the rest of the algorithms.

4   Conclusions

In this work, we have presented a benchmark of hair removal methods applied to dermosocopy images. Particularly, we have analyzed the performance of six state-of-the-art approaches using two different hair simulation datasets, created using an available dermoscopic dataset. The analysis was made by calculating nine measures of similarity, widely used, to compare the hairless reference image with that obtained with each of the hair removal algorithms. Due to the difficulty of a qualitative analysis, we decided to perform a statistical test to objectively verify the superiority, if any, of any of the methods.

This gives us a tight and complete description of whether an algorithm is capable of retrieving information without altering or blurring the surrounding texture too much. The results obtained through the statistical test of these measures lead to the conclusion that the algorithm proposed by Xie *et al.* statistically outperforms the rest of the algorithms, although not significantly all of them. As reflected in Fig. 2 and in Table 3, the Abbas' and Toossi's algorithms that produce the least suitable results. This bad behavior may be due to the fact that these algorithms do not seem to distinguish well hairs of greater thickness or dark colors.

As future work, we aim to expand the number of images used in the datasets, as well as to improve the quality obtained by the hair simulation methods in terms of its realism.

Acknowledgment. This work was partially supported by the project TIN 2016-75404-P AEI/FEDER, UE. L. Talavera-Martínez also benefited from the fellowship BES-2017-081264 conceded by the *Ministry of Economy, Industry and Competitiveness* under a program co-financed by the *European Social Fund*. We thank Dr. Mohamed Attia from the Institute For Innovation and Research, Deakin University, Australia, for providing the GAN-based simulated hair images.

References

1. European Cancer Information System. https://ecis.jrc.ec.europa.eu/index.php. Accessed 19 Nov 2018
2. Melanoma Molecular Map Project. http://www.mmmp.org/MMMP/welcome.mmmp. Accessed 28 Feb 2019
3. World Cancer Research Fund International. https://www.wcrf.org/dietandcancer/cancer-trends/worldwide-cancer-data. Accessed 1 Jan 2019
4. Abbas, Q., Celebi, M.E., García, I.F.: Hair removal methods: a comparative study for dermoscopy images. Biomed. Signal Process. Control **6**(4), 395–404 (2011)

5. Attia, M., Hossny, M., Zhou, H., Yazdabadi, A., Asadi, H., Nahavandi, S.: Realistic hair simulator for skin lesion images using conditional generative adversarial network. Preprints (2018). https://doi.org/10.20944/preprints201810.0756.v1
6. Barnston, A.G.: Correspondence among the correlation, RMSE, and Heidke forecast verification measures; refinement of the Heidke score. Weather Forecast. **7**(4), 699–709 (1992)
7. Bibiloni, P., González-Hidalgo, M., Massanet, S.: Skin hair removal in dermoscopic images using soft color morphology. In: Conference on Artificial Intelligence in Medicine in Europe, pp. 322–326. Springer (2017)
8. Egiazarian, K., Astola, J., Ponomarenko, N., Lukin, V., Battisti, F., Carli, M.: New full-reference quality metrics based on HVS. In: Proceedings of the Second International Workshop on Video Processing and Quality Metrics, vol. 4 (2006)
9. Huang, A., Kwan, S.Y., Chang, W.Y., Liu, M.Y., Chi, M.H., Chen, G.S.: A robust hair segmentation and removal approach for clinical images of skin lesions. In: 2013 35th Annual International Conference of the IEEE Engineering in Medicine and Biology Society (EMBC), pp. 3315–3318. IEEE (2013)
10. Lee, T., Ng, V., Gallagher, R., Coldman, A., McLean, D.: Dullrazor®: a software approach to hair removal from images. Comput. Biol. Med. **27**(6), 533–543 (1997)
11. Mayer, J., et al.: Systematic review of the diagnostic accuracy of dermatoscopy in detecting malignant melanoma. Med. J. Aust. **167**(4), 206–210 (1997)
12. Mendonca, T., Celebi, M., Mendonca, T., Marques, J.: PH2: a public database for the analysis of dermoscopic images. Dermoscopy Image Analysis (2015)
13. Mirzaalian, H.: Hair Sim Software. http://www2.cs.sfu.ca/~hamarneh/software/hairsim/Welcome.html. Accessed 7 Mar 2019
14. Mirzaalian, H., Lee, T.K., Hamarneh, G.: Hair enhancement in dermoscopic images using dual-channel quaternion tubularness filters and MRF-based multilabel optimization. IEEE Trans. Image Process. **23**(12), 5486–5496 (2014)
15. Rosenfeld, A., De La Torre, P.: Histogram concavity analysis as an AID in threshold selection. IEEE Trans. Syst. Man Cybern. **2**, 231–235 (1983)
16. Sheikh, H.R., Bovik, A.C.: Image information and visual quality. In: 2004 IEEE International Conference on Acoustics, Speech, and Signal Processing, vol. 3, pp. iii–709. IEEE (2004)
17. Toossi, M.T.B., Pourreza, H.R., Zare, H., Sigari, M.H., Layegh, P., Azimi, A.: An effective hair removal algorithm for dermoscopy images. Skin Res. Technol. **19**(3), 230–235 (2013)
18. Wang, Z., Bovik, A.C.: A universal image quality index. IEEE Signal Process. Lett. **9**(3), 81–84 (2002)
19. Wang, Z., Bovik, A.C.: Mean squared error: Love it or leave it? A new look at signal fidelity measures. IEEE Signal Process. Mag. **26**(1), 98–117 (2009)
20. Wang, Z., Bovik, A.C., Sheikh, H.R., Simoncelli, E.P., et al.: Image quality assessment: from error visibility to structural similarity. IEEE Trans. Image Process. **13**(4), 600–612 (2004)
21. Wang, Z., Simoncelli, E.P., Bovik, A.C.: Multiscale structural similarity for image quality assessment. In: The Thrity-Seventh Asilomar Conference on Signals, Systems & Computers, vol. 2, pp. 1398–1402. IEEE (2003)
22. World Health Organization: How common is skin cancer?. https://www.who.int/uv/faq/skincancer/en/index1.html. Accessed 28 Jan 2019
23. Xie, F.Y., Qin, S.Y., Jiang, Z.G., Meng, R.S.: PDE-based unsupervised repair of hair-occluded information in dermoscopy images of melanoma. Comput. Med. Imaging Graph. **33**(4), 275–282 (2009)

A Semi-automatic Software for Processing Real-Time Phase-Contrast MRI Data

Sidy Fall[1(✉)], Pan Liu[1], and Olivier Baledent[1,2]

[1] Facing Faces Institute/CHIMERE EA 7516, University of Picardy,
Amiens, France
sidy.fall@u-picardie.fr
[2] Medical Image Processing Department, University Hospital of Picardy,
Amiens, France

Abstract. To date, the effects cardiac and respiratory changes on cerebral blood flow dynamics are not well understood. Real-time phase contrast magnetic resonance imaging (MRI) allows to assess hemodynamic information with high temporal resolution. We have developed a software allowing to process real time phase contrast MR data. Measurements were obtained in 14 subjects and were performed in cerebral arteries and venous sinus. Our software was able to quantify blood flows in the selected vessels and to detect flow changes during inspiration and expiration periods of the cardiac cycle. The assessment of hemodynamic information with real-time flow imaging may have potential applications for the diagnosis of cardiovascular and neurovascular diseases.

Keywords: Magnetic resonance imaging · Real-time phase contrast imaging · Signal processing · Image processing

1 Introduction

The assessment of hemodynamic information of the intracranial vessels is useful in the diagnosis of a number of neuro-vascular diseases such as hydrocephalus, idiopathic intracranial hypertension [1, 2] In the field of magnetic resonance imaging (MRI), conventional cine phase-contrast (Conv-Cine PC) imaging technique may provide useful functional information for the quantification of flow velocities in vessels [3, 4]. To get the data of the entire volume of acquisition using Conv-Cine PC sequences, only a single k-line is collected between successive heartbeats (RR interval). Therefore the time required to traverse a k-space is relatively long. The flow velocities displayed in the phase images represent a weighted average over the time of the acquisition and may be insensitive to short-term variations of the flow [5]. Moreover, the acquisition time make this technique sensitive to errors introduced by heart rate variability as well as motion, such as respiration and highly pulsatile flow [6]. Although Conv-Cine PC is now the most commonly used sequence for flows quantification, velocity imaging techniques with high sampling rate are required to characterize complex flow patterns and to elucidate the effects of physiological noise on the flows measurements. Interestingly, improvements in the data acquisition rate have been achieved by optimizing the k-space sampling methods. Real-time echo-planar imaging (EPI) offers the

© Springer Nature Switzerland AG 2019
J. M. R. S. Tavares and R. M. Natal Jorge (Eds.): VipIMAGE 2019, LNCVB 34, pp. 22–28, 2019.
https://doi.org/10.1007/978-3-030-32040-9_3

possibility to traverse a k-space in a single radio-frequency excitation and therefore allows to achieve a faster sampling rate (less than 100 ms) compared to Conv-Cine PC techniques. EPI was applied to PC MRI for quantitative blood flow velocimetry and have been shown to produce accurate flow measurements [5].

Currently, a variety of commercial and non-commercial postprocessing software [7] is available for fast and semi-automatic quantification of flows from data acquired with Conv-Cine PC sequences. However, to our knowledge, no software exists to process real-time EPI data. The purpose of this work was to present a novel homemade software allowing to analyze data from real-time EPI velocity mapping and capable to determine the respiration effects on flow measurements. In vivo velocities data were obtained from fourteen healthy volunteers and were processed with this novel software.

2 Materials and Methods

Our software was developed using an Interactive Data Language (IDL) program.

2.1 Image Loading

The software use Digital Imaging and Communications in Medicine (DICOM) data, which tags are stored in a metafile header, the so-called DICOMDIR. Data are loading through the DICOMDIR file and the different series of a study can be classified using the hierarchical structure of the DICOMDIR file. After selection of all images of a

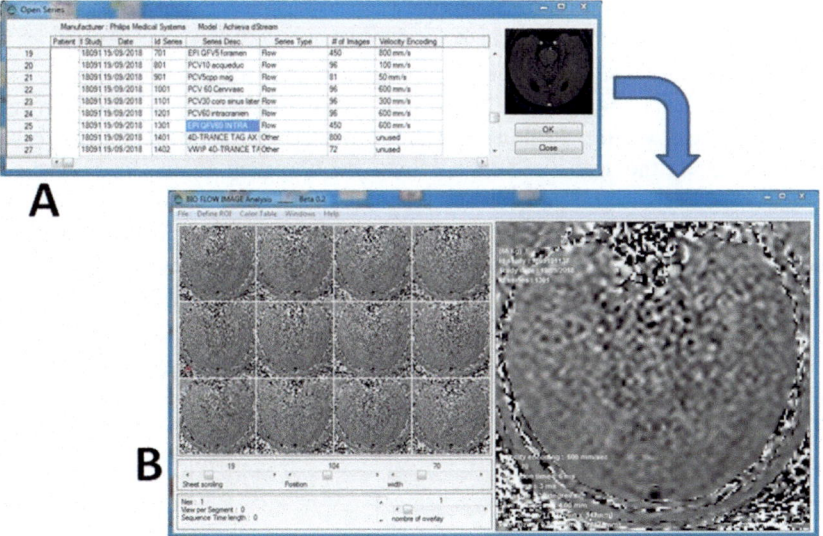

Fig. 1. A preview of the series of images (A) and the selected series can be loaded in the DICOM viewer which includes a toolbar with some functionalities (define region of interest (ROI), color table, window and help). The user may scroll through the images. Threshold filters can be applied to images to distinguish the vessels through the tissues.

series, the DICOM viewer of the software provides some basic image processing functionalities such as scrolling, windowing and zooming. Moreover, basic subject information (examination date, subject ID, series description, …) and acquisition parameters (velocity encoding, echo time, repetition time, field of view, spatial resolution, …) may be also displayed (Fig. 1).

2.2   Postprocessing

2.2.1   Segmentation of Region of Interest

The software has the ability to create segmentation based on the spectral analysis of the data. Threshold filters can be applied for noise removal or to control the grey level. The operator can isolate a ROI based on prior knowledge of anatomy. The cross-sectional area of the vessel of interest can be measured. The segmented ROI may be zoomed, allowing to carefully control its contours and shapes.

2.2.2   Visualization of Physiological Signals and Flow Curves Within the Region of Interest

The raw data of respiratory and cardiac cycles during scanning can be captured by the peripheral pulse and respiratory sensors of the scanner. These data may be processed by the software. The interface of our software presented cursors with which a bandpass filter can be applied to the recorded respiratory raw data in order to extract a respiratory waveform (Fig. 2). The temporal periods corresponding to the inspiration and expiration can be determined.

3   Flow Calculations

Functional information can be assessed through the intensity provided by the real-time EPI phase images. The software allows the user to visualize the temporal dynamic of the signal extracted within a ROI using a sliding window. The details that are relative to the fluctuations of the measured signal may be observed in sections of the signal (Fig. 2). The power spectral density of measured signal can be visualized, allowing the possibility to identify the main peaks of the signal. The software can detect the time points of the real-time EPI data that correspond to each cardiac cycle. In addition, one time averaged cardiac cycle can be created. Flows during the inspiration or expiration periods can also be calculated.

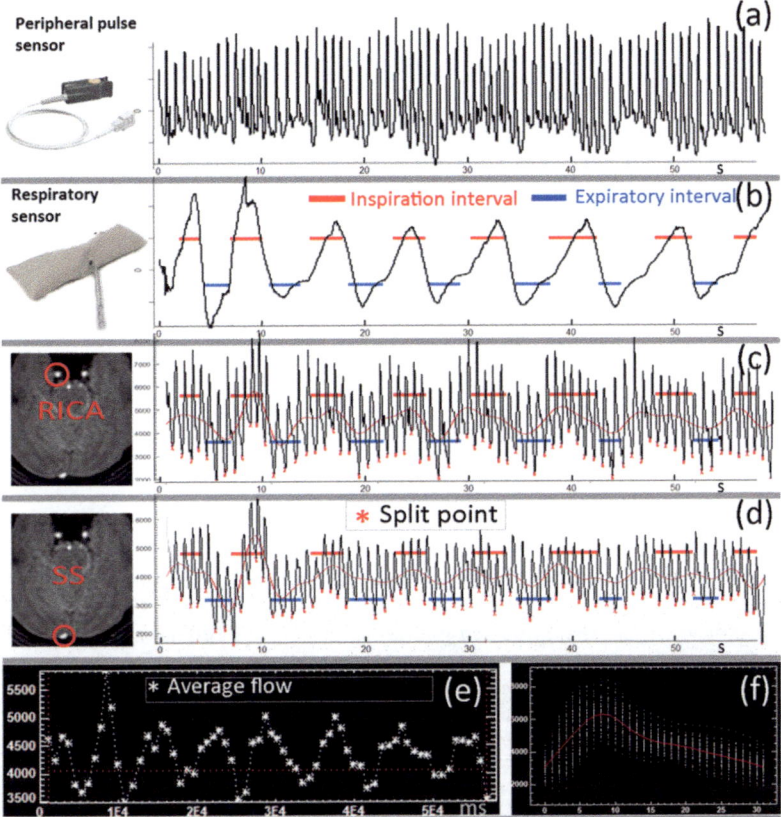

Fig. 2. Curve of the cardiac raw data (a) captured by the peripheral pulse sensor. Curve of respiration can be calculated by a band pass filtering of the respiratory signal raw data (b). As we can see in this participant, the respiratory curve is quasi periodic. Curve of the respiration (red) was superimposed on the curves of the real-time EPI data (black line) extracted in right internal carotid (RICA) (c) and the sagittal sinus (SS) (d). The red stars (split points) in (c) and (d) curves mark each cardiac cycle. The software allows to calculate the flows during the cardiac cycles (e). An average flow curve over all cardiac cycles can also be calculated (f).

4   Applications

Fourteen healthy volunteers were imaged on a Philips Achieva 3T scanner (dStream, Philips Healthcare), using a 32-elements head coil, according to an IRB-approved protocol. The two-dimensional (2D) flow data were acquired using a real-time EPI PC sequence. The imaging plane was chosen to be perpendicular to the direction of the blood flow. The acquisition parameters are summarized in the table. Physiological signals were recorded separated using respiratory and peripheral pulse sensors. Flow measurements were assessed in the cerebral sagittal sinus (SS) and internal carotid arteries. Comparison of flow measurements between the left and right internal carotid arteries was performed using a Wicoxon's signed-rank test.

Table 1. Parameters used for the acquisition of the real time echo planar phase contrast (EPI PC) data.

	EPI PC
Repetition time [TR] (ms)	6.4
Echo time [TE] (ms)	3.4
Flip angle	30°
Field of view (mm)	140
Spatial resolution (mm^2)	2×2
Slice thickness (mm)	4
Temporal resolution (ms)	110
VENC (cm/s)	60
SENSE factor	2.5
Number of images	300
Total scan time (s)	34

5   Results

The internal carotid arteries were depicted as white while the superior sagittal sinus was seen as black (Fig. 3, bottom, right). Figure 4 (bottom left) shows representative flow curves measured during the inspiration and expirations periods of a cardiac cycle.

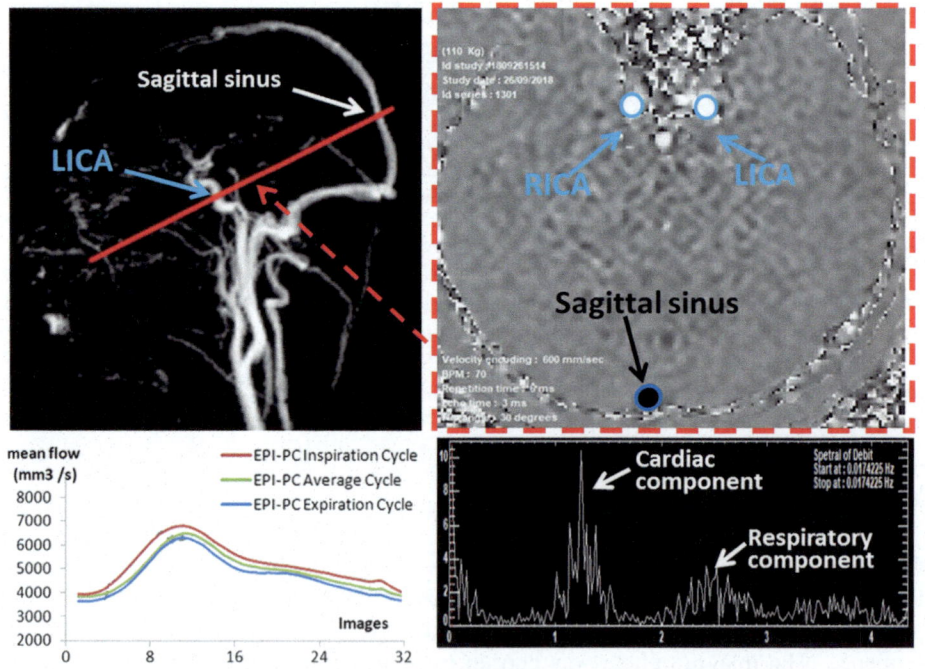

Fig. 3. Top left: Representative maximum intensity projections images (MIP) of a 3D angiogram showing the location of the acquisition plane (red line) for the flow measurements. Top right: 2D real-time PC image. Blue ROIs were superimposed on both the internal carotid arteries and sagittal sinus. Bottom left: example of mean flow curves measured in the left internal carotid artery (LICA) during the respiratory cycles. Bottom right: representative power spectrum of measured EPI PC signal in the left carotid artery. The cardiac and respiratory frequency peaks can be detected in this power spectrum.

Group mean flow in our population was 338 ± 144 ml/mn for the sagittal sinus, 382 ± 133 ml/mn for the right internal carotid artery and 337 ± 128 ml/mn for the left internal carotid artery. There was no significant difference between paired mean flows in the left and right internal carotid arteries (P > 0.37).

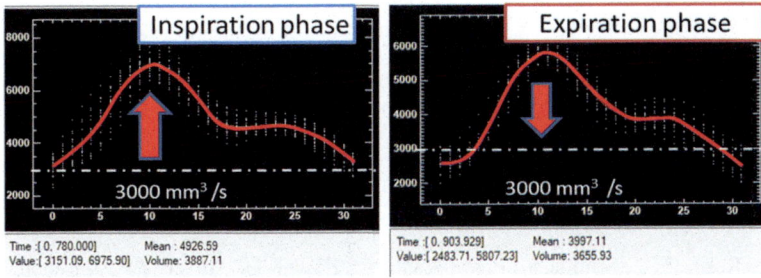

Fig. 4. Representative mean flow curves measured during the inspiration and expiration periods of a cardiac cycle.

6 Discussion

The characterization of the dynamic components of blood flow may provide insight into neurovascular physiology. The real-time EPI sequence permitted us both short acquisition time and high temporal resolution without compromising the spatial resolution of the acquired images. The effects of local variations of the blood flow and physiological noise on flow measurements can be reliably detected with this technique compared to Conv-Cine PC sequence. Our software allowed the visualization and quantification of blood flows from real-time EPI data. Although a comparative comparison of the two techniques was not performed, the blood flows measured with our software through the internal carotid arteries and the sagittal sinus in our sample of healthy participants agree with the literature data [2, 7]. In Fig. 4, we can see that the area under the curve representing the measured volume is larger during the inspiration phase than during the expiration phase. This was consistent with results from others studies that demonstrated that breathing impacts cerebral flows [10, 11]. An important clinical application of real-time velocity mapping is the characterization of stenotic flows, which can reach velocities of 5–6 m/s for severe restrictions [8]. Moreover, it has been suggested that the effects of cardiac cycles variations and the impact of breathing on cerebral flows cannot be evaluated with Conv-Cine PC MRI [9]. With the future improvement of spatial resolution of the acquisition, this real-time EPI technique is a promising tool for gaining an enhanced understanding of the role of hemodynamics in certain cerebrovascular or neurovascular diseases.

The main goal of this paper was to present a new software for processing real-time EPI data. However, to assess the accuracy of the measurements, flows quantification of in vitro data with this software should be compared to those measured with a Conv-Cine PC MRI technique.

7 Conclusion

Accelerated MRI techniques provide another approach to increase the sampling rate of the data to acquire. Advanced postprocessing tools may provide useful quantitative information in the diagnosis and treatment of diseases. Our software allowed possess real-time EPI data and to gain detailed insights into the flow characteristics and dynamic changes induced by respiration.

Acknowledgments. The authors are grateful to the staff members at the Facing Faces Institute (Amiens, France) for technical assistance, and thank the study volunteers for their participation.

References

1. Sainte-Rose, C., LaCombe, J., Pierre-Kahn, A., Renier, D., Hirsch, J.F.: Intracranial venous sinus hypertension: cause or consequence of hydrocephalus in infants? J. Neurosurg. **60**, 727–736 (1984)
2. ElSankari, S., Balédent, O., van Pesch, V., Sindic, C., de Broqueville, Q., Duprez, T.: Concomitant analysis of arterial, venous, and CSF flows using phase-contrast MRI: a quantitative comparison between MS patients and healthy controls. J. Cereb. Blood Flow Metab. Off J. Int. Soc. Cereb. Blood Flow Metab. **33**, 1314–1321 (2013)
3. Bendel, P., Buonocore, E., Bockisch, A., Besozzi, M.C.: Blood flow in the carotid arteries: quantification by using phase-sensitive MR imaging. AJR Am. J. Roentgenol. **152**, 1307–1310 (1989)
4. Pelc, L.R., Pelc, N.J., Rayhill, S.C., Castro, L.J., Glover, G.H., Herfkens, R.J., Miller, D.C., Jeffrey, R.B.: Arterial and venous blood flow: noninvasive quantitation with MR imaging. Radiology **185**, 809–812 (1992)
5. Gatehouse, P.D., Firmin, D.N., Collins, S., Longmore, D.B.: Real time blood flow imaging by spiral scan phase velocity mapping. Magn. Reson. Med. **31**, 504–512 (1994)
6. Debatin, J.F., Leung, D.A., Wildermuth, S., Botnar, R., Felblinger, J., McKinnon, G.C.: Flow quantitation with echo-planar phase-contrast velocity mapping: in vitro and in vivo evaluation. J. Magn. Reson. Imaging JMRI **5**, 656–662 (1995)
7. Balédent, O., Henry-Feugeas, M.C., Idy-Peretti, I.: Cerebrospinal fluid dynamics and relation with blood flow: a magnetic resonance study with semiautomated cerebrospinal fluid segmentation. Invest. Radiol. **36**, 368–377 (2001)
8. Thompson, R.B., McVeigh, E.R.: High temporal resolution phase contrast MRI with multiecho acquisitions. Magn. Reson. Med. **47**, 499–512 (2002)
9. Alperin, N., Lee, S.H.: PUBS: pulsatility-based segmentation of lumens conducting non-steady flow. Magn. Reson. Med. **49**, 934–944 (2003)
10. Schroth, G., Klose, U.: Cerebrospinal fluid flow. II. Physiology of respiration-related pulsations. Neuroradiology **35**, 10–15 (1992)
11. Chen, L., Beckett, A., Verma, A., Feinberg, D.A.: Dynamics of respiratory and cardiac csf motion revealed with real-time simultaneous multi-slice EPI velocity phase contrast imaging. NeuroImage **122**, 281–287 (2015)

Lower Limb Joint Angle Coordination Assessment at Sagittal Plane on Human Vertical Countermovement

C. Rodrigues[1(✉)], M. V. Correia[2], J. M. C. S. Abrantes[3], J. Nadal[4],
and M. A. B. Rodrigues[5]

[1] PRODEB – Doctoral Program of Biomedical Engineering,
FEUP, Porto, Portugal
c.rodrigues@fe.up.pt
[2] DEEC – Department of Electrical and Computer Engineering,
FEUP, Porto, Portugal
[3] MovLab – Interactions and Interfaces Lab, ULHT, Lisbon, Portugal
[4] PEB – Biomedical Engineering Program, COPPE/UFRJ, Rio de Janeiro, Brazil
[5] DES – Department of Electronic and Systems, UFPE, Recife, Brazil

Abstract. This study presents innovative analysis at the time, frequency and phase domain for lower limb joint angular coordination assessment at sagittal plane on human vertical countermovement (CM), comparing long CM on countermovement jump (CMJ) and short CM on drop jump (DJ) from 40 cm step with squat jump (SJ) in the absence of CM. Lower limb CM and muscle stretch-shortening cycle has been pointed as playing a key role on human gait efficiency as well as on run and jump performance with an open issue on objective and quantitative measures for lower limb joint angle coordination assessment at different CM in comparison with no CM condition. Case study is considered from subject specific with 20 years old, 84 kg of body mass and 1.84 m height, selected according to best performance criteria of maximum vertical jump height during CMJ, DJ and SJ from a small sample of n = 6 sports and physical education degree students with (21.5 ± 1.4) years old, (76.7 ± 9.3) kg mass and (1.79 ± 0.06) m height, with no previous injuries, specific sport abilities or training. Calibrated image system with two digital video cameras JVC GR-VL9800 operating at 100 Hz and direct linear transformation (DLT-11) was used along with Simi Motion System and Dempster adapted model with 14 segments to track 3D coordinates of joint marks and obtain joint angles, angular velocities and accelerations at sagittal plane by inverse kinematics. Entire signal analysis was implemented on complementary time, frequency and phase domains for lower limb joint angle coordination assessment at sagittal plane on human vertical jump for comparison of long, short and without CM condition. Comprehensive signal analysis allowed detection of distinct coordination at CMJ from DJ and SJ as well from untrained tested subjects to those reported on trained subjects namely with lower coordination at untrained subjects associated to inaptitude to potentiate short CM and thus presenting lower DJ performance than trained subjects.

Keywords: Lower limb · Joint angle · Coordination · Countermovement · MVJ

© Springer Nature Switzerland AG 2019
J. M. R. S. Tavares and R. M. Natal Jorge (Eds.): VipIMAGE 2019, LNCVB 34, pp. 29–40, 2019.
https://doi.org/10.1007/978-3-030-32040-9_4

1 Introduction

Muscle action constitutes the active component responsible for human movement with its study a key issue for understanding human motion. Nevertheless, multiple muscle action at each joint, multi-joint muscle action, muscle redundancy, diversity of joint configurations and movements have led to detection of unpredicted muscle action in relation to those postulated by functional anatomy, Knudson [1]. Berstein [2] postulated at the principle of equal simplicity that the central nervous system must exert control at the level of the joints or at the synergy level since it would be incredibly complex to control each and every muscle. Quoting the discussion of Helmholtz action-perception theory Gielen [3] reports several studies on stereotyped muscle activation patterns along with the ability to adapt to internal or external changes, claiming that when we flex a joint we do not consciously think about which muscles to activate, rather activation of the appropriate muscles follows directly from the intention to flex the joint. Winter [4] points out that the information on the load and characteristics of joint movements at the action to perform is essential to understand muscle action in view of the large number of factors determining the movement. Despite isometric muscle action has gathered higher research effort, muscles must undergo length changes while producing tension to develop mechanical work with isolated forms, concentric and eccentric, traditionally used to assess the basic function of the neuro-muscular system, mainly due to higher control experiments. Nevertheless, natural form of muscle action frequently involves the use of muscle stretch-shortening cycle (SSC) where the muscle is previously stretched and immediately followed by a contraction, namely at lower limb muscles increasing gait efficiency and powerful running and jumping [5]. Identification of lower limb SSC presents large challenges in relation to lower limb joint kinematics [6]. On the control of human movement Winter [4] points four levels of integration, starting at the sum of excitatory and inhibitory synaptic junctions to control individual motor units, followed by the summation of all twitches from all active motor units at the level of the tendon, the integration at each joint center of all muscle forces and passive anatomic structures crossing that joint and finally the intersegment integration of the force moments at multiple joints collaborating toward a common goal. On lower limb assessment of muscle SSC, standard maximum vertical jumps taking advantage of body weight for long countermovement jump (CMJ) and short countermovement (CM) at drop jump (DJ) have been proposed by Asmussen et al. [7] for comparison with squat jump (SJ) without CM and these protocols increasingly applied [8–10] with an open issue on the contribution of joint angle coordination to lower limb muscle SSC and MVJ performance at long and short CM in comparison with no CM condition. Several studies on MVJ coordination [9, 11] compared general variables without assessing simultaneous and sequential lower limb joint angular kinematics despite its association with higher force or velocity events, as well as entire signal analysis on complementary time, frequency and phase contribution on assessment of different CM in comparison with no CM condition.

Movements requiring higher level of force development have been associated to simultaneous segment recruitment, while lower force and higher velocity movements have been predominantly associated to sequential movement coordination as reported by Knudson [1]. Mullineaux [12] proposed on gait analysis cross-correlation of lower limb joint angular kinematics whose application to lower limb joint angular coordination during MVJ with long and short CM can provide insight on sequential or simultaneous action for comparison with no CM condition. Typical approach at kinematic analysis consists on identification of local extreme values and corresponding time instants with this method adequate at clear differences on selected parameters without considering entire signal pattern. Nevertheless, similar discrete parameters can correspond to undetected different curve signals and physiological conditions. Stergiou [13] proposed frequency domain signal processing on human movement for discrimination between patterns with similar time domain but different frequency domains. Analysis at the frequency domain of the angular displacements, velocities and accelerations of the hip, knee and ankle at the sagittal plane for CMJ, DJ and SJ can contribute to detection at frequency domain of hidden differences at time domain of lower limb angular coordination at different joints and long/short CM in comparison with no CM condition. Stergiou [13] associated phase plane of the lower limb joint angular movement to the organization of the neuromuscular system (NMS) during human gait, with the need for measures to quantify anatomic segments coordination during gait cycle, presenting central tendency and dispersion measures for circular data with mean resultant length assessing directional concentration and Rayleigh test of uniformity with null hypothesis of random phase distribution with this tools potential unveiling NMS control on lower limb joint control at long/short CM on MVJ in contrast with no CM.

The purpose of this study is thus to assess lower limb joint angle coordination as the functional link between muscle SSC and joints used to achieve highest performance at vertical flight height on MVJ with long (CMJ) and short (DJ) CM in comparison with SJ without CM. Specifically the research hypothesis are to search for objective and quantitative measures at the time, frequency and phase plane able to distinguish lower limb joint angle coordination at MVJ long and short CM from no CM condition with the arising question if those coordination differences, in case they're detected at MVJ long, short and no CM condition can be associated to untrained subjects' performance namely in comparison with elite trained subjects.

2 Materials and Methods

According to the interest of maximum performance assessment and difficulty of sample averages in representing maximum performance, attention was focused on individual study of one subject with 20 years old, 84 kg body mass and 1.84 m height. Subject was selected from a small sample of $n = 6$ sports and physical education degree students with (21.5 ± 1.4) years old, (76.7 ± 9.3) kg mass and (1.79 ± 0.06) m height, with no previous injuries, specific sport abilities or training. Selection criteria corresponds to the maximum vertical flight height assessed by higher flight time with zero ground reaction force time period on AMTI force platform model BP2416-4000 CE operating at 1000 Hz during MVJ, with long CM (CMJ), short CM (DJ) and without CM (SJ). Each subject performed a total of 9 trials on CMJ, DJ and SJ with best trial performance obtained by the same subject S1 selected for lower limb joint angle coordination assessment at sagittal plane on human vertical CM searching with selected tools for subject specific coordination strategy that could explain achieved performance, namely in relation with long and short CM in comparison with no CM application. Calibration was performed on image system with two digital video cameras JVC GR-VL9800 operating at 100 Hz on a distant parallel plane to sagittal plane at anterior and posterior position to subject medio-lateral plane using a parallelepipedal structure with known dimensions enclosing entire work volume and direct linear transformation (DLT-11). Subjects signed informed consent according to the World Medical Association Declaration of Helsinki and reflective marks were attached to each subject skin surface of palped main anatomic points at upper and lower limb joints namely right shoulder, hip, knee, ankle bone, skank and foot. Each subject performed a total of 3 trial of each MVJ without CM (SJ), with long CM (CMJ) and short CM (DJ) from 40 cm step according to the protocol defined by Asmussen et al. [7]. During each MVJ trial, ground reaction forces and force moments were acquired at 1000 Hz using AMTI BP2416-4000 CE model along with Mini Amp MAS-6 amplifiers and synchronized kinematic data of 3D marks positions captured by a pair of cameras JVC GR-VL9800 operating at 100 Hz with Simi Motion 6.1 (Simi Reality Motion Systems GmbH, Germany), Fig. 1. Contralateral lower limb performed at each MVJ approximately the same movement for which reason joint marks at right limb were considered in the analysis. 3D coordinates of joint marks were tracked, and joint angles, angular velocities and accelerations at sagittal plane obtained by inverse kinematics using Simi Motion 3D Version 7.5.280 (Simi Reality Motion Systems GmbH, Germany).

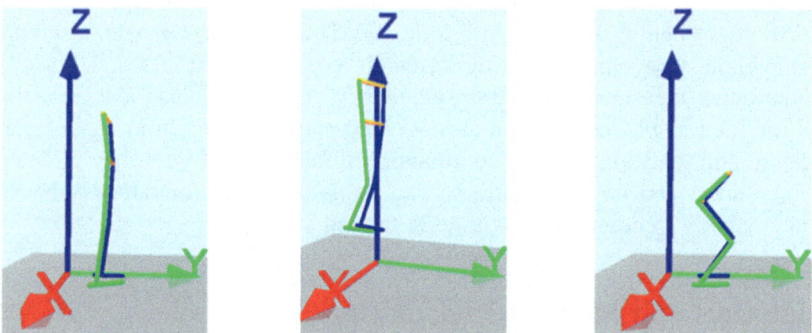

Fig. 1. Lower limb stick-figure at the start of impulse phase on CMJ, DJ and SJ with medio-lateral (X), antero-posterior (Y) and vertical axis (Z).

Hip (H), knee (K) and ankle (A) sagittal joint angles (θ), angular velocities (ω) and accelerations (α) were statistically tested on average amplitude values and variances for comparison during impulse phases at CMJ, DJ and SJ. On coordination assessment sagittal angular displacements θ_H, θ_K, θ_A, velocities ω_H, ω_K, ω_A and accelerations α_H, α_K, α_A at the hip, knee and ankle were linearly and cross-correlated at CMJ, DJ and SJ determining synchronized correlation, maximum cross-correlation and time delay of maximum correlation. Analysis at the frequency domain of angular displacements, velocities and accelerations of the hip, knee and ankle at the sagittal plane were decomposed using Fast Fourier Transform (FFT) and frequency f_{max} of maximum FFT amplitude coefficient and 90th percentile FFT convergence f_{90} compared for CMJ, DJ, SJ and H, K, A joints. Phase angles were obtained from H, K, A angles and angular velocities with directional concentration assessed by mean resultant length and Rayleigh test of uniformity applied with null hypothesis ($\rho = 0$) for each pair of MVJ and H, K, A angles at normalized phase plane to unit circle.

3   Results

CMJ presented higher variance of the angular displacement at the hip than the knee, both higher than the ankle (p < 0.05) and higher mean amplitude at the knee than the hip both higher than the ankle (p < 0.05), Fig. 2. DJ presented lower variance of the angular displacement at the hip than the knee and the ankle (p < 0.05) with similar variances at the knee and the ankle (p > 0.05) as well as similar mean amplitude at the hip and the knee (p > 0.05) both higher than the ankle. SJ presented higher variance of the angular displacement at the hip than the knee, both higher than the ankle (p < 0.05) and higher mean at the knee than the hip, both higher than the ankle (p < 0.05). When comparing each joint angular displacement at different MVJ, the hip presented higher variance at CMJ than SJ both higher than DJ (p < 0.05) with higher mean value at DJ than CMJ both higher than SJ (p < 0.05). The knee presented higher variance of the angular displacement at CMJ than DJ and SJ (p < 0.05) with similar variances at DJ and SJ (p > 0.05) as well as higher mean value at DJ than CMJ, both higher than SJ (p < 0.05). The ankle presented similar variances of the angular displacement at CMJ, DJ and SJ (p > 0.05) with higher mean value at DJ than CMJ and SJ (p < 0.05) and similar values at CMJ and SJ (p > 0.05).

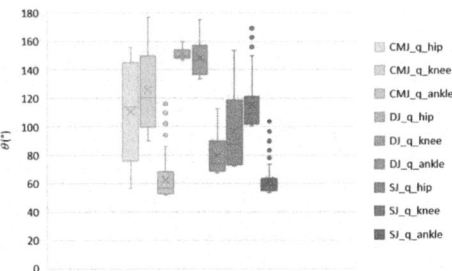

Fig. 2. Hip, knee and ankle joint angular displacement at CMJ, DJ and SJ.

CMJ presented similar variance of the angular velocity at the hip and the knee (p > 0.05) both higher than the ankle (p < 0.05), with lower mean value at the hip than the knee and the ankle (p > 0.05) as well as lower mean value at the knee than the ankle (p < 0.05), Fig. 3. DJ presented similar variability of joint angular velocity at the knee and the ankle (p > 0.05) both with higher value than the hip (p < 0.05) and lower mean joint angular velocity at the hip than the knee, both lower than the ankle (p > 0.05). SJ presented similar variance of joint angular velocity at the hip, knee and the ankle (p > 0.05) with the hip presenting higher mean value than the knee, both higher than the ankle (p > 0.05). When compared each joint angular velocity at different MVJ the hip presented similar variances at CMJ, DJ and SJ (p > 0.05) with lower mean value at CMJ than DJ (p > 0.05), both lower than SJ (p < 0.05). The knee presented similar variances of joint angular velocity at DJ when compared to CMJ and SJ (p > 0.05) with different variances at CMJ and SJ (p < 0.05) as well as lower mean value at CMJ than DJ and DJ than SJ (p > 0.05) with lower value at CMJ than SJ (p < 0.05). At the ankle DJ presented different variance of the angular velocity than CMJ and SJ (p < 0.05) with similar variances at CMJ and SJ (p > 0.05) as well as lower mean value at CMJ than DJ, both lower than SJ (p > 0.05).

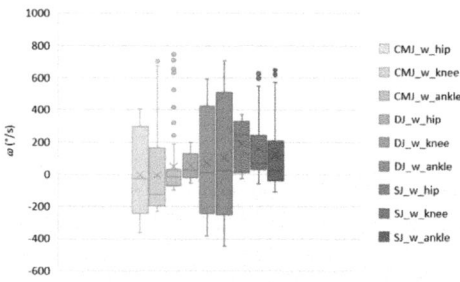

Fig. 3. Hip, knee and ankle joint angular velocity at CMJ, DJ and SJ.

CMJ presented at joint angular acceleration similar variance at the hip, knee and the ankle (p > 0.05) with lower mean value at the hip than the knee, both with lower value than the ankle (p > 0.05), Fig. 4. DJ presented at joint angular acceleration lower variance at the hip than the knee and the ankle (p < 0.05) with higher variance at the knee than the ankle (p > 0.05), as well as lower mean value at the hip than the knee and at the knee than the ankle (p > 0.05), with lower mean at the hip than the ankle (p < 0.05). SJ presented at joint angular acceleration similar variances at the hip than the knee and at the knee than the ankle (p > 0.05) with different variance at the hip than the ankle (p < 0.05) as well as lower mean value at the hip than the knee, both lower than the ankle (p > 0.05). When compared angular acceleration on each joint at different MVJ the hip presented similar variance at CMJ, DJ and SJ (p > 0.05) with lower mean value at CMJ than DJ, both lower than SJ (p > 0.05). The knee presented at joint angular acceleration different variance at DJ than CMJ and SJ (p < 0.05) with similar variance at CMJ and SJ (p > 0.05) as well as lower mean value at CMJ than SJ both lower than DJ (p > 0.05). At the ankle joint acceleration CMJ, DJ and SJ presented different variance with lower mean value at CMJ than DJ (p < 0.05) and higher mean value at DJ than SJ and at SJ than CMJ (p > 0.05).

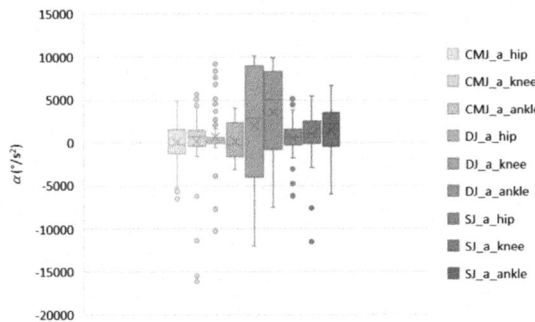

Fig. 4. Hip, knee and ankle joint angular acceleration at CMJ, DJ and SJ.

Linear correlation of joint angular displacements $r(\theta)$, angular velocities $r(\omega)$ and angular accelerations $r(\alpha)$, Table 1, present at CMJ and SJ higher $r(\theta)$ values for the hip–knee (H-K) joints and knee-ankle (K-A) joints in relation to the hip-ankle (H-A) joints with DJ presenting higher $r(\theta)$ value at K-A than H-A and H-K joints. H-K presented higher $r(\theta)$ at CMJ and SJ than DJ whereas H-A and K-A presented higher value at DJ than CMJ and SJ. at As regards to $r(\omega)$, CMJ, DJ and SJ presented higher values at H-K and K-A joints than H-A, with higher H-K, H-A and K-A $r(\omega)$ values at DJ than CMJ and SJ. Similarly $r(\alpha)$ presents at CMJ, DJ and SJ higher values on H-K and K-A joints than H-A which presents particular lower value at CMJ and negative correlation at SJ. H-K, H-A and K-A presented also higher $r(\alpha)$ at DJ than CMJ and SJ.

Table 1. Hip (H), knee (K) and ankle (A) linear correlations of joint angular displacement (θ), angular velocity (ω) and acceleration (α) at CMJ, DJ and SJ.

	$r(\theta)$			$r(\omega)$			$r(\alpha)$		
	CMJ	DJ	SJ	CMJ	DJ	SJ	CMJ	DJ	SJ
H-K	0.962	0.800	0.947	0.834	0.886	0.686	0.583	0.828	0.486
H-A	0.758	0.827	0.776	0.484	0.694	0.430	0.025	0.353	−0.100
K-A	0.857	0.954	0.932	0.813	0.930	0.928	0.650	0.722	0.629

Cross-correlations (*CCr*) of joint angular displacements (θ), angular velocities (ω) and angular accelerations (α), Table 2, presented at CMJ, DJ and SJ higher values between hip-knee joints (H-K) and knee-ankle (K-A) than at hip-ankle (H-A) joints, as well as lower time delay of maximum cross-correlation on H-K and K-A than at H-A joints, Table 3. Maximum *CCr*, was obtained for θ joint angular displacements at null time delay showing that the maximum cross-correlations of θ is detected at the natural timing of the signals with joint angular displacements synchronized for maximum correlation. Time delay of joint angular velocities $\tau(\omega)$ and accelerations $\tau(\alpha)$, Table 3, are predominantly negative, pointing for distal joint lag to proximal joint with the propagation of ω and α impulse at the proximal to distal joint direction.

Table 2. H, K, A cross-correlations for angular displacement (θ), velocity (ω) and acceleration (α) at CMJ, DJ and SJ.

	CCr(θ)			CCr(ω)			CCr(α)		
	CMJ	DJ	SJ	CMJ	DJ	SJ	CMJ	DJ	SJ
H-K	0.994	0.998	0.992	0.857	0.845	0.833	0.606	0.814	0.586
H-A	0.980	0.991	0.986	0.605	0.770	0.644	0.349	0.508	0.537
K-A	0.994	0.996	0.998	0.880	0.932	0.935	0.748	0.734	0.675

Table 3. Time delay $\tau(\theta)$, $\tau(\omega)$, $\tau(\alpha)$ of the maximum H, K, A cross-correlations.

	$\tau(\theta)$ [s]			$\tau(\omega)$ [s]			$\tau(\alpha)$ [s]		
	CMJ	DJ	SJ	CMJ	DJ	SJ	CMJ	DJ	SJ
H-K	0,000	0,000	0,000	−0,020	−0,010	−0,010	−0,010	0,000	−0,180
H-A	0,000	0,000	0,000	−0,060	−0,020	−0,020	−0,200	−0,030	−0,190
K-A	0,000	0,000	0,000	−0,020	0,000	0,000	−0,010	0,000	0,000

Cross-correlations of θ, ω and α among the joints H, K, A presented at CMJ, DJ, SJ significant correlation ($r = 0.886$) with their corresponding simple correlation, Fig. 5, which can be associated to low time delays and maximum Ccr detected at the natural timing of the signals with joint angular θ, ω and α near synchronized for maximum correlation.

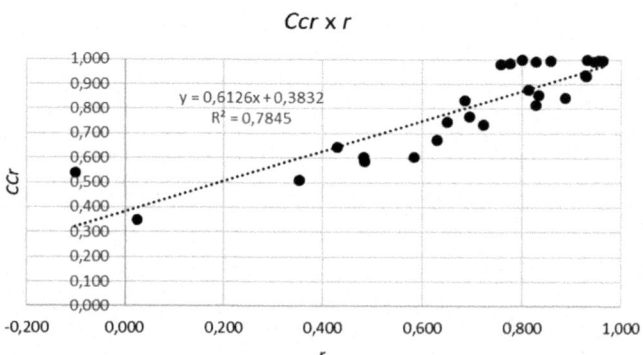

Fig. 5. Correlation CCr x r of θ, ω, α among H, K, A during CMJ, DJ, SJ.

CMJ presented for the hip, knee and the ankle joints lower frequency of the maximum amplitude coefficient for the θ, ω Fast Fourier Transform (FFT) than at SJ, both lower than at DJ, Table 4, with θ presenting higher value at the ankle (A) on SJ. CMJ and SJ presented at the joint angular acceleration α higher frequency of the maximum FFT amplitude coefficient for the knee (K) than the hip (H) and the ankle (A), with higher values at DJ for H, K, A. 90% convergence of the FFT amplitude f_{90}

from θ decomposition, Table 5, occurred at lower frequencies on CMJ and DJ than at SJ for the hip (H), knee (K) and the ankle (A), with H presenting lower value than K, both lower than A at CMJ and DJ, and lower value at K than H both lower than A at SJ. Joint angular velocity presented at CMJ, DJ and SJ lower f_{90} FFT convergence at the hip (H) than the knee (K), both lower than the ankle (A). Joint angular acceleration α presented on CMJ lower frequencies of 90% FFT convergence at the hip (H) than the knee (K), both lower than the ankle (A), with DJ and SJ presenting lower values at the ankle (A) than hip (H) both lower than the knee (K), Table 5.

Table 4. Frequency of the harmonic with maximum amplitude from θ, ω, α FFT decomposition at the hip (H), knee (K) and ankle (A) joints.

f(Hz)	θ_H	θ_K	θ_A	ω_H	ω_K	ω_A	α_H	α_K	α_A
CMJ	1,23	1,23	1,23	1,23	1,23	1,23	1,23	4,94	2,47
DJ	4,76	4,76	4,76	4,76	4,76	4,76	4,76	4,76	4,76
SJ	2,33	2,33	4,65	2,33	2,33	2,33	2,33	4,65	2,33

Table 5. Frequency of the harmonic from the 90th percentile of accumulated relative amplitude of θ, ω, α FFT decomposition at the hip (H), knee (K) and ankle (A).

f(Hz)	θ_H	θ_K	θ_A	ω_H	ω_K	ω_A	α_H	α_K	α_A
CMJ	1,23	2,47	22,22	6,17	11,11	29,63	22,22	24,69	33,33
DJ	0,00	4,76	9,52	9,52	28,57	33,33	19,05	33,33	9,52
SJ	25,58	23,26	29,71	25,58	30,23	32,56	37,21	39,53	30,23

CMJ presented at the hip (H) lower mean resultant length $\bar{r}$ and higher p significance of Rayleigh test of uniformity than at the knee (K) and the ankle (A), Table 6 and Fig. 6, pointing for H lower directional concentration at (θ, ω) than at K and A accepting null hypothesis H_0 of zero concentration ($\rho = 0$) at H and rejecting it at $p > 0.05$ for K and A. DJ presented at H and K lower $\bar{r}$ and higher p than A accepting H_0 of zero concentration ($\rho = 0$) at H, K and rejecting it at A. SJ presented higher $\bar{r}$ and lower p with higher directional concentration at (θ, ω) and rejection of H_0: $\rho = 0$.

Table 6. Mean resultant length $\bar{r}$ at unit circle normalized phase plane (θ, ω) and p significance of Rayleigh test with null hypothesis of zero concentration ($\rho = 0$)

(θ, ω)	$\bar{r}$			p		
	CMJ	DJ	SJ	CMJ	DJ	SJ
H	0,1309	0,3737	0,4257	0.2503	0.0515	$<10^{-3}$
K	0,5096	0,3044	0,6635	$<10^{-3}$	0.1430	$<10^{-3}$
A	0,8096	0,4814	0,7308	$<10^{-3}$	0.0064	$<10^{-3}$

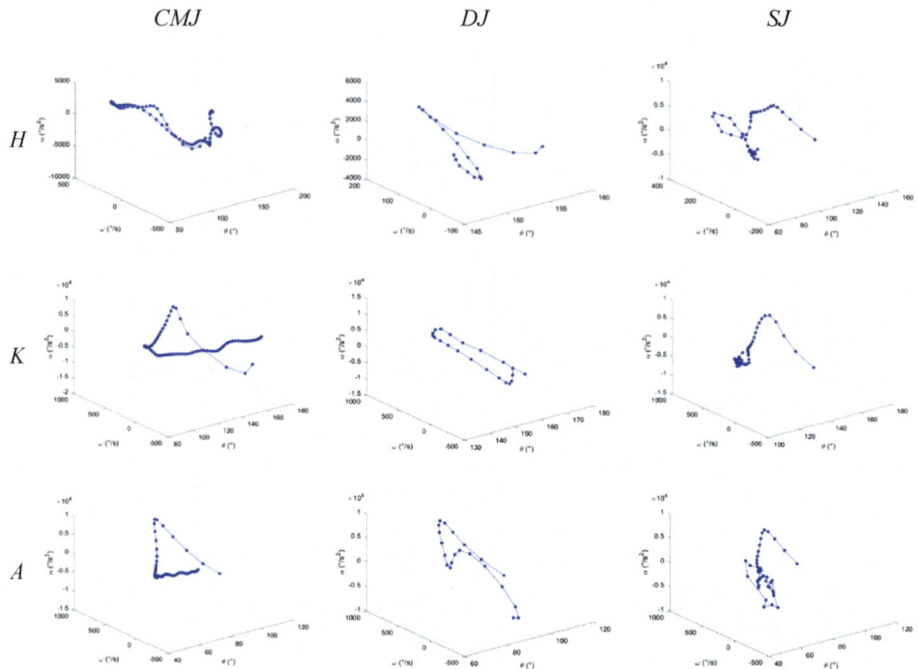

Fig. 6. Lower limb joint angular phase space (θ, ω, α) in (°, °/s, °/s²) of the hip (H), knee (K) and ankle (A) during impulse phase at CMJ, DJ and SJ.

4 Discussion

Entire signal analysis was implemented on complementary time, frequency and phase domains for lower limb joint angle coordination assessment at sagittal plane on human vertical countermovement. Computational vision was applied to track 3D movement of human external anatomical points using reflexive markers and DLT of 2D image from redundant camera system, capturing and quantifying complexity of natural human movement at lower limb vertical countermovement (CM). Inverse kinematics was performed to obtain from 3D position, lower limb joint angular displacement, angular velocity and angular acceleration of the hip, knee and ankle at sagittal plane. Entire series of the sagittal plane angular displacements, angular velocities and angular accelerations at the hip, knee and the ankle were analyzed at signal time, frequency and phase plane domains for coordination assessment of lower limb joint on standard maximum vertical jump with long CM (CMJ), short CM (SJ) and without CM (SJ).

Despite maximum vertical jump objective is to maximize vertical flight height determined by vertical velocity at takeoff, applied technique for optimization of necessary force impulse at MVJ namely with long CM (CMJ), short CM (DJ) and without CM (SJ) relies on optimization of force development and achieved velocity, jointly considered at developed mechanical power with simultaneous and sequential coordination features that must be addressed as coexisting in association with movement technique such as CM, external resistance, time for force development and subject specific motor development and condition.

Performed analysis using entire signal profile of lower limb joint angular kinematics allowed complementary approach at time, frequency and phase domain coordination assessment during impulse phase on MVJ with long CM (CMJ), short CM (DJ) and without CM (SJ) avoiding subjective discrete parameter selection and contributing to understand sequence and timing of lower limb joint actions at long and short CM for comparison without CM and associated SSC. Despite reported tests and measures were applied to subject specific best performant of MVJ small sample with long, short and without CM and the potential interest on study expand to more subjects with different motor control development and condition, these tests and measures can be complementarily applied to other studies related not only with performance but also with human gait impairment and rehabilitation.

Assessment of lower limb coordination at joint angular kinematics plays an important role since comparisons performed at angular domain do not depend on linear dimensions of anthropometric segments thus allowing direct comparison of segments coordination at different actions and subjects. On the other hand, insight at segments coordination can contribute to understand energy transference between segments and its contribution for movement synergies towards efficiency and performance. Despite focus on specific muscle action can be determinant as it happens with muscle stretch-shortening cycle on lower limb countermovement when we perform an action, we do not consciously think about muscles contraction or inhibition but about movement, with muscle action deriving from motion intention, and segment coordination synergistically contributing to action performing.

Several studies have reported highly trained performers presenting on several sports higher performance of maximum vertical jump on long CM (CMJ) than short CM (DJ), both higher than MVJ without CM (SJ), with lower limb joint coordination playing a key role for higher performance at CMJ than DJ, both higher than SJ and the natural tendency for use of CMJ preferring it to DJ and SJ in order to achieve higher performance. Lower skill or training level on the other hand is associated to lower coordination and lower performance at DJ than SJ and CMJ of reported tests, which is the case of the study sample. Best performer S1 of selected sample achieved MVJ height of 42.07 cm at CMJ, higher than SJ with 40.93 cm both higher than DJ with 30.63 cm compatible with S1 lower skill training level and reduced ability at lower limb joint coordination to exploit DJ with short CM as opposed to long CM at CMJ and SJ without CM.

5   Conclusion

In the context of conducted study on best performant subject specific lower limb joint angle coordination at sagittal plane on human vertical countermovement, lower level of impulse force associated to higher time impulse phase and muscle stretch-shortening cycle (SSC) pointed to dominant sequential of lower limb joint angular coordination at CMJ, whereas higher resistance level at ground reaction impact and impulse forces on DJ pointed to simultaneous lower limb joint angular coordination at DJ, with explosive power at intermediate force level and velocity pointing to a balance between sequential and simultaneous coordination at lower limb joint angular SJ without CM.

Acknowledgments. To EACEA, UPorto, UFRJ, UFPE and SAPIENZA Università di Roma for mobility support. To LMH – ISMAI and CRPG for trial tests and tools for data analysis.

References

1. Knudson, D.: Fundamentals of Biomechanics. Springer, New York (2007)
2. Bernstein, N.A.: The Coordination and Regulation of Movements. Pergaman Press, Oxford (1967)
3. Gielen, S.: Helmholtz: founder of the action-perception theory. In: Latash, M.L., Zatsiorsky, V.M. (eds.) Classics in Movement Science, Human Kinetics, pp. 221–243 (2001)
4. Winter, D.A.: Biomechanics and Motor Control of Human Movement. Wiley, Hoboken (2005)
5. Komi, P.V., Bosco, C.: Utilization of stored elastic energy in leg extensor muscles by men and women. J. Med. Sci. Sports **10**, 261–265 (1978)
6. Komi, P.V., Ishikawa, M., Linnamo, V.: Identification of stretch-shortening cycles in different sports. J. Physiol. **56**(1–2), 19–41 (2011)
7. Asmussen, E., Bonde-Petersen, F.: Storage of elastic energy in skeletal muscles in man. Acta Physiol. Scand. **91**(3), 385–392 (1974)
8. Bobbert, M.F., Mackay, M., Schinkelshoek, D., Huijing, P.A., van Ingen Schenau, G.J.: Biomechanical analysis of drop and countermovement jumps. Eur. J. Appl. Physiol. **54**, 566–573 (1986)
9. Aragón-Vargas, L.F., Gross, M.M.: Kinesiological factors in vertical jump performance: differences among individuals. J. Appl. Biomech. **13**, 24–44 (1997)
10. Blazevich, A.: The stretch-shortening cycle (SSC). In: Cardinale, M., Newton, R., Nosaka, K. (eds.) Strength and Conditioning – Biological Principles and Practical Applications. Wiley, West Sussex (2011)
11. Bobbert, M.F., Huijing, P.A., Schenau, V.I.: Drop jumping. I. The influence of jumping technique on the biomechanics of jumping. Med. Sci. Sports Exerc. **19**(4), 332–338 (1987)
12. Mullineaux, D.R., Barlett, R.M., Bennett, S.: Research methods and statistics in biomechanics and motor control. J. Sports Sci. **19**, 739–760 (2001)
13. Stergiou, N.: Innovative Analyses of Human Movement. Human Kinetics, Champaign (2004)

Ultrasound Speckle Noise Reduction by Radio-Frequency Data Filtering

Verónica Santo[1] and Fernando C. Monteiro[1,2]([✉])

[1] Polytechnic Institute of Bragança, Campus Santa Apolónia, Bragança, Portugal
veronica.ne.santo@gmail.com, monteiro@ipb.pt
[2] Research Centre in Digitalization and Intelligent Robotics (CeDRI),
Bragança, Portugal

Abstract. Ultrasound is a commonly used imaging modality for the examination of several pathologies due to its non-invasiveness, affordability and easiness of use. However, ultrasound images are degraded by an intrinsic artefact called 'speckle', which is the result of the constructive and destructive coherent summation of the ultrasound echoes. This paper aims to generate B-mode images out of radio-frequency (RF) data following standard procedures, a series of steps such as envelope detection, log-compression and scan conversion. Some low pass filters will be applied to RF data in order to achieve B-mode images with high quality by speckle noise reduction.

1 Introduction

Ultrasound imaging is one of the most important and cheapest instrument used for diagnostic purpose. However, the images obtained through this type of examination presents a characteristic noise type, known as speckle noise, which makes it difficult to analyse and diagnose [1–3]. Speckle noise is defined as multiplicative noise with a granular pattern formed due to coherent processing of backscattered signals from multiple distributed targets.

Over the last decades, several despeckling filters have been developed to reduce the speckle noise inherently present in ultrasound images without losing the diagnostic information. Image filtering techniques include adaptive filters, anisotropic diffusion and wavelets [2]. However, it is important to refer that the majority of the methods are applied to the B-mode images without taking advantage of the process used to obtaining these images from the RF data.

The RF signal is not usually available in the common ultrasound equipment and it is usually pre-processed (filtered and compressed) to improve its visualization. Compression is needed to reduce the dynamic range of the RF signal in order to adapted it to the dynamic range of the monitor [4].

Unfortunately, this pre-processing modifies the distribution of the RF signal. This step depends on a set of parameters such as the brightness, contrast, zoom and dynamic gain.

© Springer Nature Switzerland AG 2019
J. M. R. S. Tavares and R. M. Natal Jorge (Eds.): VipIMAGE 2019, LNCVB 34, pp. 41–48, 2019.
https://doi.org/10.1007/978-3-030-32040-9_5

Recently, new types of filters have been proposed to remove speckle noise from RF data. In [5], the authors used a low pass frequency-shift, followed by a least mean square adaptive filter. Al-Asad [6] proposed a Short Time Fourier transform applied to the envelope of each RF line before reconstruction, followed by its application to the lateral dimension of the 2D image after reconstruction.

In this paper we intend to filter the raw RF data in order to reduce noise and limit the signal to the working bandwidth. This will be done by the application of denoising filters individually to the one-dimensional RF envelopes that will constitute the B-mode image. By filtering RF data in the process of B-mode image construction, we expect to obtain images with less speckle noise.

2   From RF Data to B-Mode Images

The interaction of an acoustic wave with different tissue regions can be modelled by the backscattered radio-frequency (RF) signal. The ultrasound waves propagate through the tissue scattering and reflecting where variations in tissue density and elasticity occurs. Some of this energy returns to the transducer and it is recorded as a RF signal, as shown in Fig. 1.

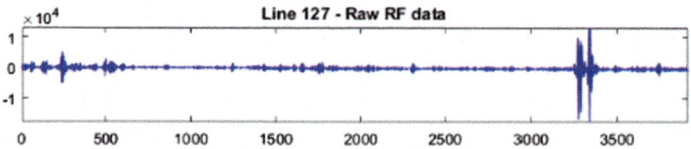

Fig. 1. Raw radio-frequency signal

There are several approaches for which useful information can be extracted from the RF signal to obtain B-mode ultrasound images. An overview of the various statistical distributions for modelling the envelope-detected RF signal can be found in [7].

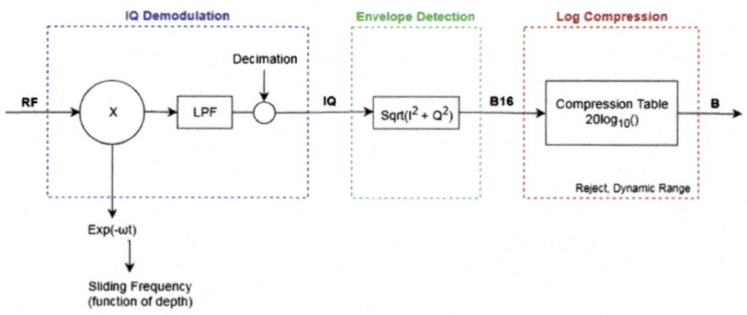

Fig. 2. From RF data to B-mode processing steps.

To obtain an ultrasound B-mode image, the RF data goes through three signal processing phases, as shown in Fig. 2: IQ Demodulation, Envelope Detection, and Log compression.

2.1   IQ Demodulation

In-phase Quadrature (IQ) demodulation the complex signal is low-pass filtered to remove the negative frequency spectrum and noise outside the desired bandwidth. The low-pass filter on the complex signal can be thought of as a filter applied to the real and imaginary part separately. With careful choice of low-pass filter, the remaining signal becomes weak for frequencies outside the pass-band for both components. This method multiplies an in-phase and quadrature-phase sinusoid with the input signal, causing signal content of that frequency to be accentuated and all other content to be reduced. The filter removes the frequencies stemming from the negative spectrum of the real RF signal, and the filter removes approximately half of the energy in the signal.

Decimation allows to reduce the sampling rate of a discrete-time signal. The Nyquist theorem states that the sampling frequency can be reduced to twice the cut-off frequency of the filter without loss of information. Low sampling rate reduces storage and computation requirements.

2.2   Envelope Detection

Since only the amplitude of the signal is of interest and not the frequency information itself, the envelope of the signal is calculated. The envelope is a non negative curve that connects the peaks (negative peaks are inverted) of the RF signal. Absolute and Hilbert transforms are commonly used methods to envelope detection. It returns the absolute or Hilbert amplitude variation of the amplitude of a time wave.

2.3   Logarithmic Compression

In order to obtain a 2D image, a series of adjacent beams (RF signals) are recorded from which a RF matrix is built. This RF matrix is then further processed, ultimately leading to the B-mode image, which is visualized on the screen.

To visualize the envelope signal of a scan line the value of each sample point is mapped to a 8 bit greyscale colour map. The result of the compression is a first B-mode image, each column of pixels corresponding to a scan line. In order to improve the contrast of the image the brightness values of the B-mode image have to be adjusted to the dynamic range of the human eye. Usually, nonlinear mapping functions are used, which keep the order of the values but reduce the difference between high peaks in the signal and lower values and therefore increasing the overall brightness of the image.

The compressed image most likely contains various image artefacts and speckles. In order to improve the image quality scan conversion has to be performed.

The RF data contains the single scan lines as single columns (See Fig. 3). Since these scan lines are usually not parallel they have to be arranged geometrically in the final image depending on the geometry of the probe as fan shaped images, as shown in Fig. 3.

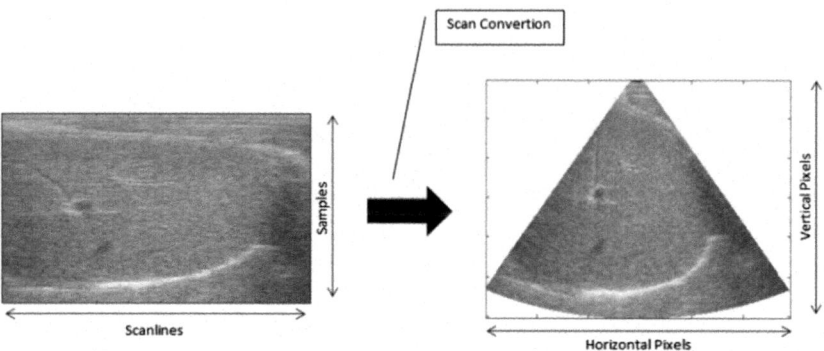

Fig. 3. Logarithmic conversion

3   Radio-Frequency Filtering Methods

Usually, techniques of image enhancement and speckle noise reduction are applied on B-mode ultrasound image [2], after the RF data goes through the phases mentioned previously.

In this work, we propose to apply several low-pass filters (LPF) before the RF signals turn into the B-mode image. In order to filter the RF signal we need to focus on the first ultrasound signal processing technique, the IQ Demodulation. In this phase it is possible to apply several LPF and test different decimation values.

3.1   Low-Pass Filters

Attenuate or reduce the amplitude of the frequencies larger than the cut-off frequency where excessive phase shift may cause oscillations. The amount of attenuation for each frequency varies from filter to filter. The different filters with different frequencies are used to remove noise. We have designed 3 classic analogue filter types to filter the noise in RF sign.

- **Bessel filter:** A Bessel filter is a type of analogue linear filter used within RF that has a maximally flat group or phase delay. Bessel low-pass filters, therefore, provide an optimum square-wave transmission behaviour. This preserves a wave shape of signals within the pass-band. However, the pass-band gain of a Bessel low-pass filter is not as flat as that of the Butterworth low-pass, and the transition from pass-band to stop-band is by far not as sharp as that of a Chebyshev low-pass filter.

- **Butterworth filter:** Butterworth filters have a maximally flat response. That means all the derivatives of the amplitude with frequency are zero at DC. The Butterworth response is a good compromise between attenuation characteristic and group delay.
- **Chebyshev filter:** The Chebyshev filter has ripples in the pass-band and the value of the ripples is a parameter that is selected as part of the filter design.

4   Results and Analysis

Figure 4 shows the results obtained by using the Bessel, Butterworth and Chebyshev filters, with different decimation of 1 and 30, and absolute and Hilbert envelopes.

By comparing the images obtained and the corresponding signal representation, we can observe that the Chebyshev filter is the one with the best results in speckle noise reduction, when used with the Hilbert Envelope. Increasing decimation from 1 to 30 produces an improvement in the results, however, for values greater than 30 an undesirable image distortion or blurring effect happens.

In order to support the visual analysis we need quantitative evaluation data. For that we evaluated speckle noise reduction with some state-of-the-art quality measures: mean square error (MSR), square root of the mean square error (RMSR), the peak-signal-to-noise radio (PSNR), signal to noise ratio (SNR), quality index and structural similarity index (MSSIM). Table 1 shows the results obtained.

Table 1. Speckling reduction according to the performance computed with state-of-the-art evaluation metrics. First six rows use *downsampling* $= 1$ and the last six rows use *downsampling* $= 30$.

Filter	Envelope	MSR	RMSE	PSNR	SNR	Quality	MSSIM
Bessel	Absolute	2145.09	46.32	17.83	11.20	0.03	0.10
Butterwoth	Absolute	954.19	30.89	21.34	15.37	0.10	0.15
Chebyshev	Absolute	1030.41	32.10	21.01	14.94	0.15	0.19
Bessel	Hilbert	950.79	30.83	21.36	15.12	0.05	0.18
Butterwoth	Hilbert	308.62	17.57	26.25	20.71	0.19	0.29
Chebyshev	Hilbert	279.38	16.71	26.68	21.04	0.29	0.37
Bessel	Absolute	1526.20	39.07	19.31	12.57	0.46	0.65
Butterwoth	Absolute	270.12	16.43	26.83	20.77	0.84	0.88
Chebyshev	Absolute	367.89	19.18	25.48	19.32	0.91	0.93
Bessel	Hilbert	643.17	25.36	23.06	16.79	0.52	0.71
Butterwoth	Hilbert	21.08	4.59	37.90	32.35	0.88	0.91
Chebyshev	Hilbert	**17.61**	**4.20**	**38.68**	**33.03**	**0.95**	**0.96**

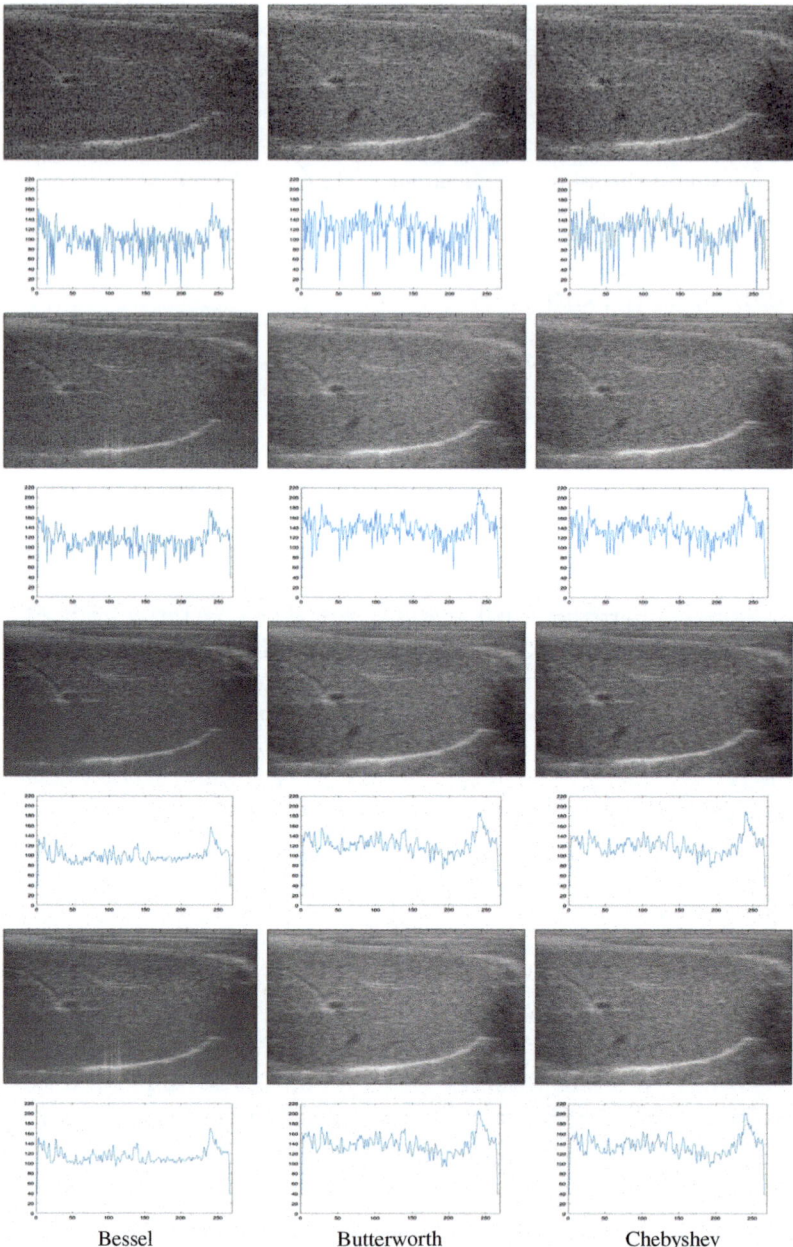

Fig. 4. Influence of Bessel, Butterworth and Chebyshev filters with different decimation and envelope. Below each result we have the profile extracted from the column 137 after denoising. First row shows the results with decimation 1 and Absolute envelope. Second row shows the results with decimation 1 and Hilbert envelope. Third row shows the results with decimation 30 and Absolute envelope. Fourth row shows the results with decimation 30 and Hilbert envelope.

From Table 1 we can see that the combination of Chebyshev filter with decimation of 30 and Hilbert envelope produces the best result in speckle noise reduction in all used quality measures. When compared to Bessel and Butterworth filter, a Chebyshev filter can achieve a sharper transition between the pass band and the stop bandwith a lower order filter. The sharp transition between the pass-band and the stop-band of a Chebyshev filter-produces smaller absolute errors and faster execution speeds than Bessel and Butterworth filters. Chebyshev filter is used where the frequency content of a signal is more important than having a constant amplitude.

Figure 5 shows the B-mode image obtained with pre-processing, without filtering; and B-mode image obtained with pre-processing and noise filtering. It is possible to observe that in the filtered Rf image the structures present in the ultrasound shows more clearly, which allows to analyze the image without losing the diagnostic information.

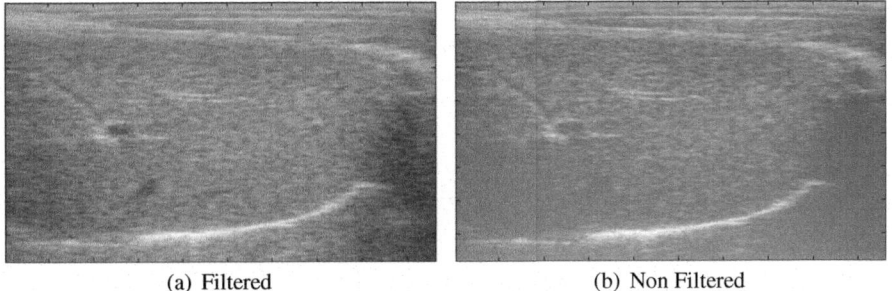

(a) Filtered   (b) Non Filtered

Fig. 5. B-mode images with and without filtering.

5   Conclusion

In this paper we proposed and approach to reduce speckle noise in ultrasound images by filtering the raw RF data, before obtain the B-mode image. This was done by using denoising filters individually to the 1D RF envelopes that will constitute the B-mode image. In the IQ demodulation phase we tested different low pass filters: Bessel, Butterworth and Chebyshev, with different decimation values and different envelope techniques. To test the effectiveness of the speckle noise reduction we evaluate the results with several state-of-the-art quality measures. From the obtained evaluation results we can conclude that filtering in RF mode, before the conversion to B-mode, is an effective method to reduce the speckle noise.

References

1. Ortiz, S.H.C., Chiu, T., Fox, M.D.: Ultrasound image enhancement: a review. Biomed. Signal Process. Control **7**(5), 419–428 (2012)
2. Monteiro, F.C., Rufino, J., Cadavez, V.: Towards a comprehensive evaluation of ultrasound speckle reduction. Lecture Notes in Computer Science, vol. 8814, pp. 141–149 (2014)
3. Joel, T., Sivakumar, R.: Despeckling of ultrasound medical images: a survey. J. Image Graph. **1**(3), 161–165 (2013)
4. Seabra, J., Sanches, J.: Modeling log-compressed ultrasound images for radio frequency signal recovery. In: International Conference of the IEEE Engineering in Medicine and Biology Society, Vancouver, Canada, pp. 426–429 (2008)
5. Wang, S., Li, C., Ding, M., Yuchi, M.: Frequency-shift low-pass filtering and least mean square adaptive filtering for ultrasound imaging. In: Progress in Biomedical Optics and Imaging Proceedings of SPIE, vol. 9790, p. 97900P (2016)
6. Al-Asad, J.F.: Despeckling the 2D medical ultrasound image through individual despeckling of the envelopes of its 1D radio frequency echo lines by STFT. J. Image Graph. **4**, 67–72 (2016)
7. Destrempes, F., Cloutier, G.: A critical review and uniformized representation of statistical distributions modeling the ultrasound echo envelope. Ultrasound Med. Biol. **36**(7), 1037–1051 (2010)

Biometric Identification Based on Forearm Vein Pattern

Ryszard S. Choraś[✉]

Department of Telecommunications, Computer Science and Electrical Engineering,
University of Technology and Life Sciences, S. Kaliskiego 7, 85-796 Bydgoszcz, Poland
choras@utp.edu.pl

Abstract. Forearm vein recognition is one of many available methods used for identification. However, forearm veins can be considered more secure compared to other biometric traits because the veins are inside the human body and therefore not easily manipulated. Veins possess several properties that make a good biometric feature for personal identification: (1) they are difficult to damage and modify; (2) they are difficult to simulate using a fake template; and (3) vein information can represent the liveness of person. In this research, the camera used contained two Charge-Coupled Devices (CCD): one used to capture visible light images and the second used to capture light in the near infrared (NIR) band of the spectrum. Features were extracted from each pair of visible and NIR images. For the visible images, feature extraction was done using the Gabor filter. For the NIR forearm images, a crossing number was used to extract properties of the veins e.g. bifurcation.

1 Introduction

Images play an important role in the identification of people across many applications. Effective biometrics systems have five key modules: A Sensor or sensors (for image/data acquisition), a Feature Extractor, a Biometric Database, a Matcher and a Decision-Maker (Fig. 1). The Sensor reads the biometric information from the user. Feature Extractor module automatically extracts features from the biometric templates. The number of features can be reduced before being used in the classification task.

Matcher module measures the similarity between extracted features from a user template and an enrolled template. More specifically, the Matcher component of the system compares feature vectors obtained from the feature extraction algorithm to produce a similarity score. This score indicates the degree of similarity between a pair of biometrics data under consideration. The Decision-Maker interprets the result. The Biometric Database maintains the templates of the enrolled users (i.e. the candidate identities).

Biometric categories can be divided into two types: (i) Identification systems and (ii) Verification systems.

Let: X_Q is input feature vector extracted from biometric data; I_k, $k \in 1, 2, \ldots, N$ be the identities of enrolled users; I_{N+1} indicate the reject case;

© Springer Nature Switzerland AG 2019
J. M. R. S. Tavares and R. M. Natal Jorge (Eds.): VipIMAGE 2019, LNCVB 34, pp. 49–61, 2019.
https://doi.org/10.1007/978-3-030-32040-9_6

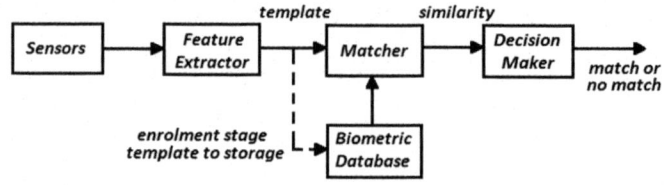

Fig. 1. Biometric system

$S(X_Q, X_I)$ is the function that measure similarity between X_Q and X_I; T_S is a predefined threshold.

(i) Identification systems. These systems give an answer to the question: *Who is it?* In other words, they are performing a one-to-many comparison.

$$X_Q = \begin{cases} I_k & \text{if } \max_k\{S(X_Q, X_I)\} \geq T_S \\ I_{N+1} & otherwise. \end{cases} \tag{1}$$

(ii) Verification systems. These systems give an answer to the question: *Is this person who they say they are?* (one-to-one comparison).

Verification can be formulated mathematically as:

$$(I, X_Q) = \begin{cases} true \text{ (a genuine user) if } S(X_q, X_I) \geq T_S \\ false \text{ (an impostor)} \qquad otherwise. \end{cases} \tag{2}$$

Desired biometric characteristics are presented in Table 1.

Table 1. Biometrics characteristics

Characteristic	Description
Robustness	Measured by the "false non-match rate"
Distinctiveness or Uniqueness	How well the biometric separates one individual from another. Measured by the "false match rate"
Permanence	How well the biometric performs over time (i.e. with an original template)
Acceptability	Describes by polling the device users
Universality	How commonly the biometric is found in humans
Performance	The accuracy of systems using the biometric
Circumvention	How easily a submitted template can be spoofed
Degree of intrusiveness	How much co-operation is required from the user to collect the biometric sample

Vein recognition is a method of biometric identification and verification, that uses pattern recognition techniques based on images of blood vessels. Blood vessel patterns (identified only on a live body) are unique to each individual. Vein recognition does not require contact during registering and authentication and is strongly immune to forgery [8,27].

To detect forearm veins and generate a vasculature map we utilize a camera containing two Charge-Coupled Devices (CCD) with two spectral filters - one is a visible light filter and the other is an IR filter. In the process of acquiring the image, the forearm of each person is set in the same way, which practically eliminates the influence of translation and rotation.

In the proposed method only reflected light is registered. The output of the camera are the captured images: typical visible light images and images captured in the near infrared (NIR) band of the spectrum.

Existing vein recognition methods can be roughly divided into two categories. The first category are methods that determine feature points within the images and match through the spatial relationships among feature points. The other category of methods combine the global features of two vein images.

Example visible and NIR forearm images shown in Fig. 2.

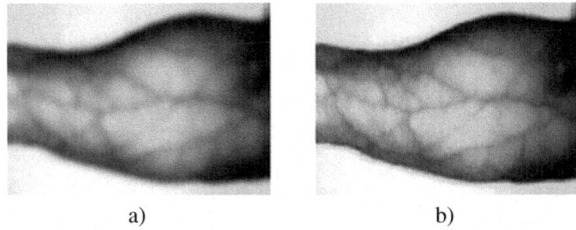

a)   b)

Fig. 2. Example visible and NIR forearm images

Human forearm vein recognition is a new biometric technology. In this paper a recognition method using feature vector derived from forearm vein images is presented.

Among the novel contributions in this work are:

- a method for the fusion of two images - visible and NIR,
- use of geometric and spectral features,
- location and extraction of features can be done easily.

The structure of the paper is as follows. In Sect. 2, information about vein image preprocessing is provided. In Sect. 3, the proposed method for feature extraction is presented. Then in Sect. 4, experimental results are provided, which are followed by conclusions and suggestions for future work in Sect. 5.

1.1 Related Works

Most related published work use vein templates for the biometrics person recognition process. Classic approaches are presented in Table 2.

Table 2. Summary of work on the vein biometric recognition

Biometric template	Methods	Image	References
Hand vein	• FFT based phase correlation	NIR	[24, 27]
	• Distance between feature points	NIR	[5, 32]
	• Matching vein triangulation and shape features	NIR	[13]
Finger vein	• Local Binary Pattern (LBP) method using a support vector machine (SVM)	NIR	[14, 27]
	• Normalized cross-correlation	NIR	[11]
	• Radon transform	NIR	[25]
	• Local derivative pattern	NIR	[15, 30]
	• Multiscale, curvelets	NIR	[10, 38]
Dorsal hand vein	• Curvelet transform	NIR	[16, 38]
	• LPP (Locality Preserving Projections)	NIR	[14, 35]
	• Shearlet transformation and Scale-invariant Feature Transformation (SIFT)	NIR	[17, 34]
Palm vein	• Local Binary Patterns (LBPs) and their variants	NIR	[20, 26]
Retina vein	• Local Binary Patterns	Visible	[19]
	• Spectral	Visible	[16, 17, 19, 21]
Ocular vein	• Texture analysis and Statistical parameters	Visible	[7, 21, 23, 29, 33]
Forearm vein	• Line Edge Map (LEM)	NIR, Visible	[10]
	• 2D-Gabor + feature extraction	NIR, Visible	[3, 4, 36]
	• Skeletonization	NIR, Visible	[3, 4, 9]
	• Shape features	NIR, Visible	[13]

We propose a novel method for personal recognition, extracting features from images in both the visible light and NIR spectrums. Based on these characteristics recognition can be realized.

2   Vein Image Processing

A gray scale digital image of size $M \times N$, $M = 2^i$, $N = 2^j$, can be mathematically defined as a matrix with entries $f(x,y)$, $x = 0, 1, \ldots, M - 1$, $y = 0, 1, \ldots, N - 1$, where the value $f(x,y)$ represents the intensity or gray level of the image at the pixel (x, y) $0 \leq f(x,n) \leq G - 1$; $G = 2^k$.

Figure 3 shows an overview of our proposed system. In the processing module, processes such as image stretching, image binarization, noise elimination and thinning are applied to extract a normalized and useful forearm image.

Each step shown in the processing module Fig. 3 will be explained in more detail below [6].

1. Image enhancement. Forearm vein image usually have poor contrast and noise. Contrast enhancement techniques can be used to improve the image [2, 28]. We used CLAHE (Contrast Limited Adaptive Histogram Equalization) [39] for this purpose.

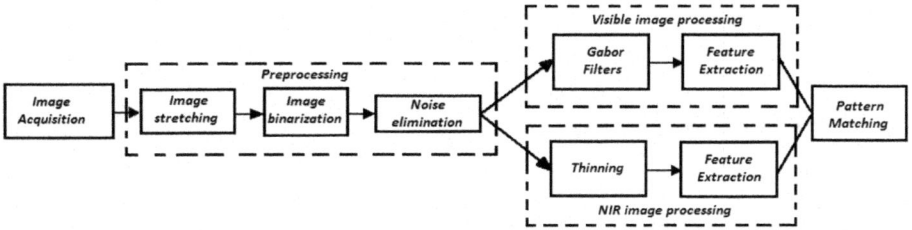

Fig. 3. Processing diagram

If h_g represents the number of pixels in an image with intensity g e.g. $f(x, y) = g$, then the probability density function is defined as $prob(f(x, y) = g) = \frac{h_g}{M \cdot N}$ for $g = 0, 1, \ldots, G - 1$ and the cumulative density function is defined as $c(f(x, y) = g) = \sum_{g=0}^{G-1} prob(f(x, y)) = g$ for $g = 0, 1, \ldots, G - 1$. The gray levels are modified as

$$\bar{g} = (max - min) \cdot c(f(x, y) = g) + min \tag{3}$$

where max and min are respectively the maximum and minimum value of image gray level.

The result after using CLAHE on the images in Fig. 2a and b are illustrated in Fig. 4a and b, respectively.

2. Image binarization. Let the initial global threshold (point between two peaks in the histogram) be th. Using th we produce two groups of pixels: the first with all pixels having intensity values $> th$, the second with pixels having values $\leq th$. Next, compute the mean intensity values $mean_1$ and $mean_2$ for the pixels within the first and second groups, respectively. Finally, a new threshold value is defined by $T = \frac{mean_1 + mean_2}{2}$.

3. Noise elimination. Median filtering is a nonlinear method based on image statistics used to remove noise. Values in the digital image are assumed to be noisy and replaced by the median value of the neighborhood pixels (the mask).

The pixels belonging to the mask are ranked in the order of their gray levels, and the median value of the group is stored to replace the noisy value. The

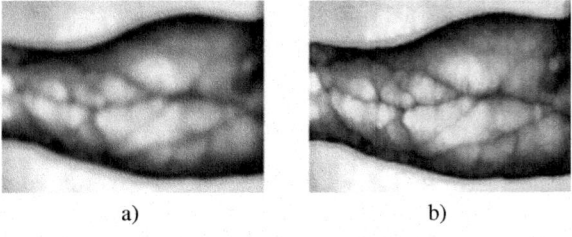

a)   b)

Fig. 4. The visible and *NIR* forearm images after applying CLAHE

2D MF for an image $f(x,y)$, where $(x,y) \in R$ is defined as

$$p(x,y) = median_{A_l}f(x,y) = median[f(m+r,n+s) \quad ; \quad (r,s) \in A_l] \quad (4)$$

where A_l is the MF window. In our case, $l = 5$, i.e. we use a 5×5 window to reduce noise.

3   Texture Feature from Gabor Wavelet Transform

The texture of an image or image segment can be characterized by a set of linear filter responses defined for different spatial frequencies, spatial localization and orientation. The Gabor filter can be written as [1,17]

$$G_{\sigma_k,\theta_i}(x,y) = \frac{1}{2\pi\sigma_k^2}e^{\left(-\frac{x^2+y^2}{2\sigma_k^2}\right)} \times \exp\{2\pi j(Wx\cos\theta_i + Wy\sin\theta_i)\} \quad (5)$$

where: $j = \sqrt{-1}$,
σ_k - the standard deviation of the Gaussian envelop,
W - the radial frequency, and
θ_i - the orientation of the Gabor filters.

In our work these parameters are shown in Table 3.

Table 3. Gabor filters parameters

Size	σ	W	θ
5×5	2, 4, 8	1	$\frac{\pi i}{6}$ where $i = 1,\ldots,6$

The result of the convolution operation of the input image with the corresponding Gabor filter is an output image calculated as

$$\Psi_{\sigma_k,\theta_i}(x,y) = f(x,y) \otimes G_{\sigma_k,\theta_i}(x,y) \quad (6)$$

where $\otimes$ denotes the 2D convolution operation and $f(x,y)$ is the input image.

The images shown in Fig. 5 are the result of the convolution operation of the input image with the above defined set of Gabor filters.

For each image in Fig. 5 we calculate two features - energy and entropy (Fig. 6) - respectively defined by

$$E_{\sigma_k,\theta_i}(x,y) = \frac{1}{M \times N}\sum_{x=0}^{M-1}\sum_{y=0}^{N-1}(\Psi_{\sigma_k,\theta_i}(x,y))^2 \quad (7)$$

$$Ent_{\sigma_k,\theta_i}(x,y) = -p_{\sigma_k,\theta_i}(x,y) \times \log_2(p_{\sigma_k,\theta_i}(x,y)) \quad (8)$$

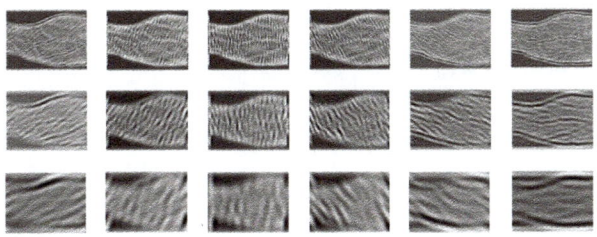

Fig. 5. Real part of the Gabor filter responses of a forearm image. Rows correspond to $\sigma_k = \{2, 4, 8\}$, and columns to $\theta_i = \{30°, 60°, 90°, 120°, 150°, 180°\}$.

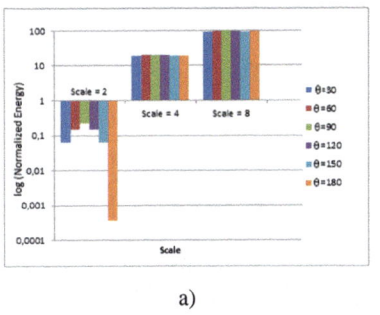

a)

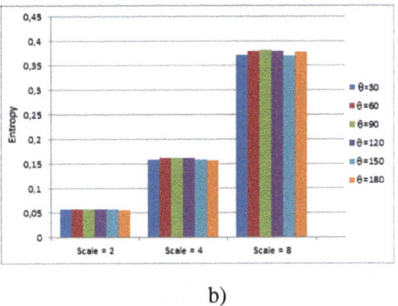

b)

Fig. 6. Energy and Entropy plots of the Gabor images

where

$$p_{\sigma_k, \theta_i}(x, y) = \frac{E_{\sigma_k, \theta_i}(x, y)}{\sum_{\sigma_k} \sum_{\theta_i} E_{\sigma_k, \theta_i}(x, y)} \tag{9}$$

In our case, the feature vector has the size $3 \times 6 \times 2 = 36$ (three scales, six orientations, two parameters) and takes the following form

$$FV_1 = \{E_{2,30}, \dots, E_{2,180}, E_{4,30}, \dots, E_{4,180}, E_{8,30}, \dots, E_{8,180},$$
$$Ent_{2,30}, \dots, E_{2,180}, Ent_{4,30}, \dots, E_{4,180}, Ent_{8,30}, \dots, Ent_{8,180}\} \tag{10}$$

4   Extraction of Geometrical Features

The thinning algorithm determines skeletal pixels by local operations. The neighbors around pixel p are enumerated as $p_1, p_2, \dots, p_8$ (Fig. 7). The quantity of such pixels that equal 1, that is $p_i = 1, i = 1, 2, \dots, 8$ pixels is $B(p)$. The number of transitions from 0 to 1 in neighbors around pixel p is $A(p)$ [37].

The point p is deleted from the image if:

(i) $2 < B(p) \le 6$
(v) $p_3 \cdot p_1 \cdot p_7 = 0$
(u) $p_1 \cdot p_7 \cdot p_5 = 0$

p(x-1,y-1)	p(x-1,y)	p(x-1,y+1)
p(x,y-1)	p(x,y)	p(x,y+1)
p(x+1,y-1)	p(x+1,y)	p(x+1,y+1)

$\equiv$

p_4	p_3	p_2
p_5	p	p_1
p_6	p_7	p_8

Fig. 7. Pixel notations

If any condition is not satisfied conditions (v) and (u) are changed to

(iv) $p_3 \cdot p_5 \cdot p_7 = 0$
(iu) $p_1 \cdot p_3 \cdot p_5 = 0$

The results of the vessel preprocessing are shown in Fig. 8.

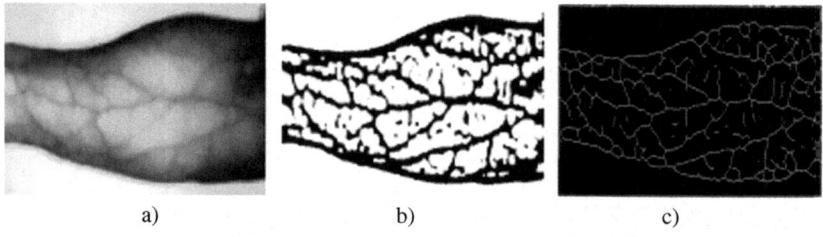

a)   b)   c)

Fig. 8. (a) The original forearm vein image; (b) the result of binarization; and (c) the skeleton of the vein region

The Crossing Number (CN) concept [12,18] is used to specify bifurcations points in the skeleton vessels image. The Crossing Number CN at a point p with a neighborhood of points p_i is obtained according to the formula:

$$CN = \frac{1}{2} \sum_{i=1}^{8} |p_i - p_{i+1}|, \ p_9 = p_1 \tag{11}$$

Properties of the point p are shown in Table 4.

The vessel topology is described by the feature vector FV_2 which defines bifurcation structure:

$$FV_2 = \{position\ of\ bifurcation\ points,\ angles\} = \{bif_1, bif_2, \ldots, bif_h\} \tag{12}$$

where

- h - the number of bifurcation points in the forearm image,
- $bif_h = \{x_h, y_h, \alpha_{1h}, \alpha_{2h}, \alpha_{3h}\}$. (x_h, y_h) are position of bifurcation points and $(\alpha_{1h}, \alpha_{2h}, \alpha_{3h})$ are the bifurcation angles with respect to the horizontal axis.

Table 4. Topological properties of p

The value of CN	Property of pixel p
0	Internal or isolate
1	End
2	Connect
3	Bifurcation
4	Cross

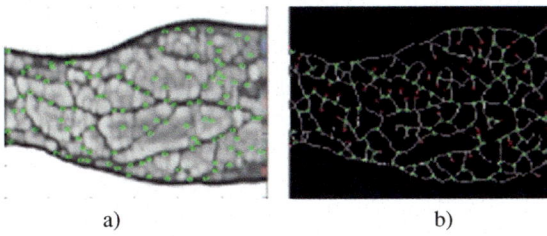

a)   b)

Fig. 9. Feature points extracted from forearm images: (a) characteristics points on original image; (b) bifurcation points on skeleton image

The normalized image of size (128×96) pixels is divided into non-overlapping blocks of (32×32) pixels. In this way we obtain 12 blocks numbered from left to right and from top to bottom. In these blocks are detected bifurcation points.

The selection of bifurcation points in individual blocks is carried out as follows:

- the coordinates (x, y) of the bifurcation point are transformed to the local coordinates of the given block $(x_{block_l}, y_{block_l})$, $l = 1, \ldots, 12$,
- $\rho = \sqrt{x_{block_l}^2 + y_{block_l}^2}$ is calculated,
- ρ-values are ranked from the smallest to the largest,
- as points of the FV_2 vector, the first t points are selected, where t is the number of points specified for the given block.

As a result, we obtain 60 bifurcation points (Fig. 9).

5   Pattern Matching

We defined the vectors of features as follows:

$$FeatVec = (FV_1, FV_2) \tag{13}$$

where FV_1 is the Gabor feature vector, and FV_2 is the bifurcation feature vector. The $FeatVec$ feature vector contains $36 + 300 = 336$ elements.

The Euclidean distance is the basis for pattern matching between the two vectors associated with Gabor features (FV_1). One vector refers to the input forearm image, the second to the template forearm image. A Euclidean distance less than a certain threshold, allows the conclusion that the input image and the template image are "very similar" or even the same.

In the case of the FV_2 vector, two bifurcation points bif_u and bif_v are considered to match, if their position and orientation are close [22, 31]:

$$D(bif_u, bif_v) = \sqrt{(x_u - x_v)^2 + (y_u - y_v)^2} < Th_1 \qquad (14)$$

$$A(bif_u, bif_v) = min(|\alpha_{u_1} - \alpha_{v_1}|, |\alpha_{u_2} - \alpha_{v_2}|, |\alpha_{u_3} - \alpha_{v_3}|) < \Theta \qquad (15)$$

If S is higher than a threshold value, then the input forearm image is matched with the template forearm image, otherwise it is considered to not be a match. S is defined as follows

$$S = \frac{H_{uv}}{\sqrt{uv}} \qquad (16)$$

where the number of matched bifurcation is H_{uv}.

6   Results

A database of forearm vein images is created for work evaluation. The biometric data we use have been collected involving 50 individuals with 4 samples per person. In this study, 200 forearms images have been captured. When recognizing visible images or recognizing NIR images, the number of true matches $50 \times C_4^2$ while the number of impostors is consistent C_{50}^2 when only one image of each forearm is selected for matching. In case fusion the visible and NIR images, there are 16 samples. 8 samples of each forearm are randomly selected for testing. The number of genuine match is $50 \times C_8^2$, and the number of impostor match is C_{50}^2.

To evaluate the effectiveness of the proposed method for forearm vein recognition we used following indicators:

- False accept rate (FAR) - the percentage of the likelihood that an impostor will be accepted by the biometric system due to large inter-user similarity,
- False reject rate (FRR) - the percentage of the likelihood that the genuine individual will be rejected by the system due to large intra-class variations,
- Genuine Acceptance Rate (GAR) obtained as $GAR = 1 - FRR$ - the percentage of the likelihood that a genuine individual is recognized as a match.

The results of our method are summarized in Table 5.

Table 5. Performance of the proposed method

Modality	False acceptance rate (FAR %)	
	0.1	1
Visible forearm	81.3	90
NIR forearm	90.4	94.1
Fusion	93	94.6

7   Conclusions

In this article we propose method to biometric identification based on forearm vein images. The proposed method uses Gabor filters to producing the feature vector based on energy and entropy parameters. Second feature vector uses bifurcation points in the skeleton vessels image. Our future work will focus on how to improve the forearm recognition by trading off the accuracy.

References

1. Choras, R.S.: Image feature extraction techniques and their applications for CBIR and biometrics systems. Int. J. Biol. Biomed. Eng. **1**(1), 6–16 (2007)
2. Choras, R.S.: Iris recognition. In: Computer Recognition Systems 3, AISC, vol. 57, pp. 637–644. Springer, Heidelberg (2009)
3. Choras, R.S.: Personal identification using forearm vein patterns. In: International Conference and Workshop on Bioinspired Intelligence (IWOBI), pp. 1–5 (2017)
4. Choras, R.S.: Biometric personal authentication using images of forearm vein patterns. In: International Conference on Signals and Systems (ICSigSys), pp. 40–43 (2017)
5. Ding, Y., Zhuang, D., Wang, K.: A study of hand vein recognition method. In: Proceedings of the IEEE International Conference on Mechatronics and Automation, pp. 2106–2110 (2005)
6. Gonzales, R.C., Woods, R.E.: Digital Image Processing. Pearson Prentice Hall, Upper Saddle River (2008)
7. Haralick, R., Shanmugam, K., Dinstein, I.: Textural features for image classification. IEEE Trans. Syst. Man Cybern. **SMC-3**(6), 610–621 (1973)
8. Jain, A.K., Flynn, P.J., Ross, A.: Handbook of Biometrics. Springer, New York (2007)
9. Kang, B.J., et al.: Multimodal biometric method based on vein and geometry of a single finger. IET Comput. Vis. **4**(3), 209–217 (2010)
10. Kirbas, C., Quek, K.: Vessel extraction techniques and algorithm: a survey. In: Proceedings of the 3rd IEEE Symposium on BioInfomratics and Bioengineering (2003)
11. Kono, M., et al.: Near-infrared finger vein patterns for personal identification. Appl. Opt. **41**(35), 7429–7436 (2002)
12. Kumar, A.: Incorporating cohort information for reliable palmprint authentication. In: Proceeding of ICVGIP, pp. 583–590 (2008)

13. Kumar, A., Prathyusha, K.: Personal authentication using hand vein triangulation and knuckle shape. IEEE Trans. Image Process. **18**, 2127–2136 (2009)
14. Lee, H.C., et al.: Finger vein recognition using weighted local binary pattern code based on a support vector machine. J. Zhejiang Univ. - Sci. C **11**(7), 514–524 (2010)
15. Lee, E.C., et al.: New finger biometric method using near infrared imaging. Sensors **11**, 2319–2333 (2011)
16. Liu, C.J., Wechsler, H.: Gabor feature based classification using the enhanced fisher linear discriminant model for face recognition. IEEE Trans. Image Process. **11**(4), 467–476 (2002)
17. Liu, D.H., Lam, K.M., Shen, L.S.: Optimal sampling of Gabor features for face recognition. Pattern Recogn. Lett. **25**(2), 267–276 (2004)
18. Maltoni, D., Maio, D., Jain, A.K., Prabhakar, S.: Handbook of Fingerprint Recognition. Springer, Heidelberg (2003)
19. Meng, X., Yin, Y., Yang, G., Xi, X.: Retinal identification based on an improved circular Gabor filter and scale invariant feature transform. Sensors **13**, 9248–9266 (2013)
20. Pierre-Olivier, L., Christophe, R., Bernadette, D.: Palm vein verification system based on SIFT matching. In: Proceedings of Third International Conference ICB, pp. 1290–1298 (2009)
21. Park, U., Jillela, R., Ross, A., Jain, A.: Periocular biometrics in the visible spectrum. IEEE Trans. Inform. Forensics Secur. **6**(1), 96–106 (2011)
22. Ranade, A., Rosenfeld, A.: Point pattern matching by relaxation. Pattern Recogn. **12**(2), 269–275 (1993)
23. Ross, A., Nandakumar, K., Jain, A.K.: Handbook of Multibiometrics. Springer, Heidelberg (2006)
24. Tanaka, T., Kubo, N.: Biometric authentication by hand vein patterns. In: Proceedings of the SICE Annual Conference, pp. 249–253 (2004)
25. Wang, D., et al.: User identication based on finger-vein patterns for consumer electronics devices. IEEE Trans. Consum. Electron. **56**(2), 799–804 (2010)
26. Wang, Y., Li, K., Cui, J.: Hand-dorsa vein recognition based on partition local binary pattern. In: IEEE 10th International Conference on Signal Processing (ICSP), pp. 1671–1674 (2010)
27. Wang, Y., Hu, J., Phillips, D.: A fingerprint orientation model based on 2D Fourier expansion (FOMFE) and its application to singular-point detection and fingerprint indexing. IEEE Trans. Pattern Anal. Mach. Intell. **29**(4), 573–585 (2007)
28. Wright, J., Yang, A., Ganesh, A., Sastry, S., Ma, Y.: Robust face recognition via sparse representation. IEEE Trans. Pattern Anal. Mach. Intell. **31**(2), 210–227 (2009)
29. Woodard, D., Pundlik, S., Lyle, J., Miller, P.: Periocular region appearance cues for biometric identification. In: Computer Vision and Pattern Recognition Workshops (CVPRW), pp. 162–169 (2010)
30. Yang, G., Xi, X., Yin, Y.: Finger vein recognition based on a personalized best bit map. Sensors **12**, 1738–1757 (2012)
31. Yoon, S., Feng, J., Jain, A.: Latent fingerprint enhancement via robust orientation field estimation. In: International Joint Conference on Biometrics (IJCB), pp. 1–8 (2011)
32. Yuksel, A., Akarun, L., Sankur, B.: Biometric identification through hand vein patterns. In: International Workshop in Emerging Techniques and Challenges for Hand-Based Biometrics (ETCHB), pp. 1–6 (2010)

33. Xueyan, L., Shuxu, G., Fengli, G., Ye, L.: Vein pattern recognitions by moment invariants. In: Proceedings of the First International Conference on Bioinformatics and Biomedical Engineering, pp. 612–615 (2007)
34. Zeng, Z., Hu, J.: Face recognition based on shearlets and principle component analysis. In: IEEE International Conference on Intelligent Networking and Collaborative Systems, pp. 697–701 (2013)
35. Zhang, D., Kong, W.K., You, J., et al.: Online palmprint identification. IEEE Trans. Pattern Anal. Mach. Intell. **25**(9), 1041–1050 (2003)
36. Zhang, H., et al.: Finger vein recognition based on Gabor filter. In: Intelligence Science and Big Data Engineering, pp. 827–834 (2013)
37. Zhang, T., Suen, C.: A fast parallel algorithm for thinning digital patterns. Commun. ACM **27**, 236–239 (1984)
38. Zhang, Z., et al.: Multiscale feature extraction of finger-vein patterns based on curvelets and local interconnection structure neural network. In: The 18th International Conference on Pattern Recognition, vol. 4, pp. 145–148 (2006)
39. Zuiderveld, K.: Contrast Limited Adaptive Histogram Equalization. Academic Press, Cambridge (1994)

A Hermite-Based Method for Bone SPECT/CT Image Fusion with Prior Segmentation

Leiner Barba-J[1,2(✉)], Lorena Vargas-Quintero[1,2],
Jose Alberto Calderon[2], and Cesar Torres Moreno[1]

[1] Optic and Computer Science Lab., Universidad Popular del Cesar,
Valledupar, Colombia
barba.leiner@unicesar.edu.co
[2] Medicina Nuclear S.A., Valledupar, Colombia

Abstract. Multimodal medical image fusion is an interesting application that has become essential in many tasks of medical diagnosis. The combination of medical images acquired from different modalities might significantly contribute to improve the diagnosis and detection process of several diseases. SPECT (Single Photon Emission Computed Tomography) and CT (Computed Tomography) are frequently fused for illness detection tasks. In this work, we have developed a novel scheme for bone SPECT/CT image fusion based on the Hermite transform (HT). It consists of a powerful tool which is able to decompose an image into a set of coefficients defined in the space of the Hermite polynomials. Three main stages are performed for the fusion process: (1) Input images are decomposed using the HT, (2) A fusion rule is applied to combine the resulting coefficients, and (3) The inverse HT is computed to obtain the fused image. Since we are interested in providing a technique focused on evaluating the bone structure, in the fusion rule we introduce prior information based on a previous segmentation obtained from the CT data. Several studies were used for performance assessment.

Keywords: Hermite transform · Image fusion · SPECT · CT

1 Introduction

Image fusion is a processing task which has found interesting utility in the field of medical applications [1]. The possibility to present in the same image space, data gathered from different modalities is of great value for disease diagnosis, treatment and therapy. One of the most common uses of image fusion for medical applications is the combination of anatomical and functional information. SPECT and CT are commonly employed together with the aim of evaluating the patient conditions. While the first modality is part of the nuclear medicine field and provides details about the metabolic behavior of the human tissues, the second is suitable to access the structural or anatomical data. Bone tissues are among the structures of the human body that have been benefited from the use of nuclear – based images such as SPECT for evaluation purposes. Several conditions and diseases can be assessed using this type of images [2]. However, the poor resolution and contrast resulting in SPECT data has made

© Springer Nature Switzerland AG 2019
J. M. R. S. Tavares and R. M. Natal Jorge (Eds.): VipIMAGE 2019, LNCVB 34, pp. 62–66, 2019.
https://doi.org/10.1007/978-3-030-32040-9_7

completely necessary its combination with other modalities with the objective of improving the evaluation process.

The image fusion task is an issue of main interest for researches. Many contributions have been made in this field and different approaches have been proposed for multimodal medical imaging [2]. Most of the approximations are based on image decomposition and transforms [3]. We can mention methods based on the wavelet, countourlet and sheartled transforms [4]. Other approaches use sparse-based methods [4], and recently machine learning techniques have played an important role for designing fusion schemes [5].

In this work, we propose a Hermite-based fusion framework applied to SPECT and CT images for bone tissues analysis. The method consists of three main steps. Firstly, the HT is applied to both input images. Posteriorly, a fusion rule is then used to perform the fusion of the resulting coefficients. Here, we used the average and classical maximum intensity value for the fusion process. Since we are only interested in the evaluation of the bone tissues, prior knowledge collected from a previous segmented structure is included into the fusion rule. Finally, the inverse HT is computed to obtain the final fused image.

The rest of the paper is organized as follows. Section 2 presents the design method and theorical foundations of the used tools. Experiments and results are discussed in Sect. 3. Conclusions are shown in Sect. 4.

2 Method

The proposed framework is introduced as follows. Let $I(x, y)$ represents an input image. Its HT is calculated as [6]

$$L_{n-m,m}(p, q) = \int \int_{-\infty}^{\infty} I(x, y) V^2(x - p, y - q) G_{n-m,m}(x - p, y - q) dx dy \quad (1)$$

where $L_{n-m,m}(p, q)$ are the Hermite coefficients, $V(x, y) = \frac{1}{\sigma\sqrt{\pi}} e^{-(x^2 + y^2)/2\sigma^2}$ corresponds to an isotropic Gaussian and $G_{n-m,m}(x, y) = \frac{1}{\sqrt{2^n (n-m)m!}} H_{n-m}(x/\sigma) H_m(y/\sigma)$ represent the normalized Hermite polynomials [7]. Here, n is the transform order. Coefficients are obtained for $n = 0, 1, 2, \ldots, N$; $m = 0, 1, \ldots, n$. Low frequency components are obtained for $n = 0$ and details components are found for $n \geq 1$. The HT can be performed by convolution between the input image and the set of filters defined by $D(x, y) = V^2(x, y) G_{n-m,m}(-x, -y)$.

On the other hand, the inverse HT can be computed as

$$I(x, y) = \sum_{(p,q)} \sum_{(n,n-m)} L_{n-m,m}(p, q) \hat{P}_{n.n-m}(x - p, y - q) \quad (2)$$

where $\hat{P}_{n-n-m} = \frac{V^2(x,y)G_{n-m,m}(x,y)}{\sum_{(p,q)} V^2(x-p,y-q)}$ corresponds to the reconstruction functions. Similarly, the inverse transformation can also be computed through the convolution operation and addition. In this approach, the HT is used without decimation which implies that resulting coefficients have the same size as original images.

Using the input CT image, a segmentation process is applied to select the region corresponding to the bone tissues. Let S denote the segmented image with $S = \{0, 1\}$. Here, region corresponding to bone tissues is represented by $S = 1$. The HT must be then applied to both input images, I^{SPECT} and I^{CT}, and the obtained coefficients are subsequently combined using the following fusion rules.

$$L^F_{n-m,m} = \begin{cases} avg\left(L^{SPECT}_{n-m,m}, L^{CT}_{n-m,m}\right) & \text{if } n = 0 \text{ and } S == 1 \\ max\left(L^{SPECT}_{n-m,m}, L^{CT}_{n-m,m}\right) & \text{if } n \geq 1 \text{ and } S == 1 \\ L^{CT}_{n-m,m} & \text{elsewhere} \end{cases} \tag{3}$$

As noted, both conditions are performed for the region defined by the segmentation of the bone tissues which ensures the algorithm focuses its analysis on these types of structures. We used a simple but an effective method based on a region growing scheme to perform the segmentation. The resulting combined coefficients, $L^F_{n-m,m}$, are posteriorly employed to compute the inverse HT which provides the final fused image I^F. Average rule, denoted as avg, was selected for combining the low frequency coefficients while the maximum rule (max) was used for combining the details Hermite coefficients.

3   Experiments and Results

Several SPECT and CT images were used for performance assessment. It is assumed that both images have been previously registered. Experiments were performed using the image intensity, however, pseudocolor images were used for visualization of SPECT and fused images.

Some image examples of the results obtained using the proposed approach is illustrated in Figs. 1 and 2. Here, the segmentation, the input images and the fusion results are illustrated. Red contour is traced to indicate the segmented region of the bone structure in the CT image which is posteriorly used as prior information in the fusion rules as shown in Eq. (3).

Quantitative analysis was also performed. Entropy (E), mutual information (MI), standard deviation (STD), edge preservation index ($Q_{AB/F}$) and the root mean square error (RMSE) were the selected metrics for evaluation. Table 1 reports the obtained results which have been average for all images.

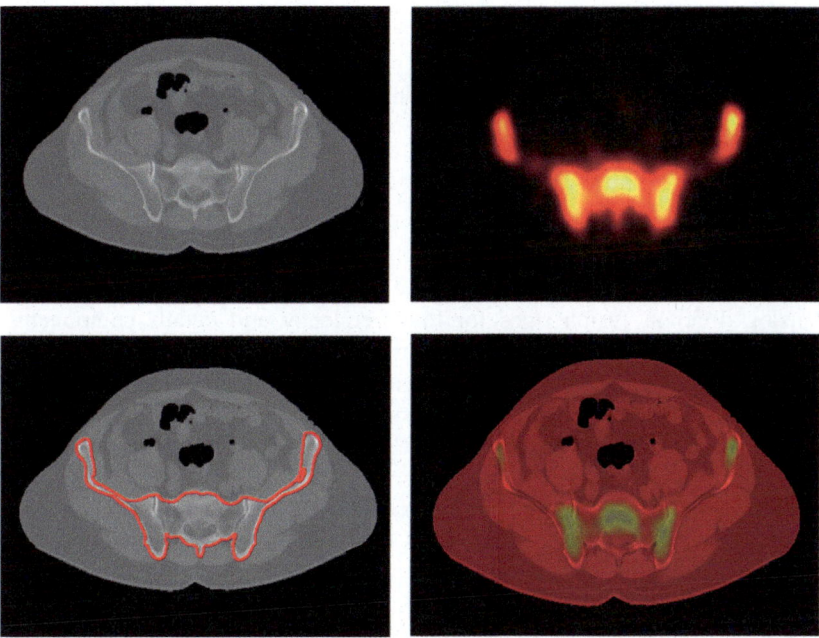

Fig. 1. Result of the fusion process obtained with the proposed method for a first image example. From left to right and top to bottom: CT, SPECT, segmented CT and fused image.

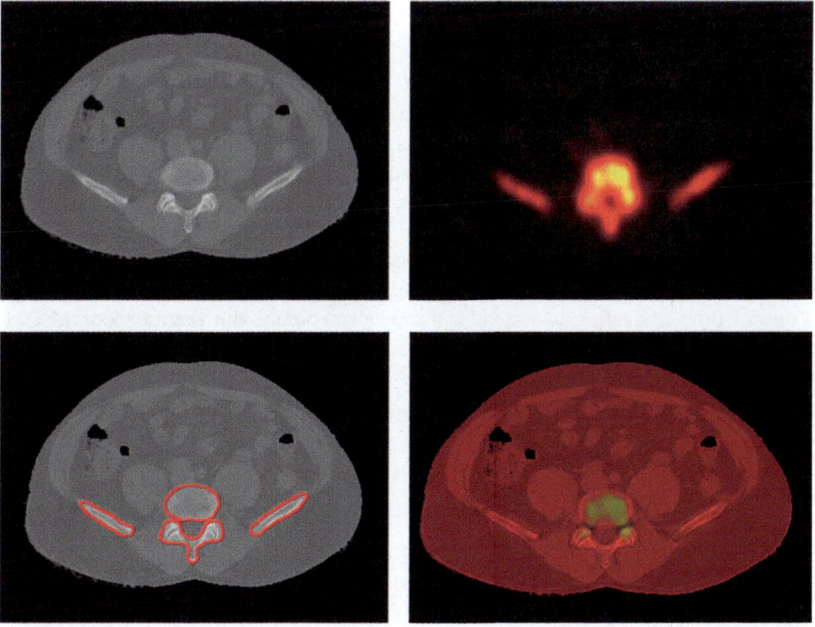

Fig. 2. Result of the fusion process obtained with the proposed method for a second image example. From left to right and top to bottom: CT, SPECT, segmented CT and fused image.

Table 1. Quantitative results using five different metrics

	E	MI	STD	$Q_{AB/F}$	RMSE
Average result	0.844	2.544	12.536	0.549	26.458

4 Conclusion

We developed an image fusion framework based on the Hermite transform applied to the analysis of SPECT and CT studies. Coefficients obtained with the HT were combined using different fusion rules for low frequency and details components. This methodology allows to highlight the bone structures in the final fused image. Including the segmentation as prior knowledge results advantageous since only the bone tissues are of interest. The presented result demonstrates that the proposed framework is suitable for the combination of anatomical and metabolic images.

Acknowledgement. This work has been sponsored by COLCIENCIAS (Colombia) through postdoctoral scholarships given to Leiner Barba-J and Lorena Vargas-Quintero.

References

1. Pappachen, A.J., Dasarathy, B.V.: Medical image fusion: a survey of the state of the art. Inf. Fusion **19**, 4–19 (2014)
2. Mariani, G., Bruselli, L., Kuwert, T., Kim, E.E., Flotats, A., Israel, O., Dondi, M., Watanabe, N.: A review on the clinical uses of SPECT/CT. Eur. J. Nucl. Med. Mol. Imaging **37**, 1959–1985 (2010)
3. El-Gamal, F.E.-Z.A., Elmogy, M., Atwan, A.: Current trends in medical image registration and fusion. Egypt. Inform. J. **17**, 99–124 (2016)
4. Ling, T., Xin, Y.: Medical image fusion based on fast finite shearlet transform and sparse representation. Comput. Math. Methods Med. (2019). Article ID 3503267
5. Zong, J., Qiu, T.: Medical image fusion based on sparse representation of classified image patches. Biomed. Signal Process. Control **34**, 195–205 (2017)
6. Silvan-Cardenas, J.L., Escalante-Ramirez, B.: The multiscale Hermite transform for local orientation analysis. IEEE Trans. Image Process. **15**(5), 1236–1253 (2006)
7. Barba-J, L., Moya-Albor, E., Escalante-Ramírez, B., Brieva, J., Venegas, E.V.: Segmentation and optical flow estimation in cardiac CT sequences based on a spatiotemporal PDM with a correction scheme and the Hermite transform. Comput. Biol. Med. **69**(1), 189–202 (2016)

Space-Adaptive Anisotropic Bivariate Laplacian Regularization for Image Restoration

Luca Calatroni[1], Alessandro Lanza[2(✉)], Monica Pragliola[2],
and Fiorella Sgallari[2]

[1] CMAP, CNRS, École Polytechnique, Institut Polytechnique de Paris,
Route de Saclay, 91128 Palaiseau, France
luca.calatroni@polytechnique.edu
[2] Department of Mathematics, University of Bologna,
Piazza di Porta San Donato 5, Bologna, Italy
{alessandro.lanza2,monica.pragliola2,fiorella.sgallari}@unibo.it

Abstract. In this paper we present a new regularization term for variational image restoration which can be regarded as a space-variant anistropic extension of the classical Total Variation (TV) regularizer. The proposed regularizer comes from the statistical assumption that the gradients of the unknown target image distribute locally according to space-variant bivariate Laplacian distributions. The high flexibility of the proposed regularizer holds the potential for the effective modelling of local image properties, in particular driving in an adaptive manner the strength and the directionality of non-linear TV-diffusion. The free parameters of the regularizer are automatically set - and, eventually, updated - based on a robust Maximumum Likelihood estimation procedure. A minimization algorithm based on the Alternating Direction Method of Multipliers is presented for the efficient numerical solution of the proposed variational model. Some experimental results are reported. They demonstrate the high-quality of restorations achievable by the proposed model, in particular with respect to classical TV-regularized models.

Keywords: Image restoration · Variational methods · ADMM · TV regularization

1 Introduction

Image restoration is the task of recovering a sharp image u starting from a given blurred and noisy observation g. In this work, we consider a degradation model of the form

$$g = Ku + e, \tag{1}$$

where $g, u \in \mathbb{R}^n$ are vectorized images, $K \in \mathbb{R}^{n \times n}$ is the linear blur operator and $e \in \mathbb{R}^n$ is an additive noise vector. A possible strategy to overcome the

© Springer Nature Switzerland AG 2019
J. M. R. S. Tavares and R. M. Natal Jorge (Eds.): VipIMAGE 2019, LNCVB 34, pp. 67–76, 2019.
https://doi.org/10.1007/978-3-030-32040-9_8

ill-posedeness of the linear system in (1) is to reformulate the problem in a well-posed *variational* form, looking for u^* estimating the desired original u, which minimizes a cost functional $\mathcal{J}(u; K, g) : \mathbb{R}^n \to \mathbb{R}$. In formula,

$$u^* \in \arg \min_{u \in \mathbb{R}^n} \left\{ \mathcal{J}(u; K, g) := \mathcal{R}(u) + \mu \mathcal{F}(u; K, g) \right\}.$$

The functionals $\mathcal{R}$ and $\mathcal{F}$ are commonly referred to as the *regularization* and the *data fidelity* term, respectively. While $\mathcal{R}$ encodes prior information on the desired image u, $\mathcal{F}$ is a data term which measures the 'distance' between the given image g and u after the action of the operator K with respect to some norm corresponding to the noise statistics in the data, cf., e.g., [10]. The regularization parameter $\mu > 0$ controls the trade-off between the two terms.

In this paper, we consider an Additive White Gaussian Noise (AWGN) corrupting the blurred image Ku, i.e. $e \sim \mathcal{N}(0, \sigma^2 I_n)$, where I_n is the n-dimensional identity matrix. It is well known that, in presence of AWGN, a suitable choice for $\mathcal{F}(u; K, g)$ is the so-called L$_2$ fidelity term, reading as,

$$\mathcal{F}(u; K, g) = \text{L}_2(u; K, g) = \frac{1}{2} \|Ku - g\|_2^2.$$

A popular choice for the regularization term $\mathcal{R}(u)$ is given by the TV semi-norm [9],

$$\mathcal{R}(u) = \text{TV}(u) = \sum_{i=1}^{n} \|(\nabla u)_i\|_2, \tag{2}$$

where $(\nabla u)_i := \left((D_h u)_i, (D_v u)_i \right)^T \in \mathbb{R}^2$ denotes the discrete gradient of image u at pixel i, with $D_h, D_v \in \mathbb{R}^{n \times n}$ linear operators representing finite difference discretizations of the first-order horizontal and vertical partial derivatives, respectively.

Coupling the L$_2$ data term with the TV regularizer leads to one of the reference variational models for image restoration, the TV-L$_2$ (or ROF) model,

$$u^* \in \arg \min_{u \in \mathbb{R}^n} \left\{ \sum_{i=1}^{n} \|(\nabla u)_i\|_2 + \frac{\mu}{2} \|Ku - g\|_2^2 \right\}.$$

The global nature of the TV-L$_2$ model (due to the presence of non-weighted sums in the model) does not allow to diversify the action of the regularizer on regions of the image presenting different properties. To overcome this issue, in [2,3,5,6], the authors have proposed space-variant regularization term based on statistical assumptions on the distribution of the ℓ_2-norm of the gradients and on the gradients themselves.

In this paper, we propose a space-variant anisotropic extension of the TV regularizer in (2) which, as it will be illustrated in Sect. 2, comes from the *a priori* assumption that the gradients of the target image u distribute locally according to a Bivariate Laplace Distribution (BLD). The proposed BLTV regularizer takes the form

$$\mathrm{BLTV}(u; \lambda^{(1)}, \lambda^{(2)}, \theta) = \sum_{i=1}^{n} \|\Lambda_i R_{\theta_i} (\nabla u)_i\|_1 \tag{3}$$

$$= \sum_{i=1}^{n} \left[\lambda_i^{(1)} |\langle r_i^{(1)}, (\nabla u)_i \rangle| + \lambda_i^{(2)} |\langle r_i^{(2)}, (\nabla u)_i \rangle| \right], \tag{4}$$

where, for every pixel i, Λ_i is a 2×2 positive definite diagonal matrix and R_{θ_i} is the rotation matrix corresponding to the angle $-\theta_i$. Mathematically,

$$\Lambda_i = \begin{pmatrix} \lambda_i^{(1)} & 0 \\ 0 & \lambda_i^{(2)} \end{pmatrix}, \quad R_{\theta_i} = \begin{pmatrix} \cos\theta_i & \sin\theta_i \\ -\sin\theta_i & \cos\theta_i \end{pmatrix} = \begin{pmatrix} r_i^{(1)} \\ r_i^{(2)} \end{pmatrix}. \tag{5}$$

We denote by $\lambda^{(1)}, \lambda^{(2)}, \theta \in \mathbb{R}^n$ the maps of the parameters defining the local distributions. Hence, the proposed BLTV-L_2 variational restoration model reads as

$$u^* \in \arg\min_{u \in \mathbb{R}^n} \left\{ \mathrm{BLTV}(u; \lambda^{(1)}, \lambda^{(2)}, \theta) + \frac{\mu}{2} \|Ku - g\|_2^2 \right\}. \tag{6}$$

The $3n$ free parameters defining the BLTV regularizer in (3)–(5) hold the potential for effectively modelling local image properties, in particular driving in an adaptive manner the strength and direction of non-linear TV-diffusion. In Figs. 1(a)–(b) the red ellipses represent TV-diffusion strengths along all possible directions at few sample pixel locations for TV and BLTV regularizers. It is evident how for classical TV such ellipses turn out to be circles (isotropy) of constant radius (space-invariance), whereas our BLTV regularizer allows for ellipses (anistropy) of different size (space-variance). In practice, such flexibility of BLTV can be (and will be) exploited to diffuse in different ways in regions exhibiting different properties: for instance, strong isotropic diffusion in homogeneous regions, strongly anisotropic diffusion in regions characterized by a dominant edge direction. We further show in Figs. 1(c)–(d) the level curves of the TV regularizer and one among the infinity of possible configuration of the level curves of the BLTV regularizer, respectively, revealing once again the flexibility of the proposed regularizer.

Together with the variational model in (6), we also propose an efficient and robust procedure for automatically estimating the parameter maps $\lambda^{(1)}$, $\lambda^{(2)}$, θ based on a Maximum Likelihood (ML) approach. The parameter maps can be updated along the iterations of the Alternating Direction Methods of Multipliers (ADMM), which is the algorithm adopted here to solve the minimization problem. Notice that the convexity of the BLTV-L_2 model ensures the convergence of the ADMM.

2 Deriving the Model via MAP

Applying the *Maximum A Posteriori* (MAP) estimation approach to image restoration consists in computing the restored image as a global maximizer of

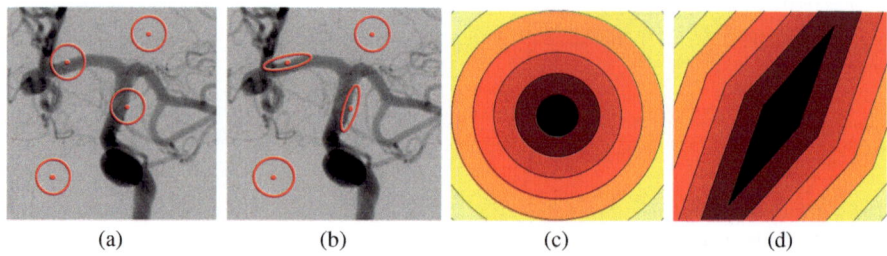

Fig. 1. Space-invariant isotropic TV-diffusion (a), space-variant anisotropic BLTV-diffusion (b), level curves of TV (c) and BLTV (d) regularization terms.

the posterior probability $\Pr(u|g; K)$ of the unknown target image u given the observation g, namely:

$$u^* \in \arg\max_{u \in \mathbb{R}^n} \Pr(u|g; K) = \arg\min_{u \in \mathbb{R}^n} \left\{ -\ln\Pr(g|u; K) - \ln\Pr(u) \right\}, \quad (7)$$

where, after applying the Bayes' rule, we drop the *evidence* term $\Pr(g)$ and extracted $-\ln$ of the objective function. The two terms $\Pr(u)$ and $\Pr(g|u; K)$ in (7) are referred to as the *prior* and the *likelihood*. The likelihood term associated with AWGN corruption takes the form

$$\Pr(g|u; K) = \prod_{i=1}^{n} \frac{1}{\sqrt{2\pi}\sigma} \exp\left(-\frac{(Ku - g)_i^2}{2\sigma^2} \right) = W \exp\left(-\frac{\|Ku - g\|_2^2}{2\sigma^2} \right), \quad (8)$$

where $\sigma > 0$ denotes the AWGN standard deviation and $W > 0$ is a normalization constant. For what concerns the prior, a common choice is to model the unknown image u as a Markov Random Field (MRF) such that the image can be characterized by its Gibbs prior distribution, whose general form is:

$$\Pr(u) = \prod_{i=1}^{n} \Pr_i(u) = \prod_{i=1}^{n} z_i \exp\left(-\alpha V_{c_i}(u) \right) = Z \exp\left(-\alpha \sum_{i=1}^{n} V_{c_i}(u) \right), \quad (9)$$

where $\alpha > 0$ is the MRF parameter, $\{c_i\}_{i=1}^n$ is the set of all cliques (a clique is a set of neighboring pixels) for the MRF, V_{c_i} is the potential function defined on the clique c_i and $Z = \prod_{i=1}^{n} z_i$ is a normalization constant. Choosing as potential function at the generic i-th pixel the magnitude of the discrete gradient at the same pixel, i.e. $V_{c_i} = \|(\nabla u)_i\|_2$ for any $i \in \{1, \ldots, n\}$, the Gibbs prior in (9) reduces to the TV prior which, plugged into (7), yields the popular TV regularizer. We remark that the TV prior corresponds to choosing

$$\Pr_i(u) = z_i \exp\left(-\alpha\|(\nabla u)_i\|_2 \right).$$

Here, we propose a space-variant anisotropic generalization of the local prior $\Pr_i(u)$ above. More in detail, we assume that the discrete gradient at any pixel i distributes according to a space-variant BLD, such that our prior reads as

$$\Pr(u) = \prod_{i=1}^{n} \Pr_i(u) = \prod_{i=1}^{n} z_i \exp\left(-\|\Lambda_i R_{\theta_i}(\nabla u)_i\|_1\right)$$

$$= Z \exp\left(-\sum_{i=1}^{n}\|\Lambda_i R_{\theta_i}(\nabla u)_i\|_1\right), \tag{10}$$

where Z is the normalization constant and $\Lambda_i, R_{\theta_i} \in \mathbb{R}^{2\times 2}$ are defined in (5).

Plugging the prior (10) and the likelihood (8) into (7) and neglecting the constant terms, the proposed BLTV-L$_2$ variational model (6) is obtained.

3   Parameter Estimation via ML

In order to make the introduction of the proposed regularizer actually useful, an efficient, robust, and automatic procedure for the estimation of the parameter maps $\lambda^{(1)}, \lambda^{(2)}, \theta \in \mathbb{R}^n$ identifying all the local BLDs has to be proposed as well. To this aim, we resort to the ML approach. Consider a set of N 2-dimensional samples $\mathcal{S} := \{s_1, \ldots, s_N\}$ drawn from a BLD with parameters $(\lambda_1, \lambda_2, \theta)$. Here, the samples play the role of image gradients at pixels of a neighborhood of radius r centered at a generic pixel i. Assuming independence of the samples, according to the definition of BLD, the likelihood function is defined by

$$\Pr(\mathcal{S}|\lambda_1, \lambda_2, \theta) = \left(\frac{\lambda_1\lambda_2}{4}\right)^N \exp\left(-\sum_{i=1}^{N}\left(\lambda_1|\langle r_1, s_i\rangle| + \lambda_2|\langle r_2, s_i\rangle|\right)\right),$$

where, clearly, r_1 and r_2 depend on θ - see (5). The goal here is to find

$$(\lambda_1^*, \lambda_2^*, \theta^*) \in \arg\max_{\lambda_1,\lambda_2,\theta} \Pr(\mathcal{S}|\lambda_1, \lambda_2, \theta) = \arg\min_{\lambda_1,\lambda_2,\theta} -\ln\Pr(\mathcal{S}|\lambda_1, \lambda_2, \theta)$$

$$= -N\ln\frac{\lambda_1\lambda_2}{4} + \sum_{i=1}^{N}\left(\lambda_1|\langle r_1, s_i\rangle| + \lambda_2|\langle r_2, s_i\rangle|\right). \tag{11}$$

Imposing a first-order optimality condition with respect to λ_1, λ_2 leads to the following closed-form estimation formulas:

$$\lambda_1 = \left(\frac{1}{N}\sum_{i=1}^{N}|\langle r_1, s_i\rangle|\right)^{-1}, \quad \lambda_2 = \left(\frac{1}{N}\sum_{i=1}^{N}|\langle r_2, s_i\rangle|\right)^{-1}. \tag{12}$$

Substituting the expressions in (12) into the objective function (11), we thus obtain the simplified minimization problem in the only variable θ:

$$\theta^* \in \arg\min_{\theta}\left\{\ln\left(\sum_{i=1}^{N}|\langle r_1, s_i\rangle|\right) + \ln\left(\sum_{i=1}^{N}|\langle r_2, s_i\rangle|\right)\right\}. \tag{13}$$

4   ADMM

In order to solve numerically the proposed image restoration model (6), we use an ADMM-based algorithm - see [1]. We first introduce two auxiliary variables $w \in \mathbb{R}^n$, $t \in \mathbb{R}^{2n}$ and rewrite the model in the equivalent linearly constrained form:

$$\{u^*, w^*, t^*\} \ \in \ \arg\min_{u,w,t} \left\{ \sum_{i=1}^{n} \|\Lambda_i R_{\theta_i} t_i\|_1 + \frac{\mu}{2} \|w\|_2^2 \right\} \tag{14}$$

$$\text{subject to}: \quad w = Ku - g, \ \ t = Du, \tag{15}$$

where the space-variant matrices Λ_i, R_{θ_i} can be estimated via the ML procedure described in Sect. 3 based only on the observed image g (i.e. as a preliminary pre-processing step) or also updated along the ADMM iterations. We define the augmented Lagrangian functional:

$$\mathcal{L}(u, w, t; \rho_w, \rho_t) := \sum_{i=1}^{n} \|\Lambda_i R_{\theta_i} t_i\|_1 + \frac{\mu}{2} \|w\|_2^2 - \rho_t^T (t - Du) + \frac{\beta_t}{2} \|t - Du\|_2^2$$

$$- \rho_w^T (w - (Ku - g)) + \frac{\beta_w}{2} \|w - (Ku - g)\|_2^2, \tag{16}$$

where $\beta_w, \beta_t > 0$ are scalar penalty parameters and $\rho_w \in \mathbb{R}^n$, $\rho_t \in \mathbb{R}^{2n}$ are the vectors of Lagrange multipliers associated with the given linear constraints. The solution $\{u^*, w^*, t^*\}$ of problem (14) is a saddle point for $\mathcal{L}$ in (16), see, e.g., [1]. Hence, we can alternate a minimization step with respect to t, u, w with a maximization step with respect to ρ_t, ρ_w. Mathematically,

$$u^{(k+1)} \ \leftarrow \ \arg\min_{u \in \mathbb{R}^n} \ \mathcal{L}(u, w^{(k)}, t^{(k)}; \rho_w^{(k)}, \rho_t^{(k)}), \tag{17}$$

$$w^{(k+1)} \ \leftarrow \ \arg\min_{r \in \mathbb{R}^n} \ \mathcal{L}(u^{(k+1)}, w, t^{(k)}; \rho_w^{(k)}, \rho_t^{(k)}), \tag{18}$$

$$t^{(k+1)} \ \leftarrow \ \arg\min_{t \in \mathbb{R}^{2n}} \ \mathcal{L}(u^{(k+1)}, w^{(k+1)}, t; \rho_w^{(k)}, \rho_t^{(k)}), \tag{19}$$

$$\rho_w^{(k+1)} \ \leftarrow \ \rho_w^{(k)} - \beta_r \left(w^{(k+1)} - (Ku^{(k+1)} - g) \right), \tag{20}$$

$$\rho_t^{(k+1)} \ \leftarrow \ \rho_t^{(k)} - \beta_t \left(t^{(k+1)} - Du^{(k+1)} \right). \tag{21}$$

The solution of the primal sub-problem (17) can be efficiently computed by means of standard linear Fast Fourier Transform (FFT) solvers. The sub-problem (18) can be solved in closed-form by following [3, Section 3]. Finally, the sub-problem (19) can be solved by computing efficiently the proximal operator of the anisotropic 1 norm, for which the proof of [2, Proposition 6.3] can be easily adapted. The regularization parameter μ is updated along the iterations so as to fulfill the global discrepancy principle as described in [4]. We refer the reader also to [5, 7] for more details on the numerical solution of the algorithm.

5   Experimental Results

In this section, we evaluate the performance of the proposed BLTV-L$_2$ restoration model compared with the baseline TV-L$_2$ model, also solved by ADMM. The stopping criteria of the ADMM for both models are defined based on the number of iterations as well as on the iterates relative change, i.e. we stop iterating as soon as

$$k \geq 1500 \quad \text{or} \quad \delta^{(k)} := \frac{\|u^{(k)} - u^{(k-1)}\|}{\|u^{(k-1)}\|} \leq 10^{-6}. \tag{22}$$

The quality of the restored images u^* is measured by means of the Improved Signal-to-Noise Ratio $\text{ISNR}(u^*, g, u) = 10 \log_{10} \frac{\|g-u\|_2^2}{\|u^*-u\|_2^2}$, with u denoting the original uncorrupted image, and of the Structural-Similarity-Index (SSIM) [11].

We consider the test image brain in Fig. 2(a) (570×430) and the test image abdomen in Fig. 4(a) (350×480) with pixel values between 0 and 255, synthetically corrupted by space-invariant blur with Gaussian kernel of parameters band $= 9$, sigma $= 2$, and by AWGN of different levels $\sigma \in \{10, 20\}$ - see, e.g., Figs. 2(b) and 4(b). The parameter maps are computed at the beginning starting from the observed image g and then updated every 300 iterations based on the current iterate. The radius of the neighborhoods used for the local parameter estimation has been set equal to 8 and 5 for the brain and abdomen test images, respectively. The reconstructions of brain for $\sigma = 20$ and of abdomen for $\sigma = 10$ via TV-L$_2$ and BLTV-L$_2$ models are shown in Figs. 2(c)–(d) and Figs. 4(c)–(d), respectively.

From a visual inspection, the restoration via BLTV-L$_2$ seems to be more neat and less cartooned than the TV reconstructions. As reported in Table 1, the ISNR and SSIM values for the two test images and for different noise levels obtained by the BLTV-L$_2$ model outperform the ones reached by the TV-L$_2$ model.

The final parameter maps computed by BLTV-L$_2$ are shown in Figs. 3 and 5.

Computational Times. We tested the joint parameter estimation + reconstruction model on a standard laptop with inbuilt MATLAB software, version 2016b. As far as the ML parameter estimation of the parameter maps procedure is concerned, we notice that the update (12) is explicit, thus very cheap, whereas the computation of θ^* in (13) requires the solution of an optimisation problem.

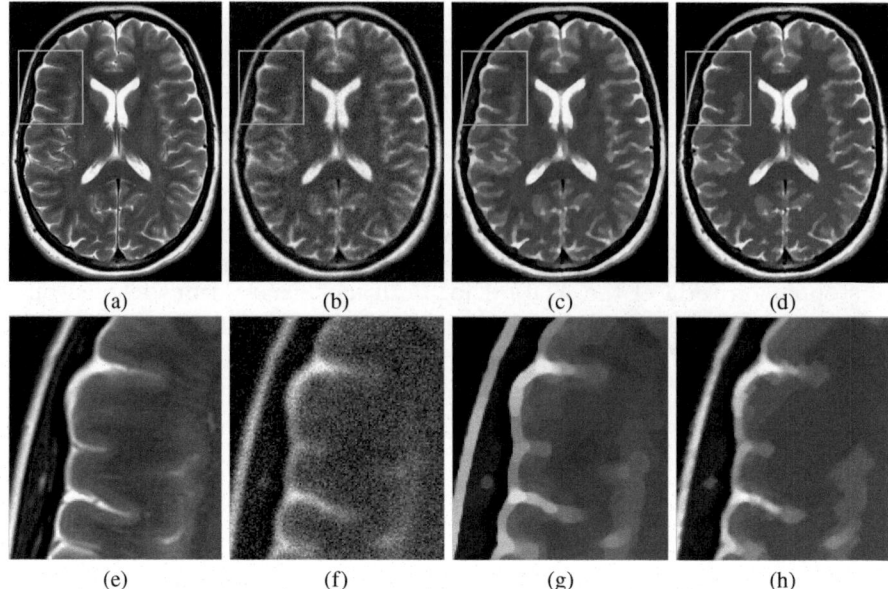

Fig. 2. *First row.* Original test image **brain** (a), observed image corrupted by Gaussian blur and AWGN with $\sigma = 20$ (b), TV reconstruction (c), BLTV reconstruction (d). *Second row.* Close-up(s) of the first row.

Table 1. Maximum ISNR/SSIM values achieved by TV-L$_2$ and BLTV-L$_2$ on **brain** and **abdomen** test images corrupted by AWGN of two different levels.

	brain				abdomen			
	$\sigma = 10$		$\sigma = 20$		$\sigma = 10$		$\sigma = 20$	
	TV-L$_2$	BLTV-L$_2$	TV-L$_2$	BLTV-L$_2$	TV-L$_2$	BLTV-L$_2$	TV-L$_2$	BLTV-L$_2$
ISNR	4.27	**5.52**	5.00	**6.67**	3.76	**4.66**	6.10	**6.77**
SSIM	0.87	**0.88**	0.83	**0.85**	0.78	**0.80**	0.74	**0.76**

We solve the problem by line-searching upon a suitable discretization of the parameter space. For Fig. 2, the ML parameter estimation procedure took 8.45 s.

The ADMM algorithmic sub-steps with automatic parameter update every 300 iterations computes the numerical solution in 381 secs for the high-resolution image in Fig. 2. A possible way to accelerate the speed of the algorithm would be the computation of the parameter maps only in terms of the given image and not along the iterations, although of course that would render a less accurate result.

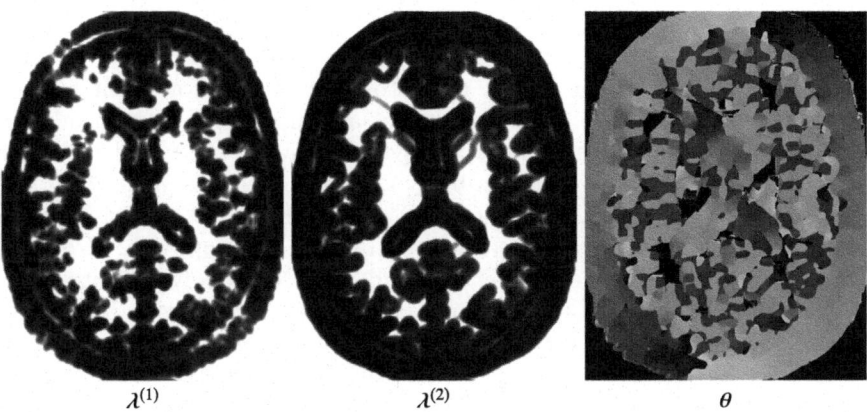

$\lambda^{(1)}$ $\lambda^{(2)}$ θ

Fig. 3. Final parameter maps for `brain` test image.

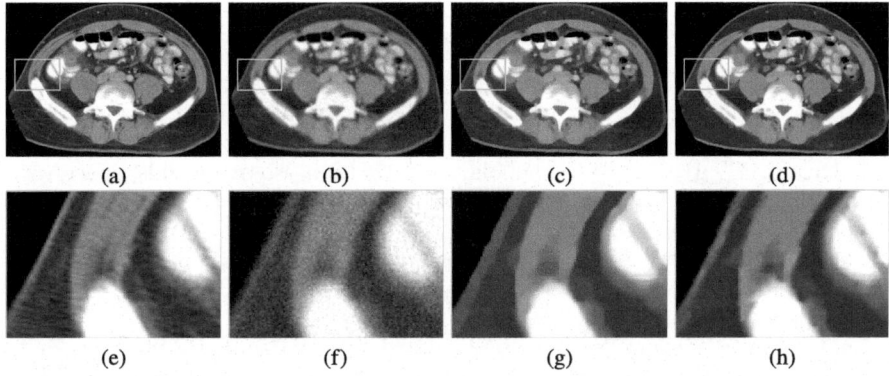

(a)   (b)   (c)   (d)

(e)   (f)   (g)   (h)

Fig. 4. *First row.* Original test image `abdomen` (a), observed image corrupted by Gaussian blur and AWGN with $\sigma = 10$ (b), TV reconstruction (c), BLTV reconstruction (d). *Second row.* Close-up(s) of the first row.

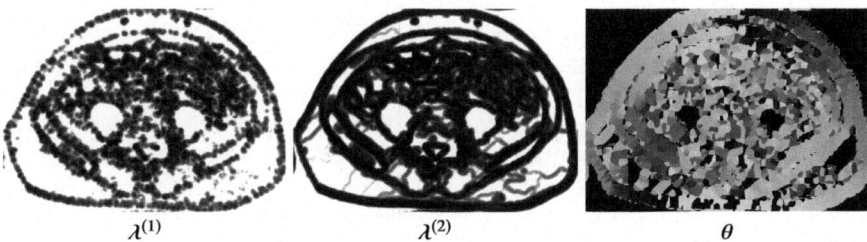

$\lambda^{(1)}$ $\lambda^{(2)}$ θ

Fig. 5. Final parameter maps for `abdomen` test image.

6   Conclusions and Future Works

We presented a new space-variant anisotropic regularization term for image restoration based on the *a priori* statistical assumption that the gradients of the unknown target image distribute locally according to space-variant bivariate Laplace distributions. The high flexibility of the proposed regularizer together with the presented ML parameters estimation procedure and ADMM-based minimization algorithm yield a very effective and efficient approach. Preliminary experiments on images corrupted by blur and AWGN strongly indicate that the proposed variational model achieves high-quality restorations and, in particular, outperforms by far the results obtained by classical TV-regularized restoration models. Coupling the proposed regularizer with other fidelity terms suitable for dealing with noises other than Gaussian - such as, e.g., Laplace, Poisson and mixed Poisson-Gaussian [8] - is a matter being studied.

References

1. Boyd, S., Parikh, N., Chu, E., Peleato, B., Eckstein, J.: Distributed optimization and statistical learning via the alternating direction method of multipliers. Found. Trends Mach. Learn. **3**, 1–122 (2011)
2. Calatroni, L., Lanza, A., Pragliola, M., Sgallari, F.: A flexible space-variant anisotropic regularization for image restoration with automated parameter selection. SIAM J. Imaging Sci. **12**, 1001–1037 (2019)
3. Calatroni, L., Lanza, A., Pragliola, M., Sgallari, F.: Adaptive parameter selection for weighted-TV image reconstruction problems. J. Phys.: Conf. Ser. NCMIP 2019 (2019, to appear)
4. He, C., Hu, C., Zhang, W., Shi, B.: A fast adaptive parameter estimation for total variation image restoration. IEEE Trans. Image Process. **23**, 4954–4967 (2014)
5. Lanza, A., Morigi, S., Pragliola, M., Sgallari, F.: Space-variant generalised Gaussian regularisation for image restoration. Comput. Methods Biomech. Biomed. Eng.: Imaging Vis., 1–14 (2018). https://doi.org/10.1080/21681163.2018.1471620
6. Lanza, A., Morigi, S., Pragliola, M., Sgallari, F.: Space-variant TV regularization for image restoration. In: Tavares, J.M.R.S., Natal Jorge, R.M. (eds.) VipIMAGE 2017. LNCVB, vol. 27, pp. 160–169. Springer, Cham (2018)
7. Lanza, A., Morigi, S., Sgallari, F.: Constrained TV_p-ℓ_2 model for image restoration. J. Sci. Comput. **68**, 64–91 (2016)
8. Lanza, A., Morigi, S., Sgallari, F., Wen, Y.W.: Image restoration with Poisson-Gaussian mixed noise. Comput. Methods Biomech. Biomed. Eng.: Imaging Vis. **2**, 12–24 (2014)
9. Rudin, L.I., Osher, S., Fatemi, E.: Nonlinear total variation based noise removal algorithms. Phys. D: Nonlinear Phenom. **60**, 259–268 (1992)
10. Stuart, A.M.: Inverse problems: a Bayesian perspective. Acta Numerica **19**, 451–559 (2010)
11. Zhou, W., Bovik, A., Sheikh, H., Simoncelli, E.: Image quality assessment: from error visibility to structural similarity. IEEE Trans. Image Process. **13**, 600–612 (2004)

Mechanical and Imaging Models-Based Image Registration

Kateřina Škardová[1], Matthias Rambausek[2], Radomír Chabiniok[3,2,4],
and Martin Genet[2,3(✉)]

[1] Department of Mathematics, Faculty of Nuclear Sciences and Physical Engineering,
Czech Technical University in Prague, Prague, Czech Republic
`katerina.solovska@fjfi.cvut.cz`
[2] Solid Mechanics Laboratory (LMS),
École Polytechnique/Institut Polytechnique de Paris/CNRS, Palaiseau, France
{`matthias.rambausek,martin.genet`}`@polytechnique.edu`
[3] M3DISIM Team, INRIA/Paris-Saclay University, Palaiseau, France
`radomir.chabiniok@inria.fr`
[4] School of Biomedical Engineering and Imaging Sciences (BMEIS),
St Thomas' Hospital, King's College London, London, UK

Abstract. Image registration plays an increasingly important role in
many fields such as biomedical or mechanical engineering. Generally
speaking, it consists in deforming a (moving) source image to match
a (fixed) template image. Many approaches have been proposed over
the years; if new model-free machine learning-based approaches are now
beginning to provide robust and accurate results, extracting motion from
images is still most commonly based on combining some statistical analysis of the images intensity and some model of the underlying deformation
as initial guess or regularizer. These approaches may be efficient even
for complex type of motion; however, any artifact in the source image
(*e.g.*, partial voluming, local decrease of signal-to-noise ratio or even
local signal void), drastically deteriorates the registration. This paper
introduces a novel approach of extracting motion from biomedical image
series, based on a model of the imaging modality. It is, to a large extent,
independent of the type of model and image data–the pre-requisite is
to incorporate biomechanical constraints into the motion of the object
(organ) of interest and being able to generate data corresponding to the
real image, *i.e.*, having an imaging model at hand. We will illustrate the
method with examples of synthetically generated 2D tagged magnetic
resonance images.

1 Introduction

Image registration plays an increasingly important role in many fields such as
biomedical (Wang and Amini 2012; Tobon-Gomez et al. 2013) or mechanical
(Bornert et al. 2012; Sutton and Hild 2015) engineering. Taking cardiology as an
example, the heart represents a particular organ in the field of medical imaging

© Springer Nature Switzerland AG 2019
J. M. R. S. Tavares and R. M. Natal Jorge (Eds.): VipIMAGE 2019, LNCVB 34, pp. 77–85, 2019.
https://doi.org/10.1007/978-3-030-32040-9_9

since *(i)* a large number of modalities and image types have been developed and the process of combining them – in which registration represents a crucial step – may maximize the amount of information for diagnostic purposes with possible outcomes in long-term management of patients' therapy (Helsen et al. 2018; Rutz et al. 2017); and *(ii)* the heart is a moving organ and registering a motion series of images allows to extract its functional characteristics (Wang and Amini 2012; Tobon-Gomez et al. 2013). While *(i)* is expected to play an increasing role in the near future when a comprehensive information across medical fields will be sought for, *(ii)* is already a very active topic nowadays, aiming at an accurate and reproducible description of heart function. It is connecting the communities of imaging, image processing, biomedical and mechanical engineering and statistical methods. Image-based motion data are paramount, for instance, in designing personalized biomechanical cardiac models (Smith et al. 2011; Krishnamurthy et al. 2013; Finsberg et al. 2018), which could be used for augmented diagnosis (Chabiniok et al. 2012; Xi et al. 2016) and treatments (Sermesant et al. 2012; Rausch et al. 2017).

Generally speaking, image registration consists in deforming a (moving) source image to match a (fixed) template image. Although new model-free machine learning-based approaches are now beginning to provide robust and accurate results (Suinesiaputra et al. 2015; Qin et al. 2018), extracting motion from images is still most commonly based on combining some statistical analysis of the images intensity and some model of the underlying deformation as initial guess or regularizer (Shi et al. 2012; Bornert et al. 2012; Tobon-Gomez et al. 2013). These approaches may be efficient even for complex type of motion; however, any artifact in the source image (*e.g.*, partial voluming, local decrease of signal-to-noise ratio or even local signal void), drastically deteriorates the registration. Mathematical, geometrical or mechanical regularization can help to alleviate the issue (Christensen et al. 1996; Veress et al. 2005; Genet et al. 2016). However, it is intrinsic to the intensity-based approach, and problems remain. For instance, radial strains are systematically underestimated in 3D tagging cardiac images (Tobon-Gomez et al. 2013; Genet et al. 2018).

This paper introduces a novel approach of extracting motion from biomedical image series, based on a model of the imaging modality. It is, to a large extent, independent of the type of image –the only pre-requisite is to be able to generate a realistic image associated to a given shape of the considered object, *i.e.*, having an imaging model at hand. We will illustrate the method with examples of synthetically generated tagged magnetic resonance images.

2   Methods

2.1   Problem Setting

$\tilde{I}_0$ & $\tilde{I}$ are two images representing the same body $\mathcal{B}$ at two instants t_0 & t:

$$\tilde{I}_0 : \begin{cases} \square_0 \to \mathbb{R} \\ \underline{X} \mapsto \tilde{I}_0\left(\underline{X}\right) \end{cases} \quad , \qquad \tilde{I} : \begin{cases} \square \to \mathbb{R} \\ \underline{x} \mapsto \tilde{I}\left(\underline{x}\right) \end{cases} \quad , \tag{1}$$

where $\square_0$ & $\square$ are the image domains at t_0 & t, which are usually identical. The domains occupied by the body $\mathcal{B}$ at t_0 & t are denoted Ω_0 & Ω, respectively. The problem is to find the smooth mapping $\underline{\Phi}$ between materials points of the reference and deformed domains:

$$\underline{\Phi} : \begin{cases} \Omega_0 \to \Omega \\ \underline{X} \mapsto \underline{x} = \underline{\Phi}(\underline{X}) \end{cases}, \tag{2}$$

where $\underline{X}$ & $\underline{x}$ denote the position of a given material point in the reference and deformed configurations. Equivalently, the problem can be formulated in terms of the smooth displacement field $\underline{U}$:

$$\underline{U} : \begin{cases} \Omega_0 \to \mathbb{R}^3 \\ \underline{X} \mapsto \underline{U}(\underline{X}) = \underline{\Phi}(\underline{X}) - \underline{X} \end{cases}. \tag{3}$$

Because of intrinsic ill-posedness, it is formulated as a minimization problem:

$$\text{find } \underline{\Phi}^{\text{sol}} = \text{argmin}_{\{\underline{\Phi}\}} \left\{ \Psi^{\text{cor}}(\underline{\Phi}) \right\}, \tag{4}$$

where Ψ^{cor} is the "correlation energy", or image similarity metric, which is in general not quadratic, but which we assume convex, at least in the neighborhood of the solution. Many approaches have been proposed to regularize this ill-posed problem (Christensen et al. 1996; Veress et al. 2005; Mansi et al. 2011; Wang and Amini 2012; Tobon-Gomez et al. 2013), which will not be discussed in details here. Specifically, details on the equilibrium gap regularization, an efficient mechanistic approach, can be found in (Claire et al. 2004; Genet et al. 2018; Lee and Genet 2019).

2.2   Intensity-Based Approaches

In intensity-based approaches, the following correlation energy is generally used:

$$\Psi^{\text{cor}}(\underline{\Phi}) = \frac{1}{2} \int_{\Omega_0} \left(\tilde{I}(\underline{\Phi}(\underline{X})) - \tilde{I}_0(\underline{X}) \right)^2 d\Omega_0. \tag{5}$$

The main drawback of this method is that it consists in mapping both images directly, so that any artifact in the images, especially partial voluming, will pollute the correlation. Here we propose to circumvent this problem by using, besides the geometrical model of the object to track, a model of the imaging process. Thus we can generate synthetic images from the model at every time step, and look for the mapping that best matches the synthetic and acquired images.

2.3   Proposed Approach

2.3.1   Imaging Model

For a body $\mathcal{B}$ occupying the spatial domain Ω_0, we define the "pure" cine-like image:

$$I_{\Omega_0}^{\text{cine}}(\underline{X}) := \mathbb{1}_{\Omega_0}(\underline{X}) = \begin{cases} 1 \text{ if } \underline{X} \in \Omega_0 \\ 0 \text{ otherwise} \end{cases} \quad \forall \underline{X} \in \square_0, \tag{6}$$

as well as a "pure" tagging-like image:

$$I_{\Omega_0}^{\text{tag}}\left(\underline{X}\right) := \mathbb{1}_{\Omega_0}\left(\underline{X}\right)\left|\sin\left(\underline{k}\cdot\left(\underline{X}-\underline{X_0}\right)\right)\right| \quad \forall\underline{X}\in\square_0, \tag{7}$$

where $\underline{k}$ and $\underline{X_0}$ are tagging pattern parameters. Tagged images are usually combined to form a grid image, so the associated model (taking $\underline{X_0}=0$) is:

$$I_{\Omega_0}^{\text{grid}}\left(\underline{X}\right) := \mathbb{1}_{\Omega_0}\left(\underline{X}\right)\sqrt[n]{\prod_{i=1}^{n}\left|\sin\left(\pi X_i/s\right)\right|} \quad \forall\underline{X}\in\square_0, \tag{8}$$

where s is the tag line distance and n the image dimension.

Once the body has moved and occupies the domain $\Omega = \Phi\left(\Omega_0\right)$, the deformed image is simply:

$$I_{\Omega_0,\Phi}\left(\underline{x}\right) = I_{\Omega_0}\left(\Phi^{-1}\left(\underline{x}\right)\right) \quad \forall\underline{x}\in\square. \tag{9}$$

Remark 1. For cine-like images, since they do not require to follow material points but only domain boundaries, we can express the deformed image solely from the deformed domain:

$$I_{\Omega_0,\Phi}^{\text{cine}}\left(\underline{x}\right) = \mathbb{1}_{\Omega_0}\left(\Phi^{-1}\left(\underline{x}\right)\right) = \mathbb{1}_{\Omega}\left(\underline{x}\right) = I_{\Omega}^{\text{cine}}\left(\underline{x}\right) \quad \forall\underline{x}\in\square. \tag{10}$$

However, it is not possible for tagging-like images, as the mapping is required to track material points.

Because MR has limited bandwidth, the frequency content of actual images is only sampled within a box window, which means, in space, that they are convoluted by a cardinal sine kernel:

$$\tilde{I}_{\Omega_0,\Phi}\left(\underline{x}\right) = \int_{\square} I_{\Omega_0,\Phi}\left(\underline{y}\right)\nu\left(\underline{y}-\underline{x}\right)d\square \quad \forall\underline{x}\in\square, \tag{11}$$

with:

$$\nu\left(\underline{z}\right) := \prod_i \text{sinc}\left(\frac{z_i}{\Delta_i}\right), \tag{12}$$

where $\underline{\Delta}$ denotes the image resolution.

2.3.2 Correlation Energy

We now seek the mapping that best matches the generated and actual images, which minimizes the following correlation energy:

$$\Psi^{\text{cor}}\left(\underline{\Phi}\right) = \frac{1}{2}\int_{\Omega_0}\left(\tilde{I}_{\Omega_0,\Phi}\left(\underline{\Phi}\left(\underline{X}\right)\right) - \tilde{I}\left(\underline{\Phi}\left(\underline{X}\right)\right)\right)^2 d\Omega_0, \tag{13}$$

where:

$$\tilde{I}_{\Omega_0,\Phi}\left(\underline{\Phi}\left(\underline{X}\right)\right) = \int_{\square} I_{\Omega_0,\Phi}\left(\underline{y}\right)\nu\left(\underline{y}-\underline{\Phi}\left(\underline{X}\right)\right)d\square \quad \forall\underline{X}\in\Omega_0. \tag{14}$$

Remark 2. In case the effect of image resolution is neglected, *i.e.*, the convolution kernel is replaced by Dirac function, this leads to:

$$\tilde{I}_{\Omega_0,\Phi}\left(\underline{\Phi}\left(\underline{X}\right)\right) = I_{\Omega_0,\Phi}\left(\underline{\Phi}\left(\underline{X}\right)\right) = I_{\Omega_0}\left(\underline{X}\right) \tag{15}$$

3 Results

Four sets of synthetic images were generated in order to test the proposed method, representing both rigid (translation & rotation) and non-rigid (compression & shear) transformations of a simple square object. All sequences were 2D, 100×100 pixels, 30 frames long. Tagging magnetic resonance image model (8) was used, with $s = 10$ pixels. For all transformations, a noise-free image was generated (*i.e.*, SNR $= +\infty$), as well as noisy images, by adding zero-mean gaussian noise with various standard deviations: 0.1 (*i.e.*, SNR $= 10$), 0.2 (SNR $= 5$) & 0.3 (SNR $= 3$). In a first step, well resolved images were used, such that the effect of image resolution and discretization can be neglected, and expression (15) can be used.

All sets of images were registered. Details on the solution procedure can be found in (Genet et al. 2018), and the code can be found at https://gitlab.inria.fr/mgenet/dolfin_dic. Figures 1, 2, 3 and 4 show the registration results, in the form of the warped mesh on top of the synthetic images (with various levels of noise). The quality of the registration is clearly seen, except for the rotation case (Fig. 2) where one node ends up outside the image. Note that this could be alleviated by using some proper mechanical regularization in order to filter out the non physical deformation patterns (Genet et al. 2018; Berberoglu et al. 2019). In other cases, even for quite noisy images (as illustrated in Fig. 3), the registration is almost perfect.

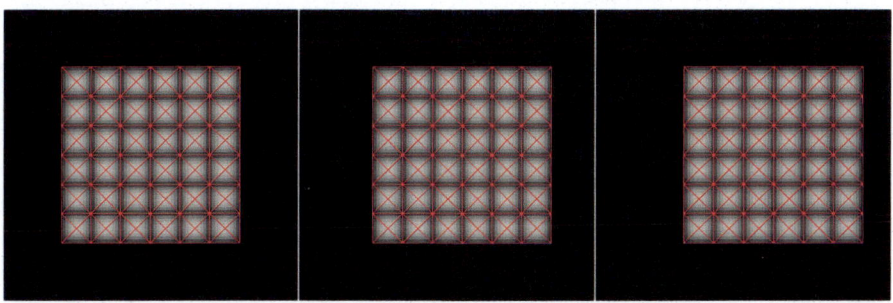

Fig. 1. Registration results on the translation case, for the noise-free images.

This is better quantified in Fig. 5, which shows, for all tested transformations and SNR, the normalized root mean square displacement error:

$$RMSE_U = \frac{\sqrt{\dfrac{1}{T} \displaystyle\int_0^T \dfrac{1}{|\Omega_0|} \int_{\Omega_0} \left\| \underline{U} - \underline{U}^{\text{ref}} \right\|^2}}{\sqrt{\dfrac{1}{T} \displaystyle\int_0^T \dfrac{1}{|\Omega_0|} \int_{\Omega_0} \left\| \underline{U}^{\text{ref}} \right\|^2}}, \tag{16}$$

Fig. 2. Registration results on the rotation case, for the SNR = 5 images. The registration fails at one node, which could be prevented by using proper mechanical regularization.

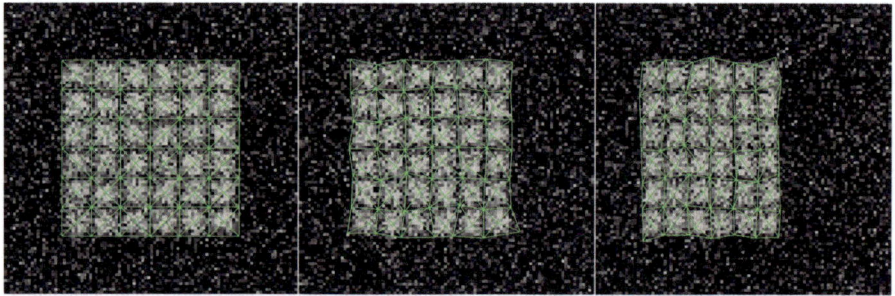

Fig. 3. Registration results on the compression case, for the SNR = 3 images.

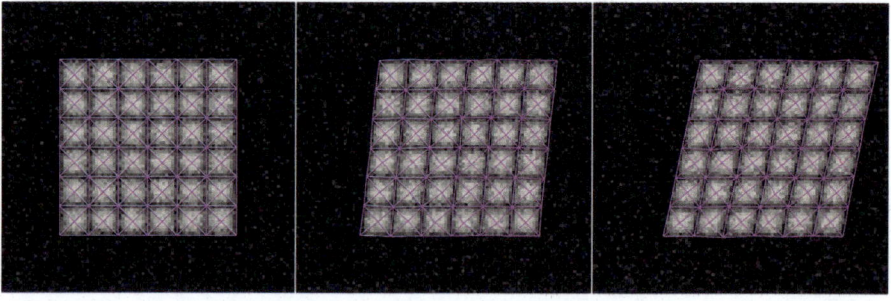

Fig. 4. Registration results on the shear case, for the SNR = 10 images.

where $\underline{U}^{\text{ref}}$ is the exact solution used to generate the synthetic images. A similar trend is found in Fig. 6, which shows the impact of SNR on the normalized root mean square image similarity error:

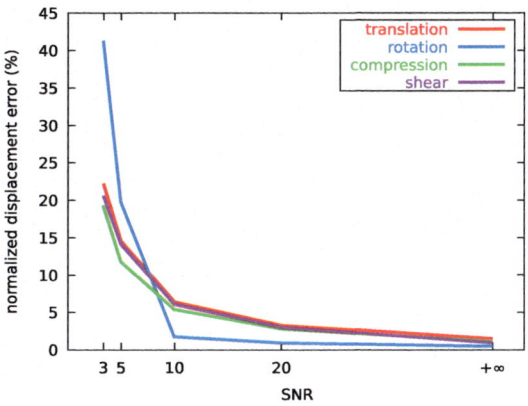

Fig. 5. Impact of SNR on the registration, in terms of normalized root mean square displacement error.

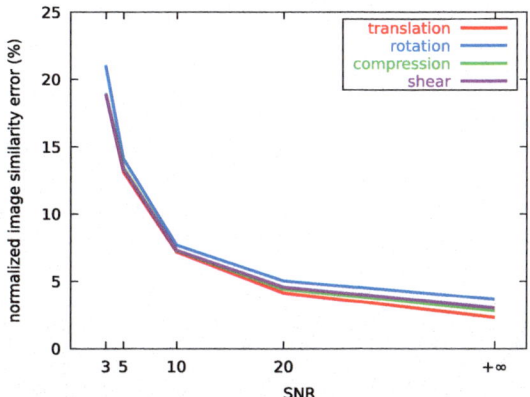

Fig. 6. Impact of SNR on the registration, in terms of normalized root mean square image similarity error.

$$RMSE_I = \frac{\sqrt{\frac{1}{T}\int_0^T \frac{1}{|\Omega_0|}\int_{\Omega_0}\left(\tilde{I}_{\Omega_0,\Phi}\left(\underline{\Phi}\left(\underline{X}\right)\right) - \tilde{I}\left(\underline{\Phi}\left(\underline{X}\right)\right)\right)^2}}{\sqrt{\frac{1}{T}\int_0^T \frac{1}{|\Omega_0|}\int_{\Omega_0}\left(\tilde{I}\left(\underline{\Phi}\left(\underline{X}\right)\right)\right)^2}}. \tag{17}$$

4   Conclusion and Perspectives

A novel image registration method has been introduced, which uses a model of the imaging device itself. Formulated and solved using the finite element method, it can naturally be used in conjunction with existing mechanics-based regularization approaches. It was tested on simple rigid and non-rigid transformations, and was shown to perform well on all of them.

The set of testing sequences will be extended in the future, notably to include the impact of image resolution. More complicated shapes and transformation will be tested as well. *In fine*, the method will be tested on real images, and could alleviate the image resolution-induced underestimation of radial strain classically found in 3D tagged magnetic resonance images.

Acknowledgements. The work was supported by the Inria-UT Southwestern Medical Center Dallas Associated team ToFMOD, and partially by the Ministry of Health of the Czech Republic (project No. NV19-08-00071).

References

Berberoglu, E., et al.: Validation of finite element image registration-based cardiac strain estimation from magnetic resonance images. In: 90th Annual Meeting of the International Association of Applied Mathematics and Mechanics (GAMM) (2019)

Bornert, M., et al.: Digital image correlation. In: Grédiac, M., Hild, F., Pineau, A. (eds.) Full-Field Measurements and Identification in Solid Mechanics. Wiley, Hoboken (2012). https://doi.org/10.1002/9781118578469.ch6

Chabiniok, R., et al.: Estimation of tissue contractility from cardiac cine- MRI using a biomechanical heart model. Biomech. Model. Mechanobiology **11**(5) (2012). https://doi.org/10.1007/s10237-011-0337-8

Christensen, G.E., Rabbitt, R.D., Miller, M.I.: Deformable templates using large deformation kinematics. IEEE Trans. Image Process. **5**(10) (1996). https://doi.org/10.1109/83.536892. A Publication of the IEEE Signal Processing Society

Claire, D., Hild, F., Roux, S.: A finite element formulation to identify damage fields: the equilibrium gap method. Int. J. Numer. Methods Eng. **61**(2) (2004). https://doi.org/10.1002/nme.1057

Finsberg, H., et al.: Efficient estimation of personalized biventricular mechanical function employing gradient-based optimization. Int. J. Numer. Methods Biomed. Eng. **34**(7) (2018). https://doi.org/10.1002/cnm.2982

Genet, M., Stoeck, C., et al.: Equilibrated warping: finite element image registration with finite strain equilibrium gap regularization. Med. Image Anal. (2018) https://doi.org/10.1016/j.media.2018.07.007

Genet, M., Stoeck, C.T., et al.: Finite element digital image correlation for cardiac strain analysis from 3D whole-heart tagging. In: 24th Annual Meeting of the International Society for Magnetic Resonance in Medicine (ISMRM2016) (2016)

Helsen, F., et al.: Advanced imaging to phenotype patients with a systemic right ventricle. J. Am. Heart Assoc. **7**(20) (2018). https://doi.org/10.1161/JAHA.118.009185

Krishnamurthy, A., et al.: Patient-specific models of cardiac biomechanics. J. Comput. Phys. **244**. https://doi.org/10.1016/j.jcp.2012.09.015

Lee, L.C., Genet, M.: Validation of equilibrated warping–image registration with mechanical regularization–on 3D ultrasound images. In: Coudière, Y. (eds.) et al.: Functional Imaging and Modeling of the Heart (FIMH), vol. 11504. Springer, Cham (2019). https://doi.org/10.1007/978-3-030-21949-9_36

Mansi, T., et al.: iLogDemons: a demons-based registration algorithm for tracking incompressible elastic biological tissues. Int. J. Comput. Vis. **92**(1) (2011). https://doi.org/10.1007/s11263-010-0405-z

Qin, C., et al.: Joint learning of motion estimation and segmentation for cardiac MR image sequences. In: International Conference on Medical Image Computing and Computer-Assisted Intervention (MICCAI) (2018)

Rausch, M.K., et al.: A virtual sizing tool for mitral valve annuloplasty. Int. J. Numer. Methods Biomed. Eng. **33**(2) (2017). https://doi.org/10.1002/cnm.2788

Rutz, T., et al.: Evolution of right ventricular size over time after tetralogy of fallot repair: a longitudinal cardiac magnetic resonance study. Eur. Heart J. - Cardiovasc. Imaging **18**(3) (2017). https://doi.org/10.1093/ehjci/jew273

Sermesant, M., et al.: Patient-specific electromechanical models of the heart for the prediction of pacing acute effects in CRT: a preliminary clinical validation. Med. Image Anal. **16**(1) (2012). https://doi.org/10.1016/j.media.2011.07.003

Shi, W., et al.: A comprehensive cardiac motion estimation framework using both untagged and 3-D tagged MR images based on nonrigid registration. IEEE Trans. Med. Imaging **31**(6) (2012). https://doi.org/10.1109/TMI.2012.2188104

Smith, N.P., et al.: euHeart: personalized and integrated cardiac care using patient-specific cardiovascular modelling. Interface Focus **1**(3) (2011). https://doi.org/10.1098/rsfs.2010.0048

Suinesiaputra, A., et al.: Big heart data: advancing health informatics through data sharing in cardiovascular imaging. IEEE J. Biomed. Health Inf. **19**(4) (2015). https://doi.org/10.1109/JBHI.2014.2370952

Sutton, M.A., Hild, F.: Recent advances and perspectives in digital image correlation. Exp. Mech. **55**(1) (2015). https://doi.org/10.1007/s11340-015-9991-6

Tobon-Gomez, C., et al.: Benchmarking framework for myocardial tracking and deformation algorithms: an open access database. Med. Image Anal. **17**(6) (2013). https://doi.org/10.1016/j.media.2013.03.008

Veress, A.I., Gullberg, G.T., Weiss, J.A.: Measurement of strain in the left ventricle during diastole with cine-MRI and deformable image registration. J. Biomech. Eng. **127**(7) (2005). https://doi.org/10.1115/1.2073677

Wang, H., Amini, A.A.: Cardiac motion and deformation recovery from MRI: a review. IEEE Trans. Med. Imaging **31**(2) (2012). https://doi.org/10.1109/TMI.2011.2171706

Xi, C., et al.: Patient-specific computational analysis of ventricular mechanics in pulmonary arterial hypertension. J. Biomech. Eng. **138**(11) (2016). https://doi.org/10.1115/1.4034559

Cross Recurrence Quantitative Analysis of Functional Magnetic Resonance Imaging

A. Lombardi[1], E. Lella[1,2], D. Diacono[1], N. Amoroso[1,2], A. Monaco[1], R. Bellotti[1,2], and S. Tangaro[1(✉)]

[1] Sezione di Bari, Istituto Nazionale di Fisica Nucleare, Bari, Italy
sonia.tangaro@ba.infn.it
[2] Dipartimento di Fisica, Università degli Studi di Bari, Bari, Italy

Abstract. In this paper, a generalized synchronization-based metric is described to assess functional connectivity in human brain. The metric is a generalized synchronization measure that considers both the amplitude and phase coupling between pairs of fMRI series. This method differs from the correlation measures used in the literature, as it is more sensitive to nonlinear coupling phenomena between time series and it is more robust against the physiological noise.

Keywords: Functional connectivity · fMRI · Cross recurrence plot · Synchronization

1 Introduction

The complex and non-linear nature of brain dynamics is widely discussed in literature. The human brain is modeled as a complex network composed of anatomical and functional subsystems that interact and allow the performance of high- and low-level cognitive functions. Such interactions are, at the state of the art, modelled through the mathematical framework of the Graph Theory, which allows to treat each subsystem as a node of a network and to identify specific topological properties concerning for example the centrality or the role of a node in the network. These techniques can be applied both to structural neuroimaging data (sMRI), which describe the brain in terms of gray matter, white matter and cerebrospinal fluid, or to functional data, which describe the dynamic brain activity through functional magnetic resonance imaging (fMRI) or electrophysiological acquisitions (EEG/MEG) [1].

However, most of the proposed models may present some limitations: for example, many approaches are static and linear as they consider the mapping of brain interactions into a specific time frame with analysis techniques that assume a linear functioning of the system.

Over the past few years, there has been an increasing interest in inferring connectivity properties from fMRI data. Functional connectivity analysis aims at assessing the strength of functional coupling between the signal responses in distinct brain areas [2]. According to the complex network framework, the anatomical regions of interest are the nodes of the network, connected by edges resulting from the adopted interregional interaction metrics. Pairwise fMRI time series connections are usually estimated

© Springer Nature Switzerland AG 2019
J. M. R. S. Tavares and R. M. Natal Jorge (Eds.): VipIMAGE 2019, LNCVB 34, pp. 86–92, 2019.
https://doi.org/10.1007/978-3-030-32040-9_10

through zero-lag correlation metrics, leading to a weighted network whose links quantify the statistical similarity between pairs of regions.

Here, we describe a novel method for quantifying functional coupling between fMRI time series and constructing functional brain networks. This method has been proven effective in detecting active sub-networks during a high-level cognitive process [3], in describing age-related functional patterns [4] and in identifying and quantifying pathological states [5].

2 Materials and Methods

In this work the mathematical details with an explanatory example are given to elucidate the synchronization index method.

A phase-space framework is used to map pairs of signals in their reconstructed phase space, i.e. a topological representation of their behavior under all possible initial conditions. This method assumes that each signal represents a projection of a higher-dimensional dynamical system evolving in time, whose trajectories are embedded into a manifold, i.e., a region of its phase space.

We use a well-known dynamic system to explain the phase-space embedding of the time series.

First, two realizations of the dynamics of the Lorenz system are generated. The Lorenz system is a simplified mathematical model for the atmospheric convection and is specified by three ordinary differential equations [6]:

$$\begin{cases} \frac{dx}{dt} = \sigma(y - x) \\ \frac{dy}{dt} = x(r - y) - y \\ \frac{dz}{dt} = xy - bz \end{cases} \tag{1}$$

where the three components x, y, z are proportional to the rate of convection, to the horizontal temperature variation, and to the vertical temperature variation. For $\sigma = 10$, $b = 8/3$, $r = 28$ the system exhibits chaotic behavior.

Figure 1 shows the time course of the three components of the system for 2000 samples and with the initial values of the three components:

$$x_1(1) = 10, y_1(1) = 10, z_1(1) = 10.$$

We then generated a second Lorenz system with the initial values of the three components: $x_2(1) = 7, y_2(1) = 10, z_2(1) = 10$.

By using the Takens's Theorem [7], it is possible to reconstruct the phase space of the system under investigation. Accordingly, the two time series $x_1(1)$ and $x_2(t)$ shown in Fig. 2 can be used to reconstruct the phase space trajectories of the two systems.

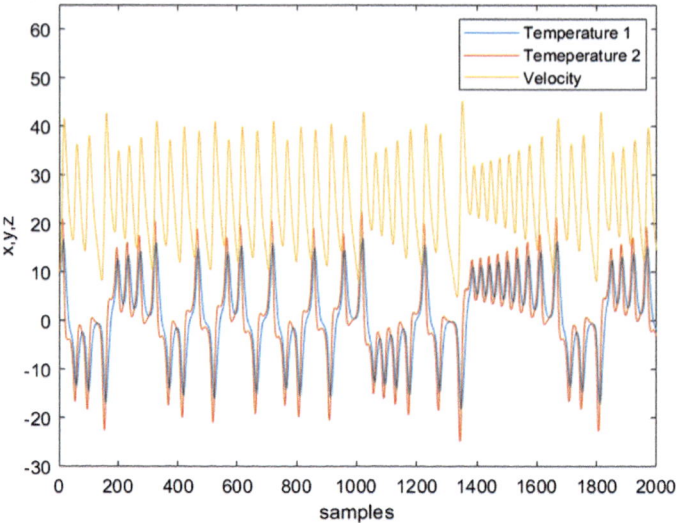

Fig. 1. Time series of the three components of the Lorenz system.

For a generic single temporal observation of a system $u(t)$, the trajectory is expressed as:

$$\vec{x}_i(t) = \left(u_i, u_{i+\tau}, \ldots, u_{i+(m-1)\tau}\right) \tag{2}$$

where m is the embedding dimension end τ is the time delay.

Both parameters must be properly selected to avoid redundancy in the phase space. The dimension m of the reconstructed phase space should be large enough to preserve the properties of the dynamical system (m $\geq$ 2D + 1, where D is the correlation dimension of the original phase space). The correct time delay τ should be chosen by determining when the samples of the time series are independent enough to be useful as coordinates of the time delayed vectors.

For the estimation of the embedded parameters m and τ several techniques have been proposed. As an example, the first local minimum of average mutual information algorithm [8] can be used to select the proper time delay. The minimum embedding dimension is usually estimated through the false nearest neighbors (FNN) algorithm [9].

After the reconstruction of the dynamic trajectories of the two systems in the phase space, it is possible to quantify their interacting behavior by projecting the phase space into the bidimensional cross recurrence plots (CRPs) [10]:

$$CR_{i,j}(\varepsilon) = \theta\left(\varepsilon - \|\vec{x}_{1i} - \vec{x}_{2j}\|\right) \; i,j = 1, \ldots, N \tag{3}$$

Where θ is the Heaviside function, ϵ is a threshold for closeness, N is the number of considered states for each system and $\| \cdot \|$ the maximum norm function.

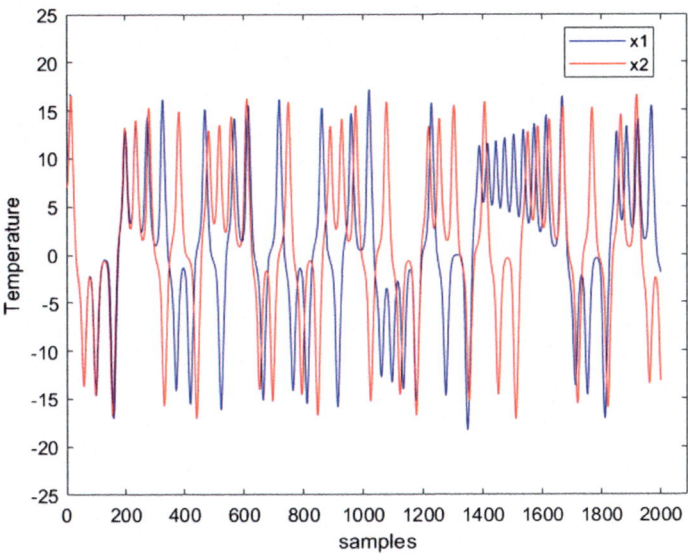

Fig. 2. Time courses of the first component of the two generated systems.

CRP is employed to reduce the dimensionality of the phase space and compare the trajectories of the interacting systems as it represents a matrix whose entries include information on the degree of closeness of each state of the first system with each state of the second system.

The value of the parameter ε must be estimated carefully, as it influences the creation of structures in the plot. The selection of an appropriate value for the threshold ε can be made by considering the influence of the observational noise that could affect the experimental measures and the minimum distance between the trajectories of the two systems. In general, choosing ε equal to few percent of the maximum phase space diameter, could ensure a sufficient number of structures in the cross recurrence plot [11], while the appearance of artifacts could be avoided by considering the signal to noise ratio for the underlying physical systems [12].

A CRP exhibits characteristic patterns that show local time relationships of the segments of the trajectories of the two interacting systems. Typical structures include single dots, diagonal lines and vertical and horizontal lines. Diagonal lines occur when the evolution of the states is similar at different times and their lengths are related to the periods during which the two systems move in similar ways remaining close to each other [13]. A CRP can also exhibit the line of synchronization (LOS), i.e., the main diagonal which implies the identity of the states of the two systems in the same time intervals. The pattern along the LOS structure can be analyzed to extract information about the synchronization of the two time series [14]. In particular, the presence of LOS suggests that the two time series are fully synchronized, while discontinuities appear when the two signals do not have the same frequency and the same phase.

We defined the synchronization time (SYNC) to quantify the mean period during which the two systems are synchronized in order to reflect the dynamical synchronization behavior of the series throughout the observation period. SYNC is proportional to the ratio of the sum of the lengths of the subsegments l_j along the LOS to the total number of samples N:

$$SYNC = \frac{1}{N} \frac{\sum_{j=1}^{N_d} l_j}{N_d} \tag{4}$$

where N_d is the total number of subsegments.

3   Results

Figure 3 shows the CRP of the two Lorenz systems for $\varepsilon = 0.9$. This threshold was chosen by using the criterion introduced in [4].

Accordingly, each time series is reconstructed from its recurrence plot following the Hirata's algorithm [15] by varying the threshold in a range of values. Then the obtained time series are correlated with the original ones and the threshold at which the maximum average correlation value is selected.

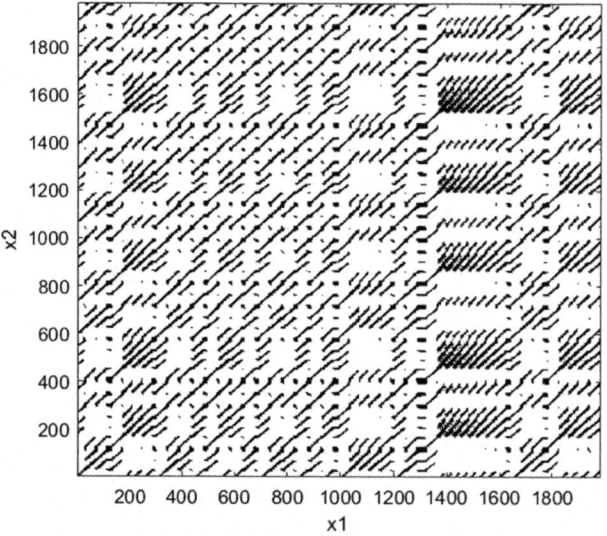

Fig. 3. CRP of the two Lorenz systems.

This threshold represents a value at which more accurate information about the original time series is retrieved, so it could also ensure a compromise between the density of the points in the plot and the removal of artifacts due to noise.

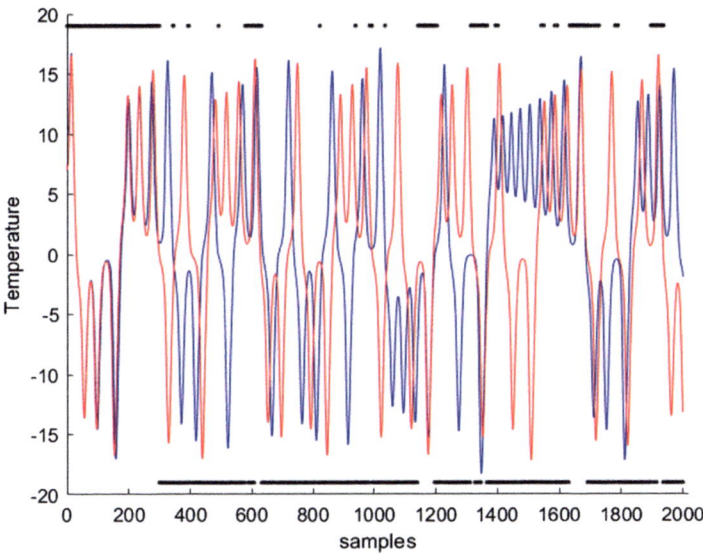

Fig. 4. Interacting behavior of the two systems represented by the LOS (positive black dots).

In Fig. 4 are displayed the two time series together with the LOS values: for sake of visualization, the presence of values in the diagonal line (unit entries) are visualized as black positive points, while the absence of values (null entries) are displayed as negative black points.

4   Discussion and Conclusion

Recently, a novel metric was introduced to assess the functional connectivity in human brain through functional magnetic resonance imaging (fMRI). Accordingly, the whole brain volume is partitioned into regions of interest (ROIs) and a phase-space framework is used to map pairs of signals of each region of interest, in their reconstructed phase space, i.e. a topological representation of their behavior under all possible initial conditions. Cross recurrence plots (CRPs) are then employed to reduce the dimensionality of the phase space and compare the trajectories of the interacting systems. The synchronization metric is then extracted from the cross recurrence to assess a coupling index of the time series.

The proposed metric is a generalized synchronization measure that takes into account both the amplitude and phase coupling between pairs of fMRI series. It differs from the correlation measures used in the literature, as it sensitive not only to the average synchronization time of two signals during the observation period, but also to their possible intermittent behavior.

In this work, some methodological procedures are elucidated to address the outlined issues by using a simple but effective example. We described all steps required to reconstruct the phase space of two interacting time series, choose the correct parameters and extract the synchronization index.

The proposed method is very promising and, in the future, may represent a robust approach to investigate the multi-scale nature of neural integration during cognitive tasks [3–5].

References

1. Bullmore, Ed., Sporns, O.: Complex brain networks: graph theoretical analysis of structural and functional systems. Nat. Rev. Neurosci. **10**(3), 186 (2009)
2. Friston, K.J.: Functional and effective connectivity: a review. Brain Connect. **1**(1), 13–36 (2011)
3. Lombardi, A., Tangaro, S., Bellotti, R., Bertolino, A., Blasi, G., Pergola, G., Taurisano, P., Guaragnella, C.: A novel synchronization-based approach for functional connectivity analysis. Complexity (2017). ISSN 1076-2787. https://doi.org/10.1155/2017/7190758
4. Lombardi, A., Amoroso, N., Diacono, D., Lella, E., Bellotti, R., Tangaro, S.: Age related topological analysis of synchronization-based functional connectivity. In: International Conference on Complex Networks and their Applications. Springer, Cham (2018)
5. Lombardi, A., Guaragnella, C., Amoroso, N., Monaco, A., Fazio, L., Taurisano, P., Pergola, G., Blasi, G., Bertolino, A., Bellotti, R., Tangaro, S.: Modelling cognitive loads in schizophrenia by means of new functional dynamic indexes. NeuroImage **195**, 150–164 (2019)
6. Sparrow, C.: The Lorenz Equations: Bifurcations, Chaos, and Strange Attractors, vol. 41. Springer, New York (2012)
7. Takens, F.: Detecting strange attractors in turbulence. In: Dynamical Systems and Turbulence, Warwick 1980, pp. 366–381. Springer (1981)
8. Fraser, A.M., Swinney, H.L.: Independent coordinates for strange attractors from mutual information. Phys. Rev. A **33**(2), 1134 (1986)
9. Kennel, M.B., Brown, R., Abarbanel, H.D.I.: Determining embedding dimension for phase-space reconstruction using a geometrical construction. Phys. Rev. A **45**(6), 3403 (1992)
10. Marwan, N., Kurths, J.: Nonlinear analysis of bivariate data with cross recurrence plots. Phys. Lett. A **302**(5), 299–307 (2002)
11. Mindlin, G.M., Gilmore, R.: Topological analysis and synthesis of chaotic time series. Phys. D: Nonlinear Phenom. **58**(1–4), 229–242 (1992)
12. Thiel, M., Romano, M.C., Kurths, J., Meucci, R., Allaria, E., Arecchi, F.T.: Influence of observational noise on the recurrence quantification analysis. Phys. D: Nonlinear Phenom. **171**(3), 138–152 (2002)
13. Marwan, N., Romano, M.C., Thiel, M., Kurths, J.: Recurrence plots for the analysis of complex systems. Phys. Rep. **438**(5), 237–329 (2007)
14. Marwan, N., Thiel, M., Nowaczyk, N.R.: Cross recurrence plot based synchronization of time series. arXiv preprint physics/0201062 (2002)
15. Hirata, Y., Horai, S., Aihara, K.: Reproduction of distance matrices and original time series from recurrence plots and their applications. Eur. Phys. J. Spec. Top. **164**(1), 13–22 (2008)

Craniofacial Morphology of Double Reed Instrument Players and Musical Performance

Miguel Pais Clemente[1]([✉]), Joaquim Mendes[2], André Moreira[3],
Afonso Pinhão Ferreira[4], and José Manuel Amarante[5]

[1] Department of Surgery and Physiology, Faculty of Medicine,
Universidade do Porto, Porto, Portugal
`miguelpaisclemente@hotmail.com`
[2] INEGI, Labiomep, Faculdade de Engenharia,
Universidade do Porto, Porto, Portugal
`jgabriel@fe.up.pt`
[3] Faculty of Dental Medicine, University of Porto, Porto, Portugal
`andre.luis.sa.moreira@gmail.com`
[4] Department of Orthodontics, Faculty of Dental Medicine,
University of Porto, Porto, Portugal
`aferreira@fmd.up.pt`
[5] Labiomep, Department of Surgery and Physiology, Faculty of Medicine,
University of Porto, Porto, Portugal
`amarante@med.up.pt`

Abstract. The craniofacial morphology can be a very important feature to take in consideration when playing a wind instrument. The embouchure mechanism and the interrelationship with the craniofacial morphology of each wind instrumentalist can be relevant on the musical performance and the presence, or absence, of pain on the cranio-cervico mandibular complex (CCMC). Lateral teleradiographies were made to three bassoon players each of them being representative of the different types of occlusion, a malocclusion Class I, a malocclusion Class II and a malocclusion Class III, in order to obtain the measurement of the mouth opening, with the mouthpiece inside the mouth of the wind instrumentalist, and correlate these data with the respective craniofacial features. The mouth opening values with the mouthpiece inside the mouth of the musician with a malocclusion Class III is higher, than in those with malocclusion Class I or II. The authors of this paper considered the law of lever to explain the embouchure of a bassoon player with a malocclusion Class III. The mouthpiece is balanced as a lever, with the upper lip being on extreme of the lever, providing a full stabilization of the bassoon mouthpiece, while the lower part of the reed will be supported on the lower jaw, that works in this case as the fulcrum. This event is in accordance with the biomechanics of the temporo-mandibular joint during musical performance of this bassoonist with a malocclusion Class III, allowing a mouth opening with 17 mm. The lateral teleradiography provides the assessment of an intra-oral view of the embouchure mechanism. This image technique is a complement on the physiological aspects that occur regarding the interaction between the mouthpiece and the craniofacial morphology.

© Springer Nature Switzerland AG 2019
J. M. R. S. Tavares and R. M. Natal Jorge (Eds.): VipIMAGE 2019, LNCVB 34, pp. 93–103, 2019.
https://doi.org/10.1007/978-3-030-32040-9_11

Keywords: Bassoon player · Craniofacial morphology · Embouchure ·
Imaging technique · Lateral teleradiography · Law of lever ·
Wind instrumentalist

1 Introduction

Musical performance can have direct implications on the general health of a practitioner, in terms of physical or psychological matters. Ackermann, highlighted the importance of developing, maintaining good communication and establishment of trust between performing artists, educators and health professionals in order to enhance a better management of their injuries [1]. Rodriguez-Lozano et al. also emphasized that the characteristics of the musical instrument and the way that it is played, can be related to the pathologies associated to occupational health of the musicians. Concerning oral-health issues, these authors stated that professional musicians generally show a propensity for bucco-dental problems [2].

The craniofacial morphology can be a very important feature to take in consideration when playing a wind instrument. The musical gesture of a single reed instrument player, a double reed instrument, or a brass player during musical performance has one of the most important anatomical area on the final outcome of music production and sound quality. The embouchure mechanism and the interrelationship with the craniofacial morphology of each wind instrumentalist can be relevant on the musical performance and the presence or absence of pain on the cranio-cervico-mandibular complex (CCMC). Some studies have related the activity of playing a wind instrument with the appearance of temporomandibular disorders (TMD) [3, 4] that can be characterized with the presence of pain in the masticatory muscles, the temporomandibular joint (TMJ) and associated structures with signs such as limitation in the mandibular range of motion, and joint noises. Nishiyama and Tsuchida, studied seventy-two non-professional players and sixty-six non-players (control group) having found that in the instrument group the frequency of subjects who felt mouthpiece pressure in the high risk of TMD group was 47.6%. The mouthpiece contact pressure was found to be a significant factor contributing to a high risk of TMD [5].

It is interesting to imagine how a musician with a malocclusion Class III, can execute the embouchure while playing a double reed instrument like a bassoon, where the temporomandibular joint biomechanics will oblige a mouth opening with the upper and lower lip being retruded over the respective upper and lower incisors. Furthermore, the obtention of different pitches, low and high, will promote respectively a protrusion or retrusion of the mandible after having the mouthpiece stabilized within the perioral structures. This means this wind instrument player (with a malocclusion Class III) will have to retrude the mandible to a higher extent when executing a higher pitch, comparing to another musician playing the same instrument with a malocclusion Class I. To study and characterize the wind instrumentalist embouchure from a clinical point of view, the dentistry field can have complementary diagnostic tools such as electromyography, piezoresistive sensors, infrared thermography and lateral teleradiography.

In the past Clemente et al., used surface electromyography on the analyses of certain muscles involved during the embouchure mechanism. It was possible to observe that the orbicular muscle of a bassoon player with a malocclusion class III had a bioelectric potential of 140–150 μV when playing a musical piece of Mozart and the anterior digastric belie presented a bioelectric potential of 39–49 μV. The activation pattern of these specific muscles are associated to the embouchure mechanism where the orbicular muscle presents a high level of hyperactivity, which can be associated to the necessity of promoting a correct lip seal between the upper lip, the mouthpiece and the lower lip, during the mandible retrusion [6]. The digastric muscles potential are representative with the mandible opening of the malocclusion Class III of the musician, which are slightly higher than a bassoon player with a malocclusion Class II of with a bioelectric potential of 31.6–31.8 μV. The bioelectric potential of the orbicular muscle of the bassoon player with a malocclusion class II of angle was of 74.6–83.1 μV, while playing the same musical Mozart piece as the bassoon player with a malocclusion class III [6].

The biomechanics of the temporomandibular joint will be different within wind instrumentalist that play the same instrument, in this particular case, a bassoon player, and this can be due to different teaching techniques, to the embouchure mechanism, to the applied pressures during the grip of the mouthpiece and surely due to the differences existing within the craniofacial morphology. Nevertheless, there can be some limitations while applying the surface electrodes of electromyography when analyzing the mechanism embouchure of wind instrumentalists.

On the other hand, the application of other techniques such as infrared thermography which does not quantify the muscle activity, but allows the assessment of the correspondent anatomical regions of interest associated to the movement during musical performance, has also been used in the characterization of the action of certain anatomical structures. This technique has been used in performing arts medicine, more specifically on the analysis of the CCMC and the embouchure mechanism. To complement this analysis, Clemente et al. introduced the Combined Acquisition Method of Image and Signal Technique (CAMIST) for temporomandibular disorders in performing arts medicine, in order to complete an exhaustive examination of a wind instrumentalist on the dentistry field the application of thermal images and piezoresistive sensors were added to the analyze the lateral teleradiographies. This provides the dentist useful information when observing a wind instrumentalist, since there is more knowledge regarding the daily activity of such important structures as the CCMC [7].

The application of lateral cephalograms has been made on orthodontic treatment planning for several years, in order to provide the analysis of the craniofacial structures. Cephalometric analysis contributes for an objective evaluation of distinctive dentoskeletal factors such as mandibular prognathism or retrognathism, maxillary prognathism or retrognathism, increased overbite, increased overjet and vertical jaw relationship among others [8–11]. There can be many advantages on applying cephalometry to the clinical examination of a wind instrument player where the existing skeletal differences will surely have an impact on the soft tissue balance of the perioral structures, while performing the embouchure. For this purpose, even though there has been an attempt on using this complementary tool on the evaluation of wind instrumentalists, it is important to understand what type of implications that can be associated to craniofacial morphology variations and musical performance.

2   Materials and Methods

This study was performed on three bassoon players selected from a sample of 50 wind instrument players. The chosen subjects are representative of the different types of occlusion, a malocclusion Class I, a malocclusion Class II and a malocclusion Class III respectively. The individuals presented the biggest discrepancy of malocclusion within the sample of the same type of wind instrument, single reed instrument, double reed instrumentalist and brass players, the lateral cephalograms allowed to report these conditions associated to musical performance.

To measure the mouth opening needed during the embouchure, it was assessed the interincisal distance in both lateral teleradiographies, in the first case this measure corresponds to the musician in maximum intercuspation, while on the other teleradiography during the embouchure. Two lines parallels to the Frankfurt's plane were traced passing at the border of the upper and lower central incisors, in order to calculate the vertical trespasse of the central incisors. In order to obtain the measurement of the mouth opening with the mouthpiece inside the mouth of the wind instrumentalist, it was taken in consideration the absolute values of the initial measurement, in the first lateral teleradiography, in maximum intercuspation, added to the mouth opening (mm) during the embouchure.

3   Results

The position of the mouthpiece inside the oral cavity of the wind instrument player with a malocclusion Class I induced a considerable higher mouth opening, since the overbite of the central incisors is of 6 mm (Fig. 1). Regarding the bassoon player with a Class II malocclusion, the value of mouth opening with the mouthpiece reaches 13 mm (Fig. 2). The mouth opening with the mouthpiece inside the mouth of the musician with the malocclusion Class III is higher (Fig. 3), than in those with malocclusion Class I or malocclusion Class II, Table 1.

The lateral teleradiography allows the dental community and the performing arts field, especially the area of wind instrument players to observe and analyze the embouchure mechanism. This work is related to bassoon players with different types of malocclusion being notorious the existing differences on the amount of mouth opening and rotation of the mandible in the clockwise direction with the inherent biomechanics of the temporomandibular joint.

Taking a special attention to the first bassoon player, it is also possible to verify that there is a tooth missing, the agenesis of the 45, which was possible to confirm during the detailed clinical history. On the other hand, this patient had already undergone a root canal treatment on the tooth 46 and the extemporaneous loss of a temporary tooth the 84. These issues are of major importance, in terms of occlusion, since the mesialization/inclination of the tooth 46 will occur and within time can certainly promote occlusal prematurity's if there is no oral rehabilitation procedure in perspective. The second patient, apparently had no major issues concerning to her dental status, apart from the malocclusion Class II, while the third patient also had the absence of the first molar.

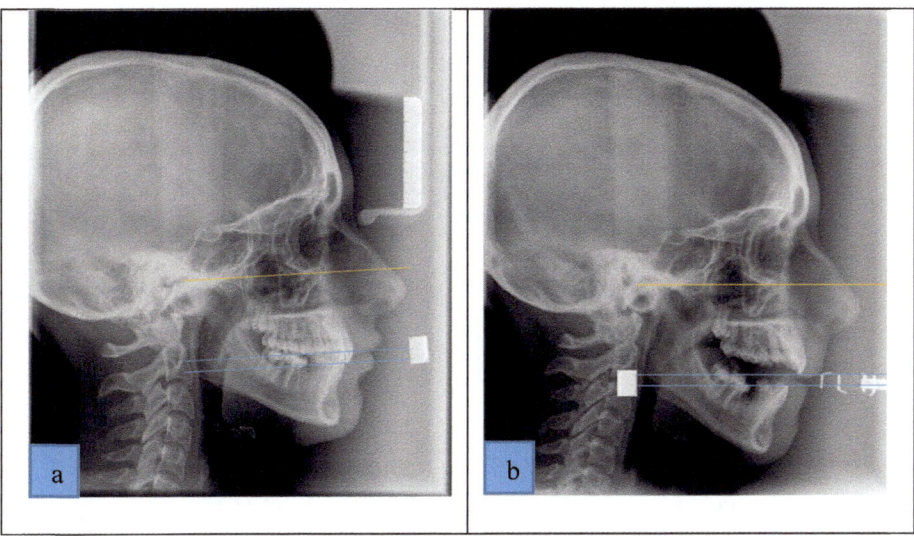

Fig. 1. (a) Lateral teleradiography of bassoon player with malocclusion Class I in maximum intercuspation, (b) Lateral teleradiography of bassoon player with malocclusion Class I during the embouchure mechanism.

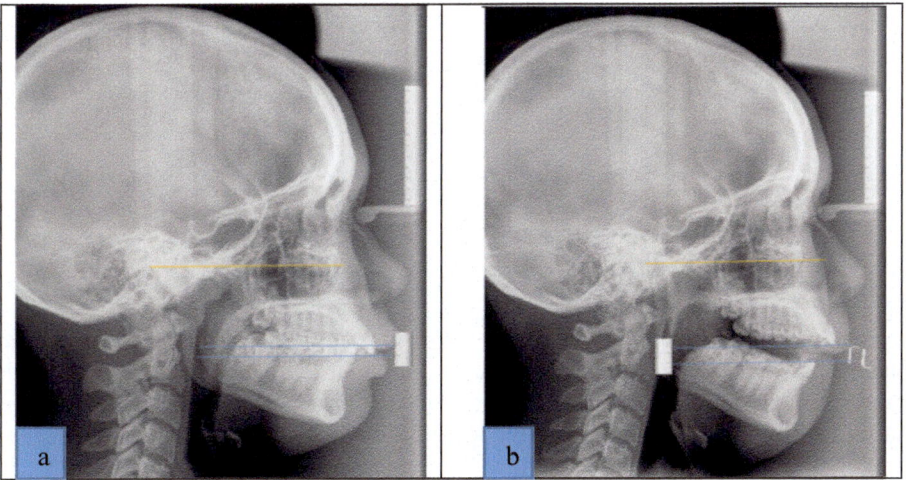

Fig. 2. Lateral teleradiography of bassoon player with malocclusion Class II in maximum intercuspation (a), and during the embouchure mechanism (b).

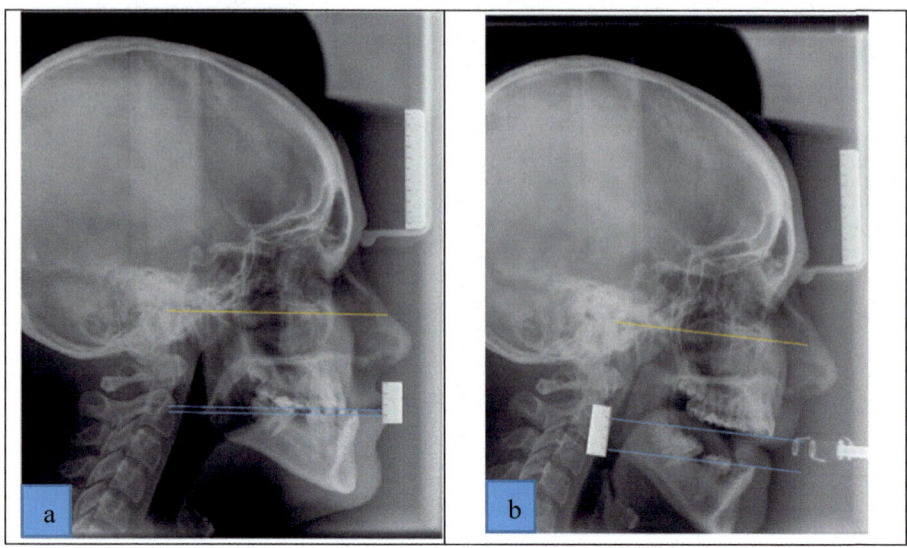

Fig. 3. Lateral teleradiography of bassoon player with malocclusion Class III in maximum intercuspation (a), and during the embouchure mechanism (b).

Table 1. Interincisal distance of the different double reed instrument players, during maximum intercuspation, with the mouthpiece inside the mouth and the corresponding value for the mouth opening.

Double reed instrument player	Maximum intercuspation	During embouchure	Total mouth opening
Class I	6 mm	6 mm	12 mm
Class II	5 mm	8 mm	13 mm
Class III	−3 mm	14 mm	17 mm

4 Discussion

The findings of our study suggest that there are specific features associated to the craniofacial morphology of wind instrumentalists in this case bassoon players that should be analyzed and considered. The embouchure mechanism of bassoon players induce high pressure on the lower lip comparing to the upper lip [12]. So the question that arises is how this is possible, since the mandible has to produce different pitches and thus adjust the position of opening/closing and protrusion/retrusion. For this purpose, it is important to have a multidisciplinary approach within the dentistry community, clinicians, engineers and music teachers who have experience with bassoon students.

The intention of this article is to highlight the intersection of distinct fields of knowledge, and allow a wider explanation of the bassoon players embouchure, taking in consideration the lateral teleradiographies, in such extreme cases as malocclusion

Class III. This situation can be analyzed by applying the law of lever to the bassoon's player embouchure. This wind instrumentalist adapts her mouthpiece inside the oral cavity having the lower incisal edge as the fulcrum. The mouthpiece is balanced between the upper lip and the instrument itself, providing a full stabilization of the bassoon mouthpiece, while the lower part of the reed will promote a higher force on the lower jaw, Fig. 4.

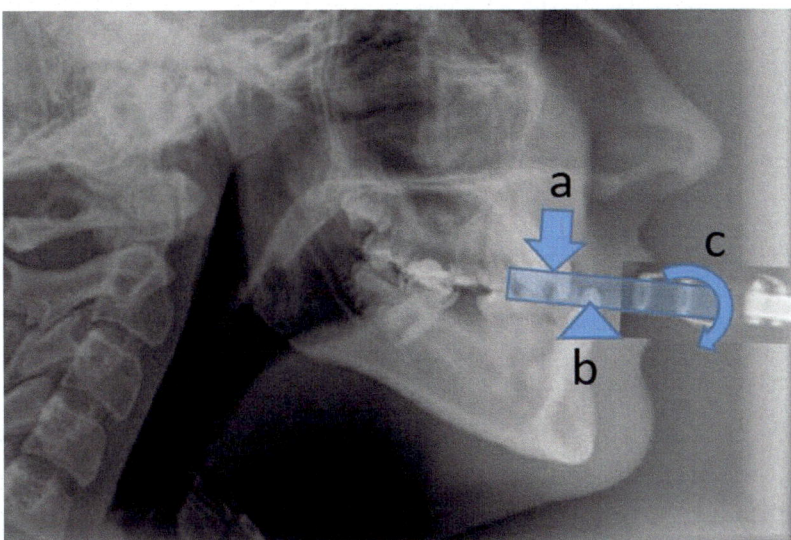

Fig. 4. Lateral teleradiography of a bassoon player with malocclusion Class III, superimposed with the law of lever applied to the embouchure. The lower incisors provide the fulcrum (b), while the upper incisors applied the balanced force in the anterior part of the double reed (a), keeping the equilibrium with the instrument momentum (c).

During musical performance the biomechanics of the temporomandibular joint of this bassoonist with a malocclusion Class III, allows a mouth to open to the maximum of 17 mm. What we can observe in the lateral teleradiography is that the pressure of the mouthpiece/reed is being supported by the lower lip. This will be proportional to the "freedom" of movement that the mandible of this wind instrumentalist will have in order to adjust anatomically to functions that usually are not convenient, like a retrusion of more than 2 mm. These particular movements can be harmful for the musician, since they can compress the bilaminar zone of the TMJ and eventually originate pain.

The lateral teleradiography provides the assessment of an intra-oral view of the embouchure mechanism. This image technique could be used as a complement of the physiological aspects regarding the mouthpiece/reed and the craniofacial morphology, which is often taught empirically by musical teachers to their students. This medical image can objectively quantify the different positions adopted by the mandible in relation to the type of occlusion the wind instrumentalist has. This research concerns bassoon players since their specificity involved with the embouchure technique is unique with the retrusion of the upper and lower lip over the respective incisal edges when stabilizing the mouthpiece. Probably, oboe players, the other double reed instrumentalists, have other playing techniques even though, as they adopt the same retrusion of the lips during the embouchure. It is generally accepted that oboe players execute higher pressures on the mouthpiece during musical performance and one simple explanation can be related to the reed which is smaller than the bassoon reed [12].

The lateral teleradiography brings light to how the equilibrium of the embouchure mechanism is performed, taking in mind the law of lever. Different type of occlusions and different pitches force the lower jaw to adapt to the right embouchure position.

Liu and Hayden, suggest the importance that physicians have in becoming familiar with the disorders associated to specific instruments in order to achieve a correct diagnosis within musicians' medical complaints. Musicians are usually prone to occupational health issues, where the most prevalent problems involve overused of muscles resulting from repetitive movements of playing and the maintenance of awkward positions [13]. For this purpose, our research intends to demonstrate with an intra-oral view by the use of the lateral teleradiography what is actually happening during the embouchure mechanism of a bassoon player. Other techniques are of common use on the analysis of certain anatomical structures within the CCMC and the existing malocclusion of the patient, like Gomes et al. 2010, found that there was a higher relative effort of the masseter and anterior temporal muscles in dolichofacial, followed by meso and brachyfacial subjects during mastication. Therefore, in the study carried out by Gomes et al. using EMG, it was concluded that vertical facial pattern influences masticatory performance, mandibular movement during mastication and the effort masticatory muscles required for chewing. Based on our findings it was possible to observe that the dolichofacial bassoon player presented the highest amount of mouth opening for the embouchure mechanism [14].

Playing a wind instrument can be considered as a parafunctional habit and with this imaging technique of the lateral teleradiography we are not analyzing the possible influence of playing a wind instrument on the occlusion of an individual, but identifying the influence that the cranio morphology can have on the embouchure mechanism itself. Parafunctional habits, such as non-nutritive sucking habits, can have an influence on the craniofacial morphology. Agarwal et al. studied a group of 415 children where it was possible to verify a correlation between a reduced intra-arch transverse diameter and non-nutritive sucking habits [15]. The possibility of observing the deciduous dentition in children with prolonged digit or pacifier habits is important in order to examine the transverse occlusal relationship [16]. The position adopted by the tongue with a low posture in the mouth related to the use of a pacifier or short frenulum linguae can be associated with posterior crossbite at the age of 4 and 5 years [17].

It is interesting to notice that these techniques such as lateral teleradiography and EMG are frequently used in the dentistry field. While the imaging technique is mainly used in orthodontics as a complementary method of diagnose in the treatment plan of a malocclusion the EMG can determine till a certain extent the degree of dysfunction of certain group of muscles, since there can be alterations in jaw clenching force in people with myogenic TMD. An EMG investigation carried out by Pires et al. 2018 in women with myogenous TMD, suggest that they exhibit a reduction in the electrical activity of the masseter muscles, with activity equaling that of the anterior temporalis muscles

[18]. The electromyographic activity in terms of muscular load can be applied when analyzing the differences between patient with painful masseter muscles and referents during the chewing of aliments with different consistencies [19]. Hagberg found no differences of the mentioned muscles during the chewing of an almond and of a gum [19]. Surface electromyography has also been used to study the muscular activity in patients undergoing myofunctional therapy, where the use of a preformed functional device in interceptive orthodontics induces a significant increase of the sEMG activity of the lower orbicularis oris (OO) muscle at rest and of the upper OO muscle during mandibular protrusion [20]. Takeuchi-Sato demonstrated that higher EMG activity could be generated between the first premolars at shorter interocclusal distances, being suggestive that different craniofacial morphologies are associated with differences in neuromuscular activity of the masticatory muscles [21].

Some of these features are important to take into consideration in the field of occlusion and orthodontics, since the maintenance of these habits for long period of time can precipitate a malocclusion with direct implications in the craniofacial morphology of an individual. Likewise, different craniofacial morphologies can have direct implications in the neuromuscular pattern of an individual or more specifically like our results demonstrated on the interincisal distance of the different double reed instrument players, during the embouchure mechanism.

According to previous cephalometric studies carried out to analyze the oral habits, malocclusion and mouth breathing pattern, where this last issue is still not consensual since there are some results which suggest that specific craniofacial patterns might not be associated with the adenoid hypertrophy or the breathing mode [22–24].

Therefore, it is recommended to visit an oral health professional for the prevention and early treatment of certain disorders related to craniofacial growth, it is fundamental to understand the interrelationship of a malocclusion and the wind instrumentalist embouchure, like this research intended to demonstrate in this particular case the double reed instrument, the bassoon.

5 Conclusions

The craniofacial morphology of distinct skeletal malocclusion will influence the amount of mouth opening during the embouchure mechanism and consequently musical performance. An increased overbite or and increased overjet will promote compensations on the position adopted by the mandible regarding the execution of the musical partiture.

The lateral teleradiography as an imaging technique that is used in orthodontics for the cephalometric analysis is useful in order to evaluate the underlying discrepancy of the occlusion and its relationship with the embouchure mechanism of the bassoon players.

These findings will help to understand that the occlusion of a wind instrumentalist should be taken in consideration at an early stage of the development of the craniofacial morphology, particularly when a child chooses his instrument.

In the cases of malocclusion there should be an efficient communication between, the dentist, the wind instrumentalist and the musical teacher. This way it will be possible to establish a treatment plan for the prevention and treatment of malocclusion in order to avoid potential discomfort or even pain when playing the wind instrument.

Acknowledgments. The authors would like to acknowledge the support of the project LAETA - UID/EMS/50022/2013.

References

1. Ackermann, B.J.: Making health care worth it: increasing value and awareness in performing arts medicine. Med. Probl. Perform. Artist. **33**(2), 146 (2018)
2. Rodriguez-Lozano, F.J., Saez-Yuguero, M.R., Bermejo-Fenoll, A.: Orofacial problems in musicians: a review of the literature. Med. Probl. Perform. Artist. **26**(3), 150–156 (2011)
3. Jang, J.Y., et al.: Clinical signs and subjective symptoms of temporomandibular disorders in instrumentalists. Yonsei Med. J. **57**(6), 1500–1507 (2016)
4. Steinmetz, A., et al.: Symptoms of craniomandibular dysfunction in professional orchestra musicians. Occup. Med. **64**(1), 17–22 (2014)
5. Nishiyama, A., Tsuchida, E.: Relationship between wind instrument playing habits and symptoms of temporomandibular disorders in non-professional musicians. Open Dent. J. **10**, 411–416 (2016)
6. Pais Clemente, M., et al.: Dental considerations and electromyographic study of orofacial muscle activity in musicians playing wind and string instruments. In: Music, Health and Happiness Conference, p 51 (2008)
7. Clemente, M.P., et al.: Combined acquisition method of image and signal technique (CAMIST) for assessment of temporomandibular disorders in performing arts medicine. Med. Probl. Perform. Artist. **33**(3), 205–212 (2018)
8. Arriola-Guillen, L.E., Flores-Mir, C.: Molar heights and incisor inclinations in adults with Class II and Class III skeletal open-bite malocclusions. Am. J. Orthod. Dentofac. Orthop. **145**(3), 325–332 (2014)
9. Hernandez-Sayago, E., et al.: Lower incisor position in different malocclusions and facial patterns. Med. Oral Patol. Oral Cir. Bucal **18**(2), e343–e350 (2013)
10. Bae, E.J., Kwon, H.J., Kwon, O.W.: Changes in longitudinal craniofacial growth in subjects with normal occlusions using the Ricketts analysis. Korean J. Orthod. **44**(2), 77–87 (2014)
11. Bavia, P.F., Vilanova, L.S., Garcia, R.C.: Craniofacial morphology affects bite force in patients with painful temporomandibular disorders. Braz. Dent. J. **27**(5), 619–624 (2016)
12. Pais Clemente, M., et al.: Wind instrumentalist embouchure and the applied forces on the perioral structures. Open Dent. J. **13**, 107–114 (2019)
13. Liu, S., Hayden, G.F.: Maladies in musicians. South. Med. J. **95**(7), 727–734 (2002)
14. Gomes, S.G., et al.: Masticatory features, EMG activity and muscle effort of subjects with different facial patterns. J. Oral Rehabil. **37**(11), 813–819 (2010)
15. Agarwal, S.S., et al.: Association between breastfeeding duration, non-nutritive sucking habits and dental arch dimensions in deciduous dentition: a cross-sectional study. Prog. Orthod. **15**, 59 (2014)
16. Bishara, S.E., et al.: Changes in the prevalence of nonnutritive sucking patterns in the first 8 years of life. Am. J. Orthod. Dentofac. Orthop. **130**(1), 31–36 (2006)
17. Melink, S., et al.: Posterior crossbite in the deciduous dentition period, its relation with sucking habits, irregular orofacial functions, and otolaryngological findings. Am. J. Orthod. Dentofac. Orthop. **138**(1), 32–40 (2010)

18. Pires, P.F., Rodrigues-Bigaton, D.: Evaluation of integral electromyographic values and median power frequency values in women with myogenous temporomandibular disorder and asymptomatic controls. J. Bodyw. Mov. Ther. **22**(3), 720–726 (2018)
19. Hagberg, C.: The amplitude distribution of electromyographic activity in painful masseter muscles during unilateral chewing. J. Oral Rehabil. **14**(6), 531–540 (1987)
20. Saccucci, M., et al.: Effects of interceptive orthodontics on orbicular muscle activity: a surface electromyographic study in children. J. Electromyogr. Kinesiol. **21**(4), 665–671 (2011)
21. Takeuchi-Sato, T., et al.: Relationships between craniofacial morphology and masticatory muscle activity during isometric contraction at different interocclusal distances. Arch. Oral Biol. **98**, 52–60 (2019)
22. Markkanen, S., et al.: Craniofacial and occlusal development in 2.5-year-old children with obstructive sleep apnoea syndrome. J. Bodyw. Mov. Ther. **41**(3), 316–321 (2019)
23. Coelho, A.R.D.P., et al.: Transverse craniofacial dimensions in Angle Class II, Division 1 malocclusion according to breathing mode. Braz. Oral Res. **24**, 70–75 (2010)
24. Feres, M.F.N., et al.: Craniofacial skeletal pattern: is it really correlated with the degree of adenoid obstruction? Dent. Press. J. Orthod. **20**(4), 68–75 (2015)

Mimetic Finite Difference Methods for Restoration of Fundus Images for Automatic Glaucoma Detection

Lola Bautista[1(✉)], Jorge Villamizar[2,3], Giovanni Calderón[2,4], Julio C. Carrillo E.[2], Juan C. Rueda[5], and José Castillo[6]

[1] Escuela de Ingeniería de Sistemas e Informática, Universidad Industrial de Santander, Bucaramanga, Colombia
lxbautis@uis.edu.co

[2] Escuela de Matemáticas, Universidad Industrial de Santander, Bucaramanga, Colombia
jorge@uis.edu.co, gcalderon@matematicas.uis.edu.co, jccarril@uis.edu.co

[3] Facultad de Ingeniería, Universidad de Los Andes, La Hechicera, Mérida 5101, Venezuela

[4] Departamento de Matemáticas, Facultad de Ciencias, Universidad de Los Andes, La Hechicera, Mérida 5101, Venezuela

[5] Centro de Prevención y Atención del Glaucoma, Bucaramanga, Colombia
jcruedaglaucoma@intercable.net.co

[6] Computational Science Research Center, San Diego State University, San Diego, CA 92182-1245, USA
jcastillo@mail.sdsu.edu

Abstract. Glaucoma is one of the leading causes of irreversible blindness in people over 40 years old. In [9] it was developed a computational tool for automatic glaucoma detection, which implements a novel method that has shown improvement in the accuracy of the detection compared to other classical methods. However, the method is sensitive to the quality of the acquired image, which is often contaminated by noise, and its quality can be poor. For this reason, automatic image restoration of the source images is needed to improve the quality of glaucoma detection. Partial differential equations to produce an image of much higher quality, enhance its sharpness, filter out the noise, extract shapes, etc. Here, we proposed the use of mimetic finite difference methods for the numerical solution of this kind of problems. The mimetic methods preserve the continuum properties of the mathematical operators often encountered in the image processing and analysis equations and ensuring better orders of convergence [5]. By ensuring these mathematical properties, the original structure of the source image is maintained, improving the diagnosis of the patient.

1 Introduction

Glaucoma is an irreversible disease that compromises the eye optic nerve, which, when untreated, often leads to blindness. Besides, its symptomatic character

© Springer Nature Switzerland AG 2019
J. M. R. S. Tavares and R. M. Natal Jorge (Eds.): VipIMAGE 2019, LNCVB 34, pp. 104–113, 2019.
https://doi.org/10.1007/978-3-030-32040-9_12

makes diagnosis difficult and requires treatment once detected. According to the World Health Organization (WHO) [14], the most common types of glaucoma are primary open-angle glaucoma (POAG) in which the iridocorneal angle is open (unobstructed) and normal in appearance without aqueous outflow is diminished [10], and angle-closure glaucoma (ACG) in which the intraocular pressure rises rapidly as a result of relatively sudden blockage of the trabecular meshwork [13].

It is estimated that by 2020 there will be approximately 80 million people with glaucoma, an increase of about 20 million since 2010 [8]. As it was reported in [9], the Colombian Ministry of Health and Social Security estimates that there are around 296,000 blind people for several causes, having glaucoma prevalence of 3.9% in people over 40 years old. This gets worse by the fact that in the country, based on the population in 2011, there are two ophthalmologists for every 100,000 patients.

The retinal fundus image is used to measure the thickness of the retinal nerve fiber layer (RFNL) to diagnose glaucoma. In Colombia, it is the preferred diagnostic technique because of its noninvasive use and portability in screening campaigns in regions with difficult accessibility [16]. The thickness of the RFNL is obtained by measuring the proportion among the size of the optic nerve (named as a disc) and the size of the excavation inside the optic nerve produced by the increased eye pressure (named cup). This parameter is known as the Cup-to-Disc Ratio (CDR) (see Fig. 1a).

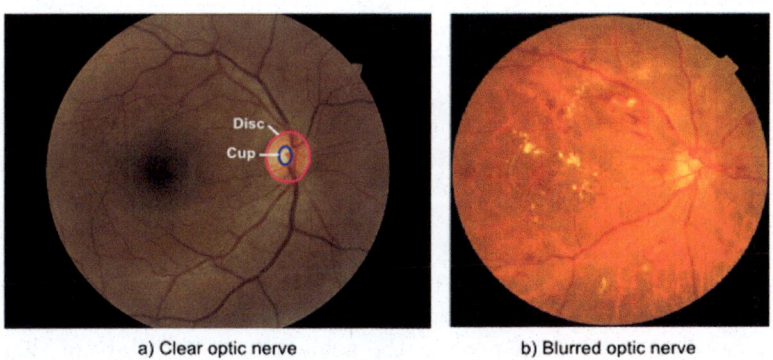

a) Clear optic nerve   b) Blurred optic nerve

Fig. 1. Fundus image.

Several works have been done for automatic glaucoma detection [12], where the main difficulty is to provide an accurate estimation of the CDR. In [9] it was developed a computational tool for automatic glaucoma detection, which implements a novel method for cup segmentation using the vessels and the cup intensities, that has shown improvement in the accuracy of the detection compared to other classical methods. However, the method is sensitive to the quality of the acquired image, due to the presence of artifacts produced either by the illumination of the camera at the moment of the shooting, the movement of the patient, or by the intrinsic features of the retina of patients with underlying diseases like diabetes or blood high pressure (see Fig. 1b). For this reason, automatic image restoration of the source images is needed to improve the quality of glaucoma detection.

Image restoration is one of the most studied problems in image processing and computer vision. Degradation in images is due to the occurrence of two phenomena that could happen simultaneously: blurring because of lenses artifacts or motion during image acquisition, and the systematic presence of noise, which has a random nature. The goal of image restoration is to remove or diminish the effects of such degradation. This problem can be stated as follows. Let $u : \mathbb{R}^2 \to \mathbb{R}$ be an original image and let u_0 be the degraded observed image of the same scene. The forward model for such observation is formulated as $u_0 = Hu + \eta$, where H is a linear operator representing the blur, and η represents white additive Gaussian noise. Given u_0, the problem is to reconstruct u. As the problem is ill-posed, it only allows us to carry out only an approximation of u.

There exist several approaches to deal with this problem, such as the energy model and the partial differential equations (PDEs) model [1], being the later the subject of our interest in this work. With this model we are able to obtain a family of functions $\{u(t, x)\}_{t \geq 0}$ representing successive smoothed versions of $u_0(x)$, preserving features such as borders and corners. The main difficulty is that smoothing has isotropic nature, which means that it does not depend on the image, and it is the same in all directions, that is, edges are not well preserved. The classical PDE's representation of smoothing is through the heat equation, which is equivalent to carrying out Gaussian linear filtering (also known as low-pass filtering).

To overcome this drawback in the diffusion coefficient, the formulation by Perona and Malik in [15, 17] establishes the advantage of an anisotropic diffusion coefficient. In spite of the practical success of the Perona-Malik model [15], it presents some issues, especially related to uniqueness and stability concerning the initial image should be no expected, i.e.; the Perona-Malik model can not ensure well-posedness. As later established, the regularizing effect of the discretization plays too much of an essential role in the solution of Perona-Malik model [3], which is possibly the key element in the achievement or failure of this model. The research towards developed after this idea have been proposed over the years. Of these works, it deserves special mention the mathematical formulation proposed by Catté et al. [7], who suggested replacing the diffusion coefficient $c(|\nabla u|^2)$ of the Perona-Malik model, by a slightly modified version, $c(|\nabla u_\sigma|^2)$, with $u_\sigma = g_\sigma * u$, where g_σ is a smooth Gaussian kernel (Gaussian distribution of mean $\mu = 0$ and variance σ^2).

In [1, 19] there have been proposed techniques to discretize this equation using finite differences methods. On this work, we deal with the design of algorithms to implement the restoration problem through mimetic finite difference methods, which preserve the continuous features of the divergence and gradient operators of the diffusion coefficient. These methods also satisfy the Green-Gauss-Stokes theorem, which is conservative, ensuring better orders of convergence [5]. The mimetic schemes preserve the continuum properties of existing mathematical operators in image processing. By guaranteeing these mathematical properties, the original structure of the source image is maintained, leading to better processing of the artifacts, to finally improve the segmentation of the disc and the cup, needed to compute the CDR parameter. In Sects. 3 and 4 of this paper, the mimetic finite difference methods are presented and applied for the restoration of fundus images for automatic glaucoma detection.

2 Model Problem

The Perona-Malik model [15] is today a widely applied nonlinear partial diffusion equation, that uses an in-homogeneous diffusivity coefficient, as a powerful and well-founded tool in multiscale image analysis, for purposes like smoothing, restoration, segmentation, filtering or detecting edges on an image u_0 with a resolution of $n_x \times n_y$ pixels. The variable t in this diffusion equation is a scale parameter that when is increased, leads to more straightforward image representations.

For purposes of increasing the scale resolution of images, the Perona-Malik model usually defines a linear diffusion filter, which dislocates edges when moving from finer to coarser scales. On the other hand, Perona-Malik established when the conduction coefficient c is chosen locally as a scalar function of the magnitude of the gradient of the brightness function (the edge detector), i.e., $c(|\nabla u(x, y, t)|^2)$, the model will not only preserve but also sharpen, the brightness edges if the diffusivity function c is chosen properly. For example, they considered

$$c(s^2) = \frac{1}{1 + s^2/\lambda^2}, \quad \lambda > 0,$$

when the main concern was to privileges wide regions over small ones. The parameter $\lambda > 0$ says how fast the function tends to zero.

Although the nonlinear model of Perona-Malik may have no weak global solution [7], still there are several methods which deal with this problem by regularizing it, both in space and time. Catté, Lions, Morell and Coll [7] found a Gaussian smoothing does not only solve this problem but reduce noise and blurs essential features such as edges and, thus, make them harder to identify [19]. In order to reduce smoothing at edges, they consider the initial-boundary problem, describing the process of filtering edges on an image,

$$u_t = \nabla \cdot \left(c(|\nabla u_\sigma|^2) \nabla u \right) \qquad \text{in } \Omega \times (0, +\infty), \qquad (1)$$
$$u(x, y, 0) = u_0(x, y) \qquad \text{on } \bar{\Omega}, \qquad (2)$$
$$\nabla u \cdot \hat{\nu} = 0 \qquad \text{on } \partial\Omega \times (0, +\infty), \qquad (3)$$

where $u(x, y, t)$ represents the filtered image of an image u_0, with $n_x \times n_y$ pixels of resolution, at a scale t; $\Omega = (1, n_x) \times (1, n_y)$ and $\bar{\Omega} = [1, n_x] \times [1, n_y]$ is the rectangular domain of the image u_0 and $\partial\Omega$ is the boundary of Ω. In contrast with the Perona-Malik model, this model considers the edge detector $|u_\sigma|^2$ with $u_\sigma = u * g_\sigma$, where g_σ a Gaussian distribution with mean $\mu = 0$ and variance σ^2. The parameter $\sigma > 0$ makes the filter insensitivity to noise at scales smaller than σ, and also a regularization parameter which guarantees well-posedness of the process [7,19]. Equations of this type have been successfully applied to processed medical images. Nevertheless, they are representative of a large scale of non-linear scale-spaces (see [4] and [19] for more details). At is mentioned in [19], the Gaussian convolution with standard deviation σ is equivalent to linear diffusion filtering ($c \equiv 1$) for some scale parameter $T = \sigma^2/2$.

Since explicit schemes, unlike explicit ones, have no restrictions on the number of scales and are globally stable, and given the good numerical results of mimetic methods in diffusion problems, this article considers an implicit discretization scheme of the nonlinear filtering model (2) based on mimetic methods of finite mimetic differences.

3   Mimetic Difference Method

The mimetic methods are based on the discretization of the classical operators of the partial differential equations (divergence, gradient and rotational) in such a way that they satisfy a discrete version of the Stokes Theorem or Green identity [6]:

$$\langle \mathbf{D}_{jd}^s v, f \rangle_Q + \langle v, \mathbf{G}_{jd}^s f \rangle_P = \langle \mathbf{B}_{jd}^s v, f \rangle_I. \tag{4}$$

In this expression, $\mathbf{D}_{jd}^s$, $\mathbf{G}_{jd}^s$ and $\mathbf{B}_{jd}^s$ are the discrete versions of their corresponding continuous: gradient (∇), divergence ($\nabla\cdot$) and boundary operator $\partial/\partial\hat{\nu}$, where s represents the mesh size and jd the spatial dimension ($j = 1$ and 2). The functionals $\langle \cdot, \cdot \rangle_A$ represent a generalized inner product of the form $\langle v, w \rangle_A \equiv v^T A w$, with weights Q, P and I; the top index T denotes the transpose of a matrix. Using the identity (4), a relationship is obtained for the boundary operator,

$$\mathbf{B}_{jd}^s = Q\mathbf{D}_{jd}^s + \left(\mathbf{G}_{jd}^s\right)^T P.$$

We will always suppose that the mimetic discretization of these operators is developed on uniform meshes and with a second order of precision in all the nodes of the mesh. Given that the discrete operators in two dimensions are defined from the one-dimensional case, and for the sake of clarity, in the following section one-dimensional case is defined in detail to make the generalization to the two-dimensional case later.

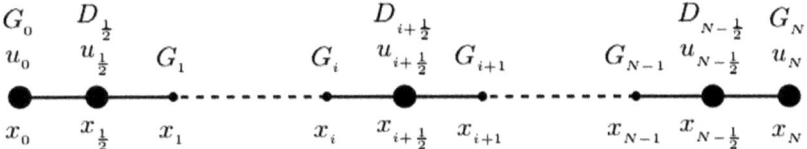

Fig. 2. one-dimensional staggered mesh.

3.1 One-Dimensional Mimetic Operators

For the discretization in a dimension, a uniform mesh is defined, whose geometry is given by the nodes x_i, with $i = 0, 1, \ldots, N$, and its cells (elements) are the intervals $[x_{i-1}, x_i]$, with size $h = x_i - x_{i-1} > 0$ for all i. The intermediate nodes (center) of the cells are given by $x_{i+1/2} = (x_i + x_{i+1})/2$, for $i = 1, \ldots, N-1$. The cell $[x_{i-1}, x_i]$ will be referred to as the cell or element Ω_i. The solution u, and the divergence operator are defined in the center of the cells, while the operator gradient in the nodes x_i that define the cells (see Fig. 2).

The discrete second-order gradient operator [6] and the second-order discrete divergence operator [2], for a uniform 1D mesh of $s = N$ cells, are given respectively by

$$\mathbf{G}_{1d}^{N} u = \frac{1}{h} \begin{bmatrix} -\frac{8}{3} & 3 & -\frac{1}{3} & 0 & \cdots & 0 \\ 0 & -1 & 1 & 0 & \cdots & 0 \\ & & \cdots\cdots\cdots & & \\ 0 & \cdots & 0 & -1 & 1 & 0 \\ 0 & \cdots & 0 & \frac{1}{3} & -3 & \frac{8}{3} \end{bmatrix} \begin{bmatrix} u_0 \\ u_{\frac{1}{2}} \\ \cdots \\ u_{N-\frac{1}{2}} \\ u_N \end{bmatrix}, \quad \mathbf{D}_{1d}^{N} v = \frac{1}{h} \begin{bmatrix} 0 & 0 & 0 & 0 & \cdots & 0 \\ -1 & 1 & 0 & 0 & \cdots & 0 \\ 0 & -1 & 1 & 0 & \cdots & 0 \\ & & \cdots\cdots\cdots & & \\ 0 & \cdots & 0 & -1 & 1 & 0 \\ 0 & \cdots & 0 & 0 & -1 & 1 \\ 0 & 0 & 0 & 0 & \cdots & 0 \end{bmatrix} \begin{bmatrix} v_0 \\ v_1 \\ v_2 \\ \cdots \\ v_{N-2} \\ v_{N-1} \\ v_N \end{bmatrix},$$

where he dimension of $\mathbf{G}_{1d}^{N}$ is $(N+1) \times (N+2)$ and the dimension of $\mathbf{D}_{1d}^{N}$ is $(N+2) \times (N+1)$. Additionally, it is consider the boundary operator $\mathbf{B}_{1d}^{N}$, whose dimension is $(N+2) \times (N+2)$ [6].

3.2 Two-Dimensional Mimetic Operators

Higher dimension mimetic discretizations can be systematically obtained for gradient and divergence operators from their one-dimensional versions, by using the Kronecker tensor product, denoted by $\otimes$ (see [18] for definition). In two dimensions, the mimetic discretization of the gradient operator on a mesh of dimension $N_x \times N_y$ (see Fig. 3) is given by

$$\mathbf{G}_{2d}^{N_x \times N_y} = \begin{bmatrix} \mathbf{I}_{N_y}^{T} \otimes \mathbf{G}_{1d}^{N_x} & \mathbf{G}_{1d}^{N_y} \otimes \mathbf{I}_{N_x}^{T} \end{bmatrix}^{T},$$

where the matrix $\mathbf{I}_i$ has dimension $(i+1) \times i$, its first and last row are null, and the rows between them are those of an identity matrix of dimension i. The submatrix on the left represents the gradient component in the x direction, while the one on the right has the derivative along the y direction. The dimensions,

$N_x + 1$ and $N_y + 1$, represent the number of nodes in the x and y components, respectively.

The mimetic discretization for the divergence operator in a mesh $N_x \times N_y$, like the one illustrated in Fig. 3, is defined by the matrix

$$\mathbf{D}_{2d}^{N_x \times N_y} = \left[\mathbf{I}_{N_y} \otimes \mathbf{D}_{1d}^{N_x} \quad \mathbf{D}_{1d}^{N_y} \otimes \mathbf{I}_{N_x} \right].$$

As for the gradient, $\mathbf{I}_i$ has the same description. The first entry of the matrix represents the approximation of the partial derivatives in the x direction, while the second entry is the approximation to the derivatives in the y direction. More details of these discretizations can be found in [18].

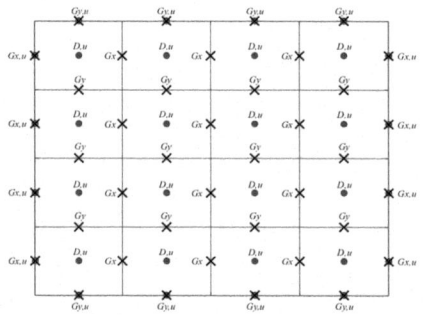

Fig. 3. Two-dimensional staggered mesh.

3.3   Model Problem: Mimetic Implementation

Notice that both the image u_0 resolution and the dimension of the two-dimensional staggered mesh agree when $n_x = N_x + 1$, $n_y = N_y + 1$ and $h = 1$. Hence, it follows from the discretization of the gradient and divergence operators in the resolution and a forward Euler method in scale (represented by the temporal spacing parameter k), the mimetic approximation for the non-linear diffusion equation (1) is given by

$$U^{n+1} = U^n + k\mathbf{D}_{2d}^{N_x \times N_y} \left(\mathbf{K}^n \mathbf{G}_{2d}^{N_x \times N_y} \right) U^{n+1}, \tag{5}$$

where U^n, following the lexicographic order, represents the approximated mimetic filtered image of u_0 at scale n in all the nodes of the staggered mesh; $\mathbf{K}^n$ is the matrix whose entries are obtained when the function c is evaluated at the estimated values of the discrete convolution associated with ∇u_σ on the nodes of the two-dimensional staggered mesh.

Since the discretized divergence operator acts on the boundary of Ω, the discretization of the Neumann boundary condition (3) is obtained from

$$\left(\mathbf{B}_{2d}^{N_x}\mathbf{G}_{2d}^{N_x \times N_y}\right)U^{n+1} = 0. \tag{6}$$

From (5) and (6), the mimetic scheme for the diffusion equation (1) together with the boundary conditions (2)–(3) is given by

$$\left(\mathbf{I} + \mathbf{B}_{2d}^{N_x}\mathbf{G}_{2d}^{N_x \times N_y} - k\mathbf{D}_{2d}^{N_x \times N_y}\mathbf{K}^n\mathbf{G}_{2d}^{N_x \times N_y}\right)U^{n+1} = U^n. \tag{7}$$

where the identity matrix $\mathbf{I}$ has dimension $(n_x + 2) \times (n_y + 2)$. Initially, is considered U^0 as the resolution of u_0 corresponding to each pixel in the two-dimensional staggered mesh.

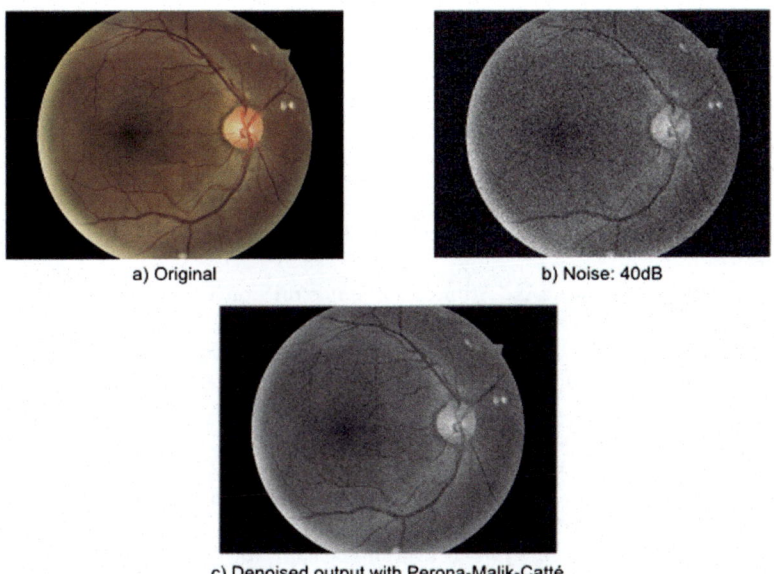

a) Original

b) Noise: 40dB

c) Denoised output with Perona-Malik-Catté

Fig. 4. Results of the Perona-Malik-Catté model

4   Numerical Experiments

Here we present the experimental setup for validating the model presented in Eq. (1) and implemented numerically following the model in Eq. (7). The input image in Fig. 4(a) was corrupted with gaussian noise at 40 dB as seen in Fig. 4(b). The parameters for the model are as follows: temporal spacing $k = 0.2$, standard deviation $\sigma = 8$, $\lambda = 0.0001$. The output of the denoised image is shown in Fig. 4(c). The *psnr* for this result is 24.78 which was reached from the first iteration. It is important to notice that the proposed model was used to study its robustness for this type of noise, however, future implementations will be used to treat artifacts given by the nature of the eye.

5   Conclusions

In this work we demonstrated that the mimetic finite difference methods have good performance on image denoising. In future works we expect to compare them against the classical finite difference methods, as well as in conditions where the image is affected by artifacts because of the nature of the patient's eye. This is a first approach for improving the preprocessing of the glaucoma images, which is going to be helpful in upcoming processing for the detection of the disease and hence, support the diagnosis for the ophthalmologists as described in [9].

Acknowledgments. G. Calderón and J.C. Carrillo E. would like to acknowledge the financial support provided by the VIE-UIS under grant No 2415.

References

1. Aubert, G., Kornprobst, P.: Mathematical Problems in Image Processing: Partial Differential Equations and the Calculus of Variations, vol. 147. Springer, New York (2006)
2. Batista, E.D., Castillo, J.E.: Mimetic schemes on non-uniform structured meshes. Electron. Trans. Numer. Anal. **34**(1), 152–162 (2009)
3. Benhamouda, B.: Parameter adaptation for nonlinear diffusion in image processing. Department of Mathematics, University of Kaiserslautern (1994)
4. Bredies, K., Lorenz, D.A.: Mathematical Image Processing. Applied and Numerical Harmonic Analysis. Birkhäuser, Cham (2018)
5. Castillo, J.E., Grone, R.D.: A matrix analysis approach to higher-order approximations for divergence and gradients satisfying a global conservation law. SIAM J. Matrix Anal. Appl. **25**(1), 128–142 (2003)
6. Castillo, J.E., Yasuda, M.: Linear systems arising for second-order mimetic divergence and gradient discretizations. J. Math. Model. Algorithms **4**(1), 67–82 (2005)
7. Catté, F., Lions, P.-L., Morel, J.-M., Coll, T.: Image selective smoothing and edge detection by nonlinear diffusion. SIAM J. Numer. Anal. **29**(1), 182–193 (1992)
8. The International Agency for the Prevention of Blindness. Glaucoma. https://www.iapb.org/knowledge/what-is-avoidable-blindness/glaucoma/. Accessed 03 Aug 2019
9. Carrillo, J., Bautista, L., Villamizar, J., Rueda, J., Sanchez, M., Rueda, D.: Glaucoma detection using fundus images of the eye. In: 2019 XXII Symposium on Image, Signal Processing and Artificial Vision (STSIVA), Bucaramanga, Colombia, April 2019
10. Kwon, Y.H., Fingert, J.H., Kuehn, M.H., Alward, W.L.: Primary open-angle glaucoma. New Engl. J. Med. **11**(360), 1113–1124 (2009)

11. Mikula, K.: Image processing with partial differential equations. In: Modern Methods in Scientific Computing and Applications, pp. 283–321. Springer (2002)
12. Nawaldgi, S.: Review of automated glaucoma detection techniques. In: 2016 International Conference on Wireless Communications, Signal Processing and Networking (WiSPNET), pp. 1435–1438. IEEE (2016)
13. American Academy of Ophthalmology: Angle-closure glaucoma. https://www.aao.org/munnerlyn-laser-surgery-center/angleclosure-glaucoma-19. Accessed 03 Aug 2019
14. World Health Organization: Blindness and vision impairment prevention. https://www.who.int/blindness/causes/priority/en/index6.html. Accessed 03 Aug 2019
15. Perona, P., Malik, J.: Scale-space and edge detection using anisotropic diffusion. IEEE Trans. Pattern Anal. Mach. Intell. **12**(7), 629–639 (1990)
16. Rueda, J.C., Lesmes, D.P., Parra, J.C., Urrea, R., Rey, J.J., Rodríguez, L.A., Wong, C.A., Galvis, V.: Valores de paquimetría en personas sanas y con glaucoma en una población colombiana. MedUNAB **10**(2), 81–85 (2007)
17. Sapiro, G.: Geometric Partial Differential Equations and Image Analysis. Cambridge University Press, Cambridge (2006)
18. Solano Feo, F., Guevara Jordan, J.M., Rojas, O., Otero, B., Rodriguez, R.: A new mimetic scheme for the acoustic wave equation. J. Comput. Appl. Math. **295**(1), 2–12 (2016)
19. Weickert, J., Romeny, B.M.T.H., Viergever, M.A., et al.: Efficient and reliable schemes for nonlinear diffusion filtering. IEEE Trans. Image Process. **7**(3), 398–410 (1998)

Segmentation, Matching, Simulation

Automatic Segmentation and Delineation of Intervertebral Discs on Lumbar Spine MR Images

Wai Teng Liu[1]($\boxtimes$) and André R. S. Marcal[2]

[1] Department of Physics and Astronomy, Faculty of Sciences,
University of Porto, Porto, Portugal
vitorliuwaiteng@yahoo.com
[2] Department of Mathematics, Faculty of Sciences,
University of Porto, Porto, Portugal

Abstract. Intervertebral discs play an important role in the human spine especially in the lumbar spine. The quantity of water that is inside the intervertebral discs and their shape are usually the criteria for diagnosis. In this paper, an automatic method for segmentation and delineation of lumbar spine and intervertebral disc is presented. The method was tested in 18 MR images with 15 good/perfect results in the automatic identification of the spine.

Keywords: Lumbar spine · Intervertebral disc · MRI ·
Automatic segmentation · Region growing

1 Introduction

Degenerative disc disease (DDD) occurs most commonly in the cervical spine or the lumbar spine, as these areas of the spine undergo the most motion and are most susceptible to wear and tear [1]. The intervertebral discs are uniquely designed as shock absorbers that protect the vertebral bone from potentially damaging loading that may arise from body weight and activation of muscle [2]. The intervertebral disc consists of a central nucleus pulposus surrounded by an annulus fibrosus. The nucleus pulposus is a pulplike gel located in the mid-to-posterior part of the disc [2]. Consisting of 70 to 90% water, the nucleus pulposus functions as a modified hydraulic shock absorber that dissipates and transfers loads between consecutive vertebrae [2]. However, the processes of degeneration and dehydration begin shortly after the age of 30 years in the human body [3]. With less water content in the discs, they tend to get thinner and become less capable for shocks absorption. Therefore, the quantity of water is a critical aspect for the intervertebral discs, not only in terms of their functional performances but also a criterion for diagnosis in degenerative disc disease [1].

In the current development of segmentation/delineation for the intervertebral disc, there are two main strategies: manual or semi-automatic and fully automatic. The former requires some local or directional parameters by manual inputs to achieve the segmentation. The fully automatic segmentation of intervertebral disc is done by designed functions or complex algorithms. Both strategies have their own benefits

© Springer Nature Switzerland AG 2019
J. M. R. S. Tavares and R. M. Natal Jorge (Eds.): VipIMAGE 2019, LNCVB 34, pp. 117–126, 2019.
https://doi.org/10.1007/978-3-030-32040-9_13

according to the need of different diagnostic purposes. In the semi-automatic strategy, the users can customize the parameters depend on different situation and delineate the intervertebral disc accurately by hand, but it is time consuming and requires an experienced human operator. On the other hand, the fully automatic strategy can achieve the segmentation without any human operation, but there might be some errors both in terms of size and shape of the discs.

The use of fully automatic strategy is a new trend that it is being improved by the contribution of researchers. In the study presented by Li et al., the segmentation was done by Fully Convolutional Network (FCN) from 3D MR image data and results in high discrimination capacity for whole intervertebral disc delineation [4]. Another study presented by Zhu et al., the segmentation was done by using Gabor Filter Band [5].

This paper presents a different method for the automatic segmentation of intervertebral discs, using image processing tools implemented in MATLAB. The proposed method starts by identifying the location of the spine within the MR image. Defining a path along the spine automatically, obtaining the information such as the signal profile and pixel location from that path, those information can then be used for delineation of the intervertebral discs. More details about the proposed method are revealed in the next section.

2   Materials and Methods

In this study, 18 sagittal T2-weighted MR images of the lumbar spine were used to perform segmentation. Those images were acquired from 8 subjects and the middle 3 slices of each subject were used. More details are presented in Table 1.

Table 1. General information of testing images.

Number of subjects	8
Number of images	24
Width of image	280 mm
Height of image	280 mm
Pixel size	0.25 mm^2
Resolution	2 pixels per mm

The segmentation and delineation are achieved by 3 main steps. First, identifying the spine region and obtaining a pathway that goes through all the vertebras and intervertebral discs. Second, using the defined pathway to locate all the intervertebral discs and obtain their signal profile. Third, delineating all the discs by using their approximate location. The concept and workflow of the proposed method are shown in Fig. 1.

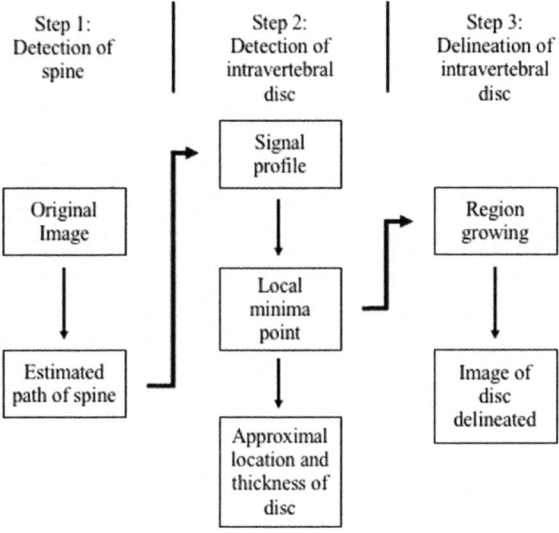

Fig. 1. The workflow of segmentation and delineation.

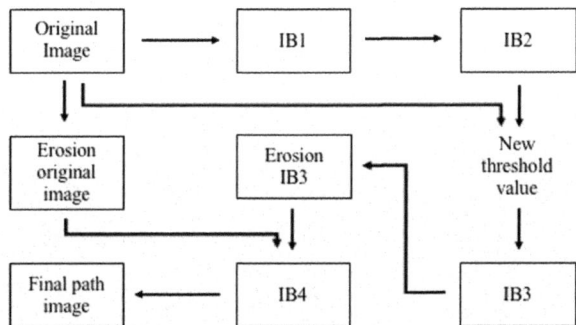

Fig. 2. The workflow of estimating the pathway of the spine.

2.1 Detection of Spine

The goal of this processing stage is to establish a path along the spine, from the top to the bottom of the image. The initial step is to create a binary image (IB1) that includes the observed region of the MRI image, using a modified version of the Otsu global thresholding [6]. Two binary morphologic operations are applied to IB1 – closing and fill holes [7], resulting in a new binary image IB2. The pixels on the original greyscale image that are on IB2 are used to establish a new threshold value, using the Otsu method [6], which is applied to the greyscale image to produce a binary image IB3. An example is presented in Fig. 3, with the original greyscale image (left) and binary image IB3 (3rd from left). A sequence of greyscale and binary erosion operations are applied to the original image and IB3 to create another binary image (IB4), as illustrated in Fig. 3 (right). Morphological hit and miss operators [7] are used to identify the largest objects in IB4 and to obtain their centroids. A scheme of the workflow followed to detect the spine is presented in Fig. 2.

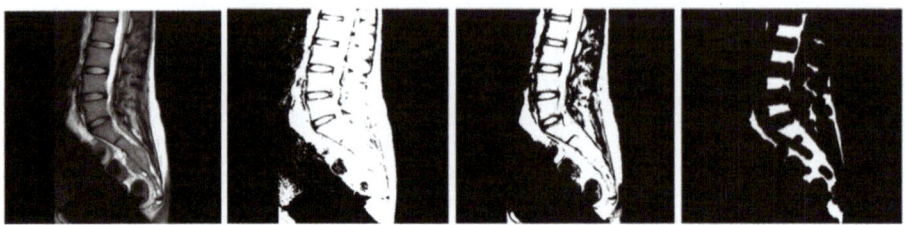

Fig. 3. An example of the steps in the detection of spine (from left to right): The original image (1st), first binary image – IB1 (2nd), IB3 (3rd) and IB4 (4th).

A validation process is performed to remove centroids close to the image corners, if available. The remaining centroids are used to establish two alternate paths from the top to the bottom of the image: piecewise linear and spline. The final path is a weighted average of both paths, with the piecewise linear contributing only to the upper part of the image. An example is presented in Fig. 4 – the initial paths (left) and the final path (right).

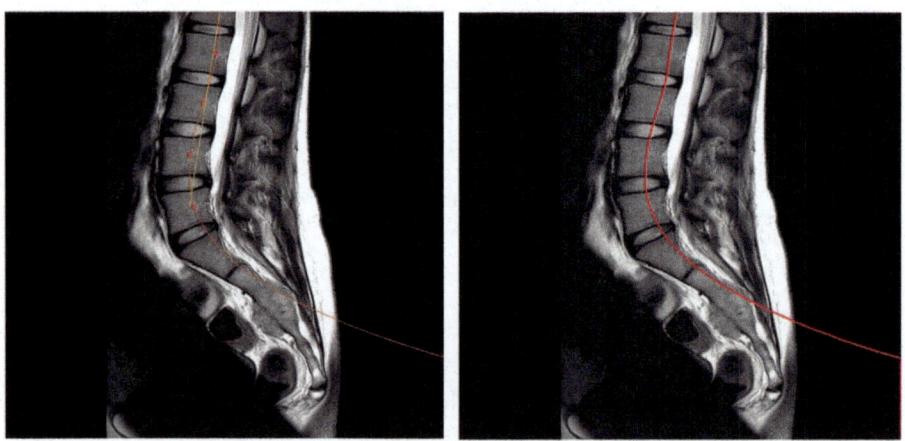

Fig. 4. Initial estimation of paths (left) and the final path (right).

2.2   Detection of Intervertebral Disc

Once the matrix information along with the path of lumbar spine is obtained, the intensity profile along the lumbar spine can then be extracted. An example of such an intensity profile is presented in Fig. 5, the brighter (darker) the pixel, the higher (lower) the signal intensity. Also, the location of spine and intervertebral discs can be identified according to the differences between the signal intensities on the profile.

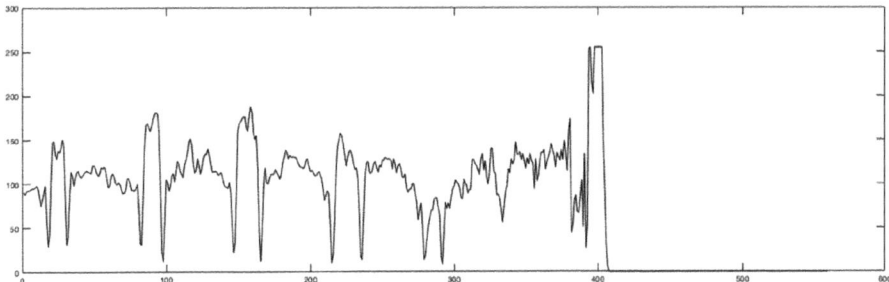

Fig. 5. Observed signal intensity profile along the estimated path of the spine.

Regarding the signal intensity of a normal lumbar spine, the regions of low signal intensities should be the spaces between vertebras and nucleus pulposus due to the fact that those regions are annulus fibrosus and they are filled with collagen fibers, which present dark values in T2 weighted images. Therefore, several valleys or local minimum points of the signal can be observed on the profile and shown periodically because of the nature of spine. These local minima of the intensity profile are useful for estimating the thickness of lumbar vertebras and discs. By setting a threshold value and comparing it with the lower part of the signal intensity along with the length of signal profile, the numbers of local minima points can be determined. The coordinate of those positions can be superposed back to the image and the annulus fibrosus between lumbar vertebras and nucleus pulposus can then be localized. An example is presented in Fig. 6.

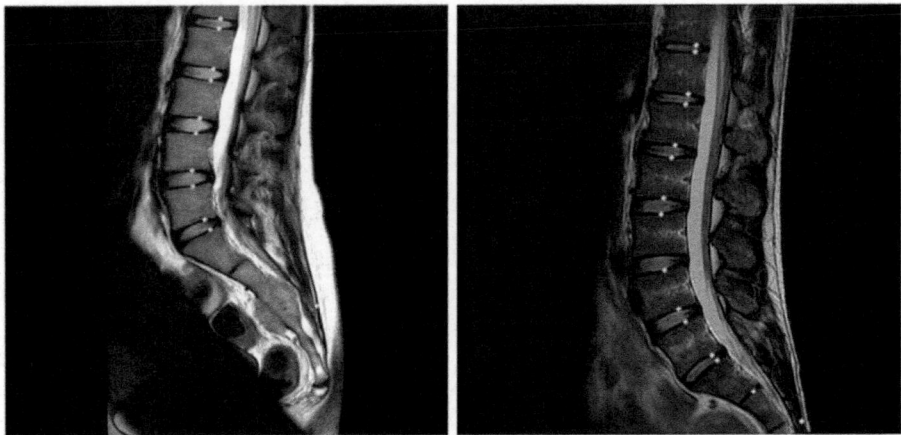

Fig. 6. The location of annulus fibrosus between vertebras and nucleus pulposus which is obtained from the points of the valley of signal intensity profile.

2.3   Delineation of Intervertebral Disc

The purpose of this study is to segment and delineate the intervertebral discs from the whole spine. Therefore, the positions or coordinates of the annulus fibrosus between lumbar vertebras and nucleus pulposus that were obtained previously can be used for this purpose. When a point in the region of annulus fibrosus is obtained, the whole region of annulus fibrosus can be identified by using region growing [8].

Region growing is an image processing tool that expands a small area of region to a larger area by appending to each seed point those neighbouring pixels that have pre-defined properties similar to the seed point [8]. The growing of region will stop when no more pixels satisfy the criteria. Thus, those coordinates that present the region of annulus fibrosus between lumbar vertebras and nucleus pulposus are used as seed points for region growing. After obtaining the whole region of annulus fibrosus, the contour of each nucleus pulposus can be finally delineated by tagging the surrounding annulus fibrosus. The coordinates of the area that surrounded by the annulus fibrosus can then be superposed back to the original image thus the location, shape, size and intensity of each nucleus pulposus can be obtained.

2.4   Evaluation

In order to verify whether this method is applicable for segmenting the spine and delineating nucleus pulposus from all the image data, some measurements were done manually before performing segmentation and compared with the values obtained from the automatic process. The first thing was to delineate the nucleus pulposus manually and compared the selected shapes with the resultant shapes of nucleus pulposus after automatic segmentation in terms of their area and size. The second was the mean signal intensity of the nucleus pulposus, comparing the signal intensities of the nucleus pulposus that were measured manually with the ones that were obtained from automatic segmentation. Both comparisons were made using a percentage scale and a statistical analysis was performed to verify the effectiveness of the purposed method.

Furthermore, a manual verification was proposed to verify the performance of the estimated pathway in each L spine image. Table 2 lists all criteria and their condition for evaluating the performance in each image in terms of whether the path goes exactly along the central line of spine and all the nucleus pulposus are detected.

3   Results

The estimated pathways of spine were obtained successfully for all test images. Table 2 presents a description of the 5 classification rates used, and the number of images labelled in each class. Among the result, there were 4 images with perfect detection, 15 images with good detection, 2 images with reasonable detection and 3 images with poor detection. One thing that should be noticed is that the estimated pathway among the images with perfect condition presented distortion in the region of T spine, although it was perfect in the region of L spine. However, as this study focused on segmentation of L spine, the investigation and results in T spine are out of scope. Table 3 presents the

Table 2. Standard and the results of the spine detector procedure.

Classification	Condition	Number of images
Perfect	Following exactly the spine through the central line of spine and all nucleus pulposus are included	4
Good	All the nucleus pulposus are detected but it does not follow the central line of spine	15
Reasonable	Missing 1 or 2 nucleus pulposus in the extra region	2
Poor	Missing several nucleus pulposus	3
Bad	Unable to detect any nucleus pulposus	0
Fail	Unable to detect any nucleus pulposus and spine	0

results of the comparison for the same slice. In the first slice of every subject, there were 1 image with perfect detection and 7 images in good detection. In the second slice of every subject, there were 2 images with perfect detection, 2 images with good detection, 1 image with reasonable detection and 3 images with poor detection. Finally, in the third slice of every subject, there were 1 image with perfect detection, 6 images with good detection and 1 image with reasonable detection.

Table 3. The result of comparison in each slice.

Slice	Perfect	Good	Reasonable	Poor	Bad	Fail	Total
1st	1	7	0	0	0	0	8
2nd	2	2	1	3	0	0	8
3rd	1	6	1	0	0	0	8

The local minima point and the respective signal intensity profile of each testing images were also obtained. The results of the signal intensity profiles in 3 images are shown in Fig. 7. The x-axis represents the position along the estimated pathway and the y-axis represents the signal intensity at the specific point of the pathway. In this way, the region of vertebras and nucleus pulposus could be discriminated by the differences between the signal intensities. Although those 3 signal intensity profiles are different, it is obvious what corresponds to vertebra or nucleus pulposus was due to the fact that the vertebra is thicker than the nucleus pulposus (indicated by arrows). Therefore, the regions of vertebra are represented by profiles that have longer lengths with the corresponding x-axis, and the region of nucleus pulposus is represented by profiles that have shorter lengths with the correlated x-axis. Also, the valleys among the profiles represent the annulus fibrosus between vertebra and nucleus pulposus, thus the thickness and the approximate location of intervertebral disc can be estimated. By using the valleys points in the signal intensity profile, the locations of the annulus fibrosus between vertebras and nucleus pulposus could be obtained and applied back to the original image.

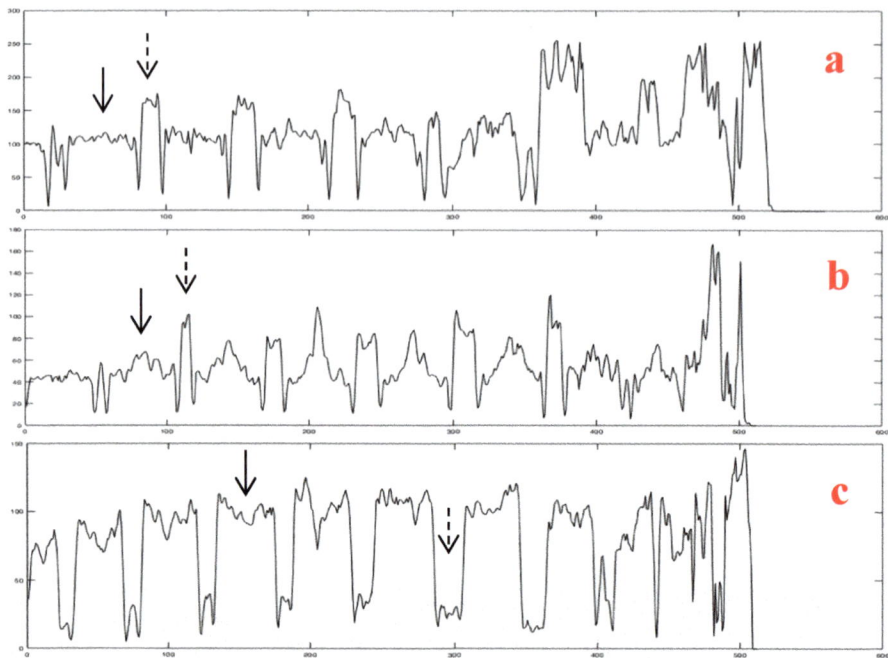

Fig. 7. Signal intensity profiles of 3 test images (a, b, c). The solid/dashed arrows indicate the regions of vertebra/nucleus pulposus respectively.

Figure 8 presents the estimated path and local minima point of the same images that were used for Fig. 7. The second row of images in Fig. 8 present the result of the estimated path of spine. Although the path on each image was not exactly along the central region of the spine, all the nucleus pulposus on each image could still be included and all the local minima points between them were thus detected successfully.

The first row of images in Fig. 8 presents the local minima points that were obtained from their respective signal intensity profile, representing the annulus fibrosus between vertebras and nucleus pulposus. These local minima points, or valleys points, can then be used as seed points for delineation of nucleus pulposus by region growing. However, as shown in image C (right), only 1 or more than 2 local minima points were detected in some intervertebral discs because of the signal intensity inside the nucleus pulposus closes to the region of annulus fibrosus or lower than it. In this case, it is difficult to estimate the thickness of intervertebral disc, but it does not affect the result of delineation of nucleus pulposus because region growing can be achieved with only 1 seed point. The effective application of the region growing algorithm is still ongoing. More fine tuning of the parameters is required, as well as comparison with manual delineation and this would lead to the estimation of discs thickness.

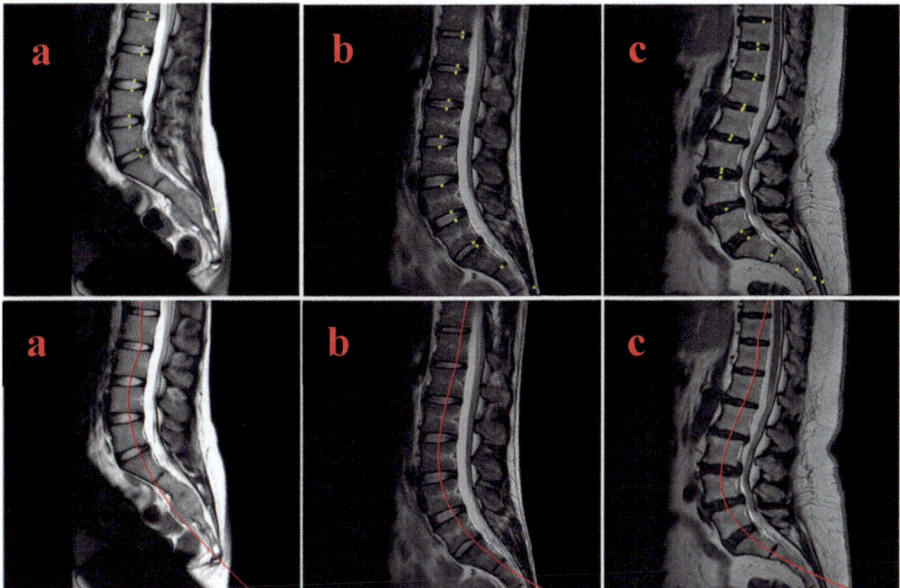

Fig. 8. The resultant local minima point (first row) and estimated path (second row) of 3 test images with respect to the signal intensity profiles in Fig. 7.

4 Conclusion

The automatic processing of MR images of spine can improve the ability to extract meaningful information with minimum human intervention. The method presented in this paper addressed the automatic identification of the alignment of spine, as well as the identification and classification of intervertebral discs. The results presented are preliminary, but provided good indications of the method's potential to accomplish these goals. The spine detection was labeled as good/perfect in 15 out of 18 images tested. Further work is required to evaluate the identification and classification of intervertebral discs. To summarize, in order to develop a robust implementation of this automatic processing for the MR image of spine, more experimental tests are required. The plan of further research includes the improvement of the proposed method for delineation of intervertebral discs and the quantification of the fluid inside each disc.

Acknowledgments. The first author wishes to thank DFA/FCUP Medical Physics MSc for the financial support. Also, we want to appreciate Ms. Lai Wai Wong who provided the information on human anatomy.

References

1. Kumar, V., Abbas, A.K., Fausto, N., Mitchell, R.N.: Robbins Basic Pathology. Elsevier Health Sciences, Amsterdam (2007)
2. Donald, A.N.: Kinesiology of the Musculoskeletal System: Foundations for Rehabilitation, 1st edn. Mosby, Maryland Heights (2002)
3. Cramer, G.D., Darby, S.A.: Clinical Anatomy of the Spine, Spine, Spinal Cord, and ANS, 3rd edn. Mosby, Maryland Heights (2013)
4. Li, X., Dou, Q., Chen, H., Chi-Wing, F., Qi, X., Belavy, D.L., Armbrecht, G., Felsenberg, D., Zheng, G., Heng, P.-A.: 3D multi-scale FCN with random modality voxel dropout learning for intervertebral disc localization and segmentation from multi-modality MR images. Med. Image Anal. **45**, 41–54 (2018)
5. Zhu, X., He, X., Wang, P., He, Q., Gao, D., Cheng, J., Baoming, W.: A method of localization and segmentation of intervertebral discs in spine MRI based on Gabor filter bank. BioMed. Eng. Online **15**, 32 (2016)
6. Otsu, N.: A threshold selection method from gray-level histograms. IEEE Trans. Syst. Man Cybern. **9**(1), 62–66 (1979)
7. MATLAB and Image Processing Toolbox Release 2017a, The MathWorks, Inc., Natick, Massachusetts, States (2017)
8. Gonzalez, R.C., Woods, R.E., Eddins, S.L.: Digital Image Processing Using MATLAB, 2nd edn. Gatesmark LLC, Knoxville (2009)

Bone Spect Image Segmentation Based on the Hermite Transform

Lorena Vargas-Quintero[1,2(✉)], Leiner Barba-J[1,2],
Jose Alberto Calderon[2], and Cesar Torres Moreno[1]

[1] Optic and Computer Science Lab., Universidad Popular del Cesar,
Valledupar, Colombia
vargas.lorena@unicesar.edu.co
[2] Medicina Nuclear S.A., Valledupar, Colombia

Abstract. Nuclear medicine image technology is currently an important modality for medical diagnosis whose applications have been extended for evaluation of a high number of human diseases. In this sense, SPECT (Single Photon Emission Computed Tomography) is one of the most used types of medical images in the field of nuclear medicine. In this work, we present a method for the segmentation of bone SPECT images. The proposed method is based on a region growing scheme implemented in the domain defined by the Hermite transform (HT) which consists of a mathematical tool that decomposes an image into a set of image coefficients representing a set of features. The method uses the coefficients obtained through the HT with the objective of designing a growing criterion that allows the segmentation of the needed region. The designed approach is focused on segmenting the bone structures in SPECT images. We evaluated the proposed scheme using several bone SPECT images. Different metrics have been also used for performance assessment.

1 Introduction

SPECT image modality is part of the nuclear medicine technologies which is commonly used in the field of radiology. This type of medical images has become essential for illness detection and treatment. It uses a set of photons generated by radioactive isotopes that have been previously injected into the human body. When the radio-pharmaceutical substance interacts with the tissues, the photons are produced, and they must be captured by a gamma camera, providing a set of image slices with the information of the organ of interest. It must be noted that with this technology we can gather metabolic information of the corresponding tissues [1]. Currently, the SPECT technology is vastly used for many applications of disease detection and diagnosis, including evaluation of the bone conditions [2]. The intensity levels coded in the SPECT images quantify the uptake of the radioactive isotope performed by the evaluated tissue, and it is an indicative of the patient conditions. SPECT, nonetheless, presents huge limitations because the spatial resolution and the contrast of the resulting images are poor.

On the other hand, quantification and assessment of the tissues of interest are tasks that might help in the diagnosis procedure, which can be carried out by using

© Springer Nature Switzerland AG 2019
J. M. R. S. Tavares and R. M. Natal Jorge (Eds.): VipIMAGE 2019, LNCVB 34, pp. 127–131, 2019.
https://doi.org/10.1007/978-3-030-32040-9_14

segmentation algorithms. The poor contrast and resolution of SPECT images hinder the correct extraction of the tissue boundaries, even for nuclear medicine physicians. In the literature, we can find several segmentation algorithms which have been designed to be applied to nuclear medicine images [3–5]. Active shape models [3], clustering-based methods [4], level sets [5] and machine learning approaches have been employed for this task [6]. Given the limitations found in this type of medical images, the segmentation problem remains open, even more when analyzing bone tissues.

In this work, we present a segmentation method based on the Hermite transform and a region growing scheme. The latter is implemented into the HT domain. Hermite coefficients are used for the design of the growing criterion. It has been demonstrated in several works, the efficiency of the HT for representing different types of image features [7]. Our method is evaluated using several SPECT image examples acquired for evaluating the bone structures.

2 Methodology

In this section, we introduce the designed scheme. Let $f(x, y)$ be an input image. The HT is calculated as [7]

$$L_{n-m,m}(p, q) = \int \int_{-\infty}^{\infty} (x, y) V^2(x - p, y - q) G_{n-m,m}(x - p, y - q) dx dy \qquad (1)$$

where $L_{n-m,m}(p, q)$ are the cartesian Hermite coefficients, $V(x, y) = \frac{1}{\sigma\sqrt{\pi}} e^{-(x^2 + y^2)/2\sigma^2}$ is a Gaussian window and $G_{n-m,m} = \frac{1}{\sqrt{2^n(n-m)m!}} H_{n-m}(x/\sigma) H_m(y/\sigma)$ are the normalized Hermite polynomials [7]. The transform order is determined by n. Coefficients must be computed for $n = 0, 1, 2, \ldots, N$ and $m = 0, 1, \ldots, n$. For $n = 0$, low frequency components are obtained, and details coefficients are calculated when $n \geq 1$. Until order $n = 2$, coefficients code features such as intensity, edges and zero crossing. These texture coefficients are then used to build the region growing scheme. Figure 1 presents cartesian Hermite coefficients up to second order applied to an image of nuclear medicine.

At the beginning, we need to provide an initial seed point. Therefore, information gathered from the neighborhood of that point is used to classify the rest of pixels. This information is evaluated using the texture coefficients obtained with the HT. The algorithm is iteratively run until a stable condition is reached which generates the segmentation of the desired bone structure. In general, the proposed method consists of two steps: (1) The HT is calculated, and (2) The region growing technique is applied using information from the Hermite coefficients. The classification criterion is configured based on the homogeneity found in the HT coefficients. The following growing rule is used.

$$C(p, q, r) = L_{m,n-m}(p, q) - L_{m,n-m}(p^r, q^r) \qquad (2)$$

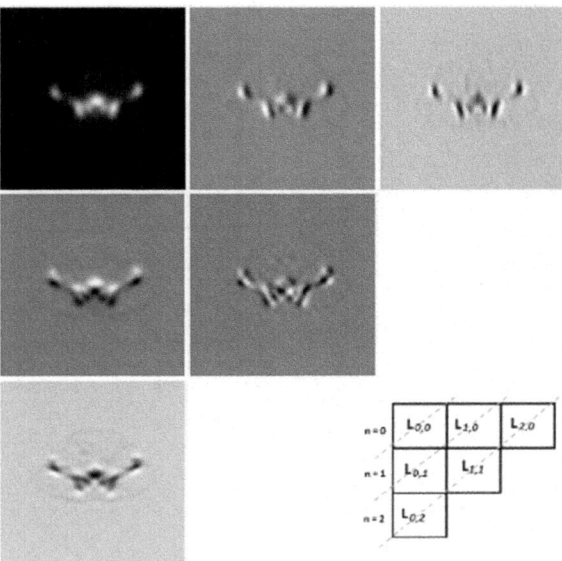

Fig. 1. Hermite transform up to order 2.

where $L_{m,n-m}(p,q)$ represents the Hermite coefficient value at point (p,q), and $L_{m,n-m}(p^r,q^r)$ corresponds to the average value computed from the neighborhood of that point. The variable C measures the homogeneity between the current point and its neighborhood. Therefore, C must be then thresholded to decide if that point corresponds to the desired object. In addition, 4 or 8 connectivity must be considered.

This method is used to provide a robust scheme that takes advantage of three types of image features.

3   Results

Examples of SPECT images were used for performance evaluation. The seed point is, until now, manually selected. The HT is computed for a second order. The growing algorithm is proved using 8 connectivity. Some results using the proposed segmentation method are illustrated in Fig. 2. It can be seen that the desired bone is satisfactory extracted.

Quantitative analysis presents the segmentation results using classical region growing and the model combining the HT and the region growing. For the evaluation we used a total of 20 images corresponding to 2 patients (Table 1).

The HT with region growing is a simple method able to achieve efficiently the contour in images of nuclear medicine. The method proposed in this work presents improved results than classical region growing. The coefficients of HT include more information of the original image which they are different texture useful in the segmentation process. The implementation scheme can be employed in the analysis of bone abnormalities.

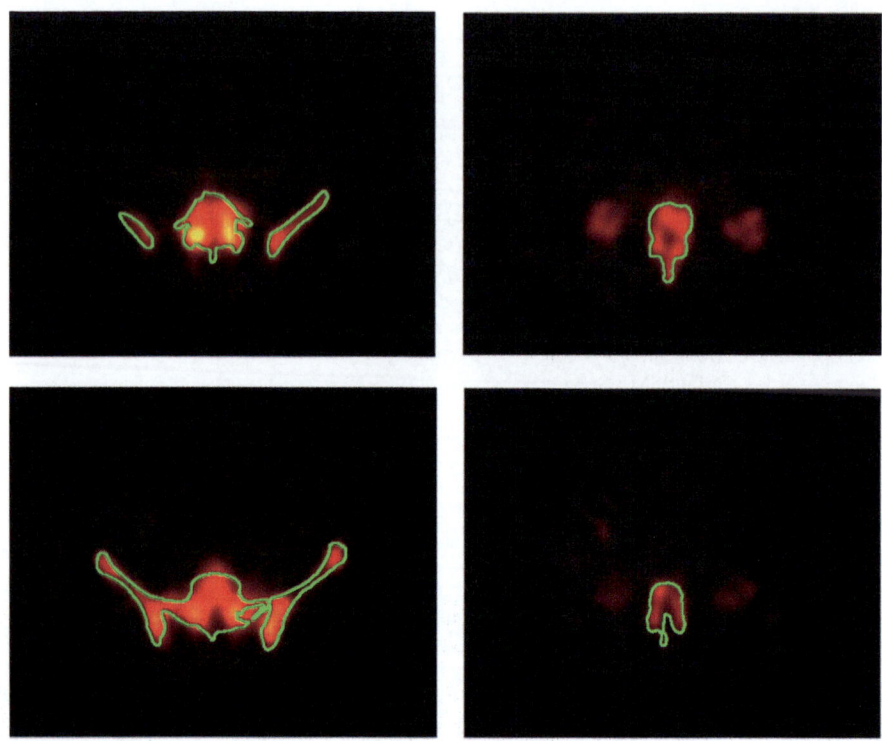

Fig. 2. Example results of the segmentation process obtained with the proposed method. The green contour corresponds to the boundary of the segmented object.

Table 1. Average point-to-curve distance obtained for 2 patients.

Patients	Images	Classical region growing	The HT with region growing
1	11	7.32 ± 3.25	3.96 ± 2.73
2	9	8.67 ± 4.50	5.01 ± 3.02

4   Conclusion

We developed an image segmentation scheme based on the Hermite transform and a region growing approach. It was applied to analysis of SPECT images. The region growing criterion is evaluated using the Hermite coefficients which allow to code several image features to improve the selection process. The method can be efficiently used for extracting the bone tissues in SPECT images.

Acknowledgments. This work has been sponsored by Colciencias.

References

1. Huellner, M., Strobel, K.: Clinical applications of SPECT/CT in imaging the extremities. Eur. J. Nucl. Med. Mol. Imaging **41**, 50–58 (2014)
2. Pelletier, M., Martineau, P.: Review of running injuries of the foot and ankle: clinical presentation and SPECT-CT imaging patterns. Am. J. Nucl. Med. Mol. Imaging **5**(4), 305–316 (2015)
3. Laading, J., McCulloch, C.: A hierarchical feature-based deformation model applied to 4D cardiac SPECT data. In: Biennial International Conference on Information Processing in Medical Imaging (1999)
4. Foster, B., et al.: A review on segmentation of positron emission tomography images. Comput. Biol. Med. **50**, 76–96 (2014)
5. Debreuve, E., Barlaud, M.: Space time segmentation using level set active contours applied to myocardial gated SPECT. In: IEEE Nuclear Science Symposium. Conference Record (1999)
6. Betancur, J., Commandeur, F.: Deep learning for prediction of obstructive disease from fast myocardial perfusion SPECT: a multicenter study. JACC cardiovasc. Imaging **11**(11), 1654–1663 (2018)
7. Silvan-Cardenas, J.L., Escalante-Ramirez, B.: The multiscale Hermite transform for local orientation analysis. IEEE Trans. Image Process. **15**(5), 1236–1253 (2006)

Image Based Search Engine - Like Using Color and Shape Features

Suhendro Y. Irianto[1(✉)], Dona Yuliawati[2], and Sri Karnila[2]

[1] Department of Informatics, Darmajaya Informatics and Business Institute,
Bandar Lampung, Indonesia
suhendro@darmajaya.ac.id
[2] Department of Information System, Darmajaya Informatics
and Business Institute, Bandar Lampung, Indonesia
{donayuliawati,Srikarnila_dj}@darmajaya.ac.id

Abstract. Content Based Image Retrieval method used to match an image query with the existing image in the database and or the image on the internet. Similarity measurements are performed using the Euclidean distance function. The image to be used is an image with JPEG format. The use of image or image feature in searching for image in a database and or internet cannot be avoided, this is because searching the image by using keyword or text is very biased and the result is far from the expectation. Search engine-like used to monitor the numbers and detect the presence of new types of fauna and flora in Indonesia. Image searching was carried out by using shape features. Search engine-like is expected also to be developed into image based search engine using CBIR method. In this work we used not less than 5,000 flora and fauna images. From the experiments, it can be concluded that the effectiveness of image retrieval is quite good in term of precision and recall.

Keywords: CBIR · Search-engine · Shape feature · Fauna · Flora

1 Introduction

Searching and surfing on the internet has become part of our daily lives, where the web browser has changed the way in searching and interacting with information. Search engine technology provides a standard interface with unlimited amount of information on the internet. Search engines can be considered as a radar that leads to a particular site or page. Once the search engines find the page (site) referred, then the search engine provides information sought by the user. Furthermore, user may then browse the site or continue to another link. Search engines are defined as a web page or web site that collects and organizes all content on the internet [1], here the user enters a query or query that cools, the search engines will provide links to the content wanted.

The problem until now still be the attention to experts in the search for a still image in a database and the internet using text or keywords, it is still far from the expectations. Text-based search techniques become impractical for two reasons: large and subjective image base sizes in interpreting images with text. To avoid this, there is an alternative in image search by applying content base image retrieval (CBIR). This CBIR technique

© Springer Nature Switzerland AG 2019
J. M. R. S. Tavares and R. M. Natal Jorge (Eds.): VipIMAGE 2019, LNCVB 34, pp. 132–137, 2019.
https://doi.org/10.1007/978-3-030-32040-9_15

is a content-based image search technique that deals with the characteristics of an image set. With the application of CBIR method, it is expected that image processing will be as easy as processing text data. Yahoo (www.yahoo.com) is one of the longest search engines since 1994, followed by Infoseek, AltaVista (www.altavista.com), AlltheWeb (www.alltheweb.com), Ask Jeeves (www.ask.com), and Google.

2 Related Works

2.1 Search Engine

Search engine basically used by user to search for information on the web, which uses the stages of searching that can be explained as follows [2] *(i) query formulation, (ii) Selection, (iii) Navigation, (iv) Query modification.* Meanwhile, many types of search engine architecture developed to date, but at first search engine architecture introduced by [3] illustration of search engine architecture can be seen in Fig. 1.

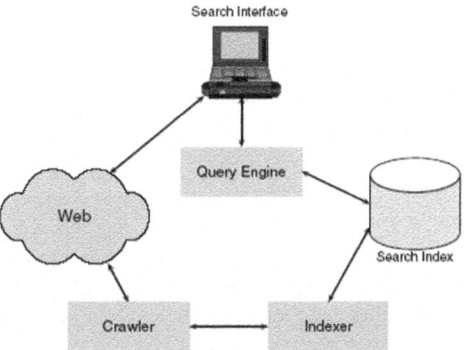

Fig. 1. Search engine architecture

2.2 Content Based Image Retrieval

Content-based image retrieval applications have been widely used in many areas of life such as biomedical, criminal, military, commerce, cultural, educational, entertainment, and agricultural fields. Many projects and research have been undertaken, such as those done by IBM (http://wwwqbic.almaden.ibm.com), Virage (www.virage.com), AltaVista, (www.altavista.com), Yahoo (www.yahoo.com), and Google (www.google.com). In CBIR to determine the resemblance of an image in a database with a reference image is to calculate its Euclidean distance. According to [4] there are four sizes to calculate the similarity, which is based on the similarity of color, shape, texture and structure. The content of an image consists of features of perception (color, shapes, textures, and structures), features of spaces, roles, views, and features of impressions and emotions.

Meanwhile, to measure the effectiveness of image searching was carried out by using precision and recall, precision is the number of similar imagery that is called divided by all retrieved images. Medium Recall is the number of invoked imagery divided by the number of images in the category [5],

$$p = \frac{a}{z} \qquad r = \frac{b}{y}$$

Where p is precision and r is recall, while a is the number of relevant or similar imagery, z is the sum of all the called images, and y the number of relevant images present in the database. While to measure the efficiency image retrieval technique is calculated using the following equation: The value of precision or recall itself is not enough to provide information, because the recall value will be made 1 by calling all the images in the database. Similarly, precision can be made high by calling multiple images.

3   Methods

3.1   Image Database

In this work we used approximately 5,000 flora and fauna images. The images are collected from the internet and from Forestry Department Office, Lampung, Indonesia.

3.2   Image Matching

In this research, the image used is image with JPEG format. In JPEG images, images are divided into blocks (8 × 8) matrices, image indexing is done by merging N blocks into a single block and each block acts as a space in a combined block. The indexing key will have 64 elements and each element has one location. Key indexing can be calculated using the following formula.

$$hi = \frac{\sum\limits_{i=1}^{N} (DCTi)^2}{N}$$

Where hi is the ke i element of the vector, and the indexing key can be explained as follows:

$H_1 = \{h_0, h_1, \ldots h_{63}\}$, while to calculate Euclidean distance can be calculated by the formula,

$$d_E^2(x, y) = \sum_{k=1}^{MN} (x^k - y^k)^2$$

M × N images sizes and can be considered as an MN dimensional euclidean space, called image space. It is natural to adopt the base e1, e2, ..., eMN to form a coordinate system of the image space, where ekN + 1 corresponds to an ideal point source with unit intensity at location (k, l). Thus an image x $= (x^1, x^2, \ldots, x^{MN})$, where x^{kN+l}

is the gray level at the $(k, l)^{th}$ pixels, is represented as a point in the image space, and x^{kN+l} with respect to $ekN + l$. Due to the lack and excess of text-based and content-based retrieval images, some researchers have tried to combine the two for image search. There are many approaches have been used to combine text or keyword and content-based techniques for shooting. A simple method used for text-based and content-based retrieval. The separate and combining the results of the retrieval has been introduced by the authors [1, 6].

They introduced annotation-based imagery search, in this system need keywords as the beginning to speed up the search by utilizing text-based search technology. Unfortunately the keyword does not always exist, so if the keyword does not exist the word system will not be efficiency. Furthermore, this system tends to be biased because no keywords are used as the beginning of an image search [7]. For an AC coefficient having a zero value, the element's feature is considered to be zero so that a vector of 64 elements will represent the pixel block texture feature. To characterize the texture features of all images, such vectors are then used as blocks to be used as indexing keys. For example there are as many as N blocks of DCT coefficients in an image, then the indexing key can be searched using the following equation:

$$H = \sum_{i=1}^{64} Ci$$

Where $Ci = \dfrac{\sum_{k=1}^{N} (category)_k}{N}$ is the to-i category between the 64 DCT coefficients.

4   Results and Discussion

4.1   Context Diagram

The system to be designed is depicted using a context diagram. Before the CBIR system designed the database of the Fauna image created with each image given index. The image is input into the CBIR system and the query image is inputted. After that the CBIR system will do searching on the image. If the image you are looking for is similar to the image database then the CBIR system will provide the same image report. The level 1 DFD design illustrates the details of the stages in the context diagram. The built system consists of two subsystems, namely: search image and add image to database. Search image function to facilitate user in search of certain image, while add image to database function to make addition of image into database, as shown in picture 3, following:

4.2   The Algorithms

The process of adding the image into the database through several stages, namely: input image, pre-processing, image segmentation and extracting process. The adding algorithm can be describe as follow, first step is initialization, follow by image query, image pre-processing, image segmentation, and feature extraction. The process of feature extraction was carried out by *Colour quantization, histogram calculation, and vectoring feature.* Image searching algorithm may be explained as Initialization, then input image query, pre-processing, image segmentation, features extraction, compare between image query dan images in the database, finally retrieve and rank similar images. From the works we calculate effectivity of image retrieval by using precision and recall parameters. Our work deploy 25 queries input in the system and demonstrate that the average of effectiveness of retrieval quite good, that is 60%. The highest precision is 100% retrieved, it justify that our algorithm is excellent. Figure 2 shows the precision and recall of image retrieval.

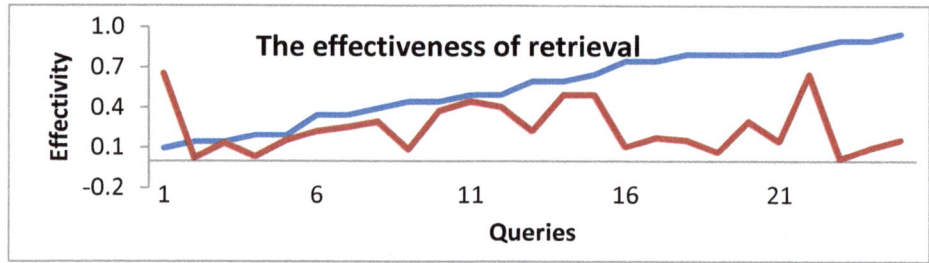

Fig. 2. The effectiveness of flora and fauna images retrieval

The application created in this work consists of the search menu there is an open query menu image preview. For the second step after the image is displayed then do extraction for color or extraction for shape. The third step of the search menu, with the search menu by color image and search by image shape, select according to the search color or shape. After a search with color/shape then the image will be displayed (Fig. 3).

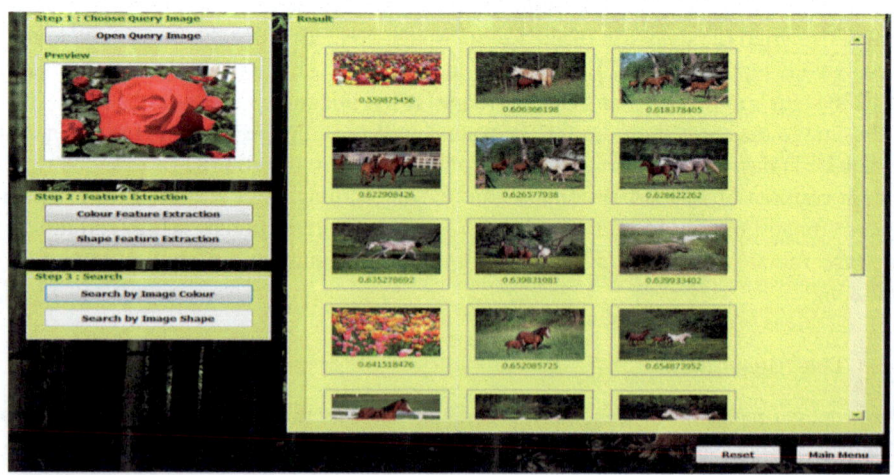

Fig. 3. Snapshot of images retrieved menu

4.3 Discussion

The results of the color search, with the HSV color type (Hue, Saturation and value) values contained from the search results of the image is the value between the comparison of the image histogram inputted with images in the database. For the form, the value contained from the search results similarity of the image is a comparison of the value of contours.

5 Conclusion

Implementation of Content Based Image Retrieval method in making search engine like image based for fauna is done through stages: system design, program flowchart and coding. Facilities provided in this prototype consists of menu image addition and image search. Menu adds the image into the data base to aims to enrich the content image in the database, while the image search menu to find images that have similarity (similarity) with the image sought (entered). From the results show that the effectiveness of retrieval in term of precision and recall is quite good.

Acknowledgment. We would like to thank to the Directorate General of Higher Education, Republic of Indonesia for supporting and funding with Hibah Produk Terapan fund. We also thank to the Research Center of Darmajaya Informatics and Business Institute for providing guiding and allowing us to use their laboratory to finish our work.

References

1. Levene, M.: An Introduction to Search Engines and Web Navigation, pp. 78–89. Wiley, Hoboken (2010)
2. Lewandowski, D.: The retrieval effectiveness of search engines on navigational queries. Aslib Proc. **63**(4), 354–363 (2011)
3. Jamali, H., Asadi, S.: Google and the scholar: the role of Google in scientists' information-seeking behavior. Online Inf. Rev. **34**(2), 282–294 (2010)
4. Wang, X., Zhang, L., Jing, F., Ma, W.: AnnoSearch: image auto-annotation by search. In: CVPR (2016)
5. Cho, W.C., Richards, D.: Improvement of precision and recall for information retrieval in a narrow domain: reuse of concepts by formal concept analysis. In: Proceedings of the 2004 IEEE/WIC/ACM International Conference on Web Intelligence, WI 2004. IEEE Computer Society, Washington, DC (2004). ISBN 0-7695-2100-2
6. Jones, G.J.F., Groves, D., Khasin, A., Lam-Adesina, A. Mellebeek, B., Way, A.: Multilingual Information Access for Text, Speech and Images. 5th Workshop of the Cross-Language Evaluation Forum, CLEF 2004. LNCS, vol. 3491, pp. 653–663. Springer, Heidelberg (2005)
7. Duan, H.Z., Zhai, C.X., Cheng, J.X., Gattani, A.: Supporting keyword search in product database: a probabilistic approach. J. VLDB Endow. **6**, 1786–1797 (2013)
8. Smeulders, A., Worring, M., Santini, S., Gupta, A., Jain, R.: Content-based image retrieval at the end of the early years. IEEE Trans. Pattern Match. Mach. Intell. **22**(12), 1349–1379 (2010)

Feature Selection

JADE-Based Feature Selection
for Non-technical Losses Detection

Clayton Reginaldo Pereira[1], Leandro Aparecido Passos[1], Douglas Rodrigues[2],
André Nunes de Souza[1], and João P. Papa[1(✉)]

[1] School of Sciences, UNESP - São Paulo State University, Bauru 17033-360, Brazil
{clayton.pereira,leandro.passos,andre.souza,joao.papa}@unesp.br
[2] Department of Computing, UFSCar - Federal University of São Carlos,
São Carlos 13565-905, Brazil
douglasrodrigues.dr@gmail.com

Abstract. Nowadays, non-technical losses, usually caused by thefts and
cheats in the energy system distribution, are among the most significant
problems an electric power company has to face. Several actions are
employed striving to contain or reduce the implications of the conducts
mentioned above, especially using automatic identification techniques.
However, selecting a proper set of features in a large dataset is essential
for successful detection rate, though it does not represent a straightforward task. This paper proposes a modification of JADE, an efficient
adaptive differential evolution algorithm, for selecting the most representative features concerning the task of computer-assisted non-technical
losses detection. Experiments on general-purpose datasets also evidence
the robustness of the proposed approach.

Keywords: Energy theft detection · Adaptive differential evolution ·
JADE · Feature selection

1 Introduction

Energy supplier companies have applied many efforts in the last years aiming to
improve the quality of distributed energy in power grids, as well as minimizing
the problem of energy loss. The latter is subject to even more attention due to the
damage implying in the companies regarding the amount of money surrounding
it [1–5].

The energy loss is often classified into two categories: (i) technical losses,
i.e., losses inherent to the transport and transmission process [6], and (ii) non-
technical losses (NTL) [7], i.e., commercial losses, which denotes the energy
that is produced and distributed by the companies but not billed due to ille-
gal interventions. The main problem regarding NTL identification concerns the
development of a proper mechanism capable of distinguishing between legal and
fraudulent customers.

Nagi et al. [8] proposed using Support Vector Machines (SVMs) [9] and
genetic SVMs [10] to tackle the aforementioned issue. Some years later, Guerrero

© Springer Nature Switzerland AG 2019
J. M. R. S. Tavares and R. M. Natal Jorge (Eds.): VipIMAGE 2019, LNCVB 34, pp. 141–156, 2019.
https://doi.org/10.1007/978-3-030-32040-9_16

et al. [11] made use of statistical techniques, text mining, and neural networks to identify NTL, while Passos et al. [12] proposed an anomaly detection model based on the Optimum-Path Forest (OPF) classifier [13,14], in which NTL is treated as outliers among samples representing regular consumers. Pereira et al. [15] proposed to optimize SVM parameters using the Social-Spider Optimization technique [16] in the context of NTL detection, and Ramos et al. [17] introduced the OPF classifier to detect non-technical losses automatically. Most recently, Papadimitriou et al. [18] presented an overview of the most current methods for electricity fraud detection based on data mining and power system analysis.

Despite the previous efforts concerning the computer-assisted classification of NTL, Nizar et al. [19] exposed that a proper selection of the features may provide faster and more accurate results. Following the same idea, Ramos et al. [20] considered using the Harmony Search [21] to identify the most representative features concerning NTL for industrial consumers. However, selecting such attributes in a dataset is not an easy task [22] due to the high number of possible combinations. Ramos et al. [23] highlighted the importance of selecting the most representative set of features when dealing with non-technical losses identification, and later on Ramos et al. [24] proposed a binary-constrained Black Hole Algorithm to handle such a task.

Recently, Zhang et al. [25] proposed JADE, an adaptive differential evolution (DE) optimization technique with an optional external archive. Such an approach presented state-of-art results regarding experiments conducted over a wide range of benchmark functions by combining a self-adaptiveness of its parameters with memory composed of past solutions. Regardless of the success obtained using JADE, as far as are concerned, there are no previous works that considered it for feature selection purposes. Therefore, the primary contribution of this work is to propose a binary version of JADE concerning the task of feature selection.

The experiments are conducted over two scenarios: (i) the binary version of JADE is compared against five well-known meta-heuristic optimization techniques to the task of feature selection, i.e., Artificial Bee Colony (ABC) [26], the Brain Storm Optimization (BSO) [27], the Cuckoo Search (CS) [28], the Firefly Algorithm (FA) [29], and the Flower Pollination Algorithm (FPA) [30]; and further (ii) eight distinct transfer functions are employed to evaluate the robustness of the binary Jade to the task of feature selection. Additionally, the proposed approach is evaluated in two scenarios: (i) NTL detection, and (ii) general-purpose classification problems.

Therefore, the main contributions of this work are twofold:

- to propose a binary-constrained JADE approach for feature selection purposes; and
- to foster the scientific community regarding NTL identification.

The remainder of the paper is organized as follows. Section 2 presents the theoretical background regarding JADE, as well as its binary version for feature selection purposes. Sections 3 and 4 present the methodology and the experimental results, respectively. Finally, conclusions and future works are stated in Sect. 5.

2 Theoretical Background

This section presents the theoretical background concerning the Adaptive Differential Evolution with Optional External Archive technique, as well as the proposed binary-constrained approach for feature selection.

2.1 Adaptive Differential Evolution with Optional External Archive

The Adaptive Differential Evolution with Optional External Archive [25] is a technique proposed to improve the optimization performance of algorithms based on the differential evolution paradigm. JADE employs an optional external archive approach, i.e., a memory composed of historical information regarding the direction of the algorithm convergence, as well as two adaptive control parameters: (i) the mutation greediness $p \in [0, 1]$, which is used to control the percentage of best solutions adopted in the mutation process, and (ii) the adaptation rate r, that is employed to update the mutation factor and the crossover probability.

Let $\mathcal{P} = \{s_1, s_2, \ldots, s_m\}$ be a population of possible solutions regarding an n-dimensional search space such that $s_i \in \Re^n$, and m stands for the number of individuals. JADE generates new solutions through a random initialization of the initial population followed by evolutionary operators, which comprise three main steps: (i) mutation, (ii) crossover, and (iii) selection. The purpose of JADE is to iteratively minimize a fitness function $f : \mathcal{P} \to \Re$ in order to find the best solution $s^* \in \mathcal{P}$. The algorithm makes use of a temporary population, hereinafter called "archived population", to keep track of older solutions that can be further used to enhance the optimization process.

The main process is implemented in Algorithm 1, where Lines 1–2 initialize the parameters μ_ψ and μ_ϕ, which denote the mean values used to calculate the crossover probability and the mutation factor, respectively[1]. The list of archived solutions $\mathcal{A}$ and the current population $\mathcal{P}$ are initialized in Lines 3–4. Further, Lines 5–20 implement the main loop, where the algorithm is executed through T epochs for convergence purposes.

Lines 6–7 initialize the sets of *successful*[2] mutation factors $\mathcal{S}_\phi$ and crossover probabilities $\mathcal{S}_\psi$. The evolutionary process takes place in the inner loop (Lines 8–14) by updating the crossover probability (Line 9) and the mutation factor (Line 10). Such step is performed by sampling from a Gaussian and a Cauchy distributions, respectively. The formulation of the Gaussian distribution is defined as follows:

$$G(\mu, \sigma) = x\sqrt{\frac{-2 \cdot \log(r)}{r}} \cdot \sigma + \mu, \tag{1}$$

where $r = x^2 + y^2$, $x \sim U(0, 1)$, $y \sim U(0, 1)$, and $r \in (0, 1]$. Similarly, one can sample from the Cauchy distribution as follows:

---

[1] The authors of the original work proposed to initialize such parameters with 0.5 [25].

[2] This term means that the new solution generated is better than the current one.

Algorithm 1. JADE algorithm.

Input: p, r, m, n, and T.
Output: global best solution s_g.
Auxiliary: t and i.

  1 $\mu_\psi = 0.5$;
  2 $\mu_\phi = 0.5$;
  3 $\mathcal{A} \leftarrow \emptyset$;
  4 Randomly initialize $\mathcal{P}$;
  5 **for** $t = 1$ **to** T **do**
  6      $\mathcal{S}_\phi \leftarrow \emptyset$;
  7      $\mathcal{S}_\psi \leftarrow \emptyset$;
  8   **for** $i = 1$ **to** m **do**
  9          $\psi = G(\mu_\psi, 0.1)$;
10          $\phi = C(\mu_\phi, 0.1)$;
           // Mutation Step
11          $v = \text{Mutation}(s_i, n, p, \phi, \mathcal{P}, \mathcal{A})$;
           // Crossover Step
12          $u = \text{Crossover}(s_i, n, \psi, v)$;
           // Selection Step
13          $\text{Selection}(s_i, u, s_g, \mathcal{A}, \psi, \phi, \mathcal{S}_\psi, \mathcal{S}_\phi)$;
14      $\mu_\psi = (1 - r) \cdot \mu_\psi + r \cdot mean_A(\mathcal{S}_\psi)$;
15      $\mu_\phi = (1 - r) \cdot \mu_\phi + r \cdot mean_L(\mathcal{S}_\phi)$;
16   **while** $|\mathcal{A}| \leq m$ **do**
17   Randomly remove solutions from $\mathcal{A}$;

$$C(\alpha, \gamma) = \frac{1}{\pi\gamma \left[1 + \left(\frac{z-\alpha}{\gamma} \right)^2 \right]}, \qquad (2)$$

where $z \sim U(0, 1)$ and $C(\alpha, \gamma) \in (0, 1]$. The core of the algorithm, i.e., mutation, crossover, and selection, are performed in Lines 11–13.

The last steps of the algorithm concern the update of μ_ψ and μ_ϕ, implemented in Lines 15–16, where $mean_A(\cdot)$ represents the standard mean and $mean_L(\cdot)$ stands for the Lehmer mean, described as follows:

$$mean_L(\mathcal{S}) = \frac{\sum_{x \in \mathcal{S}} x^2}{\sum_{x \in \mathcal{S}} x}. \qquad (3)$$

Finally, Lines 17–19 randomly remove archived solutions from $\mathcal{A}$ such that its number of elements becomes smaller or equal to the number of individuals in the current population $\mathcal{P}$.

Algorithm 2 implements the mutation process, where ϕ stands for the mutation factor. Line 1 initializes $\mathcal{P}^\star$ with the $\lfloor p.m \rfloor$ best solutions from $\mathcal{P}$, while $s^\star$ is randomly selected from $\mathcal{P}^\star$ in Line 2. Lines 3–4 randomly pick s_{r1} from $\mathcal{P}$

and s_{r2} from $\mathcal{P} \cup \mathcal{A}$, respectively, where $s_i \neq s_{r1} \neq s_{r2}$. The mutation vector v is generated in Lines 5–7 as follows:

$$v = s_i + (\phi(s^\star - s_i)) + (\phi(s_{r1} - s_{r2})). \tag{4}$$

Algorithm 2. Mutation step.

Input: s_i, n, p, ϕ, $\mathcal{P}$, and $\mathcal{A}$.
Output: mutation vector v.
Auxiliary: j.
1   $\mathcal{P}^\star \leftarrow \lfloor p.m \rfloor$ best solutions from $\mathcal{P}$.
2   $s^\star = U(1, \lfloor p.m \rfloor)$
3   $s_{r1} = U(1, \mathcal{P})$
4   $s_{r2} = U(1, \mathcal{P} \cup \mathcal{A})$
5   **for** $j = 1$ **to** n **do**
6   $\quad \lfloor \;\; v_j = s_{ij} + (\phi(s_j^\star - s_{ij})) + (\phi(s_{r1j} - s_{r2j}))$

The crossover operation is implemented in Algorithm 3. The procedure receives the crossover probability ψ, as well as the aforementioned parameters v, s_i, and n as input. A random index k is selected from the n-dimension space in Line 1. The crossover vector u is produced in the main loop, implemented in Lines 2–8, where each dimension u_j is generated as follows:

$$u_j = \begin{cases} v_j, & \text{if}(j = k) \text{ or } (\sim U(0,1) < CR) \\ s_{ij}, & \text{otherwise.} \end{cases} \tag{5}$$

Algorithm 3. Crossover step.

Input: s_i, n, ψ, and v.
Output: crossover vector u.
Auxiliary: j, k.
1   $k = U(1, n)$
2   **for** $j = 1$ **to** n **do**
3   $\quad$ **if** $(j = k)$ *or* $(\sim U(0,1) < CR)$ **then**
4   $\quad \quad \lfloor \;\; u_j = v_j$
5   $\quad$ **else**
6   $\quad \quad \lfloor \;\; u_j = s_{ij}$

The selection procedure is implemented in Algorithm 4, where Line 1 evaluates the fitness function f using the parameters obtained using both the current solution s_i and the crossover vector u. If u achieves the best result[3], s_i is added

---

[3] Notice the best result stands the lowest values since we are dealing with a minimization problem.

to $\mathcal{A}$ and its values are replaced by $\boldsymbol{u}$. CR is inserted in $\mathcal{S}_\psi$, as well as ϕ is inserted in $\mathcal{S}_\phi$. The process is implemented in Lines 2–5. Finally, Lines 6–8 update $\boldsymbol{s_g}$ with the best solution found so far.

Algorithm 4. Selection step.

Input: s_i, $\boldsymbol{u}$, s_g, $\mathcal{A}$, CR, ϕ, $\mathcal{S}_\psi$, and $\mathcal{S}_\phi$.

  1 **if** $f(s_i) > f(\boldsymbol{u})$ **then**
  2      $\mathcal{A} \leftarrow s_i$
  3      $s_i = \boldsymbol{u}$
  4      $\mathcal{S}_\psi \leftarrow CR$
  5      $\mathcal{S}_\phi \leftarrow F$
  6   **if** $f(s_g) > f(\boldsymbol{u})$ **then**
  7        $s_g = \boldsymbol{u}$

2.2 Proposed Binary-Constrained JADE

In its standard version, JADE updates each solution in the search space towards continuous-valued positions. Unlikely, in the binary version proposed for the task of feature selection, the search space is modeled as an n-dimensional boolean lattice, in which the solutions are updated across the corners of a hypercube.

Besides, as the problem is to select or not a given feature, the solution $\boldsymbol{s_i}$ is mapped onto a binary search space, where 1 corresponds that a feature will be selected to compose the new dataset and 0 otherwise. Thus, each jth feature from a possible solution $\boldsymbol{s_i}$ is mapped as follows:

$$s_i^j = \begin{cases} 1 \text{ if } T(s_i^j) > \sigma, \\ 0 \text{ otherwise} \end{cases} \tag{6}$$

where $\sigma \sim U(0,1)$ and T stands for the so-called "transfer function" (Sect. 3.2).

3 Methodology

This section presents the datasets employed in this work, as well as the setup used to conduct the experiments.

3.1 Datasets

A Brazilian electric utility has provided two private datasets, being one with $3,182$ profiles of industrial consumers and the other with $4,952$ profiles of commercial consumers, represented by eight features:

1. Demand Billed (DB): demand value of the active power considered for billing purposes, in kilowatts (kW);

2. Demand Contracted (DC): the value of the demand for continuous availability requested from the electric utility, which must be paid whether the electric power is used by the consumer or not, in kilowatts (kW);
3. Demand Measured or Maximum Demand (D_{max}): the maximum actual demand for active power, verified by measurement at fifteen-minute intervals during the billing period, in kilowatts (kW);
4. Reactive Energy (RE): energy that flows through the electric and magnetic fields of an AC system, in kilovolt-amperes reactive hours (kVArh);
5. Power Transformer (PT): the power transformer installed for the consumers, in kilovolt-amperes (kVA);
6. Power Factor (PF): the ratio between the consumed active and apparent power in a circuit. The PF indicates the efficiency of a power distribution system;
7. Installed Power (P_{inst}): the sum of the nominal power of all electrical equipment installed and ready to operate at the consumer unit, in kilowatts (kW);
8. Load Factor (LF): the ratio between the average demand ($D_{average}$) and maximum demand (D_{max}) of the consumer unit. The LF is an index that shows how the electric energy is used in a rational way.

At every 15 min, the electric utility recorded consumption data during one year. After that, such technical data was used to compute the aforementioned monthly features for both datasets. However, the company did not inform what kind of irregularity was verified in each consumer.

Moreover, to provide a more in-depth analysis regarding JADE performance to the task of feature selection, we also conducted the experiments over six public datasets:

1. German Numer[4]: a two-class dataset composed of $1,000$ samples with 24 features each;
2. Ionosphere[5]: this radar data was collected by a system in Goose Bay and consists of a phased array of 16 high-frequency antennas with a total transmitted power on the order of 6.4 kW, and have 351 instances with 34 attributes;
3. MPEG-7[6]: composed of $1,400$ binary silhouette images with 32×32 resolution. It is divided into 70 different classes (20 examples in each category) comprising shapes of animals, objects, and cartoons, to cite a few. The experiments were conducted using 180 features extracted using the Beam Angle Statistics [31] descriptor.
4. Sonar[7]: this dataset contains 111 patterns obtained by bouncing sonar signals off a metal cylinder at various angles and under various conditions, thus comprising 208 instances and 60 attributes;

---

[4] https://www.csie.ntu.edu.tw/~cjlin/libsvmtools/datasets/binary.html.
[5] https://archive.ics.uci.edu/ml/datasets/ionosphere.
[6] http://www.dabi.temple.edu/~shape/MPEG7/dataset.html.
[7] https://archive.ics.uci.edu/ml/datasets/Connectionist+Bench+(Sonar,+Mines+vs.+Rocks).

5. Splice[8]: composed of $1,000$ samples randomly selected from the original dataset of $3,190$ samples with 61 attributes;
6. SVMguide2[9]: this dataset is composed of 391 samples distributed into 3 classes with 20 features.

3.2 Experimental Setup

The experimental setup is divided into two sections, where the first compares JADE against eight well-known optimization algorithms, and the second compares the performance of the binary JADE for feature selection using eight distinct transfer functions.

Comparing JADE Against Meta-heuristic Optimization Techniques. The proposed binary JADE is compared against ABC, BSO, CS, FA, and FPA. Table 1 presents the parameters used for each optimization technique employed in this work. Besides, we have used a population of 30 agents and 60 iterations for all techniques, with such values being empirically set. The source-codes from the optimization algorithms are available on the library LibOPT [32][10]. Notice that they were implemented in C language following the guidelines provided by their references.

Table 1. Parameters employed for each optimization algorithm.

Algorithm	Parameters
ABC	$trial = 10$
BSO	$k = 3, p_{one-cluster} = 0.8$
	$p_{one-center} = 0.4, p_{two-centers} = 0.5$
CS	$\beta = 1.5, p = 0.25, \alpha = 0.8$
FA	$\alpha = 0.2, \beta_0 = 1, \gamma = 1$
FPA	$\lambda = 1.5, p = 0.8$
JADE	$c = 0.1, p = 0.05$

Binary JADE with Distinct Transfer Functions. Mirjalili and Lewis [33] proposed two families of transfer functions known as "S-shaped" and "V-shaped" functions, which define the probability of flipping each vector's element from 0 to 1 and vice versa. Tables 2 and 3 show the mentioned two families of transfer functions. It is noteworthy that many binary meta-heuristic algorithms found in the literature are implemented using a S2 transfer function [20,34,35].

---

[8] https://archive.ics.uci.edu/ml/datasets/Molecular+Biology+(Splice-junction+Gene+Sequences).
[9] https://www.csie.ntu.edu.tw/~cjlin/libsvmtools/datasets/multiclass.html.
[10] https://github.com/jppbsi/LibOPT.

Table 2. S-shaped families of transfer functions.

Name	Transfer function
S1	$T(x) = \frac{1}{1+e^{-2x}}$
S2	$T(x) = \frac{1}{1+e^{-x}}$
S3	$T(x) = \frac{1}{1+e^{\frac{-x}{2}}}$
S4	$T(x) = \frac{1}{1+e^{\frac{-x}{3}}}$

Table 3. V-shaped families of transfer functions.

Name	Transfer function
V1	$T(x) = |\operatorname{erf}(\frac{\sqrt{\pi}}{2}x)| = |\frac{\sqrt{2}}{\pi}\int_0^{\frac{\sqrt{\pi}}{2}x} e^{-t^2} dt|$
V2	$T(x) = |\tanh(x)|$
V3	$T(x) = |\frac{x}{\sqrt{(1+x^2)}}|$
V4	$T(x) = |\frac{2}{\pi}\arctan(\frac{\pi}{2}x)|$

4 Experimental Results

This section discusses the results regarding the proposed binary version of JADE employed for the task of feature selection. Aiming to evaluate the robustness of the aforementioned technique, the section is divided into two steps: (i) Sect. 4.1 compares JADE against five well-known meta-heuristic optimization techniques, and (ii) Sect. 4.2 provides a comparison of the binary JADE performance using eight distinct transfer functions. Notice the results stand for the mean accuracy and standard deviation over 15 rounds using training (50%), validating (30%), and testing sets (20%) generated randomly. Moreover, bold values stand for the best results according to the Wilcoxon signed-rank test [36] with 0.05 (5%) of significance.

4.1 JADE Against Other Meta-heuristics

This section compares JADE against five well-known meta-heuristic optimization techniques to the task of feature selection. Figures 1 and 2 depict the recognition rates, the number of selected features and the execution time over the commercial and industrial datasets, respectively. The "yellow" bar on Figs. 1a and 2a stand for the standard OPF recognition rate, i.e., without feature selection[11].

In this paper, we considered a wrapper approach, i.e., the classification rate (R) of a supervised technique is used to guide the optimization process. In this context, the primary idea is to find the subset of features that minimize $1/R$.

---

[11] The paper employs an accuracy measure proposed by Papa et al. [13], which considers unbalanced classes.

Besides, although one could use any other supervised classification technique, we opted to employ OPF [13,14] since it is parameterless, deterministic, and fast for training.

Figures 1(a) and 2(a) present the results obtained over the Brazilian electric utility datasets. JADE obtained the best statistical results in both of them, being the most accurate technique concerning commercial dataset. In Figs. 1(b) and 2(b) one can observe that the number of features is not intrinsically correlated to the best results since a different mean number of features lead toward to similar results. Moreover, Figs. 1(c) and 2(c) depict that ABC takes, in general, the twice as much computation burden as the other techniques, due to its mechanism computing a larger number of evaluations than the other techniques.

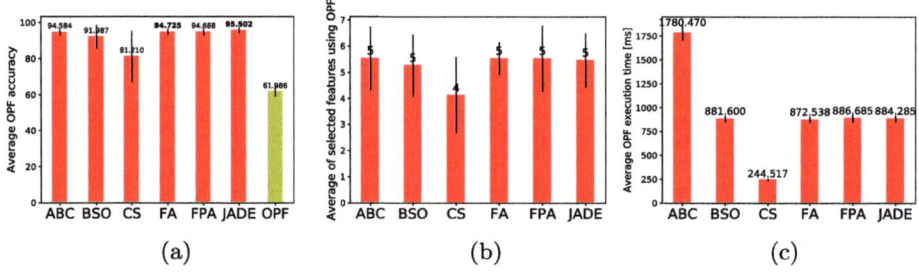

Fig. 1. Average (a) accuracy, (b) number of selected features and (c) execution time over commercial dataset.

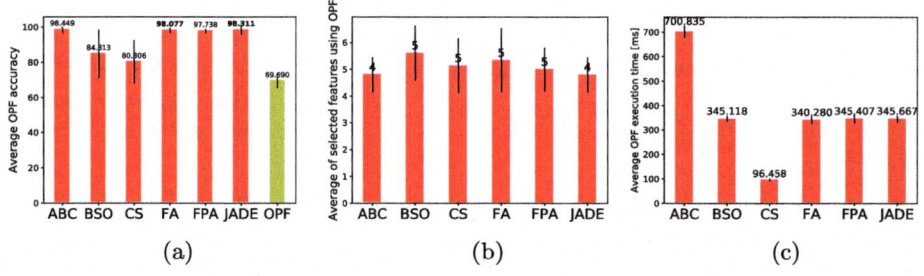

Fig. 2. Average (a) accuracy, (b) number of selected features and (c) execution time over industrial dataset.

The very same approach applied to NTL detection is now considered once again but to other five public and general-purpose datasets. The primary goal is to provide a deeper evaluation of JADE's robustness for selecting proper features. Results are presented in Table 4. Additionally, to provide a fair comparison, the results were evaluated using the Wilcoxon statistical test, as displayed in Table 5. The Wilcoxon test evaluates each pair of techniques in order to check whether they are similar to each other or not. If the p-value (parenthesis) is lesser than the desired significance level, the techniques are considered different to each other (Table 5).

Table 4. Results considering general-purpose datasets.

Dataset	Statistics	ABC	BSO	CS	FA	FPA	JADE	OPF
German Numer	Mean/Std	57.437 ± 3.52	**58.515** ± 3.72	60.421 ± 2.59	57.714 ± 2.61	56.865 ± 3.236	**58.079** ± 2.889	57.873 ± 2.451
	N° Features	14	16	14	15	14	19	24
	Runtime [ms]	84.151	42.135	11.277	41.074	41.942	45.025	-
Ionosphere	Mean/Std	**82.729** ± 5.703	**82.399** ± 5.508	**82.547** ± 5.217	**84.031** ± 5.234	81.487 ± 5.753	**82.256** ± 5.186	79.672 ± 5.549
	N° Features	21	21	20	22	23	27	34
	Runtime [ms]	13.703	6.866	1.874	6.721	6.831	7.402	-
Mpeg7	Mean/Std	**91.316** ± 1.021	**91.268** ± 1.056	91.075 ± 1.061	**91.184** ± 1.057	**91.268** ± 0.968	**91.401** ± 1.130	88.841 ± 1.389
	N° Features	121	116	114	120	119	148	180
	Runtime [ms]	516.361	260.925	66.052	248.591	255.433	298.322	-
Sonar	Mean/Std	**86.942** ± 3.296	85.304 ± 4.640	84.761 ± 4.862	**86.464** ± 5.542	84.000 ± 4.792	**86.239** ± 3.146	81.601 ± 6.346
	N° Features	39	38	40	39	40	40	60
	Runtime [ms]	5.862	3.010	0.800	2.973	2.927	3.379	-
Splice	Mean/Std	**71.059** ± 2.611	68.387 ± 2.867	66.949 ± 3.986	**69.234** ± 4.053	70.597 ± 3.287	69.929 ± 3.524	68.230 ± 3.063
	N° Features	39	39	38	41	39	49	60
	Runtime [ms]	118.077	58.154	15.807	57.206	57.779	65.367	-
SVMguide2	Mean/Std	**72.120** ± 5.364	74.317 ± 5.200	71.892 ± 4.813	**72.783** ± 4.808	72.690 ± 3.699	74.369 ± 3.460	72.362 ± 4.552
	N° Features	15	13	13	13	14	17	20
	Runtime [ms]	8.922	4.592	1.219	4.405	4.523	5.033	-

Table 5. Wilcoxon Signed rank Test considering 5% of significance: symbol '$\neq$' denotes there exists difference between the methods, and the symbol '$=$' represents the techniques are similar to each other. The values in parenthesis stand for the p-value.

Dataset	JADE/ABC	JADE/BSO	JADE/CS	JADE/FA	JADE/FPA
Commercial	$\neq$ (0.0241)	$\neq$ (0.0277)	$\neq$ (0.0038)	$=$ (0.2478)	$\neq$ (0.0280)
Industrial	$=$ (0.9528)	$\neq$ (0.0076)	$\neq$ (0.0008)	$=$ (0.5821)	$=$ (0.2475)
German Numer	$=$ (0.6701)	$=$ (0.7333)	$=$ (0.0783)	$=$ (0.8647)	$=$ (0.1914)
Ionosphere	$=$ (0.6660)	$=$ (0.8259)	$=$ (0.7004)	$=$ (0.0932)	$=$ (0.5506)
Mpeg7	$=$ (0.3664)	$=$ (0.3270)	$\neq$ (0.0235)	$=$ (0.0839)	$=$ (0.3446)
Sonar	$=$ (0.5934)	$=$ (0.8750)	$=$ (0.1258)	$=$ (0.8260)	$=$ (0.1165)
Splice	$=$ (0.2805)	$=$ (0.0535)	$=$ (0.0884)	$=$ (0.6496)	$=$ (0.6292)
SVMguide2	$=$ (0.0995)	$=$ (0.7764)	$=$ (0.0783)	$=$ (0.4603)	$=$ (0.3066)

The experimental results lead us to the following conclusions: (i) JADE was the only technique capable of achieving the best results in all datasets, according to the Wilcoxon Signed-rank Test; (ii) JADE achieved the highest accuracies on three out of eight datasets; (iii) despite ABC achieved the highest results in three out of eight datasets, JADE achieved similar statistical results consuming half of the computational burden; and (iv) considering the results without feature selection, i.e., column OPF in Figs. 1a, 2a, and Table 6, one can conclude that a proper selection of the features present a significant impact in the final results.

4.2   JADE Using Different Transfer Functions

This section provides a comparison among eight distinct transfer functions, presented in Tables 2 and 3, to the task of transforming JADE search space from real- to binary-valued features.

Regarding the Brazilian electric utility datasets, one can conclude that $S2$ is the best option for commercial profiles, as depicted in Fig. 3(a). Concerning the industrial profile, although $S2$ achieved similar results according to Wilcoxon signed-rank test, presented in Table 7, Figs. 4(a) and (c) illustrates that functions from V family present most accurate and faster results in general, being $V3$ the most accurate technique.

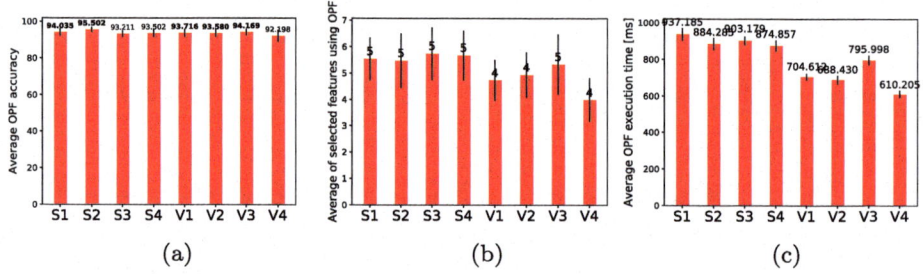

Fig. 3. Average (a) accuracy, (b) number of selected features and (c) execution time over commercial dataset considering eight distinct transfer functions.

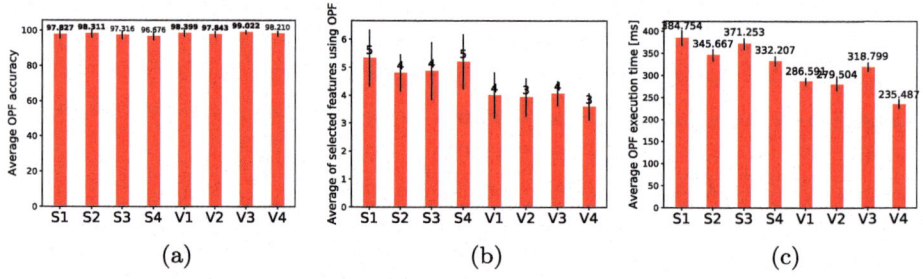

Fig. 4. Average (a) accuracy, (b) number of selected features and (c) execution time over industrial dataset considering eight distinct transfer functions.

Considering the public datasets, results presented in Table 6 expose that S family functions obtained the overall best results together with the $V3$ function, though such functions present a higher computational burden. An interesting behavior concerns Splice dataset, whose better results were achieved only with $V1$ function. Probably, such behavior is related to the complexity of DNA sequences, represented in the dataset samples.

Table 6. Execution time and accuracy over publics dataset.

Dataset	Statistics	S1	S2	S3	S4	V1	V2	V3	V4
German Numer	Mean/Std	59.19 ± 4.24	58.07 ± 2.88	58.92 ± 3.90	58.65 ± 3.83	47.67 ± 8.34	50.96 ± 7.80	60.12 ± 2.94	54.73 ± 10.62
	N° Features	18	19	13	12	2	3	14	2
	Runtime [ms]	46.350	45.021	45.293	44.504	19.583	18.402	40.196	12.883
Ionosphere	Mean/Std	84.71 ± 4.91	82.25 ± 5.18	84.01 ± 3.29	83.99 ± 5.09	84.75 ± 4.41	84.80 ± 5.27	83.94 ± 3.81	84.40 ± 4.16
	N° Features	28	27	20	19	5	4	17	3
	Runtime [ms]	6.743	7.404	6.253	6.367	4.462	4.421	5.806	3.894
Mpeg7	Mean/Std	91.50 ± 0.81	91.40 ± 1.13	91.69 ± 0.94	91.31 ± 0.84	89.78 ± 0.82	89.17 ± 0.95	91.42 ± 0.75	87.65 ± 1.34
	N° Features	148	148	107	100	21	18	114	11
	Runtime [ms]	300.591	298.322	236.421	233.994	102.917	92.533	244.162	68.969
Sonar	Mean/Std	86.71 ± 3.87	86.23 ± 3.14	84.55 ± 2.92	85.49 ± 7.09	77.74 ± 6.19	76.50 ± 9.39	82.36 ± 4.71	75.96 ± 6.38
	N° Features	49	48	36	34	9	8	38	5 1.68
	Runtime [ms]	3.350	3.379	2.750	2.835	1.937	1.933	2.757	1.688
Splice	Mean/Std	70.64 ± 2.90	69.92 ± 3.52	70.87 ± 2.86	69.73 ± 3.50	76.38 ± 7.09	61.66 ± 13.66	70.46 ± 3.65	47.63 ± 15.95
	N° Features	49	49	35	34	6	4	36	1
	Runtime [ms]	65.543	65.362	58.641	56.026	30.767	28.013	61.124	15.942
SVMguide2	Mean/Std	73.69 ± 4.37	74.36 ± 3.46	72.49 ± 5.99	69.85 ± 5.38	68.36 ± 3.91	68.12 ± 3.82	72.95 ± 6.25	63.55 ± 5.01
	N° Features	17	17	13	11	6	6	13	5
	Runtime [ms]	5.483	5.034	4.738	4.483	3.001	3.124	4.377	2.782

Table 7 presents the similarities of each function concerning the standard function $S2$. One can observe that $S1$ function presents similar results for all datasets, while $S3$ and $S4$ present similar results for 7 and 6 out of 8 datasets, respectively. Notice $V3$ represents the best values concerning the V family of functions. Moreover, $V3$ presents a similar behavior with $S2$ in six from eight datasets.

Table 7. Wilcoxon Signed rank Test considering 5% of significance: symbol '$\neq$' denotes there exists difference between the methods, and the symbol '$=$' represents the techniques are similar each other. The values in parenthesis stand for p-value.

Dataset	S2/S1	S2/S3	S2/S4	S2/V1	S2/V2	S2/V3	S2/V4
Commercial	$= (0.0691)$	$\neq (0.0016)$	$\neq (0.0106)$	$\neq (0.0054)$	$\neq (0.0268)$	$\neq (0.0303)$	$\neq (0.0076)$
Industrial	$= (0.6949)$	$= (0.1728)$	$\neq (0.0231)$	$= (0.7796)$	$= (0.4698)$	$= (0.6745)$	$= (0.5096)$
German Numer	$= (0.3343)$	$= (0.5894)$	$= (0.6496)$	$\neq (0.0012)$	$\neq (0.0106)$	$= (0.1398)$	$= (0.4603)$
Ionosphere	$= (0.4263)$	$= (0.3342)$	$= (0.3066)$	$= (0.1729)$	$= (0.2114)$	$= (0.1980)$	$= (0.2012)$
Mpeg7	$= (0.8753)$	$= (0.4954)$	$= (0.6949)$	$\neq (0.0031)$	$\neq (0.0007)$	$= (0.8261)$	$\neq (0.0007)$
Sonar	$= (0.9095)$	$= (0.3781)$	$= (0.6907)$	$\neq (0.0015)$	$\neq (0.0043)$	$\neq (0.0120)$	$\neq (0.0007)$
Splice	$= (0.6092)$	$= (0.4603)$	$= (0.9547)$	$\neq (0.0199)$	$= (0.0884)$	$= (0.8647)$	$\neq (0.0012)$
SVMguide2	$= (0.5701)$	$= (0.3343)$	$= (0.0535)$	$\neq (0.0076)$	$\neq (0.0018)$	$= (0.6092)$	$\neq (0.0010)$

5   Conclusions and Future Works

This paper introduces the differential meta-heuristic optimization algorithm JADE to the task of feature selection. Such a technique is employed for non-technical losses detection regarding commercial and industrial profiles from a Brazilian electric utility. The effectiveness of the model is confirmed using six public datasets. Moreover, eight distinct transfer functions were employed to the task of transforming JADE from the standard to a binary-constrained version.

Experimental results comparing JADE against five well-known meta-heuristic optimization techniques demonstrates the robustness of the technique, since it was the only technique capable of achieving the best results concerning all datasets, according to the Wilcoxon signed-rank test, being the most accurate in three of them. Additionally, it consumes half of the computational burden demanded by the second best technique, i.e., ABC.

Moreover, the paper led us to conclude that the S family of transfer functions are more suitable to the task of transforming JADE search space from real- to binary-valued features, although V family is capable of enhancing the results for some specific cases, i.e., Splice dataset using $V1$ transfer function. Regarding future works, we intend to propose a quaternionic version of JADE for the task of feature selection.

Acknowledgment. This study was financed in part by the Coordenação de Aperfeiçoamento de Pessoal de Nível Superior - Brasil (CAPES) - Finance Code 001. The authors acknowledge FAPESP grants 2013/07375-0, 2014/12236-1, 2016/19403-6, and 2017/02286-0, and CNPq grants 307066/2017-7 and 427968/2018-6.

References

1. Jiang, R., Tagaris, H., Lachsz, A., Jeffrey, M.: Wavelet based feature extraction and multiple classifiers for electricity fraud detection. In: IEEE/PES Transmission and Distribution Conference and Exhibition, vol. 3, pp. 2251–2256 (2002)
2. Matheus, B.R.N., Schiabel, H.: Online mammographic images database for development and comparison of cad schemes. J. Digit. Imaging **24**(3), 500–506 (2011)
3. Coleman, C.: Early detection and screening for breast cancer. Semin. Oncol. Nurs. **33**(2), 141–155 (2017)
4. He, K., Zhang, X., Ren, S., Sun, J.: Deep residual learning for image recognition. In: 2016 IEEE Conference on Computer Vision and Pattern Recognition (CVPR), pp. 770–778, June 2016
5. Howard, A.G., Zhu, M., Chen, B., Kalenichenko, D., Wang, W., Weyand, T., Andreetto, M., Adam, H.: MobileNets: efficient convolutional neural networks for mobile vision applications. CoRR, vol. abs/1704.04861, April 2017
6. Alam, M., Kabir, E., Rahman, M., Chowdhury, M.: Power sector reform in Bangladesh: electricity distribution system. Energy **29**(11), 1773–1783 (2004)
7. Patrick, O.G., Meira, J.A., Valtchev, P., State, R., Bettinger, F.: The challenge of non-technical loss detection using artificial intelligence: a survey. Int. J. Comput. Intell. Syst. **10**, 760–775 (2017)
8. Nagi, J., Yap, K.S., Tiong, S.K., Ahmed, S.K., Mohamad, M.: Nontechnical loss detection for metered customers in power utility using support vector machines. IEEE Trans. Power Deliv. **25**(2), 1162–1171 (2010)
9. Cortes, C., Vapnik, V.: Support vector networks. Mach. Learn. **20**, 273–297 (1995)
10. Nagi, J., Yap, K., Tiong, S., Ahmed, S., Mohammad, A.: Detection of abnormalities and electricity theft using genetic support vector machines. In: TENCON 2008-2008 IEEE Region 10 Conference, pp. 1–6. IEEE (2008)
11. Guerrero, J.I., León, C., Monedero, I., Biscarri, F., Biscarri, J.: Improving knowledge-based systems with statistical techniques, text mining, and neural networks for non-technical loss detection. Knowl.-Based Syst. **71**, 376–388 (2014)
12. Passos, L.A., Ramos, C.C.O., Rodrigues, D., Pereira, D.R., de Souza, A.N., da Costa, K.A.P., Papa, J.P.: Unsupervised non-technical losses identification through optimum-path forest. Electr. Power Syst. Res. **140**, 413–423 (2016)

13. Papa, J.P., Falcão, A.X., Suzuki, C.T.N.: Supervised pattern classification based on optimum-path forest. Int. J. Imaging Syst. Technol. **19**(2), 120–131 (2009)
14. Papa, J.P., Falcão, A.X., Albuquerque, V.H.C., Tavares, J.M.R.S.: Efficient supervised optimum-path forest classification for large datasets. Pattern Recogn. **45**(1), 512–520 (2012)
15. Pereira, D.R., Pazoti, M.A., Pereira, L.A.M., Rodrigues, D., Ramos, C.O., Souza, A.N., Papa, J.P.: Social-spider optimization-based support vector machines applied for energy theft detection. Comput. Electr. Eng. **49**, 25–38 (2016)
16. Yu, J., Li, V.: A social spider algorithm for global optimization. Appl. Soft Comput. **30**, 614–627 (2015)
17. Ramos, C.C.O., de Sousa, A.N., Papa, J.P., Falcã, A.X.: A new approach for non-technical losses detection based on optimum-path forest. IEEE Trans. Power Syst. **26**(1), 181–189 (2011)
18. Papadimitriou, C., Messinis, G., Vranis, D., Politopoulou, S., Hatziargyriou, N.: Non-technical losses: detection methods and regulatory aspects overview. CIRED - Open Access Proc. J. **2017**(1), 2830–2832 (2017)
19. Nizar, A.H., Zhao, J.H., Dong, Z.Y.: Customer information system data pre-processing with feature selection techniques for non-technical losses prediction in an electricity market. In: 2006 International Conference on Power System Technology, pp. 1–7, October 2006
20. Ramos, C.C.O., Souza, A.N., Chiachia, G., Falcão, A.X., Papa, J.P.: A novel algorithm for feature selection using harmony search and its application for non-technical losses detection. Comput. Electr. Eng. **37**(6), 886–894 (2011)
21. Geem, Z.W.: Music-Inspired Harmony Search Algorithm: Theory and Applications, 1st edn. Springer, Heidelberg (2009)
22. Jiang, R., Lu, R., Wang, Y., Luo, J., Shen, C., Shen, X.S.: Energy-theft detection issues for advanced metering infrastructure in smart grid. Tsinghua Sci. Technol. **19**(2), 105–120 (2014)
23. Ramos, C.C.O., Souza, A.N., Falcão, A.X., Papa, J.P.: New insights on nontechnical losses characterization through evolutionary-based feature selection. IEEE Trans. Power Deliv. **27**(1), 140–146 (2012)
24. Ramos, C.C.O., Rodrigues, D., de Souza, A.N., Papa, J.P.: On the study of commercial losses in Brazil: a binary black hole algorithm for theft characterization. IEEE Trans. Smart Grid **PP**(99), 1 (2016)
25. Zhang, J., Sanderson, A.C.: JADE: adaptive differential evolution with optional external archive. IEEE Trans. Evol. Comput. **13**(5), 945–958 (2009)
26. Karaboga, D., Basturk, B.: Artificial bee colony (ABC) optimization algorithm for solving constrained optimization problems. In: International Fuzzy Systems Association World Congress, pp. 789–798. Springer (2007)
27. Shi, Y.: Brain storm optimization algorithm. In: Proceedings of the Second International Conference on Advances in Swarm Intelligence - Volume Part I, Series ICSI 2011, pp. 303–309. Springer, Heidelberg (2011)
28. Yang, X.-S., Deb, S.: Cuckoo search via Lévy flights. In: World Congress on Nature & Biologically Inspired Computing, NaBIC 2009, pp. 210–214. IEEE (2009)
29. Yang, X.-S.: Firefly algorithm, stochastic test functions and design optimisation. Int. J. Bio-Inspir. Comput. **2**(2), 78–84 (2010)
30. Yang, S.-S., Karamanoglu, M., He, X.: Flower pollination algorithm: a novel approach for multiobjective optimization. Eng. Optim. **46**(9), 1222–1237 (2014)
31. Arica, N., Vural, F.T.Y.: BAS: a perceptual shape descriptor based on the beam angle statistics. Pattern Recogn. Lett. **24**(9–10), 1627–1639 (2003)

32. Papa, J.P., Rosa, G.H., Rodrigues, D., Yang, X.-S.: LibOPT: an open-source platform for fast prototyping soft optimization techniques, ArXiv e-prints (2017)
33. Mirjalili, S., Lewis, A.: S-shaped versus v-shaped transfer functions for binary particle swarm optimization. Swarm Evol. Comput. **9**, 1–14 (2013)
34. Rodrigues, D., Pereira, L.A.M., Papa, J.P., Ramos, C.C.O., Souza, A.N., Papa, L.P.: Optimizing feature selection through binary charged system search. In: Proceedings of 15th International Conference on Computer Analysis of Images and Patterns, pp. 377–384 (2013)
35. Rodrigues, D., Pereira, L.A.M., Nakamura, R.Y.M., Costa, K.A.P., Yang, X.S., Souza, A.N., Papa, J.P.: A wrapper approach for feature selection based on bat algorithm and optimum-path forest. Expert Syst. Appl. **41**(5), 2250–2258 (2013)
36. Wilcoxon, F.: Individual comparisons by ranking methods. Biometr. Bull. **1**(6), 80–83 (1945)

Machine Learning, Deep Learning and
Big Data

A Hybrid Approach for Breast Mass Categorization

Leandro Aparecido Passos[1], Claudio Santos[2], Clayton Reginaldo Pereira[1],
Luis Claudio Sugi Afonso[2], and João P. Papa[1(✉)]

[1] School of Sciences, UNESP - São Paulo State University, Bauru 17033-360, Brazil
{leandro.passos,clayton.pereira,joao.papa}@unesp.br
[2] Department of Computing, UFSCar - Federal University of São Carlos,
São Carlos 13565-905, Brazil
{cfsantos,sugi.luis}@ufscar.br

Abstract. Breast cancer is one of the most frequent fatal diseases among women around the world. Early diagnosis is paramount for easing such statistics, increasing the probability of successful treatment and cure. This paper proposes a hybrid approach composed of a convolutional neural network with a supervised classifier on the top capable of predicting eight specific cases of the breast tumor, being four of them malignant and four benign. The model employs the BreastNet convolution neural network to the task of mammogram images feature extraction, and it compares three distinct supervised-learning algorithms for classification purposes: (i) Optimum-Path Forest, (ii) Support Vector Machines (SVM) with Radial Basis Function, and (iii) SVM with a linear kernel. Moreover, since BreastNet is also capable of performing classification tasks, its results are further compared against the other three techniques. Experimental results demonstrate the robustness of the model, achieving 86% of accuracy over the public LAPIMO dataset.

Keywords: Breast cancer · Convolutional Neural Networks · Optimum-path forest

1 Introduction

Breast cancer is the most common cancer among women worldwide, contributing to hundreds of thousands of deaths per year [1]. Variations of the disease reflect differences in the availability of early detection and risk factors [2]. The severity of the tumor is measured, mostly, by the development stage of the cancer cells, where the mammography is employed as the primary diagnosis [3]. The evaluation is, in general, based on the appearance, structure and reproductive behavior of the affected cells. Such a diagnosis, however, requires a massive effort from a specialist, i.e., the oncologist. Therefore, a computer-based diagnostic tool powered by machine learning techniques is desirable to the task of assisting the professional in the decision-making process [4].

© Springer Nature Switzerland AG 2019
J. M. R. S. Tavares and R. M. Natal Jorge (Eds.): VipIMAGE 2019, LNCVB 34, pp. 159–168, 2019.
https://doi.org/10.1007/978-3-030-32040-9_17

In the last decades, machine learning-based approaches have been successfully employed to support medical diagnoses. One can find in the literature, for instance, dozens of works applying machine learning techniques aiming to aid early detection of Parkinson Disease [5], Diabetes [6] and Barret's Esophagus [7], among others. Regarding breast masses detection, Azar et al. [8] employed the well-known Support Vector Machines (SVM), while Karabatak et al. [9] proposed using Naïve Bayes to the task of breast cancer classification. Recently, Ribeiro et al. [10] introduced the unsupervised Optimum-Path Forest (OPF) for mammogram image segmentation.

Recently, deep learning techniques have been highlighted in the scientific community due to the outstanding results regarding a large variety of applications. The most relevant are the Restricted Boltzmann Machine-based ones [11–13], such as the Deep Belief Networks [14,15] and the Deep Boltzmann Machine [16–18], as well as the Convolutional Neural Networks (CNN) [19]. CNNs are neural networks based on the mammals' visual cortex, widely employed for image classification. Such networks execute several transformations over the input, generating feature maps from unstructured data, i.,e., set of pixels, which allows a more robust classification. Regarding breast masses detection, Sahiner et al. [20] proposed using CNNs for nodule classification purposes, to cite a few.

Although the works mentioned above achieved relevant results to the literature, they focus on detecting whether the samples represent malignant or benign tumor. However, as far as we are concerned, none of them have dealt with the problem of predicting the appropriate tumor *category*. Thus, the present work proposes a hybrid approach composed of a Convolutional Neural Network with a supervised classifier on the top to tackle the problem of breast mass categorization. Such an approach employs the BreastNet [21] to the task of feature extraction together with a supervised-learning algorithm for classification purposes. Three distinct classifiers are compared to provide an in-depth evaluation, i.e., Optimum-Path Forest [22,23], SVM with Radial Basis Function (RBF), and SVM with linear kernel. Moreover, since BreastNet is also capable of performing classification tasks, its results are further compared against the other three techniques.

Experimental results state that BreastNet itself obtained the highest overall accuracy concerning the classification task. Nevertheless, medical applications frequently demand a thorough category analysis, once it may impact directly on one's life and health. Therefore, results obtained using the linear SVM, which outperformed BreastNet classification in three out of eight categories, represent a significant contribution to the community.

Thus, the main contributions of this paper are two-fold: (i) to propose a hybrid approach for breast masses category detection, and (ii) to foster the literature regarding automated breast cancer detection. The remainder of the paper is presented as follows. Section 2 presents the proposed approach. Sections 3 and 4 present the methodology and the experiments respectively. Finally, Sect. 5 presents conclusions and future works.

2   Proposed Approach

Convolution is a widespread operation commonly used to filter images and signals, and recently to extract features as well. Given an image A and a convolution kernel C, the main idea is to obtain an output image O which highlights some properties desired during the filter design.

Let $A_{5\times5}$ be an example image (Figs. 1(a) and (b)), and $C_{3\times3}$ a *standard* convolution kernel that operates in the gray area depicted in Fig. 1(a). On the other hand, a *dilated* convolution kernel $C'_{3\times3}$ operates in a broader area, as highlighted in Fig. 1(b). Each convolution retrieves different information: C detects local information like borders while C' can disclose global features such as the content inside boundaries. BreastNet figures such skills because its blocks are generated by the combination of these two transformations, which works reasonably well for in cell images. Figure 2 shows how such an approach is designed. Using such a combination of convolutional operations, BreastNet is able to achieve state-of-art results over LAPIMO dataset [24], overcoming even some well-known CNN architectures, such as ResNet, for this particular task.

$$\begin{bmatrix} a_{11} & a_{12} & a_{13} & a_{14} & a_{15} \\ a_{21} & a_{22} & a_{23} & a_{24} & a_{25} \\ a_{31} & a_{32} & a_{33} & a_{34} & a_{35} \\ a_{41} & a_{42} & a_{43} & a_{44} & a_{45} \\ a_{51} & a_{52} & a_{53} & a_{54} & a_{55} \end{bmatrix} \qquad \begin{bmatrix} a_{11} & a_{12} & a_{13} & a_{14} & a_{15} \\ a_{21} & a_{22} & a_{23} & a_{24} & a_{25} \\ a_{31} & a_{32} & a_{33} & a_{34} & a_{35} \\ a_{41} & a_{42} & a_{43} & a_{44} & a_{45} \\ a_{51} & a_{52} & a_{53} & a_{54} & a_{55} \end{bmatrix}$$

$$\text{(a)} \qquad\qquad\qquad \text{(b)}$$

Fig. 1. Representation of: (a) regular convolution C, and (b) dilated convolution C'.

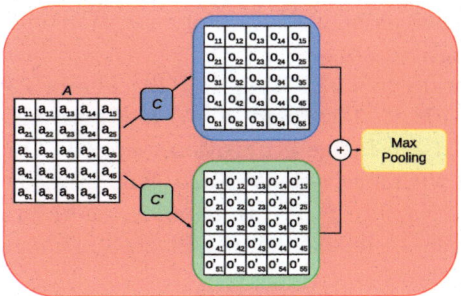

Fig. 2. BreastNet combined convolution block. Notice O and O' stand for the output of convolutions C and C', respectively.

In this paper, we make use of the full capabilities of BreastNet to distinguish among breast cancer categories. However, BreastNet itself may not be able to identify some very much specific cancer types correctly. Therefore, we propose a hybrid approach composed of BreastNet for feature learning and supervised classifiers on top of it. Figure 3 depicts the pipeline employed in this work. Notice the proposed approach is not limited to OPF and SVM classifiers only.

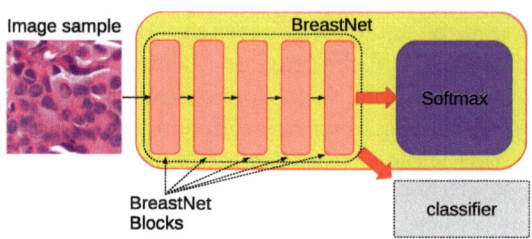

Fig. 3. BreastNet architecture. Notice this work employs the OPF, as well as the linear and RBF kernel SVM classifiers.

3   Methodology

This section presents the methodology employed to evaluate the proposed approach, as well the datasets and the experimental setup.

3.1   Datasets

The original BancoWeb Database of the LAPIMO laboratory [24] is composed of 8, 309 images generated from breast tissue biopsy slides stained with hematoxylin and eosin, prepared for histological study and labeled by specialists pathologists into four distinct categories of malignant tumors, i.e., Ductal Carcinoma (DC), Lobular Carcinoma (LC), Mucinous Carcinoma (MC), and Papillary Carcinoma (PC), and four benign tumors, i.e., Adenosis (A), Fibroadenoma (F), Tubular Adenoma (TA) and Phyllodes Tumor (PT). The images have four different magnification: 40×, 100×, 200×, and 400×. However, this work employs only the 400× images[1]. To the image acquisition step, the Olympus BX-50 system microscope was used with a relay lens with a magnification of 3.3× coupled to a Samsung digital color camera SCC-131AN and camera pixel size 6.5 μm, being the resulting images saved in a 3-channel RGB, 8-bit depth in each channel in the PNG format.

3.2   Experimental Setup

The model is trained with 75% of the dataset, i.e., 1, 820 images, while 25%, i.e., 364, are used for validation purposes. One particularity of the aforementioned

---

[1] Images are resized to 128 × 128 pixels.

dataset concerns the distribution of images available for tumor category. Ductal Carcinoma, for instance, is responsible for 788 images, while Fibroadenoma is responsible for 237, which means 56% of all available data. Such unbalanced distribution implies several constraints regarding the training and evaluation of the models.

Another critical detail relates to the amount of available data. Deep learning models are known for needing a considerable volume of instances, i.e., tens/hundreds of thousands of samples, to achieve a relevant result. To solve this issue, the total number of images were artificially increased by randomly transforming the available samples using the following rules: horizontal and/or vertical flipping, zooming from 0% up to 30%, horizontally and/or vertically shifting, considering a range of $[0, 30]\%$, and rotating in the range of $[0, 90]$ degrees. Such transformations are capable of increasing the number of samples employed from training purposes from $1,456$ images to $11,648$.

The experimental setup is divided into three main steps: the augmented training dataset is employed to feed the BreastNet network, which is in charge to learn intrinsic patterns and data structures and generates a new feature vector composed of size $32,768$ for each training and testing image. This new generated dataset is then transformed into a 50-, 100-, 300-, 500-, and 1000-feature sized samples using the Principal Component Analysis (PCA) method. Finally, the transformed data, as well as the original $32,768$-feature vectors, are employed to feed the classification algorithms, i.e., the Optimum-Path Forest, the linear-based kernel SVM, and the SVM with RBF kernel. A similar approach is observed in [5,6,25]. Notice both the linear- and radial-based SVM meta parameters were fine-tuned using a grid search.

The experiments were conducted using the LibOPF[2], a C-based library for the implementation of Optimum-Path Forest algorithm, as well as the scikit-learn[3]. The experiments were conducted using an Ubuntu 16.04 Linux machine with 64Gb of RAM running a 2x Intel® Xeon Bronze 3106 with a frequency of 1.70 GHz.

4   Experiments

This section presents the experimental results concerning the LAPIMO dataset [24] for the task of breast cancer classification. Experiments were conducted with features extracted using the BreastNet, for further feeding three supervised classifiers: Optimum-Path Forest, Support Vector Machines with linear and RBF kernels. Moreover, results are compared against a classification performed by the BreastNet itself, whose implementation is capable of performing both feature extraction and classification.

Table 1 presents the experimental results concerning the $32,768$-sized original BreastNet feature output vector, as well as five PCA-transformed feature space, i.e., employing $1,000$, 500, 300, 100, and 50 features. Notice the transformed

---

[2] https://github.com/jppbsi/LibOPF.
[3] https://scikit-learn.org/stable/index.html.

Table 1. Accuracy, F1-Measure, and computational burden over $32,768$, $1,000$, 500, 300, 100, and 50 features.

# of Features	Statistics	BreastNet	OPF	RBF-SVM	Linear-SVM
$32,768$	Accuracy	**0.8626**	0.7775	0.5824	**0.8489**
	F1-Measure	0.8598	0.7798	0.5116	0.8501
	Time (h)	1.8750	7.9108	269.2552	4.0666
$1,000$	Accuracy	–	0.6841	**0.6099**	0.3407
	F1-Measure	–	0.6890	0.5597	0.2876
	Time (h)	–	0.2333	8.6747	0.1227
500	Accuracy	–	0.7445	0.5714	0.3242
	F1-Measure	–	0.7477	0.5318	0.2747
	Time (h)	–	0.1280	3.7216	0.1966
300	Accuracy	–	0.7912	0.5220	0.3297
	F1-Measure	–	0.7893	0.4982	0.2765
	Time (h)	–	0.0808	2.0688	0.0727
100	Accuracy	–	**0.8242**	0.4478	0.3791
	F1-Measure	–	0.8194	0.3288	0.3870
	Time (h)	–	0.0297	0.7502	0.0247
50	Accuracy	–	0.7253	0.3022	0.2775
	F1-Measure	–	0.7061	0.2900	0.2941
	Time (h)	–	0.0166	0.4102	0.0130

variations are not applied to BreastNet as a classifier, since the classification is performed over the top of the network, instead of the transformed features.

BreastNet obtained the best overall results, achieving an accuracy of 0.8626. Although OPF has been outperformed by both BreastNet and the linear SVM, one may consider the accuracy of 0.8242 as a relatively good result, acknowledging it execution time that has been around 137 times faster than the Linear-SVM and 63 times faster than BreastNet. In general, OPF outperforms SVM over relative low-dimensional datasets, and in our experiments, OPF outperformed linear and RBF-SVM considering all configurations with $1,000$ samples or less.

Figure 4 depicts the confusion matrix concerning the best results obtained in each technique, i.e., 100-,$1,000$-, and $32,768$-sized feature vector for OPF, RBF-SVM, and linear-SVM, respectively. From a general point of view, Adenosis is the less constrained category. Such observation is surprising since Adenosis possesses the smallest number of samples in the dataset. Tubular Adenoma, Ductal Carcinoma, and Mucinous Carcinoma also presented high rates of accuracy for all techniques. On the other hand, Phyllodes Tumor, Lobular Carcinoma, and Papillary Carcinoma achieved relatively low results concerning every technique, which lead us to conclude the extracted features do not represent properly such categories.

Fibroadenoma presents an interesting behavior: relatively high results regarding every technique but for RBF-SVM, where not even a single sample was correctly classified. An intuitive explanation lies on the fact that the model was biased by the high number of samples from Ductal Carcinoma, misclassifying such samples.

It is noteworthy exposing the behavior of the Linear SVM, depicted in Fig. 4c. Although BreastNet obtained the best overall results, Linear SVM still achieved the highest accuracies over three out of eight tumor categories, being one of them benign, i.e., Tubular Adenoma, and two of them malign, i.e., Lobular Carcinoma and Mucinous Carcinoma. In such situations, where an accurate classification is intrinsically correlated to successful treatment, enhancing the probabilities of true positives over three categories represents a considerable contribution to the community.

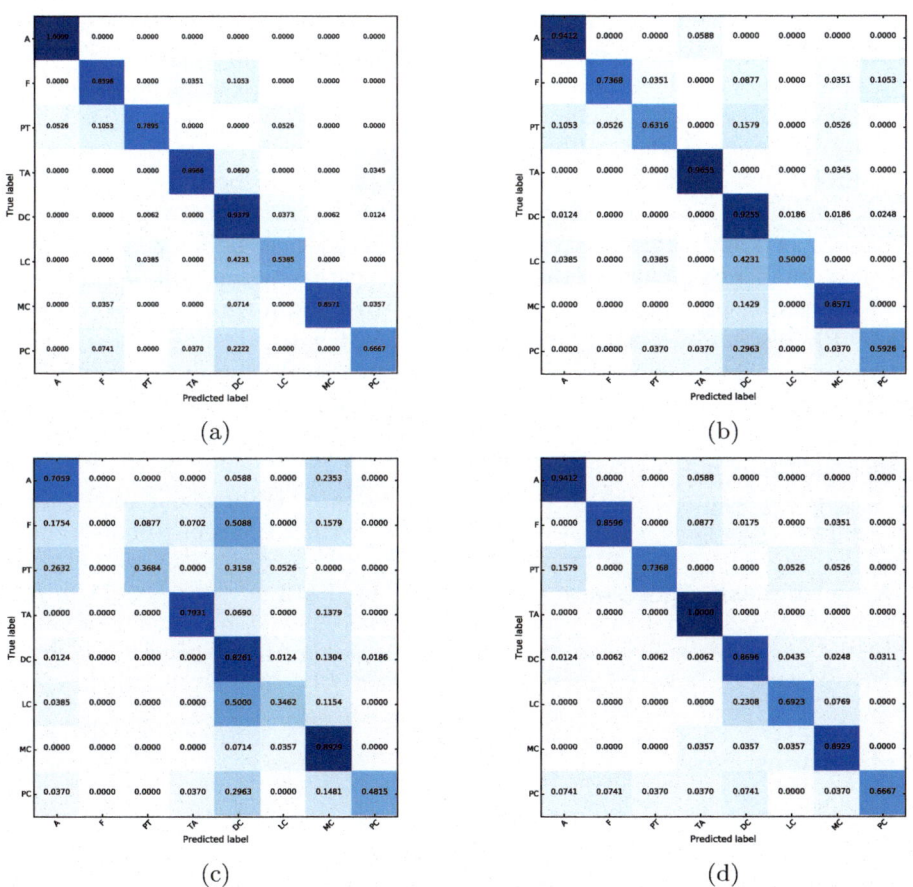

Fig. 4. Confusion matrix concerning: (a) BreastNet, (b) OPF, (c) RBF-SVM, and (d) Linear SVM.

From the experiments, one can conclude that: (i) although BreastNet and Linear-SVM obtained higher accuracies, OPF is capable of achieving relevant results demanding considerable less computational burden, (ii) we confirmed the hypothesis that OPF outperforms SVM over considerably low feature-space, since OPF obtained higher values than SVM in all results concerning the transformed vectors with $1,000$ or fewer features, and (iii) the hybrid model is capable of enhancing the probabilities of detecting determined categories of tumors, where the BreastNet itself fails on a proper classification.

5   Conclusions

This paper proposed a hybrid approach to the task of breast tumor category identification. The presented model is composed of a CNN model called BreastNet to the task of feature extraction, combined with three distinct supervised-learning algorithms on top for classification purposes. Moreover, classification performed by BreastNet is also compared against the other three algorithms.

Experiments conducted over the public LAPINE dataset demonstrated the robustness of the model, which achieved 86% of general accuracy. Additionally, it leads to conclude that, although OPF did not achieve the best results, it is suitable to the task since it obtained relative nearly best results, demanding 63 and 137 times less computational burden than BreastNet and Linear-SVM, respectively. Moreover, experiments showed that Lobular Carcinoma is the most challenging category regarding classification tasks, while Adenosis, on the other hand, presents softer constraints. Furthermore, one can also conclude that OPF presents a better behavior than SVM when dealing with relatively low-dimensional problems since it achieved better results for all cases with $1,000$ features or less. Finally, the hybrid model is capable of enhancing the probabilities of correctly classifying determined categories of tumors, where CNN-based model itself fails in classification.

Regarding future work, we intend to investigate the behavior of the model over different image magnifications, as well as distinct types of tumors.

Acknowledgment. This study was financed by FAPESP grants 2013/07375-0, 2014/12236-1, and 2016/19403-6, and CNPq grants 307066/2017-7 and 427968/2018-6. This study was financed in part by the Coordenação de Aperfeiçoamento de Pessoal de Nível Superior - Brasil (CAPES) - Finance Code 001.

References

1. Bray, F., Ferlay, J., Soerjomataram, I., Siegel, R.L., Torre, L.A., Jemal, A.: Global cancer statistics 2018: GLOBOCAN estimates of incidence and mortality worldwide for 36 cancers in 185 countries. CA: Cancer J. Clin. **68**(6), 394–424 (2018)
2. Torre, L.A., Bray, F., Siegel, R.L., Ferlay, J., Lortet-Tieulent, J., Jemal, A.: Global cancer statistics, 2012. CA: Cancer J. Clin. **65**(2), 87–108 (2015)

3. He, K., Zhang, X., Ren, S., Sun, J.: Deep residual learning for image recognition. In: 2016 IEEE Conference on Computer Vision and Pattern Recognition (CVPR), pp. 770–778, June 2016
4. Agarap, A.F.: On breast cancer detection: an application of machine learning algorithms on the Wisconsin diagnostic dataset, CoRR, vol. abs/1711.07831. http:// arxiv.org/abs/1711.07831 (2017)
5. Pereira, C.R., Passos, L.A., Lopes, R.R., Weber, S.A., Hook, C., Papa, J.P.: Parkinson's disease identification using restricted Boltzmann machines. In: International Conference on Computer Analysis of Images and Patterns, pp. 70–80. Springer (2017)
6. Khojasteh, P., Passos, L.A., Carvalho, T., Rezende, E., Aliahmad, B., Papa, J.P., Kumar, D.K.: Exudate detection in fundus images using deeply-learnable features. Comput. Biol. Med. **104**, 62–69 (2019)
7. Passos, L.A., de Souza Jr, L.A., Mendel, R., Ebigbo, A., Probst, A., Messmann, H., Palm, C., Papa, J.P.: Barrett's esophagus analysis using infinity restricted Boltzmann machines. J. Vis. Commun. Image Represent. **59**, 475–485 (2019)
8. Azar, A.T., El-Said, S.A.: Performance analysis of support vector machines classifiers in breast cancer mammography recognition. Neural Comput. Appl. **24**(5), 1163–1177 (2014)
9. Karabatak, M.: A new classifier for breast cancer detection based on Naïve Bayesian. Measurement **72**, 32–36 (2015)
10. Ribeiro, P.B., Passos, L.A., da Silva, L.A., da Costa, K.A., Papa, J.P., Romero, R.A.: Unsupervised breast masses classification through optimum-path forest. In: IEEE 28th International Symposium on Computer-Based Medical Systems, pp. 238–243. IEEE (2015)
11. Smolensky, P.: Information processing in dynamical systems: foundations of harmony theory. In: McClelland, J.L., Rumelhart, D.E., PDP Research Group (eds.) Parallel Distributed Processing: Explorations in the Microstructure of Cognition, vol. 1, pp. 194–281. MIT Press, Cambridge (1986)
12. Passos, L.A., Papa, J.P.: On the training algorithms for restricted Boltzmann machine-based models, Ph.D. dissertation, Universidade Federal de São Carlos (2018)
13. Passos, L.A., Santana, M.C., Moreira, T., Papa, J.P.: K-entropy based restricted Boltzmann machines. In: 2019 International Joint Conference on Neural Networks (IJCNN), pp. 1–8. IEEE (2019)
14. Hinton, G.E., Osindero, S., Teh, Y.-W.: A fast learning algorithm for deep belief nets. Neural Comput. **18**(7), 1527–1554 (2006)
15. Rosa, G., Papa, J.P., Costa, K., Passos, L.A., Pereira, C., Yang, X.-S.: Learning parameters in deep belief networks through firefly algorithm. In: IAPR Workshop on Artificial Neural Networks in Pattern Recognition, pp. 138–149. Springer (2016)
16. Salakhutdinov, R., Hinton, G.E.: An efficient learning procedure for deep Boltzmann machines. Neural Comput. **24**(8), 1967–2006 (2012)
17. Passos, L.A., Costa, K.A., Papa, J.P.: Deep Boltzmann machines using adaptive temperatures. In: International Conference on Computer Analysis of Images and Patterns, pp. 172–183. Springer (2017)
18. Passos, L.A., Papa, J.P.: Temperature-based deep Boltzmann machines. Neural Process. Lett. **48**(1), 95–107 (2018)
19. LeCun, Y., Bottou, L., Bengio, Y., Haffner, P., et al.: Gradient-based learning applied to document recognition. Proc. IEEE **86**(11), 2278–2324 (1998)

20. Sahiner, B., Chan, H.-P., Petrick, N., Wei, D., Helvie, M.A., Adler, D.D., Goodsitt, M.M.: Classification of mass and normal breast tissue: a convolution neural network classifier with spatial domain and texture images. IEEE Trans. Med. Imaging **15**(5), 598–610 (1996)
21. Santos, C., Pereira, C., Sugi, L.C., Passos, L.A., Papa, J.P.: Breast cancer categorization using convolutional neural networks. In: 16th International Symposium on Computer Methods in Biomechanics and Biomedical Engineering and the 4th Conference on Imaging and Visualization, CMBBE 2019 (2019)
22. Papa, J.P., Falcão, A.X., Suzuki, C.T.N.: Supervised pattern classification based on optimum-path forest. Int. J. Imaging Syst. Technol. **19**(2), 120–131 (2009)
23. Papa, J.P., Falcão, A.X., Albuquerque, V.H.C., Tavares, J.M.R.S.: Efficient supervised optimum-path forest classification for large datasets. Pattern Recogn. **45**(1), 512–520 (2012)
24. Matheus, B.R.N., Schiabel, H.: Online mammographic images database for development and comparison of cad schemes. J. Digit. Imaging **24**(3), 500–506 (2011)
25. Passos, L.A., Pereira, C.R., Rezende, E.R., Carvalho, T.J., Weber, S.A., Hook, C., Papa, J.P.: Parkinson disease identification using residual networks and optimum-path forest. In: 2018 IEEE 12th International Symposium on Applied Computational Intelligence and Informatics (SACI), pp. 000325–000330. IEEE (2018)

Comparison of Validity Indexes for Fuzzy Clusters of fMRI Data

Samuele Martinelli[1(✉)], Alberto Arturo Vergani[2], and Elisabetta Binaghi[1]

[1] Department of Theoretical and Applied Science, University of Insubria,
Varese, Italy
{smartinelli,elisabetta.binaghi}@uninsubria.it
[2] Department of Computer Science, Middlesex University, London, UK
a.vergani@mdx.ac.uk

Abstract. In computational neuroimaging, the analysis of functional Magnetic Resonance Images (fMRIs) using fuzzy clustering methods is a promising data driven approach to explore brain functional connectivity. In this complex domain, accurate evaluation procedures based on suitable indexes, able to identify optimal clustering results, are of great values strongly affecting the validity and interpretation of the overall fMRI data analysis. A large number of clustering validation indexes have been proposed in literature. This work proposes a comparison analysis of eight representative fuzzy and crisp clustering validation indexes. Salient aspects of the proposed strategy are the use of the widely adopted fuzzy c-means algorithm as underlying fuzzy clustering algorithm and the use of resting state fMRI data from the NITRC repository.

Keywords: Clustering · Clustering validation index · fMRI

1 Introduction

Data Clustering is one of the widely used methods to explore data in several domains. It utilizes only the statistical information inherent in the data without human supervision [5]. Fuzzy clustering computes degrees of membership of a single data to multiple clusters. In computational neuroimaging, the analysis of functional Magnetic Resonance Images (fMRIs) using fuzzy clustering methods is a promising data driven approach to explore brain functional connectivity. fMRI data have a complex content that regards both spatial and temporal information: the spatial ones are related to the mapping of brain regions that have common topological properties, whereas the temporal ones are referred to the detection of brain signal changes in correspondence to specific experimental times (see Fig. 1). In this context, clustering techniques find homogeneous spatio-temporal patterns without relying on any model of functional response are considered in principle more accurate than model-based methods when dealing with fMRI data analysis under complicated experimental conditions [9,17]. Clustering algorithms perform a partition of the complex fMRI content in homogeneous groups. Finding an optimized partition is a sophisticated task: not all the fMRI patterns are

© Springer Nature Switzerland AG 2019
J. M. R. S. Tavares and R. M. Natal Jorge (Eds.): VipIMAGE 2019, LNCVB 34, pp. 169–178, 2019.
https://doi.org/10.1007/978-3-030-32040-9_18

separable in distinguished crisp parcels since some of them could share common properties, as in the case of extended brain networks that vary the coactivation of different brain modules during an experimental task. Thus, the natural dynamic of the neuronal structures must be managed properly by clustering algorithms that should be able to handle both simple regularities of well-known patterns related to low-level active tasks and complex irregularities of partially-known patterns related to high-level active tasks or self-referred passive paradigms. Clustering has an important role in fMRI passive studies allowing to investigate the neurophysiological resting state that has debated biomarkers [9] and also evidence-based differences related both to gender and age [3]. In this context accurate evaluation procedures based on suitable indexes able to identify optimal (and suboptimal) clustering results are of great values strongly affecting the validity and interpretation of the overall fMRI data analysis which is still a controversial task in neuroimaging. Among the varied methods used for fMRI data clustering, fuzzy c-means [2] is certainly the most popular method [9,11,16,18]. An important issue in cluster analysis is the cluster validation aimed to measure how well the clustering results reflect the structure of the data set. For this purpose a large number of clustering validation indexes (CVIs) have been proposed in literature [4,7,13–16,19] to detect the optimal cluster number for a given dataset on the base of a balancing between the two opposite criteria of compactness within each cluster and separation between them. Several studies have been developed to investigate and compare the effectiveness of fuzzy and crisp CVIs in appropriately determining the number of clusters and measuring the goodness of clusters themselves produced by diverse algorithms [1,8]. Despite several achievements obtained, guidelines resulting from these general studies have not yet been adopted with large consensus and validation indexes are often selected basing on individual experience and/or arbitrary criteria. Critical aspects arise also in fMRI data analysis where clustering techniques are usually validated using external criteria based on prior knowledge about the data, whenever possible, or using internal different indexes depending on individual studies. The problem can be addressed by proposing comprehensive comparison studies oriented to specific clustering algorithm and specific application domains in such a way that resulting guidelines are applicable in future studies. Proceedings from these considerations, in this work we focus the attention on validation of fuzzy clustering of fMRI data and develop a comparison analysis of a set of representative fuzzy and crisp CVIs. Salient aspects of the proposed strategy are the use of the widely adopted fuzzy c-means (FCM) algorithm as underlying clustering algorithm and the use of resting state fMRI data from the NITRC repository [10]. The remaining part of the paper is organized as follows: Sect. 2 describes the clustering problem and the soft algorithm chosen to approach its solution, Sect. 3 lists the indexes used to validate the clustering results, Sect. 4 describes the general experimental procedure, the datasets used and the results obtained. Section 5 reports both the discussion of the results and the conclusions.

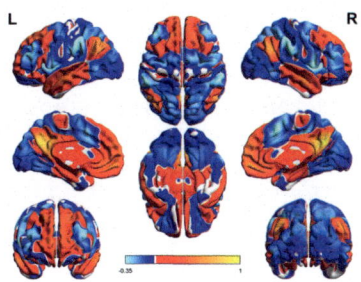

Fig. 1. This image displays resting-state functional connectivity as linear correlation for the seed region in a sample of 1,000 subjects. The seed chosen is the Precuneus (X/Y/Z MNI152 coordinates: 2 -60 30), that is the main core of the Default Mode Network (DMN), a candidate biomarker for the fMRI resting state studies. In the images, the Precuneus is in the zone with the highest functionality (yellow color).

2   Clustering Problem and Fuzzy C-Means Algorithm

The purpose of clustering is to partition a given set of data into groups (clusters) following a predefined criterion. These groups contain data that have both high similarity within clusters and high dissimilarity between the other clusters [5].

Let $X = \{x_1, x_2, \ldots, x_n\}$ a given dataset (with n elements), and let $C = \{c_1, c_2, \ldots, c_K\}$ the set of cluster, where K is the desired number of clusters. Regardless of the criterion chosen for the partition, the purpose of clustering is to develop a partition matrix of size $K \times n$ denoted as $U = [\mu_{ij}]$, with $i = 1, 2, \ldots, K$ and $j = 1, 2, \ldots, n$, where μ_{ij} is the grade of membership of point x_j to cluster c_i.

In crisp clustering, each point in the specified dataset belongs to a single cluster class. Then $\mu_{ij} = 1$ if $x_j \in c_i$, otherwise $\mu_{ij} = 0$. Instead, in fuzzy clustering, a point can be associated with more than one cluster, potentially also to all clusters, with a certain degree of membership, and the partition matrix in this case is represented as $U = [\mu_{ij}]$, where $\mu_{ij} \in [0,1]$ indicates the degree of membership of the j-th element to the i-th cluster.

The FCM algorithm proposed by Bezdek [2] is used for the data analysis in a non-supervised way in several fields. The purpose of the FCM algorithm is to create vectors called centroids that minimize the value of the function J_m that is given by the sum of the intra-cluster quadratic error. J_m it is defined as:

$$J_m = \sum_{j=1}^{n} \sum_{i=1}^{K} \mu_{ij}^m \|x_j - z_i\|^2 \tag{1}$$

where

- $m > 1$ is the exponent of the element of the fuzzy partition matrix to adjust the degree of fuzzy overlap.
- z_i is the centre of the i-th cluster.
- μ_{ij} is the degree of membership of x_j to the i-th cluster.

- $||\ldots||$ is the Euclidean norm between a point and the corresponding cluster center.

The FCM algorithm performs the following steps:

1. Randomly initialize the U matrix.
2. Calculate the cluster centroids with the following formula: $z_i = \frac{\sum_{j=1}^{n}(\mu_{ij})^m(x_j)}{\sum_{j=1}^{n}(\mu_{ij})^m}$
3. Update μ_{ij} according to the following formula: $\mu_{ij} = \frac{1}{\sum_{k=1}^{K}(\frac{||x_j - z_i||^2}{||x_j - z_k||^2})^{\frac{2}{(m-1)}}}$
4. Calculate the objective function J_m
5. Repeat steps 2–4 until J_m improves less than the prefixed threshold or until the specified maximum number of iterations is reached.

3   Cluster Validation Indexes

The use of a clustering algorithm must be complemented with the use of a validation index to detect the optimal cluster number for a given input dataset. A clustering validity index has two indicators: the compactness and the separation [12]. The compactness indicates the concentration of points that share the same cluster. The separation, evaluates the degree of isolation among clusters. A dataset is well partitioned if there is both high compactness and high separation. But often the two indicators conflict, e.g., if the compactness is high, the separation is low and *viceversa*. Therefore, a *rationale* between the two indicators is needed to design a clustering validation index.

The aim of the present work is to identify suitable CVIs for fMRI Clustering studies among a set of representative and widely used crisp and fuzzy indexes. A total of eight indexes is considered and their formal definition given below.

- The Pakhira Bandyopadhyay Maulik Index (PBMI) [13]. It evaluates the product between compactness and separation and its optimal value is towards the maximum. It is formalized as

$$PBMI(K) = \left(\frac{1}{K} \times \frac{E_1}{E_K} \times D_K\right)^2 \tag{2}$$

where K is the number of clusters used, i.e., $K = \{k', k'', \ldots, k^K\}$, the $E_K = \sum_{k=1}^{K} E_k$ holds such that the compactness is defined as crisp functional

$$J(U, Z) = E_k = \sum_{n=1}^{N} u_{nk}||x_n - z_k|| \tag{3}$$

where $U(N) = [u_{nk}]_{N \times K}$ is the binary partition matrix of the clustered data and the crisp separation is formalized as

$$D_k = \max_{k', k''} \left\{||z_{k'} - z_{k''}||\right\} \tag{4}$$

with $z_{k'} \neq z_{k''}$ (that are different centroids). Note that x_n is the n-th point in the dataset and z_k is the center of the k-th cluster. $E_1 = \sum_{n=1}^{N} ||x_n - z_1||$ z_1 is the centroid calculated on all points of the dataset

- The FPBMI is the fuzzy version of the index proposed by Pakhira et al. [13]. It evaluates the product between compactness and separation and its optimal value is towards the maximum. It is formalized similar as in the Eq. (2), except for the compactness of all clusters that it is defined as fuzzy functional, i.e.,

$$J_m(U, Z) = E_k = \sum_{k=1}^{K} \sum_{n=1}^{N} u_{nk}^m ||x_n - z_k|| \tag{5}$$

and E_1 that defined the fuzzy compactness of the cluster 1, i.e., $E_1 = \sum_{n=1}^{N} u_{n1}^m ||x_n - z_1||$. Both contain the membership value u_{nk}, where $U(N) = [u_{nk}]_{N \times K}$ is the fuzzy partition matrix of the clustered data.

- The Fukuyama Sugeno Index (FSI) [7]. It computes the difference between fuzzy compactness and fuzzy separation and its optimal value is towards the minimum., i.e.,

$$FSI(K) = \sum_{k=1}^{K} \sum_{n=1}^{N} u_{nk}^m ||x_n - z_k||^2 - \sum_{k=1}^{K} \sum_{n=1}^{N} u_{nk}^m ||z_k - \bar{z}||^2 \tag{6}$$

in which the $\bar{z}$ is the mean of all Z centroids and the u_{nk} is the membership value of the n-th point in the k-th cluster, and m is the fuzzy exponent.

- The Rezaee Lelieveldt Reider Index (RLRI) [14], also known as Compose Within and Between scattering Index (CWBI). It is the sum of compactness and separation and its optimal value is towards the minimum. In checks the average compactness and separation of fuzzy clustering by using the sum of two functions, i.e.,

$$RLR(K) = \alpha Scat(K) + Dis(K), \tag{7}$$

where α is a weighting factor equals to $Dis(K_{max})$ (the $Dis(K)$ with the maximum cluster number), and $Scat(K)$ that is the clustering compactness measure defined as

$$Scat(K) = \frac{\frac{1}{K} \sum_{k=1}^{K} ||\sigma^2(z_k)||}{||\sigma^2(X)||} \tag{8}$$

with $||x|| = (x^T \cdot x)^{1/2}$. Note that $\sigma^2(X)$ denotes the variance of all the dataset X and $\sigma^2(z_k)$ is the fuzzy variance of cluster k. The $Dis(K)$ is the clustering separation measure defined as

$$Dis(K) = \frac{D_{max}}{D_{min}} \sum_{k=1}^{K} \left[\sum_{k=1}^{K} ||z_{k'} - z_{k''}|| \right]^{-1} \tag{9}$$

with $z_{k'} \neq z_{k''}$ (different k centroids) and with Dis_{max} and Dis_{min} are the clustering separation with the maximum and minimum cluster number respectively.

- The Wang Sun Jiang Index (WSJI) [15]. It is the sum of compactness and separation and its optimal value is towards the minimum. It derived from the RLRI, adopting a linear combination of average fuzzy compactness and separation to evaluate clustering outcomes, i.e.,

$$WSJI(K) = Scat(K) + \frac{Sep(K)}{Sep(K_{max})} \tag{10}$$

where the separation $Sep(K)$ is differently defined as in Eq. 9, i.e.,

$$Dis(K) = \frac{D_{max}^2}{D_{min}^2} \sum_{k=1}^{K} \left[\sum_{k=1}^{K} ||z_{k'} - z_{k''}||^2 \right]^{-1}. \tag{11}$$

Instead, the $Scat(K)$ is the defined as in Eq. (8).

- The Xie Beni Index (XBI) [19]. It is the *ratio* between compactness and separation and its optimal value is toward the minimum. It measures the average within cluster fuzzy compactness *versus* the minimal value of the between-clusters separation, i.e.,

$$XBI(K) = \frac{\sum_{k=1}^{K} \sum_{n=1}^{N} u_{nk}^2 ||x_n - z_k||^2}{N \cdot min_{k' \neq k''} \{||z_{k'} - z_{k''}||^2\}} \tag{12}$$

with $K = \{k', k'', \ldots, k_K\}$ is the number of clusters used, N the number of data points, u_{nk} the membership values associated to the points n and a cluster k, the z_k is the centroid of a generic cluster k.

- The Davies Bouldin Index (DBI) [4]. It is the *ratio* between crisp compactness and separation and its optimal value is towards the minimum., i.e.,

$$DBI(K) = \frac{1}{K} \sum_{k=1}^{K} \max \left\{ \frac{S_{k'} + S_{k''}}{||z_{k'} - z_{k''}||} \right\} \tag{13}$$

with $k' \neq k''$ (different k centroids) and $S_{k'}$ the crisp clustering compactness of the $k' = k$-th cluster defined as

$$S_{k'} = \left(\frac{1}{N_{k'}} \sum_{x_n \in k_i} ||x_n - z_{k'}||^2 \right)^{1/2} \tag{14}$$

where $N_{k'}$ is the cardinality of the cluster k'.

- The SDBI is the soft (fuzzy) version of DBI [16]. It is the *ratio* between the fuzzy compactness and the separation and its optimal value is towards the minimum. It is defined as

$$SDBI(K) = \frac{1}{K} \sum_{k=1}^{K} \max \left\{ \frac{S_{k'}\overline{U_{k'}} + S_{k''}\overline{U_{k''}}}{||z_{k'} - z_{k''}||} \right\} \tag{15}$$

where the fuzzy compactness $S_{k'}$ is the defined as follow

$$S_{k'} = \left(\frac{1}{N} \sum_{x_n \in N} ||x_n - z_{k'}||^2 \right)^{1/2} \tag{16}$$

in which N is the cardinality of the used datasets, whereas the $\overline{U_{k'}}$ is the average of the membership values for the cluster k' (note that k' and k'' are different clusters).

4   Experiments and Results

Performances of the eight indexes introduced in Sect. 3 are evaluated using clustering results obtained by processing fMRI datasets with different configuration of FCM algorithm and comparing the optimal number of clusters indicated by the indexes with those indicated by the available ground truth.

4.1   FMRI Dataset

From the NITRC repository [10] and 1000 Functional Connectome Project, we selected the Beijing dataset with 187 healthy subjects (73M/114F; ages 18–25; all righthanded). The subjects did a resting state experimental paradigm with eyes closed. The fMRI parameters were the following: $TR = 2$, slices $= 33$ acquired with interleaved ascending procedure, time-points $= 225$, magnet $= 3$ [T]. The selection of this dataset is motivated by the specific age range and because it was just used by Biswal et al. [3] to discover resting state functional properties and their gender determinants. The brain resting state measured with fMRI has a bunch of possible biomarkers that allow researchers to build a likely ground truth (or experimental-based ground truth). The common knowledge about those biomarkers are presented in [3,9,11]. Since we want to get an empirical ground truth to validate the indexes, we defined it taking in account the acquired common knowledge about resting state fMRI biomarkers, obtaining a two classes ground truth and a four classes ground truth. The first has two labels associated to the presence/absence of regions related to the so-called Default Mode Network (DMN) [6] and the second has four labels associated to regions part of DMN and other three candidate resting networks, i.e., the Visual Network (VN), the Sensory/Motor Network (SMN) and the Other Resting Networks (ORN) (the last one encompasses all the regions that are not classified as DMN, VN or SMN).

4.2   Experiments

Two experiments have been developed by using fMRI data. In the first experiment two classes of truth are considered: what is DMN network and what is not. In the second experiment, 4 classes are considered: DMN network, VN network, SMN network and other resting networks. The FCM algorithm was configured with number of clusters $K = 2, 3, .., \sqrt{n}$ and weighting exponent $m = 1.1, 1.2, \ldots, 2.5$. To improve robustness in the evaluation, each FCM implementation was executed 200 times for each configuration and clustering result having the lowest Jm value was considered for the CVIs evaluation. In both the experiments the 8 CVIs were applied to evaluate clustering results obtained by the allowed FCM implementations distinguished by the different values of K and m parameters. To enable the quantitative comparison analysis, CVIs values were normalized taking into consideration the fact that some indexes designate the optimal number of clusters by using the maximum value, while the others the minimum value. In particular, the z-score normalization has been implemented in a positive way for the indexes that minimize their optimal value, and in a negative way for the indexes that maximize their optimal value. After normalization, the indexes indicated the number of optimal clusters with the lowest value, making them to be well comparable. Table 1 illustrates the CVIs values resulting from the evaluation of clustering fMRI dataset by FCM with $m = 2$ and i ranging from 2 to 10.

Table 1. Values of CVIs resulting from the evaluation of clustering fMRI dataset by FCM with $m = 2$ and i ranging from 2 to 10.

Index	$i = 2$	$i = 3$	$i = 4$	$i = 5$	$i = 6$	$i = 7$	$i = 8$	$i = 9$	$i = 10$
FPBMI	1.66	1.59	−1.66	−0.12	1.95	−0.44	0.62	1.62	−0.12
PBMI	1.02	0.89	−1.19	−0.47	0.69	−0.67	0.93	1.12	−0.44
FSI	0.75	0.81	−0.08	−0.53	0.07	−0.64	0.93	−0.42	−0.54
WSJI	−0.27	0.02	0.02	−0.44	0.05	−0.35	1.09	−0.95	−0.61
XBI	−0.73	−0.90	0.11	−0.46	−0.81	−0.40	−0.61	−0.63	−0.41
RLRI	−0.82	−0.98	0.65	−0.31	−0.96	−0.11	−0.57	−1.05	−0.63
DBI	−0.92	−0.97	0.93	−0.07	−1.03	0.28	−0.94	−0.29	0.46
SDBI	−0.67	−0.46	1.21	2.43	0.02	2.35	−1.44	0.61	2.30

To summarize the set of results generated and develop systematically a comparative evaluation of CVIs, we introduced a measure E defined as:

$$E = |ni - nr| \qquad (17)$$

where ni is the optimal number of clusters designated by the index, nr the number of cluster by reference. Table 2 illustrates performance of E values of the 8 CVIs, computed as average of E values obtained varying parameter m in the two experiments mentioned above.

Table 2. Mean and variance of E values for the 8 index evaluating clustering of fMRI data with 2 (Experiment 1) and 4 (Experiment 2) reference classes, the CVIs are in ascending order based on the E mean.

Index	Experiment 1		Index	Experiment 2	
	E mean	Var		E mean	Var
FSI	0.58	0.13	WSJI	0.78	0.08
RLRI	0.65	0.18	RLRI	1.35	0.17
WSJI	1.28	0.10	FSI	1.47	0.08
SDBI	1.54	1.50	SDBI	1.55	0.49
DBI	4.76	5.02	DBI	3.31	2.35
XBI	5.50	0.40	XBI	3.69	0.39
PBMI	6.22	0.42	PBMI	4.22	0.42
FPBMI	6.36	0.25	FPBMI	4.36	0.25

5   Discussion and Conclusions

In this work the performance of 8 well-known CVIs was quantitatively evaluated by using the FCM algorithm to process fMRI data. The use of the selected dataset allows to investigate the behavior of CVIs under two different levels of organizing data in two and four reference classes. The results obtained are preliminary but useful to suggests guidelines for a reliable use of cluster evaluation indexes and to contribute to a proper use of data driven, clustering techniques in the complex and more and more investigated brain function evaluation domain. Looking into the details of the results listed in Table 2, we noticed that RLRI, WSJI and FSI gained the top three positions in both the experiments even if with a different internal order. This fact leads to the conclusion that each one of them is able to both mediate between different characteristics of cluster structures and efficiently create a balance between compactness and separation. It was found also that widely used indexes such XBI, DBI and PBMI showed values considerable lower the three indexes mentioned above. The major differences between the two sets of CVIs lie in the formalization of separation component that plays an important role when dealing with clusters allocated closely as probably in case of fMRI data, and in the management of the two measures (compactness and separation) in the case of RLRI and WSJI is the sum of the two components, FSI subtraction while XBI, DBI, SDBI apply the ratio and FPBMI, PBMI the product. The novel SDBI index showed better values than crisp standard version and gained a position just below the top three positions. The XBI, PBM, FPBM, and DBI indices seem to be more suitable for contexts in which data distribution with little overlap is hypothesized, or in which cluster compactness is preferred.

Main conclusions obtained by our experimental work are consistent with results obtained in previous works [12] while considering the different experimental strategies and different domains. However caution must be exercised when applying results to other fMRI contexts taking into account the variability and complexity of these data and the different processing strategies. Future work contemplates a refinement of the metric adopted in the comparison to include other evaluation criteria and the use of a broader set of fMRI data with different levels of complexity and inter-cluster overlap, to obtain results more robust and extensible to other similar contexts.

References

1. Arbelaitz, O., Gurrutxaga, I., Muguerza, J., Pérez, J.M., Perona, I.: An extensive comparative study of cluster validity indices. Pattern Recogn. **46**(1), 243–256 (2013)
2. Bezdek, J.C., Ehrlich, R., Full, W.: FCM: the fuzzy c-means clustering algorithm. Comput. Geosci. **10**(2–3), 191–203 (1984)
3. Biswal, B.B., Mennes, M., Zuo, X., Gohel, S., Kelly, C., Smith, S.M., Beckmann, C.F., Adelstein, J.S., Buckner, R.L., Colcombe, S., et al.: Toward discovery science of human brain function. Proc. Nat. Acad. Sci. **107**(10), 4734–4739 (2010)
4. Davies, D.L., Bouldin, D.W.: A cluster separation measure. IEEE Trans. Pattern Anal. Mach. Intell. (PAMI) **1**(2), 224–227 (1979)
5. Duda, R.O., Hart, P.E., Stork, D.G.: Pattern Classification. Wiley, New York (2012)

6. Fox, M.D., Snyder, A.Z., Vincent, J.L., Corbetta, M., Van Essen, D.C., Raichle, M.E.: The human brain is intrinsically organized into dynamic, anticorrelated functional networks. Proc. Nat. Acad. Sci. **102**(27), 9673–9678 (2005)
7. Fukuyama, Y.: A new method of choosing the number of clusters for the fuzzy c-mean method. In: 1989 Proceedings of the 5th Fuzzy System Symposium, pp. 247–250 (1989)
8. Gurrutxaga, I., Muguerza, J., Arbelaitz, O., Pérez, J.M., Martín, J.I.: Towards a standard methodology to evaluate internal cluster validity indices. Pattern Recogn. Lett. **32**(3), 505–515 (2011)
9. Van Den Heuvel, M.P., Pol, H.E.H.: Exploring the brain network: a review on resting-state fMRI functional connectivity. Eur. neuropsychopharmacol. J. Eur. Coll. Neuropsychopharmacol. **20**, 519–534 (2010)
10. Kennedy, D.N., Haselgrove, C., Riehl, J., Preuss, N., Buccigrossi, R.: The nitrc image repository. Neuroimage **124**, 1069–1073 (2016)
11. Lee, M.H., Hacker, C.D., Snyder, A.Z., Corbetta, M., Zhang, D., Leuthardt, E.C., Shimony, J.S.: Clustering of resting state networks. PLoS ONE **7**(7), e40370 (2012)
12. Li, H., Zhang, S., Ding, X., Zhang, C., Dale, P.: Performance evaluation of cluster validity indices (CVIs) on Multi/Hyperspectral remote sensing datasets. Remote Sens. **8**(4), 295 (2016)
13. Pakhira, M.K., Bandyopadhyay, S., Maulik, U.: Validity index for crisp and fuzzy clusters. Pattern Recogn. **37**(3), 487–501 (2004)
14. Rezaee, R.M., Lelieveldt, B.P.F., Reiber, J.H.C.: A new cluster validity index for the fuzzy c-mean. Pattern Recogn. Lett. **19**(3–4), 237–246 (1998)
15. Sun, H., Wang, S., Jiang, Q.: FCM-based model selection algorithms for determining the number of clusters. Pattern Recogn. **37**(10), 2027–2037 (2004)
16. Vergani, A.A., Binaghi, E.: A soft Davies-Bouldin separation measure. In: 2018 IEEE International Conference on Fuzzy Systems (FUZZ-IEEE), pp. 1–8. IEEE (2018)
17. Wismuller, A.: Model-free functional MRI analysis based on unsupervised clustering. J. Biomed. Inform. **37**, 10–18 (2004)
18. Stoll, G., Meier, D., Valavanis, A., Boesiger, P., Golay, X., Kollias, S.: A new correlation-based fuzzy logic clustering algorithm for fMRI. Magn. Reson. Med. **40**(2), 249–260 (1998)
19. Xie, X.L., Beni, G.: A validity measure for fuzzy clustering. IEEE Trans. Pattern Anal. Mach. Intell. **13**(8), 841–847 (1991)

Evaluation of CNN-Based Human Pose Estimation for Body Segment Lengths Assessment

Saman Vafadar[✉], Laurent Gajny, Matthieu Boëssé, and Wafa Skalli

Arts & Métiers ParisTech, Paris, France
saman.vafadar@yahoo.com

Abstract. Human pose estimation (HPE) methods based on convolutional neural networks (CNN) have demonstrated significant progress and achieved state-of-the-art results on human pose datasets. In this study, we aimed to assess the performance of CNN-based HPE methods for measuring anthropometric data. A Vicon motion analysis system as the reference system and a stereo vision system recorded ten asymptomatic subjects standing in front of the stereo vision system in a static posture. Eight HPE methods estimated the 2D poses which were transformed to the 3D poses by using the stereo vision system. Percentage of correct keypoints, 3D error, and absolute error of the body segment lengths are the evaluation measures which were used to assess the results. Percentage of correct keypoints – the standard metric for 2D pose estimation – showed that the HPE methods could estimate the 2D body joints with a minimum accuracy of 99%. Meanwhile, the average 3D error and absolute error for the body segment lengths are 5 cm.

Keywords: Ergonomics · Anthropometry · Deep learning · Stereo vision

1 Introduction

Work on Convolutional Neural Network (ConvNet, or CNN) as a neural network model to imitate the ability of human being for pattern recognition has already begun since the late seventies [1, 2]. However, the computational costs of ConvNets had restricted its extensive use. Nowadays, the GPU-accelerated computing techniques have made the training procedure more efficient [3], resulting in the wide applications of CNNs in handwriting recognition [4], behavior recognition [5], human pose estimation [6], and medical image analysis [7].

Human Pose Estimation (HPE) methods are computer vision techniques to localize the human body joints. HPE methods, depending on the interpretation of the body structure are categorized into generative, discriminative and hybrid methods [8]. Generative methods match the image observations – or image features which are the most representative information, e.g., edges, silhouettes – with the projection of the employed human body model to the image by adjusting the body model. On the other hand, discriminative methods model the relations between the image observations and human poses [6]. For the moment, the most popular method for feature extraction is

© Springer Nature Switzerland AG 2019
J. M. R. S. Tavares and R. M. Natal Jorge (Eds.): VipIMAGE 2019, LNCVB 34, pp. 179–187, 2019.
https://doi.org/10.1007/978-3-030-32040-9_19

ConvNet. HPE methods based on ConvNets have demonstrated significant progress on challenging benchmarks (e.g. MPII [9]). The success of HPE methods based on ConvNets justifies investigation for specific applications such as anthropometry measurement.

The current applications of anthropometry measurements can be found in ergonomics. For instance, for designing fitting materials, such as the workspace and clothing to improve the safety and comfortability [10]. This study aimed to evaluate the CNN-based HPE methods for measuring anthropometric data. A stereo vision system combined with 2D HPE methods were used to recover the 3D body joint positions. Also, a marker-based motion capture system was used to assess the validity of the results.

2   Materials and Methods

Ten healthy subjects (5 males, 5 females) after informed consent participated in this study. The work has been approved by the relevant ethics committee (CPP 06036). Subjects were on average 24 years old (SD: 2, range: 21–27), mean height was 173 cm (SD: 9 cm, range: 160–187 cm), mean body mass was 64 kg (SD: 9 kg, range: 53–80 kg), and mean Body Mass Index (BMI) was 21 (SD: 2, range: 19–25).

Two calibrated and synchronized devices, a Vicon motion capture system (Vicon Motion Systems Ltd, UK) equipped with twelve Vicon Vero cameras as the reference system and a stereo vision system, were used to capture the data with the frequency of 100 Hz. The stereo vision system consisted of two GoPro Hero 7 Black cameras (GoPro, Inc., US) which recorded videos with the resolution of 1080p and linear field of view. The relative distance and angle between the two cameras were 75 cm and 15 deg, respectively, mounted on a tripod with a height of 120 cm.

Forty-eight reflective markers, according to a designed marker-set compatible with [11, 12], were attached to the subject's body segments. As shown in Fig. 1, subjects were asked to stand in a static posture in front of the stereo vision system, approximately 3.5 m away from the device, for 3 s. Table 1 shows the detail of the estimation of the morphological data using the reconstructed 3D positions of the reflective markers.

The 3 s of the videos recorded by the stereo vision system resulted in 300 frames. For the first frame of each video, a bounding box was manually defined to crop the frame in which the subject was at the center. Then, for each next frame, a dynamic bounding box was determined based on the estimated pose of the previous frame so that the height of the subject was equal to 75% of the bounding box height. Then, the cropped frames were resized to the resolution of 256×256 pixels and 368×368 pixels which are compatible with the size of the input image of the selected HPE methods. The cropped and subsequently resized images were saved to be used as the input of HPE methods. Eight HPE methods [13–20] based on convolutional neural networks for which codes were publicly available, achieving the state-of-the-art results on challenging benchmarks (e.g., MPII [9]), have been selected to estimate the 2D poses. The 2D poses consist of 16 body keypoints including upper and lower head, neck, shoulders, elbows, wrists, torso, hips, knees and ankles (Wei et al. [17] estimate

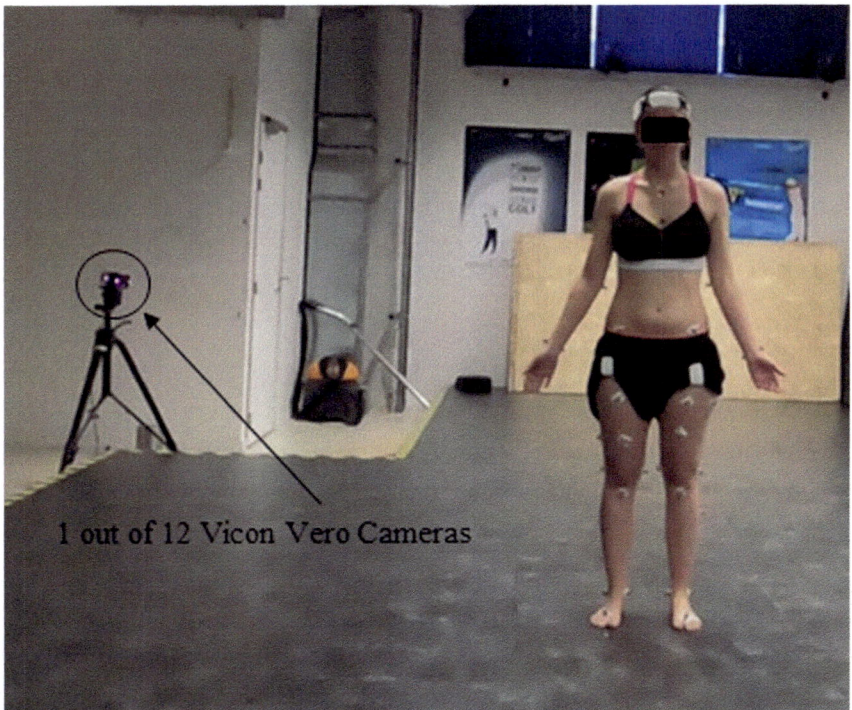

Fig. 1. One of the subjects standing in front of the stereo vision system (This image has been taken by the left camera of the stereo vision system).

14 keypoints; excluding lower head and torso). After 2D pose estimation for all the captured frames, since the subjects were standing in a static posture across the 300 frames, the mean values of the estimated body keypoints were computed. 2D to 3D pose lifting has been accomplished with the stereo vision system. After the intrinsic and extrinsic calibration of the stereo vision system, the 3D positions can be recovered using the perspective projection of the 3D point on the image planes. Herein, the retrieved 2D body keypoints were assumed to be the perspective projection of the corresponding 3D point on the left and right image plane; Thereby, the 3D positions of the body keypoints were obtained by using the linear triangulation method [21]. Hence, the morphological data were subsequently computed, as shown in Table 1. Three metrics evaluate the accuracy of the HPE methods, PCKh, 3D error and absolute error of the body segments lengths. **Percentage of Correct Keypoints** normalized with the **head** segment length (**PCKh**) defines an estimated keypoint to be correct if the distance to the corresponding reference value is less than a threshold which is a function of the head segment length – for instance PCKh@0.5 considers the 50% of the head segment length as the threshold. **3D error** measures the Euclidean distance between a reconstructed 3D keypoint and its reference value. **Absolute error** which have been used to compare the segment lengths consists of the absolute difference between the measured values by the Vicon and stereo vision system.

Table 1. The acronyms stand for: HLE = Humeral Lateral Epicondyle, HME = Humeral Medial Epicondyle, USP = Ulnar Styloid Process, RSP = Radial Styloid Process, HJC = Hip Joint Center (femoral head based on the method of Bell et al. [22]), FLE = Femoral Lateral Epicondyle, FME = Femoral Medial Epicondyle, LM = Lateral Malleolus, MM = Medial Malleolus.

Body segment	Vicon system	Stereo vision system
Forearm	$0.5 \times$ (HLE + HME) $- 0.5 \times$ (USP + RSP)	Elbow $-$ Wrist
Thigh	HJC $- 0.5 \times$ (FLE + FME)	Hip $-$ Knee
Leg	$0.5 \times$ (FLE + FME) $- 0.5 \times$ (LM + MM)	Knee $-$ Ankle

3   Results

Figure 1 shows the 2D pose estimation accuracy using the PCKh metric. The accuracy of all the selected HPE methods for 2D pose estimation was above 99% using the standard metric PCKh@0.5 [9]. Table 2 shows the 3D errors for body keypoints. The mean value of the 3D error is 5 cm. Table 3 shows the absolute error for the body segment lengths which were obtained based on the estimated 3D poses. The mean error for the lengths of the body segments, same as the 3D error, is 5 cm.

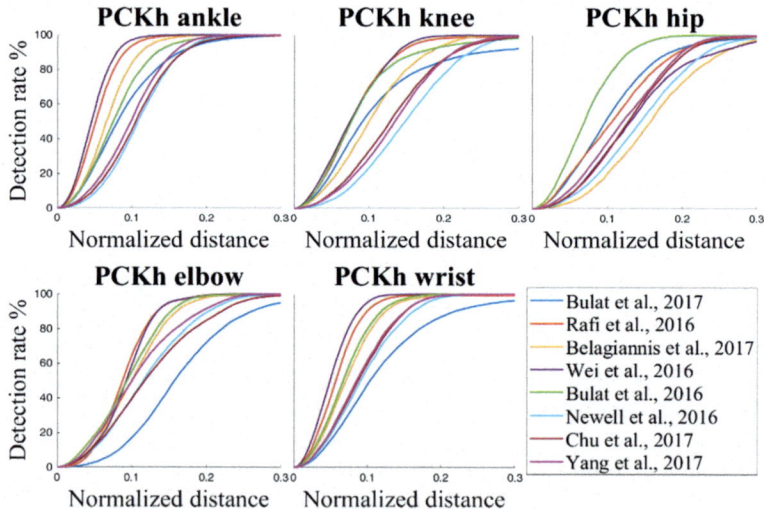

Fig. 2. Quantitative results on 2D estimated keypoints using PCKh metric.

Table 2. Quantitative results on 3D reconstructed keypoints using 3D error. Mean (min, max) values are reported in millimeter.

	Left ankle	Left knee	Left hip	Left elbow	Left wrist
Bulat et al. [13]	46 (16, 100)	43 (24, 78)	52 (25, 126)	68 (34, 140)	95 (31, 207)
Rafi et al. [15]	28 (8, 68)	48 (12, 182)	59 (30, 96)	88 (22, 598)	118 (25, 667)
Belagiannis et al. [16]	24 (8, 48)	44 (21, 81)	47 (28, 66)	34 (19, 62)	47 (10, 111)
Wei et al. [17]	23 (10, 42)	44 (19, 80)	51 (26, 101)	44 (21, 73)	60 (11, 106)
Bulat et al. [14]	32 (14, 73)	25 (7, 42)	36 (12, 64)	45 (30, 58)	36 (19, 56)
Newell et al. [18]	41 (32, 68)	64 (44, 99)	58 (44, 78)	61 (27, 122)	64 (23, 121)
Chu et al. [19]	37 (30, 61)	60 (34, 115)	50 (27, 69)	60 (37, 83)	46 (17, 127)
Yang et al. [20]	35 (24, 64)	54 (35, 83)	51 (32, 76)	52 (33, 91)	41 (14, 76)
	Right ankle	Right knee	Right hip	Right elbow	Right wrist
Bulat et al. [13]	41 (18, 79)	41 (12, 74)	49 (20, 117)	95 (31, 207)	104 (35, 288)
Rafi et al. [15]	49 (22, 126)	53 (17, 167)	58 (21, 152)	118 (25, 667)	126 (21, 790)
Belagiannis et al. [16]	29 (4, 53)	32 (16, 55)	51 (28, 67)	47 (10, 111)	51 (19, 83)
Wei et al. [17]	24 (5, 45)	29 (7, 52)	51 (33, 85)	60 (11, 106)	42 (17, 77)
Bulat et al. [14]	40 (12, 70)	43 (9, 80)	35 (13, 62)	36 (19, 56)	41 (14, 78)
Newell et al. [18]	36 (22, 61)	41 (25, 85)	45 (21, 83)	64 (23, 121)	54 (26, 93)
Chu et al. [19]	33 (19, 52)	52 (21, 85)	63 (26, 148)	46 (17, 127)	53 (26, 101)
Yang et al. [20]	32 (14, 59)	41 (17, 73)	42 (18, 100)	41 (14, 76)	42 (18, 65)

Table 3. Quantitative results on the body segment lengths using absolute error. Mean (min, max) values are reported in millimeter.

	Left forearm	Left thigh	Left leg
Bulat et al. [13]	83 (8, 195)	38 (1, 78)	34 (5, 87)
Rafi et al. [15]	63 (8, 173)	74 (14, 125)	47 (11, 142)
Belagiannis et al. [16]	32 (3, 64)	60 (11, 138)	42 (0, 140)
Wei et al. [17]	44 (3, 93)	49 (3, 130)	55 (21, 131)
Bulat et al. [14]	21 (3, 73)	30 (0, 65)	56 (24, 145)
Newell et al. [18]	58 (3, 215)	69 (15, 132)	71 (25, 123)
Chu et al. [19]	35 (3, 72)	33 (7, 87)	69 (1, 129)
Yang et al. [20]	33 (2, 78)	53 (0, 106)	54 (10, 118)
	Right forearm	Right thigh	Right leg
Bulat et al. [13]	81 (11, 147)	43 (2, 114)	59 (4, 149)
Rafi et al. [15]	37 (0, 92)	40 (2, 124)	47 (0, 115)
Belagiannis et al. [16]	30 (8, 84)	54 (1, 126)	29 (4, 59)
Wei et al. [17]	33 (3, 89)	57 (13, 110)	25 (0, 59)
Bulat et al. [14]	46 (7, 97)	51 (11, 110)	43 (17, 98)
Newell et al. [18]	37 (21, 65)	29 (4, 82)	33 (0, 83)
Chu et al. [19]	53 (18, 104)	79 (5, 165)	49 (0, 112)
Yang et al. [20]	45 (5, 87)	44 (14, 78)	40 (0, 93)

4 Discussion

In this study, with the goal of evaluation of CNN-based HPE methods for measuring body segment lengths, eight HPE methods were employed to estimate the 2D poses which were subsequently transformed to 3D poses using a stereo vision system. The body segment lengths were computed based on the estimated 3D poses. The results show the errors of 2D pose estimation, 3D reconstruction, and anthropometric data – following the selected hierarchy to compute the anthropometric data.

Table 4 shows the accuracy of the HPE methods on the MPII dataset using the PCKh metric. These results showed that the most demanding body keypoints to estimate were the ankle, knee, and wrist. However, Fig. 2, which shows the accuracy of the HPE methods on the dataset of this study, shows that the knee, hip, and elbow were the most difficult keypoints to locate accurately. Also, evaluation using PCKh@0.5 metric that measures the accuracy of all methods for all keypoints to be higher than 99% indicates that this value is comparably higher than the values reported in the literature. These two points highlight the role of the training and testing dataset. In the dataset of this study, there was no occlusion – neither self-occlusion nor occlusion by other entities – or challenging pose. Now, comparing all the HPE methods may underline that one method cannot outperform all other methods for all the body keypoints. For instance, Wei et al. [17] achieve the best performance for estimating the positions of the ankle, knee, elbow, and wrist, but its accuracy for the hip is significantly less than the other HPE methods – i.e., there is no universal method outperforming all the other methods.

The results showed that the main parameters influencing the accuracy of an HPE method based on CNNs can be, the architecture of the convolutional neural network, the training and testing dataset, and the training strategy. Size of the input image of Wei et al. [17], which is the most accurate methods for the estimation of wrist and ankle based on the results shown in Fig. 2, is 368 × 368 while for the others is 256 × 256. Thus, it strengthens the hypothesis that the resolution of the input image could also be a prominent factor in determining the accuracy of the HPE methods. Also, there are minor parameters which affect the output of the HPE methods and thereby their accuracies, such as the noise of the input image, the position, and scale of the subject inside the cropped frames.

Table 4. Quantitative results on the **MPII dataset** using the PCKh@0.5 metric.

	Ankle	Knee	Hip	Elbow	Wrist
Bulat et al. [13]	64.0	70.5	79.1	78.8	71.5
Rafi et al. [15]	73.4	80.6	86.8	86.4	81.3
Belagiannis et al. [16]	78.4	82.6	87.9	88.2	83.0
Wei et al. [17]	79.4	82.8	88.4	88.7	84.0
Bulat et al. [14]	81.9	85.7	89.4	89.9	85.3
Newell et al. [18]	83.6	87.4	90.1	91.2	87.1
Chu et al. [19]	85.0	88.0	90.6	91.9	88.1
Yang et al. [20]	85.3	88.6	91.1	91.9	88.2

3D pose recovery has been made using the linear triangulation method by assuming that the estimated 2D poses are the perspective projection of the 3D pose. However, the deviation of the 2D poses from their reference value may be exacerbated through triangulation. Figure 3 shows an explicit example that the estimated ankle joints on the left and right images are not stereo correspondent – i.e., the estimated ankles (either left or right ankle) do not refer exactly to the same anatomical point.

Table 2 shows the 3D error for the body keypoints. The mean 3D error for the estimation of body keypoints is 5 cm. In a similar study [23] which has used Microsoft Kinect™ for 3D pose estimation in a static posture, the average 3D error in standing posture has been reported to be 8 cm and 9 cm for the first- and second-generation Kinect sensor, respectively.

The 3D error, reported in Table 2, shows that the maximum errors, for the elbow and wrist using the HPE method of Rafi et al. [15], are 60 cm and 67 cm, respectively. This occurrence is because of the false 2D detections of the HPE method for several frames of the static acquisition for one of the subjects.

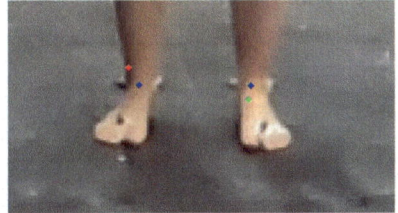 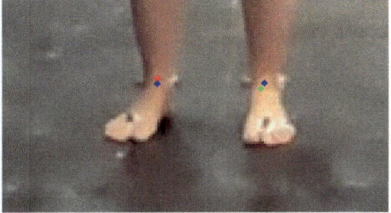

Fig. 3. 2D estimation of the ankle joints for a single frame. The red, green, and blue dots, represent the estimation of the right ankle, left ankle, and the reference values, respectively.

Table 3 shows the absolute error for the body segment lengths, while the mean error for the length of the body segments is 5 cm. A deeper look at this table shows that some methods cannot estimate the left and right body segment with the same accuracy. For instance, in the meanwhile that the 3D error for ankle and knee estimation, by Bulat et al. [13], is 4 cm, the absolute error for the left leg is 3 cm, and for the right leg is 6 cm. It also may highlight that post-processing could improve the uniformity of the results.

In conclusion, even though a stereo vision system combined with HPE methods can provide a cost-effective, easy to use, time efficient tool to measure the morphological data, the mean error is 5 cm that may not be adequate for applications in ergonomics. However, HPE methods may open new perspectives for measuring morphological data.

Acknowledgments. The authors thank the ParisTech BiomecAM chair program, on subject-specific musculoskeletal modelling and in particular Société Générale and COVEA.

References

1. Fukushima, K.: Neocognition: a self. Biol. Cybern. **202**, 193–202 (1980)
2. Fukushima, K.: Cognitron: a self-organizing multilayered neural network. Biol. Cybern. **20** (3–4), 121–136 (1975)
3. Liu, W., Wang, Z., Liu, X., Zeng, N., Liu, Y., Alsaadi, F.E.: A survey of deep neural network architectures and their applications. Neurocomputing **234**(November 2016), 11–26 (2017)
4. Wu, Y.C., Yin, F., Liu, C.L.: Improving handwritten Chinese text recognition using neural network language models and convolutional neural network shape models. Pattern Recogn. **65**(February 2016), 251–264 (2017)
5. Jain, N., Kumar, S., Kumar, A., Shamsolmoali, P., Zareapoor, M.: Hybrid deep neural networks for face emotion recognition. Pattern Recogn. Lett. **115**, 101–106 (2018)
6. Gong, W., et al.: Human pose estimation from monocular images: a comprehensive survey. Sens. (Basel) **16**(12), 1–39 (2016)
7. Litjens, G., et al.: A survey on deep learning in medical image analysis. Med. Image Anal. **42**(December 2012), 60–88 (2017)
8. Sarafianos, N., Boteanu, B., Ionescu, B., Kakadiaris, I.A.: 3D human pose estimation: a review of the literature and analysis of covariates. Comput. Vis. Image Underst. **152**, 1–20 (2016)
9. Andriluka, M., Pishchulin, L., Gehler, P., Schiele, B.: 2D human pose estimation: new benchmark and state of the art analysis. In: Proceedings of the IEEE Computer Society Conference Computer Vision and Pattern Recognition, pp. 3686–3693 (2014)
10. Bonnechère, B., et al.: Determination of the precision and accuracy of morphological measurements using the Kinect^{TM} sensor: comparison with standard stereophotogrammetry. Ergonomics **57**(4), 622–631 (2014)
11. Leardini, A., Sawacha, Z., Paolini, G., Ingrosso, S., Nativo, R., Benedetti, M.G.: A new anatomically based protocol for gait analysis in children. Gait Posture **26**(4), 560–571 (2007)
12. VICON Motion System: Nexus 2.6, documentation, full body modeling with plug-in gait, VICON documentation (2017). https://docs.vicon.com/display/Nexus26/Full+body +modeling+with+Plug-in+Gait. Accessed 15 Jan 2019
13. Bulat, A., Tzimiropoulos, G.: Binarized convolutional landmark localizers for human pose estimation and face alignment with limited resources. In: 2017 Proceedings of the IEEE International Conference on Computer Vision, pp. 3726–3734, October 2017
14. Bulat, A., Tzimiropoulos, G.: Human pose estimation via convolutional part heatmap regression. In: LNCS, vol. 9911 (2016)
15. Rafi, U., Kostrikov, I., Gall, J., Leibe, B.: An efficient convolutional network for human pose estimation. In: British Machine Vision Conference, pp. 1–11 (2016)
16. Belagiannis, V., Zisserman, A.: Recurrent human pose estimation. In: 12th IEEE International Conference on Automatic Face & Gesture Recognition (FG 2017), pp. 468–475 (2017)
17. Wei, S.-E., Ramakrishna, V., Kanade, T., Sheikh, Y.: Convolutional pose machines (2016)
18. Newell, A., Yang, K., Deng, J.: Stacked hourglass networks for human pose estimation (2016)
19. Chu, X., Yang, W., Ouyang, W., Ma, C., Yuille, A.L., Wang, X.: Multi-context attention for human pose estimation. In: CVPR 2017, pp. 1831–1840 (2017)
20. Yang, W., Li, S., Ouyang, W., Li, H., Wang, X.: Learning feature pyramids for human pose estimation. In: 2017 Proceedings of the IEEE International Conference on Computer Vision, pp. 1290–1299, October 2017

21. Heikkilä, J., Silvén, O.: A four-step camera calibration procedure with implicit image correction. In: CVPR, vol. 97, p. 1106 (1997)
22. Bell, A.L., Pedersen, D.R., Brand, R.A.: A comparison of the accuracy of several hip center location prediction methods. J. Biomech. **23**(6), 617–621 (1990)
23. Xu, X., McGorry, R.W.: The validity of the first and second generation Microsoft Kinect for identifying joint center locations during static postures. Appl. Ergon. **49**, 47–54 (2015)

White Matter, Gray Matter and Cerebrospinal Fluid Segmentation from Brain 3D MRI Using B-UNET

Tran Anh Tuan[1(✉)], Pham The Bao[2], Jin Young Kim[3],
and João Manuel R. S. Tavares[4]

[1] Faculty of Mathematics and Computer Science, University of Science,
Vietnam National University, Ho Chi Minh City, Vietnam
tatuan@hcmus.edu.vn
[2] Department of Computer Science, Saigon University, Ho Chi Minh City,
Vietnam
[3] Department of Electronic and Computer Engineering,
Chonnam National University, Gwangju, South Korea
[4] Instituto de Ciência e Inovação em Engenharia Mecânica e Engenharia
Industrial, Departamento de Engenharia Mecânica Faculdade de Engenharia,
Universidade do Porto, Porto, Portugal

Abstract. The accurate segmentation of brain tissues in Magnetic Resonance (MR) images is an important step for detection and treatment planning of brain diseases. Among other brain tissues, Gray Matter, White Matter and Cerebrospinal Fluid are commonly segmented for Alzheimer diagnosis purpose. Therefore, different algorithms for segmenting these tissues in MR image scans have been proposed over the years. Nowadays, with the trend of deep learning, many methods are trained to learn important features and extract information from the data leading to very promising segmentation results. In this work, we propose an effective approach to segment three tissues in 3D Brain MR images based on B-UNET. The method is implemented by using the Bitplane method in each convolution of the UNET model. We evaluated the proposed method using two public databases with very promising results.

Keywords: Image segmentation · Medical imaging · Deep learning · Bitplane-UNET

1 Introduction

In our aging society, many degenerative diseases of the nervous system are becoming more and more common. Neurodegenerative diseases represent a large group of neurological disorders with heterogeneous clinical and pathological effects on specific subsets of neurons in specific functional anatomic systems [1]. Many symptoms related to these diseases can be detected by using medical imaging. The most common imaging technique for the examination of brain diseases is Magnetic Resonance Imaging (MRI), which allows the visualization of the structure of the brain tissues.

© Springer Nature Switzerland AG 2019
J. M. R. S. Tavares and R. M. Natal Jorge (Eds.): VipIMAGE 2019, LNCVB 34, pp. 188–195, 2019.
https://doi.org/10.1007/978-3-030-32040-9_20

However, a crucial step in the diagnosis and treatment of brain diseases based on MRI is the accurate segmentation of the imaged brain tissues.

Many methods have been proposed for brain tissues segmentation from 3D Brain MRI [2–4]. Among other brain tissues, Gray Matter, White Matter, and Cerebrospinal Fluid are usually segmented for Alzheimer diagnosis purpose [5–8]. In the group of machine learning networks, Convolutional Neural Network (CNNs) has been successfully used to segment and classify medical images for many years [9, 10]. A CNN uses layers to transform the input data trough filters in order to obtain automatically the features for classification. There are recent approaches using CNNs for image segmentation that have shown better results than the previous ones such as UNET [11].

This article is organized in four sections. Section 2 presents the proposed method, which is based on Bitplane and UNET. Section 3 reports the experiments using two image databases: IBSR and MICCAI challenge 2018 datasets. The conclusions are presented in the last section.

2   Proposed Method

The proposed method is composed of two main steps: Skull stripping and Brain tissues segmentation, Fig. 1. From the original 3D MR image, we transform the 3D dataset into 2D slices, and implement UNET model for the image segmentation based on 2 steps, which is an extension of each neuronal convolution using features generating from the Bitplane method.

The Bitplane method [12] is based on decomposing a multilevel image into a series of binary images. The intensities of an image are based on:

$$a_{m-1}2^{m-1} + a_{m-2}2^{m-2} + \ldots + a_1 2^1 + a_0 2^0 \tag{1}$$

where a is the order bit of each pixel and m is the grey level in the input image.

We represented each input image slice by eight-bit planes as shown in Fig. 2. We realized that higher-order bits contain most of the significant commonly used visual information, and that lower-order bits contain subtle details that are not clinically relevant. In some cases, in order to reduce the effect of small intensity variations, we can use the OR operation between two bit planes: $(m - 1)^{th}$ bit and $(m - 2)^{th}$ bit planes:

$$g_i = a_{m-1} \oplus a_{m-2} \tag{2}$$

One of the CNN models commonly used in medical image segmentation is UNET [11], which is based on autoencoder. Here, we propose an approach by extending each convolution in the UNET model. With the input of convolution, multiple different features are generated based on the bit values of each pixel of the image, by using the Bitplane method, and the output is the convolution of the combination of these features. Let's consider layer i, and $L = \{K_1, \ldots, K_n\}$ as the set of output by using the Bitplane method. Hence, the number of feature maps that layer i generates is equal to the merge

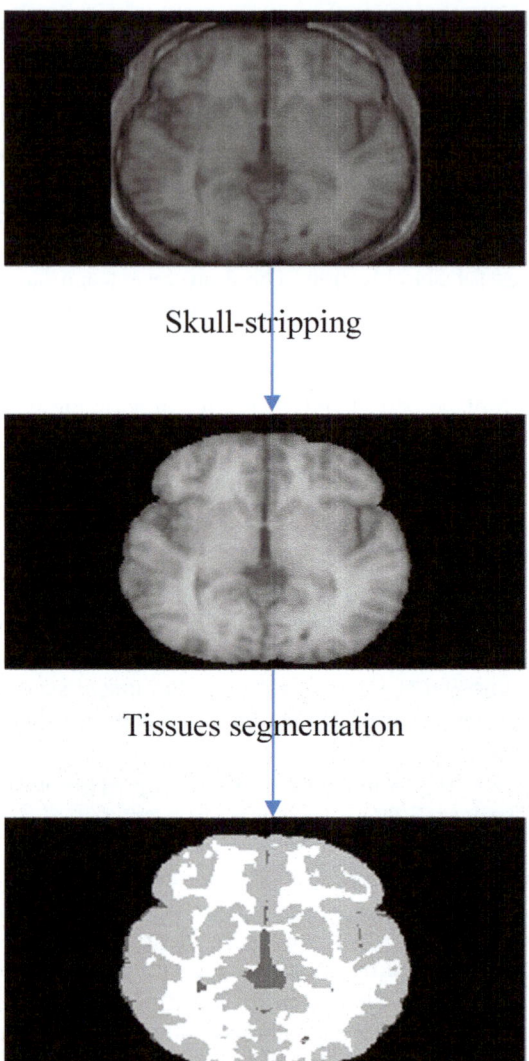

Fig. 1. The overview of the proposed method to segment three brain tissues in 3D MR images.

of the number of feature maps that the layer i generates. For example, as shown in Fig. 3, instead of using only convolution for input, multiple binary features are generated after using convolution and the output is the merge as Add operator for these binary features.

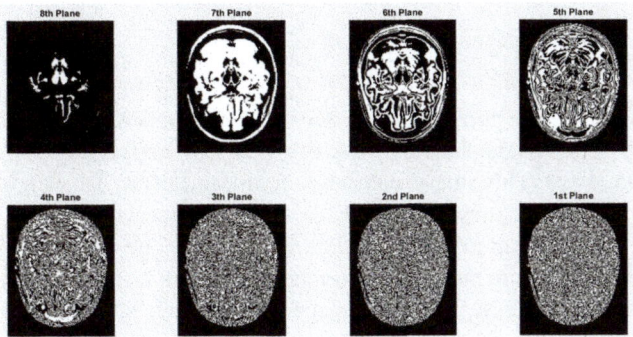

Fig. 2. Example of eight-bit planes of an image [13].

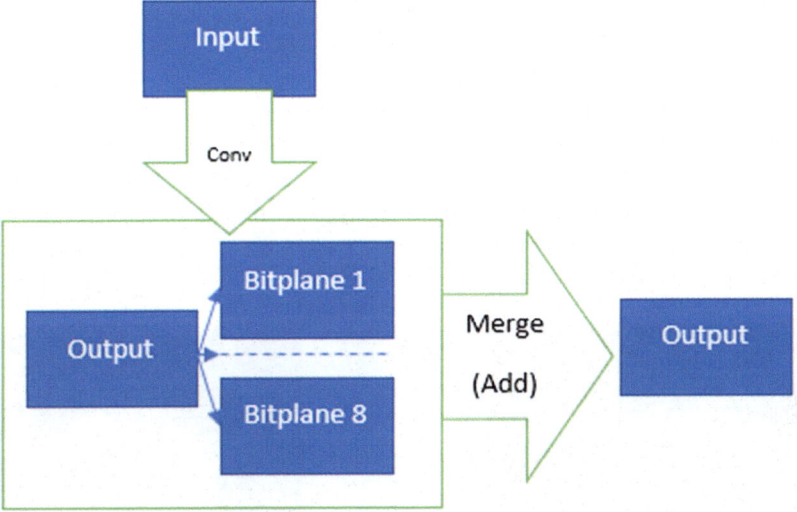

Fig. 3. Example of using the Bitplane method with a convolution.

3   Experiments and Discussion

We evaluated the proposed method on the IBSR 18 image database [14], which contains real cases and has been used in many studies regarding brain tissues segmentation in MR image scans. We also evaluated the method with a database from a MICCAI challenge [15] that is often used for comparing methods for the segmentation of gray matter, white matter, cerebrospinal fluid, and other structures on 3T MR image scans of the brain, and to assess the effect of severe pathologies on the segmentation results and brain volumetry. One of the most common metrics used for evaluating the performance of a segmentation method is Dice Similarity Coefficient [16]:

$$S(X, Y) = \frac{2|X \cap Y|}{|X| + |Y|},\tag{6}$$

where $|X|$ and $|Y|$ are the number of non zero voxels in sets X and Y, and $|X \cap Y|$ is the number of non zero voxels shared by the two sets, respectively.

IBSR 18 database: This image dataset is composed of 18 T1 weighted MRI scans with 1.5 mm of slice thickness ($256 \times 128 \times 256$). These volumes are provided after skull-stripping, normalization and bias field correction. The provided ground truth is composed of manual segmentations performed by experts using tissue labels as 0, 1, 2, 3 for background, CSF, GM, and WM, respectively. Each MRI brain volume has 128 slices of size 256×256 pixels each.

Our proposed method was implemented based on the Keras library [17] with backend Tensorflow [18] supporting convolutional network. The used optimizer was 'Adam' [19] and the loss was selected as 'binary_crossentropy' loss [20]. We evaluated the IBSR 18 database with k-folds = 3 and the obtained results are shown in Table 1. An example of a segmentation result is shown in Fig. 4.

Table 1. The Dice values obtained for the IBSR 18 dataset.

	WM	GM	CSF	Average
U-NET	0.92	0.92	0.77	0.87
B-UNET	**0.93**	0.92	**0.78**	**0.88**

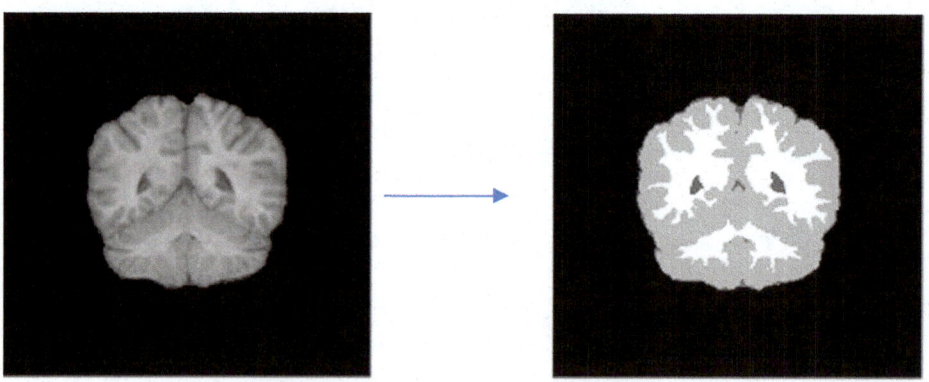

Fig. 4. Example of a segmentation result in the IBSR dataset (slice: 40 - person IBSR_01).

The Database Grand Challenge on MR Brain Segmentation proposed in MICCAI 2018 consists of 7 sets of MR brain images (T1, T1 inversion recovery, and T2-FLAIR), which are accompanied with the manual segmentation of three brain structures made by experts. All image scans have a voxel size of 0.958 mm $\times$ 0.958 mm $\times$ 3.0 mm and all are already aligned and biased field corrected using the N4ITK algorithm [21]. Here, we only used T1 and T2-FLAIR images as input for segmentation.

Our method was implemented based on a Keras library with back-end Tensorflow supporting convolutional network. We evaluated the database with k-folds = 3 obtaining the results indicated in Table 2. T. An example of a segmented image of this dataset is shown in Fig. 5.

Table 2. The Dice values obtained for MICCAI 2018 MRBRAIN dataset.

	WM	GM	CSF	Average
U-NET	0.87	0.82	0.87	0.85
B-UNET	0.87	**0.83**	**0.88**	**0.86**

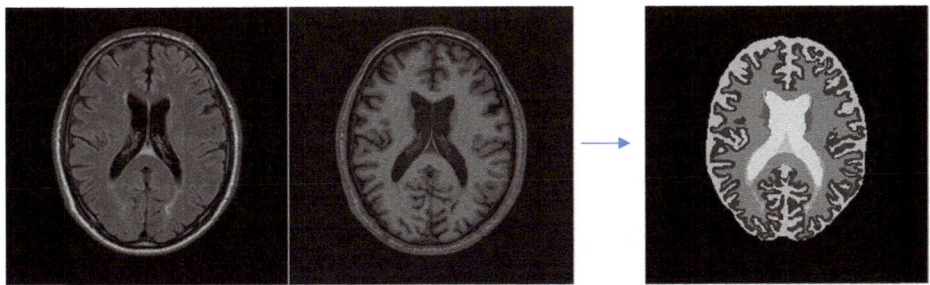

Fig. 5. Example of a segmentation result in the MRBRAIN MICCAI 2018 dataset (slice 25 person 1).

From the results obtained in the two evaluations done, we can conclude that very promising results were achieved as to the segmentation of the brain tissues. In our approach, we extended a well known architecture for 2D image segmentation (UNET) by replacing 2D convolutions with Bitplane Convolutions. From each convolution, one feature is extended multiple features that come from bit positions. The important and hidden features combined with features generated using the UNET model allows the achievement of superior segmentation results.

With our approach, the segmentation results are better than the ones obtained the original UNET because in each convolution, the original convolution method is combined with the convolution of other features to get the best classification features. However, this has high computational cost and, therefore, it improved model is slower to train than the original UNET model.

4   Conclusion

In this study, we proposed an approach to segment brain tissues in 3D MR images. The proposed solution is based on the building of more features and their combination in each convolution. Here, we implemented to generate multiple binary features by using Bitplane method in each convolution of the UNET model. The proposed method was evaluated using two public databases and obtained very promising results, basically, because the original input of each convolution is generated into multiple features before using convolution. In the future, we will concentrate on the extension of other operators in the classification model to get even better segmentation results.

Acknowledgments. This work was supported by Erasmus Mundus Program, reference IMPAKT Project 2017–2018.

João Manuel R.S. Tavares gratefully acknowledges the funding of Project NORTE-01-0145-FEDER-000022 - SciTech - Science and Technology for Competitive and Sustainable Industries, cofinanced by "Programa Operacional Regional do Norte" (NORTE2020), through "Fundo Europeu de Desenvolvimento Regional" (FEDER).

References

1. Przedborski, S., Vila, M., Jackson-Lewis, V.: Series introduction: neurodegeneration: what is it and where are we? J. Clin. Invest. **111**(1), 3–10 (2003). https://doi.org/10.1172/JCI200317522

2. Dora, L., Agrawal, S., Panda, R., Abraham, A.: State of the art methods for brain tissue segmentation: a review. IEEE Rev. Biomed. Eng. **10**, 235–249 (2017). https://doi.org/10.1109/RBME.2017.2715350

3. Akkus, Z., Galimzianova, A., Hoogi, A., Rubin, D.L., Erickson, B.J.: Deep learning for brain MRI segmentation: state of the art and future directions. J. Digit. Imaging **30**(4), 449–459 (2017). https://doi.org/10.1007/s10278-017-9983-4

4. Soni, P., Chaurasia, V.: MRI segmentation for computer-aided diagnosis of brain tumor: a review. In: Machine Intelligence and Signal Analysis. Advances in Intelligent Systems and Computing, vol. 748, pp. 375–338 (2018). https://doi.org/10.1007/978-981-13-0923-6_33

5. Gudise, S., Kande, G.B., Satya Savithri, T.: Segmentation of MR images of the brain to detect WM, GM, and CSF tissues in the presence of noise and intensity inhomogeneity. IETE J. Res. **65**(2), 250–262 (2019). https://doi.org/10.1080/03772063.2017.1409088

6. Wang, Y., Wang, Y., Zhang, Z., Xiong, Y., Zhang, Q., Yuan, C., Guo, H.: Segmentation of gray matter, white matter, and CSF with fluid and white matter suppression using MP2RAGE. J. Magn. Reson. Imaging **48**(6), 1540–1550 (2018). https://doi.org/10.1002/jmri.26014

7. Irimia, A., Maher, A.S., Rostowsky, K.A., Chowdhury, N.F., Hwang, D.H., Law, E.M.: Brain segmentation from computed tomography of healthy aging and geriatric concussion at variable spatial resolutions. Front. Neuroinform. **13**, 9 (2019). https://doi.org/10.3389/fninf.2019.00009

8. Illan, I.A., Górriz, J.M., Ramírez, J., Meyer-Base, A.: Spatial component analysis of MRI data for Alzheimer's disease diagnosis: a Bayesian network approach. Front. Comput. Neurosci. **8**, 156 (2014). https://doi.org/10.3389/fncom.2014.00156

9. Liu, X., Deng, Z., Yang, Y.: Recent progress in semantic image segmentation. Artif. Intell. Rev. **52**(2), 1089–1106 (2019). https://doi.org/10.1007/s10462-018-9641-3

10. Guo, Y., Liu, Y., Georgiou, T., Lew, M.S.: A review of semantic segmentation using deep neural networks. Int. J. Multimed. Inf. Retr. **7**(2), 87–93 (2018). https://doi.org/10.1007/s13735-017-0141-z

11. Ronneberger, O., Fischer, P., Brox, T.: U-Net: convolutional networks for biomedical image segmentation. In: Medical Image Computing and Computer-Assisted Intervention (MICCAI), vol. 9351, pp. 234–241. Springer (2015). https://doi.org/10.1007/978-3-319-24574-4_28
12. Gonzalez, R.C., Woods, R.E.: Digital Image Processing. Prentice Hall Inc., Upper Saddle River (2002)
13. Tuan, T.A., Kim, J.Y., Bao, P.T.: 3D brain magnetic resonance imaging segmentation by using bitplane and adaptive fast marching. Int. J. Imaging Syst. Technol. **28**, 223–230 (2018). https://doi.org/10.1002/ima.22273
14. Frazier, J.A., et al.: Internet brain segmentation repository (IBSR) 1.5 mm dataset. In: Child and Adolescent NeuroDevelopment Initiative (2007)
15. (2018). https://mrbrains18.isi.uu.nl. Accessed May 2019
16. Dice, L.R.: Measures of the amount of ecologic association between species. Ecology **26**(3), 297–302 (1945). https://doi.org/10.2307/1932409
17. Chollet, F., et al.: Keras (2015). https://keras.io. Accessed May 2019
18. Abadi, M., Agarwal, A., et al.: TensorFlow: large-scale machine learning on heterogeneous systems (2015). Accessed May 2019
19. Kingma, D.P., Ba, L.J.: Adam: a method for stochastic optimization. In: International Conference on Learning Representations, California (2015). arXiv:1412.6980
20. Bishop, C.M.: Pattern Recognition and Machine Learning. Springer, Boston (2006)
21. Tustison, N.J., Avants, B.B., Cook, P.A., Zheng, Y., Egan, A., Yushkevich, P.A., Gee, J.C.: N4ITK: improved N3 bias correction. IEEE Trans. Med. Imaging **29**(6), 1310–1320 (2010). https://doi.org/10.1109/TMI.2010.2046908

Computational Bioimaging and Visualization

Combining Infrared Thermography and Computational Fluid Dynamics to Optimize Whole Body Cryotherapy Protocols

Fabien Bogard[1], Sébastien Murer[1(✉)], Bastien Bouchet[2], Fabien Beaumont[1], and Guillaume Polidori[1]

[1] GRESPI, Research Group in Engineering Sciences,
University of Reims Champagne-Ardenne,
Campus Moulin de la Housse, 51100 Reims, France
{fabien.bogard,sebastien.murer}@univ-reims.fr
[2] Cryotera, 2 Rue Jules Méline, 51430 Bezannes, France

Abstract. Whole Body Cryotherapy (WBC) can be considered a therapeutic complement consisting in placing the human body in a hermetic chamber within which temperature varies between $-110\,°C$ and $-60\,°C$ over a short period of time. Despite the benefits of cryotherapy, subject safety must be ensured during the exposure to extreme cold, in the sense that the physiology of the human body should not be altered. Thus during a WBC session, accurate knowledge regarding the thermal transfer occurring at the cutaneous surface of the patient is essential. To this end, aeraulic and thermal conditions within the cryotherapy cabin are fundamental. The experimental study presented in this paper is based on the acquisition of skin temperature mappings. The derived boundary conditions are applied to the associated numerical problem which is solved using Computational Fluid Dynamics (CFD).

1 Introduction

WBC first appeared in Eastern Europe and Japan in the 1980s and is a continuation of cultural habits in cold countries. The virtues of a brief complete exposure of the body to intense cold apply to several pathological fields of the musculoskeletal system: rheumatology, traumatology, neurology, muscle recovery [6,13,14,22]. The stimulation of the sympathetic nervous system during the session and the para-sympathetic nervous system immediately after the session will allow for the initiation of a physiological reset, identified as the sources of the benefits of WBC [9,11]. The interest of whole body cryogenic chambers [9] has been demonstrated in the treatment of pain [7,21], inflammation [1,12], joint mobility [10,20], muscle recovery [10,18] and their complementarity with physiotherapy [2,8]. The durations used for the protocols vary from 120 to 240 s, depending on the studies [2,5,19,20]. However, the significance of their influence

© Springer Nature Switzerland AG 2019
J. M. R. S. Tavares and R. M. Natal Jorge (Eds.): VipIMAGE 2019, LNCVB 34, pp. 199–207, 2019.
https://doi.org/10.1007/978-3-030-32040-9_21

on some results must be underlined. Cryotherapy still suffers from the lack of scientific evidence to validate its techniques and approaches, which are described as empirical. The purpose of this study is therefore to establish a scientific basis for the mathematical modeling of WBC and fill the data void. The implementation of a predictive analytic tool requires a development phase based on the application on a test case of empirically determined evolution laws (temporal evolution of thermal losses). The results, based on iterative calculations composing the estimation tool, will be used to validate the algorithmic process. Finally, the model will be refined and calibrated to build cryotherapy protocols for target populations.

The Physiological Aspect

Body temperature varies following the energy distribution, but also as a consequence of local heat exchange coefficients. Thus, the overall temperature of the central core is about 37 °C while skin temperature at the feet level lies between 29 to 30 °C and 34 to 35 °C (daily variations are observed due to internal and external disturbances [16]). Thus, during a WBC session, skin temperatures decrease sharply to reach local minimum temperatures of around 4 °C [3]. In the human body, two types of thermoregulations can be distinguished. One is a vegetative physiological thermoregulation aimed at maintaining the internal temperature of the body at about 37 °C. The other may be regarded as "behavioral" and provides protection against sudden changes in the environment, in an attempt to lessen the physiological reactions perceived as unpleasant [15].

The Physical Aspect: Heat Exchanges

Continuous heat exchanges between the inner and outer parts of the human body involve a state of equilibrium called homeostasis. Subsequent exchanges between the human body and the environment occur through the skin surface and respiratory tract in the form of sensible and latent heat. Sensible heat is evacuated through the skin surface in three different forms of heat exchanges: convection, conduction and radiation. Besides, latent heat is consumed by sweat evaporation at the skin surface. Given the subject's position in the cryotherapy chamber, it can be reasonably assumed that conductive transfers between the sole of the protective shoes and the cabin floor are negligible.

2   Materials and Methods

Although the study is predominantly theoretical, an experimental contribution is required for the purpose of providing the conditions at the most realistic limits of the problem treated. Infrared thermography, which has already been used in the frame of similar topics, is a non-intrusive analytical method used in medicine to acquire skin temperature mappings at both the microscopic and macroscopic scales [3,4]. All bodies or solids naturally and continuously emit infrared radiations proportional to their temperature: the emitted energy thus depends on the thermal effects (global and/or local surface losses) generated by the cryotherapy session.

Experimental Protocol

A precise calibration was carried out on site, allowing parameter settings adapted to field conditions. To this end, the emissivity of the material to be visualized (the skin) must be set, as well as the reflected apparent temperature of the material and the shooting distance. The camera is placed outside the cryotherapy chamber and controlled via an interface offering extensive configuration of the acquisition parameters, such as number of images or recording duration, temperature range, emissivity or inclusion of measurement points. The subject studied (25 years old, 1,86 m, 75 kg, BFP 9%) followed all the usual guidelines of the standard procedure [3] in WBC session (safety precautions, personal protective equipment, session duration). The resulting thermal images, acquired every 30 s during the session, made it possible to quantify the temporal evolution of heat losses during exposure to extreme cold. These measurements may be carried whether locally or on a whole-body scale, depending on the nature of the desired information. Repeatability of the thermal mappings was validated on the basis of 5 identical experiments, carried out one day apart.

Numerical Protocol for Computational Fluid Dynamics (CFD) Simulation

The numerical simulation of a human being standing in the center of a whole-body cryotherapy chamber and exposed to cold dry air at a temperature of $-110\,°C$ requires a 3D model of the said subject: a 3D scanner was used in this approach. The numerical process also involves selecting relevant physical models, boundary conditions as well as the discretization of the computing domain (i.e. mesh). Our model combines two types of mesh: a surface mesh corresponding to the numerical manikin and a self-adaptive tetrahedral mesh in the vicinity of the subject for the computation of temperature within the cryotherapy chamber. A total of 15 million calculation points at each time step is used to simulate the experimental process. In this problem involving a large temperature difference between the human body and its environment, the surface temperature of the skin and the heat flow around the human body are time-dependent; unsteady computations are mandatory. The numerical solvers also make use of coupled heat transfer modes between radiation (heat radiated from the human body to the cold walls of the chamber) and natural convection (the temperature difference between the skin and the surrounding air generates a variation in air density that will generate a thermal plume). Following these computations, we are able to determine the semi-analytic expression of body heat loss during a cryotherapy session [17]. Once the global and/or local surface losses have been calculated, they are implemented as unsteady thermal boundary conditions applied to the corresponding surface.

3 Results

In this section, the results of the thermal imaging test campaign and CFD computations are presented and compared.

Experimental Results

The infrared thermography images captured every 30 s during the session are depicted in Fig. 1, and aim at bringing better understanding of the thermal phenomena occurring on the skin surface during a WBC session. As a matter of fact, comparing these images enables quantifying the temporal evolution of the skin temperature during a cryotherapy session at −110 °C. It is observed that skin temperature undergoes gradual and rapid cooling as more time is spent in the cabin. In addition, the temperature distribution is clearly uneven, with globally higher temperatures in the upper part of the body, neck, chest and upper back, as well as the surroundings of the spine. We can also notice that skin temperature is lower on the thighs and ankles.

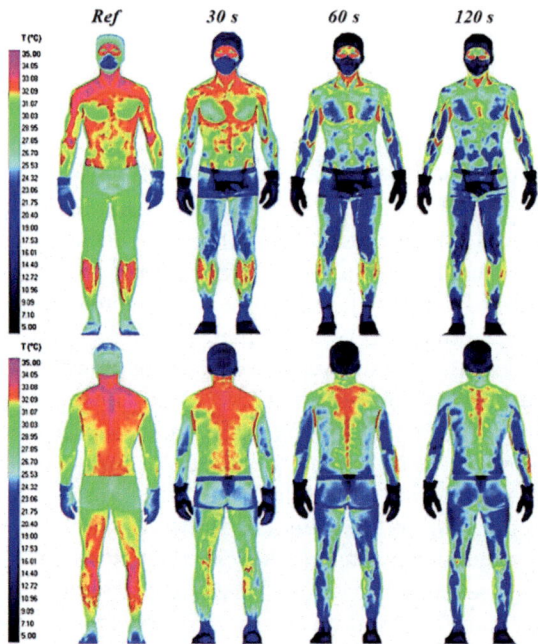

Fig. 1. Infrared thermograms before the session (Ref.), 30 s, 60 s and 120 s after the start of the cryotherapy session at −110 °C

Figure 2 summarizes the temperatures averaged on three different areas of the front side of the body. A significant overall temperature drop is observed, with marked differences depending on the area. Thus, the average temperature recorded between the reference measurement and the end of the session (after 3 min) decreases more drastically at the lower limbs (−39%) than at the trunk (∼32%).

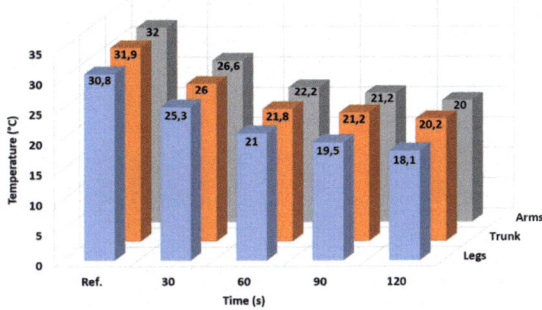

Fig. 2. Average temperature in different areas of the front side of the body, measured every 30 s after entering the cryotherapy cabin

Numerical Results
The accuracy of the theoretical model is assessed by verifying that numerical results supplied by the CFD computation lie within the margin of error of the experimental results, i.e. measurements of the mean skin temperature.

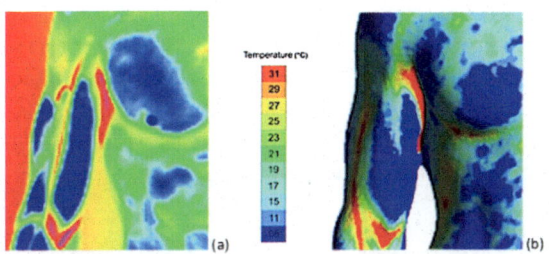

Fig. 3. Comparison in the biceps area between infrared thermogram (a) and numerical simulation (b) after 1 min 30 s of WBC at $-110\,^{\circ}$C

As seen in Fig. 3, there is good agreement in skin temperature between the numerical model prediction and the experimental thermogram. Similarly, the temperature gradient derived from both procedures is globally of the same order of magnitude.

4   Discussion

From these thermal maps, we were able to extract as much data as possible regarding skin temperature and calculate average values over 6 manually-defined regions of the lower limbs, upper limbs and trunk on the front and back sides of the body (Fig. 4). Ultimately, these values will help determine the zone-wise surface heat flux density, constituting one of the thermal boundary conditions of the problem to be modeled.

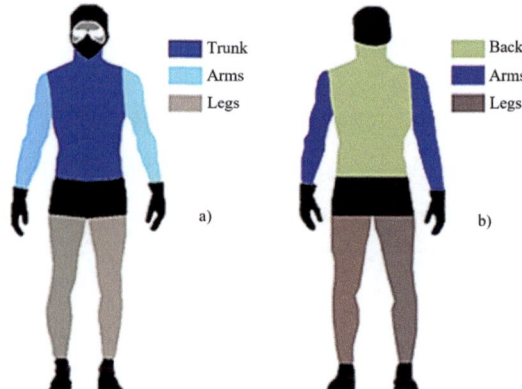

Fig. 4. Partitions of the body used for the calculation of average skin temperatures, on the front (a) and back (b) sides

During the post-treatment of the experimental results, we determined the mean skin temperature over time and compared these data with the results of the numerical simulations (Fig. 5). However, the comparison between the experimental and numerical results indicates that the partitioning of the body surface must be optimized in order to further improve the prediction of skin temperatures. In practice, this will result in a more thorough surface map containing additional areas. Thus, numerical results on the temporal evolution of skin temperatures were compared to data from infrared thermal imaging, demonstrating the predictive potential of the numerical model.

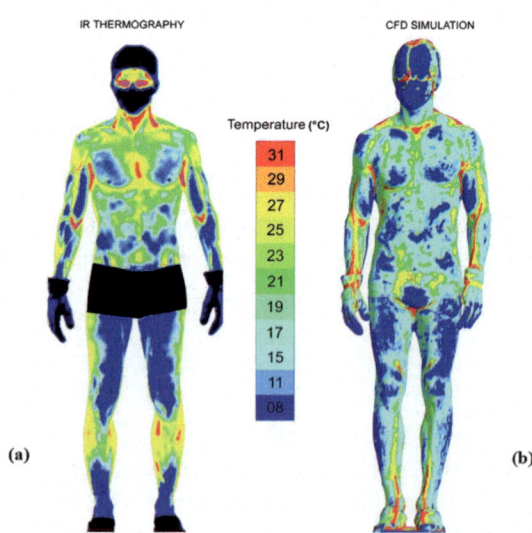

Fig. 5. Comparison between infrared thermogram (a) and temperature mapping computed using CFD code (b) at $t = 90\,s$

5 Conclusions and Prospects

The self-protection mechanisms of the human body are activated in the event of prolonged exposure to extreme temperatures, yet they are not designed to last. Therefore, sound knowledge of the thermal phenomena occurring on the patient's skin surface during a WBC session is fundamental; not to mention that of the aeraulic and thermal conditions within the cryotherapy cabin itself. With this in mind, we conducted this experimental study using infrared and digital thermal imaging along with numerical simulations based on a general-purpose CFD code. The results of infrared thermal imaging provided us with mappings of the patient's skin temperature distribution, as well as their evolution during the cryotherapy session. We were thus able to calculate average temperatures required to determine the surface density of heat fluxes per zone. This thermal parameter is converted into relevant thermal boundary conditions to be implemented in the computation code dealing with convecto-radiative heat transfers. The main advantage of mathematical modeling is that it avoids the multiplication of experiments: modeling could easily be extended to other populations, such as women, sedentary people, etc. The only parameters that would vary would be thermal time constants dependent on the thermal resistance of the human bodies being tested.

References

1. Banfi, G., Melegati, G., Barassi, A., Dogliotti, G., Melzi d'Eril, G., Dugué, B., Corsi, M.M.: Effects of whole-body cryotherapy on serum mediators of inflammation and serum muscle enzymes in athletes. J. Therm. Biol **34**(2), 55–59 (2009). https://doi.org/10.1016/J.JTHERBIO.2008.10.003
2. Burke, D.G., MacNeil, S.A., Holt, L.E., Mackinnon, N.C., Rasmussen, R.L.: The effect of hot or cold water immersion on isometric strength training. J. Strength Cond. Res. **14**(1), 21–25 (2000). https://doi.org/10.1519/00124278-200002000-00004
3. Costello, J.T., McInerney, C.D., Bleakley, C.M., Selfe, J., Donnelly, A.E.: The use of thermal imaging in assessing skin temperature following cryotherapy: a review. J. Therm. Biol **37**(2), 103–110 (2012). https://doi.org/10.1016/J.JTHERBIO.2011.11.008
4. Dębiec-Bąk, A., Gruszka, K., Sobiech, K.A., Skrzek, A.: Age dependence of thermal imaging analysis of body surface temperature in women after cryostimulation. Hum. Move. **14**(4), 299–304 (2013). https://doi.org/10.2478/humo-2013-0035
5. Fonda, B., De Nardi, M., Sarabon, N.: Effects of whole-body cryotherapy duration on thermal and cardio-vascular response. J. Therm. Biol **42**, 52–5 (2014). https://doi.org/10.1016/j.jtherbio.2014.04.001
6. Fricke, R.: Ganzkörperkältetherapie in einer kälterkammer mit temperaturen um −110 °C. Z. Phys. Med. Balneol Med. Klim **18**, 1–10 (1989)
7. Giemza, C., Matczak-Giemza, M., De Nardi, M., Ostrowska, B., Czech, P.: Effect of frequent WBC treatments on the back pain therapy in elderly men. Aging Male Off. J. Int. Soc. Study Aging Male **18**(3), 135–42 (2015). https://doi.org/10.3109/13685538.2014.949660

8. Gizińska, M., Rutkowski, R., Romanowski, W., Lewandowski, J., Straburzyńska-Lupa, A.: Effects of Whole-Body Cryotherapy in Comparison with Other Physical Modalities Used with Kinesitherapy in Rheumatoid Arthritis. BioMed research international **2015** (2015). Article ID: 409174. https://doi.org/10.1155/2015/409174

9. Hausswirth, C., Schaal, K., Le Meur, Y., Bieuzen, F., Filliard, J.R., Volondat, M., Louis, J.: Parasympathetic activity and blood catecholamine responses following a single partial-body cryostimulation and a whole-body cryostimulation. PLoS ONE **8**(8), e72658 (2013). https://doi.org/10.1371/journal.pone.0072658

10. Księżopolska-Orłowska, K., Pacholec, A., Jędryka-Góral, A., Bugajska, J., Sadura-Sieklucka, T., Kowalik, K., Pawłowska-Cyprysiak, K., Łastowiecka-Moras, E.: Complex rehabilitation and the clinical condition of working rheumatoid arthritis patients: does cryotherapy always overtop traditional rehabilitation? Disabil. Rehabil. **38**(11), 1034–40 (2016). https://doi.org/10.3109/09638288.2015.1060265

11. Louis, J., Schaal, K., Bieuzen, F., Le Meur, Y., Filliard, J.R., Volondat, M., Brisswalter, J., Hausswirth, C.: Head exposure to cold during whole-body cryostimulation: influence on thermal response and autonomic modulation. PLoS ONE **10**(4), e0124776 (2015). https://doi.org/10.1371/journal.pone.0124776

12. Lubkowska, A., Szyguła, Z., Chlubek, D., Banfi, G.: The effect of prolonged whole-body cryostimulation treatment with different amounts of sessions on chosen pro- and anti-inflammatory cytokines levels in healthy men. Scand. J. Clin. Lab. Invest. **71**(5), 419–425 (2011). https://doi.org/10.3109/00365513.2011.580859

13. Mesure, S., Catherin-Marcel, B., Bertrand, D.: La cryothérapie corps entier: littérature et perspectives de recherches. Kinésithérapie, la Revue **14**(152–153), 56–60 (2014). https://doi.org/10.1016/j.kine.2014.06.003

14. Metzger, D., Zwingmann, C., Protz, W., Jäckel, W.H.: Die Bedeutung der Ganzkörperkältetherapie im Rahmen der Rehabilitation bei Patienten mit rheumatischen Erkrankungen – Ergebnisse einer Pilotstudie. Die Rehabil. **39**(2), 93–100 (2000). https://doi.org/10.1055/s-2000-14442

15. Narçon, S.: Caractérisation des perceptions thermiques en régime transitoire - Contribution à l'étude de l'influence des interactions sensorielles sur le confort. Ph.D. thesis, École Pratique des Hautes Études, Paris (2001)

16. Parsons, K.C.: Human Thermal Environments: The Effects of Hot, Moderate, and Cold Environments on Human Health, Comfort, and Performance. CRC Press, Boca Raton (2014)

17. Polidori, G., Marreiro, A., Pron, H., Lestriez, P., Boyer, F.C., Quinart, H., Tourbah, A., Taïar, R.: Theoretical modeling of time-dependent skin temperature and heat losses during whole-body cryotherapy: a pilot study. Med. Hypotheses **96**, 11–15 (2016). https://doi.org/10.1016/j.mehy.2016.09.019

18. Pournot, H., Bieuzen, F., Louis, J., Mounier, R., Fillard, J.R., Barbiche, E., Hausswirth, C.: Time-course of changes in inflammatory response after whole-body cryotherapy multi exposures following severe exercise. PLoS ONE **6**(7), e22748 (2011). https://doi.org/10.1371/journal.pone.0022748

19. Romanowski, M., Romanowski, W., Keczmer, P., Majchrzycki, M., Samborski, W., Straburzyńska-Lupa, A.: Whole body cryotherapy in rehabilitation of patients with ankylosing spondylitis. A randomised controlled study. Physiotherapy **101**, e1294 (2015). https://doi.org/10.1016/j.physio.2015.03.1210

20. Stanek, A., Cieslar, G., Strzelczyk, J., Kasperczyk, S., Sieron-Stoltny, K., Wiczkowski, A., Birkner, E., Sieron, A.: Influence of cryogenic temperatures on inflammatory markers in patients with ankylosing spondylitis. Pol. J. Environ. Stud. **19**(1), 167–175 (2010)

21. Thomas, D.A., Maslin, B., Legler, A., Springer, E., Asgerally, A., Vadivelu, N.: Role of alternative therapies for chronic pain syndromes. Curr. Pain Headache Rep. **20**(5), 29 (2016). https://doi.org/10.1007/s11916-016-0562-z
22. Yamauchi, T.: Whole-body cryotherapy is a method of extreme cold $-175\,^{\circ}C$ treatment initially used for rheumatoid arthritis. Z. Phys. Med. Balneol Med. Klim **15**, 311 (1989)

Voxel-Based Computational Tools Help Liver Dosimetry Calculations of Multiple (*External* and *Internal*) Radiation Therapies

Paulo Ferreira[1,2(✉)], Francisco P. M. Oliveira[1], Rui Parafita[1,3],
Pedro S. Girão[2], Paulo L. Correia[2], Oriol Pares[1],
and Durval C. Costa[1]

[1] Champalimaud Centre for the Unknown, Champalimaud Foundation,
Lisbon, Portugal
paulo.ferreira@fundacaochampalimaud.pt
[2] Instituto de Telecomunicações, Instituto Superior Técnico,
Universidade de Lisboa, Lisbon, Portugal
[3] Mercurius Health, Lisbon, Portugal

Abstract. Voxel-wise absorbed dose distributions, γ-index agreement test and dose-volume histograms (DVH) for the evaluation of dose distributions have been used in stereotactic body radiotherapy (SBRT) for several years. Our research group recently adapted these concepts to liver radioembolization (RE) using Yttrium-90 (^{90}Y) glass microspheres (MS). In this work we demonstrate the application of our dosimetry tools to study the dose distributions within the tumor (PTV) and normal liver (NLV) tissues of a patient treated with multiple radiation therapies (RE after SBRT). Voxel-wise biological effective dose (BED) distributions from SBRT and RE were computed and added up. Moreover, we investigated the role of pre-treatment single-photon emission computed tomography (SPECT) in the planning dosimetry of RE.

This individual case describes a 73 years old man with pancreatic carcinoma and a single liver metastasis previously treated with 48 Gy of 10 MV SBRT photon beams in three sessions of 16 Gy. Ten months later he underwent RE due to disease progression. RE was performed according to the MS manufacturer's recommendations. Hepatic arterial angiography and pre-treatment Technetium-99 m (^{99m}Tc) macroaggregated albumin (MAA) scans (planar and SPECT) were performed in preparation of RE. Then, a ^{90}Y-MS activity of 3.239 GBq was calculated and later administered in the liver. ^{90}Y-MS positron emission tomography (PET) images were then used to verify the distribution of the MS in the liver. SPECT-MAA and PET-MS images were used, retrospectively, to calculate the voxel-wise dose distributions. Voxel-wise BED distributions of SBRT and RE were computed and added up to obtain the total BED. Cumulative dose volume histograms (DVHs) were generated to evaluate the safety of multiple radiation therapies. Finally, the voxel-wise intraclass correlation coefficient (ICC) and γ-index between pre and post-treatment dose distributions were computed. The predictive power role of the pre-treatment SPECT-MAA in the planning dosimetry of RE after SBRT was further investigated. In conclusion, the previous SBRT treatment contributed significantly to the BED in the RE PTV with no detectable contribution to the BED in the RE NLV. Therefore, higher activity might have been administered during RE keeping the NLV dose at safe

© Springer Nature Switzerland AG 2019
J. M. R. S. Tavares and R. M. Natal Jorge (Eds.): VipIMAGE 2019, LNCVB 34, pp. 208–216, 2019.
https://doi.org/10.1007/978-3-030-32040-9_22

values. Moreover, the verification PET-MS dose distribution, in this individual case, showed good agreement with the pre-treatment SPECT-MAA dose distribution (ICC = 0.73, γ-index = 98%). Consequently, SPECT-MAA could have been used in the planning dosimetry of RE.

Keywords: Stereotactic body radiotherapy (SBRT) · Radioembolization (RE) · Voxel-based dosimetry · Absorbed dose · Biological effective dose (BED) · γ-index · Dose-volume histograms (DVH)

1 Introduction

Radiation-induced liver disease (RILD) when treating liver or upper abdominal tumors with ionizing radiation [1] is a severe health problem difficult to predict.

Despite most recent advances in stereotactic body radiotherapy (SBRT), including targeting and delivery techniques, the normal liver volume (NLV) and other adjacent organs may receive significant and damaging absorbed doses. Radiation induced collateral damage to nontarget tissues continues to be a dose-limiting factor for effective treatment. Moreover, previous SBRT can theoretical be a contraindication for radioembolization (RE) with Yttrium-90 (^{90}Y) charged microspheres (MS) or for repeat SBRT series, due to increased likelihood of liver toxicity events.

To minimize liver toxicity associated with multiple radiation treatments (SBRT & RE), it is crucial to obtain precise radiation doses distributions [2]. Unfortunately, the dosimetric methods currently used in RE can only estimate the average planning tumor volume (PTV) and NLV doses. To address this issue, a computational voxel-based dosimetric methodology [3] has been developed and implemented. This work demonstrates the utility of voxel-based dosimetry to retrospectively assess absorbed dose tridimensional distributions in a patient treated with SBRT, initially, and later submitted to RE. Voxel-wise biological effective dose (BED) distributions were computed and cumulative dose volume histograms (DVHs) were generated to evaluate the safety of the multiple radiation therapies (SBRT & ^{90}Y-MS RE). In addition, the role of pre-treatment single-photon emission computed tomography (SPECT) in the planning dosimetry of RE after SBRT was investigated.

2 Materials and Methods

Patient Data

A73 years old man with pancreatic carcinoma and a single liver metastasis was treated in the liver with a total dose of 48 Gy of 10 MV SBRT photon beams in three sessions of 16 Gy. The planning X-ray computed tomography (CT) of the patient was performed in the Philips CT scanner (PHILIPS, Amsterdam, The Netherlands) set at 120 kV and 265 mA.s, with a 512 × 512 matrix and 1 mm slice thickness. During the planning CT and treatment, the patient was immobilized with a body vacuum-bag. A four-dimensional computed tomography (4D-CT) acquired with an abdominal compressor device to limit target movement was used as a respiratory control method. The calculation of the absorbed dose distributions using the SBRT technique were performed with the Eclipse treatment planning software (Varian Medical Systems, Palo Alto, CA). The three treatment sessions were performed in the Varian TrueBeam linear accelerator (Varian Medical Systems, Palo Alto, CA). Figure 1A shows the SBRT absorbed dose distribution.

Ten months later the patient underwent RE due to disease progression. RE was performed according to the recommendations of the manufacturer of the glass MS (TheraSphere, MDS Nordion, Canada). An angiographic evaluation of the hepatic vasculature was performed two weeks prior to treatment. In the same day, a pre-treatment using macro aggregate albumin (MAA) labeled with Technetium-99 m (^{99m}Tc) was applied, and planar scan and SPECT were performed to evaluate lung and extra-hepatic shunt fractions. These images were captured using the BrightView SPECT gamma camera (PHILIPS, Amsterdam, The Netherlands) equipped with Low-Energy High-Resolution (LEHR) collimators. The energy window width of 10% centered at 144 keV was defined, and 128 step-and-shoot mode projections were captured along a 360° gantry rotation and during a 20-minute time interval with a matrix of 128 × 128 pixels. The attenuation of images was corrected by CT imaging captured immediately after the SPECT, with the positron emission tomography (PET)/CT GEMINI TF (PHILIPS, Amsterdam, The Netherlands) equipment set at 140 kV and 60 mA.s, with a 512 × 512 matrix and 2 mm slice thickness. The images were reconstructed in the AutoSPECT Plus software (PHILIPS, Amsterdam, The Nether-lands) using the ordered subset expectation maximization (OSEM) algorithm with three iterations and 8 subsets, Butterworth filter with cut off 2 and order 10.

Two weeks after the treatment planning, ^{90}Y-MS were administered in a single session via the right hepatic artery. The administered ^{90}Y-MS activity of 3.239 GBq was calculated with the one compartment dosimetry model, recommended by the manufacturer of the glass MS. The prescribed absorbed dose of 120 Gy to the right liver lobe target volume of 1310 cm^3, with a 2.7% of residue waste and 6.5% lung shunt fraction was taken into account for the calculation of the ^{90}Y-MS activity to be administered in the patient's liver.

Verification ^{90}Y-MS positron emission tomography (PET) images were used to verify the distribution of the MS in the liver and, retrospectively, to calculate the voxel-wise absorbed dose distribution. The ^{90}Y-MS absorbed dose distribution is depicted in Fig. 1B.

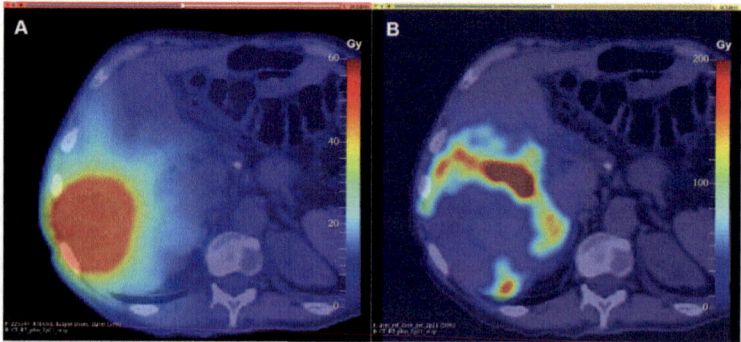

Fig. 1. Voxel-wise absorbed dose distributions superimposed on a single identical slice shown by the background CT used during the dosimetry planning for SBRT: (A) extracted from the SBRT treatment planning, and (B) from the post-RE treatment verification PET/CT ^{90}Y-MS.

Image Processing

The diagnostic, treatment-planning and post-treatment (verification) images captured from the patient were used in this work to retrospectively assess the voxel-wise absorbed dose and BED distributions following the steps shown in Fig. 2.

The SBRT plan was saved in the Eclipse software as RTDOSE, RTPLAN and RTSTRUCT DICOM files for further processing with the MATLAB R2017b (The Math Works, Inc., Massachusetts, United States) software and SlicerRT Toolkit for radiation therapy research [4]. In this regard, the PTV and liver segments defined in the SBRT planning were used to extract the NLV, by means of the segment morphology module in SlicerRT.

Regarding the RE planning, the registration of the SPECT-MAA images with the CT ones, for the correction of the attenuation effect, was assisted with the use of five physical fiducial marks that were fixed on the patient's skin [3]. The radiopaque material of the fiducial marks is visible in CT images. When the fiducial marks are loaded with ^{99m}Tc, they also become visible by SPECT imaging and then used as landmarks for the registration algorithm [4].

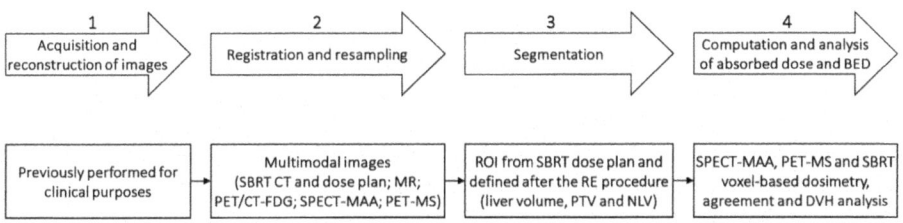

Fig. 2. Operational block diagram of the methodology used in image processing and dose computations.

The SBRT CT planning images were considered in this work as the reference image space. First, the CT images of the SBRT, SPECT-MAA and PET-MS were registered by the application of rigid transformations. Then, the same transformations were used to locate the SPECT-MAA and PET-MS images in the same space of the SBRT CT planning. During this operation, SPECT-MAA and PET-MS images were resampled to the CT voxel resolution of the SBRT CT planning, using linear interpolation, to help during the segmentation of the regions of interest (ROI) at the time of the RE.

Magnetic resonance (MR) and Fluorine-18 (^{18}F) Fluorodeoxyglucose (FDG) PET/CT diagnostic images of the patient were also selected as a complement to CT images during the segmentation task. The registration of these images was performed with reference to the physical space and voxel resolution of the SBRT CT planning.

Semi-automatic segmentation was used for the delineation of the total liver volume of the patient [5]. Manual segmentation was used for the PTV delineation. The NLV was computed in MATLAB R2017b by extracting the PTV from the total liver volume.

Computation of the Absorbed dose and BED Distributions

The SBRT absorbed dose calculations were previously done with the Eclipse treatment plan software used in the radiotherapy department. The Eclipse RTDOSE DICOM file was used in MATLAB R2017b for the computation of the BED distributions in the PTV and NLV. Previously, in MATLAB, the DoseGridScaling DICOM Tag was used for the calibration of the SBRT absorbed dose distribution.

For the computation of the RE absorbed dose distributions, we used our dosimetry methodology based on SPECT-MAA and PET-MS image voxels [3] and related work by others [6]. Voxel S values computed for ^{90}Y radioisotope, soft tissue and cubic voxels of 2.21 mm length reported in the literature [7] were used. In this regard, the SBRT dose distribution, SPECT-MAA and PET-MS images were resampled to the same cubic voxel resolution of 2.21 mm length. For the calibration of the SPECT-MAA and PET-MS activity images, the intensity of each voxel was normalized (divided) by the sum of the total intensity in the image. Then, the value of the actual ^{90}Y administered activity was multiplied by the normalized value in each voxel. Thus, the sum of all voxels values is equal to the administered activity. The convolution of the calibrated image with the 3D kernel containing the voxel S values and the application of the time integral of the ^{90}Y decay were computed as in [3].

Afterwards, voxel-wise BED distributions for both treatments (SBRT & RE) were computed, based on the SBRT and PET-MS RE absorbed dose distributions, and then added up to obtain the total BED [2, 8]. The linear quadratic model (LQM) has been reported in the literature [9] to describe the radiobiological damage to tissues by external beam radiotherapy (EBRT) and radionuclide therapies. The BED assesses the effect of absorbed doses, uniformly delivered in a few minutes in multiple fractions of SBRT. The BED takes into account the absorbed dose and dose rate. The LQM has been reformulated to radionuclide therapies, with continuously variable dose rates and non-uniform absorbed dose distributions, like in RE. The goal is to provide a way to evaluate the BED of two different radiation modalities, considering that the same BED means the same biological effect. The BED can be computed for each SBRT and RE absorbed dose voxel i, respectively, as

$$BED_i = D_i \left(1 + \frac{D_i/N}{\alpha/\beta} \right) \tag{1}$$

$$BED_i = D_i \left(1 + \frac{D_i \times T_{rep}}{\alpha/\beta \times (T_{rep} + T_{eff})} \right) \tag{2}$$

where D_i is the total absorbed dose in voxel i, N is the number of fractions in SBRT, T_{eff} is the half-life of ^{90}Y decay, T_{rep} is the half-time for repair damage, being 2.5 h for NLV and 1.5 h for PTV [2, 8], and the α/β ratio relates the intrinsic radiosensitivity α to the potential sparing capacity β. This α/β ratio is typically 2.5 Gy for the NLV and 10 Gy for PTV [2, 8].

The voxel-wise BED distributions in the PTV and NLV of both treatments were then added up to obtain the total BED. Cumulative DVH for all the absorbed dose and BED distributions were computed for further analysis.

The Role of the SPECT-MAA Planning Dosimetry in RE After SBRT

The two absorbed dose distributions obtained from SPECT-MAA and PET-MS were analyzed to investigate and understand their agreement. This was done to investigate the predictive power of the SPECT-MAA pre-treatment planning distribution and the subsequent verification by the PET-MS post-therapy distribution. The final goal was to confirm the hypothesis that SPECT-MAA dose distribution might be used for safely and more accurately planning the RE after SBRT.

The agreement of the two absorbed dose distributions (SPECT-MAA and PET-MS) was assessed by means of the computation of the voxel-wise intraclass correlation coefficient (ICC) and the passing rate (PR) of the γ-index test. The γ-index test comprises a pass/fail criterion based on the simultaneous combination of dose-difference (DD) and distance-to-agreement (DTA) test conditions. The γ-index PR outcome is computed as the percentage of voxels approved during the test. This is an evaluation tool used in EBRT for years, now adapted for RE in our work, as detailed in previous publications [3, 10].

3   Results

The voxel-wise SBRT and RE absorbed dose distributions with corresponding BED distributions are represented by the cumulative DVHs in Fig. 3. The cumulative DVH is plotted with the bin doses along the horizontal axis. The column height of a bin represents the volume receiving greater than or equal to that bin dose.

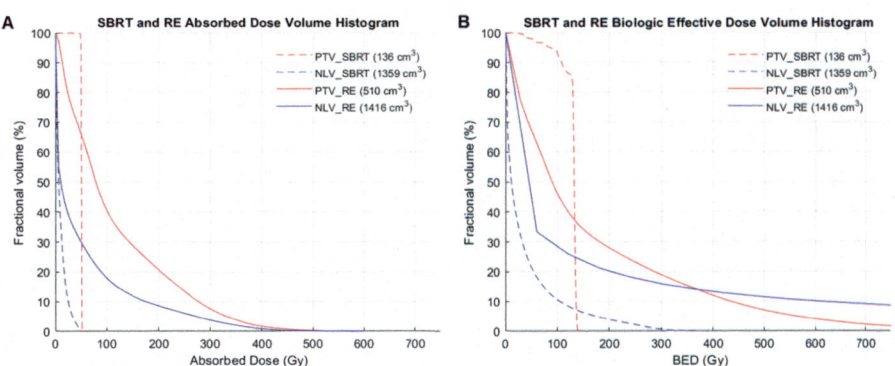

Fig. 3. DVH of the SBRT and RE: (A) absorbed dose distributions, and (B) BED distributions, computed on the PTV and NLV.

After computing the voxel-wise BED of both treatment distributions (SBRT & RE), the total voxel-wise BED distribution was obtained. The total BED distribution is represented by the DVH in Fig. 4.

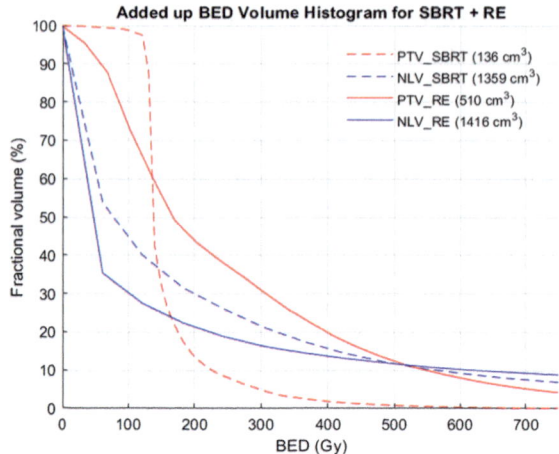

Fig. 4. DVH of the SBRT and RE BED distributions on the PTV and NLV.

Median values can be observed by the dose values at 50% of the volume of the liver ROI. The mean values are given by the area under each curve.

Table 1 summarizes the median and mean values from the DVHs in Figs. 3 and 4.

Table 1. Median and mean values in the PTVs and NLVs computed on the SBRT and RE absorbed dose and BED distributions. SBRT+RE column shows added up voxel-wise BED values.

Technique	SBRT		RE		SBRT		RE		SBRT+RE	
Liver ROI	PTV	NLV	PTV	NLV	PTV	NLV	PTV	NLV	PTV_RE	NLV_RE
Quantity	Absorbed dose		Absorbed dose		BED		BED		BED	
Median (Gy)	48	6	78	10	133	12	92	12	163	19
Mean (Gy)	48	9	115	54	126	35	171	218	246	226

Table 1 clearly demonstrates that the previous SBRT treatment contributed significantly to the PTV but there was no significant contribution to the corresponding NLV added up voxel-wise BED values.

Regarding the predictive power assessment for SPECT-MAA pre-treatment planning distribution in anticipation of the PET-MS verification distribution we studied their comparative estimated absorbed dose distributions and consequently their corresponding BED distributions. The ICC for the liver volume was 0.73. The γ-index PR was 98% in both the PTV and NLV using a combined 10%/10 mm test condition. The DVH in Fig. 5 represents the voxel-wise absorbed dose distributions estimated based on SPECT-MAA and PET-MS images.

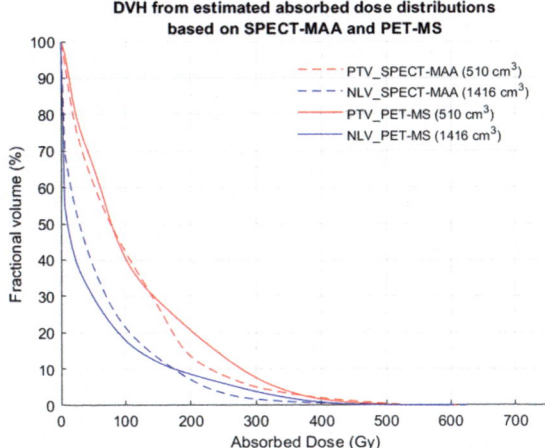

Fig. 5. DVH of the dose distributions in the PTV and NLV, estimated based on SPECT-MAA and PET-MS.

The median absorbed dose estimated and based on the SPECT-MAA was 77 Gy in the PTV (510 cm³) and 29 Gy in the NLV (1416 cm³). The mean absorbed dose was 105 Gy in the PTV (510 cm³) and 61 Gy in the NLV (1416 cm³). Regarding the absorbed doses estimated and based on the PET-MS, the median absorbed dose was 78 Gy in the PTV (510 cm³) and 10 Gy in the NLV (1416 cm³). The mean absorbed dose was 115 Gy in the PTV (510 cm³) and 54 Gy in the NLV (1416 cm³). In summary, there is good agreement between post-treatment verification PET-MS absorbed doses and the estimated absorbed doses based on the pre-treatment SPECT-MAA images. Being this true, the corresponding BED distributions are consequently also in good agreement.

4   Conclusion

Based on the absorbed dose and BED distributions we conclude, in this individual case, that the SBRT treatment contributed significantly to the BED in the RE PTV with no significant contribution to the BED in the RE NLV. Therefore, we assume that higher activity might have been administered during RE, to achieve a more effective treatment, while keeping the NLV doses at safe values. Moreover, the absorbed dose distribution estimated and based on the PET-MS images showed good agreement with the absorbed dose distribution estimated based on the pre-treatment SPECT-MAA images. Consequently, it seems that our hypothesis, SPECT-MAA may be used in the planning dosimetry of RE after SBRT, is confirmed.

Acknowledgments. We gratefully acknowledge the dedicated contribution of the entire personal from Nuclear Medicine – Radiopharmacology, Interventional Imaging and Radiation-Oncology at the Champalimaud Centre for the Unknown, Champalimaud Foundation.

PF thankfully acknowledges the Champalimaud Foundation for a recurrent PhD student grant.

References

1. Lam, M.G.E.H., et al.: Safety of 90Y radioembolization in patients who have undergone previous external beam radiation therapy. Int. J. Radiat. Oncol. Biol. Phys. **87**(2), 323–329 (2013)
2. Wang, T.-H., et al.: Combined Yttrium-90 microsphere selective internal radiation therapy and external beam radiotherapy in patients with hepatocellular carcinoma: from clinical aspects to dosimetry. PLoS ONE **13**(1), 1–14 (2018)
3. Ferreira, P.: Voxel-based dosimetry using multimodal images for patient-specific liver radioembolization with Yttrium-90 charged glass microspheres. Ph.D. thesis, Department of Electrical and Computer Engineering, Universidade de Lisboa (2019)
4. Fedorov, A., et al.: 3D Slicer as an image computing platform for the Quantitative Imaging Network. Magn. Reson. Imaging **30**(9), 1323–1341 (2012)
5. Yushkevich, P.A., et al.: ITK-SNAP: an intractive medical image segmentation tool to meet the need for expert-guided segmentation of complex medical images. IEEE Pulse **8**(4), 54–57 (2017)
6. Bolch, W.E., et al.: MIRD pamphlet no. 17: the dosimetry of nonuniform activity distributions–radionuclide S values at the voxel level. J. Nucl. Med. **40**(1), 11–36 (1999). Medical Internal Radiation Dose Committee
7. Lanconelli, N., et al.: A free database of radionuclide voxel S values for the dosimetry of nonuniform activity distributions. Phys. Med. Biol. **57**(2), 517–533 (2012)
8. Cremonesi, M., et al.: Radioembolization of hepatic lesions from a radiobiology and dosimetric perspective. Front. Oncol. Rev. **4**(210), 1–20 (2014)
9. Strigari, L., et al.: Dosimetry in nuclear medicine therapy: radiobiology application and results. Q. J. Nucl. Med. Mol. Imaging **55**(2), 205–221 (2011)
10. Low, D.A., et al.: A technique for the quantitative evaluation of dose distributions. Med. Phys. **25**(5), 656–661 (1998)

Computer Aided Diagnosis, Surgery, Therapy, and Treatment

Detection and Automatic Deletion of Staples in Images of Wound of Abdominal Surgery for m-Health Applications

Manuel González-Hidalgo[1,2(✉)], Gabriel Moyà-Alcover[1,2], Marc Munar[1,2],
Pedro Bibiloni[1,2], Andrea Craus-Miguel[2,3], Xavier González-Argenté[2,3],
and Juan José Segura-Sampedro[2,3]

[1] SCOPIA Research Group, UGiVIA Research Group,
Universitat de les Illes Balears, 07122 Palma, Spain
{manuel.gonzalez,gabriel.moya,p.bibiloni}@uib.es,
marc.munar1@estudiant.uib
[2] Balearic Islands Health Research Institute (IdISBa), 07010 Palma, Spain
acrausm@gmail.com, xavier.gonzalez@ssib.com, segusamjj@gmail.com
[3] Unidad de Cirugía General y del Aparato Digestivo,
University Hospital Son Espases, 07120 Palma, Spain

Abstract. Patients undergoing abdominal operations have to perform a post-operative follow-up during long periods of time after surgery. These face-to-face consultations have an economic cost and affect patients' quality of life. To simplify the follow-up of these patients, a smartphone application could be used. Based on an image of their surgical wound, post-operative complications may be detected. There are two cues to detect wound infection: inflammation and reddening. In this work, we aim at detecting and segmenting the wound. The first step of this process consists of locating and deleting the staples that maintain the wound closed because they distort the overall color of the image. To do this, we applied the discrete fuzzy morphological dilation operation to the whole image. Next, we applied an inpainting algorithm to restore the regions detected as staples. Finally, using fuzzy sets we determine if there are areas of the wound that present a deviation with respect to the red color, and study if the wound is reddening.

1 Introduction

Patients undergoing abdominal operations have to perform a post-operative follow-up during long periods of time after surgery. These face-to-face consultations have an economic cost and affect patients' quality of life as it can require unnecessary visits to the hospital [1]. Every surgery patient is evaluated twice

This work was supported by the project TIN 2016-75404-P AEI/FEDER, UE, and by the project PROCOE/2/2017 *Govern de les Illes Balears*.

© Springer Nature Switzerland AG 2019
J. M. R. S. Tavares and R. M. Natal Jorge (Eds.): VipIMAGE 2019, LNCVB 34, pp. 219–229, 2019.
https://doi.org/10.1007/978-3-030-32040-9_23

after the surgical procedure. The first review is at the health center after a week. A second review is performed one month later as an outpatient, at the hospital [2]. One of the main reasons for this double evaluation is to check the surgical wound and detect if it is infected. Typically, the rate of detection in these two visits is low and, when it happens, it is normally detected in the emergency department.

Telemedicine can be defined as *the use of electronic information and communication technologies to provide and support health care when distance separates participants* [3]. In recent years, the role of telemedicine in post-operative care has received significance as it has demonstrated excellent clinical outcomes, a high degree of patient satisfaction, decreased driving distance and waiting times, and cost savings to both the patient and health care systems [4]. Specifically, m-Health applications cope with medical or health issues supported by smartphones. m-Health apps are able to improve both the quality of medical services and the safety of the patients [5].

A feasible solution to change the actual follow-up of these patients is to use a smartphone application. After taking an image of their surgical wound and filling a simple questionnaire, a doctor can detect postoperative complications. In our concern, this is the first time that this objective has been addressed using image processing and analysis. The objective of this work is to locate and delete the staples that maintain the wound closed because they distort the color appearance of the image preventing the automatic detection of infections.

This article is organized as follows: Sect. 2 describes basic definitions and properties of the discrete fuzzy logical operators. Section 3 depicts the method for detecting and removing the surgery staples, and the color analysis of the inpainted image. Section 4 presents the results obtained with it. The conclusion and future work are described in Sect. 5.

2   Preliminaries and Fuzzy Morphological Operators

We will assume that the reader is familiar with the basic definitions and properties of the fuzzy, discrete or continuous, logical operators that we will use in this work, but for the sake of clarity we recall some of then. More details on these operators can be found for instance in [6]. From now on, consider L a finite chain $L = \{0, \ldots, n\}$.

Definition 1. An increasing binary operator $C : L \times L \to L$ is called a discrete conjunction if it satisfies $C(0, n) = C(n, 0) = 0$ and $C(n, n) = n$. A binary operator $I : L \times L \to L$ is called a discrete implication if it is decreasing with the first partial mapping, increasing with the second one and it satisfies $I(0, 0) = I(n, n) = n$ and $I(n, 0) = 0$. A discrete implication I is called a border implication on L if it satisfies $I(n, x) = x$ for all $x \in L$.

Given a conjunction C, the binary operator

$$I_C(a, b) = \max\{c \in L \mid C(a, c) \leq b\}$$

is an implication, which is called the *residual implication* of C.

The most well-known family of conjunctions in the discrete framework, as well as on the interval $[0, 1]$, are the *t-norms*. A conjunction T on L is called a *discrete t-norm* if it is commutative, associative and satisfies $T(x, n) = x$ for all $x \in L$. Details and properties on discrete t-norms and their residual implications can be found in [6, 7].

From now on, we will consider that A is a gray-scale image, and B is a gray-scale structuring element. In order to extend the definition of classical erosion and dilation given in [8], De Baets fuzzified the Boolean conjunction and the Boolean implication to obtain a successful fuzzification (see [9]). Recently [10], a new morphology based on the use of discrete fuzzy logical operators has been proposed. Its main difference with the non-discrete operators is that an N-dimensional gray-scale image is modelled by a function of the form $\mathbb{Z}^N \to L$, instead of a fuzzy set $\mathbb{R}^N \to [0, 1]$. This is computationally useful since, in practice, the images are stored as finite matrices and the grey values belong to a finite chain of 256 values. Thus, we introduce the following definitions.

Definition 2. The *discrete fuzzy dilation* $D_C(A, B)$ and the *discrete fuzzy erosion* $E_I(A, B)$ of A by B are the gray-scale images defined by

$$D_C(A, B)(y) = \max_x \{C(B(x - y), A(x))\}, \quad E_I(A, B)(y) = \min_x \{I(B(x - y), A(x))\}.$$

The *discrete fuzzy closing* $C_{C,\mathcal{I}}(A, B)$ and the *discrete fuzzy opening* $O_{C,\mathcal{I}}(A, B)$ of A by B are the gray-scale images defined by

$$C_{C,I}(A, B)(y) = E_I(D_C(A, B), -B)(y), \quad O_{C,I}(A, B)(y) = D_C(E_I(A, B), -B)(y).$$

Where the reflection $-B$ of a N-dimensional fuzzy set B is defined by $-B(x) = B(-x)$, for all $x \in \mathbb{Z}^N$. Among the great variety of conjunctions and discrete t-norms we have used the Łukasiewicz discrete t-norm that is $C(x, y) = T_{\mathbf{LK}}(x, y) = \max\{0, x + y - n\}$ together with their residual implication $I(x, y) = I_{\mathbf{LK}}(x, y) = \min\{n, n - x + y\}$ [11], in the experiments displayed related with the staples deletion algorithm. More information about this operators and their applications to image processing can be found in [10].

We can see the results of the k-th iteration of the discrete dilation, $D(A, B) = D_{T_{\mathbf{LK}}}(A, B)$, in Fig. 1. The structuring element considered, B, and the k-th iteration of $D(A, B)$ are, respectively:

$$B = \begin{pmatrix} 219 & 219 & 219 \\ 219 & 255 & 219 \\ 219 & 219 & 219 \end{pmatrix}, \quad \overbrace{D\Big(\dots\big(D(A, B), B\big), \dots, B\Big)}^{k \text{ times}} \tag{1}$$

It can be observed that from $k = 1$, $D(A, B)$, the staples outstand from the rest of the image. This fact will be used in the next section to design an algorithm that allows us to detect, as best as possible, the staples in the image of a wound.

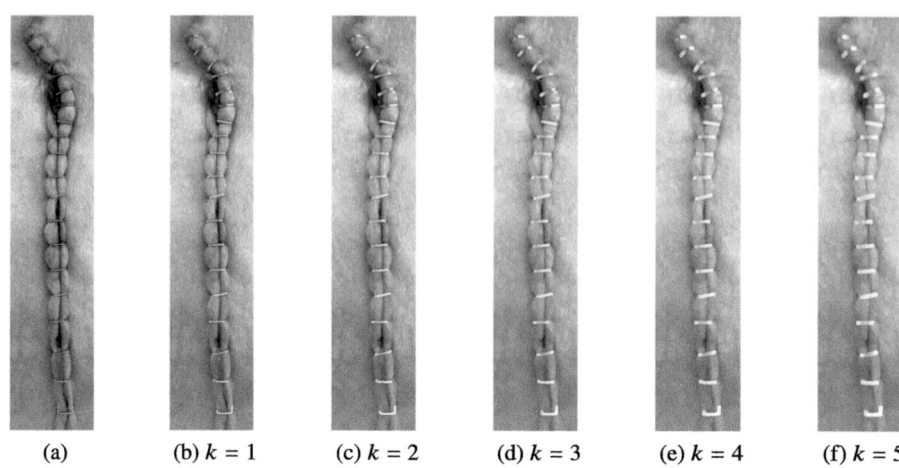

| (a) | (b) $k = 1$ | (c) $k = 2$ | (d) $k = 3$ | (e) $k = 4$ | (f) $k = 5$ |

Fig. 1. Left: original color image of a non-infected wound. The other figures show several k-th iterations of its discrete dilation. We can see how the staples outstand from their background.

Recently the operators given in Definition 2 have been extended to color images, encoded in the CIELab color space, the so-called soft color morphology [12]. The most remarkable property of the soft color morphology is that it is guaranteed that no new colors appear when the morphological operators are applied to a color image [12]. To end this section, we remark that this new soft color morphology will be the basis of the inpainting algorithm that we will use in the next section to eliminate the staples located in the wound.

3   Method

3.1   Staple Removal

The staple removal process is divided into two sequential steps. First, we detect the staples based on the technique shown in Fig. 1 by applying the discrete fuzzy morphological operators. Second, we apply an inpainting algorithm.

3.1.1   Staples Detection

Staples are metallic objects that excel in the images of surgical wounds because they typically have a high level of intensity. The designed algorithm starts from the morphological dilation over the gray-scale image, depicted in Fig. 4. As stated before, pixels that correspond to the staples are the ones with higher intensity values in the gray-scale image. We use this feature in order to select the candidates.

The images we analyze were taken by a smartphone, under uncontrolled conditions. Thus, they present irregular illumination. We try to avoid this drawback by dividing the image into $n = 5$ equally spaced horizontal regions. We create a binary mask, applying a threshold on each region, where the threshold for the n-th region is the maximum value found in that region minus a margin ϵ, where we experimentally set $\epsilon = 50$. Due to the characteristics of the morphological dilation, the results obtained have an excessive thickness. In the next step, we will refine the preliminary results.

The width of a staple in these images tends to be approximately 7 pixels. We iterate over the image columns to find vertical contours, for each contour we take its centroid selecting 3 pixels above and 3 below it, discarding the other ones. Once we have the final mask, we proceed to the next step.

3.1.2   Inpainting Algorithm

In this section, we present the inpainting algorithm for color images based on the soft color morphology operators [12]. This inpainting algorithm was used by Bibiloni et al. [13] to remove hair in dermoscopic images, and we follow the description of the algorithm done in it. The mask, obtained in previous step, was dilated to avoid undetected pixels belonging to a staple, or avoid dark areas around the staple.

In each location $x \in \mathbb{Z}^2$, the original image has either a color, denoted as $A(x) = c$ or a special symbol that means that this specific value is missing, denoted as $A(x) = \perp$. If the latter, we will say that $A(x)$ is *missing*. In this context, the dilation and erosion from Definition 2 are modified in order to obtain the soft color operators [12], and to ignore *missing* pixels. In particular, pixels that are *missing* in the input image are not taken into account to compute the maximum or minimum. Computing the maximum or minimum only over *missing* values, yields of course a *missing* value. Any other arithmetic operation with *missing* values yields a *missing* value.

In order to recover missing pixels based on nearby ones that are not missing, we introduce the following morphological transformation:

$$T(A_k) = \frac{O_{T_{\mathrm{M}}, I_{\mathrm{GD}}}(A_n, B_5) + C_{T_{\mathrm{M}}, I_{\mathrm{GD}}}(A_n, B_5)}{2} \tag{2}$$

where A_k is the original image, B_5 is a 5×5, flat, rounded structuring element; $C(x, y) = T_{\mathrm{M}}(x, y) = \min\{x, y\}$ is the minimum t-norm; $I(x, y) = I_{\mathrm{GD}}(x, y)$ its residual implication, the Gödel implication [11]. After this, the inpainting algorithm an image A is defined as the limit of the following sequence:

$$A_1 = A, \quad \forall n \in \mathbb{N}, A_{n+1}(x) = \begin{cases} A_n(x), & \text{if } A_n(x) \neq \perp, \\ T(A_n)(x), & \text{if } A_n(x) = \perp \text{ and } T(A_n)(x) \neq \perp, \\ \perp, & \text{otherwise.} \end{cases}$$

We can observe that the algorithm recovers missing pixels as soon as enough information is available. As we can see in the experiments, the iterative algorithm is able to successfully recover uniform and thin regions of the size of a staple.

3.2 Staple Removal Algorithm

Once the two main blocks of the process have been described, the whole staple removal algorithm is presented in Algorithm 1.

Algorithm 1. Staple Removal

Input: Wound image; **Output**: Image without staples

1: The image is converted into a gray-scale image.
2: The gray-scale image is used as input for the staples detector.
3: The output of the staples detector is used as a mask where the staple pixels appears in white.
4: The mask is dilated.
5: The wound image is converted into the L*a*b* color space.
6: In the L*a*b* image, we replace by ⊥ the colors in the locations indicated by the mask.
7: Missing pixels are inpainted, providing the image without staples.

Figure 2, depicts the whole algorithm designed to eliminate the staples in images of surgical wounds.

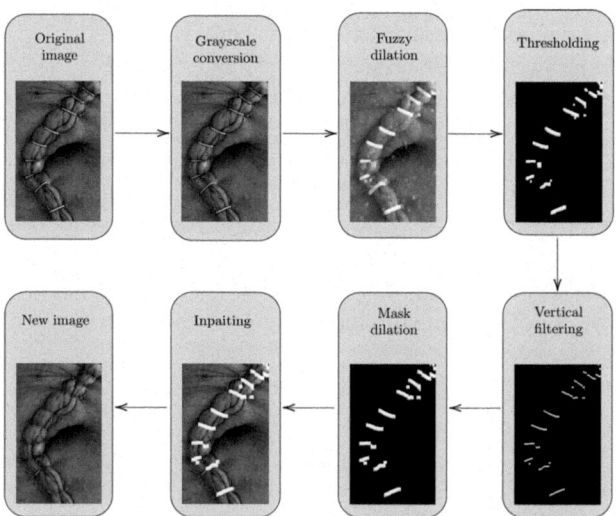

Fig. 2. Flowchart of the overall process.

3.3 Color Analysis

Once the staples have been removed, we aim to determine whether there are areas of the wound with a high proportion of red color. A color space based on the human perception of color similarity, such as HSV, HSL, or HSD, seems to be a better choice for this task. We focus on the HSV color space, whose hue

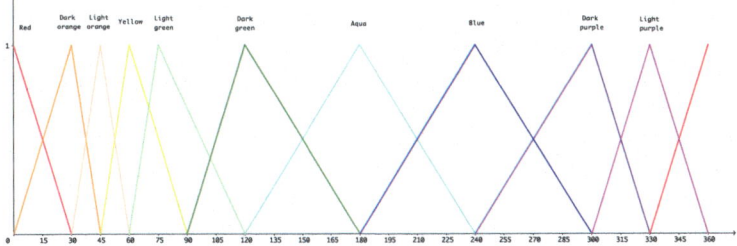

Fig. 3. Linguistic labels associated with the fuzzy sets of the H component.

(H) represents the color tone, *e.g.* red or yellow; its saturation (S) represents the amount of color; and its third component, the value (V), the amount of light.

Nevertheless, the concept of "red" is a linguistic term with a high grade of imprecision or vagueness. To deal with this vagueness we propose to use fuzzy logic, with which split the color space into linguistic labels. We have used as reference the Munsell color space, which splits into 10 segments the H component. Each one of these segments is fuzzified using a triangular function [14] to define its memberships [15,16]. The linguistic labels and the fuzzy space used on the H component used in this work are displayed in Fig. 3. As a result, we obtain the proportion of red in each of our images, and thus the presence of areas of the wound that present a high proportion of red. This can be visualized as an image, where the pixels with a maximum membership function associated with the red label are assigned to the $(1, 0, 0)$ RGB color coordinates.

4   Experimentation

In the following, we explain the preliminary results we obtained applying the process previously described. The set of images used to test our approach were

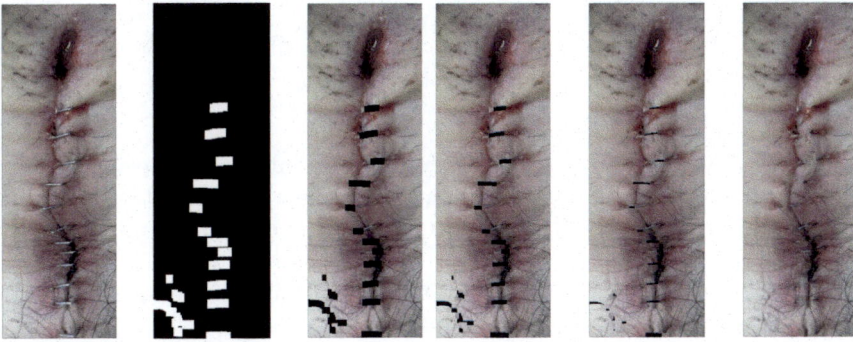

Fig. 4. Results of the iterative process of staples deletion on an image labeled as infected by the specialists.

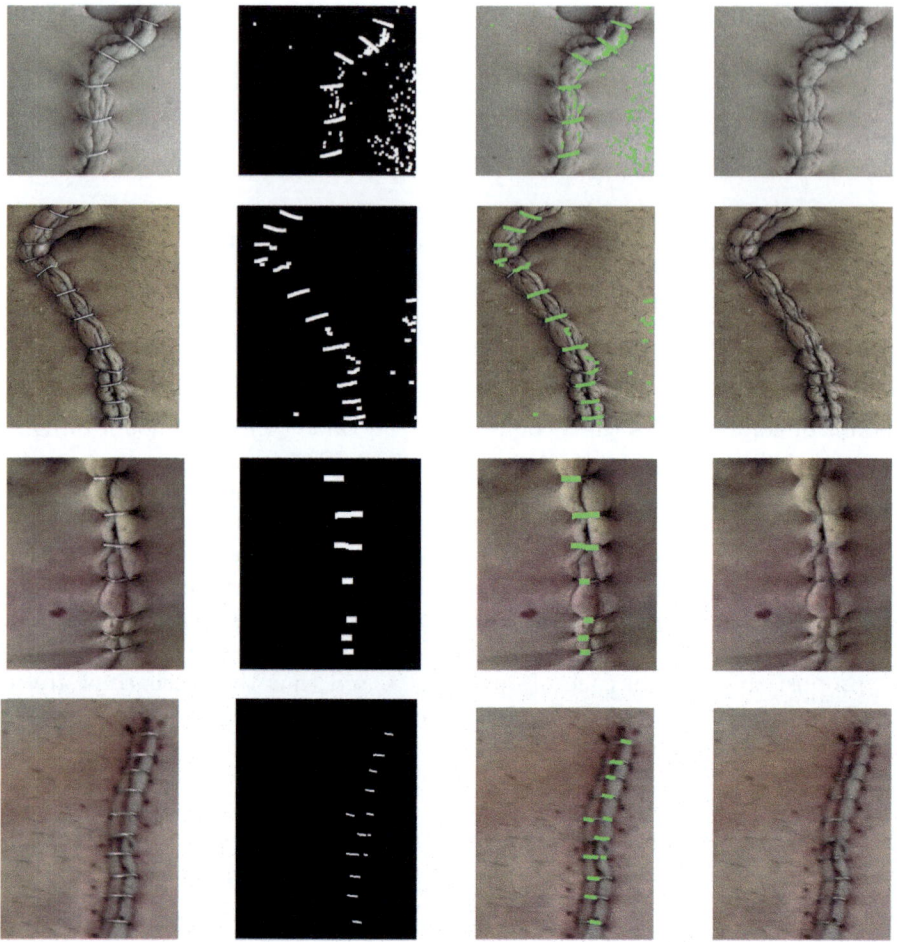

Fig. 5. Results of the application of the algorithm. The first column corresponds to the original image; the second one depicts the dilation of the obtained mask after applying the local threshold and vertical filtering; the third column shows the application of the dilated mask over the original image; and the fourth column correspond to the final result after the inpainting.

taken by the physicians from the Surgery Department of the University Hospital *Son Espases* in Palma de Mallorca, Spain. Images were taken in the clinical practice with their smartphones' embedded camera in real conditions. Thus, there was no experiment-defined control over the illumination, the capture viewpoint or the camera setup. The images depict abdominal surgical wounds, divided into two subsets: with and without infection.

In Fig. 4, we can observe the whole process of the inpainting algorithm. Starting from the detected mask, the algorithm iteratively recovers information from the local neighbourhood until the staples disappear.

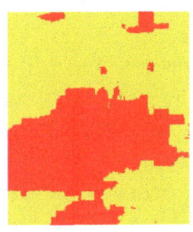

Fig. 6. From left to right, the fuzzy color image segmentation obtained for the inpainted images displayed from top to bottom in Fig. 5. As can be seen in the images labeled as infected wounds, by specialists, there are big areas with a high presence of red color.

Figure 5 depicts the whole process on different images. The first two rows correspond to examples without infection, and the two last ones correspond to images that present some degree of infection. We observe that in final result, the staples have been removed without altering the overall image appearance. The masks depicted in the second row exhibit the appearance of some noise, which appears during the thresholding step, but it does not influence the final result.

It is necessary to remember that the objective of this phase is to remove the staples and infer the missing information because it distorts the appearance of the wound area. Unfortunately, we cannot evaluate these results quantitatively because it is impossible to obtain a ground-truth of surgical wounds without staples.

Table 1. Red proportion detected in each one of the wound images displayed in the paper. Higher proportions appear in the images labeled as infected by specialists.

Wound image	Red proportion
Non-infected wound displayed in 1st row Fig. 5	0.0165
Non-infected wound displayed in 2nd row Fig. 5	0.0045
Non-infected wound displayed in Fig. 1	0.0165
Infected wound displayed in 3rd row Fig. 5	0.4002
Infected wound displayed in 4th row Fig. 5	0.4118
Infected wound displayed in Fig. 4	0.4911

In Fig. 6 we show the set of pixels detected as "red label" for each one of the images displayed in Fig. 5. As can be observed in the images corresponding to infected wounds there are big areas with a high red component. In addition, in Table 1 we can observe that in the images corresponding to infected wounds, the proportion of red pixels is larger than the images of uninfected wounds.

5 Conclusions

In this paper we presented a method to locate and delete the staples that maintain the wound closed as a first step for an m-Health application with promising benefits. After that, we estudy if the wound is reddening as a cue of wound infection. The algorithms applied are selected with the design criteria of achieving real time processing in order to be included in a smartphone application.

The method of staples detection and deletion is divided into two main steps. First, we create a binary mask of the position of the staples, based on thresholding the results obtained by the application of the discrete fuzzy morphological dilation. Second, we apply an inpainting algorithm based on a new paradigm for color image morphology: the soft color morphology.

Finally, we consider the analysis of the inpainted region in order to automatically decide whether the wound is infected or not using triangular fuzzy sets. As we have seen, all the images of areas of infected wounds have a proportion of pixels labeled as red greater than 40%, a percentage much higher than the images of uninfected wounds. This is a very clear cue of the presence of infection in some area of the wound.

The results obtained are very promising and encouraging. We need to increase the size of the dataset of wound images used. As future work, we consider the analysis of the inpainted region using others type of fuzzy sets membership functions, and to include the algorithm in a smartphone application that could be used in the clinical practice.

References

1. Segura-Sampedro, J.J., Rivero-Belenchón, I., Pino-Díaz, V., Sánchez, M.C.R., Pareja-Ciuró, F., Padillo-Ruiz, J., Jimenez-Rodriguez, R.M.: Feasibility and safety of surgical wound remote follow-up by smart phone in appendectomy: a pilot study. Ann. Med. Surg. **21**, 58–62 (2017)
2. Estarellas, N.M., Bonin-Font, F., Segura-Sampedro, J.J., Jiménez-Ramírez, A., Carrasco, P.L.N., Massot-Campos, M., Gonzalez-Argenté, F.X., Codina, G.O.: Towards a pre-diagnose of surgical wounds through the analysis of visual 3D reconstructions. In: VISIGRAPP (4: VISAPP), pp. 589–595 (2018)
3. Committee on Evaluating Clinical Applications of Telemedicine: Telemedicine: a guide to assessing telecommunications for health care. Technical report, Institute of Medicine, Washington, DC (1996)
4. Williams, A.M., Bhatti, U.F., Alam, H.B., Nikolian, V.C.: The role of telemedicine in postoperative care. mHealth **4**(5), 11 (2018)
5. Chatzipavlou, I.A., Christoforidou, S.A., Vlachopoulou, M.: A recommended guideline for the development of mHealth Apps. mHealth **2**(5), 21 (2016)
6. Mayor, G., Torrens, J.: Triangular norms in discrete settings. In: Klement, E., Mesiar, R. (eds.) Logical, Algebraic, Analytic, and Probabilistic Aspects of Triangular Norms, chap. 7, pp. 189–230. Elsevier, Amsterdam (2005)
7. Mayor, G., Suñer, J., Torrens, J.: Operations on finite settings: from triangular norms to copulas. In: Úbeda Flores, M., de Amo Artero, E., Durante, F., Fernández Sánchez, J. (eds.) Copulas and Dependence Models with Applications, pp. 157–170. Springer, Cham (2017)

8. Serra, J.: Image Analysis and Mathematical Morphology, vols. 1, 2. Academic Press, London (1982, 1988)
9. De Baets, B.: A Fuzzy Morphology: A Logical Approach, pp. 53–67. Springer, Boston (1998)
10. González-Hidalgo, M., Massanet, S.: A fuzzy mathematical morphology based on discrete t-norms: fundamentals and applications to image processing. Soft. Comput. 18, 2297–2311 (2014)
11. Baczyński, M., Jayaram, B.: Fuzzy Implications. Studies in Fuzziness and Soft Computing, vol. 231. Springer, Heidelberg (2008)
12. Bibiloni, P., González-Hidalgo, M., Massanet, S.: Soft color morphology: a fuzzy approach for multivariate images. J. Math. Imaging Vis. 61(3), 394–410 (2019)
13. Bibiloni, P., González-Hidalgo, M., Massanet, S.: Skin hair removal in dermoscopic images using soft color morphology. In: ten Teije, A., Popow, C., Holmes, J.H., Sacchi, L. (eds.) Artificial Intelligence in Medicine, pp. 322–326. Springer, Cham (2017)
14. Zadeh, L.A.: Fuzzy sets. Inf. Control 8(3), 338–353 (1965)
15. Shamir, L.: Human perception-based color segmentation using fuzzy logic. In: IPCV, vol. 2, pp. 96–502 (2006)
16. Younes, A.A., Truck, I., Akdag, H.: Image retrieval using fuzzy representation of colors. Soft. Comput. 11(3), 287–298 (2007)

Reconstruction of a Defective Finger Joint Surface and Development of an Adapted External Fixator

Lena Risse[1,2(✉)], Steven Clifford Woodcock[1,2], Gunter Kullmer[1,2], Britta Schramm[1,2], and Hans Albert Richard[1,2]

[1] Fachgruppe Angewandte Mechanik (Applied Mechanics),
Paderborn University, Pohlweg 47-49, 33098 Paderborn, Germany
risse@fam.upb.de
[2] Direct Manufacturing Research Center (DMRC),
Paderborn University, Mersinweg 3, 33098 Paderborn, Germany

Abstract. Complex injuries are not optimally treatable with standard medical methods, especially when joints or articular surfaces are affected. To achieve an intact and functional joint, the articular surface covered by hyaline cartilage must be restored geometrically correct. Since hyaline cartilage has a poor ability to regenerate, a damaged articular surface requires replacement. This article describes the CAE-assisted preoperative planning of the reconstruction of a defective proximal interphalangeal joint surface. While determining needed medical aids, optimization potentials for existing external fixators have additionally been detected. In the following, the development of an adapted finger fixator with extended pre-, intra- and postoperative adjustment possibilities will be explained. The optimized fixator therefore requires less intraoperative precision. Possible inaccuracies during the surgery can be eliminated due to the adjustment possibilities afterwards. Likewise, it is possible to respond to changing anatomical conditions during the healing process, which grants a greater chance of rehabilitation overall. The basic design and material selections for the fixator are defined through the usage of numerical methods. Subsequently, the reliability is tested by using experimental methods.

Keywords: Finger joint · Transplant · Additive manufacturing · External fixator

1 Introduction and Aim

The goal of surgery as a result of an accident or to correct congenital or acquired limitations in the functioning of the human musculoskeletal system is the complete restoration of range of motion and mobility. This ambition cannot be always fully put into practice. Standard medical methods are reaching their limits, especially in the case of complex injuries, more precisely when joints are additionally affected. The discussion about the benefits compared to the risks often leads to the fact that an optimal rehabilitation during the procedure is not given.

© Springer Nature Switzerland AG 2019
J. M. R. S. Tavares and R. M. Natal Jorge (Eds.): VipIMAGE 2019, LNCVB 34, pp. 230–238, 2019.
https://doi.org/10.1007/978-3-030-32040-9_24

The advantageous use of engineering methods and tools can contribute to risk reduction and thus obtain an increase of the achievable degree of rehabilitation. As an example of possible problems and potentials, an injury at the middle joint of a young man's middle finger is described below. With the aim of using an autologous cartilage-bone transplant a computer aided design (CAD)-assisted analysis of a suitable joint surface shows possible extraction points. Additively manufactured models were made to give the surgeon a haptic intraoperative guidance.

To ease the strain from the transplant and to passively guide motion as well as assist recovery of mobility, a two-sided external fixator is used. The success of the subsequent rehabilitation phase is strongly dependent on the placement of this fixator and thus the intraoperatively achieved precision. In order to reduce this risk or dependence, a fixator with pre-, intra- and postoperative adjustment options is developed. The fixator is intended to serve optimal adaptation to the particular patient and the injury with customizability and thus to support the recovery of joint mobility.

2 Diagnosis and Surgical Planning

A middle-aged patient suffered from an extensor defect of the proximal articular surface of the proximal interphalangeal joint (PIP joint) caused by a saw blade. Both the palmar plate and the lateral ligament attachments were not damaged. The defective extensor tendons could already be reconstructed in a previous surgery. The computer tomography (CT) data of the left hand with the injured finger joint surface of the middle finger were present and could give a first impression of the injury. The two joint ends are contracted around the finger middle joint due to the tight joint capsule, so that the extent of the defect is not exactly visible. Therefore, in a first step, the proximal phalanx and middle phalanx of the middle finger are transferred from CT data into CAD data to start an analysis and treatment plan for the injury (Fig. 1). The articular surface of the middle phalanx shows an intact articular surface, only the articular surface of the proximal phalanx is damaged by the accident. The CAD-data of the injured proximal phalanx is shown in Fig. 1a.

For the further procedure in dealing with the present joint defect different therapy options are available. If the finger joint is left without further therapy, this leads to a stiffening of the joint in the current extended position. Through a surgical intervention, an arthrodesis (joint stiffening) can alternatively be performed in a functional position (slight flexion). Both treatment alternatives would cause severe restrictions in everyday life, but not cause much risk. Another treatment alternative would be the use of an endoprosthetic partial replacement of the joint head. Literature [1–3] provides examples of successful applications of these endoprosthetic partial replacements. Due to the young age of the patient in comparison to the lifetime of a prosthesis, this solution was rejected.

Because of the desire of the patient to regain the functionality of the joint in terms of mobility and strength, a replacement of the articular surface by autologous material is a preferred treatment alternative. When using conventional planning methods, there is great uncertainty and therefore a high risk for the surgical procedure. By using CAE

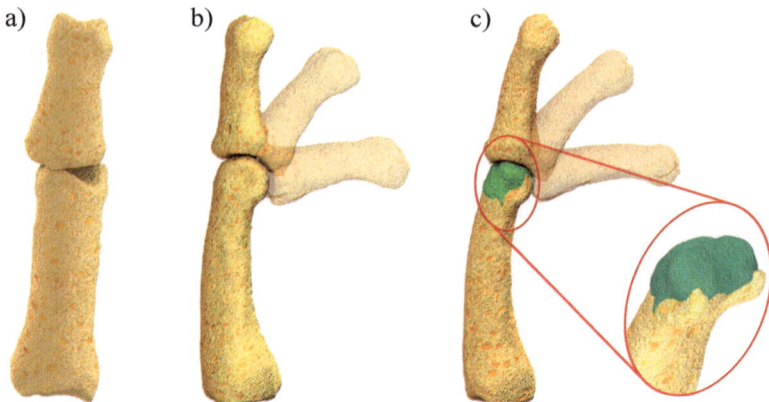

Fig. 1. Analysis of the injury (a) Present joint defect (b) Kinematic of a healthy PIP joint (c) Reconstruction of the defect.

methods, it is now possible to reconstruct the defect three-dimensionally, considering the kinematics, and then to define a suitable transplant extraction area.

The first step in the computer aided engineering (CAE)-assisted planning is to convert CT data from the affected bones into a 3D-CAD solid model. The CT images of the entire left hand are available, so that on the one hand proximal phalanx and middle phalanx of the damaged left middle finger and as a reference also the proximal phalanx and middle phalanx of the left index finger are transferred to a CAD volume model. In order to be able to reconstruct a damaged joint, it is indispensable to know the prevailing kinematics exactly. Numerous studies investigating the kinematics of the fingers and the hand, and in particular of the affected PIP joint in this case, are already available [4–6]. In [7] the joint center of the interphalangeal joints was determined. By transferring these results to the present case of application, the kinematics of the patient's finger joints are analyzed with the help of the CAD data of the healthy index finger (Fig. 1b). This made it possible to reconstruct the anatomically correct position of the defective finger bone based on the selected joint axis.

In a subsequent step, the defect is then modeled with the help of the intact counter-joint surface in the middle finger base member and the intact articular geometry of the index finger. The examination of the constructed flaw geometry is carried out by means of simulation of the joint movement in the full range of motion with included inter-ference check. Smaller adjustments due to interference lead to the reconstructed defect geometry, which acts as a "spare part that can be used optimally" and represents the pattern for the later actual graft to be taken (Fig. 1c).

Various cartilage bone donor regions of the human body are conceivable, always taking into account the avoidance of unacceptable functional restrictions to the patient. Due to the local proximity to the defect the hand is first analyzed as a possible transplant extraction region. The joints between the metacarpals and the carpal bones are amphiarthroses, whose motility is severely restricted, but which nevertheless have articular surfaces coated with hyaline cartilage. Thus, the removal of a transplant there and therefore the damage of this articular surface would hardly cause functional limitations for the patient. The exact sampling point must be chosen in a way that the graft reproduces the geometry of the original articular surface as accurately as possible. In this way, the recovery of the full range of motion of the left middle finger should be ensured. To illustrate the anatomical conditions and the planned surgical procedure, the anatomy of the human hand and of suitable graft sites are shown in Fig. 2.

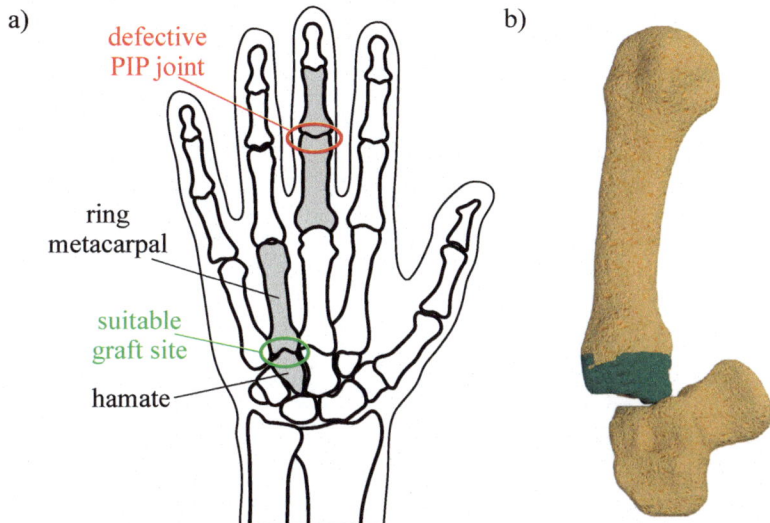

Fig. 2. Detection of possible extraction points for the cartilage bone transplant (a) anatomy of the human hand (b) CT-data of the ring metacarpal and hamate with possible cartilage bone transplant.

The CAD-assisted matching suggested that the filling of the cartilage bone defect with a graft from the base of the ring metacarpal or the hamate of the same hand would be most suitable because of the geometric similarity. The defect is to be replaced by a one-piece graft that should be as similar to the original articular surface as possible. Both factors have a positive influence on the healing process and the functionality of the joint.

Figure 2b shows the analysis of articular surface geometries of the hamate and the ring metacarpal. The contour of the articular surface of the ring metacarpal provides a great similarity with the articular surface to be replaced. Since the spatial extent of the defect is smaller than the articular surface functioning as a graft donor, the part which is geometrically most similar to the defect is detected and removed in the later surgery.

Furthermore, Fig. 2b shows the relevant articular surface of the fourth metacarpal bone with the donor site of the cartilage-bone graft marked in turquoise. This graft has high geometric matches with the missing articular surface. Examining the range of motion of the joint with the virtually-removed and defect-fitted graft shows minimal interference centered on the articular surface during flexion, and minor geometric variations in the marginal areas of the bone. During surgery, therefore, minor geometric adjustments to the graft articular surface and the graft taken are necessary.

At the end of the CAE-assisted planning process, polyamide models of the affected bones and the graft are additively manufactured. On the one hand, these are intended to provide better visualization of the steps to be performed and on the other hand, to be available as a sample during the surgical procedure after sterilization. Thus, the most important steps of the intervention can be simulated in advance on the model and possible sources of error can be eliminated.

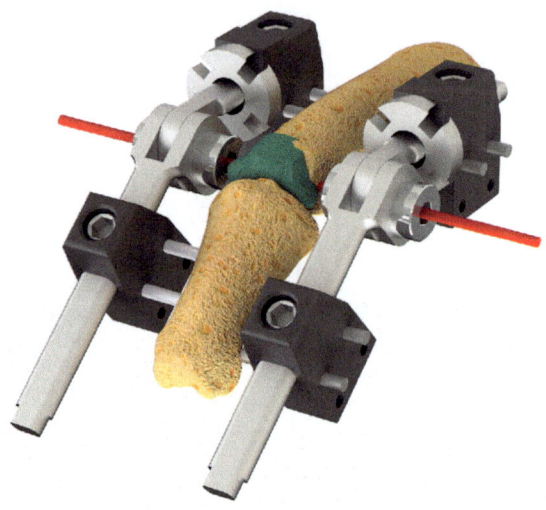

Fig. 3. Planning of external fixation.

In order to ease strain in the transplant during the healing phase, and in particular to restore the physiological joint motility and to expand the shrunken joint capsule, a gradual joint dilation is to be performed by bilateral attachment of an interlingual, unilateral, dynamic external fixator. So that the recovery of the joint mobility despite the external fixator is possible, it is necessary that the joint axes of rotation of fixator and finger joint match. This position is simulated in advance in CAD to give the surgeon a guide, such as the exact positioning must be performed as a hinge joint according to the anatomical axis of movement of the PIP joint. The result of the planning is shown in Fig. 3. The surgical procedure was carried out successfully according to the planning. A final check a few weeks after the surgery reveals successful graft healing and limited recovery of joint mobility.

3   Development of an Adapted External Finger Fixator

In the course of the preoperative planning of the finger joint medical care, it becomes clear that the chances of rehabilitation depend strongly on the achieved intraoperative precision. If the joint axes of the fixator and the finger are displaced too much, mobility is severely restricted or even impossible. In addition, during the healing process, anatomical changes can occur that can no longer be responded to. For this reason, an adapted external fixator for finger joints is being developed and shown in Fig. 4. Due to the more critical loading situation, the design of the fixator with unilateral attachment is examined. This development is characterized by the fact that there are new functions in addition to the traction possibility.

Fig. 4. Adapted external finger fixator.

The developed external fixator has a modular structure, which means that adaptation can be implemented without any great expense, for example by scaling the individual components. In this way, this fixator can be used to treat injuries in all hinge joints. Basic modules are the two fixation blocks that are connected to the bone via bone fixation elements, so-called fixator pins or Schanz screws, as well as a hinge joint. The modules are connected to each other via the load-bearing element, in this case a threaded rod. By adjusting the knurled screws, traction of the two fixation jaws from each other can be achieved.

The idea of the optimized fixator additionally allows the bone positions to be adjusted to one another at any time, as this axis is designed to be continuously adjustable and lockable via the fixation cranes in a certain angle (Fig. 5a). The distraction of the joint can be performed separately from the angular adjustment of the two bone axes. In addition, an axis angle adaptation is made possible by the rotatable and fixable center piece of the fixator. Furthermore, the location of the rotation axis can be readjusted to a certain extent (Fig. 5b). The use of the optimized finger fixator should contribute to increase the chances of rehabilitation. Adjustments that are necessary from a medical point of view or to the well-being of the patient can be carried out without much effort and at any time.

The responsibility for the reliability of this medical device is with the manufacturer, since no binding regulations exist. As a first step numerical simulations are carried out to validate the dimensions and the chosen materials for all the components.

In order to determine reliable test parameters and load cases, typical everyday stress situations of a finger joint that are shown in Fig. 6 are used to define the test conditions. The stress in the external fixator is evaluated for the fully stretched finger ($\alpha = 0°$) as well as for a moderate flexion ($\alpha = 45°$) and the full flexion ($\alpha = 105°$). The results of the FE-simulations attest that the maximum permissible stress is not reached in any of

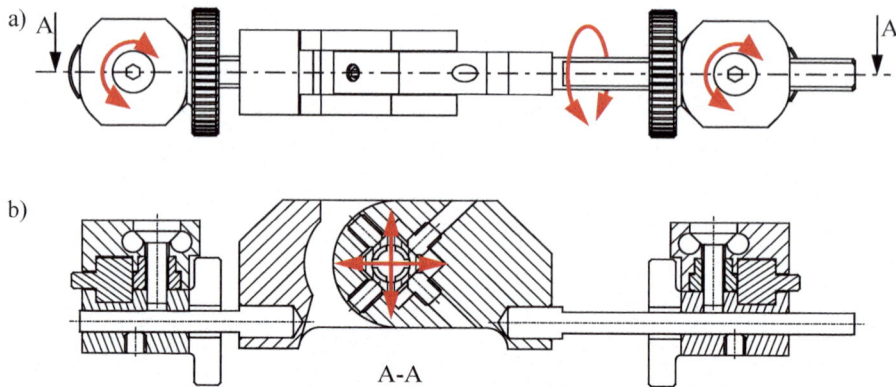

Fig. 5. New functions of the fixator (a) Adjustment of the finger axis (b) Adjustment of the fixator's rotation axis.

the components for both load cases. A reliable design could consequently be developed and prototypes can be manufactured. To validate the results of the FE simulations, experimental component tests are subsequently performed using a universal testing machine (Fig. 7a).

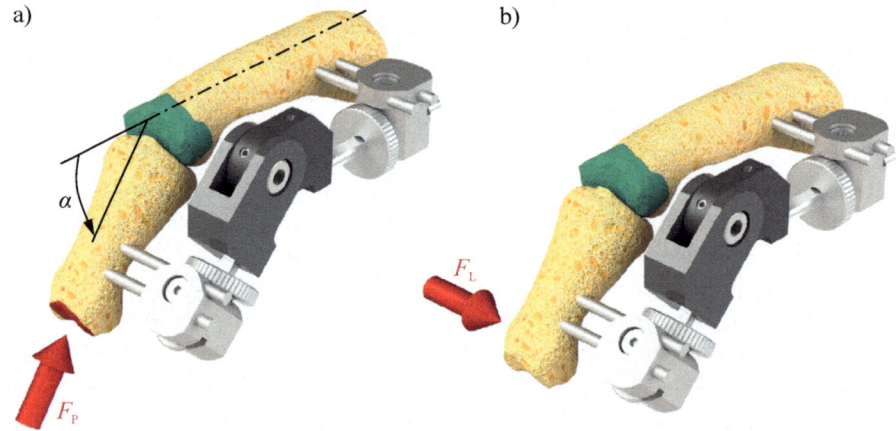

Fig. 6. Analysis of possible external fixator's load situations (a) force transmission parallel to finger axis (b) force transmission lateral to finger axis.

The load cycles for testing the fatigue properties are 50,000, determined according to [8]. The test devices designed to simulate the load cases from the numerical tests are visualized in Fig. 7. For a statistical verification of the test results, for each of the six tests (two load cases with three angular positions each) three test specimens must be used, which must survive the examinations without damage. Using [9], suitable test

a)

b)

c)

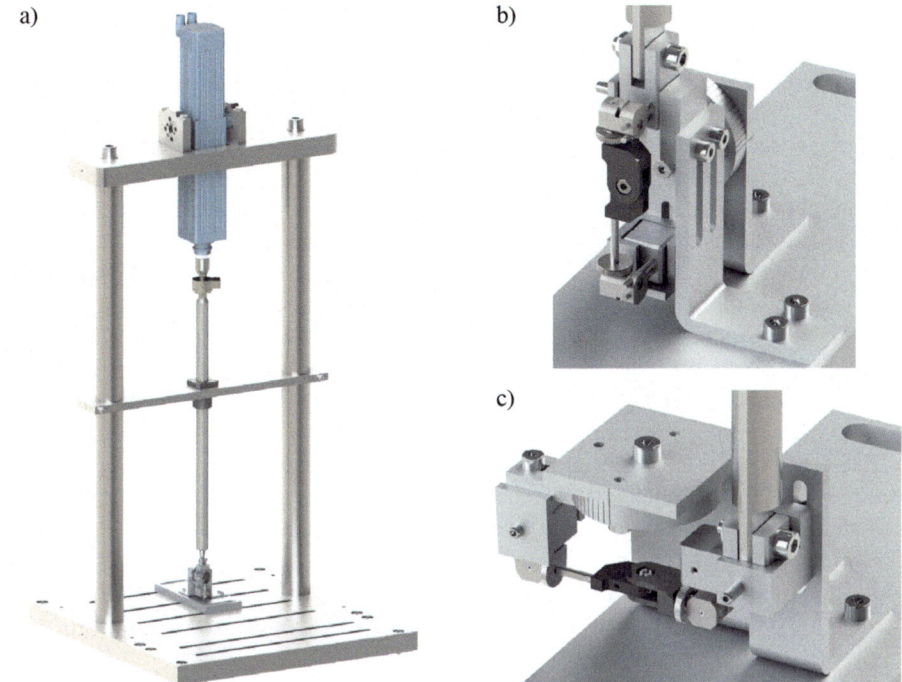

Fig. 7. Experimental tests for different loading situations (a) Universal testing machine (b) Testing device for the load case "parallel force" (c) Testing device for the load case "lateral force".

devices are developed. They have to fix the fixator and connect it to the base plate. Furthermore the force application must be enabled. As different angular positions have to be tested an angle adjustment by the testing device should be realized.

When testing the finger fixator for the load case "parallel force" (Fig. 7b), the load is transmitted parallel to the fixator axis, whereby the test specimen is fixed by two bone analogs. Using a circular disc with a round slot, the angle adjustment is enabled according to the defined load situations. In this way, the fixator can be continuously rotated and fixed. The test device for the load case "lateral force" (Fig. 7c) consists of the identical modules already described above. Since the load is transmitted perpendicular to the fixator axis, the perforated disc is connected by a Z-profile to the base plate. The experimental investigations are carried out at room temperature and ambient conditions. This is considered sufficient, as the fixator pins, which make the connection between bone and fixator, should not be checked. To connect bones and fixator standard elements are used which have already been tested for their reliability. The fixator itself has no direct interactions with the body's tissue or fluids. The tests have been successfully completed, so that the experimental validation of reliability is considered complete and a clinical study on the functionality of the adapted external fixator can follow.

4 Conclusion

The use of engineering methods, in this case especially CAD and additive manufacturing processes, offers universal potential for optimizing the achievable degree of rehabilitation. The basis for the improved planning situation is the availability of three-dimensional visualizable and modifiable data. These allow a better understanding of the situation, also without disturbing surrounding tissue. This provides the basis for analyzing and restoring the kinematics of any defect. A virtual planning of the surgical intervention leads to a better predictability and an accurate documentation of the required steps. Additively manufactured models support the planning as the surgery can be carried out in advance on the model as a test. During the procedure, they serve the haptic and visual support of the surgeon. Finally, the digital, detailed documentation of the process contributes to the fundamental enabling of complex interventions, minimizing errors and risks and thus improving the chances of recovery and the degree of rehabilitation.

The additional use of the adapted external fixator that is adjustable at any time further enhances the chances of rehabilitation. A patient-specific configuration is realized without the provision of a wide range of products. Thus, by using a single product, each patient can be individually and optimally cared for. Since the setting options are not limited to the preoperative state but are possible at any time, there is still a chance to react to subsequent changes and optimally support the healing process.

References

1. Erdogan, A., Weiss, A.P.C.: Metakarpophalangeale Arthroplastik mit der NeuFlex-Silastikprothese. Der Orthopäde **32**(9), 789–793 (2003)
2. Schindele, S.F., Hensler, S., Audigé, L., Marks, M., Herren, D.B.: A modular surface gliding implant (CapFlex-PIP) for proximal interphalangeal joint osteoarthritis: a prospective case series. J. Hand Surg. **40**(2), 334–340 (2015)
3. Schindele, S.F., Sprecher, C.M., Milz, S., Hensler, S.: Osteointegration of a modular metal-polyethylene surface gliding finger implant: a case report. Arch. Orthop. Trauma Surg. **136**(9), 1331–1335 (2016)
4. Buchner, H.J., Hines, M.J., Hemami, H.: A dynamic model for finger interphalangeal coordination. J. Biomech. **21**(6), 459–468 (1988)
5. Holguín, P.H., Rico, Á.A., Gómez, L.P., Munuera, L.M.: The coordinate movement of the interphalangeal joints – a cinematic study. Clin. Orthop. Relat. Res. **362**, 117–124 (1999)
6. Loubert, P.V., Masterson, T.J., Schroeder, M.S., Mazza, A.M.: Proximity of collateral ligament origin to the axis of rotation of the proximal interphalangeal joint of the finger. J. Orthop. Sports Phys. Ther. **337**(4), 179–185 (2007)
7. Hess, F., Fürnstahl, P., Gallo, L.M., Schweizer, A.: 3D analysis of the proximal interphalangeal joint kinematics during flexion. Comput. Math. Methods Med. **2013**(362), 138063 (2013)
8. ASTM F1541-17: Standard Specification and Test Methods for External Skeletal Fixation Devices. ASTM International, West Conshohocken, PA (2017)
9. Verein Deutscher Ingenieure: VDI-Richtlinie 2221 – Methodik zum Entwickeln und Konstruieren technischer Systeme und Produkte. Beuth Verlag, Berlin (1993)

Skin Disease Diagnosis from Photographs Using Deep Learning

Evgin Goceri[(⊠)]

Biomedical Engineering Department, Akdeniz University, Antalya, Turkey
evgin@akdeniz.edu.tr

Abstract. This work aims to study performance of different deep learning based approaches to classify skin diseases automatically from colored digital photographs. We applied recent network models, which are U-Net, Inception Version-3 (InceptionV3), Inception and Residual Network (InceptionResNetV2), VGGNet, and Residual Network (ResNet). Comparative evaluations of the results obtained by these network models indicated that automated diagnosis from digital photographs is possible with accuracy between 74% (by U-net) and 80% (by ResNet). Therefore, further studies are still required in this area to design and develop a new model by combining advantages of different network models and also to obtain higher accuracy. In addition, testing of the model should be performed with more data including more diversity to see reliability of the model.

Keywords: Automated diagnosis · Classification · Color images ·
Deep learning · Skin disease

1 Introduction

Skin diseases, which are caused by different factors such as aging, trauma, genetic or environmental factors, affect millions of people worldwide [1]. According to the report published in 2013 by the American Academy of Dermatology, 85 million Americans (i.e., one in four people of all ages) went to a dermatologist due to at least one skin disease. Also, data in the same report indicate that only direct spending of these patients is 75 million dollars (USD) for treatments [2, 3].

Traditional diagnosis of skin diseases, based on manual examination and visual evaluation, is provided by subjective decisions. Therefore, these decisions can be different according to experiences of dermatologists and their instant visual perceptions. In addition to subjectivity, the traditional approach can cause infection and pains. Also, it is a time consuming and tiring method. A symptom that has been overlooked by dermatologists, while they are working against time with intensive tempo, can lead to overlook of a new lesion in diagnosis. Moreover, if disease is infectious, diagnosis with the traditional method can be much more difficult; as lesions should be examined from a distance to prevent spreading risk of the disease. As a result of inaccurate diagnosis, treatments applied to patients become inaccurate [4]. Therefore, the importance of computer-aided analysis of images increased to overcome these problems [5].

© Springer Nature Switzerland AG 2019
J. M. R. S. Tavares and R. M. Natal Jorge (Eds.): VipIMAGE 2019, LNCVB 34, pp. 239–246, 2019.
https://doi.org/10.1007/978-3-030-32040-9_25

Traditional computerized techniques for skin disease diagnosis are usually performed using dermoscopy images and involves three steps: (1) Segmentation of lesions; (2) Extraction of hand-crafted features; (3) Classification [6]. Most of these methods are based on supervised learning. Therefore, the ground truth data and the chosen feature sets with the corresponding labels are used in the training stage of classification. A good review of these methods can be found in [7]. An important problem of these traditional techniques is lack of generalization because of different artefacts, high variations (due to various zooming effect, lighting conditions, operators, instrumentations, light reflections, dark borders, shadows or skin lines) in images, and small number of training data. Therefore, feature detection and extraction, which requires engineering skills and efficient algorithms to obtain desired features, may not be accurate and cause wrong classifications [6, 8]. Nowadays, deep learning based techniques are state-of-the-art methods (due to their ability on integration of high, middle or low level features and also enriching of these level of features with the number of layers (i.e., depth)). These techniques have the potential to classify skin diseases [9]. They have been applied with (1) different types of images (mostly dermoscopy images) and (2) different network models, also (3) to diagnose different diseases (usually melanoma) [8, 10–12].

The image data sets used in this study are photographs taken by different mobile phones or cameras. The reason to use these images is the recent advancements in mobile technologies and future applications, which can be run by doctors and patients at anywhere/anytime on smart phones. Figure 1 shows some example images used in this work.

In the literature, there are only a few works based on skin disease diagnosis from photographs [13–15]. Several methods (e.g., morphological operations [13], multi-layer perceptron [14], and logical regression [15]) except deep learning were used in these works. Recently, the authors in [16] applied a dense network and ResNet models using photographs obtained by smartphones. Their results showed efficiency of these approaches for automated diagnosis of skin diseases. However, they have handled different types of skin diseases, such as Crust, Erythema, Leukoderma, Pustule and Ulcer. In this work, we focused on five common skin diseases; (1) Acne vulgaris, (2) Hemangioma, (3) Psoriasis, (4) Rosacea, and (5) Seborrheic dermatitis.

The key advantage of deep learning based methods is automated extraction and learning of features from images. However, the main limitation for their applicability is scarcity of large public image databases. Therefore, training of these networks is usually performed with small and heavily unbalanced datasets. The data unbalancedness is a significant effect on the performance of deep learning based methods to analyze images. To solve this problem, data augmentation has been applied using flipping and rotation of images with different angles. However, increasing number of images with these approaches may not be efficient to increase performance of a network model for image classification due to unchanged variations. Also, these methods have been applied with different data sets and epochs for different skin diseases.

To find the best fit network model for computerized diagnosis of skin diseases, network architectures should be compared accurately. To make accurate and meaningful comparative analysis, evaluations should be performed with the results obtained

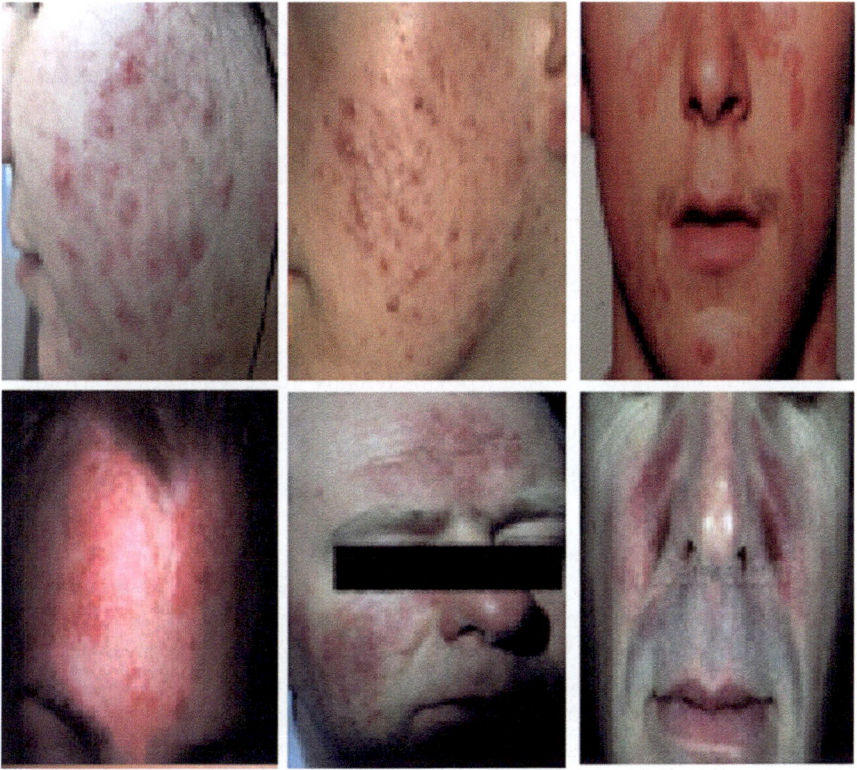

Fig. 1. Example images used in this work

using the same (1) image datasets, (2) hardware structure, and (3) evaluation metrics. Therefore, the aim in this study is to perform this comparative analysis to determine the best network model for automated diagnosis of skin diseases.

In this work, we applied those network architectures that have been proposed recently and reported with high accuracy for skin disease classification. The network models that we have focused are; U-net, InceptionV3, InceptionResNetV2, VGGNet and ResNet.

2 Applied Network Architectures

The network models applied in this work have different properties, which affect classification performance. In this section, main properties of these networks are explained.

U-Net Architecture: This architecture is based on three main stages, which are down-sampling, up-sampling steps and cross-over connections. The number of features increases in the down-sampling step. The image resolution is recovered in the up-sampling step. The reason to use the cross-over connection is to concatenate the size of feature maps equally since features might be lost in the down-sampling operation. In the U-Net architecture, up-sampling and down-sampling is performed at 5 levels. 2 convolutional layers are used at every level. The size and number of kernel at each convolutional layer is the same [17].

InceptionV3: Inception networks are based on inception blocks to handle increasing demands on image datasets and also computational power. At each layer in an inception network architecture, conceptual parameters are chosen to extract either smaller features or features that are more general. This is performed by choosing max-pooling or convolutional layer and the respective size of the receptive fields. By this way, the inception blocks allow the network to find what is more required or useful for each layer. To construct an Inception network, width of each layer is reduced by using 1×1 convolutional layers before computationally more complex convolution to reduce required processing time. This architecture is called as the Inception-V1 network [18]. Its modified versions with batch normalization (InceptionV2) [19] and additional factorization (InceptionV3) [20] have been proposed. In our work, InceptionV3 has been applied.

ResNet: Increasing the number of stacked layers of deep neural network increases error in training and validation [21, 22]. ResNet has been proposed to solve this problem and to ease training by learning residual of identity mapping [23, 24]. A ResNet is constructed by residual blocks, which comprise batch normalization layers, ReLU and convolutional layers. The residual block can shortcut (bypass) a few convolution layers at a time. These connections are aggregated with the outputs of the convolution layers. In this work, the network has been applied with 50 layers (ResNet-50).

InceptionResNetV2: The residual connection based on multiple sized convolutional filters and the Inception model are combined in this architecture. The main advantage provided by residual connections is to overcome the degradation issue that is caused by deep layers in the network. Also, the usage of residual connections reduces required time to train the network [25].

VGGNet: This architecture is designed with 3×3 filter kernels. Before a max-pooling layer, blocks of 2 to 3 convolutional layers have been used. The VGGNet decreases the number of parameters that are tied to large filters [26]. In our work, the VGGNet contains 19 weight layers, which produce results with higher performance compared to other VGGNet models containing less weighted layers [26].

3   Experimental Results

The network models applied in this work were tested on a computer with GeForce GTX 980Ti GPU, 6 GB memory, Intel Core i7-4930 K processor and 16 GB RAM memory. The batch size and epochs have been chosen as 25 and 50 respectively.

Quantitative values obtained from 800 images are shown in Fig. 2. 60% of the 800 images was selected randomly for training, 20% of the total images was used for validation and the remaining 20% of the total images was used for testing. In this work, equal number of images (160 images) have been used for each class to avoid any skew in the process and to balance the data set. These images have been obtained from public image databases [27–29]. It should be noted here that the number of images with dark skin color was not enough. Therefore, they have not been included in this study.

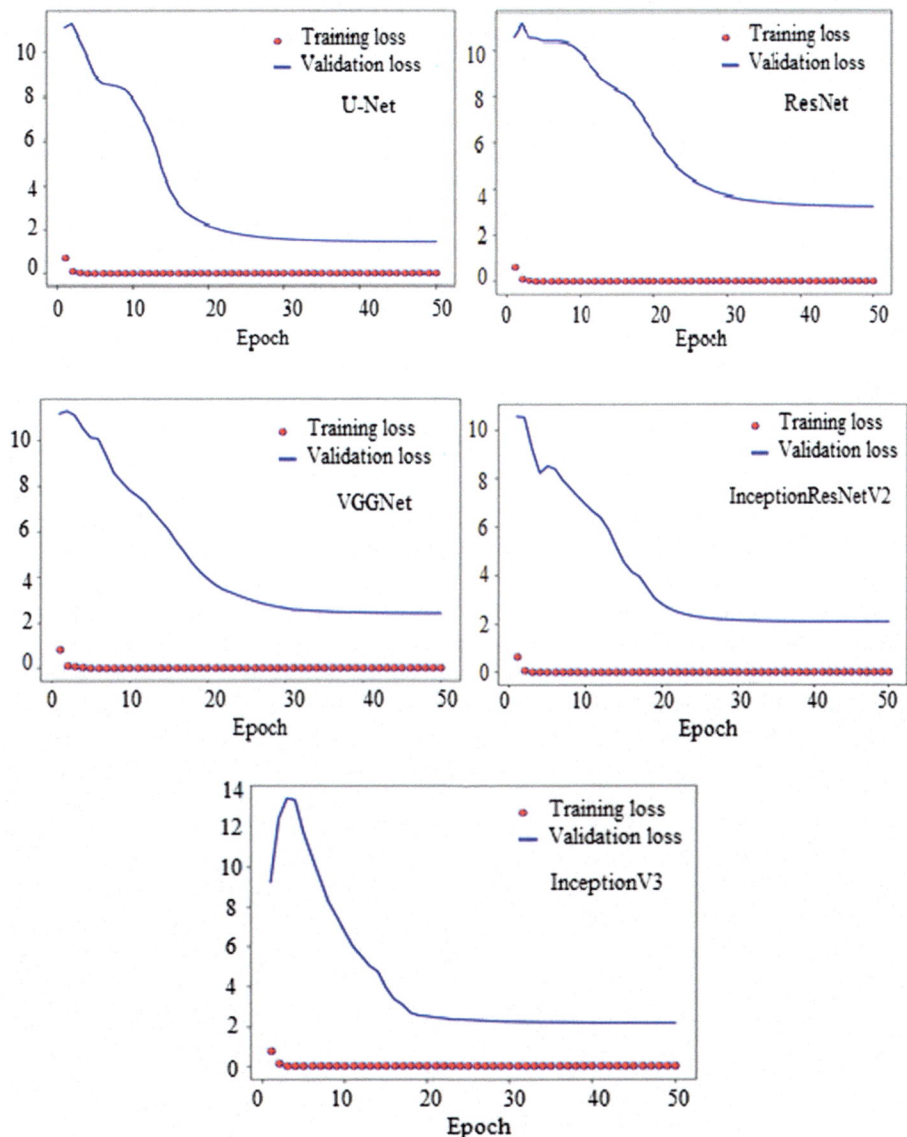

Fig. 2. Training and validation loss

According to our experimental works, validation loss values are always higher than training loss values for each network model. Also, these validation loss values are almost unchanged after about 20 epochs (Fig. 2).

Evaluations of these models have been performed in terms of training time and accuracy. These results obtained with our data sets are presented in Table 1.

Table 1. Training time and accuracy values

Method	Accuracy (%)	Training time (minute)
U-Net	74	90.44
VGGNet	75	72.77
InceptionV3	76	84.93
InceptionResNetV2	78	79.82
ResNet	80	87.68

The accuracy values obtained by the ResNet model for each disease are as follows; Acne vulgaris: 88%, Hemangioma: 75%, Psoriasis: 77%, Rosacea: 79% and Seborrheic dermatitis: 81%. For each class, the number of images used for training, testing and validation is 96,32,32 respectively.

4   Conclusion and Discussion

In this paper, we have performed comparative evaluations of five deep neural network models for automated diagnosis of five common skin diseases from digital photographs. The results obtained by our data sets indicated that the ResNet-50 performs image classification better than the other network models in terms of accuracy.

In the U-Net architecture, a general contracting network is extended with successive layers, where up-sampling operations are applied instead of pooling operations. Therefore, this architecture can concatenate higher resolution features with up-sampled features, while it is increasing the depth of the network. By this way, the resolution of the output is increased. However, to obtain high classification performance by this network, a good prior knowledge from earlier stages should be provided with efficient localization and learning since up-sampling is a sparse operation.

The ResNet-50 architecture can solve vanishing (exploding) gradient problem that inhibits convergence. Spatial resolution in outputs (feature maps) decreases when the network gets deeper and convolutional neural networks are connected fully. Also, local information is ignored when invariant features of spatial transformation are obtained for a classifier. It has been observed from the experimental results that the number of images used in the training stage should be more than 700. Also, image resolution should be about 400 dpi.

In a recent work, a dense neural network and three ResNet models (ResNet18, ResNet50, ResNet52) have been applied for automated diagnosis from photographs [16]. The authors showed that the ResNet152 model produces results with higher accuracy than the others do. Therefore, we will implement the ResNet152 structure and test with our data sets.

When those approaches applied for melanoma diagnosis from dermoscopy images are examined, we see that dense network based models have been implemented recently and their results have been compared with the results obtained by ResNet models. For instance, the authors in [10] developed deep dense deconvolution network and indicated that it is better than the ResNet. Therefore, a deep dense network architecture will also be implemented to see its performance on photographs as an extension of this work. In addition, evaluations of these networks will be performed with more images including more variations.

Acknowledgments. This work has been supported by The Scientific and Technological Research Council of Turkey (TUBITAK-118E777).

References

1. Hay, R.J., Johns, N.E., Williams, H.C., Bolliger, I.W., Dellavalle, R.P., Margolis, D.J., Marks, R., Naldi, L., Weinstock, M.A., Wulf, S.K., Michaud, C., Murray, C., Naghavi, M.: The global burden of skin disease in 2010: an analysis of the prevalence and impact of skin conditions. Invest. Dermatol. **134**, 1527–1534 (2014)
2. American Academy of Dermatology Web Site. Acne Facts and Causes. https://www.aad.org/media-resources/stats-and-facts. Accessed 30 Mar 2019
3. Lim, H.W., Collins, S.A.B., Resneck, J.S., Bolognia, J.L., Hodge, J.A., Rohrer, T.A., Van Beek, M.J., Margolis, D.J., Sober, A.J., Weinstock, M.A., Nerenz, D.R., Smith, W.B., Moyano, J.V.: The burden of skin disease in the United States. J. Am. Acad. Dermatol. **76**, 958–972 (2017)
4. Kittler, H., Pehamberger, H., Wolff, K., Binder, M.: Diagnostic accuracy of dermoscopy. Lancet Oncol. **3**, 159–165 (2002)
5. Carrera, C., Marchetti, M.A., Dusza, S.W., Argenziano, G., Braun, R.P., Halpern, A.C., Jaimes, N., Kittler, H.J., Malvehy, J., Menzies, S.W.: Validity and reliability of dermoscopic criteria used to differentiate nevi from melanoma: a web-based International dermoscopy society study. JAMA Dermatol. **152**, 798–806 (2016)
6. Oliveira, R.B., Mercedes, E.F., Ma, Z., Papa, J.P., Pereira, A.S., Tavares, J.M.R.S.: Computational methods for the image segmentation of pigmented skinlesions: a review. Comput. Methods Programs Biomed. **131**, 127–141 (2016)
7. Oliveira, R.B., Papa, J.P., Pereira, A.S., Tavares, J.M.R.S.: Computational methods for pigmented skin lesion classification in images: review and future trends. Neural Comput. Appl. **29**, 613–636 (2018)
8. Mahbod, A., Ecker, R., Ellinger, I.: Skin lesion classification using hybrid deep neural networks. arXiv preprint 1702.08434, pp. 1–5 (2017)
9. Bissoto, A., Perez, F., Ribeiro, V., Fornaciali, M., Avila, S., Valle, E.: Deep-learning ensembles for skin-lesion segmentation, analysis, classification. RECOD Titans at ISIC Challenge (2018). https://arxiv.org/abs/1808.08480
10. Hea, X., Yua, Z., Wanga, T., Leia, B., Shi, Y.: Dense deconvolution net: multi path fusion and dense deconvolution for high resolution skin lesion segmentation. Tech. Health Care **26**, 307–316 (2018)
11. Fujisawa, Y., Otomo, Y., Ogata, Y., Nakamura, Y., Fujita, R., Ishitsuka, Y., Watanabe, R., Okiyama, N., Ohara, K., Fujimoto, M.: Deep-learning-based, computer-aided classifier developed with a small dataset of clinical images surpasses board-certified dermatologists in skin tumour diagnosis. Gen. Dermatol. **180**(2), 373–381 (2019)

12. Tschandl, P., Rosendahl, C., Akay, B.N., Argenziano, G., Blum, A., Braun, R.P., Cabo, H., Gourhant, J.Y., Kreusch, J., Lallas, A., Lapins, J., Marghoob, A., Menzies, S., Neuber, N. M., Paoli, J., Rabinovitz, H.S., Rinner, C., Scope, A., Soyer, H.P., Sinz, C., Thomas, L., Zalaudek, I., Kittler, H.: Expert-level diagnosis of nonpigmented skin cancer by combined convolutional neural networks. JAMA Dermatol. **155**(1), 58–65 (2019)
13. Vasefi, F., Horita, T., Shi, K., Alhashim, M., MacKinnon, N.: Vanishing point - A smartphone application that classifies acne lesions and estimates prognosis. Int. Conf. on Imaging, Manipulation and Analysis of Bio., Cells, Tissues, California, ABD, pp. 1–6 (2016)
14. Bourouis, A., Zerdazi, A., Feham, M., Bouchachia, A.: M-Health: skin disease analysis system using smartphone's camera. Proc. Comput. Sci. **19**, 1116–1120 (2013)
15. Kim, M.: Development of a portable optical imaging system based on a smartphone and image classification using a learning algorithm. Thesis, Information and Communication Dept., DGIST (Daegu Gteongbuk Institute of Science and Technology), Kore, pp. 1–61 (2017)
16. Mishra, S., Yamasaki, T., Imaizumi, H.: Supervised classification of dermatological diseases via deep learning, pp. 1–6 (2019). arXiv:1802.03752v3
17. Ronneberger, O., Fischer, P., Brox, T.: U-net: Convolutional networks for biomedical image segmentation. In: International Conference on Medical Image Computing and Computer-Assisted Intervention, Munich, Germany, pp. 234–241 (2015)
18. Szegedy, C., Liu, W., Jia, Y., Sermanet, P., Reed, S.E., Anguelov, D., Erhan, D., Vanhoucke, V., Rabinovich, A.: Going deeper with convolutions. In: IEEE Conference on Computer Vision and Pattern Recognition (CVPR), Boston, USA, pp. 1–9 (2015)
19. Ioffe, S., Szegedy, C.: Batch normalization: accelerating deep network training by reducing internal covariate shift. In: Proceedings of The 32nd Int. Conference on Machine Learning, Lille, France, pp. 448–456 (2015)
20. Szegedy, C., Vanhoucke, V., Ioffe, S., Shlens, J., Wojna, Z.: Rethinking the inception architecture for computer vision. In: Conference on Computer Vision and Pattern Recognition (CVPR), Las Vegas, USA, pp. 2818–2826 (2016)
21. He, K., Sun, J.: Convolutional neural networks at constrained time cost. In: IEEE Conference on Computer Vision and Pattern Recognition (CVPR), Boston, USA, pp. 5353–5360 (2015)
22. Srivastava, R.K., Greff, K., Schmidhuber, J.: Highway networks. In: The ICML2015 Deep Learning Workshop, Lille, France, pp. 1–6 (2015)
23. He, K., Zhang, X., Ren, S., Sun, J.: Deep residual learning for image recognition. In: IEEE Conference on Computer Vision and Pattern Recognition (CVPR), Las Vegas, pp. 770–778 (2016)
24. Mahmood, A., Bennamoun, M., An, S., Sohel, F.A.: ResFeats: Residual Network Based Features for Image Classification. In: Conference on Image Processing, Beijing, China, pp. 1597–1601 (2017)
25. Szegedy, C., Ioffe, S., Vanhoucke, V.: Inception-v4, inception-resnet and the impact of residual connections on learning. In: 31st AAAI Conference on Artificial Intelligence, San Francisco, USA, pp. 1–12 (2017)
26. Simonyan, K., Zisserman, A.: Very deep convolutional networks for large-scale image recognition. CoRR, pp. 1–14 (2014)
27. DermWeb photo atlas. http://www.dermweb.com/photo_atlas/. Accessed 28 Mar 2019
28. DermNet image catalogue. https://www.dermnetnz.org/image-catalogue/. Accessed 28 Mar 2019)
29. American Academy of Dermatology. https://www.aad.org. Accessed 28 Mar 2019)

Toward an Affine Feature-Based Registration Method for Ground Glass Lung Nodule Tracking

Yehuda Kfir Ben Zikri[1], María Helguera[1], Nathan D. Cahill[2], David Shrier[3], and Cristian A. Linte[4(✉)]

[1] Chester F. Carlson Center for Imaging Science, Rochester Institute of Technology, Rochester, NY, USA
bz.kfir@gmail.com,mxh3658@cis.rit.edu
[2] School of Mathematical Sciences, Rochester Institute of Technology, Rochester, NY, USA
ndcsma@rit.edu
[3] Division of Radiology, University of Rochester Medical Center, Rochester, NY, USA
David_Shrier@URMC.Rochester.edu
[4] Biomedical Engineering and Chester F. Carlson Center for Imaging Science, Rochester Institute of Technology, Rochester, NY, USA
clinte@mail.rit.edu

Abstract. Lung nodule progression assessment from medical imaging is a critical biomarker for assessing the course of the disease or the patient's response to therapy. CT images are routinely used to identify the location and size and rack the progression of lung nodules. However, nodule segmentation is challenging and prone to error, due to the irregular nodule boundaries, therefore introducing error in the lung nodule quantification process. Here, we describe the development and evaluation of a feature-based affine image registration framework that enables us to register two time point thoracic CT images as a means to account for the back-ground lung tissue deformation, then use digital subtraction images to assess tumor progression/regression. We have demonstrated this method on twelve de-identified patient datasets and showed that the proposed method yielded a better than 1.5 mm registration accuracy vis-à-vis the widely accepted non-rigid image registration techniques. To demonstrate the potential clinical value of our described technique, we conducted a study in which our collaborating clinician was asked to provide an assessment of nodule progression/regression using the digital subtraction images post-registration. This assessment was consistent, yet provided more confidence, than the traditional lung nodule tracking based on visual analysis of the CT images.

Keywords: Lung nodule tracking · Affine feature-based image registration · Background tissue deformation correction · Quantitative nodule tracking

© Springer Nature Switzerland AG 2019
J. M. R. S. Tavares and R. M. Natal Jorge (Eds.): VipIMAGE 2019, LNCVB 34, pp. 247–256, 2019.
https://doi.org/10.1007/978-3-030-32040-9_26

1   Introduction

Thin-slice helical chest CT images can help identify single pulmonary nodules [3] and classify them as either part-solid (also known as sub-solid) or solid nodules [2]. When smaller than 1 cm in diameter, these nodules are typically classified as incidental, benign findings that require follow-up CT [1]. Part-solid nodules feature a "ground-glass appearance", hence commonly referred to as ground-glass opacities (GGO) nodules or ground-glass nodules (GGNs) and are characterized by hazy, increased opacities of the lung tissue that don't completely obscure pulmonary structures. Moreover, pure GGNs feature only ground-glass appearance, with no solid component.

Longitudinal analysis and tracking of nodule progression in current clinical practice resorts to visual comparison between the initial and follow-up scans, as well as the use of diameter measurements, despite their frequent inaccuracy, to quantify growth rate. As such, the radiologist relies on manual one- (1D) or two-dimensional (2D) annotations to quantify the solid portion and the whole nodule size from the axial slices that show the largest lesion diameter and the visual appearance of the margins [6]. Subsequently, the lesion volume and volumetric growth rate such as the doubling time, may be estimated according to the shape approximation of the lesion depicted in the initial and the follow-up scans displayed side-by-side, with no prior registration of the initial and follow-up images.

Although three-dimensional (3D) assessment was suggested to provide more accurate and precise nodule measurements, especially for small nodules [6], volumetric analysis is rarely used in a typical clinical workflow, as it requires segmentation of the nodule. This process is time consuming and highly subjective to intra- and inter-observer variability, especially for GGNs with nodule margins that are often blurry and not easily distinctive. As such, volumetric assessment of nodule progression that relies on either manual or automated nodule segmentation is prone to ambiguity and may mislead diagnosis, hence rendering 2D slice-based analysis as the clinical standard of care for longitudinal lung nodule analysis.

Since GGNs are non-rigid structures that grow in an irregular fashion, their estimated 2D appearance and size may also be highly misleading. Not only is a single CT slice a poor predictor of the geometry, orientation, and volume of a 3D lesion, but the different appearance of the lesion between the initial and the follow-up scan may also be highly influenced by complex deformations of the surrounding tissues between the scans. Several factors may cause significant error in the nodule assessment: intrinsic factors, such as nodule orientation relative to the chest wall or other structures, irregular nodule margins, asymmetric nodule shape and attenuation; or extrinsic factors, including patient position, changes in the parenchyma surrounding the nodules, heart rate, and respiratory motion, which significantly changes lung volume and shape [6,13]. Zheng *et al.* [13] reported that estimates at end-inspiration vs. end-expiration may lead to nodule volume detection error on the order of 12% induced by local deformations alone. Similarly, an additional study [5] reported an overall ±18% mass (volume x HU) and volume measurement fluctuations of part-solid GGNs as a result of inter-scan variability, patient position, heart rhythm, and inspiration levels. These findings suggest that true, disease-induced nodule changes might be reliably detected only if a *significant* change in lesion size occurs [4], and must be separated from both the intrinsic and extrinsic nodule changes.

Current criteria for assessing lung ground glass nodule (GGN) growth rely on visual comparison and diameter measurements from axial slices of initial and follow-up CT images that show the largest extent of the lesion, without any co-registration [6]. Volumetric analysis is rarely used clinically, as it requires segmentation of the nodule, which is highly inaccurate and could mislead diagnosis, since GGN boundaries are often blurry and not easily distinctive for segmentation. Moreover, the nodule appearance between the initial and follow-up scans is also influenced by complex deformations of the surrounding background lung tissue caused by changes in patient position, the parenchyma surrounding the nodules, heart rate, and respiratory motion, all of which significantly change lung volume and shape [13].

To objectively and accurately assess changes in GGN size and shape due to disease, the initial and follow-up CT images must first be co-registered, while accounting for any background lung tissue deformation that may influence the nodule size and geometry, not caused by the disease. To quantify the effect of extrinsic changes, such the background lung tissue deformation, on nodule geometry, and separate it from the disease-induced nodule changes, accurate registration of the initial and follow-up scans is a necessary precursor step [8,12]. Since the lung is a soft tissue organ, a rigid registration of the initial and follow-up scans will not capture the lung tissue deformation adequately.

Although deformable registration may be considered optimal, most deformable registration algorithms are highly dependent on the parameter initialization, are computationally inefficient, and pose a high risk of convergence to local rather than global minima, resulting in unrealistic deformations. As such, depending on the optimization trade-off between the similarity and regularization terms, if the registration is allowed to proceed extensively, the lesion depicted by the registered follow-up image will look similar to the lesion in the initial image, therefore compromising nodule progression assessment. As a result, the difficulty of assessing background lung tissue deformation and using it as a baseline when quantifying the disease-induced lesion changes still exists, and a reliable solution is still pending.

2 Methodology

To address this challenge, here we describe and validate a feature-based affine registration method to co-register the initial and follow-up lung CT images. This registration compensates for the back-ground lung deformation, such that the remaining differences in lesion size and shape attributable to the disease could be identified using a digitally subtracted image post-registration as suggested in [11].

2.1   Imaging Data

This study was conducted on ten pairs of chest CT datasets featuring an initial and follow-up scan. The CT image datasets were acquired on either a 16-slice Lightspeed or 64-slice VCT scanner (GE Medical Systems, Milwaukee, WI). All imaging data was retrieved following retrospective review of patient charts following informed consent granted by all patients, as approved by the Institutional Review Board. The ten datasets contained twelve lesions identified by two radiologists: one patient featured one lesion in each lung (Cases 5 and 6), while another patient featured two lesions in the same lung (Cases 11 and 12), with non-overlapping lesion ROIs. The datasets contained solitary GGNs, multiple GGNs, and solid nodules.

All scans were acquired in a single breath-hold with sub-millimeter resolution and featured lesions localized in different regions of the lung that showed different extent of progression as evaluated at different stages ranging from 0.5 months to almost 3 years. As an example, Case 9 featured a part-solid GGN with a large solid part attached to the chest wall, classified by both radiologists as a malignant juxtapleural nodule, justified by the large change detected at 32 month follow-up. Moreover, Case 2 featured a common transient part-solid nodule that showed almost complete regression in the follow-up scan. Lastly, Case 3 featured a new nodule in the follow-up scan that was not visible in the initial scan.

An expert radiologist manually segmented the nodules and also selected approximately 40 homologous fiducial landmarks on average, localized both within and in the vicinity of each lesion in both the initial and follow-up scans. Most landmarks were selected at vasculature branching points. According to the radiologists, these locations are least susceptible to non-rigid lung deformation due to breathing. The landmarks were used to quantify the TRE following each step of the registration pipeline.

2.2   Feature-Based Segmentation Overview

Following automatic segmentation and separation of the lungs from both the initial and follow-up CT scans, the registration was initialized by a centroid alignment of the lung- or lesion-centered region of interest (ROI), followed by a feature-based rigid registration. We used an approach that minimizes the chi-squared statistic between the histograms of the initial and follow-up images to identify the optimal initial rotational transform. The subsequent feature-based registration method (Fig. 1) followed a modified formulation and implementation of the iterative closest point (ICP) algorithm.

The features used for registration were edges extracted from both the initial and follow-up CT scans using monogenic filtering [9]. Unlike the ICP algorithm in which the objective function minimizes the distance between estimated corresponding points using the closed-form solution of the Euclidean distance, we built a distance map [7] of the initial image by assigning each voxel a value equal to its distance from the closest edge.

We then multiplied the distance map of the initial edge image by the transformed follow-up edge image, then computed the sum of the distances from the transformed edges to the edges in the initial image, which served as the objective function to be minimized. Moreover, we separated the edges of the lung boundaries from the edges of the lung content and defined the Energy Dissimilarity (ED) function as the weighted sum of the distances between lung boundary edges and the lung content edges from the initial and follow-up scans. To identify the optimal weighting factor alpha, we evaluated the similarity metric (normalized cross-correlation (NCC)) for several values of alpha and selected the parameter value that yielded the highest NCC. The registration was implemented on both a lung-centered and a lesion-centered ROI.

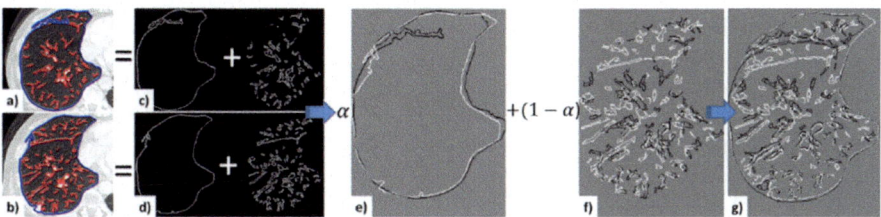

Fig. 1. Schematic of feature-based registration algorithm and computation of the similarity metric as a weighted sum between the distance map of the lung boundary and lung content.

2.3   Intensity-Based Deformable Registration

The result of the affine feature-based registration served as input for the last step of the registration pipeline—the deformable registration. Moreover, in the effort to quantify any potential shortcomings of the affine registration or improvements provided by an additional deformable registration step, the intensity-based deformable registration also served as reference against which the affine feature-based registration was assessed. We used a high-degree of freedom diffeomorphic transformation method proposed in [10], with several pre-processing steps recommended by the authors, including the rescaling and truncation of the image intensities, followed by de-noising. This technique is referred to as the Greedy Symmetric Normalization (SyN) registration and is based on the LDDMM algorithm available via the ANTS (Advanced Normalization Tools) open source package integrated within ITK. This registration is initiated via a global affine registration on the lungs masks, rather than on the CT data. The mutual information (MI) similarity measure is used on four-level image pyramid together with gradient descent successive optimization. The successive optimization objective is controlling the affine registration parameters successively.

3   Results

We tested the developed registration algorithm on twelve image datasets consisting of initial and follow images of patients at various stage of disease. To assess registration accuracy, we computed the target registration error (TRE) across a set of 30–50 homologous fiducial markers selected by a collaborating radiologist in both the initial and follow-up scans.

We compared the TRE achieved using the proposed registration to the TRE achieved using the ANTS Symmetric Normalization deformable registration method deemed optimal by the EMPIRE10 lung registration challenge [10]. In addition, we used the residual fiducial registration error (FRE) across the homologous fiducial landmark datasets selected from the initial and follow-up images as a registration accuracy baseline control, as it assesses the homology of the landmarks based on a rigid least squared fit registration.

Table 1. Target registration error (TRE in mm) across 12 patient datasets for affine, deformable, and baseline registration performed on the lung-centered ROI.

Registration method	Mean ± Std. Dev. TRE (mm)	Median TRE (mm)
Affine registration	1.8 ± 1.6	1.4
Deformable registration	1.2 ± 1.2	0.8
Least square fit FRE (Control)	1.6 ± 1.1	1.3

Our accuracy study showed differences on the order of 0.5 mm between the TRE achieved using feature-based affine vs. deformable registration (Tables 1, 2 and Fig. 2), however with significant computing performance improvements (Tables 3 and 4) using the feature-based affine registration.

Table 2. Target registration error (TRE in mm) across 12 patient datasets for affine, deformable, and baseline registration performed on the lesion-centered ROI.

Registration method	Mean ± Std. Dev. TRE (mm)	Median TRE (mm)
Affine registration	1.5 ± 1.2	1.1
Deformable registration	1.2 ± 1.2	0.8
Least square fit FRE (Control)	1.5 ± 1.2	1.2

To visualize the registration results, we generated subtraction images i.e., initial (fixed) image minus the registered follow-up (moving) image from the lesion

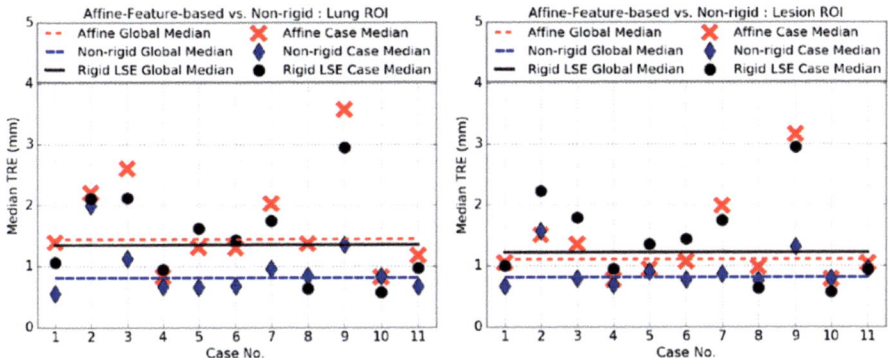

Fig. 2. Target registration error (TRE) comparison across 12 patient datasets between the feature-based affine registration, deformable registration, and baseline control (rigid least squared fit) registration for both the lung-centered ROI and lesion-centered ROI.

Table 3. Performance (Mean ± Std. Dev. in minutes) comparison of affine and deformable intensity vs. feature-based registration performed on lung-centered ROI ($357 \times 248 \times 455$).

Registration method	Intensity-based registration (mins)	Feature-based registration (mins)
Affine registration	119.0 ± 10.3	8.0 ± 10.0
Deformable registration	245.0 ± 40.1	N/A

Table 4. Performance (minutes) of rigid, affine, and deformable intensity- and feature-based registration performed on lesion-centered ROI.

Registration method	Intensity-based registration (mins)	Feature-based registration (mins)
Affine registration	1.3 ± 2.3	0.7 ± 1.5
Deformable registration	17.2 ± 10.2	N/A

mid-slices after each registration. As mentioned earlier, subtraction images were shown to be efficient when assessing nodule growth-rate and reducing variability according to [11]. As shown in Fig. 3, both the affine and deformable registrations yielded similar visual results, with no significant visual differences.

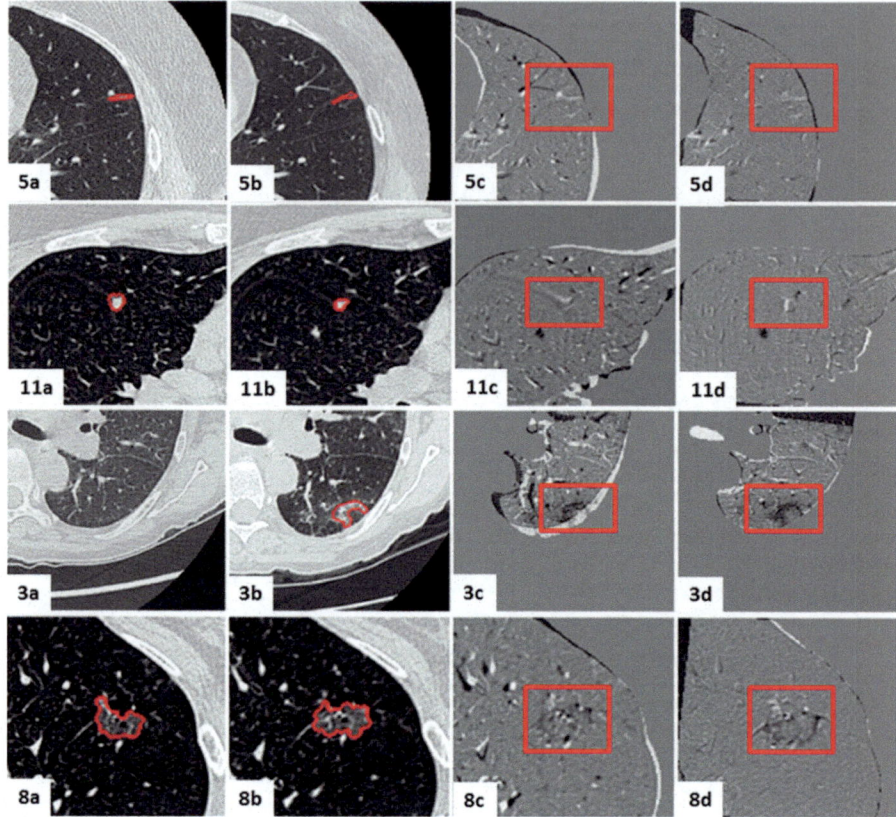

Fig. 3. Visual assessment of nodule changes: (a) initial image; (b) follow-up image; (c) digital subtraction image after centroid alignment; (d) digital subtraction image after feature-based affine registration. Four cases are showcased here: Case 5—no nodule change; Case 11—moderate nodule change; Case 3—new nodule appearing in follow-up image; and Case 8—severe nodule change.

4   Discussion and Conclusion

We described and validated a feature-based affine registration method designed to co-register initial and follow-up lung CT images to correct for the background lung tissue deformation to help objectively assess lung nodule changes induced by disease. Our study showed less than 1 mm difference between the registration accuracy achieved using the feature-based affine registration, deformable registration, and baseline control registration, with significant performance improvement when using the affine registration. Moreover, the qualitative visual assessment conducted by two radiologists confirmed similar conclusions about the nodule changes using the digital subtraction image post-registration as the current standard of care assessment method.

Acknowledgements. Research reported in this publication was supported by the National Institute of General Medical Sciences of the National Institutes of Health under Award No. R35GM128877 and by the Office of Advanced Cyber infrastructure of the National Science Foundation under Award No. 1808530.

References

1. Fischbach, F., Knollmann, F., Griesshaber, V., Freund, T., Akkol, E., Felix, R.: Detection of pulmonary nodules by multislice computed tomography: improved detection rate with reduced slice thickness. Eur. Radiol. **13**, 2378–2383 (2003)
2. Hansell, D.M., Bankier, A., MacMahon, H., McLoud, T.C., Müller, N.L., Remy, J.: Fleischner society: glossary of terms for thoracic imaging. Radiology **246**(3), 697–722 (2008)
3. Henschke, C.I., McCauley, D.I., Yankelevitz, D.F., Naidich, D.P., McGuinness, G., Miettinen, O.S., Libby, D.M., Pasmantier, M.W., Koizumi, J., Altorki, N.K., Smith, J.P.: Early lung cancer action project: overall design and findings from baseline screening. Lancet **354**, 99–105 (1999)
4. Kakinuma, R., Ashizawa, K., Kuriyama, K., Fukushima, A., Ishikawa, H., Kamiya, H., Koizumi, N., Maruyama, Y., Minami, K., Nitta, N., Oda, S., Oshiro, Y., Kusumoto, M., Murayama, S., Murata, K., Muramatsu, Y., Moriyama, N.: Measurement of focal ground-glass opacity diameters on CT images. Interobserver agreement in regard to identifying increases in the size of ground-glass opacities. Acad. Radiol. **19**, 389–394 (2012)
5. Kim, H., Park, C.M., Lee, S.M., Lee, H.J.: Measurement variability of volume and mass in nodules with a solid portion less than or equal to 5 mm. Radiology **269**, 585–593 (2013)
6. Ko, J.P., Berman, E.J., Kaur, M., Babb, J.S., Bomsztyk, E., Greenberg, A.K., Naidich, D.P., Rusinek, H.: Pulmonary nodules: growth rate assessment in patients by using serial CT and three-dimensional volumetry. Radiology **262**, 662–671 (2012)
7. Maurer, C.R., Qi, R., Raghavan, V.: A linear time algorithm for computing exact euclidean distance transforms of binary images in arbitrary dimensions. IEEE Transact. Pattern Anal. Mach. Intell. **25**(2), 265–270 (2003)
8. Oliveira, F.P.M., Tavares, J.M.R.S.: Medical image registration: a review. Comput. Methods Biomech. Biomed. Eng. Imaging Vis. **17**, 73–93 (2014)
9. Rajpoot, K., Grau, V., Noble, J.A.: Local-phase based 3-D boundary detection using MONOGENIC signal and its application to real-time 3-D echocardiography images. In: Proceedings of IEEE International Symposium on Biomedical Imaging: From Nano to Macro 2009, pp. 783–786 (2009)
10. Song, G., Tustison, N., Avants, B., Gee, J.C.: Lung CT image registration using diffeomorphic transformation models. In: Medical Image Analysis for the Clinic: A Grand Challenge, pp. 23–32 (2010)
11. Staring, M., Pluim, J.P.W., de Hoop, B., Klein, S., van Ginneken, B., Gietema, H., Nossent, G., Schaefer-Prokop, C., van de Vorst, S., Prokop, M.: Image subtraction facilitates assessment of volume and density change in ground-glass opacities in chest CT. Invest. Radiol. **44**, 61–66 (2009)

12. Viergever, M.A., Maintz, J.B., Klein, S., Murphy, K., Staring, M., Pluim, J.P.W.:
    A survey of medical image registration – under review. Med. Image Anal. **33**,
    140–144 (2016)
13. Zheng, Y., Kambhamettu, C., Bauer, T., Steiner, K.: Accurate estimation of pul-
    monary nodule's growth rate in CT images with nonrigid registration and pre-
    cise nodule detection and segmentation. In: 2009 IEEE Computer Society Confer-
    ence on Computer Vision and Pattern Recognition Workshops, pp. 101–108. IEEE
    (2009)

Computer Aided Effective Prediction of Complete Responders After Radiochemotherapy Based on Tumor Regression Grade Estimated by MR Imaging

Chiara Losquadro[1], Silvia Conforto[1], Maurizio Schmid[1],
Gaetano Giunta[1(✉)], Marco Rengo[2], Damiano Caruso[2],
and Andrea Laghi[2]

[1] Department of Engineering, Applied Electronics Section,
Roma Tre University, Rome, Italy
{chiara.losquadro, silvia.conforto, maurizio.schmid,
gaetano.giunta}@uniroma3.it
[2] Department of Radiologic, Oncologic, and Anatomo-Pathologic Sciences,
La Sapienza University, Rome, Italy
{marco.rengo, damiano.caruso, andrea.laghi}@uniroma1.it

Abstract. The aim of this work is to implement an automatic method to predict and classify complete responders (CRs) patients, affected by rectal cancer and treated with neoadjuvant radiochemotherapy (RCT), by exploiting the tumor regression grade (MR-TRG) estimated by magnetic resonance imaging. For the purpose of the study, a total of 65 patients were enrolled and the magnetic resonance (MR) examinations to calculate TRG were performed using a 3.0 T scanner. By processing and testing patients' data, the algorithm allows to determine the optimum threshold dividing CRs patients from patients that are considered non responders. The prediction accuracy of the classifier was investigated by using cross-validation statistical analysis in order to automatically determine the best testing rule. After collecting the outcomes of the performed cross-validation, the obtained results show the percentages of correct instances and misclassified patients. The automatic classification of CRs appears to be feasible and can be considered as a helpful method to predict CRs assisting clinicians to predict disease prognoses and patient survival prospects in order to provide treatments' customization.

Keywords: Automatic prediction · Rectal cancer · Magnetic resonance imaging · Computer-aided prognosis · Automatic classification

1 Introduction

Rectal cancer is one of the most common diseases in Western countries with its high rate of mortality: during the last ten years, colorectal cancer was recognized as the third most commonly diagnosed cancer in men and the second in women. This trend is due to the increasing exposure to risk factors for colorectal cancer, including unhealthy diet, obesity, and smoking [1].

© Springer Nature Switzerland AG 2019
J. M. R. S. Tavares and R. M. Natal Jorge (Eds.): VipIMAGE 2019, LNCVB 34, pp. 257–266, 2019.
https://doi.org/10.1007/978-3-030-32040-9_27

Colon/rectum or colorectal is considered as a fundamental part of the gastrointestinal (GI) or digestive system: the colon is also called large intestine and it starts from small intestine connecting the rectum and it is involved to perform important functions (e.g. water, mineral and salts absorption, excretion of refusal substances) [2, 3].

Magnetic Resonance Imaging (MRI) is considered the most accurate imaging modality to detect and stage rectal cancer diseases [4]: with its high resolution, it has the ability to distinguish normal rectal tissue from tumor tissue exploiting signal intensity differences based on T2-weighted sequences [5]. For such a purpose, feature extraction could be also used [6, 7].

The provided treatments for rectal cancer cure may include surgical, radiologic, and oncologic treatment; the treatment of choice in patients with rectal cancer is considered the radiation therapy also called neoadjuvant radiotherapy [8].

In particular, recently published studies [9] underlined and demonstrated that preoperative radiotherapy is superior to postoperative therapy and then it has increasingly been used in the management of patients.

The goal of this neoadjuvant radiotherapy, with or without chemotherapy, is to act on the tumor before the surgery treatment and it is significantly associated with less local recurrence and it improves long-term survival [10, 11]; furthermore, it ensures less invasive surgery, because it allows to reduce the tumor size and the depth of tumor invasion, inducing a development of fibrosis [9]. The reduction of tumor may lead to a complete clinical response with the consequent absence of residual primary tumor clinically detectable or to a complete pathologic response defined as the absence of viable tumor cells after a complete pathologic examination of the resected specimen: these situations can be observed in 10% to 30% of patients that have been treated by neoadjuvant radiochemotherapy (RCT) [12].

Patients who obtained a complete response were treated with different clinical treatments including a less invasive surgery or wait-and-see approaches; on the other hand, for patients which don't demonstrate a complete response, surgery may be mandatory.

To follow-up the impact of RCT on patients, the use of MRI is required. It allows to restage the tumor after the therapy, but it is difficult to distinguish fibrosis from residual tumor because of the similarity of T2 signal intensity [13].

The assessment of response to therapy is performed using an important predictor of patient's outcome [14–17], the histopathological tumor regression grade (TRG), defined as the ratio between fibrosis and residual tumor.

According to the above-mentioned reasons, identification of complete responders (CRs) after RCT is very important: in CRs patients, a treatment customization may be provided, surgery may be deferred and an active surveillance plan can be realized [18, 19].

The aim of our study was to propose, analyze, optimize, and implement an algorithm to effectively classify patients as CRs or non-responders (NRs), taking advantages of the investigated MR-TRG, including its assessment on prediction accuracy.

2   Materials and Methods

2.1   Patients

The study involved a total of 65 patients with histologically-confirmed rectal adeno-carcinoma; with a view to be enrolled in the study, all patients signed a written informed consent. The characteristics of patient's population are summarized in Table 1 as reported in a previous study [20]. The involved patients received biopsy for histologic analysis and an MRI study to stage rectal cancer disease. After two weeks of MRI staging, patients started the neoadjuvant RCT treatment and a repeated MR study was performed after RCT to restage the tumor one week before the surgery. From the subsequent follow-up, we know that, over the 65 considered patients, 23 patients were considered ex-post as CRs, while 42 patients as NRs.

Table 1. Summarized characteristics of patients enrolled in the study

Characteristics	Total	CRs	NRs
Sex			
Male	44 (66.7%)	14 (60.9%)	29 (69.1%)
Female	21 (33.3%)	9 (39.1%)	13 (30.9%)
Age y ($\pm$ SD)	64.8 ($\pm$ 8.43)	62.5 ($\pm$ 6.7)	65.3 ($\pm$ 9.2)
T stage			
2	1 (1.5%)	1 (4.3%)	1 (2.3%)
3	56 (86.2%)	19 (82.6%)	36 (85.7%)
4	8 (12.3%)	3 (13.1%)	5 (12%)
N stage			
0	12 (18.4%)	5 (21.7%)	8 (19.1%)
1	26 (40%)	6 (26.1%)	18 (42.8%)
2	27 (41.6%)	12 (52.2%)	16 (38.1%)
Overall stage			
II	12 (18.4%)	5 (21.7%)	8 (19.1%)
IIIA	1 (1.5%)	0 (0%)	1 (2.3%)
IIIB	27 (41.6%)	7 (30.4%)	19 (45.3%)
IIIC	25 (38.5%)	11 (47.9%)	14 (33.3%)
Tumor dimension			
$\leq$ 5 cm	43 (66.2%)	18 (78.3%)	26 (61.9%)
< 5 cm	22 (33.8%)	5 (21.7%)	16 (38.1%)
Distance from anal verge			
$\leq$ 5 cm	37 (56.9%)	14 (60.9%)	23 (54.7%)
> 5 $\leq$ 8 cm	15 (23.1%)	2 (8.7%)	11 (26.2%)
> 8 cm	13 (20%)	7 (30.4%)	8 (19.1%)

Data taken from [20].

2.2   MR-TRG Computation

For the purpose of this study, we considered TRG estimated from MRI, i.e. the MR-TRG. All the MR examinations were performed using a 3.0 T scanner (Discovery MR750, General Electrics, Milwaukee, Wisconsin, USA) provided with a phased-array coil and a slice thickness of 4 mm; MR-TRG was extracted analyzing T2-weighted images acquired after neoadjuvant RCT as described in the previous study [20].

In order to calculate MR-TRG, an automatic quantification of fibrosis was performed using a K-means algorithm [21]; before the automatic analysis the entire tumor volume was manually contouring using T2-weighted images and the total tumor volume in cubic millimeters was provided. The dataset of the contoured tumor was processed with the K-means algorithm, so pixels of MR images were clustered into two partitions on the basis of their median signal intensity, where high signal intensity represents the residual tumor, while low signal intensity represents fibrosis as shown in Fig. 1.

The algorithm provided two outputs, obtained from all the disease image slices of the tumor region, namely the volume of low signal intensity pixels identified as fibrosis, and the volume of high signal intensity pixels identified as residual tumor, expressed in cubic millimeters. Then the percentage of fibrosis voxels (integrating all the pixels of each image slice) was calculated as (Fibrosis volume/Tumor volume) × 100 and on the basis of this percentage, just representing the MR-TRG value of the examined patient.

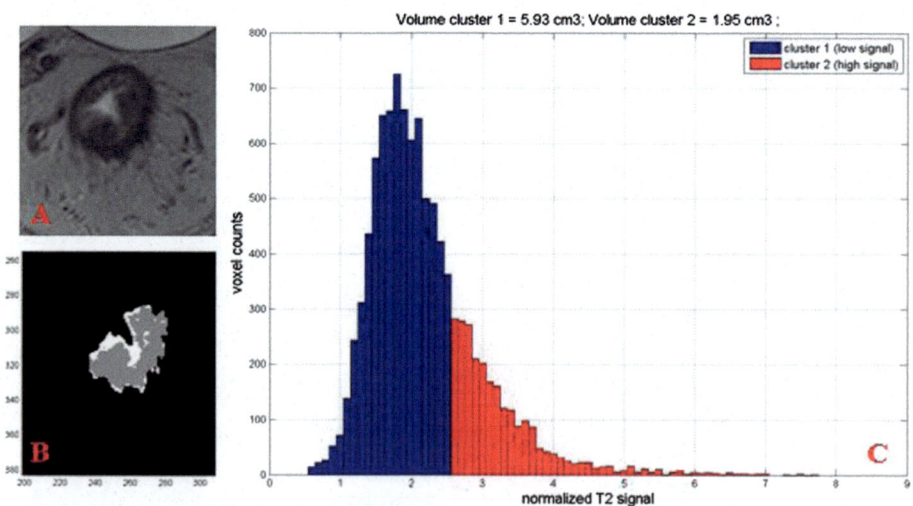

Fig. 1. (A) a native T2-weighted image; (B) Tumor image extracted and clustered: grey pixels represent the low signal intensity, identifying fibrosis and the white ones the high signal intensity, identifying residual tumor; (C) Histogram of the distribution of pixels considering signal intensity. Integrating all the tumor slices, the volume clusters are finally computed.

2.3   Prognostic Prediction of Complete Responders

The algorithm to process and test the data was implemented and developed in MATLAB software (The MathWorks Inc., Natick, Massachusetts, United States) and a schema of its implementation is reported in Fig. 2.

The purpose of this study is to develop a classifier that allows to correctly predict and identify CRs by separating them from NRs.

The data of the 65 patients involved in our study were collected and, giving the importance of MR-TRG, it was chosen as the most informative predictor. MR-TRG after the RCT treatment was considered. A value of MR-TRG higher than a given threshold likely denotes CRs, while lower values are representative of NRs.

One goal of this work is just to set a proper threshold to decide whether the patients have to be prognostically considered as responders or not.

The threshold decision about the prognostic value of involved patients could be suffering from the period after the therapy: patients were followed up for thirty months, with physical examination, blood tests, and yearly whole body computed tomography at intervals of three months to prevent the presence of local recurrences or distant metastases. During the follow-up period, a patient classified as NR could successfully turn into CR, due to the therapy effects over time; conversely, a patient classified as CR may not turn into a NR reversing his healthy conditions. The results presented in [20] are derived from one unique set for which the same patients are used for training (to set the threshold) and validation (to evaluate testing performance). Conversely, it is widely recognized that to perform an effective test from a statistical viewpoint, the patients should be partitioned into two *non-overlapping* subsets, namely one training set and one (separated) validation set. The training set is used to automatically detect the MR-TRG threshold which correctly classify the CRs, by reducing misclassified patients. As the following step, the validation set is used to test the model and then the numbers of correct and incorrect predictions are calculated. The results represent the so-called confusion matrix, that shows the correct and wrong decisions, often normalized to the number of patients of the validation set (test size).

3   Results and Discussion on Data Processing and Validation

3.1   Test Optimization and Statistical Validation

Two partitions of complementary patients' subsets are respectively used for training and validation steps. The training set of patients (with known follow-up) allows to set the threshold of the MR-TRG value as the one that minimizes the amount of errors of wrong classification (i.e. actual CR classified as NR, or actual NR classified as CR) or, equivalently, maximizes the number of correct decisions (i.e. actual CR classified as CR, or actual NR classified as NR).

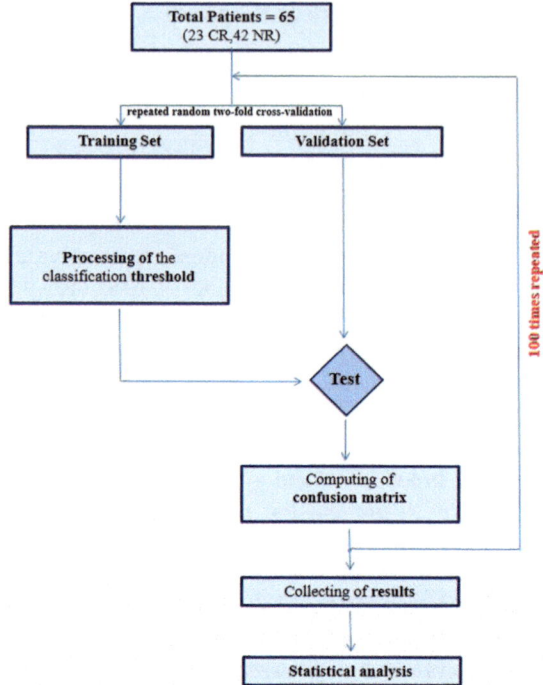

Fig. 2. The validation process of the implemented algorithm

In our attempts, 30 patients have been randomly chosen as the training set among all the 65 patients. Indeed, in order to avoid a systematic bias due to the fact that the total number of true CRs are about one third of the total (according to ex-post statistics), the 30 patients have been randomly chosen as 10 actual CRs and 20 actual NRs.

As a consequence, the validation set has consisted of the remaining 35 patients. The performance of the test, whose threshold is fixed according to the 30 training patients, can be therefore validated by matching 'the decision obtained with the pre-fixed threshold and the knowledge of the follow-up of the remaining 35 patients.

In order to enhance the statistical significance of validation, a repeated random two-fold cross-validation [22] has been implemented. The two-fold cross-validation has allowed to split the dataset into two partitions: the two-fold has been chosen such that both training set and validation set were large enough to be statistically representative. For the purpose of this study, the usually chosen 5-fold cross-validation or 10-fold cross-validation would not perform appropriate results because the dataset would split within too small partitions to estimate significant statistics.

In practice, as depicted in Fig. 2, we have reiterated many times the training and validation procedures by independent random extractions of 30 training patients (always 10 true CRs and 20 true NRs), used to fix new MR-TRG thresholds for each run, while the remaining 35 patients have been used as further validation trials. In each iteration, of course, the set of 35 patients is validated by using the threshold fixed according to the complementary 30 patients. Note that the thresholds may numerically differ in each iteration.

The process has been repeated 100 times because 100 cycles have been proven to be enough in our attempts to reach a reasonable convergence of estimated error percentages; in addition, using 100 different independent partitions, the equivalent number of instances increases and it is considered very useful in a medical domain, where the patients' number is often restricted.

The number of correct and wrong predictions was collected in a confusion matrix that takes all 100 runs into account.

After collecting the results of all the 100 iterations, a statistical analysis has been performed, based on an equivalent set of 3500 (= 35•100) validation patients.

Previous studies have approached the problem of classification, in particular van Stiphout et al. [23] developed and validated an accurate predictive model for pathologic complete response after chemoradiotherapy for rectal cancer, based on PET-CT data. For the aim of their study, a binary classification from the machine learning field was used: the support vector machine (SVM). Similarly to our study, van Stiphout et al. validated the model using a validation group in order to estimate the accuracy of the predictions. A significant difference from our work is the dataset dimension: our dataset is very limited. An additional study [24] also aimed to predict the response to the treatment after neo-adjuvant radiochemotherapy of rectal cancer: the work used the early tumor regression grade incorporating it on a radiobiologically based predictive index to apply it to the patients likewise considering tumor volume. The basis of our work concerns a previous study from Rengo et al. [20]: they evaluated an automatic method to discriminate between favorable and unfavorable pathologic response groups. A limitation of the work is the relatively small study population and that they didn't validate the method on a control group, conversely to our explained work.

3.2 Numerical Validation of Results

The performance of the classification algorithm was evaluated by calculating and adding all the confusion matrices, in which the correct and wrong decisions of the predictor clearly appear. The number of the correct predictions and the number of incorrect predictions has been extracted and expressed as median and as mean ± standard deviation (SD).

The overall percentage error of the classification is determined as:

$$100 \bullet \sum \text{incorrect predictions}/(\sum \text{incorrect predictions} + \sum \text{correct predictions}).$$

The results show that the total correct instances are 3062 over 3500. Instead, the total incorrect instances are 438 over 3500. The total percentage error of the classification was also determined, and it results to be the 12,5% of cases. The numerical results are summarized in Table 2.

Table 2. Results of statistical analysis over (repeated) 35 patients and percentages

	Average	Median	Standard deviation
Correct predictions	30.62 (87.5%)	31 (88.5%)	±1.55 (4.4%)
Wrong predictions	4.38 (12.5%)	4 (11.5%)	±1.55 (4.4%)

4 Conclusion

This study provides useful information about the utility of computer-aided prediction proving to be a helpful support for clinicians predicting patient's disease outcome.

Classifier's building for predicting treatment response can lead to improve personalized treatment, to defer surgery or to realize surveillance plans for the benefit of patients. The results encourage a deeper investigation regarding the identification of multiple clinical and histopathologic factors that are relevant in improving classification accuracy to easier determine treatments after RCT.

In this paper, we have presented, optimized, and statistically analyzed a computer-aided method to predict and classify patients, affected by rectal cancer and treated with neoadjuvant RCT, as complete responders, by exploiting the MR-TRG index estimated by magnetic resonance imaging. The accomplished analysis has shown the encouraging results that allow to assist clinicians to early identify CRs with a reasonable classification accuracy. Further developments in this area will come about interactions between computer and imaging scientists, clinicians, oncologists, radiologists, and pathologists. In particular, the method presented in this paper could also be improved by considering more parameters and imaging feature in addition to MR-TRG (such as the effect of tumor volume [24]) and extending the study to a bigger population.

Acknowledgments. Medical images were acquired at La Sapienza University Hospital in Rome (Italy) with the consent for research purposes of patients.

References

1. Torre, L.A., Bray, F., Siegel, R.L., Ferlay, J., Lortet-Tieulent, J., Jemal, A.: Global cancer statistics. CA Cancer J. Clin. **65**(2), 87–108 (2015)
2. Maier, A., Fuchsjager, M.: Preoperative staging of rectal cancer. Eur. J. Radiol. **47**(2), 89–97 (2003)
3. Ashiya: Notes on the Structure and Functions of Large Intestine of Human Body (2013). http://www.preservearticles.com/notes/notes-on-the-structure-and-functions-of-large-intestine-of-human-body/5312
4. Krestin, G.P., Steinbrich, W., Friedmann, G.: Recurrent rectal cancer: diagnosis with MR imaging versus CT. Radiology **168**(2), 307–311 (1988)
5. de Lange, E.E., Fechner, R.E., Edge, S.B., Spaulding, C.A.: Rectal carcinoma treated by preoperative irradiation: MR imaging and histopathologic correlation. Am. J. Roentgenol. **158**(2), 287–292 (1992)

6. Soomro, M.H., Giunta, G., Laghi, A., Caruso, D., Ciolini, M., De Marchis, C., et al: Haralick's texture analysis applied to colorectal T2-weighted MRI: a preliminary study of significance for cancer evolution. In: Proceedings of the 13th IASTED International Conference Biomedical Engineering, pp. 16–19 (2017)
7. Caruso, D., Zerunian, M., Ciolina, M., De Santis, D., Rengo, M., Soomro, M.H., et al.: Haralick's texture features for the prediction of response to therapy in colorectal cancer: a preliminary study. Radiol. Med. (Torino) **123**(3), 161–167 (2018)
8. Iafrate, F., Laghi, A., Paolantonio, P., Rengo, M., Mercantini, P., Ferri, M., et al.: Preoperative staging of rectal cancer with MR imaging: correlation with surgical and histopathologic findings. Radiographics **26**, 701–714 (2006)
9. Sauer, R., Becker, H., Hohenberger, W., Rödel, C., Wittekind, C., Fietkau, R., et al.: Preoperative versus postoperative chemoradiotherapy for rectal cancer. N. Engl. J. Med. **351**, 1731–1740 (2004)
10. Kapiteijn, E., Marijnen, C.A., Nagtegaal, I.D., Putter, H., Steup, W.H., Wiggers, T., et al.: Preoperative radiotherapy combined with total mesorectal excision for resectable rectal cancer. N. Engl. J. Med. **345**(9), 638–646 (2001)
11. Cedermark, B., Dahlberg, M., Glimelius, B., Påhlman, L., Rutqvist, L.E., Wilking, N.: Improved survival with preoperative radiotherapy in resectable rectal cancer. Swedish rectal cancer trial. N. Engl. J. Med. **336**, 980–987 (1997)
12. Habr-Gama, A., Perez, R.O., Nadalin, W., Sabbaga, J., Ribeiro Jr., U., Silva e Sousa Jr., A. H., et al.: Operative vs nonoperative treatment for stage 0 distal rectal cancer following chemoradiation therapy: long term results. Ann. Surg. **240**(4), 711–717 (2004)
13. Beets-Tan, R.G., Lambregts, D.M., Maas, M., Bipat, S., Barbaro, B., Caseiro-Alves, F., Curvo-Semedo, L., et al.: Magnetic resonance imaging for the clinical management of rectal cancer patients: recommendations from the 2012 European Society of Gastrointestinal and Abdominal Radiology (ESGAR) consensus meeting. Eur. Radiol. **23**(9), 2522–2531 (2013)
14. Abdul-Jalil, K.I., Sheehan, K.M., Kehoe, J., Cummins, R., O'Grady, A., McNamara, D.A., et al.: The prognostic value of tumour regression grade following neoadjuvant chemoradiation therapy for rectal cancer. Colorectal Dis. **16**(1), O16–O25 (2014)
15. Park, Y.J., Oh, B.R., Lim, S.W., Huh, J.W., Joo, J.K., Kim, Y.J., et al.: Clinical significance of tumor regression grade in rectal cancer with preoperative chemoradiotherapy. J. Korean Soc. Coloproctol. **26**(4), 279–286 (2010)
16. Rodel, C., Martus, P., Papadoupolos, T., Fuzesi, L., Klimpfinger, M., Fietkau, R., et al.: Prognostic significance of tumor regression after preoperative chemoradiotherapy for rectal cancer. J. Clin. Oncol. **23**(34), 8688–8696 (2005)
17. Suarez, J., Vera, R., Balen, E., Gomez, M., Arias, F., Lera, J.M., et al.: Pathologic response assessed by Mandard grade is a better prognostic factor than down staging for disease-free survival after preoperative radiochemotherapy for advanced rectal cancer. Colorectal Dis. **10**(6), 563–568 (2008)
18. Nahas, S.C., Rizkallah Nahas, C.S., Sparapan Marques, C.F., Ribeiro Jr., U., Cotti, G.C., Imperiale, A.R., et al.: Pathologic complete response in rectal cancer: can we detect it? Lessons learned from a proposed randomized trial of watch-and-wait treatment of rectal cancer. Dis. Colon Rectum **59**(4), 255–263 (2016)
19. Habr-Gama, A., Sabbaga, J., Gama-Rodrigues, J., Sao Juliao, G.P., Proscurshim, I., Bailao Aguilar, P., et al.: Watch and wait approach following extended neoadjuvant chemoradiation for distal rectal cancer: are we getting closer to anal cancer management? Dis. Colon Rectum **56**(10), 1109–1117 (2013)

20. Rengo, M., Picchia, S., Marzi, S., Bellini, D., Caruso, D., Caterino, M., et al.: Magnetic resonance tumor regression grade (MR-TRG) to assess pathological complete response following neoadjuvant radiochemotherapy in locally advanced rectal cancer. Oncotarget **8** (70), 114746–114755 (2017)
21. Soomro, M.H., Giunta, G., Laghi, A., Caruso, D., Ciolina, M., De Marchis, C., et al.: Segmenting MR images by level-set algorithms for perspective colorectal cancer diagnosis. In: European Congress on Computational Methods in Applied Sciences and Engineering, pp. 396–406. Springer, Cham (2017)
22. Kuhn, M., Johnson, K.: Applied Predictive Modeling. Springer, New York (2013)
23. van Stiphout, R.G.P.M., Lammering, G., Buijsen, J., Janssen, M.H.M., Gambacorta, A., Slagmolen, P., et al.: Development and external validation of a predictive model for pathological complete response of rectal cancer patients including sequential PET-CT imaging. Radiother. Oncol. **98**, 126–133 (2011)
24. Fiorino, C., Gumina, C., Passoni, P., Palmisano, A., Broggi, S., Cattaneo, G.M., et al.: A TCP-based early regression index predicts the pathological response in neo-adjuvant radio-chemotherapy of rectal cancer. Radiother. Oncol. **128**(3), 564–568 (2018)

Prediction of the Child's Head Growth in the First Year of Life

Wojciech Wolański[1(✉)], Edyta Kawlewska[1], Dawid Larysz[2], Marek Gzik[1], Joanna Gorwa[3], and Robert Michnik[1]

[1] Department of Biomechatronics, Faculty of Biomedical Engineering,
Silesian University of Technology, Zabrze, Poland
{wojciech.wolanski,edyta.kawlewska,marek.gzik,robert.michnik}@polsl.pl
[2] Department of Radiotherapy, Maria Sklodowska-Curie Memorial Cancer Center
and Institute of Oncology, Gliwice, Poland
dawilar@gmail.com
[3] Department of Biomechanics, Faculty of Physical Education,
Sport and Rehabilitation, Poznan University of Physical Education, Poznań, Poland
sarbi@poczta.onet.pl

Abstract. The article presents statistical research aiming at the formulation of functions of the child's head growth in the first year of age. The research group consisted of 129 infants, from 0 to 12 months of age, whose heads were subjected to computer tomography tests within a regular diagnosis of neurological and oncological diseases and other conditions. On the basis of medical images, 3D models of infants' skulls were developed. They enabled determination of the coordinates of anatomical points. The craniometric database was subjected to a statistical analysis in order to develop a head growth model. Five directional vectors of the head growth were formulated in particular directions and three functions were proposed. The degree of the functions' compatibility with empirical data was tested. The best-fitting model was selected for each vector by means of a nonlinear estimation method.

Keywords: Head · Statistics · Infant · Growth function

1 Introduction

Infant skull is composed of several major bones that are held together by fibrous material called sutures. In the first two years of life the skull growth is significant and its course can run properly only with non-ossified sutures [6,18]. Better understanding of the head growth mechanisms, particularly in the period of its most dynamic development, could constitute a significant source of information. Thanks to such knowledge, it would be possible to predict a shape of the head, for instance in the case of genetic defects (such as craniosynostosis). Available standards of development of healthy infants and children most often refer to the progression of mass, growth or the head circumference. A typical characteristic of

© Springer Nature Switzerland AG 2019
J. M. R. S. Tavares and R. M. Natal Jorge (Eds.): VipIMAGE 2019, LNCVB 34, pp. 267–275, 2019.
https://doi.org/10.1007/978-3-030-32040-9_28

growth is a variable curve including the periods of quick growth ("growth erup-
tion") and stagnation [9,10,16]. A seasonal speed of growth is also noticeable,
which is higher in the spring and summer months [3]. Extensive anthropometric
research on healthy children has been being conducted by the world statistical
centres and organizations, such as the WHO (World Health Organization) or the
NCHS (National Centre for Health Statistics). Growth charts have been created
[17] which define standards for particular parameters in the scope of age, sex,
weight, growth, etc. On the grounds of such charts, it is possible to initially
define regularity of the children's growth as well as to predict the changes of
individual parameters.

Skull morphometry This article focuses on model and statistical research
which aimed to formulate the function of the infant's head growth in the first
year of age on the basis of craniometric data.

2   Materials and Methods

A test group consisted of 129 infants of a regular head structure (Table 1). Computer tomography was done during the diagnosis of neurological, oncological and
other diseases.

Table 1. Specification of children group with regular skull shape

Age (months)	1	2	3	4	5	6	7	8	9	10	11	12	Σ
Number	6	12	9	11	9	6	16	13	13	11	14	9	**129**

In anthropometric measurements CT images were used, which were obtained
from the Upper Silesian Children's Health Centre in Katowice. CT examinations
were performed with a 16-slice scanner (Aquilion S16, Toshiba, Tokyo, Japan).
Standard vendor diagnostic protocols for head CT were modified according to
ALARA principle and patients age with following parameters: spiral mode, 80–
120 kV tube voltage, 80–150 mAs tube current, gantry rotation time 0.5 s, FOV
240. The whole head from gonion to vertex was included in the field of view,
together with soft tissue of the face. Scanner gantry was not tilted. The reconstructed slice thickness was 0.5 mm, with an increment of 0.5 mm. The children
in which the feed and wrap technique was not working were examined in general
anaesthesia. Regional Ethics Board approval was received and all parents gave
informed consent for CT scan. On the basis of medical images obtained using
Mimics (Materialise) software, 3-D skull models were created and characteristic
anatomical points were determined (Fig. 3). The coordinates obtained in such a
way constituted the craniometric database which later on served the purpose of
determining the functions of the infant's head growth.

Standard check-up examinations of infants consider the evaluation of the
child's head circumference to be an indicator of regular development. Taking

into consideration the above, first of all the function of the child's head growth was developed on the basis of average circumferences (Fig. 1) and brain's volumes (Fig. 2) measured in particular age groups.

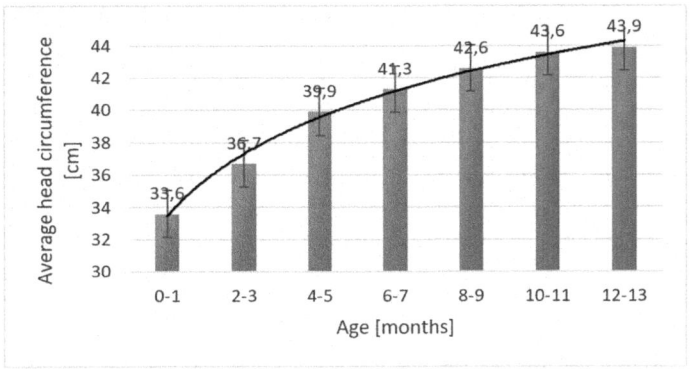

Fig. 1. Changes in the child's head circumference in the first year of life

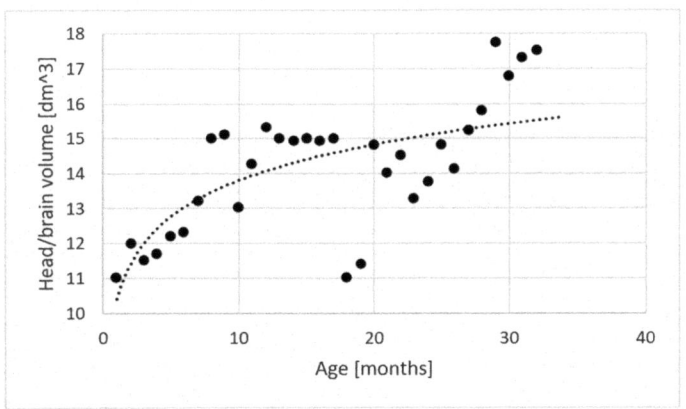

Fig. 2. Head (brain) volume in the first year of life

Each age group (two-month diversification) is characterised by a different growth intensity rate of the child's head. On the basis of the conducted measurements of the circumference of physical models of a regular head (variables of a normal distribution) the researchers developed a statistical model of the infant's head growth for each age group. The result of the correlation between age and the child's head circumference has been presented in Eq. (1). The correlation coefficient approximating to 1 shows that the dependence formulated on the basis of classic curves correlates very well with the head circumference data.

$$y = 5,5451 ln(x) + 33,446;\ R^2 = 0,9912 \tag{1}$$

where: y – head circumference [cm], x – child's age [months.], R – correlation coefficient.

The determined growth curve corresponds well to the change of the head circumference in each age group, however, it is insufficient to predict the shape changes which occur in all three directions of Euclidean space. In order to obtain the simulation of the changes in a three-dimensional shape of the infant's head, it is necessary to determine functions which describe the head growth in three main directions: anterior-posterior, vertical and lateral. During the development of statistical models describing the head growth, identification of proper anthropometric points constituted a major challenge. For such anthropometric points the growth indices were developed. Further statistical analysis took into consideration previously selected distances (segments) between the points on the skull which showed a strong correlation with the child's development. Point basion 'ba' was adopted as a reference point for the analysis of the growth or the change of position of other points. Its location on the circumference of the foramen magnum of the occipital bone guarantees the invariability of its position. A 3D model of the skull with a marked reference system as well as anthropometric points subject to a statistical analysis are presented in Fig. 3.

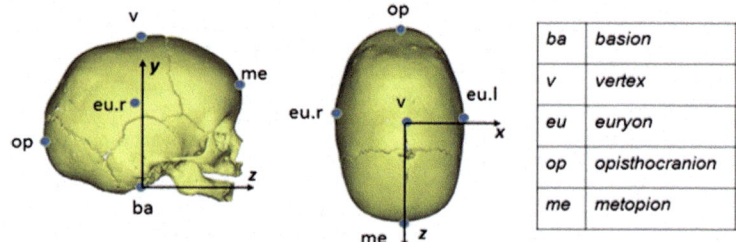

Fig. 3. The model of the infant skull with marked anatomical points and a reference system: (a) a side view, (b) a top view

The distances between selected anatomical points on the infants' skulls constituted a set of variables on the basis of which the curves of growth were determined. The directions of the skull growth were represented by localized vectors which began at a basion point and their ends determined other anthropometric points. The statistical analysis encompassed five vectors (Table 2).

All vectors representing the directions of the skull growth were organized in types encompassing the whole range of the child's age under examination (under the first year of age). Next, each set of variables was matched with a statistical model. In search of the best vector-fitting model a method of a nonlinear estimation was applied. In the process of selection of the estimator's curves, various combinations and specifications of the model were examined taking into

Table 2. Vectors representing directions of skull growth

Vector	Initial point	Terminal point	Direction
$\overrightarrow{w1}$ (eu.r-ba)	Basion	Euryon right	Lateral $(-x)$
$\overrightarrow{w2}$ (eu.l-ba)	Basion	Euryon left	Lateral $(+x)$
$\overrightarrow{w3}$ (br-ba)	Basion	Bregma	Vertical (y)
$\overrightarrow{w4}$ (me-ba)	Basion	Metopion	Anterior $(+z)$
$\overrightarrow{w5}$ (op-ba)	Basion	Opisthocranion	Posterior $(-z)$

consideration the distribution of variables, the degree of their asymmetry and concentration coefficient (kurtosis). Kurtosis in a normal distribution always equals three $(K = 3)$. The higher the kurtosis value (K), the higher concentration (the diagram is higher and slimmer). In the set of variables examined, the concentration around the mean is lower, and the dispersion is higher than in the case of a normal distribution $(K < 3$ – the diagram is wider than in a normal distribution, Fig. 4). That is the reason why matching the curves of growth was done for various models on the basis of the analysis and evaluation of their fitting to the database. A model having the best-fitting parameters was adopted as a representative one.

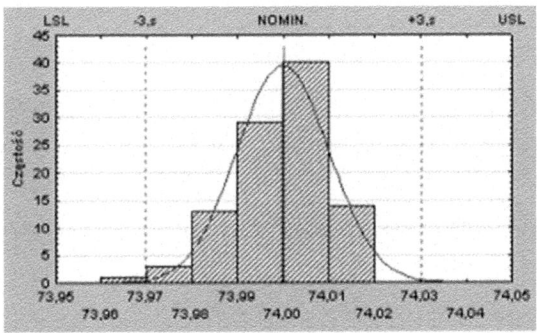

Fig. 4. Distribution of variable vector representing the infant skull growth

In order to develop the curves of growth, all available data were used, i.e. from the birth to 13-th month of age. Due to the distribution of variables different from a normal distribution, a nonlinear estimation was used in the selection of a statistical model. Three statistical models were proposed to represent the function of growth:

$$\textbf{FUNCTION\_1 } y = a - \frac{a}{1 + \left(\dfrac{x}{b}\right)^c} \tag{2}$$

$$\textbf{FUNCTION\_2 } y = a + b \cdot arctan(c \cdot x) \tag{3}$$

$$\textbf{FUNCTION\_3} \ y = \frac{a}{1 + (b \cdot e^{c \cdot x})} \tag{4}$$

where:
a, b, c – parameters of the model,
x – child's age adequate to the growth value [months],
y – growth vector (from $\overrightarrow{w1}$ to $\overrightarrow{w5}$).

Each of the functions of the proposed models (power FUNCTION_1, cyclometric FUNCTION_2 or exponential/logarithmic FUNCTION_3) corresponds to variability of vectors representing the directions of the skull growth. The selection of a model will be determined only by the degree of correlation.

3   Results

The accuracy assessment of the results of the statistical analysis was conducted on the grounds of the difference between empirical and predicted data. The lower the value of the rest, the higher the correlation coefficient and thereby better fitting of the model to the variables. Table 3 presents the determined parameters of the statistical models made for particular directions of growth and the degree of their fitting to the data observed.

Table 3. Vectors representing directions of skull growth

Vector (y)	FUNCTION_1				FUNCTION_2				FUNCTION_3			
	a	b	c	R	a	b	c	R	a	b	c	R
$\overrightarrow{w1}$	97,561	0,0502	0,3869	0,891	68,2395	11,1018	0,21501	0,890	84,5958	0,24426	−0,5697	0,990
$\overrightarrow{w2}$	100,95	0,0451	0,3312	0,892	69,4531	12,0247	0,38917	0,861	85,7784	0,24533	−0,5492	0,992
$\overrightarrow{w3}$	171,78	0,5669	0,2434	0,973	87,647	22,513	0,2601	0,976	117,058	0,33485	−0,2831	0,975
$\overrightarrow{w4}$	205,29	32,867	0,1594	0,863	73,5354	18,1381	0,18592	0,986	96,7176	0,31695	−0,2103	0,986
$\overrightarrow{w5}$	101,02	0,3806	0,7022	0,923	66,956	20,793	0,2619	0,920	91,5608	0,42379	−0,4422	0,987

On the basis of the statistical analysis it can be observed that FUNCTION_3 shows better fitting to the data describing the head growth. The mean coefficient of R correlation, which equals 0.986, is higher than the rest. The functions obtained for particular directions of growth make it possible to predict the skull shape in a common vector space with regard to the analyzed child's age. The diagrams below present the course of changes of all vectors in the whole period of the child's age analysis (under one year of age) (Fig. 5). On the grounds of own research, the following general guidelines were determined. They refer to the head growth forecasting and thereby the brain development.

- average head circumference at birth equals approx. 34 cm,
- in the first year of age the head circumference increases on average by approx. 1 cm per month,

- the most rapid growth occurs within the first six months: during the first month,
- based on the vectors values it was stated that the rate of growth in anterior and lateral directions is similar and equals about 30 mm per year,
- mass/volume of the head/brain doubles within four to six months, it trebles in the first year of age, after the fourth year it considerably slows down.

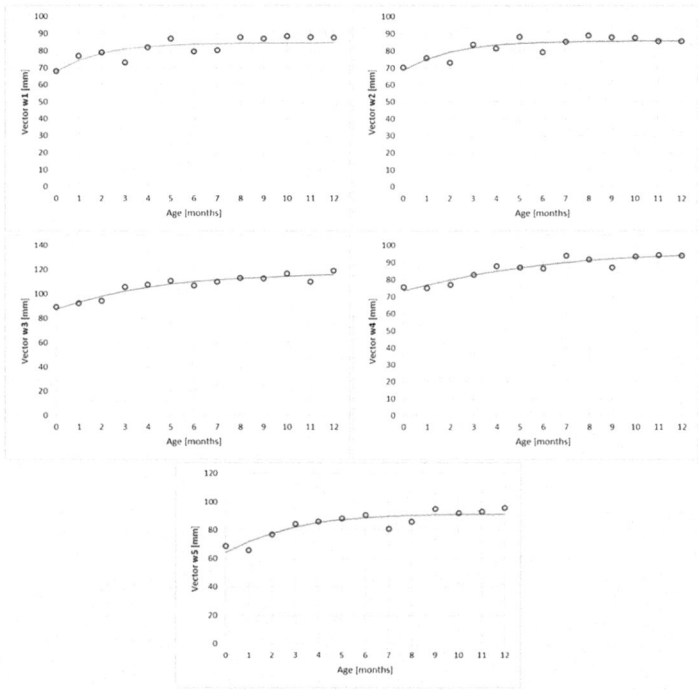

Fig. 5. Functions of growth of the infant skull in each direction

The developed statistical models of growth enable understanding of the changes of the skull shape in the course of the child's regular development. Forecasting of growth of the infant's head which takes into account the growth function makes it possible to re-create a new geometry of the skull by means of linear interpolation of all or a few selected anatomical points.

4   Discussion

There are a few factors which influence regular head growth in infants. They may act independently and at different stages of the child's development. Taking into consideration the time of their occurrence, they can be classed as prenatal and postnatal, and depending on the type they are divided into biological and environmental factors. At the time of birth, it is the factors of the intrauterine environment that are first of all visible in the scope of the changes of growth parameters, whereas the influence of biological (genetic) factors is noticed in a further period of development. The child's head tends to develop quickly within the first 18 months, and then this tendency decreases with age [2,4,14].

Statistical models coupled with mathematical methods make it possible to find connections between the observed real phenomena as well as lead to better understanding of the mechanisms occurring between the variables analyzed. As can be seen in various researches, both linear and non-linear models were applied to interpret the data. In order to analyze hormones concentration, authors Wang et al. (2000) used non-parametric estimators of the two-dimensional random variable density function for many data obtained from dependent trials. Walijski and Yee (2006) applied non-parametric regression to assess the relation between the profile of the systolic and diastolic blood pressure changes and the Body Mass Index (BMI). Chamidah (2012) used a model of local polynomial regression to estimate the curve of growth in children under 5 years of age on the basis of their height and bodily mass.

Statistical models can be also applied to multi-dimensional aspects, for instance, in prediction of shapes or reconstruction of an unknown shape of objects on the basis of geometrical data [1,5–8,11–13,15]. The statistical analysis enables understanding of general dependence between variable coordinates as well as facilitates selection of the type of a parametric model.

5   Conclusion

This article presents the results of statistical research which aimed to develop functions of the infant's head growth enabling forecasting of the skull shape. The developed model of the infant's head growth makes it possible to predict already at the stage of preoperative planning the effects of various treatment methods and head shape defects. In order to obtain an objective (model) function of the head growth, a statistical analysis was conducted on the basis of the anthropometric data of skulls with regular anatomical structure. The growth functions obtained in such a way ensure a correct long-term forecasting of the skull shape in children.

References

1. Dvoracek, L., Skolnick, G., Nguyen, D.C., Naidoo, S.D., Smyth, M.D., Woo, A.S., Patel, K.B.: Comparison of traditional versus normative cephalic index in patients with sagittal synostosis: measure of scaphocephaly and postoperative outcome. Plast. Reconstr. Surg. **136**(3), 541–548 (2015)
2. Fujimura, M., Seryu, J.I.: Velocity of head growth during the perinatal period. Arch. Dis. Child. **52**, 105 (1977)
3. Gelander, L., Karlberg, J., Albertsson-Wikland, K.: Seasonality in lower leg length velocity in prepubertal children. Acta Paediatr. **83**, 1249 (1994)

4. Hall, J.G., Froster-Iskenius, U.G., Allanson, J.E.: Head circumference (occipitofrontal circumference, OFC). Handbook of Physical Measurements, p. 72. Oxford University Press, New York (2007)
5. Herlin, C.H., Largey, A., de Matteï, C.H., Daurès, J.P., Bigorre, M., Captier, G.: Modeling of the human fetal skull base growth: interest in new volumetrics morphometric tools. Early Human Dev. **87**(4), 239–245 (2011)
6. Kawlewska, E., Wolański, W., Larysz, D., Gzik-Zroska, B., Joszko, K., Gzik, M., Gruszczyńska, K.: Statistical analysis of cranial measurements - determination of indices for assessing skull shape in patients with isolated craniosynostosis. In: Gzik, M., Tkacz, E., Paszenda, Z., Piętka, E. (eds.) Advances in Intelligent Systems and Computing, Innovations in Biomedical Engineering, vol. 526, pp. 132–144. Springer, Heidelberg (2017). https://doi.org/10.1007/978-3-319-47154-9_16
7. Lamecker, H., Zachow, S., Haberl, H., Stiller, M.: Medical applic models. In: Weber, S., et al., (ed.) Fortschritt-Berichte VDI. Computer Aided Surgery around the Head, vol. 17, No. 258, p. 61 (2005)
8. Lamecker, H., Zöckler, M., Haberl, H., Zachow, S., Hege, H.C.: Statistical shape modeling for craniosynostosis planning. In: Proceedings of Advanced Digital Technology in Head and Neck Reconstruction, Banff (2005)
9. Lampl, M., Johnson, M.L., Frongillo Jr., E.A.: Mixed distribution analysis identifies saltation and stasis growth. Ann. Hum. Biol. **28**, 403 (2001)
10. Lampl, M., Veldhuis, J.D., Johnson, M.L.: Saltation and stasis: a model of human growth. Science **258**, 801 (1992)
11. Larysz, D., Larysz, P., Filipek, J., Wolański, W., Gzik, M., Kawlewska, E.: Morphometric analysis of the skull shape in children with hydrocephalus. In: Proceedings of VIPIMAGE 2013-IV ECCOMAS on Computational vision and medical image processing IV, Funchal, pp. 337–340 (2013)
12. Likus, W., Bajor, G., Gruszczyńska, K., et al.: Cephalic index in the first three years of life: study of children with normal brain development based on computed tomography. Sci. World J. **2014**(502836), 6 (2014). https://doi.org/10.1155/2014/502836
13. Likus, W., Bajor, G., Gruszczyńska, K., Baron, J., Markowski, J.: Nasal region dimensions in children: a CT study and clinical implications. BioMed Res, Int (2014)
14. Swaiman, K.F.: Neurologic examination of the term and preterm infant. In: Swaiman, K.F., Ashwal, S., Ferriero, D.M., (eds.) Pediatric Neurology: Principles and Practice, 4th ed, Mosby, St. Louis, p. 48 (2006)
15. Tejszerska, D., Wolański, W., Larysz, D., Gzik, M., Sacha, E.: Morphological analysis of the skull shape in craniosynostosis. Acta Bioeng. Biomech. **13**(1), 35–40 (2011)
16. Thalange, N.K., Foster, P.J., Gill, M.S., et al.: Model of normal prepubertal growth. Arch. Dis. Child. **75**, 427 (1996)
17. WHO Child Growth Standards, Department of Nutrition for Health and Development, World Health Organization (2007) ISBN 9789241547185
18. Wolański, W., Larysz, D., Gzik, M., Kawlewska, E.: Modeling and biomechanical analysis of craniosynostosis correction with the use of finite element method. Int. J. Numer. Meth. Biomed. Engng. **29**, 916–925 (2013). https://doi.org/10.1002/cnm.2506

Medical Imaging

Influence of Mutual Rotation
of Polarizing Filters on Light Intensity
Measured with Collagen Fibres

Michaela Turčanová[1(✉)], Martin Hrtoň[2], Petr Dvořák[3], and Jiří Burša[1]

[1] Institute of Solid Mechanics, Mechatronics and Biomechanics,
Brno University of Technology, Brno, Czech Republic
`Michaela.Turcanova@vutbr.com`, `bursa@fme.vutbr.cz`
[2] Central European Institute of Technology,
Brno University of Technology, Brno, Czech Republic
`martin.hrton@ceitec.vutbr.cz`
[3] Institute of Physical Engineering, Brno University of Technology,
Brno, Czech Republic
`dvorak.p@fme.vutbr.cz`

Abstract. The most common technique for observation of collagen fibres and evaluation of their directionality is polarized light microscopy. In the microscope, there are two polarizing filters, polarizer and analyzer, that can rotate individually. The default position of the polarizer is perpendicular to the analyzer. When the light passes through a birefringent specimen, the plane of the polarized light is rotated. This leads to a 90° periodic change in light intensity passing through the sample and both filters. If the orientation of the polarizer differs from the perpendicular direction to the analyzer, the period of light intensity changes to 180°; one of the original maxima increases and the other decreases. This change is very useful for the determination of directions of collagen fibres by an automatic algorithm.

Keywords: Polarized light microscopy · Collagen fibers · Birefrigence · Light intensity

1 Introduction

Collagen, the most abundant protein in the human body, is a major load-bearing component in different organs and tissues (e.g. skin, tendons, blood vessels). In arteries its load-bearing capability applies especially at high blood pressures [1]. Collagen fibres are arranged here in families with a dominant direction and more or less significant dispersion. However, unlike elastin fibres, less than 10% collagen fibres are straightened at physiological pressures [2]. Collagen fibres are 100–1000× stiffer than elastin fibres [3] and changes in their arrangement alter substantially the mechanical properties of the tissue. The orientation of collagen fibres in the vessel wall, their dispersion and waviness are the main

© Springer Nature Switzerland AG 2019
J. M. R. S. Tavares and R. M. Natal Jorge (Eds.): VipIMAGE 2019, LNCVB 34, pp. 279–285, 2019.
https://doi.org/10.1007/978-3-030-32040-9_29

characteristics and input parameters into structural constitutive models which are becoming more and more useful in stress analyses of e.g. diseased arteries [4].

Existing methods used for the detection of directions of collagen fibres in soft tissues can be divided into two main groups: manual and automatic methods.

Manual methods have a lot of disadvantages, e.g. long duration and hence small numbers of evaluated samples, operator fatigue and his/her loss of concentration, or different fibre intensities. This may result in inaccurate evaluation (e.g. underestimation of fibre dispersion or determination of more dominant directions than existing in reality); a large sample size and large number of measured points can result in the loss of operator's orientation and thus an incorrect recording of the angle. Also intrinsic operator's tendency to repeat first the successful position from the previous measurements may induce overestimation of dominant directions while limited sensitivity of human eye can make the results operator dependent. In contrast, limited number of analysed points can lead to mismatch between dispersion and waviness.

On the contrary, automatic methods are very fast and free of human errors. Unfortunately, various limitations, e.g. high expenses, inhibit from their wider use [5–8]. The easiest and cheapest method - polarized light microscopy (PLM) – is broadly used in manual evaluation [9–13] but also an automated algorithm for its exploitation was created recently in our group [14]. It suffers, however, from some limitations originating from the very fundamentals of the method; they are analyzed below and a proposal how they can be overcome is presented.

The light emanating from a lamp vibrates (as described by the theory of electromagnetism) always perpendicularly to the direction of its propagation, i.e. disorderly (chaotically) in all transversal directions. When passing through the polarizer (the first filter in the light path) only the part of light with the electric intensity vector vibrating in one direction can pass, while the rest of the light is absorbed. In a standard setting the second filter in the light path, the analyzer, is oriented perpendicularly to the first one so that no light can pass both of them. When a birefringent sample is located in the light trajectory between both filters, it turns the direction of polarized light and consequently a portion of light can pass all the set. Its output intensity depends on the birefringence of the sample and in case of a fibrous material also on its orientation. When the orientation of fibres coincides with one of the filters, no light can pass and the fibres appear dark; under other angles the intensity increases and achieves its maximum under 45°.

For detection of collagen fibre directions, we can use its two abilities - birefringence and diattenuation. The birefringence is due to the internal structure of the fibres themselves (see [15]) and can be enhanced up to seven times by the Picro Sirius Red dye (PSR) (see [16]). This causes the polarized light to divide into two beams, so called ordinary and extraordinary (see Fig. 1). Each of these beams is rotated by another angle and some portion of it can pass through the analyzer. The intensity of the light visible in the lens is given by superposition of these two beams and thus varies according to a sinusoidal function with 90° periodicity (see Fig. 2).

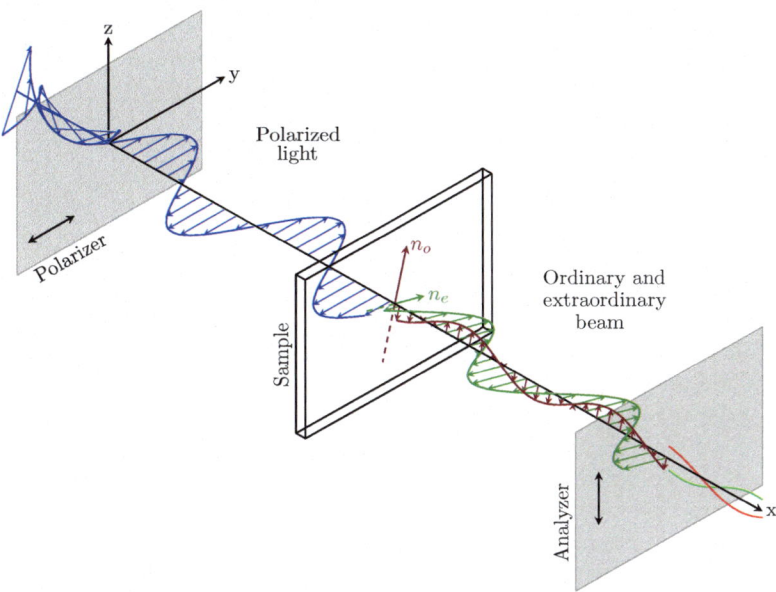

Fig. 1. Schematic of the transmission of polarized light through a birefringent sample in the polarization microscope. The light passes through the polarizer only in the same direction as its orientation. The birefringence of the sample causes the beam to be divided into ordinary (n_o) and extraordinary (n_e) beam. Therefore, the light passing through the analyzer must have the appropriate orientation. Taken and modified from [17].

Fig. 2. Micrographs showing intensity changes of an optically anisotropic structure (porcine Achilles tendon) depending on the rotation of this structure relative to the polarizer and the analyzer ($\delta = 0°$ means perpendicular filters). If the collagen fibres are oriented in the direction of the polarizer or analyzer (x or y-axis), the light beam cannot pass through these two filters. In any other arrangement, the structure appears lighter and the sample reaches its maximum intensity at 45°.

Diattenuation is the total anisotropic attenuation of light that is caused by absorption and scattering. The intensity of the transmitted light depends on the polarization state of the incident light, i.e. the intensity of the transmitted light is maximal for the light polarized in a certain direction and minimal for the light polarized in the corresponding orthogonal direction [18].

The behaviour of an optically anisotropic structure under polarized light with perpendicular filters is known and commonly used. The local fibre orientation at a point is equal to the extinction angle (minimum light intensity) at this point;

however, the same holds also for the perpendicular direction of the fibre. This intrinsic 90° periodicity of the light intensity can be overcome when both filters are not perpendicular; this mutual turn of the filters can cause increase of the intensity at one maximum and decrease at the other one. To analyse this effect, several measurements with different polarizer angles and different wavelengths were performed in this study.

2   Materials and Methods

The preparation of histological sections was done at the St. Anne's University Hospital in Brno. $18 \times 4\,\text{mm}^2$ rectangle specimens were cut out of the Achilles tendon. Then all samples were fixed in 10% formaldehyde at room temperature for 24 h and afterwards embedded in wax. Furthermore, the tendon samples were cut into slices of $5\,\mu\text{m}$ thickness; the section plane was parallel to the preferred direction of the collagen fibres which is evident with this tissue. In the last phase, the slices were dyed with PSR (0.1%).

Samples from histological sections were placed on a rotating table in a Padim-Drexx polarization microscope and imaged with 5MP microcam BresserOptik, GmbH. The sample was centred and both polarizing filters set to the perpendicular position. Then the dependence of diattenuation on the wavelength was tested to choose the most suitable wavelength corresponding to the collagen highest optical anisotropy.

3   Results

The biggest difference between the light intensities of two mutually perpendicular fibre positions was at the wavelength of 550 nm, in parallel orientation of polarizers, see Fig. 3.

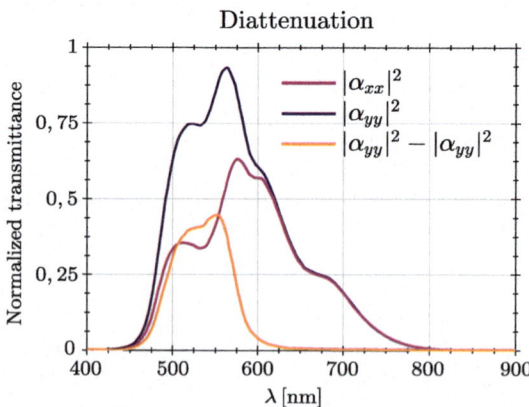

Fig. 3. Spectral dependence of transmittance in both principal directions of anisotropy (measured directly with aligned filters) and their difference. The collagen fibres are apparently strongly anisotropic, and this anisotropy is most pronounced for wavelengths of 500–570 nm.

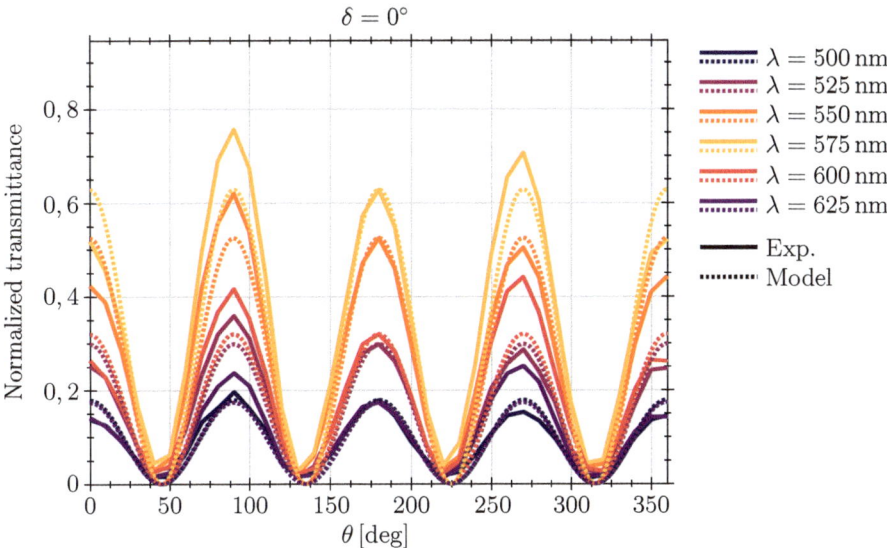

Fig. 4. Behaviour of light intensity at different wavelengths λ and specimen rotation θ with perpendicular polarizers ($\delta = 0°$).

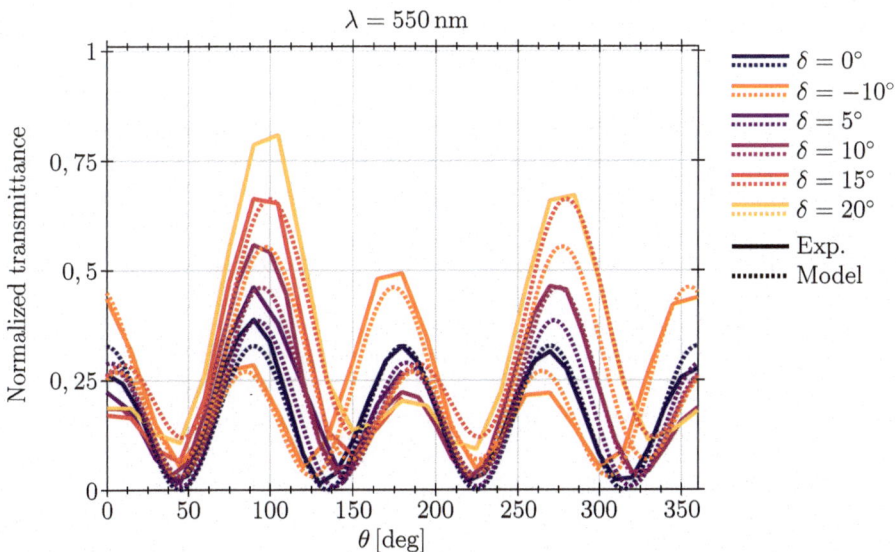

Fig. 5. Comparison of measurements with the polarizers being perpendicular ($\delta = 0°$) and deflected by different δ angles.

Measurements with the perpendicular polarizers and changing specimen orientation angle θ brought almost sinusoidal light intensity at a randomly chosen point of the birefringent sample. In Fig. 4 you can see comparison of several wavelengths. It is not possible to distinguish unambiguously from the graph with $\delta = 0°$ (perpendicular filters) it is not possible to distinguish unambiguously between mutually perpendicular orientations of the fibres; thus the range of evaluation is limited to $90°$. In contrast, the measurement has shown that the deflection of the polarizers from the perpendicular adjustment results in a change in the dependence of intensity of the passing polarized light on the fibre angle. When increasing the polarizer deflection, one of the original maxima increases and the other decreases. The higher one of both maxima is decisive for fibre angle detection and its next minimum determines the angle of extinction (rather that one of the lowest intensity) of the fibres and thus the orientation of the collagen fibres in the investigated point. Figure 5 represents a comparison of the measurements for one chosen wavelength ($\lambda = 550$ nm) with perpendicular filters and with their various deflection angles. It shows that with increasing δ angle the difference between two adjacent maxima increases (theoretically up to $\delta = 45°$) and enables us to distinguish between the mutually perpendicular orientations of the fibres.

4   Conclusion

We have found that by deflecting (turning) the polarizers from the perpendicular direction, we can determine the extinction angle, or fibre orientation in their whole range of $180°$. With the increasing deflection angle δ the difference between two maxima increases and enables us extend the range of evaluation of fibre directions to $180°$. This impact of polarizer rotation on light intensity is crucial for us in the upcoming algorithm for automatic detection of directions of collagen fibres in soft tissues. Although many other effective methods for determination of collagen fibre directions exist, their application is rather rare due to an expensive equipment needed. As the presented approach offers a cheap and generally applicable alternative, it could contribute to enhancement of knowledge on collagen fibres orientation and dispersion in different soft tissues. Lack of this information represents a crucial limitation for applications of structure-based constitutive descriptions in computational modelling of these tissues.

Acknowledgements. This work was supported by Czech Science Foundation project No. 18-13663S.

References

1. Fung, Y.C.: Biomechanics: mechanical properties of living tissues (2013)
2. Armentano, R.L., Levenson, J., Barra, J.G., Fischer, E.I., Breitbart, G.J., Pichel, R.H., Simon, A.: Assessment of elastin and collagen contribution to aortic elasticity in conscious dogs. Am. J. Physiol. Heart Circ. Physiol. **260**, H1870–H1877 (1991)
3. Wagenseil, J.E., Mecham, R.P.: Elastin in large artery stiffness and hypertension. J. Cardiovasc. Transl. Res. **5**, 264–273 (2012)
4. Humphrey, J.D., Canham, P.B.: Structure, mechanical properties, and mechanics of intracranial saccular aneurysms. J. Elast. Phys. Sci. Solids **61**, 49–81 (2000)

5. Akyildiz, A.C., Speelman, L., Gijsen, F.J.H.: Mechanical properties of human atherosclerotic intima tissue. J. Biomech. **47**, 773–783 (2014)
6. Chai, C.K., Akyildiz, A.C., Speelman, L., Gijsen, F.J.H., Oomens, C.W.J., van Sambeek, M.R.H.M., van der Lugt, A., Baaijens, F.P.T.: Local anisotropic mechanical properties of human carotid atherosclerotic plaques-characterisation by micro-indentation and inverse finite element analysis. J. Mech. Behav. Biomed. Mater. **43**, 59–68 (2015)
7. Hoffman, A.H., Teng, Z., Zheng, J., Wu, Z., Woodard, P.K., Billiar, K.L., Wang, L., Tang, D.: Stiffness properties of adventitia, media, and full thickness human atherosclerotic carotid arteries in the axial and circumferential directions. J. Biomech. Eng. **139**, 124501 (2017)
8. Akyildiz, A.C., Chai, C.K., Oomens, C.W.J., van der Lugt, A., Baaijens, F.P.T., Strijkers, G.J., Gijsen, F.J.H.: 3D fiber orientation in atherosclerotic carotid plaques. J. Struct. Biol. **200**, 28–35 (2017)
9. Smith, J.F.H., Canham, P.B., Starkey, J.: Orientation of collagen in the tunica adventitia of the human cerebral artery measured with polarized light and the universal stage. J. Ultrastruct. Res. **77**, 133–145 (1981)
10. Canham, P.B., Finlay, H.M., Dixon, J.G., Boughner, D.R., Chen, A.: Measurements from light and polarised light microscopy of human coronary arteries fixed at distending pressure. Cardiovas. Res. **23**, 973–982 (1989)
11. Canham, P.B., Finlay, H.M., Dixon, J.G., Ferguson, S.E.: Layered collagen fabric of cerebral aneurysms quantitatively assessed by the universal stage and polarized light microscopy. Anat. Rec. **231**, 579–592 (1991)
12. Finlay, H.M., McCullough, L., Canham, P.B.: Three-dimensional collagen organization of human brain arteries at different transmural pressures. J. Vasc. Res. **32**, 301–312 (1995)
13. Schriefl, A.J., Reinisch, A.J., Sankaran, S., Pierce, D.M., Holzapfel, G.A.: Quantitative assessment of collagen fibre orientations from two-dimensional images of soft biological tissues. J. R. Soc. Interface **9**, 3081–3093 (2012)
14. Novak, K., Polzer, S., Tichy, M., Bursa, J.: Automatic evaluation of collagen fiber directions from polarized light microscopy images. Microsc. Microanal. **21**, 863–875 (2015)
15. Wolman, M., Kasten, F.H.: Polarized light microscopy in the study of the molecular structure of collagen and reticulin. Histochemistry **85**, 41–49 (1986)
16. Junqueira, L.C.U., Bignolas, G., Brentani, R.R.: Picrosirius staining plus polarization microscopy, a specific method for collagen detection in tissue sections. Histochem. J. **11**, 447–455 (1979)
17. Example: Polarizing microscope. Available via DIALOG, 28 March 2019. http://www.texample.net/tikz/examples/polarizing-microscope/
18. Chenault, D.B., Chipman, R.A.: Measurements of linear diattenuation and linear retardance spectra with a rotating sample spectropolarimeter. Appl. Opt. **32**, 3513–3519 (1993)

Lung Nodule Segmentation Based on Convolutional Neural Networks Using Multi-orientation and Patchwise Mechanisms

Paulo H. J. Amorim[1], Thiago F. de Moraes[1], Jorge V. L. da Silva[1], and Helio Pedrini[2(✉)]

[1] Tridimensional Technology Division,
Center for Information Technology Renato Archer, Campinas, SP 13069-901, Brazil
[2] Institute of Computing, University of Campinas, Campinas, SP 13083-852, Brazil
helio@ic.unicamp.br

Abstract. Image segmentation is used in several knowledge domains, such as medicine, biology, remote sensing, industrial automation, surveillance and security. More specifically, image segmentation plays a crucial role in various medical imaging applications, as an important part of clinical diagnosis. Deep learning techniques have recently benefited medical image segmentation and classification tasks. In this work, we have explored the use of Convolutional Neural Networks (CNN) for lung nodule segmentation using multi-orientation and patchwise mechanisms. Experiments conducted on the public LIDC-IRI dataset demonstrate that our results were able to reduce the number of false negatives, which is important in this task. High segmentation rates were achieved when compared to medical specialists.

1 Introduction

Lung cancer is one of the most aggressive forms of the pulmonary disease. In Brazil, for instance, it is estimated that 31,270 new cases occurred only in 2018. In the United States, 1,688,780 new cases were reported in 2017 [26].

Early detection of lung nodules is crucial to increase patient's survival rate. The detection of the lung nodule malignancy by radiologists depend of their clinical experience.

A common clinical diagnostic tool is the use of computed tomography (CT), where radiologists usually measure certain characteristics, such as size [22], texture [10] and density [14] of possible nodules. Before performing such measurements, it is necessary to select the lesion using an image segmentation technique.

Image segmentation techniques aim to partition an image into regions or objects [12,16,25,27]. They can be classified into three different approaches: (i) manual segmentation, where the radiologist indicates each pixel of the lesion manually, (ii) semi-automatic segmentation, where some parameters are defined

© Springer Nature Switzerland AG 2019
J. M. R. S. Tavares and R. M. Natal Jorge (Eds.): VipIMAGE 2019, LNCVB 34, pp. 286–295, 2019.
https://doi.org/10.1007/978-3-030-32040-9_30

by the user and the algorithm performs the selection of the lesion, and (iii) automatic segmentation methods, where there is no need to configure any parameters, just trigger an algorithm or intelligent system.

The main objective of this work is to perform an automatic segmentation of pulmonary nodules using convolutional neural networks (CNN) based on multi-orientation and patchwise mechanisms. For false negative reduction, a region growing algorithm was used after the union of three volumes containing the probability of occurring a lesion. Experimental results on the public LIDC-IRI dataset demonstrated that the proposed method was able to achieve segmentation levels comparable to a medical specialist.

As contribution, the developed approach is intended to be included in InVesalius [3], an open-source framework, in order to improve medical diagnosis and patient treatment.

The text is organized as follows. Section 2 presents concepts and work related to the topic under investigation. Section 3 describes the proposed methodology for lung segmentation. Experimental results are presented and discussed in Sect. 4. Finally, Sect. 5 concludes the work and presents directions for future work.

2   Background

The segmentation of lung nodules from computed tomography (CT) images provides relevant information for diagnosis, reducing manual and subjective intervention by human experts.

Although several lung nodule segmentation approaches have been developed in recent years [2,7,10,18,29], the achievement of accurate and robust segmentation results is still a challenging task due to the large volumes of CT images and data variability.

In recent years, deep neural networks [20] have been applied in several areas, for instance, robotics [28], speech synthesis [4,33] and medicine [8,21,23]. More specifically for segmentation of pulmonary nodules, these networks have been applied in the works proposed by Wang et al. [30] and Kamal et al. [17].

A deep neural network, compared to traditional artificial neural networks, is differentiated by a number of hidden layers and their respective neurons, in addition to the way they are interconnected. Another characteristic is the excessive amount of data needed to perform the training stage.

Similarly to traditional artificial neural networks, there are several categories of deep neural networks, such as deep belief networks (DBN) [15], generative adversarial networks (GAN) [13] and convolutional neural networks (CNN) [19]. The convolutional neural networks are distinguished by their ability to apply convolution filters in the data in order to extract more feature.

A CNN architecture designed to biomedical image segmentation is the U-Net. The general idea of the U-Net is to contract the input image using max-pooling (2×2) along with convolution (3×3) and expand using up-convolution (2×2) followed by convolution (3×3). At the end of each convolution, an activation

layer uses a rectified linear unit (ReLU). The last layer consists of a convolution
(1×1) to map each component to its respective class [24].

3   Lung Nodule Segmentation Method

The main steps of the lung segmentation proposed in this work are illustrated
in Fig. 1. We considered computed tomography (CT) images of lung with nodules of different sizes. Initially, pixel intensity and contrast of the images were
normalized to the lung window, with values of -600 HU and 1500 HU [6]. The
images and the ground truth were stacked and interpolated using the cubic spline
interpolation [9]. Then, both volumes were resliced in the axial (XZ), coronal
(XY) and sagittal (YZ) orientations.

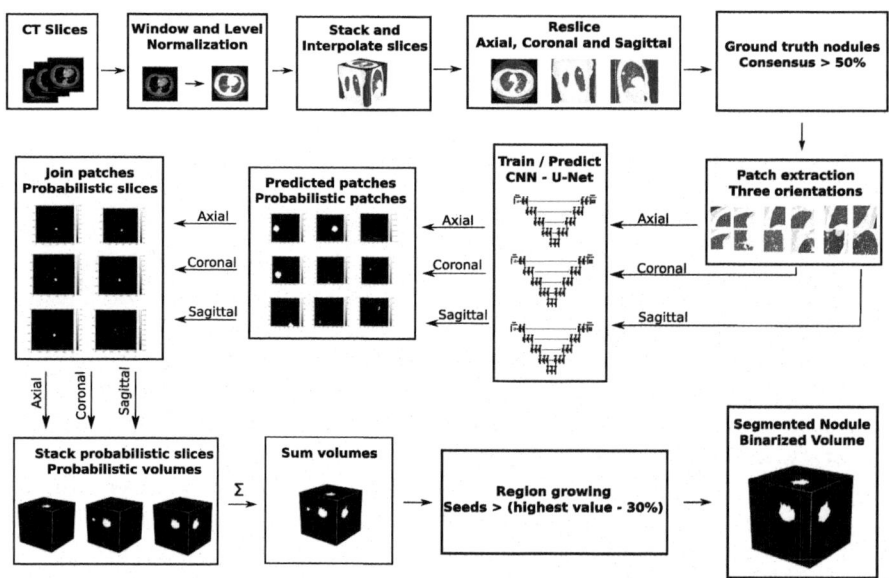

Fig. 1. Diagram of the proposed lung image segmentation approach.

The ground truth was used to select the pixels that had more than 50% consensus between at least two radiologists (Fig. 2(a)). In order to avoid inconsistencies in the training phase, nodules recognized by only one radiologist (Fig. 2(b))
were discarded.

From the images at each of the three orientations (axial, coronal and sagittal), we extracted 128×128-pixel resolution patches in steps of 64 pixels. Three
instances of the U-Net convolutional neural network architecture were employed,
each one for training and predicting orientation patches. The input image dimensions and network weights were changed in relation to the original paper [24].
These new values can be observed in Fig. 3.

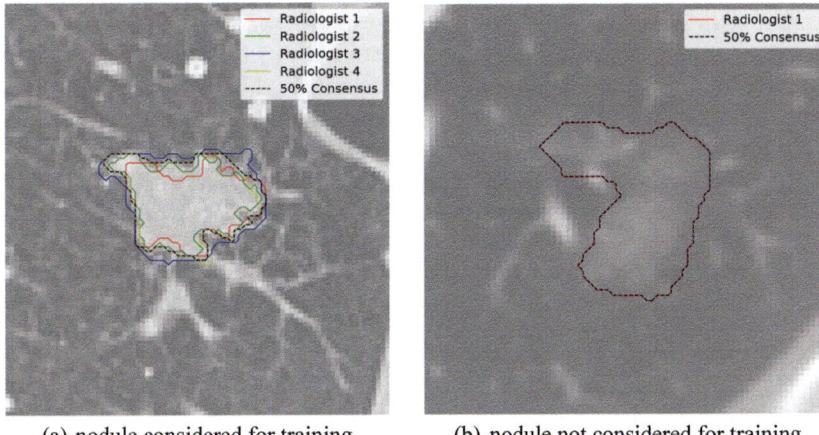

(a) nodule considered for training   (b) nodule not considered for training

Fig. 2. Example of nodules found by 4 radiologists and found by 1 radiologist. The dotted line marks 50% of the consensus of the pixels marked by the radiologists.

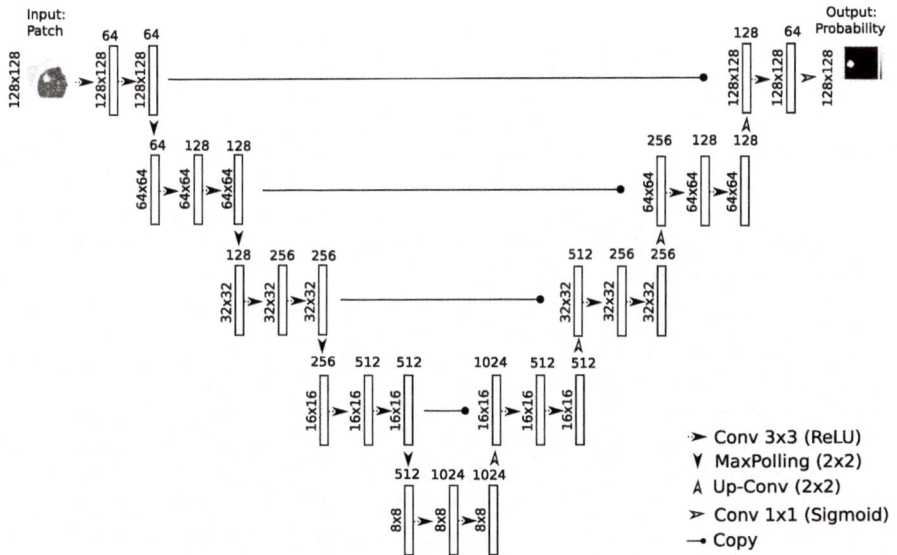

Fig. 3. Modified U-Net architecture for lung segmentation.

After each activation layer, batch normalization was used. In the last layer, the sigmoid activator was employed. In addition, Adam optimizer was used with learning rate of 0.0001. In the training step, data augmentation techniques generated different versions of the images by randomly applying the following geometric transformation: rotation (1–180°), shear (1–60%), zoom (1–30%), horizontal and vertical flip.

In the prediction step, the output of the network is a set of patches, whose pixels indicate the probability of belonging to a nodule. These patches are joined to form a probability image in both orientations. Three volumes are formed from the stacking of these probability images. The volumes are then summed, generating a single volume. In order to reduce false negatives, all pixels with values that satisfy Eq. 1 are selected from the resulting volume. The pixels are used as seeds in the application of the region growing algorithm [1].

$$seeds = p(x, y, z) > HighestValue(p) - 0.30 * HighestValue(p) \qquad (1)$$

4 Experimental Results

This section describes and discusses the experimental results obtained with the proposed lung segmentation method based on convolutional neural networks.

4.1 Lung Dataset

In our experiments, we used the public LIDC-IRI dataset [5] to evaluate the effectiveness of our lung segmentation method. This dataset consists of a set of 1,018 computed tomography examinations acquired from 1,010 patients, resulting in 7,371 lesions recognized by at least one radiologist.

The acquisition process was performed by different equipments from seven manufacturers. The lesion segmentation was conducted by four radiologists. In addition to the images, there is an XML file that describes some meta-data, such as subtlety, internal structure, calcification, sphericity, margin, lobulation, spiculation, radiographic solidity and malignancy.

4.2 Training and Evaluation

The dataset, after the pre-processing stage, was divided into the proportion of 80% for training and 20% for validation and test. The patchwise strategy with 128×128 size patches was chosen in order to not overload the video card memory and to be able to use a batch of 64 ground-truth images during the training. The training process was executed for 200 epochs, using 10-fold cross-validation. The total number of patches in the training stage for the axial, coronal and sagittal orientations were approximately 20,000, 26,000 and 27,000, respectively.

All codes in this work were implemented using Python programming language, as well as Keras, TensorFlow, Numpy, Scipy and pyLIDC libraries. A video card Geforce Titan Xp 12 GB GDDR5X was used on a computer Intel Core i7 processor, 16 GB DDR3 RAM, 1 TB HD Seagate SATA 7200 rpm. The execution time was approximately 60 h to train all three networks.

The following metrics were used in the validation step: Sørensen-Dice coefficient (DSC) (Eq. 2) and intersection over union (IoU) or Jaccard coefficient

(Eq. 3). In both equations, P is the result of the prediction and G is the ground truth, which must be binary images.

$$DSC = \frac{2 \mid P \cap G \mid}{\mid P \cap G \mid + \mid P \cup G \mid} \tag{2}$$

$$IoU = \frac{\mid P \cap G \mid}{\mid P \cup G \mid} \tag{3}$$

From our experiments, it was possible to obtain 83% in Sørensen-Dice coefficient and 76% intersection over union after sum three volumes. For axial orientation only, it was obtained 74% in Dice coefficient, whereas 63% and 68% for coronal and sagittal orientations, respectively. Our approach to summing all the three probability volumes from each orientation and then using as seed the highest values in the region growing algorithm allowed to reduce the amount of false positive rate by 50% average and increased the Dice coefficient.

Figure 4 presents the obtained Free-response Receiver Operating Characteristic (FROC) curve, whereas Table 1 presents a comparison of some segmented nodules after the sum of volumes. Additionally, Table 2 presents a comparison against other methods available in the literature.

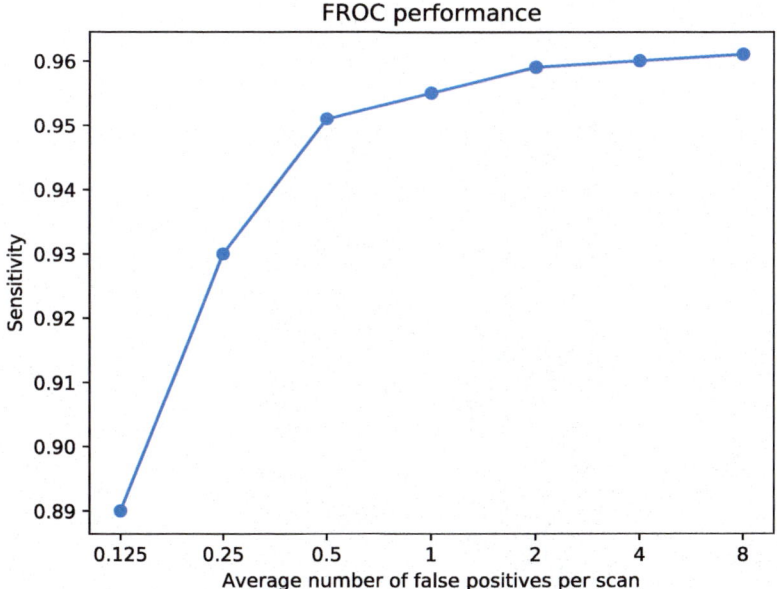

Fig. 4. Free-response Receiver Operating Characteristic (FROC) curve.

Table 1. Examples of input, ground-truth and output images, as well as their respective intersection over union (IoU) and Dice coefficients.

Input	Ground Truth	Predicted	IoU	DSC
			0.897	0.946
			0.885	0.939
			0.874	0.933
			0.866	0.928
			0.859	0.924
			0.774	0.873
			0.773	0.872
			0.771	0.870

Table 2. Comparison of the results achieved with the proposed method (highlighted in bold) against other methods found in the literature.

Method	Strategy	Dice (%)
Farag et al. [11]	Level sets	60
Ye et al. [31]	Graph cut	74
Kamal et al. [17]	Deep learning U-Net 3D	74
Ye et al. [32]	Shape based + SVM	76
Wang et al. [29]	Multi-view deep convolutional neural networks	77
Wang et al. [30]	Deep learning CF-CNN	80
Khosravan and Bagci [18]	Single task learning	82
Proposed method	**U-Net with multi-orientation and patchwise**	**83**
Khosravan and Bagci [18]	Multi-task learning	86

5 Conclusions and Future Work

In this work, we proposed and analyzed a method for segmenting lung nodules in computed tomography images using convolutional neural networks with multi-orientation and patchwise approaches. Our experimental results demonstrated the effectiveness of our method, achieving high Sørensen-Dice coefficient rates.

As directions for future work, we intend to explore the segmentation of other human body parts, as well as incorporate the developed lung nodule segmentation approach into InVesalius [3], an open source medical imaging software.

Acknowledgements. The authors are thankful to CNPq (grants #300047/2019-3 and #305169/2015-7) and São Paulo Research Foundation (grant FAPESP #2014/12236-1) for their financial support, as well for the NVidia GPU Grant Program for a Titan Xp donation.

References

1. Adams, R., Bischof, L.: Seeded region growing. IEEE Transact. Pattern Anal. Mach. Intell. **16**(6), 641–647 (1994)
2. Alilou, M., Beig, N., Orooji, M., Rajiah, P., Velcheti, V., Rakshit, S., Reddy, N., Yang, M., Jacono, F., Gilkeson, R.C.: An integrated segmentation and shape-based classification scheme for distinguishing adenocarcinomas from granulomas on lung CT. Med. Phys. **44**(7), 3556–3569 (2017)
3. Amorim, P., Moraes, T., Silva, J., Pedrini, H.: InVesalius: an interactive rendering framework for health care support. In: International Symposium on Visual Computing, pp. 45–54. Springer, Heidelberg (2015)

4. Arik, S.Ö., Chrzanowski, M., Coates, A., Diamos, G., Gibiansky, A., Kang, Y., Li, X., Miller, J., Ng, A., Raiman, J.: Deep voice: real-time neural text-to-speech. In: 34th International Conference on Machine Learning, vol. 70, pp. 195–204. JMLR.org (2017)
5. Armato, S.G., McLennan, G., Bidaut, L., McNitt-Gray, M.F., Meyer, C.R., Reeves, A.P., Zhao, B., Aberle, D.R., Henschke, C.I., Hoffman, E.A.: The lUng Image Database Consortium (LIDC) and Image Database Resource Initiative (IDRI): a completed reference database of lung nodules on CT scans. Med. Phys. **38**(2), 915–931 (2011)
6. Bankier, A.A., MacMahon, H., Goo, J.M., Rubin, G.D., Schaefer-Prokop, C.M., Naidich, D.P.: Recommendations for measuring pulmonary nodules at CT: a statement from the fleischner society. Radiology **285**(2), 584–600 (2017)
7. Bobadilla, J.C.M., Pedrini, H.: Lung nodule classification based on deep convolutional neural networks. In: 21st Iberoamerican Congress on Pattern Recognition, pp. 117–124. Springer, Lima (2016)
8. Brebisson, A., Montana, G.: Deep neural networks for anatomical brain segmentation. In: IEEE Conference on Computer Vision and Pattern Recognition Workshops, pp. 20–28 (2015)
9. De Boor, C., De Boor, C., Mathématicien, E.U., De Boor, C., De Boor, C.: A Practical Guide to Splines, vol. 27. Springer, New York (1978)
10. Dennie, C., Thornhill, R., Sethi-Virmani, V., Souza, C.A., Bayanati, H., Gupta, A., Maziak, D.: Role of quantitative computed tomography texture analysis in the differentiation of primary lung cancer and granulomatous nodules. Quant. Imaging Med. Surg. **6**(1), 6 (2016)
11. Farag, A.A., El Munim, H.E.A., Graham, J.H., Farag, A.A.: A novel approach for lung nodules segmentation in chest CT using level sets. IEEE Transact. Image Process. **22**(12), 5202–5213 (2013)
12. Gonzalez, R., Woods, R.: Digital Image Processing. Pearson Education, London (2011)
13. Goodfellow, I., Pouget-Abadie, J., Mirza, M., Xu, B., Warde-Farley, D., Ozair, S., Courville, A., Bengio, Y.: Generative adversarial nets. In: Advances in Neural Information Processing Systems, pp. 2672–2680 (2014)
14. Heidinger, B.H., Anderson, K.R., Nemec, U., Costa, D.B., Gangadharan, S.P., VanderLaan, P.A., Bankier, A.A.: Lung adenocarcinoma manifesting as pure groundglass nodules: correlating CT size, volume, density, and roundness with histopathologic invasion and size. J. Thorac. Oncol. **12**(8), 1288–1298 (2017)
15. Hinton, G.E.: Deep belief networks. Scholarpedia **4**(5), 5947 (2009)
16. Junqueira Amorim, P.H., de Moraes, T.F., Rezende, R.A., Silva, J.V.L., Pedrini, H.: Medical imaging for three-dimensional computer-aided models. In: 3D Printing and Biofabrication, pp. 185–207 (2017)
17. Kamal, U., Rafi, A.M., Hoque, R., Hasan, M.: Lung Cancer Tumor Region Segmentation Using Recurrent 3D-DenseUNet. arXiv preprint arXiv:1812.01951 (2018)
18. Khosravan, N., Bagci, U.: Semi-supervised multi-task learning for lung cancer diagnosis. In: 2018 40th Annual International Conference of the IEEE Engineering in Medicine and Biology Society (EMBC), pp. 710–713. IEEE (2018)
19. Krizhevsky, A., Sutskever, I., Hinton, G.E.: Imagenet classification with deep convolutional neural networks. In: Advances in Neural Information Processing Systems, pp. 1097–1105 (2012)
20. LeCun, Y., Bengio, Y., Hinton, G.: Deep learning. Nature **521**(7553), 436 (2015)
21. Li, W., Jia, F., Hu, Q.: Automatic segmentation of liver tumor in CT images with deep convolutional neural networks. J. Comput. Commun. **3**(11), 146 (2015)

22. Margerie-Mellon, C., Heidinger, B.H., Bankier, A.A.: 2D or 3D measurements of pulmonary nodules: preliminary answers and more open questions. J. Thorac. Dis. **10**(2), 547 (2018)
23. Moeskops, P., Wolterink, J.M., van der Velden, B.H., Gilhuijs, K.G., Leiner, T., Viergever, M.A., Išgum, I.: Deep learning for multi-task medical image segmentation in multiple modalities. In: International Conference on Medical Image Computing and Computer-Assisted Intervention, pp. 478–486. Springer, Heidelberg (2016)
24. Ronneberger, O., Fischer, P., Brox, T.: U-Net: convolutional networks for biomedical image segmentation. In: International Conference on Medical Image Computing and Computer-Assisted Intervention, pp. 234–241. Springer, Heidelberg (2015)
25. Schwartz, W.R., Pedrini, H.: Color textured image segmentation based on spatial dependence using 3D co-occurrence matrices and markov random fields. In: 15th International Conference in Central Europe on Computer Graphics, Visualization and Computer Vision, pp. 81–87. Plzen, Czech Republic (2007)
26. Siegel, R.L., Miller, K.D., Jemal, A.: Cancer statistics. CA Cancer J. Clin. **66**(1), 7–30 (2017)
27. Silva, R.D., Minetto, R., Schwartz, W.R., Pedrini, H.: Satellite image segmentation using wavelet transforms based on color and texture features. In: International Symposium on Visual Computing, pp. 113–122. Springer, Heidelberg (2008)
28. Sünderhauf, N., Brock, O., Scheirer, W., Hadsell, R., Fox, D., Leitner, J., Upcroft, B., Abbeel, P., Burgard, W., Milford, M.: The limits and potentials of deep learning for robotics. Int. J. Robot. Res. **37**(4–5), 405–420 (2018)
29. Wang, S., Zhou, M., Gevaert, O., Tang, Z., Dong, D., Liu, Z., Tian, J.: A multi-view deep convolutional neural networks for lung nodule segmentation. In: 2017 39th Annual International Conference of the IEEE Engineering in Medicine and Biology Society, pp. 1752–1755. IEEE (2017)
30. Wang, S., Zhou, M., Liu, Z., Liu, Z., Gu, D., Zang, Y., Dong, D., Gevaert, O., Tian, J.: Central focused convolutional neural networks: developing a data-driven model for lung nodule segmentation. Med. Image Anal. **40**, 172–183 (2017)
31. Ye, X., Beddoe, G., Slabaugh, G.: Automatic graph cut segmentation of lesions in CT using mean shift superpixels. J. Biomed. Imaging **2010**, 19 (2010)
32. Ye, X., Lin, X., Dehmeshki, J., Slabaugh, G., Beddoe, G.: Shape-based computer-aided detection of lung nodules in thoracic CT images. IEEE Transact. Biomed. Eng. **56**(7), 1810–1820 (2009)
33. Zhang, X.L., Wu, J.: Deep belief networks based voice activity detection. IEEE Transact. Audio Speech Lang. Process. **21**(4), 697–710 (2013)

Alzheimer's Disease and the Volume of Hippocampal Subfields: Cluster Analysis Using Neuroimaging Data

C. G. Fonseca Pachi[1,2(✉)], L. Torralbo[1], and J. F. Yamamoto[3]

[1] Laboratory of Psychiatric Neuroimaging (LIM-21), Instituto de Psiquiatria,
Hospital das Clinicas HCFMUSP, Faculdade de Medicina,
Universidade de Sao Paulo, Paulo, SP, Brazil
claricgfp@gmail.com
[2] Senac Santo Amaro University Center, São Paulo, SP, Brazil
[3] Information Technology Center, Hospital das Clínicas HCFMUSP,
Faculdade de Medicina, Universidade de São Paulo, São Paulo, SP, Brazil

Abstract. This study analyzes the presence of volumetric changes in subfields of the hippocampus using structural MR image from a large database of volunteers, selected and screened in three distinct groups of elderly subjects: diagnosed with Alzheimer's disease (AD) or cognitive impairment mild (MCI) and control group (CG). Images were processed using the FreeSurfer (FS) software methodology specifically developed for automated segmentation of these hippocampal subfields and this selected dataset has based a dissertation research about aging. Statistical tools allowed evaluating the presence of correlation between age and volumetric changes of hippocampal subfields in the advancing age. In addition, clustering analysis showed important similarities between the volumes of the hippocampal subfields of these subjects.

Keywords: Hippocampal subfields · Clusters analysis · Volumetry

1 Introduction

Magnetic resonance imaging (MRI) is widely used for mapping the anatomical changes in the human brain. The hippocampus is composed of several distinct and specialized subfields with different cognitive functions and due to its particular susceptibility to neurodegenerative processes; it is one of the main brain structures of interest for studies on aging.

Usually, the presence of atrophy in these hippocampal subfields is related to the emergence of diseases on advanced ages [3, 6, 17]. Hippocampal sclerosis, for example, is characterized by neuronal loss and gliosis (white matter alteration) specifically in CA1 [5, 10]. On the other hand, some studies on Alzheimer's disease (AD) report that there may be greater neurodegeneration in the subiculum early during the disease course, while other subfields such as the GD and CA3 would be relatively preserved [21]. In recent years, some automated methods of hippocampal segmentation have been proposed. Most of these methods use pre-segmented models (atlas) that are aligned to the input image in the segmentation process. FreeSurfer is an open-source

© Springer Nature Switzerland AG 2019
J. M. R. S. Tavares and R. M. Natal Jorge (Eds.): VipIMAGE 2019, LNCVB 34, pp. 296–306, 2019.
https://doi.org/10.1007/978-3-030-32040-9_31

software package for the analysis and visualization of structural and functional neuroimaging data from cross-sectional or longitudinal studies. It is the most commonly used computational software for automatic segmentation of hippocampal subfields. FreeSurfer can be used to extract the cortical envelope from a T1 MRI. It also registers the individual cortex surfaces to surface-based anatomical atlases (https://surfer.nmr.mgh.harvard.edu/fswiki/FreeSurferMethodsCitation). Recently, the FreeSurfer (FS version 5.3) received improvements that allowed the segmentation of hippocampal subregions to cranial magnetic resonance data acquired from the 1.5 or 3T equipment [22]. This work intends to analyze a retrospective database of structural MR images acquired in 123 elderly volunteers and evaluate the volume of their hippocampal subfields. Thereby we find significant relationships between the chronological age of the subjects, the diagnostic classification and the increase and decrease of volume among them. Further, we propose a hierarchical agglomerative clustering to investigate similarity within the formed subfields clusters and classifying assumptions associated with process imaging and interrelating explanatory variables.

Technology has a deep impact in studies of brain structure and the multivariate statistics could improve methods of dimensionality reduction and of pattern-searching algorithms for high-dimensional spaces. Widely used for exploratory and classificatory purposes, cluster analysis presents itself as a powerful tool for identifying and relating elements of a database. In structural analysis of MR images, the cluster analysis allowed to reveal relationship between the observations that could not be possible only from the individual observations and the conventional statistical methods.

2   Methods

2.1   MRI Acquisition

The data for this experiment was made available by the Laboratório de Neuroimagem em Psiquiatria (LIM21) from the Hospital das Clínicas. This lab provided MRI data sets acquired in elderly individuals over the course of a Projeto temático funded by FAPESP (Process: 2012/50329-6), approved by Comitê de Ética do HCFMUSP.

Structural MRI data were acquired in the Magnetic Resonance Sector of the Institute of Radiology (InRad) and the Heart Institute (InCor) of the Hospital das Clinicas of the Medical School of the University of São Paulo (HC-FMUSP). The protocol of the Philips Achieva 3T Apparatus was used, in addition to a volumetric T1 sequence (Fluid-Attenuated Investment Recovery - FLAIR) and also a T2 * sequence with TE = 30 ms, TR = 2000 ms, flip angle = 80°, FOV = 240 × 240, shear thickness = 4 mm, matrix = 80 × 80 [20].

2.2   Automated Segmentation of the Hippocampus Subfields Using FreeSurfer

All T1-weighted images were on the same workstation and they had processed for automatic cortical reconstruction and subcortical volumetric segmentation using FS image analysis software version 5.3 (http://surfer.nmr.mgh.harvard.edu/).

Image processing included the correction of the movement [15, 16] and the removal of non-cerebral tissue through a mixed procedure that utilized both, signal strength and SB connectivity, and surface geometry from its deformation followed by automated spatial registration [18]. Then, the design of the pial and SB surfaces was done by automatic segmentation of SB, SC and LCR volumetric structures [8, 9], intensity normalization [20] and the delineation (tessellation) of the boundaries between SC and SB.

The topology was automatically corrected [8, 19], ensuring that the surface has the topological properties of a sphere and surface deformation follow gradients of intensity to establish the boundaries between SC/SB and SC/CSF at sites where the highest gradient of intensity sets the transition to the other tissue class [7]. The subdivision of the subfields of the hippocampus was automated in the FreeSurfer (version 6.0), using a computational atlas of the hippocampal formation with the MR data (Fig. 1).

INPUT (1 mm T1)   FREESURFER 5.3   FREESURFER 6.0

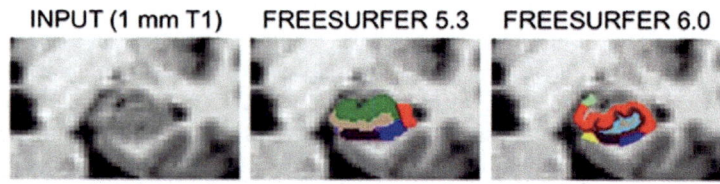

Fig. 1. Coronal T1-weighted image 1 mm scanning and the corresponding segmentation by FreeSurfer 5.3 and FreeSurfer 6.0 (https://surfer.nmr.mgh.havard.edu/fswiki/HippocampalSub fields, 2017).

This atlas was drawn using 15 samples of high-resolution autopsies (isotropic resolution 0.13 mm) from custom hardware resulting images manually segmented into 13 different sub-regions of the hippocampus. After, we segmented all the images by an algorithm that modified RM contrast variables [13].

The boundaries of the CA1, CA2, CA3, and subfields are blurred and to distinguish them we also used an algorithm to improve the quality of image through prior information about the thickness and intensity of the pyramid layer, combined with the location of these subfields [13] (Fig. 2).

The visual inspection of the images was done in the FreeSurfer package, as directed by its developers [11] and the quality control considered the final visualization of these maps of the hippocampal subfields. Finally, these maps were evaluated through superimposed on T1-weighted images.

This quality control protocol was developed by the ENIGMA consortium, an international neuroimaging enterprise which LIM-21 maintains partnership [12].

The protocol is running on the Shell software (http://www.tldp.org/LDP/Bash-Beginners-Guide/html/), using MatLab (www.mathworks.com/products/matlab/) for the production of the figures and the R (https://www.r-project.org/) to determine which hippocampal subfields images are outliers.

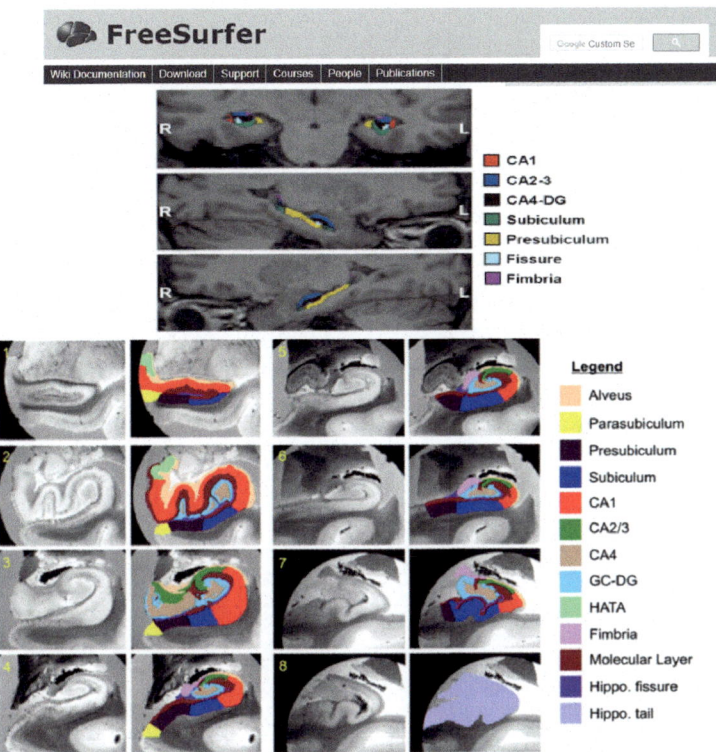

Fig. 2. Above: shows the subdivision of the sub-regions of the hippocampus with the mask overlapping of each region in the structural image T1. Below: shows sagittal sections of the hippocampus medial to lateral. Right: shows cortical sections represented as the anterior or posterior poles of the hippocampus [14].

2.3   Statistical Analysis

One hundred twenty-three subjects (123) between 58 and 90 years of age took part in the study, which had formed three groups of elderly people: 36 patients with Alzheimer's disease (AD), 44 patients with mild cognitive impairment (MCI) and 43 subjects without cognitive deficits (CG) (Fig. 3).

Statistical analyses were carried out using StatSoft Inc. (2011), STATISTICA (data analysis system) version 10 software (www.statsoft.com) and Matlab, version 7.14.0. 739 (R2012a, The MathWorks Inc., Natick, MA).

Volumes of the hippocampal subfields were normalized using the intracranial volume of each individual, thus composing a proportional relation between each subfield and the total intracranial volume of each patient, defined by:

$$srh_i = \frac{a_i}{vT_i} \; i = 1, 2, \ldots \ldots, 123,$$

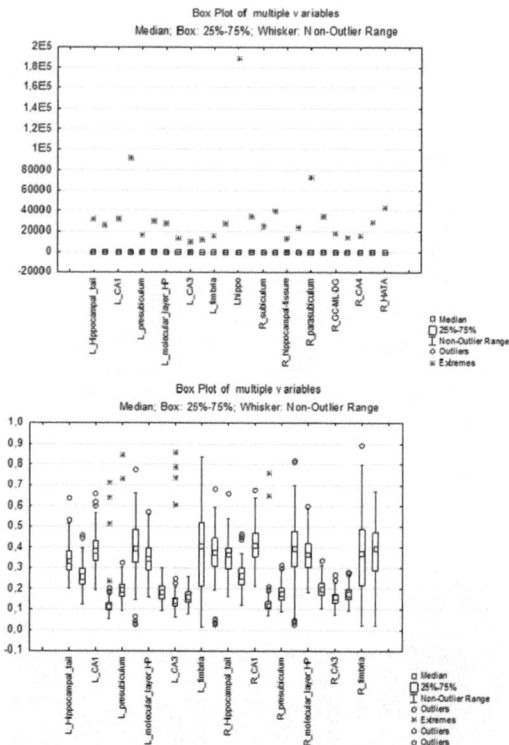

Fig. 3. Above: Box Plot shows outliers in hippocampal subfields in both hemispheres (n = 123). Below: After exclusion of the extreme value (AD paciente - subject 20), the Grubbs test confirmed that there were no more statistically significant outliers (n = 122).

Where,

$$srh_i = normalized\ volume\ of\ the\ hippocampal\ subfield\ of\ each\ individual\ i,$$

$$a_i = volume\ of\ the\ hippocampal\ subfield\ of\ each\ individual\ i,$$

$$vT_i = total\ intracranial\ volume\ of\ each\ individual\ i.$$

The volumes of each subfield had inspected and an outlier (subject 20) removed to ensure a 95% confidence interval. Then, the Grubbs test was used to corroborate that there is not extreme or atypical data in spite of the inherent variability in this finally sample (122 subjects).

Shapiro-Wilk normality test and Levene homogeneity test had made in all data set and they had considered statistically significant at p < 0.05. Mann-Whitney and t-student tests compared the differences in subfields volumes between right and left hemisphere in all groups of subjects. Pearson correlation between age and volume in the left and right hemisphere analyzed the subfields volumes trend in time.

Finally, Cluster Analysis conducted to explore the relationship between hip-pocampal subfield volumes and the previous diagnosis used to rank the groups (AD, MCI, and Control). The Euclidean distance had used to measure dissimilarity and the agglomerative method chosen to maximize the distance between the groups was Complete Linkage.

The Euclidean distance between two points p and q is the length of the line segment connecting them $(\overline{pq})$ and is given by Pythagorean formula:

$$d(pq) = d(qp) = \sqrt{\sum_{i=1}^{n} (q_i - p_i)^2}.$$

Complete linkage method defines the distance between two classes as the greatest distance that could be obtained if we select one element from each group and measure the distance between these elements. Thus, each cluster is formed when all the dis-similarities ('links') between pairs of objects in the cluster are less than a particular level. The complete linkage function is described by the following expression:

$$D(C_1, C_2) = max_{p \in C_1, q \in C_2}(d(p, q))$$

Where:

$$D(C_1, C_2) = distance\ between\ elements\ p \in C_1\ and\ q \in C_2$$
$$C_1\ and\ C_2 = two\ sets\ of\ elements\ (clusters)$$

Complete linkage clustering avoids the chaining phenomenon usually formed via Single linkage clustering where the elements may be forced to form a cluster even that the elements may be very distant to each other. Due this particularity, the Single linkage method tends to find compact clusters of approximately equal diameters.

3 Results

Although t-student test had showed no statistically differences to age into patients groups, Fig. 4 shows that the subjects with Alzheimer's disease present a slight neg-ative asymmetry when compared to the others.

Comparing volume measurements of the hippocampal subfield of each hemisphere groups we did not find any statistically significant difference between the MCI and Control Groups, only the MCI and DA groups showed significant differences (p < 0.05) in Subiculum, Molecular_layer_HP, GC-ML-DG, and CA3, only in the left hemisphere.

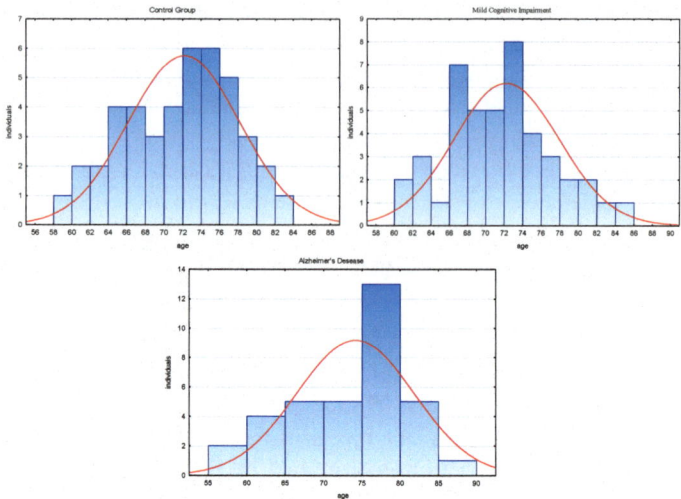

Fig. 4. Subject age distribution according previous diagnosis: we can see a similar pattern where 70% of the subjects in the control and MCI groups are above 69 years. In the AD group, we found 71% of individuals above this age.

However, comparing the DA and Control groups we found statistical significance in all subfields on the left hemisphere, except for Hippocampal-fissure (p = 0.362). In the right hemisphere, only the subfields Hippocampal_tail, Subiculum, CA1, Presubiculum, Molecular_layer_HP, and CA4 presented statistical significance (p < 0.05).

Pearson's correlations highlighted the trend of decreasing volume of hippocampal subfields with aging, although they had been all considered weak correlations.

Multivariate analysis showed the same pattern of similarity between left and right hippocampal subfields in the control group. However, when we examined the clusters in the AD group we saw the displacement of some subfields to other levels of similarity between the hemispheres.

We can observe a change in the grouping pattern of the G3 and CA3 subfields in the CCL and DA groups when compared to the control group. In addition, although the CA1 maintained the same grouping, it is possible to note that in the AD group there is a greater distance between theirs similarities (Fig. 5).

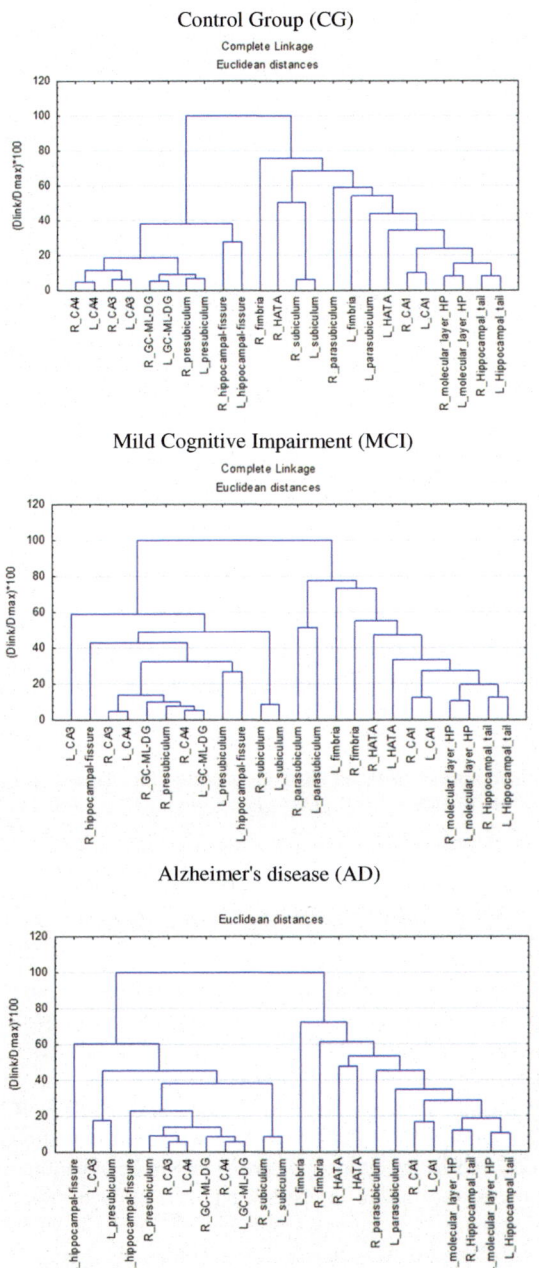

Fig. 5. Dendrograms present the clusters of hippocampal subfields in both hemispheres for each diagnostic group.

4 Discussion

This study investigated hippocampal subfields volumes using advanced automated hippocampal segmentation and it had found a different pattern among three distinct groups of elderly subjects.

A growing number of studies are employing the multivariate statistical techniques approaches to deal with the data generated by neuroimaging investigations, and cluster analysis should provide a good insight to compare groups and variables. Next step will study other methods of clustering to understand the effect of biomarkers in each patient group and thus, seek meaningful relationships that affect the volume of hippocampal subfields [3].

5 Conclusion

Studies are in progress to extend the dataset including newly processed images and to research multivariate statistical tools to analyze information on molecular processes in the brain and cerebrospinal fluid biomarkers [1–4].

Next steps are concerning about clarifying the hidden impact of both, preprocessing steps and statistical methods to reduce across-study heterogeneity.

This context has showed the importance of multidisciplinary works and it justifies the initiatives of several researchers from different subject field to improve the routines of laboratories.

Acknowledgments. Laboratory of Psychiatric Neuroimaging (LIM-21), Instituto de Psiquiatria, Hospital das Clinicas HCFMUSP, Faculdade de Medicina, Universidade de Sao Paulo, Sao Paulo, SP, BR.

References

1. Alves, T.C.T.F., Scazufca, M., Silveira, P.S., Duran, F.L.S., Duran, J.H.T., Vallada, H.P., Andrei, A., Wajngarten, M., Menezes, P.R., Busatto, G.F.: Subtle gray matter changes in temporo-parietal cortex associated with cardiovascular risk factors. J. Alzheimer's Dis. **27**, 575–589 (2011)
2. Alves, T.C.T.F., Rays, J., Junior, R.F., Wajngarten, M., Telles, R.M.S., Duran, F.L.S., Meneghetti, J.C., Robilotta, C.C., Prando, S., De Castro, C.C., Buchpiguel, C.A., Busatto, G. F.: Association between major depressive symptoms in heart failure and impaired regional cerebral blood flow in the medial temporal region: a study using 99mTc-HMPAO single photon emission computerized tomography (SPECT). Psychol. Med. **36**, 597–608 (2006)
3. Busatto, G.F., Buchpiguel, C.A., Britto, L.R.G., Tumas, V.: Neurociência Translacional da Doença de Alzheimer: Estudo pré-clínicos e clínicos do peptídeo amiloide e outros biomarcadores (2012)
4. Busatto, G.F., Garrido, G.E.J., Almeida, O.P., Castro, C.C., Camargo, H.P., Cid, C.G.: A voxel-based morphometry study of temporal lobe gray matter reductions in Alzheimer's disease. Neurobiol. Aging **24**, 221–231 (2003)

5. Crespel, A., Coubes, P., Rousset, M.C., Corinne, B., Rougier, A., Rodouin, G.: Inflammatory reactions in human medial temporal lobe epilepsy with hippocampal sclerosis. Brain Res. **952**(2), 159–169 (2002)
6. Ferreira, L.R.K., Busatto, G.F.: Resting-state functional connectivity in normal brain aging. Neurosci. Biobehav. Rev. **37**, 384–400 (2013)
7. Fischl, B., Dale, A.M.: Measuring the thickness of the human cerebral cortex from magnetic resonance images. Proc. Natl. Acad. Sci. U.S.A. **97**(20), 11050–11055 (2000)
8. Fischl, B., Liu, A., Dale, A.M.: Automated manifold surgery: constructing geometrically accurate and topologically correct models of the human cerebral cortex. IEEE Trans. Med. Imaging **20**(1), 70–80 (2001)
9. Fischl, B., Salat, D.H., Busa, E., Albert, M., Dieterich, M., Haselgrove, C., der Kouwe, A., Killiany, R., Kennedy, D., Klaveness, S., Montillo, A., Makris, N., Rosen, B., Dale, A.M.: Whole brain segmentation: automated labeling of neuroanatomical structures in the human brain. Neuron **33**(3), 341–355 (2002)
10. Flores, R., Joie, R.L., Gael, C.: Structural imaging of hippocampal subfields in healthy aging and Alzheimer's disease. Neuroscience **309**, 29–50 (2015)
11. FreeSurfer (2017). https://surfer.nmr.mgh.harvard.edu/fswiki/HippocampalSubfields
12. Hibar, D.P., Westlye, L.T., Erp, T.G.M., Rasmussen, J., Leonardo, C.D., Faskowitz, J., Haukvik, U.K., Hartberg, C.B., Doan, N.T., Agartz, I., Dale, A.M., Gruber, O., Krämer, B., Trost, S., Liberg, B., Abé, C., Ekman, C.J., Ingvar, M., Landén, M., Fears, S.C., Freimer, N. B., Bearden, C.E., the Costa Rica/Colombia Consortium for Genetic Investigation of Bipolar Endophenotypes, Sprooten, E., Glahn, D.C., Pearlson, G.D., Emsell, L., Kenney, J., Scanlon, C., McDonald, C., Cannon, D.M., Almeida, J., Versace, A., Caseras, X., Lawrence, N.S., Phillips, M.L., Dima, D., Delvecchio, G., Frangou, S., Satterthwaite, T.D., Wolf, D., Houenou, J., Henry, C., Malt, U.F., Bøen, E., Elvsåshagen, T., Young, A.H., Lloyd, A.J., Goodwin, G.M., Mackay, C.E., Bourne, C., Bilderbeck, A., Abramovic, L., Boks, M.P., van Haren, N.E.M., Ophoff, R.A., Kahn, R.S., Bauer, M., Pfennig, A., Alda, M., Hajek, T., Mwangi, B., Soares, J.C., Nickson, T., Dimitrova, R., Sussmann, J.E., Hagenaars, S., Whalley, H.C., McIntosh, A.M., Thompson, P.M., Andreassen, O.A.: Subcortical volumetric abnormalities in bipolar disorder. Mol. Psychiatry **21**, 1710–1716 (2016)
13. Iglesias, J.E., Augustinack, J.C., Nguyen, K., Player, C.M., Player, A., Wright, M., Roy, N., Frosch, M.P., McKee, A.C., Wald, L.L., Fischl, B., Leemput, K.V.: A computational atlas of the hippocampal formation using ex vivo, ultra-high resolution MRI: application to adaptive segmentation of in vivo MRI. NeuroImage **115**, 117–137 (2015)
14. Pizzi, S.D., Franciotti, R., Bubbico, G., Thomas, A., Onofrj, M., Bonanni, L.: Atrophy of hippocampal subfields and adjacent extrahippocampal structures in dementia with Lewy bodies and Alzheimer's disease. Neurobiol. Aging **40**, 103–109 (2016)
15. Reuter, M.R., Diana, H., Fischl, B.: Highly accurate inverse consistent registration: a robust approach. NeuroImage **53**(4), 1181–1196 (2010)
16. Reuter, M., Schmansky, N.J., Rosas, H.D., Fischl, B.: Within-subject template estimation for unbiased longitudinal image analysis. Neuroimage. **61**(4), 1402–1418 (2012)
17. Rosas, H.D., Liu, A.K., Hersch, S., Glessner, M., Ferrante, R.J., Salat, D.H., van der Kouwe, A., Jenkins, B.G., Dale, A.M., Fischl, B.: Regional and progressive thinning of the cortical ribbon in Huntington's disease. Neurology **58**(5), 695–701 (2002)
18. Ségonne, F., Dale, A.M., Busa, E., Glessner, M., Salat, D., Hahn, H.K., Fischl, B.: A hybrid approach to the skull stripping problem in MRI. Neuroimage **22**(3), 1060–1075 (2004)

19. Ségonne, F., Pacheco, J., Fischl, B.: Geometrically accurate topology-correction of cortical surfaces using nonseparating loops. IEEE Trans. Med. Imaging **26**(4), 518–529 (2007)
20. Sled, J.G., Zijdenbos, A.P., Evans, A.C.: A nonparametric method for automatic correction of intensity nonuniformity in MRI data. IEEE Trans. Med. Imaging **17**(1), 87–97 (1998)
21. Small, S.A., Scott, A.S., Buxton, R.B., Witter, M.P., Barnes, C.A.: A pathophysiological framework of hippocampal dysfunction in ageing and disease. Nat. Rev. Neurosci. **12**(10), 585–601 (2011)
22. Wisse, L.E., Biessels, G.J., Geerlings, M.I.: A critical appraisal of the hippocampal subfield segmentation package in FreeSurfer. Front. Aging Neurosci. **6**, 261 (2014). https://doi.org/10.3389/fnagi.2014.00261

Automated Spinal Midline Delineation on Biplanar X-Rays Using Mask R-CNN

Zixin Yang[1], Wafa Skalli[1], Claudio Vergari[1], Elsa D. Angelini[2,3], and Laurent Gajny[1(✉)]

[1] Institut de Biomécanique Humaine Georges Charpak,
Arts et Métiers Institute of Technology, 151 boulevard de l'Hôpital,
75013 Paris, France
laurent.gajny@ensam.eu
[2] LTCI, Telecom ParisTech, Department Image-Data-Signal, Paris, France
[3] ITMAT Data Science Group, NIHR Imperial BRC, Imperial College London,
London, UK

Abstract. Manually annotating medical images with few landmarks to initialize 3D shape models is a common practice. For instance, when reconstructing the 3D spine from biplanar X-rays, the spinal midline, passing through vertebrae body centers (VBCs) and endplate midpoints, is required. This paper presents an automated spinal midline delineation method on frontal and sagittal views by using Mask R-CNN. The network detects all vertebrae from C7 to L5, followed by vertebrae segmentation and classification at the same time. After postprocessing to discard outliers, the vertebrae mask centers were regarded as VBCs to get the spine midline by polynomial fitting. Evaluation of the spinal midline on 136 images used root mean square error (RMSE) with respect to manual ground-truth. The RMSE ± standard error values of predicted spinal midlines (C7–L5) were 1.11 mm ± 0.67 mm on frontal views and 1.92 mm ± 1.38 mm on sagittal views. The proposed method is capable of delineating spinal midlines on patients with different spine deformity degrees.

Keywords: Biplanar X-rays · Spine 3D reconstruction · Mask R-CNN · Spinal midline

1 Introduction

Scoliosis such as adolescent idiopathic scoliosis (AIS) is a three-dimensional (3D) local and global deformation of the spine. Clinical parameters, like the Cobb angle and vertebrae axial rotation, are essential both for diagnosis [1], treatment planning [2] and decision follow up [7].

The EOS system (EOS Imaging, Paris, France) allows diagnosing scoliosis, by taking radiographs in frontal and sagittal views in standing position in order to perform 3D reconstruction. In comparison with computerized tomography (CT) and magnetic resonance imaging (MRI), it has advantages in terms of low radiation dose and accurate deformity assessment in standing position [3–5]. Semi-automatic 3D reconstruction methods have been introduced and, for instance, the method in [6] is used in daily

© Springer Nature Switzerland AG 2019
J. M. R. S. Tavares and R. M. Natal Jorge (Eds.): VipIMAGE 2019, LNCVB 34, pp. 307–316, 2019.
https://doi.org/10.1007/978-3-030-32040-9_32

routine to measure clinical parameters. However, it requires supervision and training to adjust anatomical features of a parametric statistical shape model. Vertebral occlusions, due to projections of soft tissues, organs, braces, air cavities and other bones, lead to additional manual adjustments (meantime over 12 min [7]), which limits their use in the clinical workflow.

Recently, research works have aimed to automated 3D spine reconstruction [8–10]. In general, the first step is to initialize a simplified statistical shape model [6] thanks to several landmarks such as the spinal midline, passing through endplates or vertebral body centers. In [8], they use manual inputs while in [9] they rely on a fully convolutional neural network (CNN) combined with an additional differentiable spatial to numerical (DSNT) layer to predict the locations of landmarks. In [10], a coarse localization based on the image intensity distribution of columns and rows is used, followed by a detailed statistical shape model (SSM) for landmarks position prediction and CNN patch-based regression models to correct the landmarks.

In this work, contributing to automatic statistical shape model initialization, we present a method to delineate the spinal midline from vertebra C7 to vertebra L5 from both sagittal and frontal views. The method uses Mask R-CNN [12, 13] to segment and identify vertebrae. A post-processing method is introduced to remove outliers. Mask centers are regarded as vertebrae centers and polynomial fitting is applied to get the spinal midline.

2   Materials and Methods

2.1   Database

A database of 136 biplanar X-rays of 92 asymptomatic subjects and 44 scoliotic patients (Cobb Angle = $34.7° \pm 19.9°$) has been collected retrospectively. Every subject underwent the EOS™ ultra-low dose system (EOS imaging, Paris, France) under a protocol validated by the Ethical Committee (C.P.P. Ile de France VI). 3D reconstructions were performed by trained experts using the method from [6] for T1–L5 and from [14] for cervical spine reconstruction (C3–C7). Vertebral bodies from digitally reconstructed spines were backprojected on frontal and sagittal views (Fig. 1). Vertebrae masks from these projections were regarded as our manual ground-truth references.

2.2   Pre-processing

Images from sagittal and frontal views were automatically cropped based on the backprojected vertebral bodies to only contain vertebrae from C7 to L5, and resized to 512×216 pixels. In the resizing process, original ratios were preserved. Both images were preprocessed by adaptive noise-removal filtering, median filtering and contrast-limited adaptive histogram equalization to decrease noise and enhance contrast [17]. Each pair of planar radiographs were combined into a single image with a fixed size of 512×512 pixels, in which the left half consisted of the sagittal projection whereas the right half included the frontal projection (Fig. 2a).

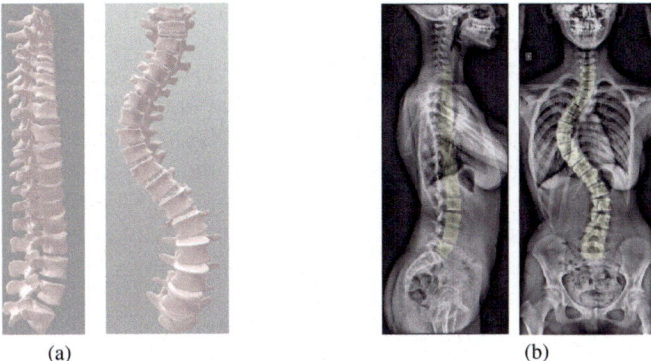

Fig. 1. Ground-truth generation: (a) A digitally reconstructed 3D spine; (b) Backprojections of vertebrae bodies on sagittal and frontal view images.

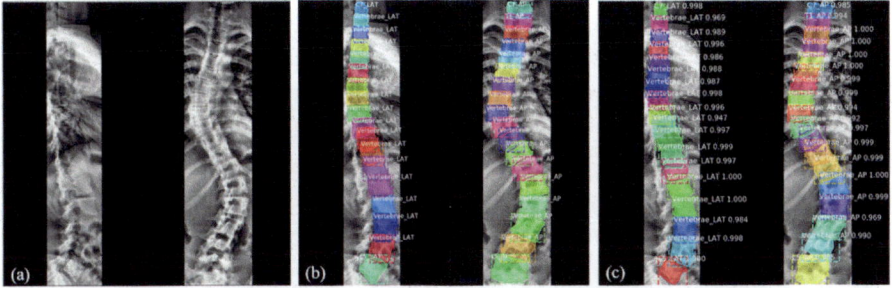

Fig. 2. Inputs and outputs of Mask R-CNN: (a) Input image; (b) Input training labels with vertebrae bounding boxes, masks and class labels. (c) Output predicted bounding boxes, masks, class labels and scores. On the sagittal view, vertebrae were divided into C7, T1–L4, and L5. On the frontal view, vertebrae were divided into C7, T1, T2–L4 and L5.

2.3 Mask R-CNN

Mask R-CNN [11] was chosen in this work to perform the spinal midline delineation. It is a two-stage detector that here localizes vertebrae firstly by predicting bounding boxes containing objects of interest, followed by joint segmentation and class labeling in the second stage. Mask R-CNN was proposed in 2017 as an extension of Faster R-CNN [19] for semantic segmentation. This was achieved by adding a branch to the network for predicting segmentation masks on region of interests. The loss function for training this network was:

$$L = L_{cls} + L_{box} + L_{mask} \tag{1}$$

This is a multi-task loss function combining the loss of classification (L_{cls}), localization (L_{box}) and segmentation mask (L_{mask}). We have used Mask R-CNN based on the implementation by Matterport Inc. [12] released under an MIT License.

Some classes needed to be defined. C7, L5 from sagittal view, and C7, T1, L5 from frontal view were assigned to individual classes. On the sagittal view, vertebrae in T1–L4 (16 vertebrae) were assigned to the same class. On the frontal view, T2–L4 (15 vertebrae) were assigned to the same class. In total, in both views, 18 vertebrae had to be detected, segmented and classified into 7 classes (Fig. 2b). The data was split for 5-fold cross-validation without overlap for training and testing with a ratio of 4:1. Augmentation techniques included randomized rotations ±3° and contrast normalization [18] to generalize unseen data. Mask R-CNN predicted vertebrae bounding boxes and masks with their class names and class probability scores (Fig. 2c).

2.4   Post-processing and Evaluation

Firstly, we discarded sagittal masks and frontal masks whose class scores were below 0.65 (Fig. 3a). Secondly, for each view, each mask center (x_c, y_c) was calculated. On each view, predicted masks were fitted by a sixth order polynomial function:

$$X_m = f_m(Y_m),$$

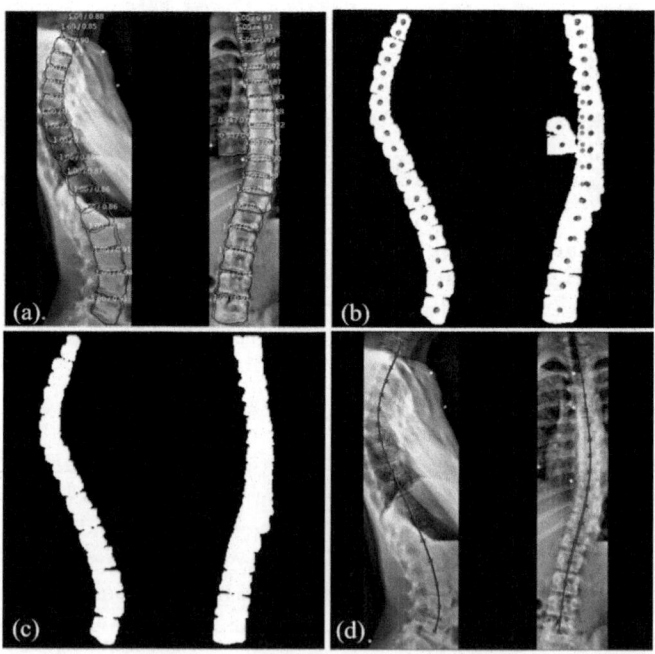

Fig. 3. Post-processing workflow. (a) Select predicted masks with class scores above 0.65. Ground truth contours (red) and predicted contours (blue) are shown. (b) Ground truth mask centers are in red and predicted mask centers are in blue. Each predicted mask center has a corresponding point (green) predicted by the polynomial fit of predicted masks. (c) After removal of predicted masks with a horizontal distance between its blue and green points above 15 pixels. (d) Final spinal midlines fitted by a 6th-order polynomial function. Ground truth in red and predicted in blue.

where (X_m, Y_m) represents locations of all pixels within the mth predicted mask. New horizontal locations of each mask centers x'_c were calculated by $x'_c = f_m(y_c)$ (green points in Fig. 3b). Then, the horizontal distance d was calculated as $d = |x'_c - x_c|$. Masks were discarded if $d > 15$ *pixels* (Fig. 3c). Finally, remaining masks centers were fitted on each view by another sixth order polynomial function to obtain the final spinal midline on frontal and sagittal views from C7 to L5 (Fig. 3d).

The segmentation performance was evaluated with the Dice similarity coefficient [14]. To evaluate the identification task, we defined correct detection when a vertebra was assigned the right class name and when its Dice value was above 0.5. Based on this definition, the precision and accuracy for identification were calculated. The first one calculated the percentage of correct predictions among all predictions while the later one calculated the percentage of ground-truth vertebrae that were correctly detected.

We used root mean square error (RMSE) to evaluate the distance between the predicted and ground-truth spinal midlines:

$$RMSE(pixel) = \sqrt{\frac{1}{N}\sum_{i=1}^{N}(Ref_i - Pred_i)^2} * ratio_{resize} * ratio_{mm} \qquad (2)$$

We sampled $N = 170$ points from the ground-truth spinal midline between the C7 and the L5 body centers. *Ref* stands for the horizontal locations of sampled ground-truth points. (*Pred*) stands for the horizontal locations of the predicted points at the same vertical locations. RMSE values were transformed into millimeter by adjusting for the resize ratio $ratio_{resize}$ and multiplying by the pixel to physical size (mm) ratio $ratio_{mm}$.

Similarly, we used Eq. 2 to evaluate the spinal midline RMSE in sub-regions (e.g. T1–T4) after truncating all midline (C7–L5).

3   Experimental Results

The experiments were performed on a PC with Intel Core i7 2.8 GHz CPU, 16 GB memory, and NVIDIA GeForce GTX 1050 GPU, based on Python. We used ResNet101 as the backbone architecture.

3.1   Segmentation and Identification Evaluation

We report segmentation and identification performances in Table 1. Vertebrae L5 has the highest average Dice scores: 0.92 in frontal view and 0.91 in sagittal view. Accuracy (the percentage of detected vertebrae) and precision (the percentage of correct predictions among all predictions) of vertebrae C7 had lower value than other classes in both views.

Table 1. Dice similarity coefficient: mean ± SD. Identification: accuracy and precision.

Name	Dice	Accuracy	Precision
C7 (Frontal)	0.88 ± 0.06	0.970	0.936
T1 (Frontal)	0.88 ± 0.05	0.993	0.957
T2–L4 (Frontal)	0.91 ± 0.05	0.992	0.999
L5 (Frontal)	0.92 ± 0.03	1.0	0.965
C7 (Sagittal)	0.87 ± 0.06	0.963	0.942
T1–L4 (Sagittal)	0.88 ± 0.08	0.976	0.992
L5 (Sagittal)	0.91 ± 0.03	0.993	0.971

3.2   Spinal Midline Evaluation

Table 2 reports RMSE (mm) values from the frontal views and Table 3 from the sagittal views. On the frontal views, we got an average RMSE of 1.11 mm and 0.67 mm standard error in the C7–L5 region. T5–T8, T9–T12 and L1–L5 on frontal views have lower RMSE than other regions. In comparison, spinal midline delineation of C7–L5 on sagittal views (Table 3) had larger errors than on frontal views, with an average RMSE of 1.92 mm and a standard error of 1.38 mm. In T1–T4 sagittal views, 39 subjects (28.68%) had RMSE values above 3 mm, which is almost four times the proportion seen in frontal views.

Table 2. RMSE (mm) on frontal views: mean ± SD (Max) for vertebrae body centers location and spinal midline delineation from 136 subjects. Number (%) of subjects with RMSE > 3 mm.

Frontal	RMSE (mm)	>3 mm
C7	1.55 ± 1.23 (6.52)	18 (13.2%)
T1–T4	1.18 ± 1.31 (9.42)	10 (7.35%)
T5–T8	0.87 ± 0.85 (4.27)	5 (3.68%)
T9–T12	0.70 ± 0.78 (7.45)	2 (1.47%)
L1–L4	0.90 ± 0.66 (4.23)	2 (1.47%)
L5	1.41 ± 1.24 (8.84)	10 (7.35%)
All	1.11 ± 0.67 (5.02)	3 (2.21%)

Table 3. RMSE (mm) on sagittal views: mean ± SD (Max) for vertebrae body centers location and spinal mid-line delineation from 136 subjects. Number (%) of subjects with RMSE > 3 mm.

Sagittal	RMSE (mm)	>3 mm
C7	1.91 ± 4.82 (36.05)	14 (10.29%)
T1–T4	2.60 ± 2.36 (13.99)	39 (28.68%)
T5–T8	1.58 ± 1.42 (13.90)	9 (6.62%)
T9–T12	1.40 ± 1.79 (13.04)	14 (10.29%)
L1–L4	1.02 ± 0.79 (13.04)	5 (3.68%)
L5	1.33 ± 1.94 (20)	9 (6.62%)
All	1.92 ± 1.38 (9.44)	16 (11.76%)

4 Discussion

Results showed that the proposed method was capable of automatically delineating spinal midline from vertebrae C7 to L5 on X-rays radiographs from frontal and sagittal views acquired on either asymptomatic or scoliosis subjects (Fig. 4).

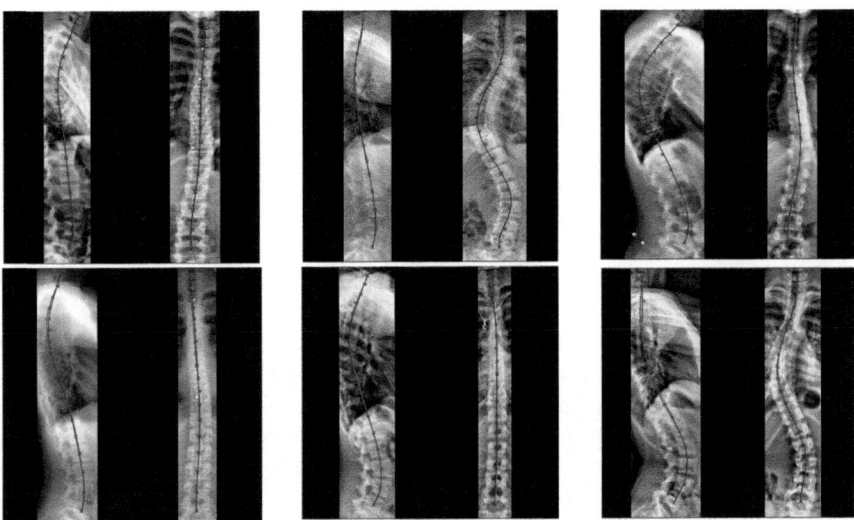

Fig. 4. Spinal midline delineation examples. Ground-truth in red, prediction in blue.

In the method, although reconstructions from C3 to L5 were available, we chose to detect vertebrae from C7–L5 within seven classes: C7, T1–L4 (16 vertebrae), L5 from sagittal view, and C7, T1, T2–L4 (15 vertebrae), L5 from frontal views. This decision was based on properties of the spine projections on biplanar X-rays and the mechanism of Mask R-CNN. On the frontal views, we chose C7 as the starting vertebra for the spinal midline because vertebrae C3–C6 are likely to be occluded by patients' head. Our preliminary experiments confirmed that the network was unable to combine correctly spatial location information from vertebrae in C3–C6. Due to similar appearances, vertebrae T2–L4 were likely to also be wrongly classified but ended up being correctly segmented. Quite distinct from T2–L4, vertebrae C7, vertebrae T1 and vertebrae L5 were correctly segmented and identified as individual classes. In sagittal views, T1–T4 are commonly occluded, as confirmed by the high T1–T4 RMSE value in Table 3. For this reason, we did not separate T1 as an individual class on the sagittal view. C3–C6 are clearly visible on sagittal views, but we did not study them in this work, keeping the same field of view on the two views. In our experiments, increasing the field of view had negative impacts on the outputs.

Mean Dice values of each class, all above 0.87, compare well with [16] where they studied only frontal views using a smaller dataset (35 images). Our mean RMSE (SD) 1.11 mm ± 0.67 mm on frontal views and 1.92 mm ± 1.38 mm on sagittal views are also comparable with the reconstruction results in [10], with 1.6 (1.3) mm for mean 3D Euclidean distance (SD) errors of VBC landmark locations.

Performances of our method was inevitably influenced by vertebral occlusion, especially in the thoracic region on sagittal views. In our dataset, we had many images with strong occlusions. Overlay projections of the shoulder, ribcages and hands contributed to the high RMSE errors in T1–T4 (sagittal views), shown in Table 2. As the vertebra C7 was likely occluded by the patient's head in frontal view and the shoulder in sagittal views, it has relative high errors in Table 2. Signals from organs, local tissues and air cavities were sometimes mistaken for vertebrae.

From visual examination, spinal midlines with RMSE < 3 mm will not likely cause any problem for the statistical shape model initialization. Overall, on frontal views, only 2.21% of spinal midlines with RMSE > 3 mm might need minor corrections. Influenced by vertebrae occlusion on the T1–T4 sagittal views, 11.76% of spinal midlines had RMSE values above 3 mm. However, the errors obtained on T1–T4 spinal midline on sagittal view could be decreased by regressing a statistical shape model in which greater weights would be assigned to the most reliable spinal midline segments.

Recent papers exploiting convolutional networks can be divided in two categories. The first category predicts spinal landmarks [9, 15] or segments vertebrae [16] to measure directly some clinical parameters, mainly the Cobb angle, on landmarks or masks. The second category predicts landmarks and uses them to initialize a statistical shape model [10]. This enables full spine reconstruction, and the evaluation of more clinical parameters. In [16], vertebrae were localized using a series of traditional imaging processing steps. However, the processing steps were tuned on a specific dataset, thus prone to failure on other cohorts. In comparison, the proposed method using Mask R-CNN has demonstrated great robustness and is one of the most popular architecture for segmentation in computer vision. We exploited its potential in vertebrae identification and segmentation on biplanar X-rays in frontal and sagittal views. The main difficulty of applying the network to biplanar X-rays was the presence of outliers which were successfully removed by the post-processing method.

5   Conclusion

We presented an automated spinal midline delineation method on biplanar X-rays images using a Mask R-CNN. The results suggested that we can obtain robust and accurate spinal midlines from C7 to L5, especially in frontal views. The spinal midlines in sagittal views were less accurate due to occlusions by the shoulder and ribcage. The method should contribute to the automated initialization of a statistical shape model, the first step in current spine reconstruction methods. Future works will focus on the detection of additional landmarks that are required for the accurate adjustment of a 3D spine model on biplanar X-rays images.

Acknowledgments. The authors thank the ParisTech BiomecAM chair program, on subject-specific musculoskeletal modelling and in particular Société Générale and COVEA. The authors would also like to thank François Girinon for having initiated this work.

References

1. Skalli, W., Vergari, C., Ebermeyer, E., Courtois, I., Drevelle, X., Kohler, R., Abelin-Genevois, K., Dubousset, J.: Early detection of progressive adolescent idiopathic scoliosis: a severity index. Spine **42**(11), 823–830 (2017)
2. Vergari, C., Courtois, I., Ebermeyer, E., Bouloussa, H., Vialle, R., Skalli, W.: Experimental validation of a patient-specific model of orthotic action in adolescent idiopathic scoliosis. Eur. Spine J. **25**(10), 3049–3055 (2016)
3. Brenner, D.J., Hall, E.J.: Computed tomography—an increasing source of radiation exposure. N. Engl. J. Med. **357**(22), 2277–2284 (2007)
4. Yazici, M., Acaroglu, E.R., Alanay, A., Deviren, V., Cila, A., Surat, A.: Measurement of vertebral rotation in standing versus supine position in adolescent idiopathic scoliosis. J. Pediatr. Orthop. **21**(2), 252–256 (2001)
5. Dubousset, J., Charpak, G., Dorion, I., Skalli, W., Lavaste, F., Deguise, J., Kalifa, G., Ferey, S.: A new 2D and 3D imaging approach to musculoskeletal physiology and pathology with low-dose radiation and the standing position: the EOS system. Bull. de l'Academie nationale de medecine **189**(2), 287–297 (2005)
6. Humbert, L., De Guise, J.A., Aubert, B., Godbout, B., Skalli, W.: 3D reconstruction of the spine from biplanar X-rays using parametric models based on transversal and longitudinal inferences. Med. Eng. Phys. **31**(6), 681–687 (2009)
7. Ilharreborde, B., Steffen, J.S., Nectoux, E., Vital, J.M., Mazda, K., Skalli, W., Obeid, I.: Angle measurement reproducibility using EOS three-dimensional reconstructions in adolescent idiopathic scoliosis treated by posterior instrumentation. Spine **36**(20), E1306–E1313 (2011)
8. Gajny, L., Ebrahimi, S., Vergari, C., Angelini, E., Skalli, W.: Quasi-automatic 3D reconstruction of the full spine from low dose biplanar X-rays based on statistical inferences and image analysis. Eur. Spine J. **28**(4), 658–664 (2019)
9. Galbusera, F., Niemeyer, F., Wilke, H.J., Bassani, T., Casaroli, G., Anania, C., Costa, F., Brayda-Bruno, M., Sconfienza, L.M.: Fully automated radiological analysis of spinal disorders and deformities: a deep learning approach. Eur. Spine J. **28**, 1–10 (2019)
10. Aubert, B., Vazquez, C., Cresson, T., Parent, S., De, J.G.: Towards automated 3D spine reconstruction from biplanar radiographs using CNN for statistical spine model fitting. IEEE Trans. Med. Imaging (2019)
11. He, K., Gkioxari, G., Dollár, P., Girshick, R.: Mask R-CNN. In: Proceedings of the IEEE International Conference on Computer Vision, pp. 2961–2969 (2017)
12. Abdulla, W.: Mask R-CNN for object detection and instance segmentation on keras and tensorflow (2017). https://github.com/matterport/Mask_RCNN
13. Dice, L.R.: Measures of the amount of ecologic association between species. Ecology **26**(3), 297–302 (1945)
14. Rousseau, M.A., Laporte, S., Chavary-Bernier, E., Lazennec, J.Y., Skalli, W.: Reproducibility of measuring the shape and three-dimensional position of cervical vertebrae in upright position using the EOS stereoradiography system. Spine **32**(23), 2569–2572 (2007)
15. Wu, H., Bailey, C., Rasoulinejad, P., Li, S.: Automated comprehensive Adolescent Idiopathic Scoliosis assessment using MVC-Net. Med. Image Anal. **48**, 1–11 (2018)

16. Horng, M.-H., Kuok, C.-P., Fu, M.-J., Lin, C.-J., Sun, Y.-N.: Cobb angle measurement of spine from X-ray images using convolutional neural network. Comput. Math. Methods Med. **2019**, article ID 6357171, 18 p. (2019). https://doi.org/10.1155/2019/6357171
17. Ebrahimi, S., Gajny, L., Skalli, W., Angelini, E.: Vertebral corners detection on sagittal X-rays based on shape modelling, random forest classifiers and dedicated visual features. Comput. Methods Biomech. Biomed. Eng.: Imaging Vis. **7**(2), 132–144 (2019)
18. Jung, A.: imgaug (2017). https://github.com/aleju/imgaug
19. Ren, S., He, K., Girshick, R., Sun, J.: Faster R-CNN: towards real-time object detection with region proposal networks. In: Advances in Neural Information Processing Systems, pp. 91–99 (2015)

Breast Skin Temperature Evaluation in Lactating and Non-lactating Women by Thermography: An Exploratory Study

Ana Gouveia[1,2(✉)], Luís Pires[3], Nuno Garcia[4], Ana Barbosa[1],
Ana Jesus[3], Nuno Pombo[4], Marta Soares[5],
and José Martinez-de-Oliveira[1]

[1] Faculdade de Ciências da Saúde, Universidade da Beira Interior,
Av. Infante D. Henrique, 6200-506 Covilhã, Portugal
anagouveia@fcsaude.ubi.pt
[2] Instituto de Biofísica e Engenharia Biomédica, Faculdade de Ciências,
Universidade de Lisboa, 1749-016 Lisbon, Portugal
[3] Center for Mechanical and Aerospace Science and Technologies,
Universidade da Beira Interior, Calçada do Lameiro, 6200-001 Covilhã, Portugal
[4] Instituto de Telecomunicações, Universidade da Beira Interior,
Covilhã, Portugal
[5] Serviço de Obstetrícia e Ginecologia, Centro Hospitalar Universitário da Cova
da Beira, Covilhã, Portugal

Abstract. During pregnancy and lactation, woman breasts feel changes like blood flow increasement, associated with a higher breast temperature. We performed an exploratory study of the breast skin temperature of lactating and non-lactating women based on thermography, with a qualitative analysis of the temperature patterns and a quantitative evaluation of the differences. Frontal breast thermograms of four non-lactating young women and four women with well-established lactation were acquired and analyzed. Qualitative analysis of the images obtained show some evidence of the existence of a characteristic skin temperature pattern for lactating women. Quantitative differences between thermograms were also noticed, especially when considering dispersion metrics: lactating women present higher breast skin temperature gradients and amplitudes. Results obtained, especially based on central tendency metrics, should be interpreted with caution because some of the acquisition conditions for non-lactating women may lead to some bias on the results. Further investigation will be performed to quantify breast skin temperature gradient and be able to classify images based in the breast skin temperature pattern.

Keywords: Infrared thermography · Lactation · Breastfeeding · Breast skin temperature

1 Introduction

Breastfeeding is an essential process to provide infants optimal nutrients for their development and an important protection for mothers [1]. During pregnancy and lactation, woman breasts feel many changes [2], like extension and branching of the ductal

© Springer Nature Switzerland AG 2019
J. M. R. S. Tavares and R. M. Natal Jorge (Eds.): VipIMAGE 2019, LNCVB 34, pp. 317–322, 2019.
https://doi.org/10.1007/978-3-030-32040-9_33

system, influenced by hormones (e.g., oestrogen, progesterone, prolactin), and blood flow increasement. This blood flow variation is associated with a higher metabolic activity and breast temperature, the same authors says, which remains during the lactation period until about 2 weeks after stopping lactation [3].

Literature focused in the evaluation of breast skin temperature is scarce. Kimura and Matsuoka [4] studied the changes of breast skin temperature during breastfeeding. Breast skin temperature was measured during breastfeeding in 11 women, from 1–2 days until 7–8 weeks postpartum, by probes attached on breasts skin. Other authors have studied breast temperature using thermography in lactating women with lactation problems, like engorgement [5] and fissures [6]. Heberle, Ichisato, and Nohama [7] studied different alterations in puerperal breast based in thermography, as well as clinical examination and pressure algometry, in engorgement or mastitis situations. However, no research was found studying temperature skin lactating breasts without problems and based in thermography, as well as evaluating temperature differences of breast skin between non-lactating and lactating women. In this way, the aims of this work are (1) the qualitative study of skin temperature patterns of lactating and non-lactating women and (2) the quantitative evaluation of breast skin temperature differences between the same groups.

2 Materials and Methods

Thermographic acquisitions were performed using a Flir camera, E50 model, equipped with an uncooled microbolometer Focal Plane Array detector. It has a thermal sensitivity of 0.05 °C, at 30 °C, and it is sensitive in the spectral range of 7.5 to 13 μm. The focus mode is exclusively manual and the image resolution is 240 × 180 pixels.

Thermograms of the frontal and lateral area of the breasts were obtained, with arms up and down, for four non-lactating young women (20 years old) and four women with well-established lactation (39 to 46 years old), before breastfeeding their infants (5 and 32 months old). Acquisitions were carried out in different locations but with similar thermal conditions, with an ambient temperature of about 20 °C and relative humidity of 50%. Care was taken to ensure that there weren't relevant radiative sources within the rooms where acquisitions took place. The distance between the camera ant the imaged women area was approximately the same in all acquisitions (1 m). Acquisitions were preceded by a 10 min thermal acclimatization period in which they were naked above the waist.

For an adequate temperature evaluation and comparison of images, we only considered frontal images (and with arms down) because in these conditions we can observe a larger and more irrigated area of the breast. We started by extracting two regions of interest, one for each breast, in each frontal thermogram. Due to amorphous nature of infrared images and their unclear anatomical limits, this can be a difficult task, and most authors prefer non-automatic methods [8]. We performed a manual region of interest (ROI) extraction in thermograms using visual inspection of correspondent breast photography whenever it was possible. The breast ROI defined was the mammary fold below the axillar level as shown in Fig. 1.

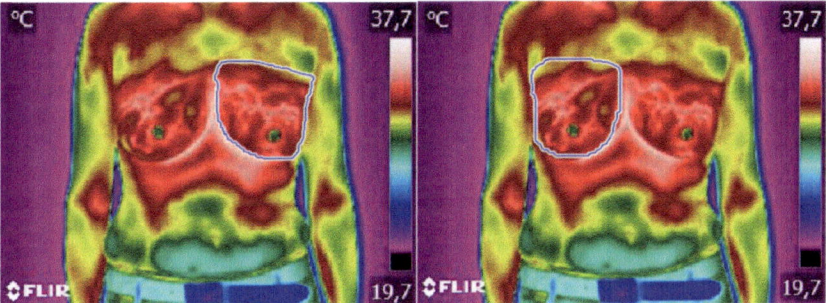

Fig. 1. Breast ROI definition for left breast (left image) and right breast (right image) (woman #6).

Computations were performed for the eight women mentioned and considering both breasts separately.

3 Results

Thermograms of the breast frontal view and with arms down, obtained for the eight women, are showed in Fig. 2.

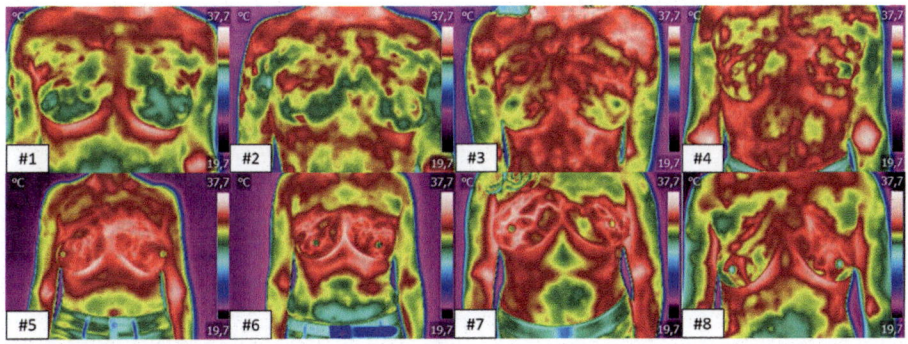

Fig. 2. Thermograms of four non-lactating (upper images - #1 to #4) and four lactating women (below images - #5 to #8).

From visual inspection and although each woman has her specific skin temperature distribution, some interesting differences are highlighted from visual inspection: breasts in non-lactating women have areas with more homogenous temperatures (specially cases #1 and #2); breasts in lactating women seem to have a common pattern with very hot well-defined lines (colored white in images). Another interesting point is the difference between right and left breasts in lactating women, specially #5, #7 and #8. In these cases, the breast with more hot line concentration agrees with the one that produces more milk or it is the infant favorite – left breast for #5 and #8, right breast for

#7 - according women report. In case of woman #6, it wasn't reported any difference in milk production or baby preference.

Histograms and boxplots for each breast and each woman were performed (Figs. 3 and 4).

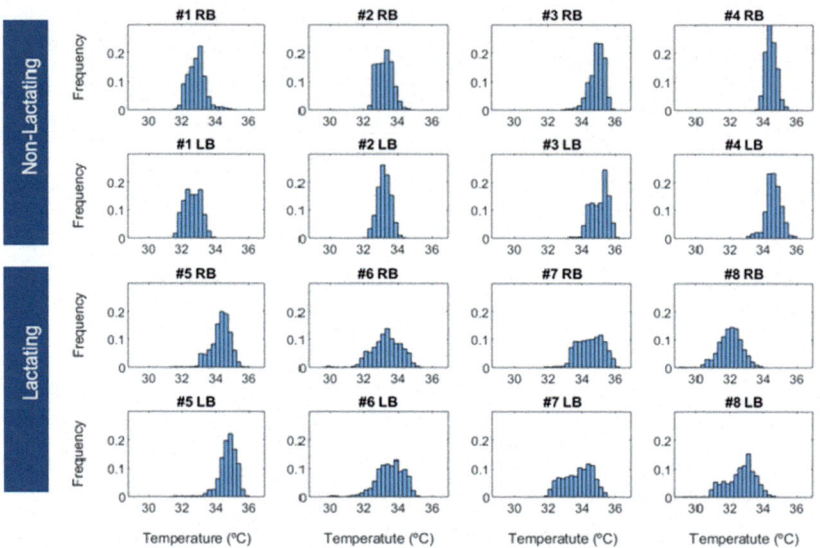

Fig. 3. Histograms of ROI thermograms for right breast (RB) and left breast (LB) for non-lactating (#1 to #4) and lactating (#5 to #8) women. YY axis refers to relative frequency.

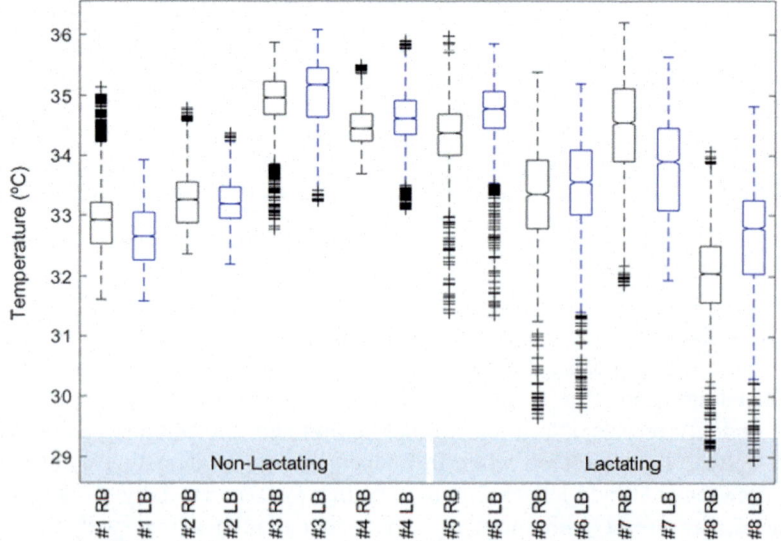

Fig. 4. Boxplots of ROI thermograms for right breast (RB) and left breast (LB) for non-lactating (#1 to #4) and lactating (#5 to #8) women.

In general, histograms of non-lactating women are sharper than those of lactating women, which relative frequency values are more spread, which is confirmed by boxplots that show clearly more dispersion in lactating temperature values then in non-lactating. On another hand, there is no evidence of any tendency for higher temperatures for one of the groups.

4 Discussion

Qualitative analysis of the images obtained show some evidence of the existence of a characteristic skin temperature pattern for lactating women. Even considering individual characteristics and milk production differences due to infant's needs (from 5 to 32 months), there are similarities between thermal images of lactating women and there are differences when comparing to thermal images of non-lactating cases. Another interesting observation was the prominence breast skin temperature pattern for one of the breasts of three lactating women that reported a milk production imbalance or infant preference. We also noticed some quantitative differences between thermograms, especially when considering dispersion metrics: lactating women present higher breast skin temperature gradients and amplitudes. However, our results do not reveal a higher breast skin temperature for lactating women, like some authors justify physiologically [2] or claim by their own results [4]. Some other conditions besides the lactation occurrence can affect skin temperature [9, 10] and can justify our results: menstrual phase of lactating women, insufficient time to climatization of non-lactating women and stress during acquisition, etc. Although the thermal and humidity conditions, acquisition context for non-lactating women could add some stress that eventually affected results. More care must be taken to choose the place and time to acquisitions, for decreasing stress as lower as possible.

A challenging step of the analysis was the limits definition of ROI. Several trials were performed – larger and smaller region, from images with arms up and down – but these options did not change much the results. The possibility of visual inspection of breast photography taken simultaneously was very important.

5 Conclusion

This paper proposes a qualitative and a quantitative study of breast skin temperature patterns of non-lactating and well-established lactating women. An exploratory study is presented.

Our experiments revealed the existence of a characteristic skin temperature pattern for lactating women. In addition, quantitative differences between lactating and non-lactating thermograms were observed, especially when considering dispersion metrics. However, these results should be interpreted with caution because some of the acquisition conditions for non-lactating women may lead to some bias on the results.

Further investigation is required to better understand the characteristics and differences of skin temperature patterns of lactating and non-lactating women. Effort must be made to improve the protocol and create better acquisition conditions as we earlier mentioned. Nevertheless, the main focus will be in image processing and classification, to quantify breast skin temperature gradient and be able to classify images based in the breast skin temperature pattern.

References

1. Lawrence, R.A., Lawrence, R.: Breastfeeding: A Guide for the Medical Profession. Mosby Inc., St. Louis (2006)
2. Geddes, D.T.: Inside the lactating breast: the latest anatomy research. J. Midwifery Women's Health **52**(6), 556–563 (2007)
3. Thoresen, M., Wesche, J.: Doppler measurements of changes in human mammary and uterine blood flow during pregnancy and lactation. Acta Obstet. Gynecol. Scand. **67**(8), 741–745 (1988)
4. Kimura, C., Matsuoka, M.: Changes in breast skin temperature during the course of breastfeeding. J. Hum. Lact. **23**(1), 60–69 (2007)
5. dos Santos Heberle, A.B., de Moura, M.A.M., de Souza, M.A., Nohama, P.: Assessment of techniques of massage and pumping in the treatment of breast engorgement by thermography. Rev. Lat. Am. Enfermagem **22**(2), 277–285 2014
6. di Migueli, M.F., da Luz, S.C.T., dos Santos, K.M., da Silva Honório, G.J., da Roza, T.: Thermography study of nipple-areola complex in immediate puerperas. Man. Ther. Posturol. Rehabil. J. **14**(48), 2236–5435 (2016)
7. dos Santos Heberle, A.B., Ichisato, S.M.T., Nohama, P.: Breast evaluation during lactation using thermography and pressure algometry. Acta Paul. Enferm. **28**(3), 256–263 (2015)
8. Borchartt, T.B., Conci, A., Lima, R.C.F., Resmini, R., Sanchez, A.: Breast thermography from an image processing viewpoint: a survey. Signal Process. **93**(10), 2785–2803 (2013)
9. Chudecka, M., Lubkowska, A.: Thermal maps of young women and men. Infrared Phys. Technol. **69**, 81–87 (2015)
10. Fernández-Cuevas, I., et al.: Classification of factors influencing the use of infrared thermography in humans: a review. Infrared Phys. Technol. **71**, 28–55 (2015)

Automatic Measurement of the Volume of Brain Ventricles in Preterm Infants from 3D Ultrasound Datasets

Lionel C. Gontard[1]([✉]), Joaquin Pizarro[1],
Isabel Benavente-Fernández[2], and Simón P. Lubián-López[2]

[1] Department of Informatics, ESI, University of Cádiz, 11519 Puerto Real, Spain
lionel.cervera@uca.es
[2] Neonatology Department, Puerta Del Mar University Hospital, Cadiz, Spain

Abstract. In this work we evaluate and validate the lateral ventricular brain volume obtained using transfontanellar three-dimensional ultrasound (3D US). The goal is to improve diagnosis of excessive dilation of ventricles after germinal matrix- intraventricular hemorrhage in the preterm infant, a complication known as posthemorrhagic ventricular dilatation, with a computer-aid method that is not operator-dependent. To this end, a software has been developed for automatic segmentation of 3D US using conventional image processing techniques. It significantly reduces the processing time compared to the state-of-the-art methods. The results are discussed with respect to the measurements obtained using manual methods of segmentation that are considered as the gold standard.

Keywords: Transfontanellar 3D US · 3D image processing · PHVD

1 Introduction

Very low birth weight preterm infants (VLBW), defined as those with a birth weight equal or less than 1500 g and/or a gestational age equal or less than 32 weeks present an increased risk of *germinal matrix- intraventricular* hemorrhage (GM-IVH). One third of those that with a severe form of GM-IVH develop posthemorrhagic ventricular dilatation (PHVD). This complication can evolve generating increased intraventricular pressure, hydrocephalus that requires the surgical placement of an intraventricular catheter to allow CSF removal. Despite the negative impact on neuro developmental outcome [1, 2] there is still no clear international consensus about the best time to intervene or what are the best measures to guide the diagnostic-therapeutic action.

In routine clinical practice, ventricular growth is monitored by US over time (days or weeks) and in the case of PHVD therapeutic intervention is carried out [3]. The most generalized approach to monitoring the progression of ventricular growth is based on 2D US linear measurements of the ventricles [4]. Nevertheless, these type of linear measurements have great limitations since they do not reflect the complexity of the ventricular system. They are also operator-dependent and inexact, so it is necessary to search for more exact measures of the ventricular size.

© Springer Nature Switzerland AG 2019
J. M. R. S. Tavares and R. M. Natal Jorge (Eds.): VipIMAGE 2019, LNCVB 34, pp. 323–329, 2019.
https://doi.org/10.1007/978-3-030-32040-9_34

3D ultrasound (3D US) represents a very important leap in the quality of the study of PHVD [5, 6]. However, to incorporate 3D US for diagnosis of PHVD in a clinical setting, an accurate and efficient segmentation algorithm is crucial to extract information from the lateral ventricles of US 3D images. The segmentation of organs from 3D US images poses many challenges, such as noise, low contrast of soft tissues and less details of the image of the structures. Most often, 3D US requires intensive manual segmentation by a trained physician. Some studies have quantified 3D US ventricular volumes in neonates using manual segmentation, which implies a high processing time and inter-subject variability [7–9].

Automatic processing of these datasets is therefore essential. The processing of the three-dimensional volume of data generated with 3D US generally consists of segmenting each of the 2D sections of the volume. The most common segmentation techniques employ thresholding methods, region growth, and state-of-the-art techniques that combine deformable surfaces, STAPLE, majority-of-vote (MV), and single-phase level set (LS) [10, 11]. Recently, automatic segmentation of the neonatal ventricular system of premature brains from 3D US images has been demonstrated [12, 13]. They represent state-of-the-art in automatic segmentation from 3D US and use a combination of the phase congruence map, the multiple atlas initialization technique, the atlas selection strategy and the evolution of multiphase geodetic level sets (MGLS) combined with a previous derived spatial form of multiple presegmented atlases. The experimental results with 30 images of patients showed a good correlation with reference to the manually obtained volumes. The disadvantages of this methods are that (1), the algorithm performs image processing of 2D slices from the 3D volume one by one; (2), its effectiveness depends on the correspondence of the 3D US data obtained with those stored in an atlas, and therefore, it may not be exact with respect to the data used in other laboratories or with different acquisition conditions; (3), it demands a high computation time before the measurement is obtained; (4), and finally, the method requires the robotization of the ultrasound machine.

In this work, we evaluate an alternative algorithm that can segment the ventricular volume automatically from freehand 3D US datasets. Because it does not rely on atlases of data, it is simple, more general and fast. The method can help diagnose PHVD, and we evaluate its reliability comparing the results with manual methods.

2   Materials and Methods

We have studied the progression of ventricular growth in a preterm neonate over a month before and after surgical intervention. Twenty-three transfontanellar 3D US images of the brain were acquired through the 4D option in the 3D/4D Voluson i portable ultrasound system (GE Healthcare, Milwaukee, WI, USA). First, the transducer (SVNA5-8B, 5–8 MHz), using a center frequency of 6.5 MHz, was situated in the third coronal plane, and the scan angle was settled to 90°. With the transducer fixed in that position, the beam would move from anterior to posterior planes and from side to side through lateral planes. Volumes were saved, and exported to *.nrrd format, widely used in medical imaging and easier to import in other programming packages.

The algorithm for the automatic segmentation of the ventricles is summarized in Fig. 1. It consists of three steps: (i) the 3D US image is preprocessed for noise reduction using a median 3D filter; (ii) the filtered dataset is segmented for darker intensities (low density in the brain) using a global threshold calculated iteratively over the histogram of the 3D volume. In each iteration segmented blobs are classified according to geometrical parameters; (iii) after calibration, the segmented volumes of the left and right ventricle (LV and RV) are converted from voxels into milliliters (ml). Finally, the volumes were rendered using an isosurface for visualization.

```
PSEUDOCODE

Thresh = 1-by-N vector containing N threshold values using Otsu's method
M = 3D dataset
Mresult = Empty array with size M
Mf = median filter (M)
number_object_detected=0
While N > 2 and number_object_detected < 2
    • Segmented_volume = thresh(N-1) < Mf < thesh(N)
    • objects = measures a set of properties for each connected component (object) in the 3-D
      volumetric binary image segmented_volume (volume, Principal Axis Length, orientation)
    • objects = sort objects by its volume in descending order.
    • K = 1|
    • Nobject = 0
    • while k < 6 (only the greatest object are taken into account)
        ▪ If objects(k) is considered as ventricle
            o   Increment Nobject
            o   Add objects(k) to Mresult
        ▪ endIf
        ▪ k = k+1
    • endWhile
    • number_object_detected = Nobject
    • N = N-1;
endWhile
```

Fig. 1. Pseudocode for the automatic segmentation of the 3D US images.

For validation of the automatic segmentation, a trained operator carried out the manual delineation of the ventricular contour of the datasets. For this task, we used the computer-aided analysis of virtual organs (VOCAL) software (4D View® Version 10. x, GE Healthcare, Austria). This software tool allows the performance of volume measurements, by rotating the organ or structure of interest around a fixed axis, while 2D US contours are manually delineated on each plane. A rotation angle of 15° for each contour plane was selected leading to 12 rotation steps. (see Reference [4] for more details).

3 Results

Figure 2 shows the results of the application of the algorithm for automatic segmentation on 3 transfontanellar 3D US representative of a temporal series of 23. Each of the rows show two orthoslices (coronal view and sagital view) of the brain. The segmented ventricular matrix is contoured with green colour. From day 1 to day 11, the ventricles become progressively enlarged due to cerebrospinal fluid (CSF) accumulation. Of note, the density of the CSF is low and display dark intensities in the 3D US. On day 11 (middle row) surgical intervention is carried out (intraventricular reservoir placed) allowing CSF extraction leading to ventricular volume reduction.

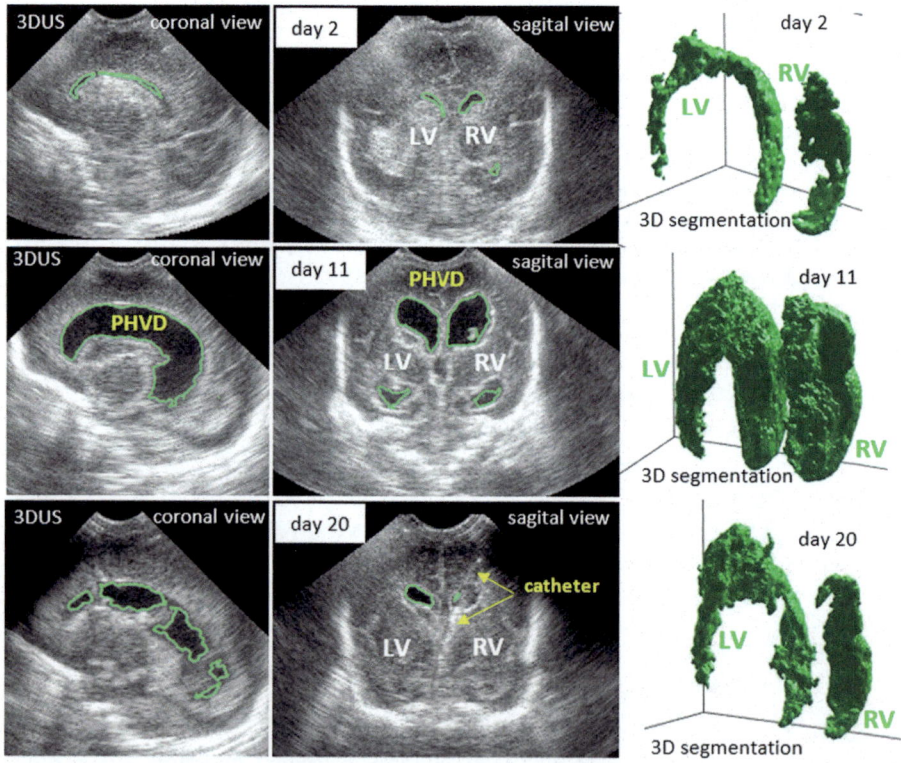

Fig. 2. Three representative 3D US out of a temporal series of 23 3D US of the same preterm infant. The rightmost column shows the 3D isosurface visualization of the volumes of LV and RV calculated using the automatic 3D segmentation algorithm described in this work. Each row shows two orthoslices of the brain. In the front (coronal) view the anterior horns of the LV and RV are visible. The top row was acquired shortly after the start of the monitorization. In the case of PHVD, the ventricles can grow in excess and are the cause of abnormal neurological development of the infant. One of the strategies to reduce ventricular volume is to place an intraventricular catheter to allow CSF removal. The bottom row was acquired towards the end of the monitorization in day 20, and the intraventricular reservoir is visible.

Figure 3 shows the evolution plots of the ventricular volume (LV and RV) of the preterm infant over time. Each graph displays two curves: one represents the volume of the ventricle measured automatically with the algorithm described in this work and the second one the volume of the ventricle measured manually using VOCAL.

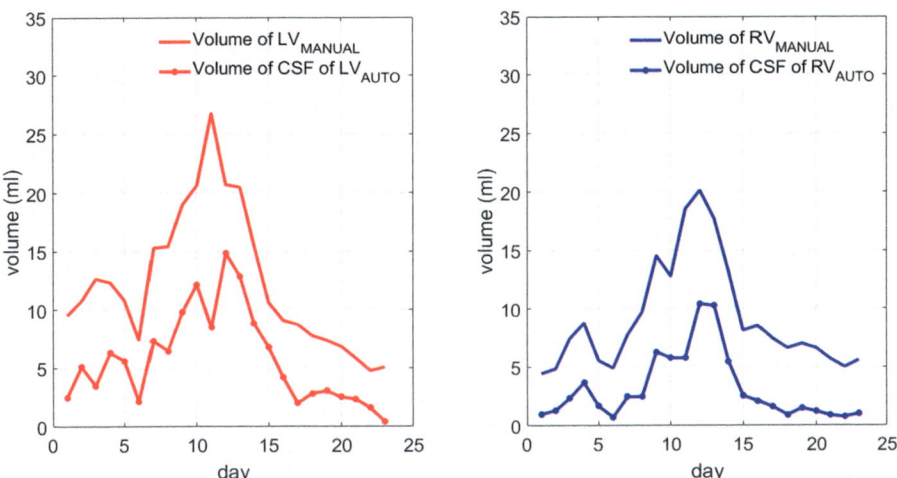

Fig. 3. Temporal series of 23 days of the volumes of the LV and RV of a preterm neonate with PHVD. The curves compare the measurements obtained using manual segmentation and the automatic segmentation proposed in this work. Manual segmentation estimates the total volume of the ventricles while the automatic segmentation only measures the liquid content of the ventricles.

4   Discussion

We have developed a simple algorithm for the automatic ventricular segmentation on 3D US images that displays a high degree of correlation with manual segmentation. The curves in Fig. 2 using VOCAL and the automatic segmentation are very similar in shape, with several coincident peaks and the same growth and decay. In both curves, the volume decreases from day 11 after a ventricular reservoir was inserted allowing CSF extraction.

The volumes using VOCAL are invariably higher than those measured automatically. The main reason is that the algorithm shown here measures only the volume of liquid in the ventricles. Because the ventricles include also the choroid plexus (an intraventricular structure that produces CSF) and can be partially filled with blood clots, the measure of CSF is smaller than the whole volume. This is evident if one looks at the coronal view of day 20 (Fig. 2). The volume of the ventricle contains 4 smaller sections filled with CSF.

The automatic measurements of CSF are more unprecise, with larger oscillations. This is particulary true on the left ventricle, in which in day 11 the curves diverge. A plausible explanation is that the insertion of the catheter adds pressure and changes the average density (and the intensities measured with US). Because the algorithm is based on global thresholding the intensities, the measurements are distorted. This effects is clear in the sagital view of day 20, the catheter (an elongated shape with bright intensity) difficults the measurement of liquid.

5   Conclusion

Evaluations both *in vitro* and *in vivo* show that the manual method VOCAL is considered the 3D US gold standard method for volumetric measurements. However, VOCAL has the following limitations: (1) difficulties in identifying the edges of some structures, and (2) the time needed to perform the measurement can vary from 2 to 10 min [7]. Therefore, automatic segmentation is a desirable tool for medical diagnose from 3D US because it can be fast and independent of inter-subject variability.

The automatic segmentation procedure described here, which measures the CSF contained in the ventricles of infants from 3D US datasets, is a potential tool for monitoring the upcoming of PHVD. Although there are differences with the measurements obtained manually, the measurements provided by the automatic segmentation are more "objective" in terms of reproducibility. The algorithm can miss bits of the ventricles with low amount of CSF, therefore underestimating the result, hence, further research to validate its accuracy and assess its robustness will be performed applying it to a bigger study population. Compared with the current state-of-the-art methods this algorithm has important advantages like simplicity, generality and speed.

Acknowledgments. LC Gontard acknowledges financial support from the program ASECTI, 'Contratos de Acceso al Sistema Español de Ciencia, Tecnologia e Innovacion'.

References

1. Volpe, J.J.: Brain injury in premature infants: a complex amalgam of destructive and developmental disturbances. Lancet Neurol. **8**, 110–124 (2009)
2. Synnes, A.R., Anson, S., Arkesteijn, A., Butt, A., Grunau, R.E., Rogers, M., Whitfield, M. F.: School entry age outcomes for infants with birth weight ≤ 800 grams. J. Pediatr. **157**, 989–994 (2010)
3. Brouwer, M.J., de Vries, L.S., Pistorius, L., Rademaker, K.J., Groenendaal, F., Benders, M. J.: Ultrasound measurements of the lateral ventricles in neonates: why, how and when? A systematic review. Acta Paediatr. **99**, 1298–1306 (2010)
4. Benavente-Fernandez, I., Lubián-Gutierrez, M., Jimenez-Gomez, G., Lechuga-Sancho, A. M., Lubián-López, S.P., Neonatal Neurology Foundation (Fundación Nene): Ultrasound lineal measurements predict ventricular volume in posthaemorrhagic ventricular dilatation in preterm infants, Acta Paediatr, **106**, 211–17 (2017)
5. Riccabona, M.: Potential role of 3DUS in infants and children. Pediatr. Radiol. **41**(Suppl 1), S228–S237 (2011)

6. Chang, H.H., Larson, J., Blencowe, H., Spong, C.Y., Howson, C.P., Cairns-Smith, S., Lackritz, E.M., Lee, S.K., Mason, E., Serazin, A.C., Walani, S., Simpson, J.L., Lawn, J.E.: Preventing preterm births: analysis of trends and potential reductions with interventions in 39 countries with very high human development index. Lancet **381**, 223–234 (2013)
7. Kusanovic, J.P., Nien, J.K., Goncalves, L.F., Espinoza, J., Lee, W., Balasubramaniam, M., Soto, E., Erez, O., Romero, R.: The use of inversion mode and 3D manual segmentation in volume measurement of fetal fluid-filled structures: comparison with virtual organ computer-aided analysis (VOCAL). Ultrasound Obstet. Gynecol. **31**, 177–186 (2008)
8. Haratz, K.K., Oliveira, P.S., Rolo, L.C., Nardozza, L.M., Milani, H.F., Barreto, E.Q., Araujo Júnior, E., Ajzen, S.A., Moron, A.F.: Fetal cerebral ventricle volumetry: comparison between 3D ultrasound and magnetic resonance imaging in fetuses with ventriculomegaly. J. Matern. Fetal Neonatal Med. **24**, 1384–1391 (2011)
9. Kishimoto, J., de Ribaupierre, S., Lee, D.S., Mehta, R., St Lawrence, K., Fenster, A.: 3D ultrasound system to investigate intraventricular hemorrhage in preterm neonates. Phys. Med. Biol. **58**, 7513–7526 (2013)
10. Noble, J.A., Boukerroui, D.: Ultrasound image segmentation: a survey. IEEE Trans. Med. Imaging **25**, 987–1010 (2006)
11. Mozaffari, M.H., Lee, W.: 3D Ultrasound image segmentation: A Survey (2016). arXiv preprint arXiv:1611.09811
12. Qiu, W., Chen, Y., Kishimoto, J., de Ribaupierre, S., Chiu, B., Fenster, A., Yuan, J.: Automatic segmentation approach to extracting neonatal cerebral ventricles from 3D ultrasound images. Med. Image Anal. **35**, 181–191 (2017)
13. Boucher, M.-A., Lippé, S., Dupont, C., Knoth, I.S., Lopez, G., Shams, R., El-Jalbout, R., Damphousse, A., Kadoury, S.: Computer-aided lateral ventricular and brain volume measurements in 3D ultrasound for assessing growth trajectories in newborns and neonates. Phys. Med. Biol. **63**, 225012 (2018)

The Application of Infrared Thermography as a Quantitative Sensory Measure of the DC/TMD

Miguel Pais Clemente[1(✉)], Carlos Faria[2],
Francisco Azevedo Coutinho[2], Joaquim Mendes[3],
João Correia Pinto[4], and José Manuel Amarante[5]

[1] DMD, Department of Surgery and Physiology, Faculty of Medicine Porto,
Porto, Portugal
miguelpaisclemente@hotmail.com

[2] Stomatology Resident, Stomatology Department of Centro Hospitalar
Universitário de São João; Surgery and Physiology Department - Faculty of
Medicine, University of Porto, Porto, Portugal
carlosafaria@gmail.com, franciscoac@outlook.com

[3] INEGI, Labiomep, Faculdade de Engenharia, Universidade Do Porto,
Porto, Portugal
jgabriel@fe.up.pt

[4] Department of Centro Hospitalar Universitário de São João, Porto, Portugal
jcpl950@gmail.com

[5] Labiomep, Department of Surgery and Physiology, Faculty of Medicine,
University of Porto, Porto, Portugal
amarante@med.up.pt

Abstract. The most common internal derangement of the temporomandibular joint is the displacement of the articular disc. Infrared thermography can be a useful method to avoid the establishment of chronic pain of the masticatory muscles. An early diagnosis and evaluation of the infrared thermal patterns can monitor the patients' symptoms associated to the regions of interest (ROIs). The clinical relevance of detecting these muscular conditions and dysfunctions are important to determine the location of pain in order to avoid the progression of this pathological condition, where sensitization mechanisms can lead to the evolution and chronicity of pain on the cranio-cervical-mandibular complex (CCMC). The results of this pilot study, highlight the possibility of using infrared thermography as a quantitative sensory measure of the diagnostic criteria for temporomandibular disorders (DC/TMD), since there is a neuro-thermal pain pattern associated to specific ROIs of the CCMC.

Keywords: Infrared thermography · Imaging modality · Myalgia masticatory muscles · Temporomandibular disorders · Diagnostic criteria for temporomandibular disorders

© Springer Nature Switzerland AG 2019
J. M. R. S. Tavares and R. M. Natal Jorge (Eds.): VipIMAGE 2019, LNCVB 34, pp. 330–340, 2019.
https://doi.org/10.1007/978-3-030-32040-9_35

1 Introduction

Infrared thermography, as a medical imaging technique related to certain health issues can be used as complementary method of diagnosis and treatment evaluation [1–5]. Infrared thermography intends to monitor and assess the skin temperature adjacent do specific anatomical sites which are denominated Regions Of Interest (ROIs). This modality provides accurate measurements of surface temperature in inflammatory conditions or muscular hyperactivity, where the skin temperature assessed of the ROI is compared with the contralateral ROI, provided by the infrared thermograms. The temperature differential of 0.3 °C between each region of interest is the reference value in terms of asymmetry to determine if the anatomical site can be considered pathological, or associated with the presence of an eventual symptomatology [6]. Nevertheless, infrared thermography by itself will not provide a diagnose, this will be always the clinician's responsibility. This technique can be a quantitative method for measuring the anatomic-physiological aspects of the musculoskeletal system. Since temporomandibular disorders can be considered as a sub group of musculoskeletal disorders, the infrared thermography is applicable.

The term temporomandibular disorder (TMD) involves a group of clinical signs such as joint noises, deviation during function, limitations in the range of motion and symptoms with pain on the temporomandibular joint, the masticatory muscles and the associated structures [7, 8]. Since the etiology of TMD is multifactorial, it is imperative to provide an accurate clinical examination, with a detailed clinical history of the patient, where the differential diagnosis of orofacial pain is very important [9, 10]. In order to understand these facts, there is a useful tool for TMD assessment, which is the Diagnostic Criteria for Temporomandibular Disorders (DC/TMD), providing a diagnose of both, physical and psychological aspects associated to the TMD identification [11–14].

2 Materials and Methods

Six patients who underwent the consultation of occlusion and temporomandibular disorders of Department of Stomatology of Centro Hospitalar de São João, were included in the study. The Inclusion criteria was a diagnosis of myalgia according to the (DC/TMD). All subjects gave their informed consent before participating in the study. This investigation is in accordance with the revised Helsinki Declaration (1983).

The physical examination of the patient was performed following the Axis I of the (DC/TMD), the physical diagnosis involved the TMJ and the adjacent musculature being done by an oral medicine specialist. The clinical diagnose in the present study was made according to the guidelines of the (DC/TMD). While the Axis II of the (DC/TMD) was distributed to the patients in order to achieve their psychological assessment.

In terms of infrared thermography, the images capture protocol is based in the 'Glamorgan Protocol', which defines the conditions to observe before and during the examination. The images were captured in a room with controlled temperature and humidity, absence of air flow and absence of incident lightning over the subject. The thermal camera was a FLIR i7 (Wilsonville, Oregon, USA), operating in the long-range infrared spectrum (7.5 to 13 μm), with a focal plane array size of 140 × 140, a Noise Equivalent Temperature Difference (NETD) of <70 mK at 25 °C and measurement uncertainty of ±2% of the overall reading. The camera was switched on 15 min before the image capture to avoid start-up drifting, the emissivity value was set to 0.98 and the rainbow medical scale was chosen to define the display colors. The patient was previously advised to not eat a heavy meal, drink alcohol or smoking, since this may interfere with the thermal measurements. In addition, sports or physical activity two hours before examination should also be avoided, since it can also induce side effects on the temperature absolute values measurements. At the appointment, the patient should not have any make up, oil or ointment over the face skin. Before the images were taken the participant was acclimatized for 10 min to the room temperature. Infrared lateral right side and lateral left side thermograms of the cranio-cervical-mandibular complex (CCMC) were performed at a distance of 1.5 m from the patient, with the object of interest occupying 2/3 of the image.

3   Results

3.1   TMD Symptoms Results

The results of this study suggest a high correlation between the skin temperature assessed by infrared thermography and the presence of myalgia of the masticatory muscles according to the (DC/TMD), as it can be seen by the example of a patient infrared thermogram, with right and left lateral view of the CCMC (Figs. 1, 2 and 3 (a)–(b)). The temperature correspondent to the ROIs of the masseter muscle on the right lateral view is of 35.0 °C when comparing to the contralateral region with 34.4 °C. Regarding the ROIs associated to the anterior temporal muscle the temperature differential is also of 0.6 °C (Table 1), being the temperature of 35 °C for the right view thermogram and 34.4 °C for the left side view. In this particular case, this could be associated to patient's posterior crossbite on the right side together with the masticatory preference for the same side. The clinical examination with the DC/TMD confirmed the thermographic results of patient n°1, as it can be observed in Fig. 1 (a)–(b), and the Table 4 summarizes the prevalence of the temperature asymmetry of the subjects involved in this study. In order to exemplify the thermal pattern characteristics, the thermographic results present the lateral thermograms of the most significant differential temperatures.

3.2   Thermographic Results

The infrared thermal patterns presented a significant association with the patient examination according to the DC/TMD. Patient number two has bruxism and this parafunctional habit can be associated to the asymmetry of 0.3 °C correspondent to the ROIs of the anterior temporal muscle (Fig. 2 (a)–(b) and Table 2). Regarding patient number three, the clinical history examination reports that she was retired at the age of 47 years due to a health problem. Now that she is 57 years old and reveals a context of

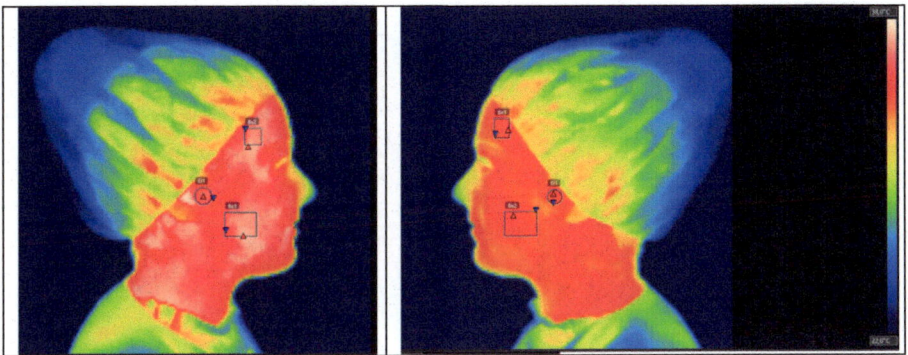

Fig. 1. Infrared thermograms of the CCMC, right (a) and left (b) views with the ROIs correspondent to the anterior temporal muscle, the temporomandibular joint and the masseter muscle – patient n°1.

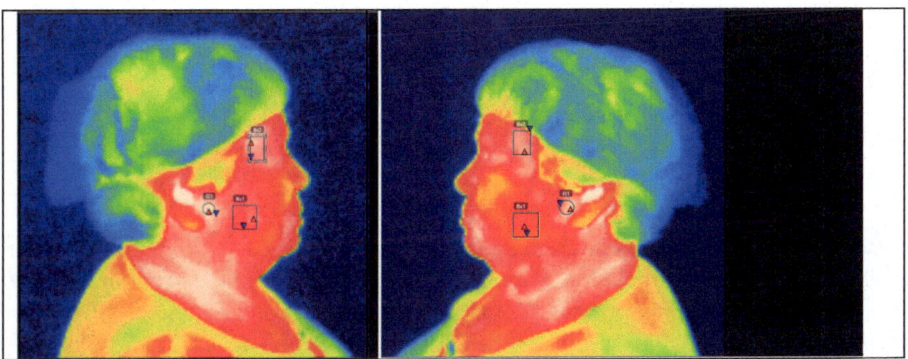

Fig. 2. Infrared thermograms of the CCMC, right (a) and left (b) view with the ROIs corresponding to the anterior temporal muscle, the temporomandibular joint and the masseter muscle – patient n°2.

depression with psychological symptoms and clinical symptomatology of pain in the masticatory can be associated to this context, where the thermal images are in accordance with the presence of pain, on the right jaw with the asymmetry of 0.7 °C of the anterior temporal muscle and 0.2 °C of the masseter muscle, Fig. 3 (a)–(b) and Table 3. Another significant asymmetric temperature was noticed in patient number four regarding the ROIs of the right masseter muscle, with a differential temperature of 0.6 °C. The findings of the DC/TMD of this patient presented bilateral masseter myalgia, however the pain site was described by the subject as being more intense on the right masseter muscle (Table 4). It is also possible to observe the data correspondent to patient number five and six, where there is no significant differential temperature between the different ROIs. While in the first case, the DC/TMD results indicate the presence of myalgia of the right masseter muscle, which can be associated to the patient complaint of an oral rehabilitation with removable partial dentures

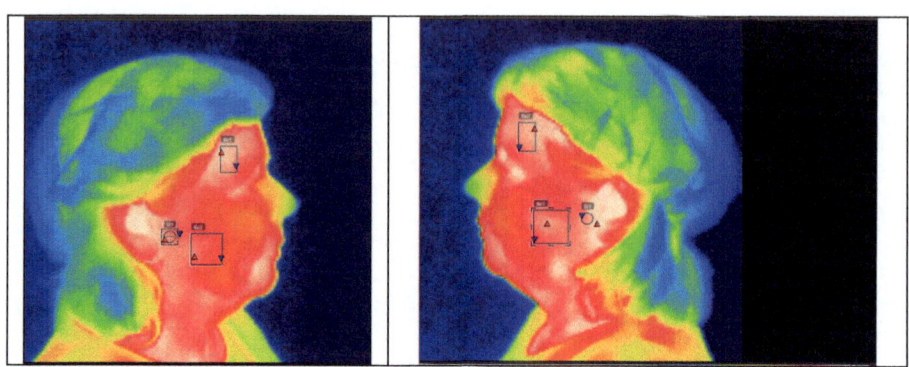

Fig. 3. Infrared thermograms of the CCMC, right (a) and left (b) view with the ROIs corresponding to the anterior temporal muscle, the temporomandibular joint and the masseter muscle – patient n°3.

Table 1. Temperature data of the lateral thermograms of the CCMC – patient n°1.

ROIs ASSESSED	Right lateral view (°C)	Left lateral view (°C)
Temporal muscle	35.0	34.4
Temporomandibular joint	34.4	34.4
Masseter muscle	35.0	34.4

maladjusted. The occlusal interferences and unbalanced muscle coordination can be the origin of the pain, nevertheless not enough to represent a significant asymmetry of the masseter or anterior temporal muscle. On the other hand, patient number six evidence an asymmetry of 0.1 °C on the masseter muscle and the DC/TMD results are in agreement with the patients' complaint of bilateral masseter muscle pain, however the pain is reported to occur in the morning which can be associated to a parafunctional habit of bruxism during sleep time (Table 4). This patient had a difference of 0.1 °C, between the left masseter muscle (33.1 °C) and the contralateral (33 °C). In this particular case, the absolute temperature value is lower on the left masseter muscle compared to the right masseter muscle, however the temperature asymmetry is only of 0.1 °C, a very low value to be conclusive and determinant in terms of thermographic results.

Table 2. Temperature data of the lateral thermograms of the CCMC – patient n°2.

ROIs ASSESSED	Right lateral view (°C)	Left Lateral view (°C)
Temporal muscle	33.0	32.7
Temporomandibular joint	34.1	33.8
Masseter muscle	33.7	33.8

Table 3. Temperature data of the lateral thermograms of the CCMC - patient n°3.

ROIs ASSESSED	Right Lateral view (°C)	Left Lateral view (°C)
Temporal muscle	34.5	33.8
Temporomandibular joint	33.8	34.0
Masseter muscle	33.2	33.4

Table 4. ROIs corresponding to affected/pathological sites with myalgia of the masticatory muscles (in bold).

	ROIs ASSESSED	Temperature asymmetry (°C)
Patient 1	Temporal muscle	**0.6**
	Temporomandibular joint	0
	Masseter muscle	**0.6**
Patient 2	Temporal muscle	**0.3**
	Temporomandibular joint	**0.3**
	Masseter muscle	0.1
Patient 3	Temporal muscle	**0.7**
	Temporomandibular joint	0.2
	Masseter muscle	0.2
Patient 4	Temporal muscle	0
	Temporomandibular joint	1.2
	Masseter muscle	**0.6**
Patient 5	Temporal muscle	0.1
	Temporomandibular joint	0
	Masseter muscle	0.1
Patient 6	Temporal muscle	0.2
	Temporomandibular joint	0.2
	Masseter Muscle	0.1

4 Discussion

The use of infrared thermography as complementary method of diagnostic to identify temporomandibular disorders can be effective [15, 16], nevertheless there are authors referring that literature is still lacking in number of studies regarding the reliability of this technique in the diagnosis of TMD [17]. Therefore, it is important to understand some limitations of this imaging modality, and highlight the advantages of infrared thermography namely: non-ionizing radiation (thus presenting no harm to the patient), non-invasive, no consumables are necessary, no maintenance, and portable.

Some patients of our study group presented bruxism, which is a parafunctional habit of the muscles of the masticatory system, involving the elevator muscles of the mandible such as the masseter and the temporal muscles, leading to the occurrence of teeth clenching. This activity may lead to disturbances in muscle blood flow on the supra-cited muscles, with the activation and sensitization of nociceptors of the central nervous system causing muscle pain. Thus, the patients that were referred to the Department of Stomatology of the University Hospital of São João, with bruxism presented symptomatology of the masseter muscles bilaterally. In these case where the myalgia of the masseter muscles appeared in both sides, it could be expected that the thermograms did not present asymmetrical patterns of these ROI. However, this did not occur in the ROIs of the anterior temporal muscle which can be related to the bruxism pattern, in accordance to our results. Nevertheless, it should be highlighted that this small group of patients has the pathology associated to the muscles of the masticatory system, according to the clinical examination and DC/TMD. Eventually, in clinical cases of functional disorders of the temporomandibular joint, with dislocation of the disc (with or without reduction), arthralgia or osteoarthritis there could appear compensatory patterns of muscle activation, which can represent a higher prevalence on mobility restriction of the mandible. In terms of thermographic findings, this could eventually be traduced in significant asymmetrical thermograms of the CCMC, which was not the case of these patients when observing the ROIs of the masticatory muscles.

Since pain is a subjective feeling, associated to a pathophysiological process that occurs in the central nervous system, and linked to an emotional experience of each particular individual, it is very difficult to quantify the pain as an objective measurement. The implementation of diagnostic tools such as the DC/TMD tend to assess and diagnose pain in a more accurate procedure. Other criteria have been implemented in Pain centers, such as the visual analog scale (VAS) during clinical examination. Schiffman, E. *et al.*, refer the fact that adjunctive tests including electronic diagnostic instruments evidence an incremental validity for a positive diagnosis of TMD that should be considered for clinical use [11]. The results of our research, even though it is still a pilot study, intends to highlight the possibility of using infrared thermography as a quantitative sensory measure of the DC/TMD, since there is a neuro-thermal pain pattern associated to specific ROIs of the CCMC.

The present findings suggest this imaging modality as a complementary method of diagnosis can eventually be used in the future as an Infrared Pain Index – "IPI", where the index of thermal scale asymmetry corresponds to the pain level of each individual. The relationship between the "IPI" and the pathological region should be validated in upcoming investigations taking in consideration the asymmetry threshold of 0.3 °C for the pathological region of the CCMC, where the highest temperature usually corresponds to the affected area. However, arthrogenic TMD were not included in this work, the patients involved in our study represent a group of Myogenic TMD that mention pain arising from the muscles. In case of analyzing arthralgia of the TMJ, it could be possible to compare the Infrared imaging technique with magnetic resonance imaging (MRI), since this is the gold standard for the evaluation of TMD associated to arthralgia [18, 19]. Vogl et al., showed in a group of 546 female and 248 male patients, who were clinically diagnosed with TMD, that the MRI has a significant correlation on the anatomical findings of the articular disc. The results of these authors indicate that anterior disc displacement with reduction showed a specificity of 88% and a sensitivity of 78%, whereas anterior disc displacement without reduction showed a specificity of 84% and a sensitivity of 73% [20].

The most common internal derangement of the TMJ is the displacement of the articular disc, which can be diagnosed with MRI [19]. This image modality allows a qualitative and morphologic study of the articular disc, where the alterations of this anatomical site can be associated to an increase mechanical load of the TMJ, among other factors. The specificity of each condition originating TMD is important to categorize between arthralgia and myalgia, since the etiology and the treatment plan can be different according to each particular case. The MRI has been well documented as a complementary method of diagnose for the TMJ arthralgia, in addition the infrared thermography can be a useful method to avoid the establishment of chronic pain of the masticatory muscles. An early diagnosis and evaluation of the infrared thermal patterns can monitor the patients' symptoms associated to the ROIs. The clinical relevance of detecting these muscular conditions and dysfunctions are important to determine the location of pain in order to avoid the progression of this pathological condition, where sensitization mechanisms can lead to the evolution and chronicity of pain on the CCMC.

In this study it is highlighted the need for evaluating complementary diagnostic tests, like infrared thermography in TMD. The implementation of rigorous protocols that are being conducted in the Department of Stomatology of the University of Hospital São João with the analysis of the correspondent thermograms of the CCMC intend to contribute to the validation of this technique. In the future, a larger sample should be analyzed with eventually the inclusion of patients that present both conditions, myalgia and arthralgia.

Sawada et al. referred that the evaluation of muscular pain of the TMD has mainly been performed by visual inspection and palpation, where an objective quantification is difficult [21]. Therefore, this study intends to demonstrate and identify patients with an appropriate imaging method such as infrared thermography in the evaluation of myalgia of the masticatory muscles. Whereas Sawada et al. study intended to quantitatively assess the myalgia of the masticatory muscles in patients with TMD, where the mean apparent diffusion coefficient on diffusion weighted MRI, showed values of the masticatory muscles of the pain side significantly higher than those of the no-pain side [21].

Based on our findings, although we have to highlight the limitation of the small sample of our research, it was possible to observe a relationship between anxiety and TMD, which has been associated in previous studies as Bertoli et al. described [22]. These authors also correlate that adolescents with a type of malocclusion Class II had a higher prevalence of myofascial pain than adolescents with Class I ($p < 0.015$), the same was found regarding adolescents that presented a malocclusion Class III when compared to Class I ($p < 0.004$) [22]. In fact, one of the patients of our study, that presented higher values of thermal asymmetry (0.6 °C), had a malocclusion Class II with a posterior crossbite on the side corresponding to the myalgia of the masseter muscle which was in consonance with the correspondent thermograms.

Ohrbach et al., highlighted the association between oral parafunctional behaviors as significant predictors of TMD [23]. A study carried out by Chen et al. showed that TMD subjects presented a higher level of comorbid pain conditions comparing with the control group, and emphasized the importance of integrating bodily pain assessment and psychological assessment in the evaluation of TMD patients [24].

On the other hand, Meloto et al. when analyzing 260 participants with first-onset TMD patients, using the DC/TMD protocol, concluded that the clinical measures available for clinicians to have access when TMD first develops are useful in calculating the risk of developing persistent TMD [25]. Whereas psychological measures being important predictors are not taken in consideration to add meaningfully predictive capacity of clinical measures. Meloto *et al.* reinforce that there are practical implications when the TMD first develops and that clinicians should use clinical measures in treatment planning and for monitoring an intervention [25]. For this purpose, our research advocates the implementation of infrared thermography as a quantitative sensory measure on the supervision of the treatment plan. It is therefore important to analyze the multivariate symptoms that can be related with risk of first onset TMD [26].

The therapeutic itself, should be at a first stage the most conservative as possible, allowing the patient to have awareness of some of the parafunctional habits that can be associated to the etiology of the TMD myalgia, with patient education and self-care advice, physical therapy, acupuncture, transcutaneous electrical nerve stimulation (TENS), intra-oral devices such as an occlusal splint, and psychological advising [27–30]. Lindfors *et al.* had most of the 150 patients involved in their study - 78%, reporting that the information about the underlying cause of their symptoms made them more involved in the treatment, while 72% of the participants stated that the jaw exercises were effective in reducing symptoms associated to masticatory myofascial pain [31].

Infrared thermal imaging can be an adjunctive tool for the screening of TMD with myalgia of the masticatory muscles and some authors also start to apply this modality in the evaluation of the implemented treatment namely with occlusal splints or low level laser therapy in myofascial pain syndrome [32, 33].

5   Conclusions

The findings of this study lead to the conclusion that an evaluation of TMD patients with myalgia of the masticatory muscles can have on the infrared thermography a reliable complementary diagnosis technique. Nevertheless, there are limitations in this study mainly regarding the sample size.

Medical history and the standardized clinical examinations for TMD, such as the DC/TMD should be considered to be the standard reference point. In the near future, it can be desirable to include biometrics signals to a clinical protocol involving the assessment of TMD where infrared thermography can be considered a fundamental imaging modality as it provides quantitative sensory measures on specific anatomical sites correspondent to the ROIs of the masticatory muscles with myalgia without any harm to the patient.

References

1. Ring, E.F.: Progress in the measurement of human body temperature. IEEE Eng. Med. Biol. Mag.: Q. Mag. Eng. Med. Biol. Soc. **17**(4), 19–24 (1998)
2. Friedman, M.S.: The use of thermography in sympathetically maintained pain. Iowa Orthop. J. **14**, 141–147 (1994)

3. Jiang, L.J., Ng, E.Y., Yeo, A.C., Wu, S., Pan, F., Yau, W.Y., et al.: A perspective on medical infrared imaging. J. Med. Eng. Technol. **29**(6), 257–267 (2005)
4. Merla, A., Ciuffolo, F., D'Attilio, M., Tecco, S., Festa, F., De Michele, G., et al.: Functional infrared imaging in the diagnosis of the myofascial pain. In: Conference Proceedings : Annual International Conference of the IEEE Engineering in Medicine and Biology Society IEEE Engineering in Medicine and Biology Society Annual Conference, vol. 2, pp. 1188–1191 (2004)
5. Dworkin, S.F., LeResche, L.: Research diagnostic criteria for temporomandibular disorders: review, criteria, examinations and specifications, critique. J. Craniomandib. Disord.: Facial Oral Pain **6**(4), 301–355 (1992)
6. Canavan, D., Gratt, B.M.: Electronic thermography for the assessment of mild and moderate temporomandibular joint dysfunction. Oral Surg. Oral Med. Oral Pathol. Oral Radiol. Endod. **79**(6), 778–786 (1995)
7. Dworkin, S.F.: Research diagnostic criteria for temporomandibular disorders: current status & future relevance. J. Oral Rehabil. **37**(10), 734–743 (2010)
8. National institutes of health technology assessment conference statement: Management of temporomandibular disorders, 29 April–1 May 1996. Oral Surgery, Oral Medicine, Oral Pathology, Oral Radiology, and Endodontology. **83**(1), 177–183 (1997)
9. Gavish, A., Halachmi, M., Winocur, E., Gazit, E.: Oral habits and their association with signs and symptoms of temporomandibular disorders in adolescent girls. J. Oral Rehabil. **27**(1), 22–32 (2000)
10. Okeson, J.P.: Management of Temporomandibular Disorders and Occlusion, vol. 69. Elsevier, St Louis (2008)
11. Schiffman, E., Ohrbach, R., Truelove, E., Look, J., Anderson, G., Goulet, J.P., et al.: Diagnostic criteria for temporomandibular disorders (DC/TMD) for clinical and research applications: recommendations of the international RDC/TMD consortium network* and orofacial pain special interest Groupdagger. J. Oral Facial Pain Headache **28**(1), 6–27 (2014)
12. Ohrbach, R.: Diagnostic Criteria for Temporomandibular Disorders: Assessment Instruments. Version 15 May 2016. Faria, C., Coutinho, F.A., Resende, T., Ferreira, H., Gonçalves, M., Gomes, R., Gomes, D., Pinto, J.C.: Critérios de Diagnóstico para Disfunção Temporomandibular: Portuguese Version September 2017
13. Schiffman, E., Ohrbach, R.: Executive summary of the diagnostic criteria for temporo-mandibular disorders for clinical and research applications. J. Am. Dent. Assoc. (1939) **147**(6), 438–445 (2016)
14. Per Alstergren, G.: Guidelines for DC/TMD Training and Calibration. International RDC-TMD Consortium (2016)
15. Clemente, M.P., Mendes, J.G., Vardasca, R., Ferreira, A.P., Amarante, J.M.: Combined Acquisition Method of Image and Signal Technique (CAMIST) for assessment of temporomandibular disorders in performing arts medicine. Med. Prob. Perform. Artist. **33**(3), 205–212 (2018)
16. Wozniak, K., Szyszka-Sommerfeld, L., Trybek, G., Piatkowska, D.: Assessment of the sensitivity, specificity, and accuracy of thermography in identifying patients with TMD. Med. Sci. Monit.: Int. Med. J. Exp. Clin. Res. **21**, 1485–1493 (2015)
17. de Melo, D.P., Bento, P.M., Peixoto, L.R., Martins, S., Martins, C.C.: Is infrared thermography effective in the diagnosis of temporomandibular disorders? A Syst. Rev. Oral Surg. Oral Med. Oral Pathol. Oral Radiol. **127**(2), 185–192 (2019)
18. Laskin, D., Greene, C., Hylander, W.: Temporomandibular Disorders: An Evidence-Based Approach to Diagnosis and Treatment. pp. 1492–1493 (2006)

19. Whyte, A.M., McNamara, D., Rosenberg, I., Whyte, A.W.: Magnetic resonance imaging in the evaluation of temporomandibular joint disc displacement–a review of 144 cases. Int. J. Oral Maxillofac. Surg. **35**(8), 696–703 (2006)

20. Vogl, T.J., Lauer, H.C., Lehnert, T., Naguib, N.N., Ottl, P., Filmann, N., et al.: The value of MRI in patients with temporomandibular joint dysfunction: correlation of MRI and clinical findings. Eur. J. Radiol. **85**(4), 714–719 (2016)

21. Sawada, E., Kaneda, T., Sakai, O., Kawashima, Y., Ito, K., Hirahara, N., et al.: Increased apparent diffusion coefficient values of masticatory muscles on diffusion-weighted magnetic resonance imaging in patients with temporomandibular joint disorder and unilateral pain. J. Oral Maxillofac. Surg.: Official J. Am. Assoc. Oral Maxillofa. Surr. (2019)

22. de Paiva Bertoli, F.M., Bruzamolin, C.D., de Almeida Kranz, G.O., Losso, E.M., Brancher, J.A., de Souza, J.F.: Anxiety and malocclusion are associated with temporomandibular disorders in adolescents diagnosed by RDC/TMD. Cross-Sectional Study J. Oral Rehabil. **45** (10), 747–755 (2018)

23. Ohrbach, R., Bair, E., Fillingim, R., Gonzalez, Y., Gordon, S.M., Lim, P.-F., et al.: Clinical orofacial characteristics associated with risk of first-onset TMD: the OPPERA prospective cohort study. J. Pain. **14**, T33–T50 (2013)

24. Chen, H., Slade, G., Lim, P.F., Miller, V., Maixner, W., Diatchenko, L.: Relationship between temporomandibular disorders, widespread palpation tenderness, and multiple pain conditions: a case-control study. J. Pain **13**(10), 1016–1027 (2012)

25. Meloto, C.B., Slade, G.D., Lichtenwalter, R.N., Bair, E., Rathnayaka, N., Diatchenko, L., et al.: Clinical predictors of persistent temporomandibular disorder in people with first-onset temporomandibular disorder: a prospective case-control study. J. Am. Den. Assoc. (1939) **150**(7), 572–581 (2019). e10

26. Ohrbach, R., Bair, E., Fillingim, R.B., Gonzalez, Y., Gordon, S.M., Lim, P.F., Ribeiro-Dasilva, M., Diatchenko, L., Dubner, R., Greenspan, J.D., Knott, C., Maixner, W.: Clinical orofacial characteristics associated with risk of first on set TMD: the OPPERA prospective cohort study. J. Pain **14**(12), 33–50 (2013)

27. Durham, J., Al-Baghdadi, M., Baad-Hansen, L., Breckons, M., Goulet, J.P., Lobbezoo, F., et al.: Self-management programmes in temporomandibular disorders: results from an international Delphi process. J. Oral Rehabil. **43**(12), 929–936 (2016)

28. Shimada, A., Ishigaki, S., Matsuka, Y., Komiyama, O., Torisu, T., Oono, Y., et al.: Effects of exercise therapy on painful temporomandibular disorders. J. Oral Rehabil. **46**(5), 475–481 (2019)

29. Ferreira, A.P., Costa, D.R., Oliveira, A.I., Carvalho, E.A., Conti, P.C., Costa, Y.M., et al.: Short-term transcutaneous electrical nerve stimulation reduces pain and improves the masticatory muscle activity in temporomandibular disorder patients: a randomized controlled trial. J. Appl. Oral Sci. Rev. FOB **25**(2), 112–120 (2017)

30. Greene, C.S., Menchel, H.F.: The use of oral appliances in the management of temporomandibular disorders. Oral Maxillofac. Surg. Clin. North Am. **30**(3), 265–277 (2018)

31. Lindfors, E., Magnusson, T., Ernberg, M.: Patients' experiences of therapeutic jaw exercises in the treatment of masticatory myofascial pain-a postal questionnaire study. J. Oral Rehabil. (2019)

32. Clemente, M.P., Mendes, J., Moreira, A., Vardasca, R., Ferreira, A.P., Amarante, J.M.: Wind instrumentalists and temporomandibular disorder: from diagnosis to treatment. Dent. J. **6**(3), 41 (2018)

33. Altindis, T., Gungormus, M.: Thermographic evaluation of occlusal splint and low level laser therapy in myofascial pain syndrome. Complement. Ther. Med. **44**, 277–281 (2019)

Tracking and Analysis of Movement

GRIDDS - A Gait Recognition Image and Depth Dataset

João Ferreira Nunes[1,2](✉)(iD), Pedro Miguel Moreira[1](iD),
and João Manuel R. S. Tavares[2](iD)

[1] ARC4DigiT - Centro de Investigação Aplicada para a Transformação Digital,
Instituto Politécnico de Viana do Castelo, Viana do Castelo, Portugal
{joao.nunes,pmoreira}@estg.ipvc.pt
[2] Instituto de Ciência e Inovação em Engenharia Mecânica e Engenharia Industrial,
Departamento de Engenharia Mecânica, Faculdade de Enhenharia,
Universidade do Porto, Porto, Portugal
tavares@fe.up.pt

Abstract. Several approaches based on human gait have been proposed
in the literature, either for medical research reasons, smart surveillance,
human-machine interaction, or other purposes, whose validation highly
depends on the access to common input data through available datasets,
enabling a coherent performance comparison. The advent of depth sensors leveraged the emergence of novel approaches and, consequently, the
usage of new datasets. In this work we present the GRIDDS - A Gait
Recognition Image and Depth Dataset, a new and publicly available
gait depth-based dataset that can be used mostly for person and gender
recognition purposes.

Keywords: Gait dataset · Person recognition · Gender recognition ·
RGB-D sensors · GRIDDS

1 Introduction

Human gait and its underlying dynamics can reveal relevant information for a
manifold of applications. For example, gait can be used either as an indicator
of a person's health condition [7,18], or to reveal their state of mind [12], or,
in another context of use, it may work as a biometric feature, enabling the
identification of individuals that are under observation, based on their individual
walking styles. In the latest example, when compared to other classical biometric
traits, like finger-print, face, iris or retina, gait reveals to be more advantageous
in some aspects, since gait can be captured at a distance, through non-intrusive
technologies, without the implicit need of the individuals' collaboration [9,17].

The great majority of the dedicated work to this field of research is based on
image sequences and on their intrinsic features in order to extract gait characteristics [2,8,19]. However, during the last decade, the dissemination and availability
of low cost RGB-D sensors like Microsoft Kinect, Intel RealSense or Asus Xtion,

© Springer Nature Switzerland AG 2019
J. M. R. S. Tavares and R. M. Natal Jorge (Eds.): VipIMAGE 2019, LNCVB 34, pp. 343–352, 2019.
https://doi.org/10.1007/978-3-030-32040-9_36

among others, prompted the development of highly-improved 3D vision systems, as well as the development of new approaches and novel applications. These sensors, with a built-in infrared camera providing additionally depth information, are capable to track in real time at least one human figure and also to extract the coordinates of a set of points, which mostly correspond to human body joints, forming a schematic representation of the human skeleton, without using any kind of markers attached to the human body. Thus, in order to derive gait features, some of the more recently introduced approaches are using the estimated skeleton structure by means of the depth data, [20–22].

At the time that we conducted some preliminary work based on human gait for person and gender recognition, we noticed that the number of publicly available datasets including depth data was reduced, as demonstrated in [14]. We also concluded that the existing datasets either did not provide all the sensors' collectable data (color images, depth information, joints' coordinates, etc.), or that they had been acquired with an older version of the sensor that we had (Microsoft Kinect v1 versus Microsoft Kinect v2), varying in data resolution and precision, as well as in the total number of tracked joints. For that reason, we decided to develop a new dataset: the GRIDDS (Gait Recognition Image and Depth Dataset), whose potential applicability is in person and gender recognition.

2   Related Datasets

Most of the published work that is based on human gait uses image sequences of people walking in order to extract a set of features so that gait can be properly characterized. These approaches brought the need for public video-based datasets of people walking under different conditions (e.g.: indoor/outdoor environments, single/multiple views, clothing and carrying variations, etc.). The usage of common input data, available in such datasets, enables a coherent comparison of different approaches and gives an insight into the capabilities of respective methods. A detailed description of some of the must used video-based datasets can be found in [19] and in [15].

More recently, new gait-based methods have been proposed. These methods use the depth information beyond the objects represented in images, both provided by the RGB-D sensors. Evidently the validation of these new approaches also benefits from the access to common input data by means of public datasets. However, the majority of the existing datasets do not include depth information, or if any, they have been developed with a previous version of the RGB-D sensor that we had at the time, with lower performances in people tracking [23]. Thus, it became necessary to create new datasets that would also include depth information, and, depending on the sensor used, include also some additional data, like the joints' coordinates of the human figures detected and tracked in the scene. In fact, the body-skeleton stream provided by some of the depth sensors has been proven to be significantly accurate [3,4] and it has been used in robust gait-based recognition systems, as evidenced by the number of devoted published studies in the recent years [6,16]. A review of existing depth-based datasets has been presented in [5], and [14].

When compared to other public depth-based datasets, GRIDDS has the advantage that it was acquired using the latest version of Kinect sensor, thus tracking a bigger number of joints, with grater data resolution and precision. In addition, it includes a greater variety of available data (*color* images, *depth* images and *depth* data, *infrared* images, *joints' coordinates* and the corresponding *timestamps* of each captured frame).

3   GRIDDS

The GRIDDS - Gait Recognition Image and Depth Dataset was recorded at the Polytechnic Institute of Viana do Castelo (IPVC) facilities, in June of 2018. The dataset is publicly available online at [13], and its usage is allowed according to the instructions described on the same web page.

3.1   Participants

For the development of this dataset we had the collaboration of 35 volunteers, among students, teachers and staff from the IPVC. A written informed consent was obtained from all subjects prior to their participation. Besides the recorded walking sessions, some additional data was also collected, including the participant's age, height and gender. This information is also available online and it is summarized in Table 1.

Table 1. Description of the 35 volunteers in numbers, according to their gender, age and height.

	Gender	Age		Height	
		Mean	SD	Mean	SD
♂	11	29.2	9.7	178.2	6.4
♀	24	39.0	10.4	163.0	7.4
Total	35	35.9	11.2	167.8	10.0

3.2   Depth Sensor and Data Specifications

The sensor used to collect the data was the Kinect v2 (also known as the Xbox One Kinect), manufactured by Microsoft. The collected data modalities included the color, depth, infrared and body streams, and their corresponding timestamps for each captured frame. As stated by the sensor specifications, all data modalities were acquired at an approximately frame rate of 30 fps, varying in their content, but also in their format and resolution: both the depth and infrared images have a resolution of 512 × 424 pixels, while the color images have a resolution of 1920 × 1080 pixels. The body data, which was inferred by the sensor's SDK, consists in 2D and 3D coordinates of 25 body points, which mostly correspond to human body joints that are detected and tracked in the scene. The 2D coordinates correspond to the body joints' coordinates on the color images, having its origin (x = 0, y = 0) located at the top-left corner of every image and each unit corresponds to one pixel. The 3D coordinates are referred to the coordinate system used by the Kinect, whose origin (x = 0, y = 0, z = 0) is located at the center of the sensor, where each unit value corresponds to one meter. The timestamps for every captured stream are expressed in the time unit returned by the Kinect—in order to estimate the time passed between two frames (f_i and f_k, where $k \geq i$), we can refer to the Eq. (1).

$$time_{\langle f_i, f_k \rangle} = (timestamp_{f\,k} - timestamp_{f\,i})/10000000 \tag{1}$$

3.3 Environment and Data Acquisition Description

The recording sessions occurred in a controlled indoor environment, with a static background and with both natural and artificial lighting. Two trajectories were defined, in a straight line across the room: one starting from the left side of the room to the right side, and the other on the opposite direction. The sensor, supported on a tripod, was fixed at 1.8 m high, perpendicular to the defined trajectories (see Fig. 1). The gray triangle represents the sensor's range, according to its technical specifications provided by the manufacturer.

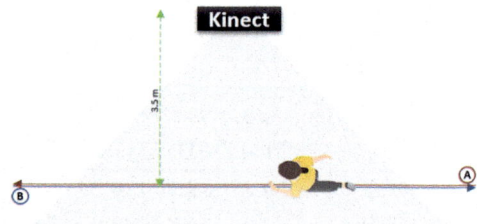

Fig. 1. A graphical representation of the environment where the recording sessions occurred. Letters A and B correspond to the two defined trajectories.

Each one of the 35 volunteers completed 5 walking sequences per trajectory, at a distance of approximately 3.5 m from the sensor, resulting in 10 sequences per participant, and a total of 350 recorded sequences.

3.4 Data Availability

The dataset is composed by 35 folders (one per participant), each one containing the following collected data: *color* images, *depth* images and *depth* data, *infrared* images, *joints' coordinates* and the corresponding *timestamps* of each captured frame of the ten recorded sessions per participant. Additionally, we included the *body silhouettes* images, cropped, facing all to the same side, and normalized in size, with a resolution of 80 × 120 pixels. The information that is made available inside of each folder, is in either one of the following two formats:

- `vvv_ss_stream_nnn.fmt`, for the *color*, *depth*, *silhouette* and *infrared* streams;
- `vvv_ss_stream.fmt`, for the *timestamp* and *joints' coordinates* streams;

where *vvv* corresponds to the volunteers' id, *ss* to the session number, *stream* to the different available streams, *nnn* to the frame number and *fmt* to the different file formats (PNG or CSV). For example, the file named `003_09_depth_021.csv` corresponds to the '*depth*' stream from the volunteer number '003', captured during session number '09', at the frame number '21', saved in the 'CSV' file format. All image files are in the PNG format, varying only in the bit-depth color information, where the color images are in 24-bit, depth images in gray-scale 16-bit, body silhouettes in 1-bit and the infrared images in 16-bit. The depth data files (which are in CSV format) have the same resolution as the depth images, however, in this case, each cell contains the distance between the Kinect device and the objects in front of the device, in millimeters. The coordinates files are also in CSV format and have a resolution of $6 \times N_{frames} \times N_{joints}$, where N_{frames} is the number of captured frames, N_{joints} is the number of tracked joints (25 joints) and the 6 columns correspond to: the frame number, the 3D coordinates and the 2D coordinates of the joints, both in meters. Figure 2 illustrates some of the extracted data during a recorded session.

Furthermore, we include the `body-viwer` tool[1] that we developed to visualize the gait sequences with a graphical representation of all tracked joints and '*bones*', as well as a representation of the angles between hip-knee-ankle joints and shoulder-elbow-wrist joints from the body side closest to the sensor.

3.5 Summary

In conclusion, GRIDDS can be briefly described in form of a table (see Table 2), based on set of characteristics, namely: (i) its applicability, referring to potential application fields (person recognition (PR), gender recognition (GR)) (ii) number of subjects that participate in the recorded sessions, indicating also their distribution in respect to gender (iii) type of sensor used and how it was placed on scene (iv) number and type of defined trajectories (v) number of sequences recorded per participant (vi) list of the collected data; (vii) list of additional data provided (e.g.: source-code or applications to manipulate data, documentation, etc.).

---

[1] Available at https://github.com/joaofnunes/gridds.

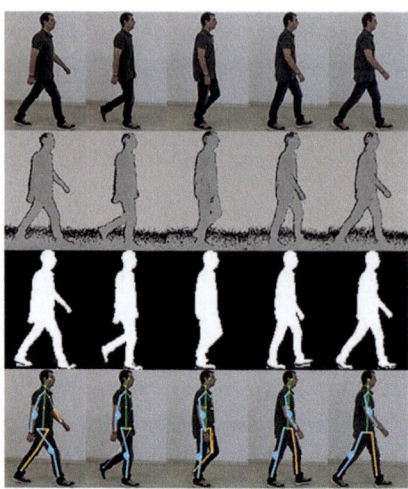

Fig. 2. Examples of normalized and aligned captured streams. First row: color images; Second row: depth images; Third row: body silhouettes; Fourth row: skeleton representation on top of color images.

Table 2. Proposed framework to summarily describe our dataset.

Identification	Year	Applicability[a]	#subjects	#sensors	#trajectories[b]	#sequences/subject[b]	Collected data[c]	Extra data
GRIDDS	2018	PR+GR	35 11♂+24♀	kinect v2 fixed at 1.8 m high	2 S	10 (5S+5S)	C, D, SK, T, DS, S	start-end frame/gait cycle + matlab scripts

[a]PR: Person Recognition; GR: Gender Recognition.
[b]S: Side (Left-to-Right and/or Right-to-Left).
[c]C: RGB data; D: Depth data; SK: 3D Skeleton Coordinates;
T: Time; S: Silhouettes; DS: Depth Silhouettes

4 Application Examples

In order to demonstrate some of the potential usefulness of GRIDDS, we have conducted two sample applications. The first one consists in extracting valid sequences of gait cycles, based on the joints' coordinates. The second is a demonstration of developing some state-of-the-art gait image representations, which are commonly used for person recognition purposes.

4.1 Gait Cycle Detection and Validation

Human gait is considered to be as a periodic activity and a single gait cycle can be regarded as the time passed between two identical events that occurred during the human walking sequence. The proposed method consists in detecting 'valid' sequences of one or more gait cycles by ensuring an effective feet side alternation. The method is exclusively based on the joints' coordinates, and regardless the availability of different approaches to extract joints' coordinates, for practical reasons we have decided to work with the coordinates provided by the sensor's SDK. Typically, there are two techniques for estimating gait cycles: the double support method, based on the local maxima, in which both legs are farthest from each other; and the mid-stance method, based on local minima, when both legs are in the rest (standing) position at the minimum distance from each other. In our proposal we used the double support phase to determine gait cycles. This validation can reveal to be quite useful, since the Kinect's tracking system may tend to confuse the left and right sides of the body joints, particularly when the sensor is placed perpendicular to the defined trajectories.

The first step consists in identifying all the double support positions of the walking sequence (i.e., when both feet are at a maximum distance from each other), knowing that three consecutive double support positions represent a gait cycle. Therefore, the Euclidean distance between ankle joints is computed, and local maxima (peaks) of the computed distance are identified, with a minimum separation criterion between peaks. Figure 3 illustrates the Euclidean distance signal between ankle joints (left side), and the same distance signal smoothed with a moving average filter with a fixed window length, determined heuristically (right side).

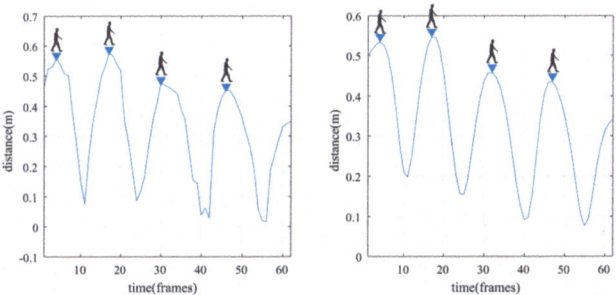

Fig. 3. Peaks detection in the Euclidean distance between the ankle joints (left side), and a moving average filter applied to the same signal (right side).

Then, in order to verify that the side of the ahead ankle joint at each peak has been alternating (left-right-left or reverse), a characters sequence is build indicating the side of the ahead ankle joint at each detected peak (either 'L' for left side or 'R' for right side). Whenever a "LRL" or "RLR" characters sequence is detected, it means that a valid gait cycle has been detected.

4.2   Gait Cycle Representations

In this experiment the depth images in form of silhouettes were used to create different state-of-the-art gait image representations, which are commonly used for person recognition purposes [8, 10, 11, 24]. Firstly some additional image processing operations were needed, specifically: images' flipping, ensuring that the human silhouettes were all facing to the same side; and then the implementation of some basic morphological operations (e.g.: dilation and erosion). Then, following each of the selected five representations' descriptions, we developed the code necessary to generate each of the gait image representation, which is also available to download. Figure 4 illustrates the selected representations: Motion-Energy Image [2], Motion-History Image [2], Gait Energy Image [8], Active Energy Image [24] and Gait Entropy Image [1].

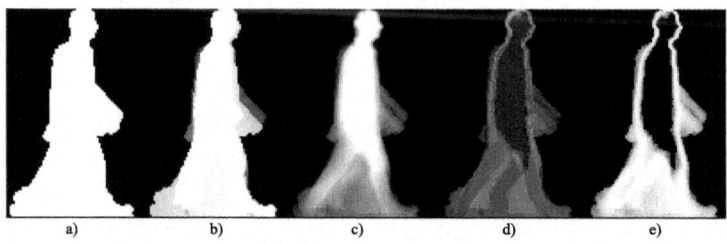

Fig. 4. Five different gait image representations: (a) Motion-Energy Image [2], (b) Motion-History Image [2], (c) Gait Energy Image [8], (d) Active Energy Image [24], and (e) Gait Entropy Image [1].

All referred representations convert a sequence of silhouettes into a two-dimensional image, varying in the way the resulting image is computed. The first representation, Motion Energy Image, is a binary image describing where the motion has occurred in an image sequence. The second, Motion History Image, is a scalar-valued image where the intensity is a function of recency of motion. Both representations, together, can be considered as a two component version of a temporal template, a vector-valued image where each component of each pixel is a function of the motion at that pixel location. The third representation, Gait Energy Image, is one of the most adopted model free representations, representing gait in a single gray scale image obtained by averaging the silhouettes of a complete gait cycle. The fourth representation, Active Energy Image, has the advantage of reducing the influence of carrying or clothing conditions. By calculating the difference between two adjacent silhouettes in the gait sequence, it aims to extract only the active regions. The last representation, Gait Entropy Image, captures mostly motion information, and that is why it is also robust to covariate condition changes that affect appearance (e.g.: clothing or carrying).

5   Conclusions and Future Work

The emergence of low cost depth sensors gave a new impetus to human motion studies. This was mainly due to the fact that these sensors enabled a relative accurate, real time, and markerless tracking, even without the collaboration or awareness of the people under study. At the same time, there was also a need for public data for the validation and performance comparison of newly proposed methodologies. Given the lack of public datasets containing more information, we decided to create our own data and make it public for the scientific community. Thus, we have developed a new depth-based gait dataset, publicly available, especially devoted for person and gender recognition purposes: the GRIDDS - A Gait Recognition Image and Depth Dataset. The dataset was acquired using the Microsoft Kinect v2 depth sensor, and unlike other datasets, it makes available all the collected data. Despite its contribution to the scientific community, we have identified some limitations of the dataset, which we hope to overcome some of them in a timely manner. One of them is related to the nonexistence of covariates. We expect to acquire more sessions in the near future, repeating some of the volunteers, thereby ensuring at least a variation in time and clothing. Another limitation is related to the reduced number of trajectories proposed, and for this reason we plan to gather new sessions in which participants move in front of the sensor, towards to it. And finally, another limitation identified is the scenario used for the recordings, however in this aspect we do not have much room for maneuver due to the limitation of the power supply to the sensor.

References

1. Bashir, K., Xiang, T., Gong, S.: Gait recognition using gait entropy image. In: Proceedings of the 3rd International Conference on Imaging for Crime Detection and Prevention, ICDP. IET (2009)
2. Bobick, A.F., Davis, J.W.: The recognition of human movement using temporal templates. IEEE Trans. Pattern Anal. Mach. Intell. **23**(3), 257–267 (2001)
3. Bonnechère, B., Jansen, B., Salvia, P., Bouzahouene, H., Omelina, L., Moiseev, F., Sholukha, V., Cornelis, J., Rooze, M., Jan, S.V.S.: Validity and reliability of the Kinect within functional assessment activities: comparison with standard stereophotogrammetry. Gait Posture **39**(1), 593–598 (2014)
4. Clark, R.A., Pua, Y.H., Fortin, K., Ritchie, C., Webster, K.E., Denehy, L., Bryant, A.L.: Validity of the Microsoft Kinect for assessment of postural control. Gait Posture **36**(3), 372–377 (2012)
5. Firman, M.: RGBD datasets: past, present and future. In: Proceedings of the IEEE Conference on Computer Vision and Pattern Recognition - Workshops. IEEE (2016)
6. Gabel, M., Gilad-Bachrach, R., Renshaw, E., Schuster, A.: Full body gait analysis with Kinect. In: Proceedings of the Annual International Conference of the IEEE Engineering in Medicine and Biology Society, pp. 1964–1967. IEEE (2012)
7. Givon, U., Zeilig, G., Achiron, A.: Gait analysis in multiple sclerosis: characterization of temporal–spatial parameters using GAITRite functional ambulation system. Gait Posture **29**(1), 138–142 (2009)
8. Han, J., Bhanu, B.: Individual recognition using gait energy image. IEEE Trans. Pattern Anal. Mach. Intell. **28**(2), 316–322 (2006)
9. Kale, A., Sundaresan, A., Rajagopalan, A.N., Cuntoor, N.P., Roy-Chowdhury, A.K., Kruger, V., Chellappa, R.: Identification of humans using gait. IEEE Trans. Image Process. **13**(9), 1163–1173 (2004)

10. Lam, T., Cheung, K.H., Liu, J.N.K.: Gait flow image: a silhouette-based gait representation for human identification. Pattern Recogn. **44**(4), 973–987 (2011)
11. Liu, Z., Sarkar, S.: Simplest representation yet for gait recognition: averaged silhouette. In: Proceedings of the International Conference on Pattern Recognition. IEEE (2004)
12. Michalak, J., Troje, N.F., Fischer, J., Vollmar, P., Heidenreich, T., Schulte, D.: Embodiment of sadness and depression—gait patterns associated with dysphoric mood. Psychosom. Med. **71**(5), 580–587 (2009)
13. Nunes, J.F., Moreira, P.M., Tavares, J.M.R.S.: GRIDDS - a gait recognition image and depth dataset (2018). http://gridds.ipvc.pt
14. Nunes, J.F., Moreira, P.M., Tavares, J.M.R.S.: Benchmark RGB-D gait databases: a systematic review. In: VII ECCOMAS Thematic Conference on Computational Vision and Medical Image Processing, VipIMAGE 2019 (2019)
15. Poppe, R.: A survey on vision-based human action recognition. Image Vis. Comput. **28**(6), 976–990 (2010)
16. Preis, J., Kessel, M., Werner, M., Linnhoff-Popien, C.: Gait recognition with kinect. In: Proceedings of the International Workshop on Kinect in Pervasive Computing (2012)
17. Reid, D.A., Samangooei, S., Chen, C., Nixon, M.S., Ross, A.: Soft biometrics for surveillance: an overview. In: Handbook of Statistics - Machine Learning: Theory and Applications, pp. 327–352. Elsevier (2013)
18. Salarian, A., Russmann, H., Vingerhoets, F.J.G., Dehollain, C., Blanc, Y., Burkhard, P.R., Aminian, K.: Gait assessment in Parkinson's disease: toward an ambulatory system for long-term monitoring. IEEE Trans. Biomed. Eng. **51**(8), 1434–1443 (2004)
19. Sarkar, S., Phillips, P.J., Liu, Z., Vega, I.R., Grother, P.J., Bowyer, K.W.: The humanID gait challenge problem: data sets, performance, and analysis. IEEE Trans. Pattern Anal. Mach. Intell. **27**(2), 162–177 (2005)
20. Shotton, J., Fitzgibbon, A., Cook, M., Sharp, T., Finocchio, M., Moore, R., Kipman, A., Blake, A.: Real-time human pose recognition in parts from single depth images. In: Proceedings of the IEEE Conference on Computer Vision and Pattern Recognition, CVPR, pp. 1297–1304. IEEE Computer Society (2011)
21. Wang, J., Liu, Z., Wu, Y., Yuan, J.: Mining actionlet ensemble for action recognition with depth cameras. In: Proceedings of the IEEE Conference on Computer Vision and Pattern Recognition, CVPR. IEEE (2012)
22. Wei, X., Zhang, P., Chai, J.: Accurate realtime full-body motion capture using a single depth camera. ACM Trans. Graph. - Proc. ACM SIGGRAPH Asia **31**(6), 188:1–188:12 (2012)
23. Zennaro, S., Munaro, M., Milani, S., Zanuttigh, P., Bernardi, A., Ghidoni, S., Menegatti, E.: Performance evaluation of the 1st and 2nd generation Kinect for multimedia applications. In: Proceedings of the IEEE International Conference on Multimedia and Expo, pp. 1–6. IEEE, June 2015
24. Zhang, E., Zhao, Y., Xiong, W.: Active energy image plus 2DLPP for gait recognition. Signal Process. **90**(7), 2295–2302 (2010)

Validation of Whole-Body COM Movement from 3D Anthropometric Image with Dynamic Data at Different Human Standard MVJ

C. Rodrigues[1(✉)], M. V. Correia[2], J. M. C. S. Abrantes[3], J. Nadal[4],
and M. A. B. Rodrigues[5]

[1] PRODEB – Doctoral Program of Biomedical Engineering, FEUP,
Porto, Portugal
c.rodrigues@fe.up.pt
[2] DEEC – Department of Electrical and Computer Engineering, FEUP,
Porto, Portugal
[3] MovLab – Interactions and Interfaces Lab, ULHT, Lisbon, Portugal
[4] PEB – Biomedical Engineering Program, COPPE/UFRJ, Rio de Janeiro, Brazil
[5] DES – Department of Electronic and Systems, UFPE, Recife, Brazil

Abstract. This study presents and applies noninvasive subject specific validation of human whole-body (WB) center of mass (COM) kinematic from 3D anthropometric multibody model using dynamic data from ground reaction forces during impulse phase at standard maximum vertical jump (MVJ) with long countermovement (CM) on countermovement jump (CMJ) and short CM on drop jump (DJ) for comparison with MVJ without CM on squat jump (SJ), assessing lower limb CM contribution and muscle stretch-shortening cycle (SSC) at WB COM vertical impulse. A small group of $n = 6$ sports and physical education degree students with (21.5 ± 1.4) years old, without previous injuries, specific sport abilities or train were weighed (76.7 ± 9.3) kg and their height measured (1.79 ± 0.06) m. Adhesive reflective marks were attached at main upper and lower limb joints. Each subject performed a total of 3 trial at each MVJ, CMJ, DJ and SJ. During trial tests kinematics of anatomical points were registered with two JVC GR-VL9800 digital video cameras at 100 Hz and ground reaction forces with AMTI platform model BP2416-4000 CE operating at 1000 Hz. WB COM kinematics was determined using calibrated SIMI motion tracking of joint reflective marks and Dempster model selecting vertical WB COM displacement Δz_k according to higher amplitude and research interest on MVJ WB COM movement for CM and SSC assessment. Dynamic of WB COM vertical displacement Δz_d was determined from double time integration of COM vertical acceleration. Comparison of kinematic Δz_k and dynamic Δz_d was statistically tested on average and variance at each MVJ type, Δz_k with Δz_d and on root mean square-error (RMSE) during impulse phase. Results present similar variability of Δz_k and Δz_d at each MVJ $p > 0.05$, with mean values discriminating CMJ different means $p < 0.05$ from DJ and SJ with equal means $p > 0.05$, pointing dynamic data as suitable for validation of WB COM movement from 3D anthropometric image as well as for detection of different RMSE at each type of MVJ, its influence on assessment of CM and SSC and improve accuracy on kinematic, dynamic measurements and models.

© Springer Nature Switzerland AG 2019
J. M. R. S. Tavares and R. M. Natal Jorge (Eds.): VipIMAGE 2019, LNCVB 34, pp. 353–365, 2019.
https://doi.org/10.1007/978-3-030-32040-9_37

Keywords: Validation · Whole body COM · 3D anthropometric · Dynamic · MVJ

1 Introduction

Development of higher precision equipment has led to the emergence of increased accuracy at human movement analysis and noninvasive assessment of natural phenomena contributing for objective and quantified study of human movement efficiency and performance, Payton et al. [1]. Muscle stretch-shortening cycle (SSC) constitutes one of these phenomena whose research interest dates to first studies on *ex-vivo* muscle action performed by Hill [2]. Despite isolated forms of muscle action, isometric, concentric and eccentric, have been traditionally used to assess the basic function of the neuromuscular system with isometric action receiving the greatest research effort, natural form of muscle function involves the use of SSC muscle actions in which the muscle is previously stretched and immediately followed by a contraction with impact at lower limb muscles on gait efficiency and powerful running and jumping [3]. Muscle SSC has been assessed noninvasively due to its large expression and accessibility on lower limb concentric and eccentric action at natural human movement using body weight for center of mass (COM) downward acceleration and muscle stretch achieving higher intensity of muscle contraction action and upward acceleration on maximum vertical jump (MVJ), with short countermovement (CM) on drop jump (DJ), long CM on countermovement jump (CMJ) and without CM on squat jump (SJ) addressing the contribution and the mechanisms of CM and SSC on vertical impulse of whole body COM on MVJ [4–7]. Regarding whole body COM location on MVJ two methods have been mainly applied. The first method is based on 3D anthropometric image of skin adhesive marks and biomechanical model as determined by Dempster [8]. Second method is based on inverse dynamics with COM velocity obtained from time integration of COM acceleration from resultant external force and COM displacement obtained from time integration of COM velocity. Despite calibration systems and simultaneous synchronized acquisition of kinematic and dynamic signals, arising question is based on resulting differences at both methods and the need of validation of whole-body COM location obtained with 3D anthropometric image with dynamic data at different human standard MVJ. This issue is determinant due to the nature of whole-body COM which cannot be directly observed during human movement, changing relative position on human body and plays a central role on assessment of CM and SSC contribution at COM vertical impulse on human standard MVJ.

2 Materials and Methods

AMTI force platform model BP2416-4000 CE operating at 1000 Hz was previously calibrated applying vertical and orthogonal transversal calibration loads obtaining calibration report and 6×6 sensitivity matrix relating applied forces and force moments with measured forces and force moments. Excitation voltages and gains of coupled amplifiers Mini Amp MAS-6 to force platform were adjusted according to the movement under analysis, considering the high intensity of vertical forces recorded at the impulsion and reception phases during the maximum vertical jump (MVJ). Adopted procedure prevents signal saturation of registered forces and force moments while ensuring use of the maximum dynamic amplitude of the available scale for higher resolution of the registered values. According to the symmetry on sagittal plane of human body movement during MVJ, two digital video cameras JVC GR-VL9800 were used for image acquisition at 100 Hz on a distant parallel plane to sagittal plane at anterior and posterior position to subject medio-lateral plane, with camera system calibration performed at each trial set using a parallelepipedal structure with known dimensions enclosing entire work volume, left (anterior) and right (posterior) camera views, Fig. 1.

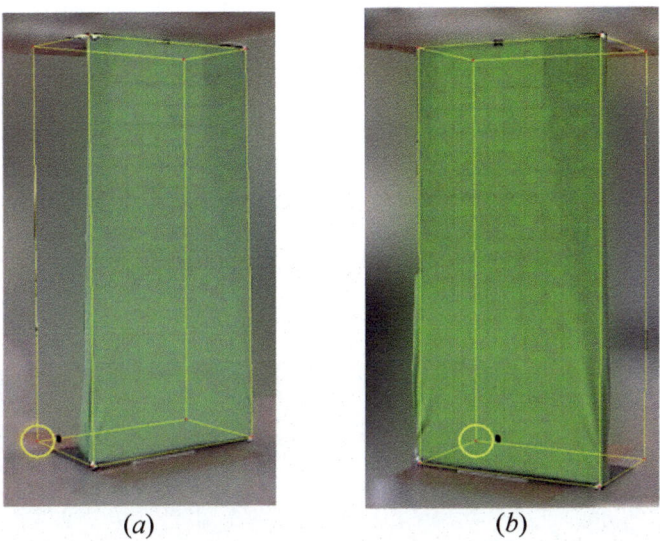

(a)   (b)

Fig. 1. (a) Left (posterior) and (b) right (anterior) camera views during image system calibration.

3D calibration system was performed on Simi Motion 6.1 (Simi Reality Motion Systems GmbH, Germany) using DLT-11 (Direct Linear Transformation) algorithm and X, Y, Z known world coordinates of 8 points from calibration structure associating each calibration point of inserted X, Y, Z world coordinates to corresponding x, y coordinates of selected point at the image space on each camera determining a_{ij} calibration constants from Eq. 1, for $i = 1,...,8$ the calibration points and $j = 1, ..., 11$ the calibration constants as proposed by Addel-Aziz et al. [9],

$$x_i = \frac{a_{i1}X + a_{i2}Y + a_{i3}Z + a_{i4}}{a_{i9}X + a_{i10}Y + a_{i11}Z + 1}; \quad y_i = \frac{a_{i5}X + a_{i6}Y + a_{i7}Z + a_{i8}}{a_{9i}X + a_{i10}Y + a_{i11}Z + 1} \quad (1)$$

Since the number of equations (8 points × 2 cameras) at the reconstruction process is larger than the number of variables, iterative least square solution of collinearity equations (ILSSC) is applied with linearization of Eq. 1 by Taylor series expansion in the proximity of a starting point as proposed by Pedotti et al. [10]. New point at the parameter space is computed using Eq. 1 and registered coordinates minimizing the errors. New point then becomes the next starting point with iterations proceeding until the parameters update is small than defined tolerance within the maximum of six iterations. Space intersection is then performed minimizing perpendicular distance connecting the two lines joining perspective centers of the two cameras to the projections of the point on the sensor as proposed by Borghese et al. [11]. Experimental trial set is composed by in vivo subject specific multibody model of $n = 6$ sports students' degree with (21.5 ± 1.4) years old, without specific training or exercise skills. After signing informed consent according to the World Medical Association Declaration of Helsinki subjects were weighed (76.7 ± 9.3) kg and their height measured (1.79 ± 0.06) m. Reflective marks were attached to each subject skin surface palped anatomic points at main upper and lower limb joints namely right shoulder, hip, knee, ankle bone, skank and foot. Kinematic data from 3D anthropometric image and dynamic data from ground reaction forces (GRF) were timely synchronized based on the identification of impact time instant of a dropping golf ball with the force platform producing a sharp peak at vertical GRF register and the frames corresponding to the mean time instant of the ball contact with ground at both camera images, Fig. 2, setting up those time instants as offsets.

   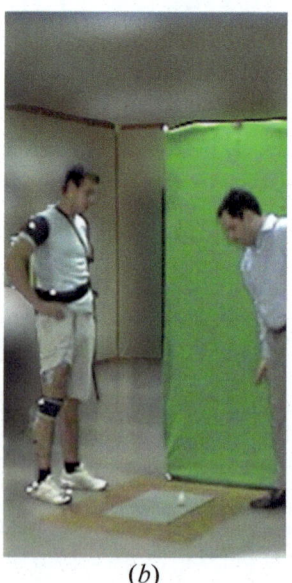

(a)   (b)

Fig. 2. (a) Left (posterior) and (b) right (anterior) camera views during synchronization of image system and force platform.

Each subject performed 3 trial of each MVJ without countermovement (SJ – Squat Jump), with long countermovement (CMJ – Countermovement Jump) and short countermovement (DJ – Drop Jump) according to the protocol defined by Asmussen et al. [12]. During each MVJ trial, ground reaction forces and force moments were acquired at 1000 Hz using AMTI BP2416-4000 CE model along with Mini Amp MAS-6 amplifiers and kinematic data of 3D marks positions captured by a pair of cameras JVC GR-VL9800 operating at 100 Hz. 3D coordinates of joint marks were tracked using Simi Motion 3D Version 7.5.280 (Simi Reality Motion Systems GmbH, Germany) and whole-body COM 3D movement coordinates obtained using a fourteen-segment Dempster [8] adapted model, Fig. 3.

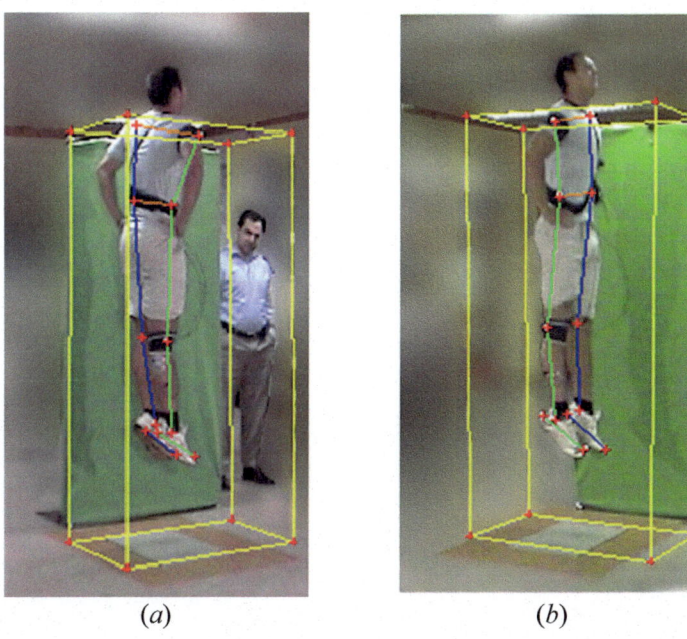

(a) (b)

Fig. 3. (a) Left (posterior) and (b) right (anterior) camera views during trial tests on maximum vertical jump.

Acquired components of GRF with force platform during MVJ trials, Fig. 4, including antero-posterior (*GRFx*), medio-lateral (*GRFy*) and vertical (*GRFz*) force components were time delimited selecting impulse phase, starting with $dGRFz/dt < 0$ and $GRFz < mg$ at CMJ, $dGRFz/dt > 0$ and $GRFz > 0$ at DJ, and $dGRFz/dt > 0$ GRFz > mg at SJ, with m the mass of each subject and $g = 9.81$ m/s^2 the acceleration of gravity, ending with $GRFz \sim 0$ ($GRFz \leq 2\%$ mg) for each subject SJ, CMJ and DJ MVJ trial corresponding to maximum flight time measured by the time interval with $GRFz \sim 0$. Best performance trial was selected for each MVJ with higher flight time on CMJ, DJ and SJ corresponding to the same S1 subject for comparison of whole-body COM vertical displacement obtained by kinematic (Δz_k) and dynamic (Δz_d).

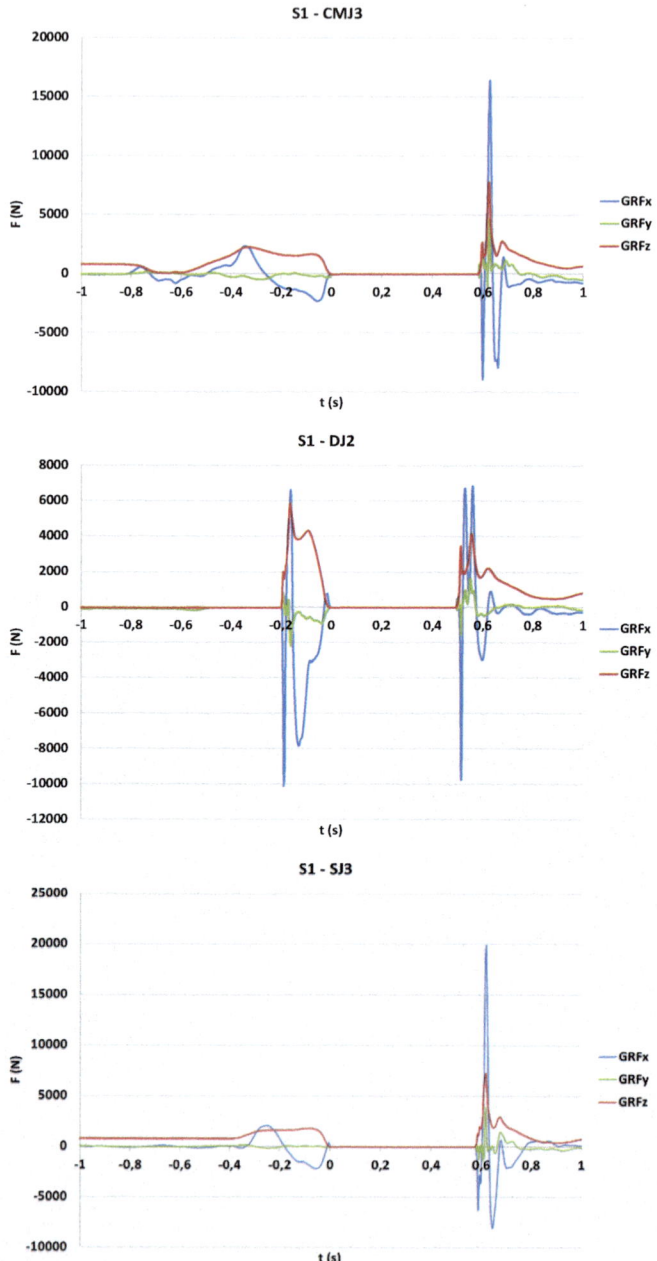

Fig. 4. Antero-posterior (*GRFx*), medio-lateral (*GRFy*) and vertical (*GRFz*) ground reaction force (GRF) components for S1 at CMJ3, DJ2 and SJ3 trials.

The same procedure was applied with time segmentation during MVJ of the impulse phase determined according to GRF time profiles on kinematic signal from 3D anthropometric data and whole-body COM obtained by the adapted Dempster model [8], Fig. 5.

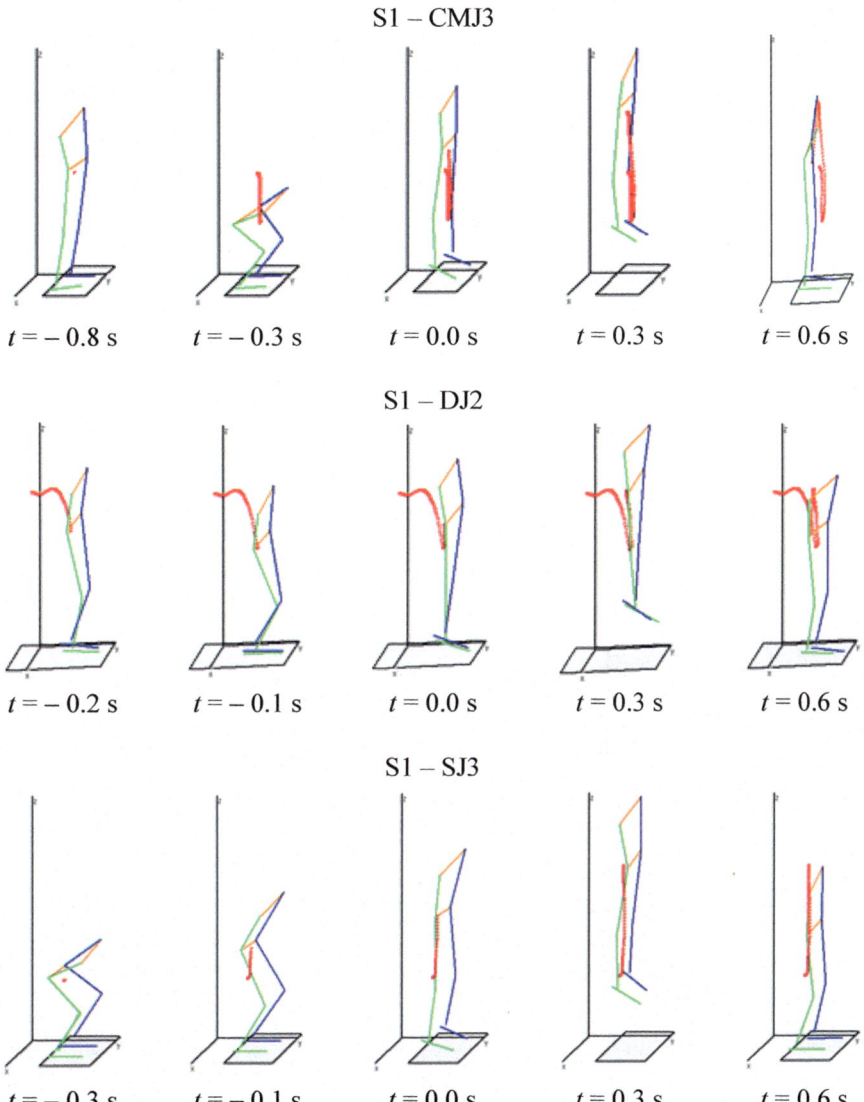

Fig. 5. Stick-figures from 3D anthropometric data and whole-body center of mass (COM) at selected time instants during one trial of S1 performed tests at CMJ3, DJ2 and SJ3.

Vertical ground reaction force ($GRFz$) was selected according to higher impulse contribution and higher COM vertical excursion on MVJ. Vertical resultant force (RFz) was obtained from $GRFz$ subtracting subject weight $Fg = m\,g$ assessed by the force platform on orthostatic standing position and whole body COM vertical acceleration a_z obtained from $GRFz$ dividing it by the subject mass obtained from $m = Fg/g$ with $g = 9.81$ m/s^2 the gravitic acceleration. Vertical velocity v_z of whole-body COM was obtained from time integration of a_z and initial vertical velocity v_{0z}. Vertical position z of whole-body COM was obtained from time integration of v_z and initial vertical position z_0 of the COM, Fig. 6. Whole-body COM vertical displacement obtained from dynamic method Δz_d was resampled at 100 Hz for comparison with whole-body COM vertical displacement Δz_k from kinematic method.

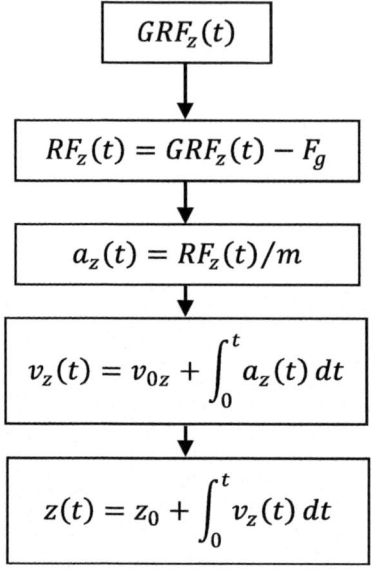

Fig. 6. Block diagram of whole-body center of mass (COM) vertical acceleration a_z, velocity v_z and position z obtained from vertical ground reaction force $GRFz$.

Whole-body COM vertical displacement Δz was statistically compared on mean values and dispersion between different MVJ and kinematic (Δz_k) with dynamic (Δz_d) using also mean square-error MSE during impulse phase Δt for validation of kinematic with dynamic measures, Eq. 2,

$$D_{\Delta z} = \sqrt{\frac{1}{\Delta t} \int_0^{\Delta t} [\Delta z_k(t) - \Delta z_d(t)]^2 dt} \tag{2}$$

3   Results

Representative examples of S1 trial whole-body vertical displacement obtained by kinematic (Δz_k) and dynamic (Δz_d) methods during impulse phase at different MVJ, CMJ, DJ and SJ are presented at Fig. 7, with $t = 0$ s the take-off time instant.

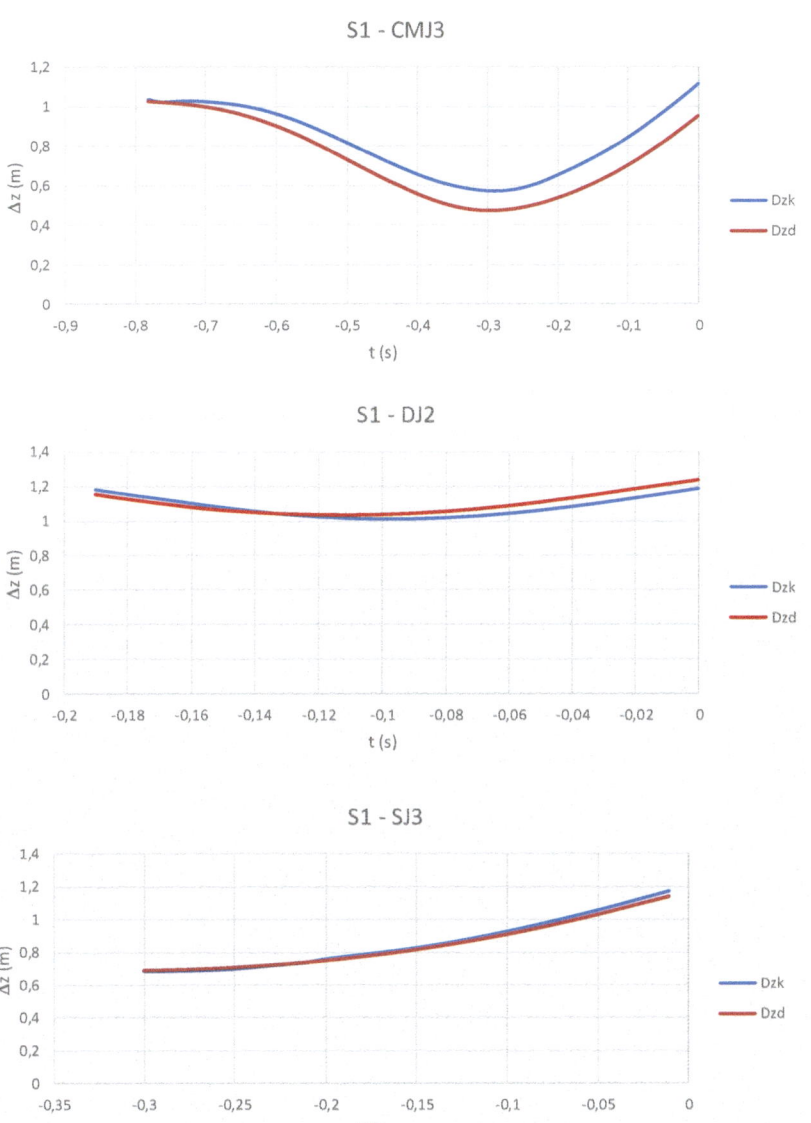

Fig. 7. Center of mass (COM) vertical displacement during impulsion phase on kinematic (Δz_k) and dynamic (Δz_d) methods for S1 at CMJ3, DJ2 and SJ3 trials.

Whole-body COM vertical displacement presented for each S1 selected MVJ trial CMJ3, SJ3 and DJ2 similar variability by kinematic method (Δz_k) when compared with dynamic (Δz_d) method using Levene statistical test ($p > 0.05$), Table 1 and Fig. 8. Mean values of Δz_k and Δz_d presented at CMJ3 statistical significative differences ($p < 0.05$) with higher mean Δz_k than Δz_d, without statistical significative differences ($p < 0.05$) at DJ2 and SJ3. When comparing Δz_k, CMJ3 and SJ3 presented similar mean values and variabilities ($p > 0.05$), with different mean values and variabilities ($p < 0.05$) at CMJ3 and SJ3 when compared to DJ2, with higher variability at CMJ3 and SJ3 than DJ2 and lower mean values at CMJ3 and SJ3 than DJ2. As regards to Δz_d mean values and variabilities CMJ3, SJ3 and DJ2 presented statistical significative differences ($p < 0.05$), with lower mean value at CMJ3 than SJ3 both lower than DJ2 and higher variability at CMJ3 and SJ3 than DJ2.

Table 1. Mean ($\bar{x}$) and standard deviation (*std.*) of whole-body COM vertical displacement obtained by kinematic (Δz_k) and dynamic (Δz_d) methods at CMJ, SJ, DJ.

	CMJ3 ($\bar{x} \pm std.$)	SJ3 ($\bar{x} \pm std.$)	DJ2 ($\bar{x} \pm std.$)
Δz_k (m)	0.822 ± 0.171	0.858 ± 0.152	1.078 ± 0.059
Δz_d (m)	0.730 ± 0.192	0.848 ± 0.141	1.100 ± 0.061

Root mean square-error (RMSE) of whole-body COM kinematic (Δz_k) and dynamic (Δz_d) vertical displacement during MVJ impulse phase, Eq. 2, presented higher value 0.076 m at CMJ3 than DJ2 with 0.031 m, both higher than SJ3 with 0.014 m in agreement with vertical displacement profiles, Fig. 7, presenting increasing $D_{\Delta z}$ values from start time instant to take-off instant $t = 0$ s at each MVJ.

4   Discussion

Whole-body COM vertical displacement during impulse phase on long lower limb countermovement (CMJ), short countermovement (DJ) and without countermovement (SJ) was assessed using kinematic method with adapted Dempster model from 3D anthropometric image (Δz_k) and dynamic (Δz_d) method from simultaneous and synchronized ground reaction force measured with force platform and double time integration of COM vertical acceleration a_z. Standard MVJ protocol has been applied making use of whole-body weight for natural long (CMJ) and short (DJ) lower limb countermovement (CM) assessment with corresponding lower limb muscle stretch-shortening cycle (SSC) effect on whole-body vertical impulse for comparison with vertical impulse without countermovement (SJ). Most frequent approaches on assessment of CM effect at MVJ, CMJ and DJ for comparison with MVJ without CM at SJ are exclusively kinematic using anthropometric marks or dynamic with acquired ground reaction forces and simultaneous kinematic and dynamic analysis applied mainly on inverse methods for estimation of internal actions with residual forces playing a key role on agreement of F with ma at Newton's second law and the need for

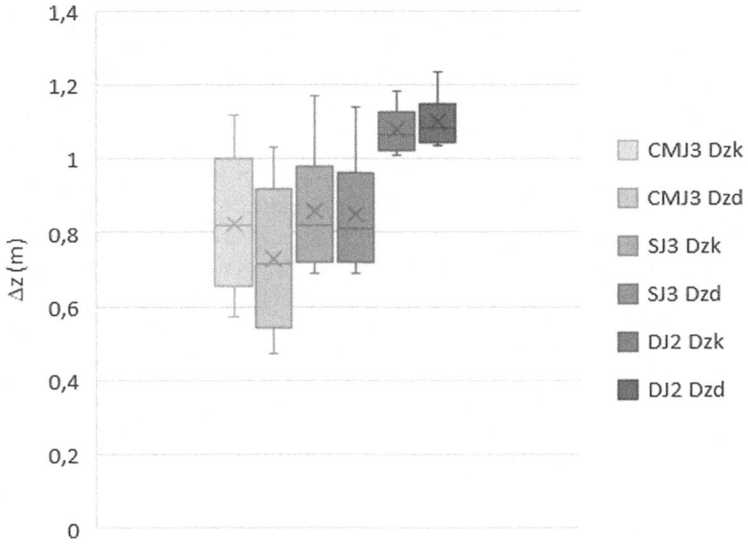

Fig. 8. Whole-body COM vertical displacement obtained by kinematic (Δz_k) and dynamic (Δz_d) methods at CMJ, SJ and DJ MVJ.

assessment of residual forces accounting for differences at kinematic and dynamic data. With this purpose in mind attained results allowed detection of similar variability of COM vertical displacement by kinematic Δz_k and dynamic Δz_d method at S1 CMJ3, SJ3 and DJ2 trials pointing for the ability of dynamic and kinematic methods to detect the same variability at each form of selected MVJ. Mean Δz_k and Δz_d presented comparable values, differentiating registered behavior at CMJ in relation to DJ and SJ, with statistical significative differences ($p < 0.05$) at CMJ3 and higher mean at Δz_k than Δz_d, without statistical significative differences ($p < 0.05$) at DJ2 and SJ3. These results are consistent with mean square-error of whole-body COM kinematic (Δz_k) and dynamic (Δz_d) vertical displacement during MVJ impulse phase presenting higher value at CMJ than DJ both higher than SJ. Consistency between statistical tests of mean and dispersion with mean square-error of whole-body COM kinematic (Δz_k) and dynamic (Δz_d) vertical displacement during MVJ impulse phase point to the potential of the validation of whole-body COM movement from 3D anthropometric image with dynamic data at different human standard MVJ. This validation of whole-body COM vertical displacement using complementary methods plays furthermore importance due to the potential sources of error on each method and the key role assessment of whole-body COM vertical displacement on evaluation of the lower limb vertical counter-movement and associated SSC muscle contribution on MVJ. Since whole-body COM relies on each body segment mass, position and mass distribution within it, with COM not directly observed, constantly moving in relation to body segments the possibility of validation of its position based on complementary kinematic and dynamic methods constitutes an important contribution to the assessment and reduction of the necessary residual forces to match kinematic and dynamic methods as well to the improvement in the accuracy of the estimation of the CM and associated lower limb muscle SSC for the impulse of whole-body COM.

When compared separately whole-body COM kinematic (Δz_k) and dynamic (Δz_d) vertical displacement presented during MVJ impulse phase higher variance at CMJ than SJ $p > 0.05$ at Δz_k and $p < 0.05$ at Δz_d all higher than DJ $p < 0.05$. Also compared separately whole-body COM kinematic (Δz_k) and dynamic (Δz_d) vertical displacement presented during MVJ impulse phase lower mean value at CMJ than DJ $p > 0.05$ at Δz_k and $p < 0.05$ at Δz_d all lower than SJ. These consistent behavior on comparison of mean values and variances whole-body COM kinematic (Δz_k) and dynamic (Δz_d) vertical displacement presented during MVJ impulse phase at CMJ, SJ and DJ point for the acceptance on the congruence hypotheses of the validation of whole-body COM movement from 3D anthropometric image with dynamic data at different human standard MVJ. When considering a detailed analysis of the evolution of the mean square-error of COM vertical kinematic and dynamic displacement $D_{\Delta z}$ during impulse phase this presented higher value at CMJ when compared to DJ both with higher value than SJ roughly increasing from start of impulse phase to the take-off time instant. This increase is associated to the start with integration constants at dynamic process v_{0z} and z_0 from kinematics but also to the uncertainty errors accumulated during kinematic and dynamic measures. These errors can be associated to uncertainty on kinematic measures of 3D anthropometric image, integration errors from dynamic measures, synchronization or calibration errors reinforcing the need and utility for validation of whole-body COM movement from 3D anthropometric image with dynamic data at different human standard MVJ.

5   Conclusion

Validation of whole-body COM movement from 3D anthropometric image with dynamic data at different human standard MVJ presented as a valid approach to assess and reduce errors from kinematic measures of 3D anthropometric image and dynamic measures, synchronization or calibration errors contributing to increase accuracy on the estimation of lower limb countermovement and corresponding muscle SSC contribution for the impulse of whole-body COM on MVJ. Several methods including linear impulse, flight time and kinematic anthropometric models have been used to assess the contribution of lower limb countermovement and corresponding muscle SSC for the impulse of whole-body COM on MVJ without accounting for differences of results between the different methods, the effect of applied method on attained results at different MVJ and the accuracy and validation of results at each method.

Presented techniques can contribute to development of better subject specific anthropometric models overcoming subjective parameter estimation as well as to improvement of kinematic and dynamic complementary methods, reducing magnitude of necessary accounting residual forces to match kinematic and dynamic methods for accurate estimation of whole body COM displacement and the improvement for contribution assessment of natural phenomena such as lower limb muscle SSC on human MVJ but also at running performance and gait efficiency.

Acknowledgments. To EACEA, UPorto, UFRJ, UFPE and SAPIENZA Università di Roma for mobility support. To LMH – ISMAI and CRPG for trial tests and tools for data analysis.

References

1. Payton, C.J., Bartlet, R.M.: Biomechanical Evaluation of Movement in Sport and Exercise: the British Association of Sport and Exercise Science Guide. Routledge Taylor & Francis Group, Abingdon Oxon (2008)
2. Hill, A.V.: The maximum work and mechanical efficiency of human muscles, and their most economical speed. J. Physiol. **56**(1–2), 19–41 (1992)
3. Komi, P.V., Ishikawa, M., Linnamo, V.: Identification of stretch-shortening cycles in different sports. J. Physiol. **56**(1–2), 19–41 (2011)
4. Komi, P.V., Bosco, C.: Utilization of stored elastic energy in leg extensor muscles by men and women. J. Med. Sci. Sports **10**, 261–265 (1978)
5. Bobbert, M.F., Mackay, M., Schinkelshoek, D., Huijing, P.A., van Ingen Schenau, G.J.: Biomechanical analysis of drop and countermovement jumps. Eur. J. Appl. Physiol. **54**, 566–573 (1986)
6. Aragón-Vargas, L.F., Gross, M.M.: Kinesiological factors in vertical jump performance: differences among individuals. J. Appl. Biomech. **13**, 24–44 (1997)
7. Blazevich, A.: The stretch-shortening cycle (SSC). In: Cardinale, M., Newton, R., Nosaka, K. (eds.) Strength and Conditioning – Biological Principles and Practical Applications. Wiley, West Sussex (2011)
8. Dempster, W.T.: Space requirements of the seated operator: geometrical, kinematic, and mechanical aspects of the body with special reference to the limbs (Wright Air Development Center Technical report no. 55-159). Wright-Patterson Air Force Base, WADC, Dayton, OH (1955)
9. Addel-Aziz, Y.L., Karara, H.M.: Direct linear transformation from comparator coordinates into object space coordinates in close-range photogrammetry. In: Proceedings of the Symposium on Close Range Photogrammetry, pp. 1–18 (1971)
10. Pedotti, A., Ferrigno, G.: Optoelectronic-based system. In: Allard, P., Stokes, I.A.F., Blanchi, J.P. (eds.) Three-Dimensional Analysis of Human Movement, Human Kinetics, pp. 57–78 (1995)
11. Borghese, N.A., Ferrigno, G.: An algorithm for 3D automatic movement detection by means of standard TV cameras. IEEE Trans. Biomed. Eng. BME **22**, 259–264 (1990)
12. Asmussen, E., Bonde-Petersen, F.: Storage of elastic energy in skeletal muscles in man. Acta Physiol. Scand. **91**(3), 385–392 (1974)

Benchmark RGB-D Gait Datasets:
A Systematic Review

João Ferreira Nunes[1,2(✉)] 🆔, Pedro Miguel Moreira[1] 🆔,
and João Manuel R. S. Tavares[2] 🆔

[1] ARC4DigiT - Centro de Investigação Aplicada para a Transformação Digital,
Instituto Politécnico de Viana do Castelo, Viana do Castelo, Portugal
{joao.nunes,pmoreira}@estg.ipvc.pt
[2] Instituto de Ciência e Inovação em Engenharia Mecânica e Engenharia Industrial,
Departamento de Engenharia Mecânica, Faculdade de Enhenharia, Universidade do
Porto, Porto, Portugal
tavares@fe.up.pt

Abstract. Human motion analysis has proven to be a great source of
information for a wide range of applications. Several approaches for a
detailed and accurate motion analysis have been proposed in the literature, as well as an almost proportional number of dedicated datasets.
The relatively recent arrival of depth sensors contributed to an increasing interest in this research area and also to the emergence of a new
type of motion datasets. This work focuses on a systematic review of
publicly available depth-based datasets, encompassing human gait data
which is used for person recognition and/or classification purposes. We
have conducted this systematic review using the Scopus database. The
herein presented survey, which to the best of our knowledge is the first
one dedicated to this type of datasets, is intended to inform and aid
researchers on the selection of the most suitable datasets to develop, test
and compare their algorithms.

Keywords: Gait datasets · Depth sensors · Systematic review

1 Introduction

The analysis of human motions has been a very active research topic, with a
manifold of potential applications. Some of them have already been discussed in
reviews like [21,23,28,31]. Particularly, gait, as a specific type of human movement, has been used for a wide range of applications, either as a data source
or simply as a way to interact with them. For example, in healthcare, gait data
has been used (i) for geriatrics assistance (elderly fall prevention [16,20], mobility assessment, and home monitoring [9]); (ii) to conduct orthopedic studies
[4,29]; and (iii) to neurological and chronic disorders assessment [10,13,33]).
In the sports domain, gait patterns have been used to assist athletes so they
can perform better and safer [14,27]). In smart surveillance systems, gait sig-
natures are used as a new type of biometric authentication [6,7]). And also, in

© Springer Nature Switzerland AG 2019
J. M. R. S. Tavares and R. M. Natal Jorge (Eds.): VipIMAGE 2019, LNCVB 34, pp. 366–372, 2019.
https://doi.org/10.1007/978-3-030-32040-9_38

human-computer interaction, where gait has been used to interact in video-game environments [3] or to animate virtual characters [15].

Several methodologies for a detailed and accurate gait analysis have been proposed, including computer vision-based, inertial sensors-based or other based approaches. Simultaneously, dedicated gait datasets have appeared almost in a proportional way. The relatively recent arrival of depth sensors, like Microsoft Kinect, Intel RealSense or Asus Xtion, capable of tracking humans in real-time without the need to ware special suits or markers, contributed to an increasing interest in this research area leading also to the emergence of a new type of motion datasets. A detailed description of a widely used depth sensor functionality is given in [35]. These depth-based datasets, besides the RGB data, also include raw depth data and in same cases also the 3D coordinates of a set of points that in general, correspond to human body joints.

Regardless the methodologies nor the technologies used, it is very important for the scientific community to use common input data, enabling coherent comparisons of performances and results. For that reason, and focusing specifically on the human gait, we have conducted this systematic review, whose primary goal is to identify all the existing, freely available, depth-based datasets containing human gait information, whose applicability encompasses person recognition and/or classification purposes. Consequently, we are assisting researchers by presenting an updated framework, easy to analyse, useful to identify existing datasets, and suitable to compare them, avoiding, eventually, the creation of new (and sometimes redundant) datasets.

2   Related Work

Several human motion-based datasets reviews can be found in the literature. Some examples are presented in [18, 22, 25, 30]. It is evident the diversity between datasets in terms of their applicability, the acquisition environment conditions, the number of participants, the number of sequences, etc. However, none of those reviews have been presented in a simple form, easy to analyse and to compare datasets. In most of the reviews each dataset is described without any formal organization or structure. Moreover, in the examples previously identified, the revised datasets do not include depth data. Nevertheless, in [11], it is presented a revision of depth-based datasets within eight categories: semantics, object pose estimation, camera tracking, scene reconstruction, object tracking, human actions, human faces and human identification. The latter encompasses four datasets, in which two of them are also present in our study.

3   Systematic Review

The protocol that was used to locate, gather and evaluate the datasets under study is described in this Section. The review took place in the first quarter of 2019 and it was conducted using the Scopus database. The criteria defined for the selection of articles were as follows: Domain (Gait Analysis); Purpose (Person Recognition and/or Gender Classification); and Dataset (Depth-Based).

Thus, the searched terms used in the Scopus database were: ("gait recognition" OR "gait identification" OR "person recognition" OR "person identification" OR "gender identification" OR "gender recognition") AND ("depth sensor" OR "RGBD" OR "RGB-D" OR "Kinect" OR "RealSense" OR "Xtion" OR "ToF") AND ("dataset" OR "database" OR "data set"). These terms were searched in the title, abstract and keywords of the indexed articles in Scopus database and were refined by: Publication types = (ALL) AND Languages = (ENGLISH). This search resulted in 58 articles.

Every resulting article was analysed and both self-constructed and referenced depth-based gait datasets were included in the first set of articles. From among the 58 articles retrieved, we were able to identify 10 freely available datasets, which are described in Sect. 4. A newly created dataset that was developed by us was also included in the list, thus resulting in a total of 11 datasets reviewed.

4   Framework for Datasets Comparison

For a simple and easy way to analyse and also to compare datasets we decided to present them in form of a table, describing each dataset based on a set of generic features. In this Section the features used to describe each dataset are listed, and then the selected datasets are identified. The proposed framework is presented in Table 1.

4.1   Features Description

The selected features were the following:

- *applicability*: the context for which the dataset was created;
- *subjects*: number of participants, and if possible, their description regarding gender and age;
- *sensor*: number and type of sensors used, and how they were placed on scene;
- *trajectories*: number of different defined trajectories;
- *sequences/subject*: number of sequences performed per subject;
- *covariates*: list of existing variations between sequences (e.g.: walking styles, clothing, etc.);
- *collected data*: list of data that was collected and made available to download;
- *additional data supplied*: list of additional data that was made accessible, like code/applications to manipulate data.

A few other features were initially included in the datasets description structure proposal, however, given that their values were constant between all datasets, we have decided to omit them. Those features included information about the environment where the sessions took place (indoor laboratories), about the frame rate of used sensors (30 fps) and about whether the participants were aware that they were being filmed (all participants were conscious that they were being observed).

Table 1. The proposed framework used to describe each dataset.

Identification	Year	Applicability[a]	#subjects	#sensors	#trajectories[b]	#sequences/subject[b]	Covariates[c]	Collected data[d]	Extra data
BVH MoCap Databse [12]	2010	PR	6 + 4 static + moving sessions	Kinect v1 static or in motion	5 (2S+F+ B+MS)	IN	SP	SK	c# code + matlab scripts
Depth-Based Gait Dataset [8]	2013	PR	29	2×kinect v1 fixed at 2.5 m high	2(F+B)	4 (2F+2B)	VP, OC, FR, WC (NW+FW)	D, SK, T	–
DGait [5]	2012	GR	55 36♂+19♀	Kinect v1 fixed at 2 m high	8 (F+B +2S+2D45 +2D-45)	11 (2F+B+ 4S+2D45 +2D-45)	VP	C, D, SK	Start-end frame/gait cycle
GRIDDS [24]	2018	PR + GR	35 11♂+24♀	Kinect v2 fixed at 1.8 m high	2S	10 (5S+5S)	–	C, D, SK, T, DS, S	Start-end frame/gait cycle + matlab scripts
Kinect Gait Biometry Dataset [1]	2014	PR	164 17-35 years old	2×kinect v1 in motion	1SC	5SC		SK	–
RGB-D Person Re-identification Dataset [2]	2012	PR	79	Kinect v1 fixed	2(F+B)	4 (3F+1B)	T, C, WC (SW+NW)	C, SK, S, 3DM, EF	Matlab scripts
SAIVT-DGD [26]	2011	PR	15	Kinect v1 fixed	1F	20F	WC (NW+FW), CC (BC+SC+ FC) + S	D, DS, 3DV	Matlab scripts + documentation
SDUgait [32]	2015	PR	52 28♂+24♀	2×kinect v2 fixed at 1 m high	Kinect1: 5 F+B+S+ D45+A Kinect2: 5 F+2S+ D-45+A	5	VP	DS, SK	C# code
SZU RGB-D Gait Dataset [34]	2013	PR	99	ASUS Xtion PRO LIVE fixed at 0,8 m high	4 (2S+2D30)	8 (4S+4D30)	VP	D, DS	–
TUM-GAID Database [17]	2012	PR	305	Kinect v1 fixed at 1.9 m high	2S	10[e]	CC, S T[f], C[f]	C, D, A	–
UPCV Gait Dataset [19]	2013	GR	30 15♂+15♀ 23-55 years old	Kinect v1 fixed at 2 m high	1S	5S	–	SK	Matlab scripts

a   PR: Person Recognition; GR: Gender Recognition.

b   F: Frontal; B: Backwards; S: Side (Left-to-Right and/or Right-to-Left); Dxx: Diagonal at xx Degrees; SC: Semi-Circular; MS: Moving Sensor; A: Arbitrary; IN: Irregular Number.

c   VP: Viewpoint; WC: Walking Conditions (SW: Slow Walk, NW: Normal Walk, FW: Fast Walk); C: Clothing; S: Shoes; FR: Frame Rate; OC: Occlusions; CC: Carrying Conditions (BC: Back Carrying, SC: Side Carrying, FC: Front Carrying); T: Time; SP: Sensor Position.

d   SK: 3D Skeleton Coordinates; T: Time; C: RGB data; D: Depth data; A: Audio; S: Silhouettes; 3DM: 3D Mesh; EF: Estimated Floor; 3DV: 3D Volumes; DS: Depth Silhouettes.

e   10 sequences for 273 subjects and 20 sequences for 32 subjects.

f   Only applied to 32 subjects.

Selected Datasets. The datasets that were part of our study, whose selection procedure was explained in Sect. 3, and that consequently are present in our proposed framework are the following: Depth-Based Gait Dataset [8], DGait [5], GRIDDS, [12], Kinect Gait Biometry Dataset [1], RGB-D Person Re-identification Dataset [2], SAIVT-DGD [26], SDUgait [32], SZU RGB-D Gait Dataset [34], TUM-GAID Database [17] and UPCV Gait Dataset [19].

5   Conclusions

The importance of human gait is quite evident, considering its wide range of application domains. Systems that are able to recognize humans and to classify human attributes like gender, age or mood may have a great impact in our society. In the last decade, the dissemination and availability of RGBD sensors (depth + image) prompted the development of new methods and the availability of datasets including depth information beyond the objects represented in the images. This work, to the best of our knowledge, is the first systematic review dedicated to depth-based gait datasets, whose purpose is person recognition and/or classification, covering a total of 11 datasets, of which 72% were acquired between 2011 and 2015, using the first version of the Kinect sensor, whilst 18% were acquired between 2016 and 2018, using the second version of Kinect. Concerning the original context that they were created for, 81% aimed to recognize persons by their gait, while for the remaining 19% their goal was to classify people's gender. As an outcome of this work, we introduce our perspective on how these datasets can be classified and compared and a logical, structured and feature oriented classification framework is presented.

A   Datasets' URLs

- BVH MoCap Databse https://bit.ly/2kqdqtx
- Depth-Based Gait Dataset http://www.facweb.iitkgp.ac.in/~shamik/Gait/Dataset1.html
- DGait http://www.cvc.uab.es/DGaitDB
- GRIDDS http://gridds.ipvc.pt
- Kinect Gait Biometry Dataset https://bit.ly/2QbDu6U
- RGB-D Person Re-identification Dataset https://bit.ly/2HLXZU7
- SAIVT-DGD https://research.qut.edu.au/saivt/databases/saivt-dgd-database
- SDUgait https://sites.google.com/site/sdugait
- SZU RGB-D Gait Dataset http://yushiqi.cn
- TUM-GAID Database https://www.mmk.ei.tum.de/en/misc/tum-gaid-database
- UPCV Gait dataset http://www.upcv.upatras.gr/personal/kastaniotis/datasets.html

References

1. Andersson, V.O., Araújo, R.M.: Person identification using anthropometric and gait data from kinect sensor. In: Proceedings of the 29th Association for the Advancement of Artificial Intelligence, pp. 425–431 (2015)
2. Barbosa, I.B., Cristani, M., Bue, A.D., Bazzani, L., Murino, V.: Re-identification with RGB-d sensors. In: Proceedings of the 12th international conference on Computer Vision - Volume Part I, pp. 433–442. Springer, Heidelberg (2012)

3. Bloom, V., Argyriou, V., Makris, D.: Hierarchical transfer learning for online recognition of compound actions. Comput. Vis. Image Underst. **144**, 62–72 (2016)
4. Bonnechère, B., Jansen, B., Salvia, P., Bouzahouene, H., Omelina, L., Moiseev, F., Sholukha, V., Cornelis, J., Rooze, M., Jan, S.V.S.: Validity and reliability of the kinect within functional assessment activities: comparison with standard stereophotogrammetry. Gait Posture **39**(1), 593–598 (2014)
5. Borràs, R., Lapedriza, À., Igual, L.: Depth information in human gait analysis: an experimental study on gender recognition. In: Proceedings of the 9th International Conference on Image Analysis and Recognition - Volume Part II, ICIAR, pp. 98–105. Springer, Heidelberg (2012)
6. Bouchrika, I., Carter, J.N., Nixon, M.S.: Towards automated visual surveillance using gait for identity recognition and tracking across multiple non-intersecting cameras. Multimed. Tools Appl. **75**(2), 1201–1221 (2016)
7. Boulgouris, N.V., Hatzinakos, D., Plataniotis, K.N.: Gait recognition: a challenging signal processing technology for biometric identification. IEEE Signal Process. Mag. **22**(6), 78–90 (2005)
8. Chattopadhyay, P., Sural, S., Mukherjee, J.: Frontal gait recognition from occluded scenes. Pattern Recogn. Lett. **63**, 9–15 (2015)
9. Cheng, H., Liu, Z., Zhao, Y., Ye, G.: Real world activity summary for senior home monitoring. In: Proceedings of the IEEE International Conference on Multimedia and Expo, ICME, pp. 1–4 (2011)
10. Dobson, F., Morris, M.E., Baker, R., Graham, H.K.: Gait classification in children with cerebral palsy: a systematic review. Gait Posture **25**(1), 140–152 (2007)
11. Firman, M.: RGBD datasets: Past, present and future. In: Proceedings of the IEEE Conference on Computer Vision and Pattern Recognition - Workshops. IEEE (2016)
12. Galinska, K., Luboch, P., Kluwak, K., Bieganski, M.: A database of elementary human movements collected with RGB-d type camera. In: Proceedings of the 6th IEEE International Conference on Cognitive Infocommunications. IEEE (2015)
13. Galna, B., Barry, G., Jackson, D., Mhiripiri, D., Olivier, P., Rochester, L.: Accuracy of the microsoft kinect sensor for measuring movement in people with parkinson's disease. Gait Posture **39**(4), 1062–1068 (2014)
14. Gouwanda, D., Senanayake, S.M.N.A.: Emerging trends of body-mounted sensors in sports and human gait analysis. In: Proceedings of the 4th International Conference on Biomedical Engineering, pp. 715–718. Springer, Heidelberg (2008)
15. Gross, R., Shi, J.: The CMU motion of body (MoBo) database. Technical report CMU-RI-TR-01-18, Carnegie Mellon University, Pittsburgh, PA (2001)
16. Hausdorff, J.M., Rios, D.A., Edelberg, H.K.: Gait variability and fall risk in community-living older adults: a 1-year prospective study. Arch. Phys. Med. Rehabil. **82**(8), 1050–1056 (2001)
17. Hofmann, M., Geiger, J., Bachmann, S., Schuller, B., Rigoll, G.: The TUM gait from audio, image and depth (GAID) database - multimodal recognition of subjects and traits. J. Vis. Commun. Image Represent. **25**(1), 195–206 (2014)
18. Ji, X., Liu, H.: Advances in view-invariant human motion analysis: a review. IEEE Trans. Syst. Man, Cybern. Part C: Appl. Rev. **40**(1), 13–24 (2010)
19. Kastaniotis, D., Theodorakopoulos, I., Economou, G., Fotopoulos, S.: Gait-based gender recognition using pose information for real time applications. In: Proceedings of the 18th International Conference on Digital Signal Processing. IEEE (2013)
20. Kwolek, B., Kepski, M.: Improving fall detection by the use of depth sensor and accelerometer. Neurocomputing **168**, 637–645 (2015)

21. Moeslund, T.B., Hilton, A., Krüger, V.: A survey of advances in vision-based human motion capture and analysis. Comput. Vis. Image Underst. **104**(2), 90–126 (2006)
22. Nixon, M.S., Tan, T., Chellappa, R.: Human Identification Based on Gait. Springer, US (2006)
23. Nunes, J.F., Moreira, P.M., Tavares, J.M.R.S.: Human motion analysis and simulation tools: a survey. In: Miranda, F., Abreu, C. (eds.) Handbook of Research on Computational Simulation and Modeling in Engineering, pp. 359–388. IGI Global, Hershey (2016)
24. Nunes, J.F., Moreira, P.M., Tavares, J.M.R.S.: GRIDDS - a gait recognition image and depth dataset. In: VipIMAGE 2019/VII ECCOMAS Thematic Conference on Computational Vision and Medical Image Processing (2019)
25. Phillips, P.J., Sarkar, S., Vega, I.R., Grother, P.J., Bowyer, K.W.: The gait identification challenge problem: data sets and baseline algorithm. In: Proceedings of the 16th International Conference on Pattern Recognition. IEEE (2002)
26. Sivapalan, S., Chen, D., Denman, S., Sridharan, S., Fookes, C.: Gait energy volumes and frontal gait recognition using depth images. In: Proceedings of the International Joint Conference on Biometrics, pp. 1–6 (2011)
27. Stasi, S.L.D., Logerstedt, D., Gardinier, E.S., Snyder-Mackler, L.: Gait patterns differ between ACL-reconstructed athletes who pass return-to-sport criteria and those who fail. Am. J. Sports Med. **41**(6), 1310–1318 (2013)
28. Vasconcelos, M.J.M., Tavares, J.M.R.S.: Human motion analysis: methodologies and applications. In: Proceedings of the 8th International Symposium on Computer Methods in Biomechanics and Biomedical Engineering. CMBBE (2008)
29. Vernon, S., Paterson, K., Bower, K., McGinley, J., Miller, K., Pua, Y.H., Clark, R.A.: Quantifying individual components of the timed up and go using the kinect in people living with stroke. Neurorehabil. Neural Repair **29**(1), 48–53 (2014)
30. Wang, J., She, M., Nahavandi, S., Kouzani, A.: A review of vision-based gait recognition methods for human identification. In: Proceedings of the International Conference on Digital Image Computing: Techniques and Applications, DICTA, pp. 320–327. IEEE (2010)
31. Wang, L., Hu, W., Tan, T.: Recent developments in human motion analysis. Pattern Recogn. **36**(3), 585–601 (2003)
32. Wang, Y., Sun, J., Li, J., Zhao, D.: Gait recognition based on 3D skeleton joints captured by kinect. In: Proceedings of the IEEE International Conference on Image Processing, ICIP, pp. 3151–3155. IEEE (2016)
33. Webster, D., Celik, O.: Systematic review of kinect applications in elderly care and stroke rehabilitation. J. Neuroeng. Rehabil. **11**(1), 108 (2014)
34. Yu, S., Wang, Q., Huang, Y.: A large RGB-D gait dataset and the baseline algorithm. In: Biometric Recognition, pp. 417–424. Springer (2013)
35. Zhang, Z.: Microsoft kinect sensor and its effect. IEEE Multimed. **19**(2), 4–10 (2012)

Software Development for Image Processing and Analysis

Processing Thermographic Images for the Pre Diagnosis of Breast Cancer

Diannys Granadillo$^{(\boxtimes)}$, Yaileth Morales, Eberto Benjumea, and Cesar Torres

Optics and Informatics Group, SIPIB, Popular University of Cesar, Valledupar, Cesar, Colombia
ddgranadillo@unicesar.edu.co

Abstract. Breast cancer is a disease that begins when the cells of the breast begin to grow uncontrollably. It can not be predicted, because the factors that mainly affect the suffering of this pathology are genetic, they can be suffered by both women and men [1]. It is the most common cancer among women worldwide and accounts for 16% of all female cancers. In Colombia, it is estimated that around 8600 cases are reported every year and 2,660 women die due to this [2]. Temperature variations in the skin are organic indicators of several types of cancer, including carcinoma-type breast cancer. The most widely used method for the detection of breast cancer is mammography, a radiation that provides an anatomical image of the breast, with which atypical formations can be observed in the tissue. Mammography, "the use of X-rays, which are ionizing radiation, can cause problems, you can not take more than three or four X-ray tests per year, pregnant women can not have mammograms." For this reason, other diagnostic alternatives have been sought. In this scenario is where infrared thermography appears, technique that allows to evaluate the behavior and the physiognomy of tissues using the temperature they have as a base, diseased tissues such as those affected by some type of cancer can present different temperature levels. If they are compared with those that are healthy. When taking this into account, breast cancer is a disease that can be characterized, detected and analyzed by thermographic images.

In this work an algorithm was developed to detect breast cancer through the segmentation and processing of thermographic images of the breast. The digital process is based on the identification of the area of interest, the segmentation of the color and the quantitative discrimination by color tonality. It is contained in a graphical interface to facilitate its use.

Keywords: Breast cancer · Thermographic image · Segmentation · Histogram · K-means · Image processing

1 Introduction

Breast cancer is a disease that starts when the cells in the breast grow uncontrollably. It is not predicted, because the factors that mainly affect the suffering of this pathology are genetic and can be suffered by both women and men [1]. According to the World Health Organization (WHO), the most frequent cancer among women is breast, which

© Springer Nature Switzerland AG 2019
J. M. R. S. Tavares and R. M. Natal Jorge (Eds.): VipIMAGE 2019, LNCVB 34, pp. 375–387, 2019.
https://doi.org/10.1007/978-3-030-32040-9_39

worldwide represents 16% of all female cancers. It is estimated that 1.38 million new cases are detected every year. This occurs more frequently in developed countries, but has a greater impact on the population of low and middle income countries [2]. Therefore, it is very important to find useful tools for its detection at an early stage. An adequate diagnosis allows the specialist to make a decision about the treatment that should be administered to the patient according to the type of anomaly found.

Medical images constitute a fundamental tool during the analysis and study of different pathologies [3–5]. Mammograms and ultrasound are the most used techniques for the diagnosis of breast cancer. However, this technique presents many difficulties when applied to young women and can also be aggressive for patients, because of the radiation it has. Therefore, thermography is an alternative that overcomes the disadvantages of another type of clinical imaging tool. This modality of imaging is applied to breast tissue at a primary diagnostic stage and is important because it is a non-invasive, low-cost technique and does not use ionizing radiation [3]. Infrared thermography presents a functional analysis of the breast. This technique can be used to detect and analyze the cancer that develops in these tissues. The correct interpretation of an image obtained by thermography can help in the early detection of cancer. Cancerous tumors, specifically those of carcinoma-type breast cancer, have a higher metabolic heat generation because the cells reproduce faster than in normal tissue. These temperature variations can be captured through infrared thermography " [6]. The use of this technology is very useful in the evaluation of vascular problems or some type of cancer. The analysis is performed using temperature differences in a given region and taking into account the symmetry of the breasts. In general, these quantitative assessments are given by the temperature difference of 1 °C that are associated with problems such as cancer and angiogenesis. A range of variation of the temperature difference may correspond to the diagnosis of benign tumors [7]. In thermographic images, many of the evaluations developed by specialists require segmentation methods and techniques for breast analysis. These methods are used for the identification of problems such as tumors or abnormalities. In general, this analysis requires some type of manual involvement by physicians, making these tasks tedious, time-consuming and associated with operator dependence [8, 9]. For this reason, it is necessary and beneficial to develop robust techniques for the analysis of images of thermographies of the breast. Image processing techniques have been proposed for automatic and semi-automatic segmentation of breast thermograms [10, 11]. These applications are taken into account as diagnostic tools by specialists for the identification of problems in the breast. The analysis of segmentation techniques in thermographic images has been developed by researchers using several approaches. Among the works carried out are mathematical operators, statistical methods, neural networks and vector support [12, 13]. Many authors have made their contributions on these techniques on thermography. However, applications in segmentation and analysis in the detection of breast cancer using thermograms still cover a large field of work.

In this research, a method for pre-diagnosis of breast cancer was developed through segmentation algorithm and thermographic image analysis in order to contribute to the early detection of cancer pathologies in the breast; for this, a database with existing thermographic images was used, courtesy of Visual Lab-DMR of patients with breast cancer, identifying through a thermal map variations in the skin that indicate the presence or absence of cancer and finally developed a digital tool to support the specialist in health, for the analysis and early identification of anomalies, which interrelates the acquisition of thermographic images, with the different algorithms based on the color parameters (Thermal Map) and Evolution of the thermographic image. The general tools to be used are composed of a conventional personal computer (PC) and the MATLAB® technical programming platform.

2 Methodology

This work was carried out on the Matlab platform and the thermographic images are hosted on the web by Visual Lab-DMR [14]. The analyzed images are taken under the rainbow color palette. The development of this research was composed of three phases:

1. Characterization of breast cancer lesions.
2. Tests on mammary thermographic images.
3. Development of a computer tool for the pre-diagnosis of breast cancer.

The characterization of breast cancer lesions by means of an algorithm was aimed at identifying, through the thermal map, variations in the temperature of the breast that would indicate presence or absence of cancer or some anomaly. Taking into account that the red areas in the images indicate the presence of temperatures higher than the normal maximum, we sought to select these areas since they are indicative of an anomaly. Then, the images were examined by algorithms of selection of region of interest and clustering by K-means; used to segment colors with the aim of establishing the validity of other authors' conclusions and perform analyzes of the thermal map by using histograms. The stages of this process are shown in Fig. 1.

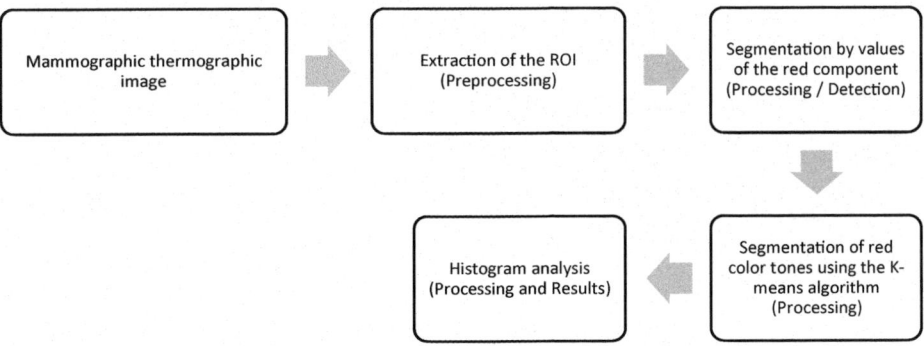

Fig. 1. Stages of breast cancer characterization

2.1 Acquisition of the Thermal Image

The lesion images were taken from a thermographic image database available on the web for students, researchers or anyone interested in collaborating in the area of thermography [14]. Due to the fact that thermographic cameras with adequate characteristics in terms of thermal graphic resolution have commercial prices in excess of 20 million Colombian pesos, it was not possible to create an image database of their own. It must be taken into account that the thermographic images to be examined must be acquired under the rainbow palette and by configuring the range of temperature measurements in such a way that the values above 30 °C are represented under the red color [15], given that the Normal range of temperature in the part of the thorax is between 29.5 °C and 30.5 °C [16].

2.2   Development of the Base of Thermographic Images

Since the base of images provided by Visual Lab-DMR (database of thermographic images available on the web) [14], it contains more than 100 images and not all are suitable for the processing that will be done with the algorithm designed in this work. Ten images of healthy patients and 10 images of sick patients were chosen to perform the analysis and the respective characterization. It must be taken into account that the website provides data on patients such as age, race, diseases, diagnosis by mammography classified as healthy or sick, bad habits, among other characteristics that may be important when making a diagnosis or medical follow-up (Figs. 2 and 3).

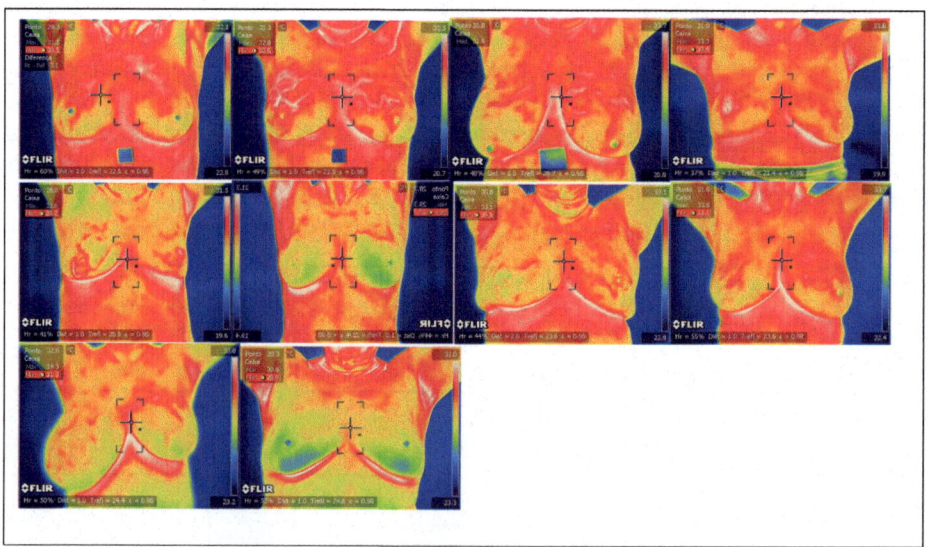

Fig. 2. Thermographic images of sick patients

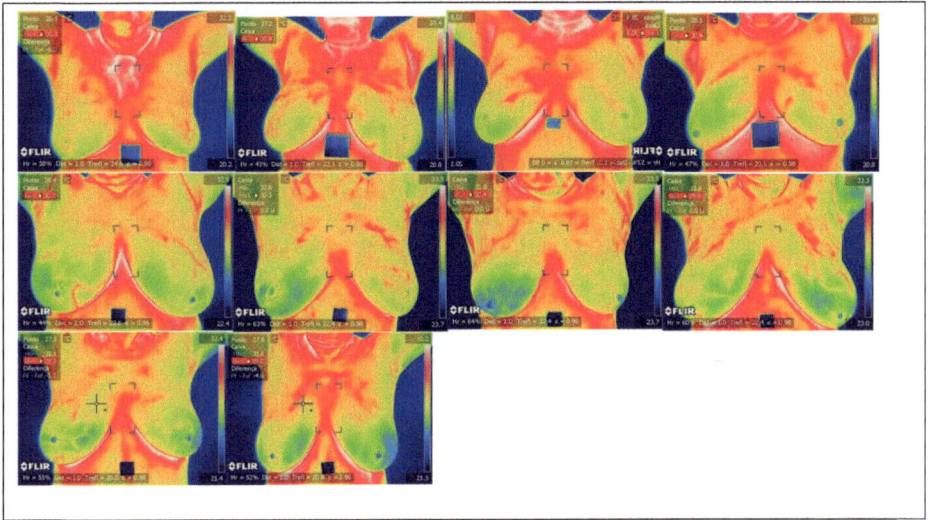

Fig. 3. Thermographic images of healthy patients

2.3   Characterization of Breast Cancer

- **Extraction of the region of interest (ROI)**

The computer interface allows to load any type of thermographic image in image format with the exception of the radiometric image format is 2. Since many times in thermographic images appear elements that are not of interest or that can generate false positives, the interface allows the clipping in the image of the desired area for the study. Figure 4 illustrates this process. Figure 4a shows a thermal image courtesy of Visual Lab-DMR [14]. The manual selection is made as can be seen in Fig. 4b, and finally the original image is cut according to this selection obtaining the image that appears in Fig. 4c.

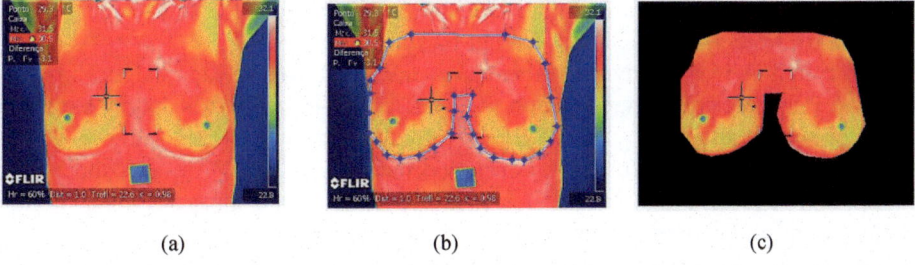

      (a)   (b)   (c)

Fig. 4. (a) Thermography of a patient with breast cancer, (b) manual selection of the region of interest (ROI), (c) trimming done in the thermography following the selection of the user.

- **Segmentation by values of the red component**

In view that lesions with breast cancer, due to the acquisition protocol, exhibit red colors in the thermographies, it is segmented based on the existing values of red components. By analyzing the histograms of several thermographic images of both healthy patients and patients, it was observed that the values of the red component in sick patients are above 129, this characteristic was observed through the use of histograms such as shown in Fig. 5, the red component starts to show minimum values from 128 and in some images at 130; and higher values in the range of 234 to 235 as seen in the image.

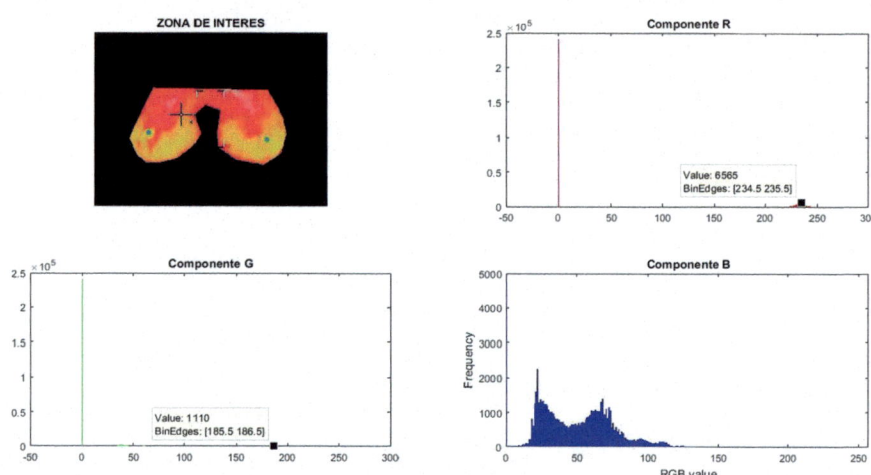

Fig. 5. Histogram of the red, green and blue components of the region of interest

- **Segmentation of red color tones using the K-means algorithm**

Performing at characterization of breast lesions by means of repetitive patterns such as borders, or by range of values as shown in the histogram, some algorithms were performed that, although they did not yield the expected results, served to rule out detection methods. In the case of edge detection, the methods of Canny, Sobel, Prewitt and Roberts were used. As shown in Fig. 6.

Some authors in their investigations have concluded that the canny algorithm turns out to be very efficient in the detection of edges, and in this case it is not the exception as can be seen in the images shown. However, after applying this algorithm to several thermographic images, it could be concluded that in this type of lesions there is no evidence of a pattern that can be taken as repetitive in the images, which is why segmentation was carried out using the Clustering algorithm K-means.

First, a segmentation of the red tones was carried out in order to establish discrimination parameters according to the values of the color components of the image. The image obtained in the segmentation by values of the red component is taken as input to the K-means algorithm after changing its color space to CIE L * a * b. The objective is to divide the image into four according to the average of the values of its

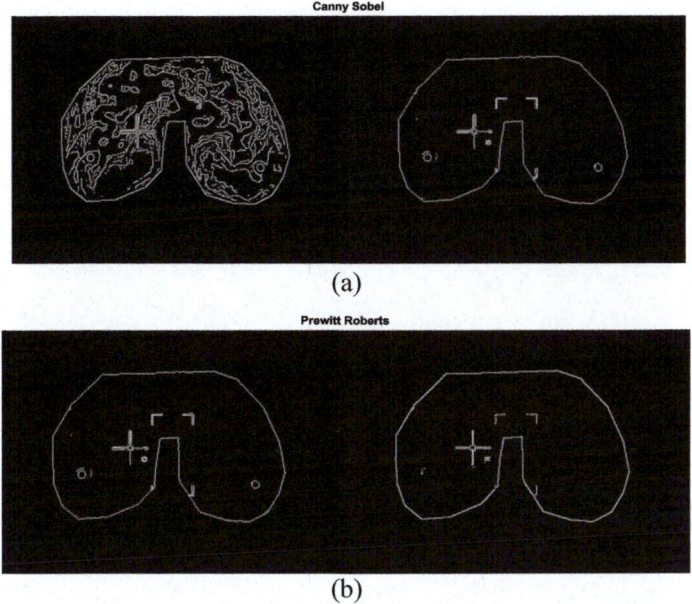

Fig. 6. (a) Edge detection by the Canny and Sobel algorithms, (b) edge detection with Prewitt and Roberts

components a and b. At this stage of the process, one of the images obtained in the segmentation contains the area with the highest value of the red component or, failing that, the breast cancer lesion. The implementation of the K-means algorithm was performed using the MatLab sample code for color segmentation using the K-means grouping [17]. This is shown in Fig. 7.

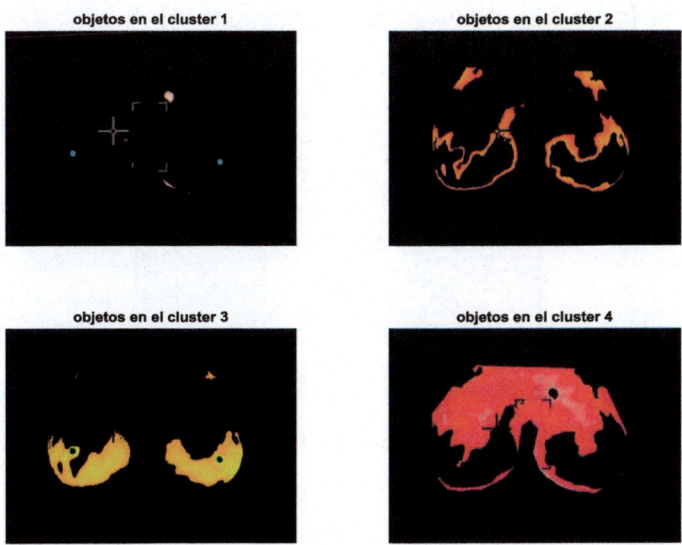

Fig. 7. Clustering of colors by means of K-means in the thermal image

With this process it is not yet possible to know if the thermography in question belongs to the group of patients or healthy, for which only the red component of the image is extracted, which is what we really want to analyze, given that this is where they show the highest temperatures. See Fig. 8.

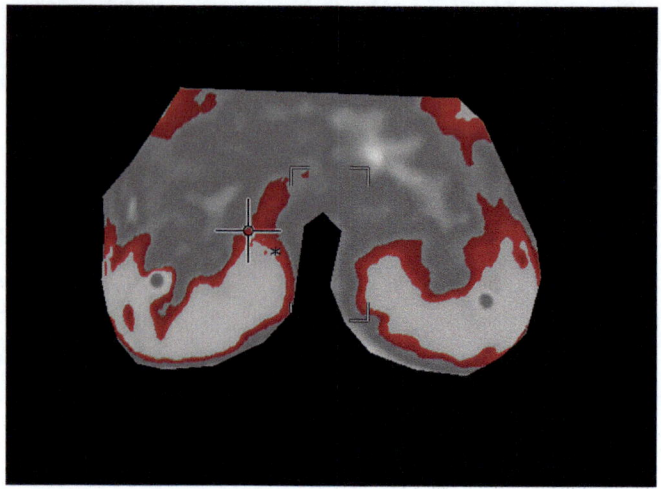

Fig. 8. Grayscale image showing only the red component.

Based on this image, a comparison of the values contained in the red component is made, and in the algorithm it is indicated that now only the values that are above 129 are shown. With this it is already possible to know if the image is a healthy or sick patient, since if it shows some red values on the grayscale image indicates that the image is of a sick patient, otherwise, if the image is only in grayscale this belongs to a healthy patient. In this case, the image that is being worked belongs to a sick patient, as can be seen in Fig. 9.

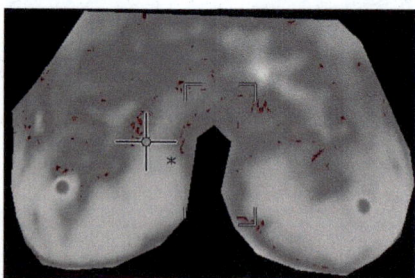

Fig. 9. Final image of the segmentation by k-means

3 Results

3.1 Development of a Computer Tool for the Pre-diagnosis of Breast Cáncer

The designed interface allows the acquisition of the image from the computer with the button load image, the manual selection of the study area or region of interest, by means of the buttons selection of ROI and Cutout of the panel. (See Fig. 10).

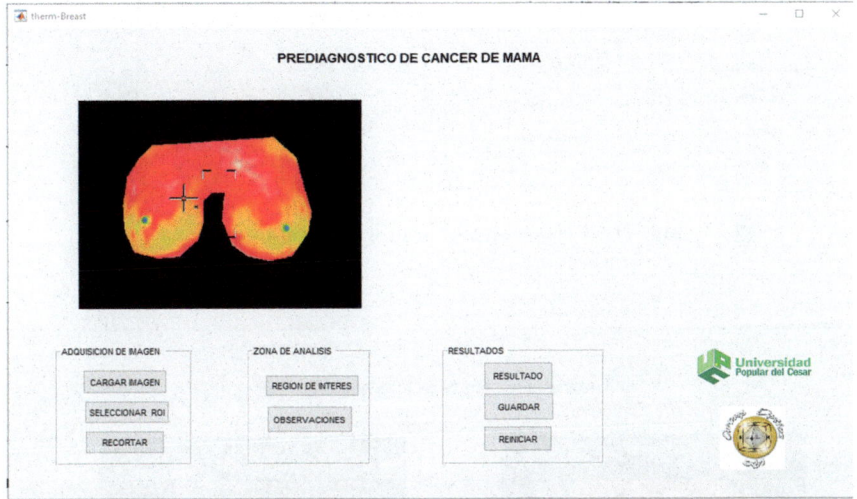

Fig. 10. Trimming of region of interest.

Then process the image taking only the areas of the study area that have red components, and this process is shown with the region button of interest. See Fig. 11.

Finally, when clicking on the Results button, the software segments the image, taking those 27 areas whose values show a higher average of red values. Identifying the type of image that is being analyzed (sick or healthy). See Fig. 12.

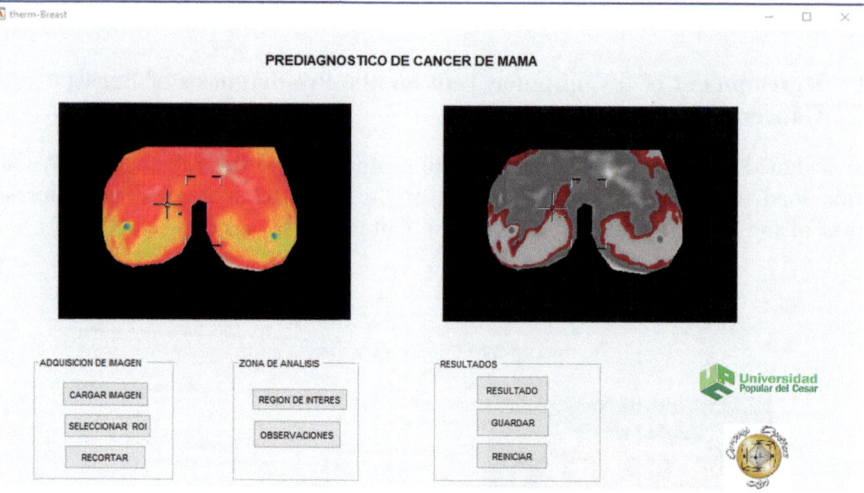

Fig. 11. Cropped image with the region of interest.

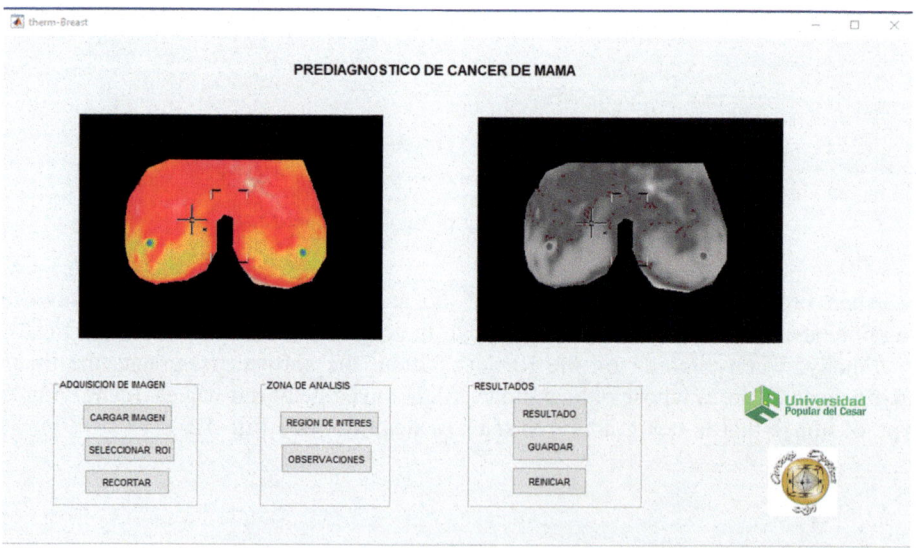

Fig. 12. Final image with breast cancer result.

3.2   Tests in Healthy Patients

The software was tested in 10 images of healthy patients, to avoid extending too much, the thermographic image result of a patient with a healthy diagnosis will be shown below in the Visual Lab-DMR image database. The interface for the case of healthy patients should show the final image completely gray as shown in Fig. 13.

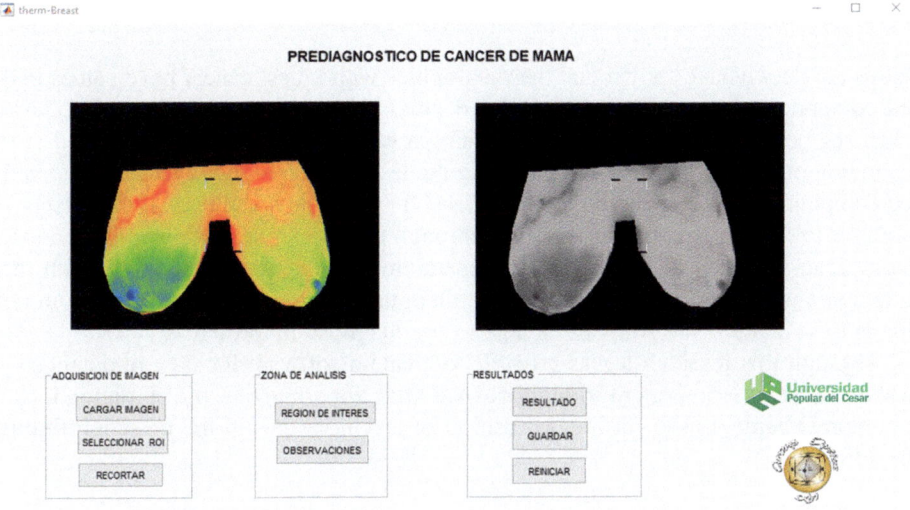

Fig. 13. Thermal imaging of a patient with a healthy diagnosis.

3.3   Tests in Sick Patients

In the case of patients with diagnosed patients, the interface should show in the gray final image some points that indicate that the temperature exceeds the normal in that area, so it is considered that there is an anomaly. This can be seen in Fig. 14.

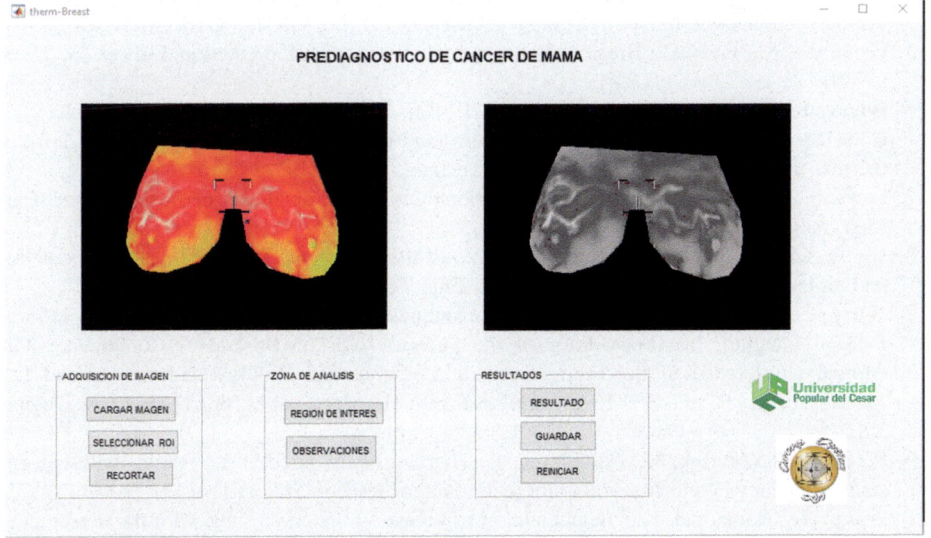

Fig. 14. Detection in thermographic image of patient with sick diagnosis.

4   Conclusions

In this article we have shown that thermographies with breast cancer have values in the red component above 129 under the RGB color space on a scale of 0 to 255. Also, when segmenting a thermography containing a carcinoma of the type specified at the beginning of the work, the zone containing the lesion has a higher average value in the red component with respect to the other areas. It should be noted that all these facts are valid for thermographic images acquired through the use of the rainbow palette and the appropriate selection of the range of temperature measurement. Even tests with thermographic images of Angiosarcoma and inflammatory breast cancer are less common, this is to validate if the software is able to pre-diagnose its presence.

Through this research it was possible to create a software for the pre-diagnosis of breast cancer, it is important to highlight the vital importance of the configurations of the thermal camera used for the acquisition of lesion images in the detection through the software.

References

1. Robles, S., Galanis, E.: El cáncer de mama en América Latina y el Caribe. Revista panamericana de salud pública 12(2), 141–143 (2002)
2. El universal: 2.600 mujeres en Colombia mueren al año de cáncer de seno (2018). www.eluniversal.com.co. https://www.eluniversal.com.co/salud/2600-mujeres-en-colombia-mueren-al-ano-de-cancer-de-seno-208783-OSEU311258
3. Ogihara, H., Hamamoto, Y., Fujita, Y., Goto, A., Nishikawa, J., Sakaida, I.: Development of a gastric cancer diagnostic support system with a pattern recognition method using a hyperspectral camera. J. Sens. (2016)
4. Lehmann, T., Tavakoli, M., Usmani, N., Sloboda, R.: Force-sensor-based estimation of needle tip deflection in brachytherapy. J. Sens (2013). https://doi.org/10.1155/2013/263153
5. Gautherie, M., Gros, C.: Breast thermography and cancer risk prediction. Cancer 45, 51–56 (1980)
6. Bagavathiappan, S., Philip, J., Jayakumar, T., Raj, B., Rao, P., Varalakshmi, M., Mohan, V.: Correlation between plantar foot temperature and diabetic neuropathy by using an infrared thermal imaging technique. J. Diab. Sci. Technol. 4, 1386–1392 (2010)
7. Lawson, R.: Implications of surface temperatures in the diagnosis of breast cancer. Can. Med. Assoc. J. 75, 309–310 (1956)
8. Ng, E., Chen, Y., Ung, L.: Computerized breast thermography: study of image segmentation and temperature cyclic variations. J. Med. Eng. Technol. 25, 12–16 (2001)
9. Herry, C., Frize, M.: Digital processing techniques for the assessment of pain with infrared thermal imaging. In: Proceedings of the Second Joint EMBS-BMES Conference: 24th Annual Conference of the Engineering in Medicine and Biology and the Annual Fall Meeting of the Biomedical Engineering Society, Houston, vol. 2, pp. 1157–1158, October 2002
10. Schaefer, G., Závišek, M., Nakashima, T.: Thermography based breast cancer analysis using statistical features and fuzzy classification. Pattern Recogn. 42, 1133–1137 (2002)
11. Prasad, K., Rajagopal, K.: "Segmentation of breast thermogram images for the detection of breast cancer". A projection profile approach. J. Image Graph 3, 47–51 (2015)

12. Garduño, M., Vega, S., Morales, L., Osornio, R.: Supportive noninvasive tool for the diagnosis of breast cancer using a thermographic camera as sensor. Sensors **17**, 1–21 (2017)
13. Rajendra, A., Ng, E., Jen-Hong, T., Vinitha, S.: Thermography based breast cancer detection using texture features and support vector machine. J. Med. Syst. **36**, 1503–1510 (2012)
14. Visual Lab-DMR: Banco de imágenes mastológicas (2012). http://visual.ic.uff.br/dmi/prontuario/home.php
15. Herman, C.: Emerging technologies for the detection of melanoma: achieving better outcomes. CCID **5**, 195–212 (2012)
16. Verduzco, J., Cetina, W.: Vista de Sistema de monitoreo de temperatura de los senos humanos en la detección temprana del cáncer de seno. RICS Revista Iberoamericana de las Ciencias de la Salud (2015). https://www.rics.org.mx/index.php/RICS/article/view/24/51
17. MathWorks: Segmentación basada en color mediante clustering K-means - MATLAB & Simulink Example. MathWorks América Latina (2019). https://la.mathworks.com/help/images/examples/color-based-segmentation-using-k-means-clustering.html?searchHighlight=kmeans&s_tid=doc_srchtitle

Computational Vision

Automatic Classification of Pterygium-Non Pterygium Images Using Deep Learning

Yadini Perez Lopez and Luis Rojas Aguilera$^{(\boxtimes)}$

SIDIA Institute of Science and Technology, Manaus, Brazil
yadini.l@samsung.com, luis.rojas@sidia.com

Abstract. Pterygium is an ocular disease caused by the invasion of a fibro-vascular tissue onto the cornea region. Several researches has been developed for automatic detection of pterygium in eyes images. In those researches, color and shape information of pterygium were explored using Digital Image Processing techniques and Machine Learning algorithms such as Artificial Neural Networks (ANN) and Support Vector Machine (SVM). More recently, Deep Learning techniques were applied for implementing a system for diagnosing multiple ocular diseases including pterygium, however no study have been developed on using Deep Learning focused on ptyregium detection only. We present a method for automatic classification of pterygium - non pterygium images using Convolutional Neural Networks (CNN). A dataset of positive (pterygium) and negative (non pterygium) images, previously used in early researches, was employed in order to train and test a CNN model with one convolutional layer. The images were studied in two color formats, RGB and grayscale. The best result in the pterygium – non pterygium image classification task was attained using RGB format, getting an Area Under the ROC curve of 99.4%. The results obtained overcome the results found in literature.

Keywords: Classification · Convolutional Neural Network · Deep learning · Pterygium

1 Introduction

Pterygium is a degenerative wing-shaped ocular surface lesion that in most of the cases it expands from the nasal side of the eye towards the cornea. There is no consensus regarding the pathogenesis of pterygia, nevertheless, most studies support the concept that ultraviolet (UV) light radiation plays a major role in the development of pterygium. Countries closer to the equator have shown the higher rates of incidence [1]. "Besides being a cosmetic issue, pterygium is a clinical condition that affects vision by blocking the visual axis, causing tear film instability, or inducing corneal astigmatism" [2].

There is no proven method to prevent this condition, but limiting sun light exposure and avoiding remain in dusty environments could help to slowdown progression [3].

The treatment of pterygium depends on stage of the disease. In early-stage patients may be prescribed with eye drops for lubrication and the use of sunglasses. If the lesion grows, surgical intervention becomes more compelling [4].

© Springer Nature Switzerland AG 2019
J. M. R. S. Tavares and R. M. Natal Jorge (Eds.): VipIMAGE 2019, LNCVB 34, pp. 391–400, 2019.
https://doi.org/10.1007/978-3-030-32040-9_40

Currently, in most of the health care centers around the world, pterygium diagnosis exam is accomplished by manually inspection of the patient's eyes, here an eye image is manually marked by the doctor for grading [5]. This exam is straightforward but it must be performed by qualified personnel, since the diagnosis accuracy relies on the expertise of the doctor. Here, an automated pterygium detection method could be used for supporting doctor's diagnosis, especially in low-income localities or rural areas where hospitals struggle with medical material and specialist deficiency.

Until 2018, several researches have been implemented aiming the automatic detection of pterygium. In most of works authors used Digital Image Processing (DIP) and/or traditional machine learning techniques for segmenting pterygium. More recently, Zhan et al. [6], used Convolutional Neural Network (CNN) for implementing a diagnostic system for multiple ocular disease, among which is pterygium. CNN is a type of deep neural network widely used for analysis, classification and object detection in images [7]. For the best of our knowledge, deep learning techniques have not been applied for exclusively detecting pterygium in eyes images.

This work presents a method for automatic classification of pterygium – non pterygium images using CNN and two different image color formats: Red, Green and Blue (RGB) and grayscale.

2 Related Works

For the best of our knowledge, automatic identification of pterygium has not been widely explored and only few works have been published on this topic. In early researches, authors used digital image processing techniques for pterygium identification in images [8–10].

In Gao et al. [8], was presented a color information based method for filtering out pterygium from cornea images, aiming to improve the accuracy of a cortical cataract grading system. The pterygium detector achieved a true positive rate of 66.67% and true negatives rate of 80.33%. Here, the missed cases were due to blurred and small size pterygium presented in some images. Authors proposed including the iris region in the pterygium detection for improving the method's results [8].

Mesquita and Figuereido et al. [9] developed an algorithm for automatically measuring the advancement of the pterygium in already diagnosed eyes. The proposal consisted in segmenting the iris using the Circular Hough Transform (CHT). Then, authors implemented a second segmentation for detecting the pterygium in the iris, this time, using a region growing algorithm based on Otsu's thresholding method. Authors reached a hit rate of 63.4% for pterygium segmentation on correctly iris segmented images. On the other hand, authors claimed that the missed cases mainly were due to poor lighting, small advancement of the pterygium over the iris, and in cases where the consistency of the pterygium gradually decreases and merges with the iris [9].

More recently, machine learning techniques were applied in order to perform pterygium detection in anterior segmented images [5]. In 2017, Zaki et al. [5] used machine learning techniques for implementing a system for automatic pterygium detection composed by four modules: image preprocessing, corneal segmentation, feature extraction and classification (pterygium or non-pterygium). Preprocessing (module 1) consisted in applying HSV-Sigmoid enhancement aiming to improve the image's contrast. Corneal segmentation (module 2) was implemented using 3SFD method and Otsu thresholding. Four features were extracted in module 3: comprising circularity, Haralick's circularity, eccentricity, and solidity. This features were used in the final module: Classification, where a Support Vector Machine (SVM) and Artificial Neural Network (ANN) with different parameter configurations were implemented and used for testing the proposed detection system. In this research, authors used four different datasets: UBIRIS [11], MILES [12], Brazilian pterygium (BP) from Professor Rafael Mesquita, Center of Informatics, Federal University of Pernambuco, Recife, Brazil [9] and Australian pterygium (AP) from Professor Lawrence Hirst of the Australian Pterygium Centre [13]. The best results for pterygium detection were obtained with the SVM classifier with an RBF kernel of standard deviation equal to 1, reaching an area under ROC curve (AUC) of 0.956 [5].

In 2018, Zhan et al. [6] implemented a system for diagnosing multiple ocular diseases including pterygium. Authors used a dataset of ophthalmic medical images labeled with five classes (normal eyes, pterygium, keratitis, subconjunctival hemorrhage and cataracts). The diagnosis framework proposed comprised four stages, disease identification (accuracy reached of 93%), localization of the anatomical parts and foci of the eye (accuracy reached using images under natural light without fluorescein sodium eye drops of 82% and 90% using images under cobalt blue light or natural light with fluorescein sodium eye drops), classification of the specific condition of each anatomical part or focus using the results from the second stage (accuracy of 79%–98%), and finally, the system provides treatment advice according to medical experience and artificial intelligence, which is merely involved with pterygium (accuracy over 95%) [6].

Zaki et al. [5] and Zhan et al. [6] presented very promising results for pterygium detection. Although Zhan et al. [6] applied CNN for attaining this goal, the proposed system was not exclusively created for pterygium detection. On the other hand, Zaki et al. [5] worked only with pterygium disease, however deep learning techniques were not used in this work.

This works presents a new method for automatic classification of pterygium – non pterygium images using CNN. The results obtained in this research overcome the results found in the literature.

3   Materials and Methods

3.1   Dataset Description

The Image Dataset used in this research corresponds with an union of UBIRIS, MILES, BP and AP datasets used in Zaki et al. [5]. We employed this image database because we had the opportunity to access it and we wanted to have a baseline for comparing our results using deep learning with results obtained using traditional machine learning for pterygium classification task. The mentioned datasets where created by capturing images from the frontal view of human eyes using a digital camera and a smartphone, involving two eyes conditions: eyes with and without pterygium. UBIRIS and MILES included together 2,692 non-pterygium images. Meanwhile, BP and AP comprehend 325 pterygium images [8]. Table 1 summarizes relevant information about the mentioned datasets, presenting different characteristics in regard to quantity of images and image's spatial dimensions.

Table 1. Characteristics of the original images datasets used.

	Datasets				
	UBIRIS	MILES	BP	AP	Dataset
Total images	1,877	815	58	267	3,017
Spatial dimensions	200 × 150	1747 × 1180	Multiple	4064 × 2704	150 × 150

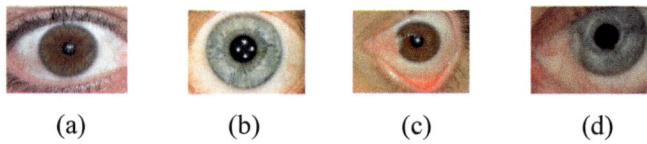

(a)   (b)   (c)   (d)

Fig. 1. Images from UBIRIS (a), MILES (b), BP (c) and AP (d) datsets.

It is worth mentioning that there are differences in the way that eyes were shown in the captures; e.g. in UBIRIS dataset, eyes are normally opened, whereas, in BP eyes are forced open, see Fig. 1(a) and (c).

We resized every image from Image Dataset to a spatial dimension of 150 × 150 pixels aiming to reduce the complexity and memory demand for training the CNN model. No information was lost when resizing the original images.

3.2   Proposed Method

The proposed methodology is composed by two main stages: Preprocessing the images and Image classification. Each stage is composed by different steps. Figure 2 shows the proposed methodology in a block diagram. We study the images in RGB and grayscale color formats.

3.3   Image Preprocessing

The preprocessing stage includes two steps: Obtaining the grayscale image version and Image normalization. The first step is optional and is applied only when grayscale images are analyzed. Here, is important to highlight that we considered working with grayscale images since pterygium can appear in different colors, including red, pink, white, yellow or gray [14]. Because of this, color information might be not outstanding. Moreover, working with one channel images (grayscale format) reduces the weights of the CNN model compared to working with three channels (RGB format).

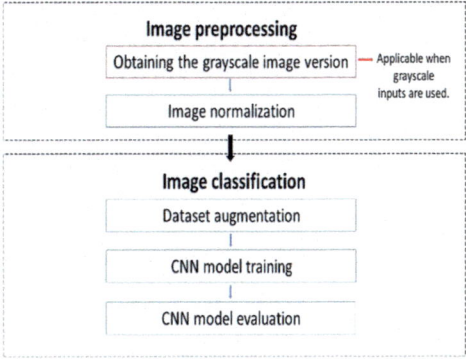

Fig. 2. Proposed methodology.

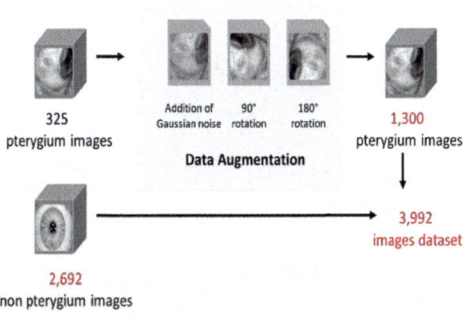

Fig. 3. Procedure to increase the amount of positive images.

The RGB to grayscale conversion was implemented according to Eq. 1, where the resulting image (Ig) is obtained through a weighted sum of the R, G, and B components.

$$Ig = 0.2989 * R + 0.5870 * G + 0.1140 * B \qquad (1)$$

The second step (Image or data normalization) was implemented by dividing images by 255.

3.4   Image Classification

The first step of the Image classification stage consisted in applying data augmentation. This is a commonly used technique when working with deep models, since for training those, large and balanced datasets are required in order to obtain appropriated generalization rates [7, 15]. The Image dataset (we will call it Original dataset from here) used in this research, included only 3,017 samples in total. Moreover, the main problem of the Original dataset is that is not balanced. This means that the quantity of samples per classes (pterygium and non pterygium) was not equal, presenting 325 positive images and 2,692 negative samples. Aiming to balance the Original dataset we implemented data augmentation by doing three different transformations in each positive image: addition of Gaussian noise, rotation in 90 and 180. Therefore, the quantity of pterygium images increased up to 1,300 samples. Figure 3 illustrates the data augmentation procedure implemented for obtaining a more balanced dataset, called Augmented dataset.

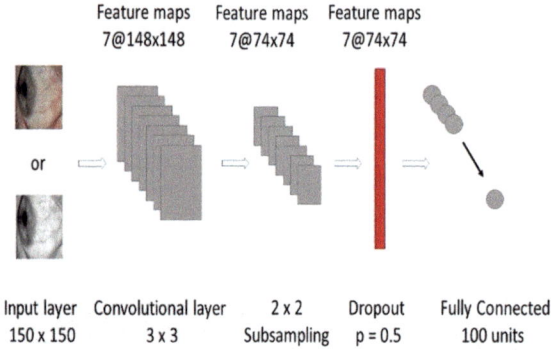

Fig. 4. Illustration of the CNN model's architecture.

The second step in the Image classification module consisted in training one layer CNN model using the images in RGB and grayscale format. We did not train deeper architectures since in an early simulation with two layers CNN model, the results obtained were not promising. This may be due to the fact that, even after data augmentation, the Augmented dataset remains small when it comes to the application of deep learning techniques. The proposed CNN model's architecture is shown in Fig. 4.

Model was trained using the following optimization parameters: Medium Quadratic Error as *error function*, nesterov momentum as *weights update function*, with parameter *update_momentum*, starting by 0.9 and increasing up to 0.999, and *learning rate* starting by 0.03 and decreasing to 0.0001. Was defined 1,000 *epochs* for training.

For evaluating the performance of the model (CNN model evaluation step) were used five metrics: ROC and Precision- Recall curves, precision, recall and accuracy. This metrics were calculated using scikit-learn library[1].

The computer used in this work was a NVIDIA Intel(R) Xeon(R) CPU E5-1650, 3.350 Hz, with RAM of 16 GB.

---

[1] http://scikit-learn.org/stable/modules/classes.html#module-sklearn.metrics.

4 Results

4.1 Dataset Description

The spatial resolution transformation applied to the image Dataset did not have an impact on the information screened in the images.

4.2 Proposed Method

Image normalization and data augmentation contributed to reach better hit rates in the training process of the model. Without applying data augmentation, model was not able to learn.

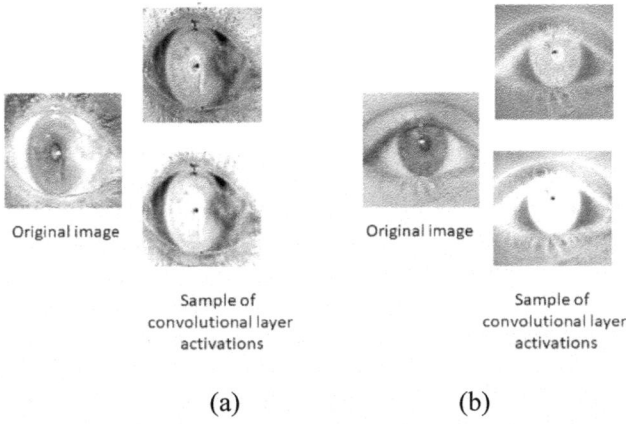

Original image   Original image

Sample of   Sample of
convolutional layer   convolutional layer
activations   activations

(a)   (b)

Fig. 5. Original grayscale image and activations samples of the convolutional layer for pterygium (a) and non-pterygium (b) cases respectively.

The output of convolutional layers (activations) reveals several features of the input image. Shallow layers identify more visually understandable features like edges, and deeper layers detect more abstract and complex features [15]. Figure 5 shows the activations of the proposed model's convolutional layer for a positive (a) and negative (b) image. Here it is possible to observe the differences between these activations. The positive image, Fig. 5(a), clearly shows the non-circular shape of the iris due to the presence of pterygium. The opposite happens in Fig. 5(b), for a negative image, here, the iris preserves a circled shape.

As mentioned in section *Proposed Method*, the proposed CNN model was trained using RGB and grayscale images format. Table 2 shows the results (precision, recall and accuracy) attained for pterygium and non pterygium classification using the mentioned color formats. The best results were reached using RGB format. Here, a precision of 0.998 was reached, meaning that the method presents a discrete amount of false positive[2] classification cases. Nevertheless, the recall obtained, 0.882, reveals that the proposed method presents a higher false negative[3] rate.

Figure 6 shows the Precision-Recall (a) and ROC (b) curves of the model using RGB and grayscale format.

Table 2. precision, recall and accuracy attained for RGB and grayscale pterygium and non-pterygium image classification using the proposed CNN model.

Images format	Metrics		
	Precision	Recall	Accuracy
RGB	0.998	0.882	0.935
Grayscale	0.993	0.765	0.869

5   Discussion

In this research was presented a method for automatic classification of pterygium – non pterygium images using deep learning. For accomplishing this goal, was proposed a CNN classifier, which was trained and tested using the same image dataset from Zaki et al. [5]. We worked with the images in two color formats, RGB and grayscale. The best results were reached using colored images (RGB), accuracy of 93.5%, sensitivity (recall) of 88.2% and AUC of 99.4%.

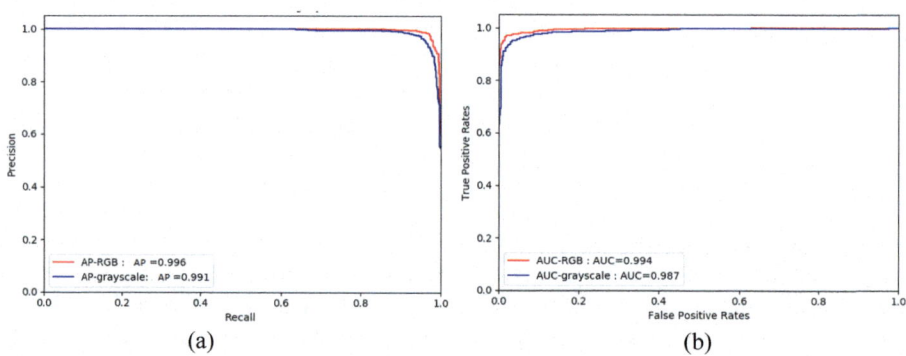

Fig. 6. Performance of the proposed CNN model for RGB and Grayscale images. Precision-Recall (a) and ROC (b) curves.

In Zaki et al. [5] authors presented very promising results for pterygium screening. The proposed method was composed by four stages, firstly was performed an image preprocessing using HSV-Sigmoid enhancement, then corneal segmentation was

---

[2] False Positive: Non-pterygium image mistakenly classified as Pterygium.

[3] False Negative: Pterygium image mistakenly classified as Non-pterygium.

applied, finally a SVM classifier was applied for analyzing features extracted from the previously segmented images. Authors reported an AUC of 95.6% and a sensitivity of 88.7%.

In Zhan et al. [6] was presented a system for diagnosing multiple ocular diseases including pterygium. An accuracy of 93% was attained for pterygium identification.

We do not compare our results with Zhan et al. [6] since the researches conditions are different (was not used the same dataset and the quantity of classes also was different).

Our proposal presented higher AUC compared with Zaki et al. [5] 3.8%. Moreover, we present a solution with a simpler pipeline, this could mean lower consumption of computational resources.

This work differs from previously developed researches founded in literature by the following points:

- Utilization of Convolutional Neural Networks for exclusively classification (pterygium or non-pterygium) of eyes images.
- Studio of (pterygium - non pterygium) images in two different color formats, RGB and grayscale.

6   Conclusion

This work presents a method for automatic classification of eyes images as positive (pterygium) or negative (non-pterygium) using Convolutional Neural Networks. The model was trained and tested using two different color formats separately, RGB and grayscale. The best results were reached using colored images (RGB). The proposed CNN model has one convolutional layer and can classify an eye image as positive or negative for pterygium presence with an accuracy of 93.5%, a recall of 88.2%, a precision of 99.8% and an AUC of 99.4%. The results attained could be improved augmenting the depth of the model and consequently augmenting the image dataset size, aiming to identify more complex features of the eyes images.

Acknowledgments. We would like to thank Professor Lawrence Hirst of the Australia Pterygium Center for granting access to the use of the pterygium databases and to Professor Rafael for the Brazilian pterygium database. We also thank the other database providers cited in the article.

References

1. Todorovic, D., Vulovic, T., Sreckovic, S., Jovanovic, S., Janicijevic, K., Todorovic, Z.: Updates on the treatment of pterygium. Serbian J. Exp. Clin. Res. **17**(3), 257–262 (2016)
2. Hossain, P.: Pterygium Surgery, Autumn edn. Royal College of Ophthalmologists, London (2011)
3. Hall, A.: Understanding and managing pterygium. Community Eye Health **29**(95), 54–56 (2016)
4. Hellem, A.: Pterygium: what is "surfer's eye"? (2017). https://www.allaboutvision.com/conditions/pterygium.htm. Accessed 10 Aug 2018

5. Zaki, W.M.D.W., Daud, M.M., Abdani, S.R., Hussain, A., Mutalib, H.A.: Automated pterygium detection method of anterior segment photographed images. Comput. Methods Programs Biomed. J. **154**, 71–78 (2017)
6. Zhang, K., et al.: Works citing "an interpretable and expandable deep learning diagnostic system for multiple ocular diseases: qualitative study". J. Med. Internet Res. **20**, 11 (2018)
7. Goodfellow, I., Bengio, Y., Courville, A.: Deep Learning. MIT Press (2016). http://www.deeplearningbook.org
8. Gao, X., et al.: Automatic pterygium detection on cornea images to enhance computer-aided cortical cataract grading system. Presented at the 34th Annual International Conference of the IEEE EMBS, San Diego, California USA (2012)
9. Mesquita, R.G., Figueiredo, E.M.N.: An algorithm for measuring pterygium's progress in already diagnosed eyes. Presented at the IEEE International Conference on Acoustics, Speech and Signal Processing (ICASSP), Kyoto, Japan (2012)
10. Buitelaar, P., Cimiano, P.: Frontiers in Artificial Intelligence and Applications, vol. 167. IOS Press, Amsterdam (2008)
11. Proenc, H., Alexandre, L.A.: UBIRIS: a noisy iris image database. In: International Conference on Image Analysis and Processing (2005)
12. Miles, J.: A selection of sample iris photos (grouped by illumination type), 18 November 2015. https://drive.google.com/drive/u/0/folders/0B5OBp4zckpLnYkpBcWlubC0tcTA
13. Lawrence, H.: The Australian pterigium institute. http://pterygium.info/en/
14. American Academy of Ophthalmology 2018: Advanced ophthalmology Inc. (2015) https://www.aao.org/topic-detail/pterygium-europe#top
15. Mitchell, T.M.: Machine Learning. McGraw-Hill Inc, New York (1997)

A Tool for Building Multi-purpose and Multi-pose Synthetic Data Sets

Marco Ruiz$^{(\boxtimes)}$, Jefferson Fontinele, Ricardo Perrone, Marcelo Santos, and Luciano Oliveira

Intelligent Vision Research Lab, Federal University of Bahia, Salvador, Brazil
{marco.ruiz,jeffersonfs,perrones,marceloms,lrebouca}@ufba.br

Abstract. Modern computer vision methods typically require expensive data acquisition and accurate manual labeling. In this work, we instead leverage the recent progress in computer graphics to propose a novel approach of designing and generating large scale multi-purpose image data sets from 3D object models directly, captured from multiple categorized camera viewpoints and controlled environmental conditions. The set of rendered images provide data for geometric computer vision problems such as depth estimation, camera pose estimation, 3D box estimation, 3D reconstruction, camera calibration, and also pixel-perfect ground truth for scene understanding problems, such as: semantic and instance segmentation, object detection, just to cite a few. In this paper, we also survey the most well-known synthetic data sets used in computer vision tasks, pointing out the relevance of rendering images for training deep neural networks. When compared to similar tools, our generator contains a wide set of features easy to extend, besides allowing for building sets of images in the MSCOCO format, so ready for deep learning works. To the best of our knowledge, the proposed tool is the first one to generate large-scale, multi-pose, synthetic data sets automatically, allowing for training and evaluation of supervised methods for all of the covered features.

Keywords: Synthetic data · 3D rendering · Multi-purpose · Multi-pose · Tool

1 Introduction

Results of convolutional neural network (CNN) methods based on supervised learning strongly depend on large-scale training data sets. However, the annotation of thousands of images and intrinsic aspects are a tedious and huge task which demands time in high precision human observation, even more for pixel-wise annotations or 3D information. To cope with these issues, works like the MSCOCO [9] make use of any massive collaborative annotation mechanism. Nevertheless, this solution poses some challenges to work synergistically that affects the standardization of the full process, promoting inaccuracies in the categorization process, the location of bounding boxes and poor boundary pixel annotation.

© Springer Nature Switzerland AG 2019
J. M. R. S. Tavares and R. M. Natal Jorge (Eds.): VipIMAGE 2019, LNCVB 34, pp. 401–410, 2019.
https://doi.org/10.1007/978-3-030-32040-9_41

In effect, these particularities increase the divergence on labeling thousands of images, thus demanding a detailed verification process at the final stage. On the other hand, rendered data sets can be automatically annotated, but mostly have been specially designed for representing a few of the more common aspects of real-world problems, *e.g.*, pose, depth, cluttering, lighting, etc; without allowing to extend features or examples of a new non-common object class. The cost and time to collect and label images are less when images are rendered from 3D models. Furthermore, synthetic data provides access to a reliable set of data for research, while not compromising on the principle of reproducibility [1].

Particularly with respect to data sets for deep learning methods, we highlight four issues: (i) the scarcity of training images with accurate annotations from different categorized viewpoints; (ii) the lack of powerful features specifically linked to 3D tasks; (iii) an impediment for rapid deployment of detection systems of less common objects, and; (iv) the mammoth time of image annotation. Recently, a successful research direction to overcome these four issues is to train CNNs from built scenarios with rendered synthetic objects [4]. In this paper, we propose to overcome these concerns by making publicly available a software tool that automatically generates multi-purpose and multi-pose images from 3D models. To expose our approach, we compare it at the feature level with other works and then present five examples of data sets generated with our tool, such as the *Car poses* exhibited in Fig. 1.

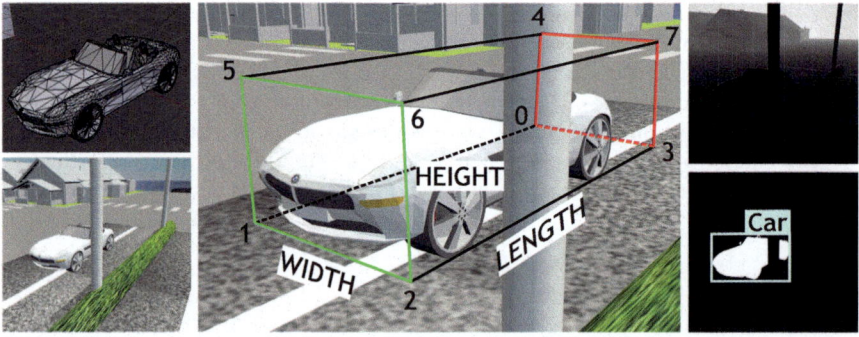

Fig. 1. The *Car poses* data set was generated with our tool by placing 3D car models in a virtual environment, then rendered from several camera viewpoints. Each snapshot consists of a 3D model (top left) and its corresponding RGB rendered image (bottom left), 3D bounding box coordinates and real size of cars (center), depth data in meters (top right), 2D bounding box coordinates and pixel-wise segmentation (bottom right), 3D poses of all target objects (not shown), among other annotations. Best viewed in color.

2   Related Work

Can synthetic images be trained to represent inherent characteristics of real scenes?. Mainly focused on basic computer vision problems, this discussion has been frequently addressed during the last years [12,18,22]. The use of existing 3D models has been advocated in the past [4] and remains an appealing strategy [8]. Results expose that CNNs are capable of extracting discriminative features from synthesized images when evaluated on real scenarios, demonstrating competent and even better results compared to methods purely trained with real images. Training on synthetic rendered images represents an alternative solution for feeding networks with trustworthy data with perfect object semantic annotations [8]. In this way, the problem of domain bias [1] (sometimes referred as domain gap [5]) between synthetic and real-world images has been conquered by using photorealistic scenarios [2,20] combined with 3D representations [20], or by transferring style from real-world scenarios to rendered images [1]. Our virtual environment enables users to freely rich the scene with realistic 2D and 3D representations.

Tools or methods for building synthetic data sets: a software library whose purpose is quite similar to ours is introduced in [19]. The main goal is to allow the computer vision community for easily extending or generating data sets from 3D models, accordingly with different parameters like lighting, pose, and texture, and image metadata like image label, object outline, etc; finally outputting 2D bounding boxes automatically labeled, and a caption that users manually input to summarize the scene. To demonstrate the software's potential, the authors exposed two data sets that together with the code for data sets generation, are publicly available. Similarly, the work in [8] extracts a set of poses and class discriminant features from synthetic 3D object models, to build a viewpoint-independent detector. The approach they follow to extract multiple rendered views of synthetic 3D models is pretty similar to our method for automatic labeling. The latter generated a data set from 58 synthetic models containing 3D annotations similar to us, except the automatic annotation of depth, segmentation masks, and real object dimensions. The authors focused explicitly on the 2D/3D object detection application and did not make available the generated data set nor the code to build it.

Following a truly virtual approach, CARLA [3] and AirSim [16] bring a virtual environment capable of simulating a dynamic world, including physical phenomena like rain and other weather conditions. These simulators provide an open-source platform to build algorithms for autonomous vehicles, encompassing the features to design and develop by hand convenient hiper-realistic data sets automatically annotated. Similarly, the MINOS simulator [15] is designed to support the development of multi-sensory models for goal-directed navigation in indoor environments. MINOS is used to benchmark deep-learning-based navigation methods, showing that current deep reinforcement learning approaches fail in large realistic environments, also demonstrating that multi-modality is beneficial in learning to navigate cluttered scenes. A wide set of images can be rendered from the virtual indoor scene, including depth, GPS positions, and segmentation mask annotations. However, the simulator is not available to categorize camera poses nor other features for applications that aim to exploit the 3D domain of objects.

SURREAL [21] is not just a large-scale data set of people synthetically-generated rendered from 3D sequences of human motion capture data, but a publicly available software library to create new images or video sequences from a pre-configured virtual scene. Similar to our tool, the library allows to design some assets in the scene e.g, camera position, lighting, and textures, also using a database of images to compose the background for the rendered frames.

Synthetic data sets: much of the data sets are thought primarily for evaluating new CNN methods or aiming to release new benchmarks either for 3D context learning applications [17], visual odometry [6], 3D detection or reconstruction [6, 22], SLAM [6], or viewpoint estimation [18]. Despite coming from simulated environments, tightly intertwined problems are not addressed and consequently other context-relevant features are no longer explored. On the other hand, a small portion of the works create data for more general purposes, for instance for the context of driving scenarios [5, 7, 14], to wide range of 3D object reconstruction problems [23], or very specialized for particular tasks such as optical flow [2], 3D scene understanding [23], or multiple tasks related with visual perception [13]. These latter manifest a greater effort to exploit the automatic annotation from virtual environments, delivering a larger number of features than other researchers can benefit from similar applications.

To show a comprehensive resume of the literature review, we group all discussed works here, in Table 1; highlighting the features that are most common among them and meeting the scope of computer vision applications that our tool could address.

Contributions

Rather than releasing a static data set, we propose a software tool intentionally designed to support the process of building well-planned data sets. This approach contains a wide set of features including a configurable scenario and discretization parameters for rendering images automatically labeled and formatted for supervised learning applications (especially for the training process). As part of this work, very relevant works are reviewed and compared from a feature-oriented strategy. According to the literature review, our work has several key strengths in comparison to others publicly available.

Here, we are focused on how to improve the process of generating a new data set or even expand an existing one. Our initial motivation came from the scarce availability of labeled data designed to support traffic surveillance problems. In particular, the majority of existing data is not well-suited to represent complex traffic scene and driving scenarios such as *top-down* traffic monitoring, law enforce, anomalies detection, 3D object detection, and segmentation, etc. In any application, pan-tilt-zoom cameras pose an important challenge related to the camera viewpoints that most of the published data sets do not

Table 1. Comparison of synthetic image data sets and tools.

Data set/ Method	# frames	2D Box	3D Box in pxs	Mask	Disparity/ Depth	3D Model	Pose Categ.	Calibrat. Matrices	Optical flow	Real size
[2]	+1k	✗	✗	✓	✓	✗	✗	✗	✓	✗
[5]	+21k	✓	✓	✓	✓	✗	✓	✓	✓	✗
[14]	+213k	✓	✗	✓	✓	✗	✓	✓	✓	✗
[17]	+45k	✗	✗	✗	✓	✓	✗	✗	✗	✗
[15]	+45k	✗	✗	✓	✓	✗	✗	✗	✗	✗
[21]	+6M	✗	✗	✓	✓	✗	✓	✗	✓	✗
[11]	+35k	✗	✗	✗	✓	✓	✓	✓	✓	✗
[6]	N/A	✗	✗	✗	✓	✓	✓	✗	✗	✗
[7]	200k	✓	✓	✓	✓	✗	✓	✗	✗	✗
[13]	+250k	✓	✓	✓	✓	✓	✗	✗	✓	✗
[16]	Unlimited	✓	✗	✓	✓	✓	✓	✓	✗	✗
[3]	Unlimited	✓	✗	✓	✓	✓	✗	✓	✗	✓
[18]	N/A	✗	✗	✗	✗	✓	✓	✗	✗	✗
[10]	14k	✓	✓	✓	✗	✓	✓	✓	✗	✗
[23]	+30k	✓	✓	✓	✗	✓	✓	✗	✗	✗
[22]	+90k	✓	✗	✓	✗	✓	✓	✓	✗	✗
[8]	N/A	✓	✓	✗	✗	✓	✓	✗	✗	✗
[20]	60k	✓	✓	✓	✓	✓	✓	✗	✗	✓
[19]	Unlimited	✓	✗	✗	✗	✓	✓	✗	✗	✗
Ours	Unlimited	✓	✓	✓	✓	✓	✓	✓	✗	✓

address. Furthermore, expanding a data set by applying the same techniques of acquisition/annotation may be prohibitively very high in terms of cost and time. From the above, our main contribution is a software tool to design and automatically generate multi-purpose synthetic object data sets from multiple camera viewpoints and environmental conditions. The data contains features for 3D context learning and geometric computer vision problems such as depth estimation, pose estimation, 3D box estimation, 3D reconstruction, camera calibration, and also a pixel-perfect ground truth for scene understanding problems such as scene-level and instance-level semantic segmentation, and object detection (see Table 1).

Different of other tools [8,19] and simulators [3,16], our approach is implemented to give the final user flexibility on easily adjust normalized parameters (e.g. pose, lighting, blurring, backgrounds, output type, viewpoints, and number of rendered images) while improving the interoperability with other data sets. Thus, all outputs can be automatically annotated and stored in two optional JavaScript Object Notation (JSON)-style notations: Native style or MSCOCO [9] style. By applying a parameterized procedure to automatically generate images, we face the practical difficulty of gathering thousands of real examples at the same time that guaranteeing a reasonable level of variability of object poses. This kind of data is well-suited for traffic monitoring scenarios, where the camera can move in several directions, capturing uncountable poses to be detected and recognized. Code and data sets are available in https://github. com/IvisionLab/traffic-analysis/tree/master/synthetic-dataset-generator.

3   Building Data Sets

The pipeline for the data rendering process is depicted in Fig. 2, consisting of three basic steps: place the models in the scene, set the discretization parameters, and execute the rendering script. To accomplish it, a virtual environment was designed in Blender[1] and some Python scripts were integrated for parameters setting, in the same way as other research works of similar purposes [11, 21].

Components of the virtual scene: any 2D or 3D object allowed by Blender can be imported into the scene. The scenario can be modeled at the user's disposition, for instance by adding multiple lighting sources (*e.g.* sun, lamps, spots, hemis), new image processing parameters (*e.g.* blurring, edge highlighting, shadows) or full 3D environments (*e.g.* neighborhoods, rooms). The blender scene already contains some flat objects and one 3D environment of a neighborhood, with the objective of contextualizing the objects in random and fixed backgrounds. Furthermore, a node was created to apply a blending step "smoothen" on the RGB image before rendering, just as [4]. Extending new nodes graphically is simple and directly alters the output files of the rendered images.

Placing the models: if perfect poses annotation is a concern, so all the objects must be roughly aligned to have orientation 0 (front face facing to the camera), and each model must be centered on the point (0,0) of the scene from its geometric centroid.

Setting parameters: the viewpoint is parameterized as a tuple (α, φ, d) of camera rotation parameters, where α is the azimuth, φ the elevation, and d the distance to the camera. Three principal parameters of discretization should be defined to create the animation of the camera: the range of distances $D(d)$, the range of elevations $E(\varphi)$, and the range of azimuth rotation $A(\alpha)$, with $[\alpha, \varphi, d] \in \mathbb{R}$ (see Fig. 2(2)). These ranges are split into discretized steps. Thereby, the number of rendered frames directly depends on set the parameters chosen, and the added number of 3D models of the category to be built (target objects) as $NumberOfFrames = NumberOfModels * NumberOfPoses$, where the number of poses results of the multiplication of the of the three arrays cardinality, which conforms the viewpoint tuple, as $NumberOfPoses = |D(d)| * |E(\varphi)| * |A(\alpha)|$.

Execute rendering: here is when the poses are conceived by animating a sphere who drives the camera over several 3D positions according to α, φ and d values. Each *pose* constitutes a key combination of these three parameters. Then, the 3D target models are stored in OBJ format. From these models, the coordinates of their 3D boxes in (x, y, z) are extracted directly from the 3D environment virtually calibrated. The eight vertices come for designing a tight 3D box faces fitted around each 3D model in a standardized manner. For each of the poses obtained from the animation, the following is performed:

---

[1] Free open-source 3D software https://www.blender.org.

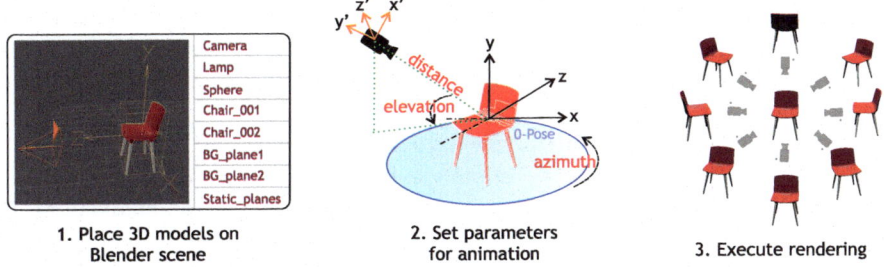

1. Place 3D models on Blender scene   **2. Set parameters for animation**   **3. Execute rendering**

Fig. 2. Overview of our data set generation approach. The first stage is the 3D model selection and placing in the scene. The second stage is the discretization parameter setting for the animation process. The last stage is to run the rendering for automatic labeling.

Fig. 3. Examples of rendered images and segmentation masks of the Bike poses and Hairdryer poses data sets.

- Calibration matrices: intrinsic and extrinsic calibration matrices are computed from the scene. First, the 3×3 matrix K called the camera matrix, and finally the extrinsic pose parameters $R|t$, which both compose the calibration 4×4 matrix P calculated by multiplying the camera matrix K by the Rotation R matrix and the Translation t vector, like this: $P = K * [R|t]$. Any 3D point ev of the calibrated scene can be transformed accurately to the pixel space by multiplying it by the camera matrix, as follows: $projectedPoint = P * ev$.
- 3D Scene Layout: all 3D boxes are labeled vertex to vertex, following the same labeling pattern. In the same way, the faces of the 3D boxes are labeled as Top, Bottom, Left, Right, Back or Front. These 3D positions correspond to the feature's location in the image after backprojection onto the object geometry.
- Render images: at this moment, all the necessary parameters for the production of the data set are defined, hence the rendering process is ordered. Here, RGB images, binary masks and depth maps are stored in folders. From the binary segmentation masks, the 2D boxes that circumscribe the target objects, and the segmentation polygons are extracted following the MSCOCO

style. As aforementioned, the tool allows to generate data sets in native format, one proposed by us, or in MS COCO data set format. The structure of the files and JSON files is different for each case.

Putting in test our generator: five data sets are generated for demonstration: the *Car poses* with thousands of camera viewpoints from 100 full-annotated vehicles. The other four data sets are similar for the categories: bike, chair, boat, and hairdryer. The *Car poses* is the biggest with a hundred models of cars and vans of different models, additionally annotated with real dimensions (length, height, and width) obtained online from car manufacturers and some pre-annotated from [12]. For consistency in the size of objects, a length of 1.8 m was manually set for the 3D cars as in [12]. Table 2 summarizes the main characteristics of the data sets created, while Fig. 3 illustrates examples of some of their rendered images, in which the fine-grained labeling is noteworthy. All of these five data sets were rendered in approximately 13 h using an Intel Core i7 quad-core CPU @ 2.8 GHz accelerated by Nvidia GeForce GTX 1070, and a RAM of 12 GB.

Table 2. Comparison of the five data sets generated with our tool.

Parameter	Car poses	Bike poses	Chair poses	Boat poses	Hairdryer poses
Azimuth	0–360° in 5° steps	0–340° in 20° steps	0–170° in 10° steps	0–300° in 10° steps	0–180° in 20° steps
Elevation	0–90° in 10° steps	0–20° in 20° steps	0–80° in 10° steps	0–60° in 10° steps	10–30° in 20° steps
Distance	5, 7 [m]	4, 6, 8 [m]	4, 5 [m]	7, 8, 9 [m]	2, 3, 4 [m]
# Poses	2,736	108	324	651	60
# 3D target Models	100	6	6	10	6
# RGB images	273,600	648	1,944	6,510	360

4 Discussion and Conclusion

It was proposed a generator tool of synthetic data sets whose characteristics are well suited for fine-tuning CNNs, or pre-training or training from scratch. Our generator allows for building a large labeled set of multi-viewpoint rendered images to facilitate user specific experiments. As little human effort is involved in this process, it can scale very well.

The use of a modeling software can better deal with inherent difficulties on annotation than traditional ways using real images, once the modeling process already embeds it in a more precise manner. For instance, there is no divergence on labeling each object model or even annotate all its pixels, because the modeling process already demands the designer to name each 3D object model and its boundary. Hence, all elements of the virtual environment are accurate annotated and labeled before the rendering process. Consequently, there is a significant reduction of time and costs compared to conventional annotation over thousands of real images by hand-craft processes.

When comparing our approach with the most similar [8,13,19,21,22], the ours has a larger number of characteristics that makes it more applicable to different scenarios. It is vital to recognize that, although our work contains several annotations than others miss, some others works are better fitted for the specific scenarios for which they were carefully designed, such as the Virtual KITTI [5] and Falling Things [20] data sets which contain more hyper-realistic scenarios for autonomous vehicles applications. In the same way, the simulators as CARLA [3] and AirSim [16] provide hiper-realistic interactive environments well suited for many computer vision problems, although considerable knowledge of game engines and computer graphics is already needed for extending new features such as 3D box annotations or camera viewpoints categorization. According to the literature review, there are many other works using rendered images for computer vision applications, nevertheless, most of them do not release the code for creating or extending the data. Adding animation to the scene, optical flow annotations, and test the data sets on a diversity of applications, will be the future direction of this work.

References

1. Atapour-Abarghouei, A., Breckon, T.P.: Real-time monocular depth estimation using synthetic data with domain adaptation via image style transfer. In: Proceedings of the IEEE Conference on Computer Vision and Pattern Recognition, pp. 2800–2810 (2018)
2. Butler, D.J., Wulff, J., Stanley, G.B., Black, M.J.: A naturalistic open source movie for optical flow evaluation. In: Fitzgibbon, A., et al. (eds.) European Conference on Computer Vision (ECCV), Part IV, LNCS 7577, pp. 611–625. Springer, Berlin (2012)
3. Dosovitskiy, A., Ros, G., Codevilla, F., Lopez, A., Koltun, V.: CARLA: an open urban driving simulator. In: Proceedings of the 1st Annual Conference on Robot Learning, pp. 1–16 (2017)
4. Dwibedi, D., Misra, I., Hebert, M.: Cut, paste and learn: Surprisingly easy synthesis for instance detection. In: The IEEE International Conference on Computer Vision (ICCV), pp. 1310–1319 (2017)
5. Gaidon, A., Wang, Q., Cabon, Y., Vig, E.: Virtualworlds as proxy for multi-object tracking analysis. In: 2016 IEEE Conference on Computer Vision and Pattern Recognition (CVPR), pp. 4340–4349 (2016)
6. Handa, A., Whelan, T., McDonald, J., Davison, A.: A benchmark for RGB-D visual odometry, 3D reconstruction and SLAM. In: IEEE International Conference on Robotics and Automation (ICRA), pp. 1524–1531 (2014)
7. Johnson-Roberson, M., Barto, C., Mehta, R., Sridhar, S.N., Rosaen, K., Vasudevan, R.: Driving in the matrix: can virtual worlds replace human-generated annotations for real world tasks? In: IEEE International Conference on Robotics and Automation (ICRA), pp. 746–753 (2017)
8. Liebelt, J., Schmid, C., Schertler, K.: Viewpoint-independent object class detection using 3D feature maps. In: 2008 IEEE Conference on Computer Vision and Pattern Recognition, pp. 1–8 (2008). https://doi.org/10.1109/CVPR.2008.4587614
9. Lin, T., Maire, M., Belongie, S.J., Bourdev, L.D., Girshick, R.B., Hays, J., Perona, P., Ramanan, D., Dollár, P., Zitnick, C.L.: Microsoft COCO: common objects in context. CoRR **abs/1405.0312**, pp. 740–755 (2014)

10. Matzen, K., Snavely, N.: NYC3DCars: a dataset of 3D vehicles in geographic context. In: International Conference on Computer Vision (ICCV), pp. 761–768 (2013)
11. Mayer, N., Ilg, E., Hausser, P., Fischer, P., Cremers, D., Dosovitskiy, A., Brox, T.: A large dataset to train convolutional networks for disparity, optical flow, and scene flow estimation. In: Proceedings of the IEEE Conference on Computer Vision and Pattern Recognition, pp. 4040–4048 (2016)
12. Pepik, B., Stark, M., Gehler, P., Schiele, B.: Teaching 3D geometry to deformable part models. In: 2012 IEEE Conference on Computer Vision and Pattern Recognition, pp. 3362–3369 (2012). https://doi.org/10.1109/CVPR.2012.6248075
13. Richter, S.R., Hayder, Z., Koltun, V.: Playing for benchmarks. In: International Conference on Computer Vision (ICCV), pp. 2232–2241 (2017)
14. Ros, G., Sellart, L., Materzynska, J., Vazquez, D., Lopez, A.M.: The synthia dataset: a large collection of synthetic images for semantic segmentation of urban scenes. In: 2016 IEEE Conference on Computer Vision and Pattern Recognition (CVPR), pp. 3234–3243 (2016). https://doi.org/10.1109/CVPR.2016.352
15. Savva, M., Chang, A.X., Dosovitskiy, A., Funkhouser, T., Koltun, V.: MINOS: multimodal indoor simulator for navigation in complex environments. arXiv:1712.03931 (2017)
16. Shah, S., Dey, D., Lovett, C., Kapoor, A.: AirSim: high-fidelity visual and physical simulation for autonomous vehicles. In: Field and Service Robotics (2017)
17. Song, S., Yu, F., Zeng, A., Chang, A.X., Savva, M., Funkhouser, T.: Semantic scene completion from a single depth image. IEEE Conference on Computer Vision and Pattern Recognition, pp. 190–198 (2017)
18. Su, H., Qi, C.R., Li, Y., Guibas, L.J.: Render for CNN: viewpoint estimation in images using CNNs trained with rendered 3D model views. In: 2015 IEEE International Conference on Computer Vision (ICCV), pp. 2686–2694 (2015). https://doi.org/10.1109/ICCV.2015.308
19. Sun, B., Peng, X., Saenko, K.: Generating large scale image datasets from 3D cad models. In: CVPR 2015 Workshop on the Future of Datasets in Vision (2015)
20. Tremblay, J., To, T., Birchfield, S.: Falling things: a synthetic dataset for 3D object detection and pose estimation. In: Proceedings of the IEEE Conference on Computer Vision and Pattern Recognition Workshops, pp. 2038–2041 (2018)
21. Varol, G., Romero, J., Martin, X., Mahmood, N., Black, M.J., Laptev, I., Schmid, C.: Learning from synthetic humans. In: Conference on Computer Vision and Pattern Recognition (CVPR), pp. 4627–4635 (2017)
22. Xiang, Y., Kim, W., Chen, W., Ji, J., Choy, C., Su, H., Mottaghi, R., Guibas, L., Savarese, S.: ObjectNet3D: a large scale database for 3D object recognition. In: European Conference Computer Vision (ECCV), pp. 160–176 (2016)
23. Xiang, Y., Mottaghi, R., Savarese, S.: Beyond pascal: a benchmark for 3D object detection in the wild. In: IEEE Winter Conference on Applications of Computer Vision, pp. 75–82 (2014). https://doi.org/10.1109/WACV.2014.6836101

Extracting Clothing Features for Blind People Using Image Processing and Machine Learning Techniques: First Insights

Daniel Rocha[1(✉)], Vítor Carvalho[1,2(✉)], Filomena Soares[1(✉)],
and Eva Oliveira[2(✉)]

[1] Algoritmi R&D, University of Minho, Guimarães, Portugal
id8057@alunos.uminho.pt, fsoares@dei.uminho.pt
[2] 2Ai Lab, School of Technology, IPCA, Barcelos, Portugal
{vcarvalho, eoliveira}@ipca.pt

Abstract. Vision is one of the senses that dominates the life of humans. It allows them to know and have the perception of the world around them, while giving meaning for objects, concepts and ideas, and tastes. How to dress and the style we prefer for different occasions is part of one's identity. Blind people do not have this sense, and dressing can be a difficult and stressful task. With the advance of technology it is important to minimize all the limitations of a blind person in the management of garments. Not knowing the colors, the type of pattern, or even the state of the garments make this a daily challenge in which nowadays resources are not the best. Thus, the approach of this project is to address this issue of extracting the basic characteristics and conditions of the garment (in good conditions, dirty or wrinkly) in order to help the blind.

Keywords: Clothes recognition · Blind people · Image processing · Machine learning

1 Introduction

Visual impairment is a type of sensory impairment and refers to the severe and irreversible loss or diminution of visual ability in both eyes, not being susceptible to improvement or correction with the use of lenses and/or clinical or surgical treatment. Visual loss can be sudden and severe or the result of a gradual deterioration, where objects at great distances become increasingly difficult to see. Vision impairment thus encompasses all conditions in which vision impairment exists.

The individual who is born with the sense of sight, later losing it, stores visual memories, can remember images, lights and colors he/she has known, and this is very important for his/her re-adaptation. On the other hand, whoever is born without the capacity of seeing, can never form or possess visual memories.

Despite the high technology already available, there are still some gaps, in particular, the aspect of aesthetics and image that are little explored.

© Springer Nature Switzerland AG 2019
J. M. R. S. Tavares and R. M. Natal Jorge (Eds.): VipIMAGE 2019, LNCVB 34, pp. 411–418, 2019.
https://doi.org/10.1007/978-3-030-32040-9_42

It is in the search of these characteristics that comes the motivation for this project, namely, to enable people with visual impairment to feel equally well with what they wear, functional and without needing help. The objective of this paper is to recognize the techniques used for image recognition, retrieval and suggestions of clothing items.

The scope of the research follows the previous work of the team [1–3], as through image processing techniques, it is possible to help the blind to choose the clothing and to manage the wardrobe.

In this sense, the overall objective of the project is to build a robust algorithm that will recognize and take into account the following elementary features:

- Type of clothes;
- The season of the year and weather;
- Suggesting combinations of clothing pieces;
- Identification of the cloth pattern;
- Presence of stains;
- Modifications in the state of the garment.

This paper is focused on the last three points mentioned above.

This paper in divided in five sections. Section 2 describes the related work; in Sect. 3 we present the system goals; Sect. 4 describes the experiments and machine learning approaches. Finally, Sect. 5 concludes with the final remarks.

2   Related Work

Image processing and machine learning developments have reached important proportions in terms of retrieval and recognition of clothes. In this way, this section presents some solutions to identify the highlights and the gaps that can be overcome.

Pal and Pal [4] review the techniques of segmentation. Theses researchers noted that there are challenging issues such as developing a unified approach to image segmentation that can be applied to all types of images. However, there is no universally accepted method, as each specific case needs its own techniques [4].

In the identification of garments, Yamaguchi et al. [5, 6], propose an approach to clothes analysis. The analysis approach consists of two main steps: retrieve similar images from the database of analyzed images and use extracted images and tags to analyze the image to be examined. The proposed clothing analysis solution is able to classify 53 different categories of fashion photo clothes. The method is able to separately segment each piece of clothing by exploiting super-pixels and model of flexible pieces for human pose estimation [5, 6].

Wazarkar and Keshavamurthy [7] propose the classification of fashion images, incorporating the concepts of linear convolution and corresponding points using local characteristics. First, the linear convolution is used to obtain the corresponding representative image. Then the corresponding points are identified with the help of local resources using Scale Invariant Feature Transform (SIFT) [7].

Manfredi et al. [8] propose a complete system for clothing segmentation and color classification from images taken from online fashion stores. As elements for segmentation of the garment it is used the Random Forest classifier and later the Gaussian mixture model (GMM) [8].

Liang et al. [9] proposed a clothes co-parsing (CCP) structure to jointly analyze a batch of clothing images and produce an accurate annotation of clothing items. The system consists of two sequential phases of inference on a set of clothing images, the segmentation of images to extract regions of distinguishable clothing and co-labeling regions to recognize various clothing items [9].

Yuan et al. [10] developed a computer-based prototype to match a pair of images of two clothes to standard and color. The researchers in the proposed method for pattern detection achieved 85% accuracy being robust for clothing with complex patterns of texture, various colors and variations of rotation and illumination. For pattern matching, the major errors occurred with images with very similar texture patterns. Regarding the color matching in 100 pairs of images, only one pair does not match correctly due to background distraction. To deal with complex texture patterns and lighting changes, they combined techniques using Radon transform, wavelet features, and co-occurrence matrix for pattern matching [10].

Yang et al. propose [11] a system for pattern recognition. This system identifies 11 clothing colors and recognizes 4 categories of clothing patterns (checkered, patterned and irregular). The following methods are used to extract the pattern recognition characteristics: Radon Signature, statistical descriptor (STA), and scale invariant feature transform (SIFT). The combination of the previous methods is later used by classifiers, the Support Vector Machine Algorithm (SVM) for pattern recognition [11].

Although there has been a great effort to develop systems to aid visual impaired, there is still no solution capable of covering all the questions proposed in the scope of this project, namely an automatic system to extract the basic characteristics of the garment and its condition (in good conditions, dirty or wrinkly) for the visually impaired.

Image processing is essential for extraction of characteristics into an image. Dalal and Triggs [12] have introduced the Histogram of Oriented Gradients (HOG) features to classify humans in images [12]. HOG descriptors are mainly used to describe the structural shape and appearance of an object in an image, making them excellent descriptors for object classification.

In addition to computational vision algorithms, it is still necessary to include automatic learning algorithms that support decision making.

Automatic learning is divided into two types, unsupervised learning and supervised learning. In unsupervised learning, the data are grouped on the basis of patterns, that is, learning is based on observation. In this group the category of Clustering algorithms is inserted. On the other hand, in supervised learning a predictive model is created based on existing data, that is, learning is carried out based on existing examples [13]. The SVM is designed as an efficient tool for solving classification, regression and related problems that involves highly dimensional and complex data [14].

3   System Goals

This work has as main objective the development of a system, capable of supporting the visually impaired in performing the tests of inspection, identification and combination of garments.

In order to achieve this goal, specific objectives were set, in particular:

- Development of an algorithm for image acquisition, processing and analysis;
- Based on the previous algorithm, it is intended to develop software that is intuitive, easy to use, respecting accessibility rules;
- Development of artificial intelligence algorithms for combinations, suggestions and learning of the User's preferences;
- Building a database for storing all the information that will support artificial intelligence algorithms.

In pursuit of the objectives the following research question arises:

- How can artificial intelligence make the inspection, identification, combination and management of clothing for a blind person?

Based in techniques of image processing and machine learning, we have performed some experiments that will be described in the next section.

4   Experiments

This section describes the implementation of algorithms HOG, SVM and Neuronal Networks in preliminary tests.

The algorithms are implemented over Python programming language due to its power and wide field in data science and computer vision.

To implement the algorithms, we collected some images from the online catalog of Salsa store [15]. To simulate the stains in the clothes, brown circles were placed upon the image to represent coffee stains.

We use dlib that is a modern C++ toolkit containing machine learning algorithms and tools for creating complex software in C++ to solve real world problems [16].

The detector implementation has been created with the following steps:

- Gather training data;
- Perform the labeling of the data;
- Training and parameterization of the detector;
- Evaluate and testing the detector.

After editing the images, using the imglab tool that belongs to the dlib, the labeling of the stain location was done in 7 images.

Later, we parameterized the model and performed the training. After completing the training, 4 images with coffee stains were supplied (Fig. 1) and through a box our model signaled the zone of dirt.

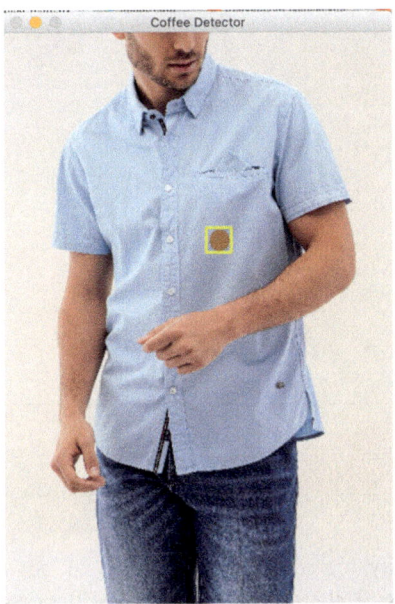

Fig. 1. SVM classification – detection of dirty in the garment

This model with only 7 training images provided, was able to achieve a precision of 100%. However, with different stain formats, texture and over pattern the accuracy would certainly not be the same.

Through the construction of a neural network, we intend to identify different types of clothes, like jeans, and coat. Fashion MNIST is a database containing 70,000 individual clothing images, characterized by 10 different classes, such as T-shirt/top, Trouser, Pullover, Dress, Coat, Sandal, Shirt, Sneaker, Bag and Ankle boot [17]. In the creation of the neural network 60,000 images of the database were used to train the network and 10,000 images to evaluate the accuracy of the model.

For the creation of our model, we used Keras that is a high level framework that makes TensorFlow easier to use, allowing to build and train deep learning models.

In this way, we have adapted an example of a basic classification using Keras and TensorFlow [18].

The model has train images and train labels, which are the data used during the training. In order to construct the neural network, a configuration of the model layers was carried out, to be compiled later.

The network consists of a two-layer sequence designated as densely-connected, or fully-connected. The first Dense layer has 128 nodes (neurons). The second is a 10-node softmax layer that returns a vector with 10 probability values, totaling the sum of 1. The probability that the current image belongs to one of 10 is indicated by the value of each node.

The cost function and accuracy metrics are shown throughout the model trains and this model achieves an accuracy of about 86% in the training data, Fig. 2.

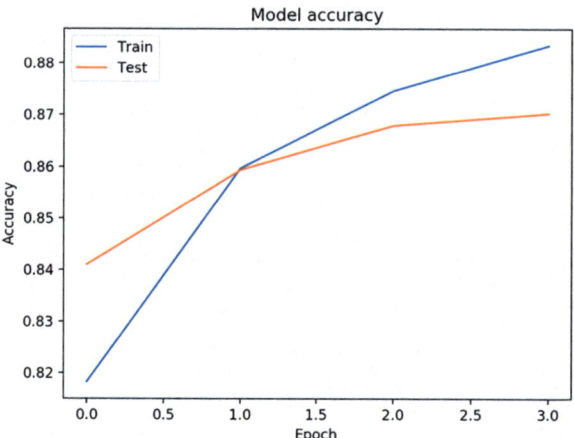

Fig. 2. Model accuracy distribution of train and test.

In the test data, we evaluated the performance of our model. However, as it is possible to observe in Fig. 2, the accuracy in the test data is smaller than the accuracy in the training data, which reflects the occurrence of overfitting. That is, our model works worse on new data than on training phase.

After the training of the model, four images were provided to predict the corresponding label, Fig. 3.

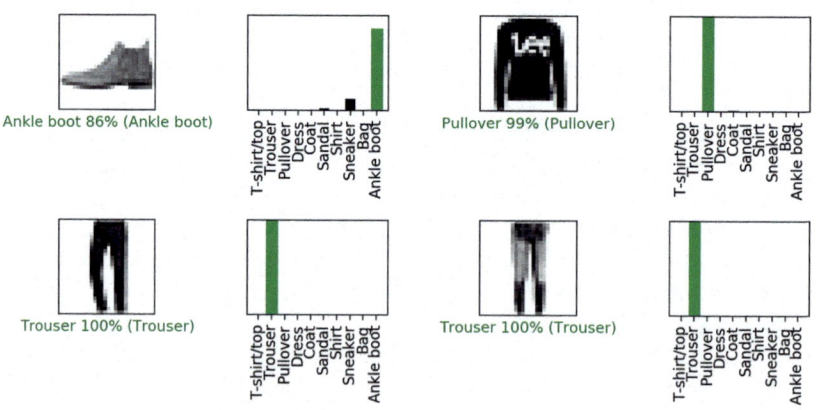

Fig. 3. Predicted labels by model.

Figure 3 shows through the green bar the confidence, in percentage, of the classifications.

The construction of this neuronal network could be improved changing the dense layers and the epoch of the training. Even the good results achieved, the overfitting needs to be avoided in order to improve the accurate of our test data.

5   Final Remarks

This paper discusses the implementation of basic steps inside of our objective, that is the identification and classification of type the clothes.

For simple object detection scenarios in controlled environments, the dlib toolkit is very well suited to deliver fast and reliable results. The advantage of this method is the little training data that is required.

The neural network performed with Keras and TensorFlow have shown that it is an advantage to recognizing and learning from the database, with optimizations of the layers, achieving good results.

Next step of the research considers exploring other techniques with robust algorithms to achieve the final purpose of this project; deep learning approaches can be a possibility.

This system proposes a concept for significantly improving the daily life of the blind, in particular, allowing them to recognize clothe features as stains and wrinkles.

Acknowledgments. This work has the support of Association of the Blind and Amblyopes of Portugal (ACAPO) and Association of Support for the Visually Impaired of Braga, Portugal (AADVDB). Their considerations gave (and still give) this project the first insights to a viable solution for the blind people community.

References

1. Rocha, D., Carvalho, V., Oliveira, E., et al.: MyEyes-automatic combination system of clothing parts to blind people: first insights. In: 2017 IEEE 5th International Conference Serious Games Applications for Health (SeGAH), pp. 1–5 (2017)
2. Rocha, D., Carvalho, V., Oliveira, E.: MyEyes - automatic combination system of clothing parts to blind people: prototype validation. In: Sensordevices' 2017 Conference. Rome, Italy, 10–14 September 2017
3. Rocha, D., Carvalho, V., Gonçalves, J., et al.: Development of an automatic combination system of clothing parts for blind people: MyEyes. Sensors Transducers **219**, 26–33 (2018)
4. Pal, N.R., Pal, S.K.: A review on image segmentation techniques. Pattern Recognit. **26**, 1277–1294 (1993). https://doi.org/10.1016/0031-3203(93)90135-J
5. Yamaguchi, K., Kiapour, M.H., Ortiz, L.E., Berg, T.L.: Parsing clothing in fashion photographs. In: 2012 IEEE Conference Computer Vision Pattern Recognition (CVPR), pp. 3570–3577 (2012)
6. Yamaguchi, K., Kiapour, M.H., Ortiz, L.E., Berg, T.L.: Retrieving similar styles to parse clothing. IEEE Trans. Pattern Anal. Mach. Intell. **37**, 1028–1040 (2015). https://doi.org/10.1109/TPAMI.2014.2353624
7. Wazarkar, S., Keshavamurthy, B.N.: Fashion image classification using matching points with linear convolution. Multimed. Tools Appl. **77**, 25941–25958 (2018). https://doi.org/10.1007/s11042-018-5829-4
8. Manfredi, M., Grana, C., Calderara, S., Cucchiara, R.: A complete system for garment segmentation and color classification. Mach. Vis. Appl. **25**, 955–969 (2014). https://doi.org/10.1007/s00138-013-0580-3

9. Liang, X., Lin, L., Yang, W., et al.: Clothes Co-parsing via joint image segmentation and labeling with application to clothing retrieval. IEEE Trans. Multimed. **18**, 1175 (2016). https://doi.org/10.1109/TMM.2016.2542983

10. Yuan, S., Tian, Y., Arditi, A.: Clothing matching for visually impaired persons. Technol. Disabil. **23**, 75–85 (2011). https://doi.org/10.3233/TAD-2011-0313

11. Yang, X., Yuan, S., Tian, Y.: Assistive clothing pattern recognition for visually impaired people. IEEE Trans. Hum.-Mach. Syst. **44**, 234–243 (2014). https://doi.org/10.1109/THMS. 2014.2302814

12. Dalal, N., Triggs, B.: Histograms of oriented gradients for human detection. In: Proceedings - 2005 IEEE Computer Society Conference on Computer Vision and Pattern Recognition, CVPR 2005, pp 886–893 (2005)

13. Sharma, A., Sharma, A.: Machine learning: a review of techniques of machine learning (2018)

14. Muñoz, A., Moguerza, J.M., Venturini, G.A.M.: Support vector machines (2019)

15. Salsa Jeans ® | Jeans, Roupa e Acessórios para Mulher e Homem. https://www.salsajeans. com/pt/?gclid=CjwKCAjwscDpBRBnEiwAnQ0HQH6hls7uDEkOZMqwWww5E68wLSve KTqnIN4OYskqR5VFZtkpzwS-LRoCrEsQAvD_BwE. Accessed 19 Jul 2019

16. Dlib C++ library. http://dlib.net/. Accessed 11 Jul 2019

17. Fashion MNIST | Kaggle. https://www.kaggle.com/zalando-research/fashionmnist. Accessed 15 Jul 2019

18. Train your first neural network: basic classification | TensorFlow Core | TensorFlow. https:// www.tensorflow.org/tutorials/keras/basic_classification. Accessed 21 Jul 2019

HARM - The Human Action Recognition Module

Brolin Fernandes, Gunish Alag$^{(\boxtimes)}$, and Saumya Kumaar

University of Twente, Enschede, The Netherlands
g.alag@student.utwente.nl

Abstract. We present **HARM**, a three stage human action recognition algorithm for static images. The proposed methodology incorporates pose estimation, pose classification and scene classification to develop a complete understanding of humans in images with respect to their environments. The module utilizes a blend of modern state-of-the-art pose estimation and image classification algorithms, thereby resulting in accurate classifications of human poses and actions in static images.

1 Introduction

Computer vision has come a long way in assisting robotic systems with several tasks like autonomous driving, surveillance etc. Even without any sensor data fusion, image data can explicitly provide with valuable information about the environment, provided we use appropriate algorithms for feature extraction. The foundations that arise from image processing are extensively applied to a variety of fields like nanotechnology, space physics, botanical analysis etc. A similar important example of machine vision applications is human action recognition, which plays a vital role in crowd monitoring and surveillance systems. Recognizing actions of human beings has been addressed with a number of algorithms with various approaches (Fig. 1).

Initially, several temporal domain based classification techniques were proposed, wherein all the frames in an observed sequence were summarized into a single representation. The above suggested dimensionality reduction methodologies were adopted in [24,25] like PCA, IsoMap and Locality Preserving Projections (LPP) but they did not guarantee a decent discrimination between classes. An improvement was suggested in [26], where the authors introduced spatio-temporal discriminant embedding, even though the proposed techniques were computationally expensive. A lot of authors have used k-NN classifiers in conjunction with dimensionality reduction. For instance, Wang *et al.* [27] either use the minimum mean frame-wise distance in an embedded space, or a frame-order preserving variant. However, learning a discriminative metric for the same was tedious. In line with the above issue, discriminative classifiers like SVMs and RVM (Relevance Vector Machines) were studied and experimented in human action recognition by Oikonomopoulos *et al.* in [28]. Still, absolute labeling was

© Springer Nature Switzerland AG 2019
J. M. R. S. Tavares and R. M. Natal Jorge (Eds.): VipIMAGE 2019, LNCVB 34, pp. 419–433, 2019.
https://doi.org/10.1007/978-3-030-32040-9_43

Fig. 1. An example of the output of HARM.

not the optimal solution as humans are indeed capable of performing multiple actions at the same time, which implies that probabilistic or discriminative dynamics models are likely to give better estimates. Hidden Markov Models (HMM) use states that correspond to different phases in the performance of an action. They model state transition probabilities and observation probabilities. Feng *et al.* [29] use static HMMs, where key-poses correspond to states. Weinland *et al.* [30] constructed a code-book based on a discrimination inspired search. However, the independence assumptions of HMMs indicate that the observations in time are uncorrelated, which is unfortunately not very often the case. Discriminative models like CRFs (Conditional Random Fields) that use multiple overlapping features can overcome this issue as also corroborated by Sminchisescu *et al.* in [31]. Post establishment of superiority of CRFs in human action recognition, multiple variations of CRFs were suggested in literature to further improve the performance. Wang *et al.* [32] proposed factorial CRFs whereas Zhang *et al.* [33] suggested to use hidden CRFs for accurate human action classification. Shi *et al.* [34] suggested to use a semi-Markov modeling approach for both action segmentation and recognition.

Although the above mentioned methods provide a reasonable accuracy in understanding human behaviors, there was still a lot of scope for improvement in this field of research. With the advent and publicity of CNNs (Convolutional Neural Networks) in [18], they have been at the core of almost all state-of-the-art machine vision solutions. Zhang *et al.* [35] experimented self-regulated adaptation schemes which can reposition observations adaptively to facilitate accurate action recognition. Furthermore, they combine the scheme with a CNN-based classification approach, thereby making it end-to-end trainable. However, the dataset they have used is RGB-D format, and they use depth maps to evaluate actions, thereby making their algorithms non-desirable for real-time applications. Saggese *et al.* [36] learn skeletal representation for action recognition utilizing a custom-dataset. Other depth map based approaches have been suggested in [37,38], without any trade-offs between accuracy and speed. Another interesting methodology commonly found in literature for action classification is using inertial sensors [39, 40]. This implies that the subjects are supposed to have wearable tech which provide the data for recognition algorithms, thereby making them sub-optimal for real-world deployment.

As mentioned earlier, this particular field still has a lot of scope of improvement, as human actions can have a diverse range. Furthermore, real-time considerations are minimal. The literature review indicates a common understanding that actions and poses of human beings are not correlated, which is usually not the case, and if we intend to design a surveillance system, we need to be absolutely certain about the correlation between pose and action in images. For instance, $shop - lifting$ and $sitting$ are practically unlikely to be simultaneous. Hence, this calls for an algorithm that can accurately estimate the pose, classify the pose and then classify the action/scene with respect to the latter data and image content. In this paper, we present **HARM**, a human activity identification system for static images, which can accurately estimate the pose and actions of humans. The primary contributions of this paper are as follows:

- An experimental blend of pose estimation, pose classification and action classification into a single module for semantic understanding of human behaviors.
- An end-to-end trainable module that requires no pre-processing suitable for offline surveillance processes.

Furthermore, we use the keyword *scene* classification as opposed to *action*, as the latter word could be ambiguous, in the sense that *walking* might be an action, but $walking - the - dog$ might be more particularly related to the overall image environment.

1.1 Datasets

Several human action datasets have been suggested across literature. The Weizmann human action dataset [41] contains 10 actions, INRIA XMAS multi-view dataset comprises of 14 [42], whereas UCF sports action [43] and Hollywood human action datasets [43, 44] encompass 9 activities each. As it can be observed, 10–14 actions are insufficient to accurately model a generic environment. So it becomes increasingly clear that we need a dataset which encompasses a larger proportion of human actions, and hence, we chose Stanford 40 dataset [9] for action classification, which has 40 very prominent day-to-day human activities and MPII Human Pose Dataset [7], which has 23,000 samples. The dataset for pose classification is custom collected and will be released very soon.

2   Supplementary Material

The final implementation scripts for singular images and webcam feed, trained models and visualization scripts can be found at this GitHub repository.

3   Methodology

The methodology adopted in this research is a 3-step process, which is as follows:

- Pose Estimation
- Pose Classification
- Scene Classification

We emphasize every section individually in this paper. A visualization of the overview is presented in Fig. 2 for easier understanding.

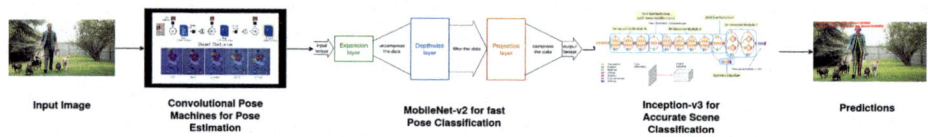

Input Image **Convolutional Pose Machines for Pose Estimation** **MobileNet-v2 for fast Pose Classification** **Inception-v3 for Accurate Scene Classification** **Predictions**

Fig. 2. The diagram represents the overall flow of the proposed HARM model in the research. The algorithms are executed in series, thereby resulting in a very deep interwoven network. Understanding human actions accurately requires immense data and fine-tuned models, which are inherently deep and hence would require significant computation power.

3.1   Pose Estimation

In this research, we employ Convolutional Pose Machines [1] for efficient pose estimation in singular images. CPMs are heavily based on the articulated inference pose machines are described in [2]. The fundamental difference between CPMs [1] and Pose Machines [2] is that in CPMs, prediction and feature computation modules have been replaced by a deep hierarchical convolutional architecture. Once the hyperparameters of this network are optimized, it can be shown that it learns the same features as Pose Machines [2]. Furthermore, convolutional architectures have the added advantage of being end-to-end trainable. We now describe the architecture in detail.

(1) Keypoint Localization Using Local Image Evidence: The initial stage of a CPM attempts to predict part beliefs from only local image evidence. This evidence is *local* because the receptive field of the first stage is constrained around a local patch of the concerned pixels. For this research, we adopted the fully convolutional architecture mentioned in [3], for an efficient receptive field. This structure consists of five convolutional layers followed by two 1×1 convolutional layers and the input images are cropped to size 368×368. The mathematical background of this assumption has been omitted from this paper as it is beyond the scope but can be referenced from Section 4.2 in [1].

(2) Sequential Prediction with Learned Spatial Context Features: Due to the large variances in configuration and appearance of the lower half of the kinematic model of the human skeleton, the accuracies of landmark detection in such regions are much lower as compared to the upper sectors like head and shoulders. The landscape of the belief maps around a part location, albeit noisy, can, however, be very informative.

A predictor in subsequent stages $(g_{t>1})$ could use the spatial information $(\psi_{t>1}(.))$ of the noisy belief maps in and around the image position z and increase the prediction rate using the fact that parts occur in geometric configurations. The second stage of CPM is a classifier g_2 that accepts the input image features X_z^2 and belief-computed features using a computation function ψ for every part of the previous stage. An interesting thing to note about the feature function ψ is that it encodes the landscape of belief maps in a spatial region around the location z of different parts. But for a CPM it is not needed to have an explicit function that computes the visual features, and ψ is defined instead as the receptive field of the predictor.

A receptive field at the output layer of the second stage network, which is essentially large enough to permit grasping of complex parametric representations and long-range correlations between different parts, assists in the design of the network. It has been empirically proven in [1] that such large-sized receptive fields are actually effective in learning complex models. The experimentation is carried out on smaller normalized images though rather than the best setting. Next, the following convolutional layers in the later stages allow the classifier to choose between the dominant features, which is analogous to Principal Component Analysis. In the first stage, the image is examined only locally with the help of a smaller receptive field. In the second stage, the receptive field is increased manifold. There are many ways this could be done. We could either increase the kernel size of the convolutional filters at the risk of encountering vanishing gradients during training, or we could add pooling layers but at the expense of precision. For this research, we do not compromise with precision, and hence the optimal choice for us was to increase the number of convolutional layers to achieve large receptive field on the 8×down-scaled heat-maps. We observed that the stride-8 network performs as well as a stride-4 one even at high precision region, while it makes us easier to achieve larger receptive fields.

(3) Learning in Convolutional Pose Machines: The resulting large number of layers in the CPM architecture makes it prone to a commonly faced problem of vanishing gradients [4–6], where the magnitudes of the back-propagated gradients decrease. Fortunately, the sequential arrangement of CPM naturally avoids this issue.

The target cost function to be minimized at the output of each stage is given by:

$$\sum_{p=1}^{P+1} \sum_{z \in Z} ||b_T^p(z) - b_*^p(z)||_2^2 \tag{1}$$

All the stages are jointly trained in the network using standard stochastic gradient descent. Image feature sharing occurs across all stages from the corresponding convolutional layers.

3.2   Pose Classification

For pose classification, we employ use of the state-of-the-art mobile architecture MobileNetV2 [14] which is based on an inverted residual structure. This arrangement has shortcut connections between the thin bottleneck layers. The intermediate expansion layer uses lightweight depth-wise convolutions to filter features as a source of non-linearity.

(1) Architecture: The detailed model can be seen in Table 1. MobileNetV2 contains initial fully convolutional layers with 32 filters, succeeded by 19 bottleneck layers. ReLU6 is used a non-linearity because of its robustness with low-precision computational resources [15]. Kernel size is fixed to 3×3 and dropouts and batch normalization are used during training.

Table 1. Bottleneck residual block transforming from k to k' with stride s.

Input	Operator	Output
h × w × k	1×1 conv2d, ReLU6	h × w × (tk)
h × w × (tk)	3×3 dwise s = s, ReLU6	$\frac{h}{s} \times \frac{w}{s} \times$ (tk)
$\frac{h}{s} \times \frac{w}{s} \times$ (tk)	linear 1×1 conv2d	$\frac{h}{s} \times \frac{w}{s} \times$ k'

Except for the first layer, a constant expansion rate is use throughout the network. Expansion rates between 5–10 were observed to result nearly identical performance measures with smaller networks being better of at lower values of expansion rates and vice-versa for larger networks.

An expansion factor of 6 is applied to the size of the input tensors. Furthermore, the tunable hyperparameters considered in this section were input image resolution and width multiplier. They were adjusted according to our requirements for trade-offs between accuracy/performance. The primary network has a computation cost of 300M multiply-adds and uses 3.4M parameters.

Table 2 describes the a sequence of 1 or more identical layers, that are repeated n times. All layers belonging to one sequence have exactly the same number of outputs. All spatial convolutional operations have a 3×3 kernel size, s is the stride and t is the expansion factor. The bottleneck algorithm can be further referenced from [8].

The most remarkable difference between [14] and [15] is that for multipliers less than one, in [14] width multipliers are applied to all layers except the very last convolutional layer. This improves performance for smaller models, and hence MobileNetV2 performs better than MobileNetV1.

Table 2. MobileNetV2

Input	Operator	t	c	n	s
$224^2 \times 3$	conv2d	–	32	1	2
$112^2 \times 32$	bottleneck	1	16	1	1
$112^2 \times 16$	bottleneck	6	24	2	2
$56^2 \times 24$	bottleneck	6	32	3	2
$28^2 \times 32$	bottleneck	6	64	4	2
$14^2 \times 64$	bottleneck	6	96	3	1
$14^2 \times 96$	bottleneck	6	160	3	2
$7^2 \times 160$	bottleneck	6	320	1	1
$7^2 \times 320$	bottleneck	–	1280	1	1
$7^2 \times 1280$	bottleneck	–	–	1	–
$1 \times 1 \times 1280$	bottleneck	–	k	–	

3.3 Scene Classification

Action (or scene) classification is achieved using the popular Inception-v3 architecture [16]. This network has been chosen as it is 42 layers deep and still has only 2.5 times the computation cost as compared to GoogLeNet [17] and much better performance than VGGNet [19]. Although, VGGNet has the added advantage of simple architecture, this actually comes at a cost of classification rate.

(1) Architecture: Here we specify the Inception-v3 architecture and the layout is mentioned in the table below (Table 3):

Table 3. Modified Inception-V3 architecture used for this research, with concatenating feature vectors from 5×5 and 3×3 convolutional outputs. *Vector Concat* is the module where the 'inception' aspect comes into the picture. At these stages in the model, the output features from all 5×5, 3×3 and 1×1 convolutional operations are merged.

Type	Stride	Input size
conv	$3 \times 3/2$	$229 \times 229 \times 3$
conv	$3 \times 3/1$	$149 \times 149 \times 32$
conv padded	$3 \times 3/1$	$147 \times 147 \times 64$
pool	$3 \times 3/2$	$73 \times 73 \times 64$
conv	$3 \times 3/1$	$71 \times 71 \times 80$
conv	$3 \times 3/2$	$35 \times 35 \times 192$
conv	$3 \times 3/1$	$229 \times 229 \times 3$
3×inception	Vector Concat	$35 \times 35 \times 288$
5×inception	Vector Concat	$17 \times 17 \times 768$
2×inception	Vector Concat	$8 \times 8 \times 1280$
pool	8×8	$8 \times 8 \times 2048$
linear	Logits	$1 \times 1 \times 2048$
softmax	Classifier	$1 \times 1 \times 40$

The traditional 7×7 convolutions have been factorized into three different 3×3 based on the ideas suggested in [16]. For the Inception part of the network, we have 3 traditional inception modules at the 35×35 with 288 filters each. This is further reduced to a 17×17 grid with 768 filters using a technique called grid reduction (as described in [16]). This is followed by 5 instances of the factorized inception modules.

4   Datasets, Training and Testing Setup

Pose estimation models were trained on the MPII Human Pose Dataset [7] for 1000 epochs. Pose classification dataset was collected manually, while Stanford 40 dataset [9] was used for understanding human actions in static images with respect to their environments.

MobileNet-V2 was trained for 4000 training samples (Fig. 3) on a custom dataset. Stanford 40 dataset has a total of $9,523$ samples out of which they were split in a 90–10% ratio for training-testing respectively. Inception-v3 architecture was trained on this dataset for 20,000 training steps, with a final test accuracy of 93.5%.

Fig. 3. Training samples for pose classification

All models were trained on a system with Ubuntu 18.04, 8 GB DDR4 RAM, i5-8240 Intel Processor and NVIDIA GTX 1050Ti (768 Cores with 4 GB DDR5 VRAM) GPU.

5   Evaluation and Results

The evaluation criteria for the research presented in this article is manifold. We provide metrics for each of the three stages of the algorithms. We publicly release the source code and details on the architecture, learning parameters, design decisions and data augmentation to ensure re-producibility, as already mentioned previously in the Supplementary Material section.

The evaluation for pose estimation is done using PCKh metric [7] (Fig. 3). Here the error tolerance is normalized with respect to the head-size of every individual target. The results below are shown on the MPII Human Pose Dataset

[7], as it contains more than $28,000$ training samples. The scores of CPM achieve state-of-the-art performance at 87.95%, which is more than 6% higher than the next competitor.

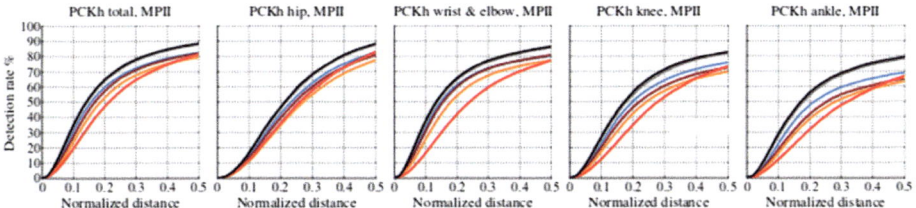

Fig. 4. Quantitative analysis of CPM using the PCK metric. The complete evaluation can be referenced from [1]

Next, we evaluate the pose classification algorithms. We first establish a CNN-baseline and compare it with multiple machine learning algorithms and modern deep perceptual architectures.

The feature vectors for these ML structures consist of angular orientations between the successive skeletal key-points. We take into account 3 consecutive key-points near the hips and near the shoulders to estimate an approximate orientation of the human body. Next, we evaluate the angular difference between the leg-hip structure and the ground, which assists the system to differentiate *lying-down* from upright and sitting.

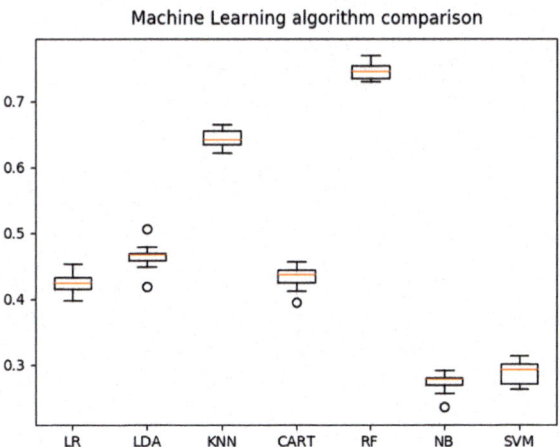

Fig. 5. Quantitative analysis of ML algorithms for pose classification. As we can see that random forest seems to outperform all other architectures, but still does not provide desirable accuracy.

Figure 4 shows the performance of ML algorithms with respect to pose classification. However, there could be many variations within a single class, like for example sitting and lying down, and hence algorithms that require hard-coded feature engineering, might be sub-optimal. This can be overcome by using deep convolutional models, where feature extraction is automatic, which can be again, confirmed from Table 4.

Classification has multiple metrics. The ones utilized in this research have been are $Precision = \frac{TP}{TP+FP}$, $Recall = \frac{TP}{TP+FN}$, $F1 = (1 + \beta^2)\frac{P*R}{\beta^2*P+R}$ and $Accuracy = \frac{TP+TN}{TP+FP+TN+FN}$, where, TP, TN, FP and FN are *true positives*, *true negatives*, *false positives* and *false negatives* respectively. P denotes precision and R denotes recall metric. A decent value of β is 0.9 as suggested in [5]. These metrics have been tabulated below (Table 2) for comparison (Fig. 5).

Table 4. A mathematical comparison of algorithms for pose classification. The Inception-v3 architecture seems to outperform all the mentioned classifiers, which could be attributed to the fact that Inception-v3 takes into account the outcomes of all 5×5, 3×3 and 1×1 convolutions.

Architectures	Evaluation metrics and performances			
	Accuracy	*Precision*	*Recall*	F1
CNN BaseLine	0.8844	0.8231	0.8542	0.8754
AlexNet [18]	0.8987	0.8564	0.8765	0.8896
ResNet-50 [20]	0.9458	0.8996	0.8945	0.9456
VGG-16 [19]	0.9584	0.9231	0.9227	0.9418
VGG-19 [19]	0.9655	0.8833	0.9285	0.9321
MobileNet-v1 [15]	0.9287	0.8845	0.8561	0.9333
MobileNet-v2 [8]	0.9354	0.9121	0.9145	0.9444
Inception-v3 [16]	0.9856	0.9213	0.9554	0.9645

As we can see that for pose classification (a ternary classification problem), Inception-v3 gives the best accuracy. However, this task has only 3 classes as mentioned previously, with significant inter-class variations. Therefore, using a huge network for a 3-class classification problem seems to be sub-optimal in terms of evaluation speed (FPS). Hence, we choose a smaller model (MobileNet-v2) for this task which again gives close enough accuracy but is way better in terms of real-time performance, as also suggested in [8] (Fig. 6).

Choosing Inception-v3 for scene classification makes sense because unlike pose classification, we have 40 classes here. Furthermore, image scenarios can have multiple intra-class variations which can significantly affect the performance of smaller models like MobileNet. Hence, it would be sub-optimal to go for speed rather than accuracy for scene classification, as we need the system to accurately predict the relative semantic dynamics of images. This is the underlying reason for choosing a deep perceptual network.

Fig. 6. Quantitative analysis

Though accuracy can be evaluated for a multi-class classifier depending upon the correctness of the outputs, rest of the metrics are defined for binary classification usually. Hence, in order to evaluate F1, Recall and Precision, we evaluated these for every class instance separately and tabulated the average values for each. For example, we evaluate the three metrics for *smoking* vs all others, and we get the results for *smoking* category, and then we average all the results (Table 5).

Table 5. A mathematical comparison of algorithms for scene classification. The Inception-v3 architecture seems to outperform all the mentioned classifiers once again.

Architectures	Evaluation metrics and performances			
	Accuracy	*Precision*	*Recall*	F1
CNN BaseLine	0.7865	0.7536	0.7123	0.7755
AlexNet [18]	0.8131	0.7737	0.7548	0.8177
ResNet-50 [20]	0.9252	0.8317	0.8229	0.9187
VGG-16 [19]	0.8528	0.7911	0.7224	0.9275
VGG-19 [19]	0.8455	0.7833	0.7785	0.9144
OverFeat [21]	0.9187	0.8564	0.7912	0.9433
MobileNet-v2 [8]	0.8879	0.8798	0.8725	0.8844
Inception-v3 [16]	0.9356	0.9213	0.9054	0.9645

The CNN-base lines for pose and scene classification were derived from [18], with a simple sequential architecture. The model consists of 2 convolutional layers with max-pooling operations, flattening layer followed by 2 fully connected layers for feature-representation.

6   Discussion

HARM implementation attempts to recognize and identify pose variations and actions of humans in static images. The overall inference time for a single image is 1.423 s on an average. This clearly indicates that the module is yet to be optimized for real-time performance.

Furthermore, HARM needs to incorporate the case where multiple people are present in images. This case was tackled using state-of-the-art person detection algorithms [10–13], but there were inconsistencies in outputs. The actions which are related to close proximity of humans, such as smoking, applauding, playing guitar etc. can be easily identified as the scene associated with the human is within vicinity. However, other actions such as cleaning dishes, feeding horses etc. require the system to semantically understand the regions around the human beings as well, which are not incorporated in generic person/object detection algorithms. Human detection algorithms localize humans, disregarding other visual features present in the image. So the research presented in this article could be used as a starting point to create a more generic activity recognition module.

HARM is not yet customized for understanding human actions in videos. This is another feature that could be included in further developments of the algorithm. Understanding video feeds requires temporal abstractions along-with perceptual features to accurately recognize and predict the actions of people. This can be considered as a specific application of generic image captioning [22, 23], where the objective is to develop a semantic understanding of the image and the actions visualized in it.

References

1. Wei, S.-E., Ramakrishna, V., Kanade, T., Sheikh, Y.: Convolutional pose machines. In: Proceedings of the IEEE Conference on Computer Vision and Pattern Recognition, pp. 4724–4732 (2016)
2. Ramakrishna, V., Munoz, D., Hebert, M., Bagnell, J.A., Sheikh, Y.: Pose machines: articulated pose estimation via inference machines. In: European Conference on Computer Vision, pp. 33–47. Springer, Cham (2014)
3. Long, J., Shelhamer, E., Darrell, T.: Fully convolutional networks for semantic segmentation. In: Proceedings of the IEEE Conference on Computer Vision and Pattern Recognition, pp. 3431–3440 (2015)
4. Bengio, Y., Simard, P., Frasconi, P.: Learning long-term dependencies with gradient descent is difficult. IEEE Trans. Neural Netw. 5(2), 157–166 (1994)
5. Bradley, D.M.: Learning in modular systems. No. CMU-RI-TR-09-26. Carnegie-Mellon University, Pittsburgh, PA Robotics Institute (2010)
6. Glorot, X., Bengio, Y.: Understanding the difficulty of training deep feedforward neural networks. In: Proceedings of the Thirteenth International Conference on Artificial Intelligence and Statistics, pp. 249–256 (2010)

7. Andriluka, M., Pishchulin, L., Gehler, P., Schiele, B.: 2D human pose estimation: new benchmark and state of the art analysis. In: Proceedings of the IEEE Conference on computer Vision and Pattern Recognition, pp. 3686–3693 (2014)

8. Sandler, M., Howard, A., Zhu, M., Zhmoginov, A., Chen, L.-C.: MobileNetv2: inverted residuals and linear bottlenecks. In: Proceedings of the IEEE Conference on Computer Vision and Pattern Recognition, pp. 4510–4520 (2018)

9. Yao, B., Jiang, X., Khosla, A., Lin, A.L., Guibas, L.J., Fei-Fei, L.: Human action recognition by learning bases of action attributes and parts. In: International Conference on Computer Vision (ICCV), Barcelona, Spain, 6–13 November 2011

10. Khraief, C., Benzarti, F., Amiri, H.: Multi person detection and tracking based on hierarchical level-set method. In: Tenth International Conference on Machine Vision (ICMV 2017), vol. 10696, p. 106960G. International Society for Optics and Photonics (2018)

11. Zhang, S., Wen, L., Bian, X., Lei, Z., Li, S.Z.: Single-shot refinement neural network for object detection. In: Proceedings of the IEEE Conference on Computer Vision and Pattern Recognition, pp. 4203–4212 (2018)

12. Zhang, Z., Qiao, S., Xie, C., Shen, W., Wang, B., Yuille, A.L.: Single-shot object detection with enriched semantics. In: Proceedings of the IEEE Conference on Computer Vision and Pattern Recognition, pp. 5813–5821 (2018)

13. Womg, A., Shafiee, M.J., Li, F., Chwyl, B.: Tiny SSS: a tiny single-shot detection deep convolutional neural network for real-time embedded object detection. In: 2018 15th Conference on Computer and Robot Vision (CRV), pp. 95–101. IEEE (2018)

14. Sandler, Mark, Andrew Howard, Menglong Zhu, Andrey Zhmoginov, and Liang-Chieh Chen. "Mobilenetv2: Inverted residuals and linear bottlenecks." In Proceedings of the IEEE Conference on Computer Vision and Pattern Recognition, pp. 4510–4520. 2018

15. Howard, A.G., Zhu, M., Chen, B., Kalenichenko, D., Wang, W., Weyand, T., Andreetto, M., Adam, H.: MobileNets: efficient convolutional neural networks for mobile vision applications. arXiv preprint arXiv:1704.04861 (2017)

16. Szegedy, C., Vanhoucke, V., Ioffe, S., Shlens, J., Wojna, Z.: Rethinking the inception architecture for computer vision. In: Proceedings of the IEEE Conference on Computer Vision and Pattern Recognition, pp. 2818–2826 (2016)

17. Szegedy, C., Liu, W., Jia, Y., Sermanet, P., Reed, S., Anguelov, D., Erhan, D., Vanhoucke, V., Rabinovich, A.: Going deeper with convolutions. In Proceedings of the IEEE Conference on Computer Vision and Pattern Recognition, pp. 1–9 (2015)

18. Krizhevsky, A., Sutskever, I., Hinton, G.E.: Imagenet classification with deep convolutional neural networks. In: Advances in Neural Information Processing Systems, pp. 1097–1105 (2012)

19. Simonyan, K., Zisserman, A.: Very deep convolutional networks for large-scale image recognition. arXiv preprint arXiv:1409.1556 (2014)

20. He, K., Zhang, X., Ren, S., Sun, J.: Deep residual learning for image recognition. In: Proceedings of the IEEE Conference on Computer Vision and Pattern Recognition, pp. 770–778 (2016)

21. Sermanet, P., Eigen, D., Zhang, X., Mathieu, M., Fergus, R., LeCun, Y.: Overfeat: integrated recognition, localization and detection using convolutional networks. arXiv preprint arXiv:1312.6229 (2013)

22. Vinyals, O., Toshev, A., Bengio, S., Erhan, D.: Show and tell: a neural image caption generator. In: Proceedings of the IEEE Conference on Computer Vision and Pattern Recognition, pp. 3156–3164 (2015)

23. Xu, K., Ba, J., Kiros, R., Cho, K., Courville, A., Salakhudinov, R., Zemel, R., Bengio Y.: Show, attend and tell: neural image caption generation with visual attention. In: International Conference on Machine Learning, pp. 2048–2057 (2015)
24. Masoud, O., Papanikolopoulos, N.: A method for human action recognition. Image Vis. Comput. **21**(8), 729–743 (2003)
25. Blackburn, J., Ribeiro, E.: Human motion recognition using isomap and dynamic time warping. In: Workshop on Human Motion, pp. 285–298. Springer, Berlin (2007)
26. Jia, K., Yeung, D.-Y.: Human action recognition using local spatio-temporal discriminant embedding. In: 2008 IEEE Conference on Computer Vision and Pattern Recognition, pp. 1–8. IEEE (2008)
27. Wang, L., Suter, D.: Learning and matching of dynamic shape manifolds for human action recognition. IEEE Trans. Image Process. **16**(6), 1646–1661 (2007)
28. Oikonomopoulos, A., Patras, I., Pantic, M.: Spatiotemporal salient points for visual recognition of human actions. IEEE Trans. Syst. Man Cybern. Part B (Cybern.) **36**(3), 710–719 (2005)
29. Feng, X., Perona, P.: Human action recognition by sequence of movelet codewords. In: Proceedings of First International Symposium on 3D Data Processing Visualization and Transmission, pp. 717–721. IEEE (2002)
30. Weinland, D., Boyer, E., Ronfard, R.: Action recognition from arbitrary views using 3D exemplars (2007)
31. Sminchisescu, C., Kanaujia, A., Metaxas, D.: Conditional models for contextual human motion recognition. Comput. Vis. Image Underst. **104**(2–3), 210–220 (2006)
32. Wang, L., Suter, D.: Recognizing human activities from silhouettes: motion subspace and factorial discriminative graphical model. In: 2007 IEEE Conference on Computer Vision and Pattern Recognition, pp. 1–8. IEEE (2007)
33. Zhang, J., Gong, S.: Action categorization with modified hidden conditional random field. Pattern Recogn. **43**(1), 197–203 (2010)
34. Shi, Q., Wang, L., Cheng, L., Smola, A.: Discriminative human action segmentation and recognition using semi-Markov model. In: 2008 IEEE Conference on Computer Vision and Pattern Recognition, pp. 1–8. IEEE (2008)
35. Zhang, P., Lan, C., Xing, J., Zeng, W., Xue, J., Zheng, N.: View adaptive neural networks for high performance skeleton-based human action recognition. IEEE Trans. Pattern Anal. Mach. Intell. (2019)
36. Saggese, A., Strisciuglio, N., Vento, M., Petkov, N.: Learning skeleton representations for human action recognition. Pattern Recogn. Lett. **118**, 23–31 (2019)
37. Jalal, A., Kamal, S., Azurdia-Meza, C.A.: Depth maps-based human segmentation and action recognition using full-body plus body color cues via recognizer engine. J. Electr. Eng. Technol. **14**(1), 455–461 (2019)
38. Weiyao, X., Muqing, W., Min, Z., Yifeng, L., Bo, L., Ting, X.: Human action recognition using multilevel depth motion maps. IEEE Access **7**, 41811–41822 (2019)
39. Imran, J., Raman, B.: Evaluating fusion of RGB-D and inertial sensors for multimodal human action recognition. J. Ambient Intell. Humanized Comput. 1–20 (2019)
40. Nazir, S., Yousaf, M.H., Nebel, J.-C., Velastin, S.A.: Dynamic spatio-temporal bag of expressions (D-STBoE) model for human action recognition. Sensors **19**(12), 2790 (2019)
41. Blank, M., Gorelick, L., Shechtman, E., Irani, M., Basri, R.: Actions as space-time shapes. In: Tenth IEEE International Conference on Computer Vision (ICCV 2005) Volume 1, vol. 2, pp. 1395–1402. IEEE (2005)

42. Weinland, D., Ronfard, R., Boyer, E.: Free viewpoint action recognition using motion history volumes. Comput. Vis. Image Underst. **104**(2–3), 249–257 (2006)
43. Rodriguez, M.D., Ahmed, J., Shah, M.: Action MACH a spatio-temporal maximum average correlation height filter for action recognition. In: CVPR, vol. 1, no. 1, p. 6 (2008)
44. Laptev, I., Marszałek, M., Schmid, C., Rozenfeld, B.: Learning realistic human actions from movies, June 2008

Industrial Applications

Automatic Extraction of Marbling Measures Using Image Analysis, for Evaluation of Beef Quality

C. M. R. Caridade[1]($\boxtimes$), C. D. Pereira[2], A. F. Pires[2], N. G. Marnotes[2], and J. F. Viegas[2]

[1] Polytechnic of Coimbra - Institute of Engineering, Coimbra, Portugal
caridade@isec.pt
[2] Polytechnic of Coimbra - College of Agriculture, Coimbra, Portugal
{cpereira,jviegas}@esac.pt, arona_pires@hotmail.com.pt,
nataligarcia.mar@gmail.com

Abstract. In beef, the marbling flecks are determinant regarding to flavor and tenderness. Beef cuts with higher levels of marbling are likely to be tenderer, flavor full and juicier than cuts with lower marbling levels [1]. Therefore, the USDA (United States Department of Agriculture) quality grade standards are based on subjective evaluation of the relative degree of visible intramuscular fat [2]. There are 10 official marbling classes: devoid (D) practically devoid (PD), traces (TR), slight (SL), small (SM), modest (MT), moderate (MD) slightly abundant (SA), moderately abundant (MA) and abundant (AB). So, there is a growing interest in developing methods and techniques for beef quality classification using digital images. This paper presents an automatic methodology, based on image processing techniques, to identify and to locate the beef in the image and to calculate the marbling measures, in order to evaluate the beef quality. The tests realized showed that it is possible to use image analysis in color photographs of beef steaks to automatically extract the marbling measures and to identify the beef quality. However, to develop more accurate algorithms, larger training and testing sets must be used.

1 Introduction

Marbling is the definitive factor in establishing the quality of the meat. It refers to the white streaks and flecks of fat within the lean sections of meat, not considering the fat that surrounds the meat which can be trimmed off (Fig. 1). Marbling refers to the intramuscular fat and is one of the main factors influencing the flavour of meat, and in most cases, the more marbling, the better the quality. Marbling fat melts at a temperature lower than other fats, resulting in increased tenderness and flavor.

It is reported that marbling affects taste [4,5], to have a significant relationship to consumer acceptance of beef steaks [6] and that marbling measures taken on one muscle can be predictive of marbling in other muscles of the same carcass

© Springer Nature Switzerland AG 2019
J. M. R. S. Tavares and R. M. Natal Jorge (Eds.): VipIMAGE 2019, LNCVB 34, pp. 437–446, 2019.
https://doi.org/10.1007/978-3-030-32040-9_44

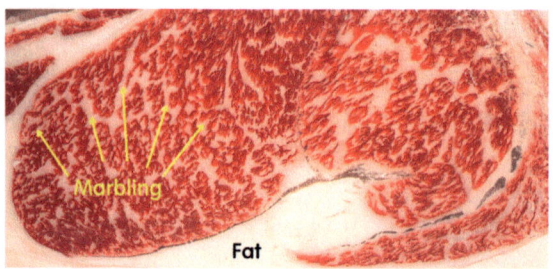

Fig. 1. Marbling flecks in beef steak [3].

[7]. Many studies have been done in recent years on the relationship between the quality of steaks and their degree of marbling. However, marbling evaluation is normally done by trained assessors or by comparison to patterns. Although well trained assessors are able to accurately classify the meat, there is a growing interest in the development of automatic measurement systems such as the use of digital images as a test for beef quality classification [8,9]. The purpose of this work was to develop an automatic methodology, based on image processing techniques, to identify and to locate the beef in the image and to calculate the marbling measures in view of beef quality evaluation.

2 Methodology

Direct photographs were taken on a smooth and homogenous background picking up the beef steak. The images were stored in jpeg format with 3072×4608 pixels size. Labels with the type of piece and place of collection are also sometimes added to the photograph to facilitate their identification. Figure 2 shows examples of some beef steaks images.

Images were processed with algorithms developed in Matlab Version 7 software (The Math-work, Natick, MA, USA). The method developed has four separate stages:

1. Transformation of the RGB image in HSV image;
2. Identification of the binary images of HSV channels;
3. Pre-processing in the HSV channels;
4. Detection of the fat and the muscle area.

2.1 Transformation of the RGB Image in HSV Image

The original image in RGB system can be converted to HSV system and then in three HSV separate channels. The separation of the HSV channels is used to analyse the components of the colors of each portion of an image [10].

In Fig. 3 it is possible to see the test RGB image I_4 (top), the HSV corresponding image (center) and the image representation of the 3 different channels

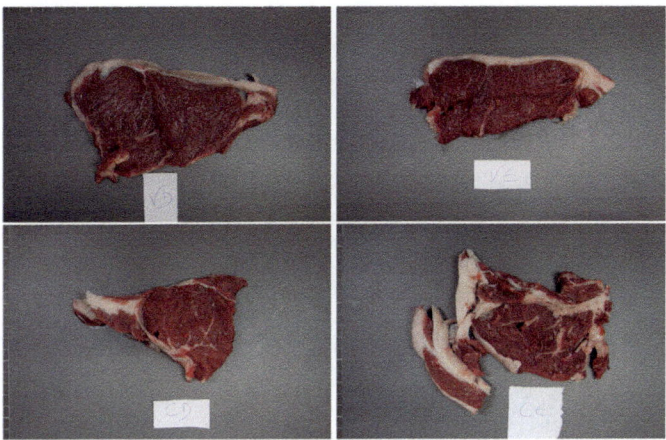

Fig. 2. Examples of beef steaks images obtained from direct photographs: I_1 (top left); I_2 (top right); I_3 (bottom left); I_4 (bottom right).

(bottom). The hue, saturation and value channels images are greyscale images in which the only colors are shades of grey represented by $H(x,y)$, $S(x,y)$ and $V(x,y)$ respectively.

2.2 Identification of the Binary Images of HSV Channels

The HSV channels in greyscale are converted to a binary image using segmentation by thresholding operation with the threshold value obtained by the Otsu method [11]. The creation of a new binary image was defined using values between the found threshold (T) and 255 (white color in binary image). A threshold selection method from all grey images is given by Eq. 1,

$$B(x,y) = \begin{cases} 1, & \text{if } T \leq HSV(x,y) \leq 255 \\ 0, & \text{if } HSV(x,y) \leq T \end{cases} \tag{1}$$

where T is the Otsu's threshold value and $HSV(x,y)$ is represented by $H(x,y)$, $S(x,y)$ and $V(x,y)$ for the hue, saturation and value channels respectively. The binary images defined between these values for each of the three channels are represented in Fig. 4.

In these images we can see more clearly that the fat (internal and external) is more visible in the value channel whereas in the saturation channel the definition of the muscles is the most highlighted region. In this way, these two channels are the most important for the detection of fat and muscles in meat.

2.3 Pre-processing in the HSV Channels

The next step is to select the interest area (beef and fat area), using the binary images of the three channels of the HSV image. To eliminate the label present

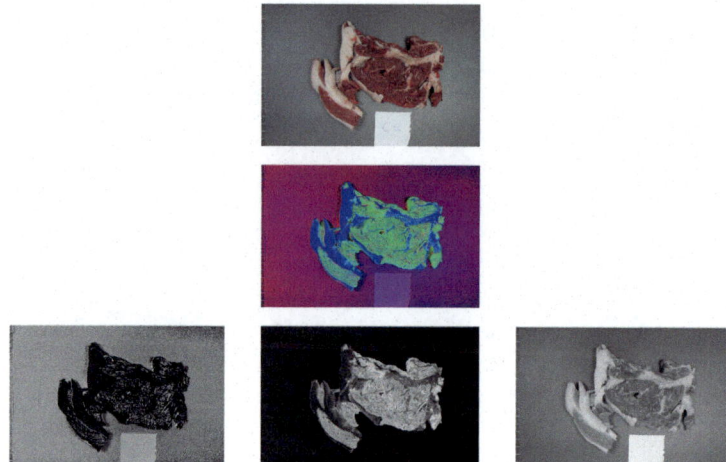

Fig. 3. Image representation I_4: RGB (top); HSV (center); hue, saturation and value channels (bottom).

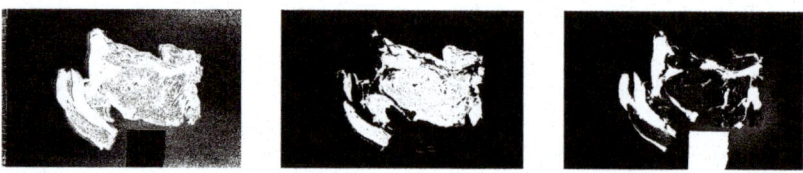

Fig. 4. Binary image I_4: hue channel with $T = 0.3098$ (left); saturation channel with $T = 0.3255$ (center); value channel with $T = 0.5882$ (right).

in the image (small white rectangle in binary images - Fig. 4 right) the value channel was used. In this channel the label is well delimited in white colour on the black background. The use of the binary value channel image of the Fig. 4 was necessary to remove the noise. A morphological operator erosion was applied to the value channel image (Fig. 5 top left), using as structuring element a disk of 20 pixels radius [10]. The resulting image has fewer objects and the label present in this image is more clearly defined (Fig. 5 top right). Then all the objects in the image were identified and those whose area is less than 5% of the total area of the beef (small objects) were deleted as shown in Fig. 5 below in the left side.

Finally, all objects still in the image were identified through their bounding box (yellow rectangles in Fig. 5 below right). The bounding box of the label will be the one whose difference between the area of the bounding box and the area of the object is less than 0.05% (represented in red). This process automatically identifies the label object and then the label position in the image. Finally the label is eliminated in all channels as shown in Fig. 6.

After elimination of the label on each HSV channel of the image, a pre-processing is performed which consists of eliminating noise and undesirable

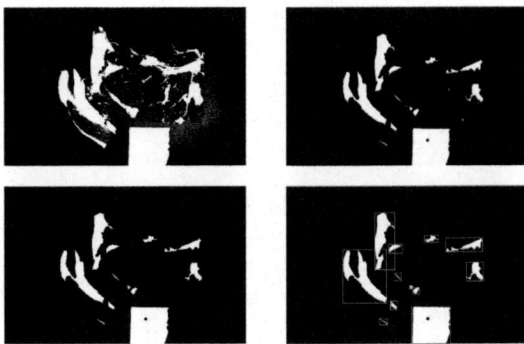

Fig. 5. Automatic detection of the label position in the value channel image: binary value channel (top left), noise removal (top right), small objects removal (bottom left), bounding box objects (bottom right).

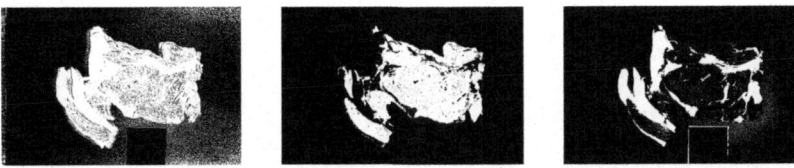

Fig. 6. Elimination of the label in the three HSV channels: hue channel (left), saturation channel (center) and value channel (right).

objects, highlighting the region of the muscles and fat that are to be identified in the beef image (Fig. 7). In the value channel 3, erosion with disc structuring element of 2 pixels radius was applied 3 consecutive times to eliminate the noise. Finally, objects with an area less than 0.5% of the total area defined in the value image are eliminated. In the saturation channel the morphological close operator was applied with a disk-shaped structuring element with 2 pixels radius. Then all objects whose areas are less than 0.1% of the area defined in the saturation image were removed. Channel hue image is not used at this stage for the detection of fat and muscle in beef steak.

2.4   Detection of the Fat and the Muscle Area

Now, in the value channel image, it is possible to identify the external fat of the beef steak. The objects (external fat) present in the value channel image after pre-processing (Fig. 7 right) are identified and segmented. In Fig. 8, at the center, are highlighted the contours of these objects (in yellow) over the original image. At the right side only the cut out objects of the original image are visualized.

The beef muscle is identified in the saturation channel after the pre-processing applied to this channel. In Fig. 8, mid row, the muscle is defined as blue color over the original image (center) and cropped from the original image (right). With respect to marbling flecks (internal fat), they are identified by the image

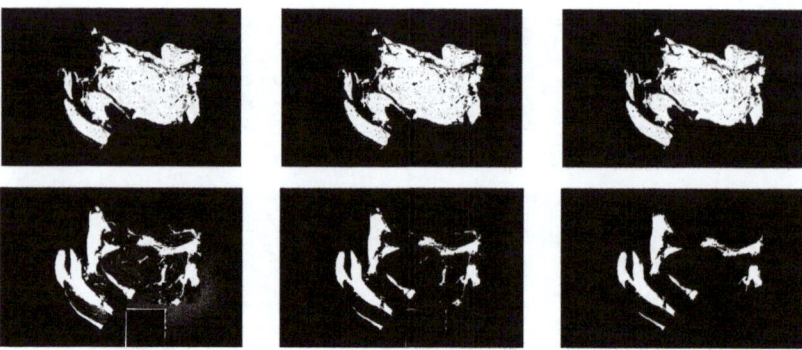

Fig. 7. The saturation (top) and value (bottom) channels after applying the pre-processing techniques: without label (left); without noise (center) and without small objects (right).

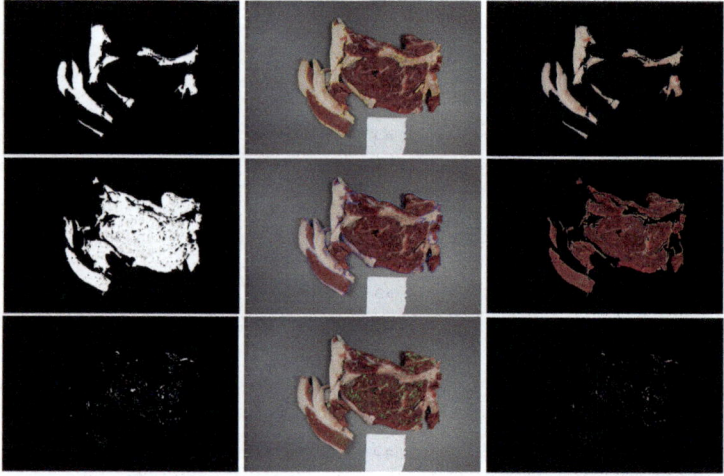

Fig. 8. Detection and segmentation of the external fat (top), muscle (center) and internal fat (bottom). Binary image after pre-processing (left); segmentation over original image (center); original image segmented (right) - I_4

obtained subtracting the external fat (Fig. 8 top row left) to the muscle (Fig. 8 mid row left) image. Then, all the objects present in this image are filled and finally the muscle image is subtracted again from the resulting image.

In Figs. 9, 10 and 11 are presented the results obtained for the images I_1, I_2 and I_3 (Fig. 2) respectively.

3   Results

To specify the area of muscle, the external fat and the marbling flecks, some features are automatically calculated in the digital image. A range of marbling

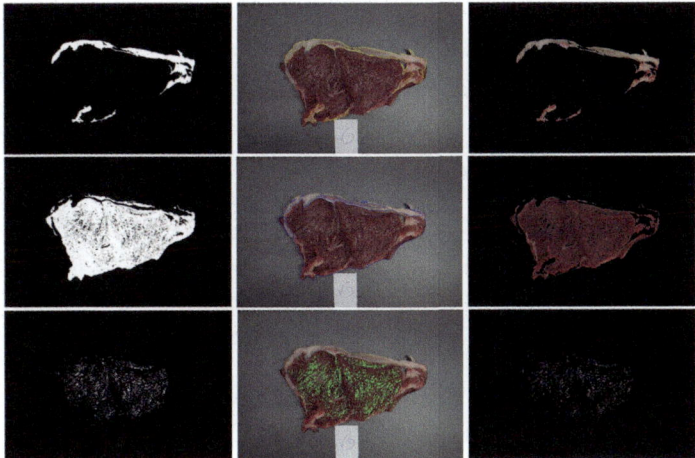

Fig. 9. Detection and segmentation of the external fat (top), muscle (center) and internal fat (bottom). Binary image after pre-processing (left); segmentation over original image (center); original image segmented (right) - I_1.

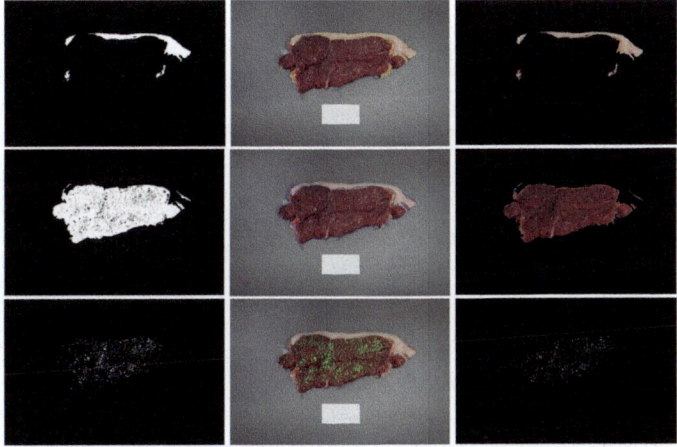

Fig. 10. Detection and segmentation of the external fat (top), muscle (center) and internal fat (bottom). Binary image after pre-processing (left); segmentation over original image (center); original image segmented (right) - I_2.

measures was calculated. In Table 1 the values obtained for the image I_4 are represented. In this case, the beef has 1, 940, 561 pixels and the fat 598, 801 pixels. So this beef has 30.8% fat. Most of this fat is external fat (89%) while internal fat (rib eye fat) represents only 11%. The internal fat is defined by the amount and size of the marble. Thus, in this example there are 5 large marbled areas with a total area of 17, 261 pixels (26.3%) and 3, 300 small marbled areas with a total area of 48, 359 pixels (73.7%).

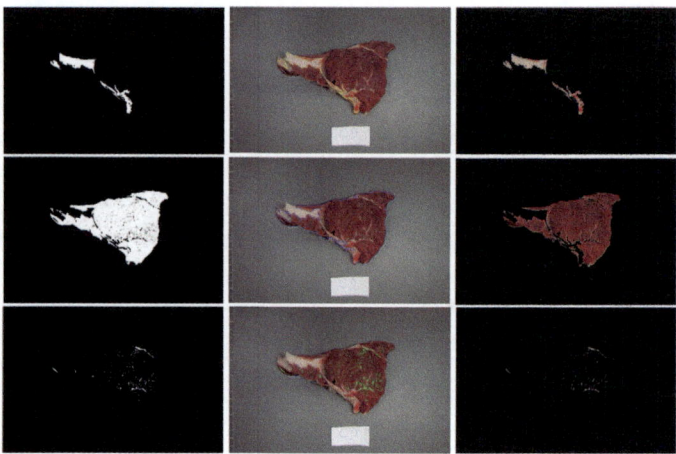

Fig. 11. Detection and segmentation of the external fat (top), muscle (center) and internal fat (bottom). Binary image after pre-processing (left); segmentation over original image (center); original image segmented (right) - I_3.

Table 1. Beef features for image I_4.

Beef features	Definition	Value (in pixels)
Beef area	Beef area with external fat	$1,940,561$
Rib eye area	Beef area without external fat	$1,407,380$
Rib eye fat area	Fat area in rib eye	$65,620$
Largest marbling area	Large marbled area	$17,261$
Small marbling area	Small marbled area	$48,359$
External fat area	Area of external fat	$533,181$
Total fat area	Area of total fat	$598,801$

With these features some measures were made based in the procedure proposed by Kuchida [12] and developed by Nakahashi [13]. A range of marbling measures was calculated:

- The fat area ratio (FAR) was calculated by dividing all pixels of the fat rib eye image by the rib eye area.

$$FAR = \frac{\text{Rib eye fat area}}{\text{Rib eye area}}, \tag{2}$$

- Coarseness of minimum marbling particles (CmM) was calculated by dividing small marbling area (with areas $\leq 0.1\%$ of the rib eye area) by the rib eye fat area.

$$CmM = \frac{\text{Small marbling area}}{\text{Rib eye fat area}}, \tag{3}$$

- Coarseness of maximum marbling particles (CMM) was calculated by dividing pixels of the largest marbling particle by all pixels of the rib eye fat area. The largest marbling particles have areas with more than 0.1% of the rib eye area.

$$CMM = \frac{\text{Largest marbling area}}{\text{Rib eye fat area}},$$ (4)

- Number of small flecks (NSF) was the number of marbling small flecks [3].

$$NSF = \text{Number of small marbling flecks}.$$ (5)

Table 2 shows the results obtained by applying these measures to the proposed examples.

Table 2. Measures based in [12] for images I_1, I_2, I_3 and I_4

Measures	Value - I_1	Value - I_2	Value - I_3	Value - I_4
FAR	8.0%	7.2%	3.6%	4.7%
CmM	0.0%	83.2%	80.6%	73.7%
CMM	100.0%	16.8%	19.4%	26.3%
NSF	5514	3923	1793	3300

4   Conclusions

This study shows that it is possible to use image analysis in color photographs of beef steaks to automatically extract the marbling measures and to identify the beef quality. However, to develop more accurate algorithms, larger training and testing sets must be used. In future we intend to evaluate and compare the marbling measured automatically using image analysis with the conventional AOAC procedure and near Infrared analysis (NIR) to quantify the marbling particle fineness. It will also be investigated the relationship between the fineness of marbling and sensory evaluation.

Acknowledgements. This project is supported by national funds through the ministry of Agriculture and Rural Development and co-financed by the European Agricultural Fund for Rural Development (EAFRD), through the partnership agreement Portugal2020 - PDR, under the project PDR2020-101-030748: Valor Jarmelista.

References

1. Lepper-Blilie, A.N., Berg, E.P., Buchanan, D.S., Berg, P.T.: Effects of post-mortem aging time and type of aging on palatability of low marbled beef loins. Meat Sci. **112**, 63–68 (2016)

2. United States Department of Agriculture. https://www.usda.gov/media/blog/2013/01/28/whats-your-beef-prime-choice-or-select
3. A lesson in steak cuts. https://www.afoodieworld.com/jenpaolini/7277-a-lesson-in-steak-cuts
4. Wood, J., Richardson, R., Nute, G., Fisher, A., Campo, A., Kasapidou, E., Enser, M.: Effects of fatty acids on meat quality: a review. Meat Sci. **66**(1), 21–32 (2004)
5. Calkins, C., Hodgen, J.: A fresh look at meat flavour. Meat Sci. **77**(1), 63–80 (2007)
6. Platter, W.J., Tatum, J.D., Belk, K.E., Chapman, P.L., Scanga, J.A., Smith, G.C.: Relationship of consumer sensory ratings, marbling score, and shear force value to consumer acceptance of beef strip loin steaks. J. Anim. Sci. **81**, 2741–2750 (2003)
7. Konarska, M., Kuchida, K., Tarr, G., Polkinhorne, R.J.: Relationships between marbling measures across principal muscles. Meat Sci. **123**, 67–78 (2016)
8. Lee, B., Yoon, S., Choi, Y.M.: Comparison of marbling fleck characteristics between beef marbling grades and its effect on sensory quality characteristics in high-marbled Hanwoo steer. Meat Sci. **152**, 109–115 (2019)
9. Schulz, L., Sundrum, A.: Assessing marbling scores of beef at the 10th rib vs. 12th rib of longissimus thoracis in the slaughter line using camera grading technology in Germany. Meat Sci. **152**, 116–120 (2019)
10. Gonzalez, R.C., Woods, R.E.: Digital Image Processing, 2nd edn. Prentice Hall International Inc., Upper Saddle River (2002)
11. Otsu, N.: A threshold selection method from gray-level histograms. IEEE Trans. Syst. Man Cybern. **9**(1), 62–69 (1979)
12. Kuchida, K., Osawa, T., Hori, T., Kodaka, H., Maruyama, S.: Evaluation and genetics of carcass cross section of beef carcass by computer image analysis. Jpn. Dobutsu Iden Ikushu Kenkyu **34**, 45–52 (2006)
13. Nakahashi, N., Maruyama, S., Seki, S., Hidaka, S., Kuchida, K.: Relationships between monounsaturated fatty acids of marbling flecks and image analysis traits in longissimus muscle for Japanese Black steers (2014)

Virtual Reality

Immersive Multiple Object Tracking Using Virtual Reality
Pilot Study on Athletes and Non Sport-Practicing Subjects

Astrid Kibleur[1]([✉]), Morgan Le Chénéchal[1], Nicolas Benguigui[2],
and Jonas Chatel-Goldman[1]

[1] Open Mind Innovation, Caen, France
{astrid,morgan,jonas}@omind.me
[2] Université de Caen Normandie, Caen, France
nicolas.benguigui@unicaen.fr

Abstract. Virtual Reality (VR) offers new ways to propose visual immersive stimuli to subjects and is particularly relevant in sports studies. Multiple Object Tracking (MOT) is highly relevant to measure visuomotor coordination performance but it is currently mainly based on flat display, which partly restricts depth perception and natural visual exploration. In this pilot study, we proposed and tested a protocol using the MOT task in a VR environment. In this experiment, the immersive level of the task was split in three conditions: 'far' from the objects to track, 'close' and 'inside' (ie, the objects were moving around the subject in this specific condition). This protocol was tested in two participants groups: an athlete group and a non athlete group. Our results show that there are effects of sport practice on the task performance, but that these effects were reduced in the 'inside' condition. This suggests that the 'inside' condition, that we introduced here for the MOT task, requires skills that are not necessarily linked to the sport practice.

1 Introduction

Multiple object tracking (MOT), is a task classically used in sports studies [1,7]. In this task, the subject has to follow visual targets that move amongst distractors. The goal is to focus on the target tracks and positions. In sport studies, it was used to show that basketball athletes perform better than non athletes in this task when there are at least 3 targets to track and this effect was related to a reduced activation in task related brain areas (athletes needed less focused attention to track targets) [6]. To make the MOT more immersive than in 2D, 3D MOT (displayed on a 3D TV) was introduced and it was shown to be a good tool to evaluate perceptual capacity of professional athletes after a concussion [1]. Virtual reality (VR) provides an ideal way of measuring movements in 3D because it is ecological and immersive, hence, the movements done in a VR environment are much closer to real world movements than when stimuli are

© Springer Nature Switzerland AG 2019
J. M. R. S. Tavares and R. M. Natal Jorge (Eds.): VipIMAGE 2019, LNCVB 34, pp. 449–453, 2019.
https://doi.org/10.1007/978-3-030-32040-9_45

presented on a 2D screen as is classically done. Therefore, it seems adequate for sport studies. MOT has been successfully used in VR for its entertainment effect to distract children from stress and pain in pediatric nephrology [4]. However, to our knowledge, MOT in VR were never used in sports studies. Moreover, we introduce a novel condition that becomes available within a VR environment where the subject is at the center of the box in which objects are moving: tracking occurs all around the subject instead of only in front of him as classically done (see Neurotracker software [2]). Here, we propose a protocol of MOT in VR with 3 different levels of immersion and we tested this protocol in subjects with different level of sport practice. The purpose was to test if our design could discriminate both population (sport-practicing and non sport-practicing groups) and study the behavior of the subjects in this task. Then, building on this pilot study, we will refine the design and work on a larger sample size to measure visuo-motor performance (through multiple object tracking ability) and their relation to sport practice.

2   Materials and Methods

22 participants (21.7 ± 1.3 yo, 7 women, 16 right handed) were included in this experiment: 11 were athletes and 10 were non high level sport-practicing subjects. Participants were presented with a VR MOT task using the HTC Vive Pro[1].

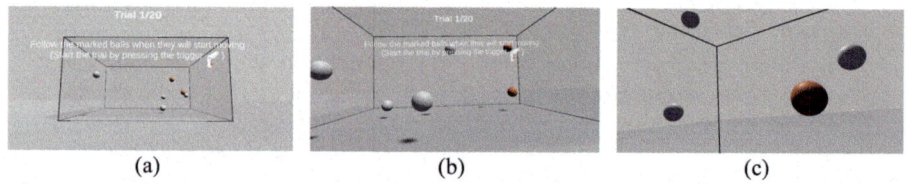

| (a) | (b) | (c) |

Fig. 1. The three conditions used in our MOT task: (a) far, (b) close, and (c) inside the box.

The goal was to track 2 targets amongst 8 spheres. The 2 target spheres were highlighted in orange at the beginning of each trial then they turned back to grey (as the 6 other spheres) and the 8 spheres started to move in a defined parallelepiped (see Fig. 1) with a constant velocity. At some point (defined by the total length of the spheres' path kept constant in each trial at 18 m) they stopped, and the participant was asked to point the 2 target spheres. The task was split in 3 blocks with 3 different immersion conditions: 'far' (at 3 m) from the box with the moving spheres, 'close' (at 0.6 m) to the box and 'inside' (in the center of) the box. The 3 immersion conditions' order was randomized across participants. In each immersion condition block, 20 trials were presented

---

[1] https://www.vive.com/fr/product/vive-pro/.

with an adaptive difficulty: the speed of the moving spheres was increased and decreased by $0.2\,\text{m.s}^{-1}$ after respectively a successful trial (2 spheres accurately detected) and a failed trial, bounded between 0.4 and $3.6\,\text{m.s}^{-1}$. Moreover, two consecutive successful trials made the speed increased twice faster. The average distance between the two targets across all trials was bounded between 2.5 and 3.5 m.

After each trial, the participant was asked to quote his performance, meaning, on a visual analog scale (graduated from 0 to 2), he indicated how many spheres he thought he pointed accurately. This was used to compute a metacognitive sensitivity score: the metacognitive Phi (described in [3]).

The comparisons between the professional sport-practicing group and the non sport-practicing group were performed using permutation t-tests. We also performed some correlation analyses (Pearson's R) between the weekly amount of sport practice and the metrics from the task. The main performance metric used here is the sphere's speed. From the speed values in each trials in one session, we could compute 'speed curves' and fit sigmoid function to these curves to study the subject's progression dynamic.

3   Results

The metacognitive sensitivity in this task was slightly higher in non sport-practicing group (ϕ nopro) than in sport-practicing group (ϕ pro): $\phi pro = 0.54 \pm 0.13; \phi nopro = 0.62 \pm 0.10; p = 0.079$. When looking at the mean of the spheres speed across trials for each immersion conditions, there were no significant differences between the 2 groups. However, when looking at the correlations between the mean speed and the weekly hours of sport practice, there was a positive correlation in the 'far' immersion condition ($R = 0.39, p = 0.080$), meaning that the more sport the participant practiced every week, the better he was in the task in the 'far' immersion condition. When looking at the speed curves fits, we extracted the half maximum trial number (HMTN) which corresponds to the number of trials that the participant needed to reach a speed equal to half of his speed range. This value indicates the number of repetition that was necessary for a participant to reach half of his maximum performances in the task. This metric was different in the 2 groups in the 'far' and the 'close' immersion conditions. In the 'far' condition: HMTN nopro $= 5.4 \pm 1.6$ trials, HMTN pro $= 4.7 \pm 0.4$ trials, $p = 0.1$. In the 'close' condition: HMTN nopro $= 5.1 \pm 0.8$ trials, HMTN pro $= 4.7 \pm 0.4, p = 0.06$.

Finally, when studying the correlations between the HMTN and the weekly sport practice, we found an anti-correlation ($R = -0.42, p = 0.052$) between the weekly number of hours of sport practice and the HMTN in the 'far' condition, indicating that the more the participant practiced sport weekly, the less trials he required to improve his performance in this task (Fig. 2).

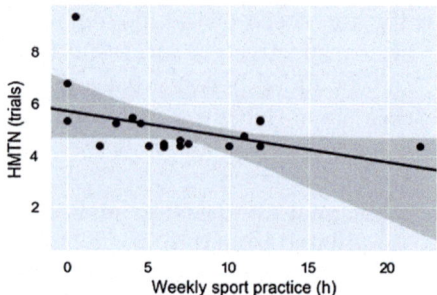

Fig. 2. Correlation between the weekly sport practice (in hours) and the half maximum trial number (HMTN) in the 'far' immersion condition ($R = -0.42, p = 0.05$). The HMTN represents the dynamics of the acquisition curve, the higher this value is, the more trials the subject needed to reach its maximum performance and reciprocally.

4   Discussion and Conclusion

In this pilot study, we tested an experimental design of multiple object tracking in virtual reality. We found that there was effects of sport practice on the participants' performance in this task. More precisely, the effects that we measured were specific to the 'far' and 'close' immersion conditions where we found that the amount of sport practice was related to higher performance in the 'far' condition and quicker adaptation to the 'far' and 'close' task conditions. The fact that we did not see clear effects in the 'inside' condition might be related to several effects: first, the cohort size is clearly a major limitation in this study as there were only 10 versus 11 participants in this new task design. Furthermore, the observed effects are small and this might be linked to the fact that we used only 2 targets, as it was shown in [5,6], that athletes and non athletes do not differ in the MOT when only 2 targets are used. We might also hypothesize that the performance in the 'inside' situation might be related to other factors than sport practice, as for instance, the use of immersive video games, the type of sport practiced which may modulate how well a subject tracks objects behind him or the novelty of virtual reality in immersion which might defocus the subject attention because of scene exploration (and multiple object tracking task are very dependent on attentional level).

References

1. Chermann, J.F., Romeas, T., Marty, F., Faubert, J.: Perceptual-cognitive three-dimensional multiple-object tracking task can help the monitoring of sport-related concussion. BMJ Open Sport Exerc. Med. **4**(1), e000384 (2018)
2. Corbin-Berrigan, L.A., Kowalski, K., Faubert, J., Christie, B., Gagnon, I.: Three-dimensional multiple object tracking in the pediatric population: the neurotracker and its promising role in the management of mild traumatic brain injury. NeuroReport **29**(7), 559–563 (2018)

3. Fleming, S.M., Lau, H.C.: How to measure metacognition. Front. Human Neurosci. **8**, 443 (2014)
4. Piskorz, J., Czub, M.: Effectiveness of a virtual reality intervention to minimize pediatric stress and pain intensity during venipuncture. J. Spec. Pediatr. Nurs. **23**(1), e12201 (2018)
5. Qiu, F., Pi, Y., Liu, K., Li, X., Zhang, J., Wu, Y.: Influence of sports expertise level on attention in multiple object tracking. PeerJ **6**, e5732 (2018)
6. Qiu, F., Pi, Y., Liu, K., Zhu, H., Li, X., Zhang, J., Wu, Y.: Neural efficiency in basketball players is associated with bidirectional reductions in cortical activation and deactivation during multiple-object tracking task performance. Biol. psychol. **144**, 28–36 (2019)
7. Romeas, T., Guldner, A., Faubert, J.: 3d-multiple object tracking training task improves passing decision-making accuracy in soccer players. Psychol. Sport Exerc. **22**, 1–9 (2016)

Using Augmented Reality in Patients with Autism: A Systematic Review

Anabela Marto[1,2], Henrique A. Almeida[1(✉)],
and Alexandrino Gonçalves[1]

[1] School of Technology and Management, Computer Science
and Communication Research Centre (CIIC),
Polytechnic Institute of Leiria, Leiria, Portugal
`henrique.almeida@ipleiria.pt`
[2] INESC-TEC, Porto, Portugal

Abstract. Augmented Reality (AR) technology has been used for a wide range of areas, uncovering diverse benefits regarding to its usage. Focusing on medicine, AR is being used for the last few decades in several different tasks – from surgery support to treatment of phobias and other cognitive or physiological disorders. This study intends to focus on the usage of AR in one specific pathology: the rehabilitation of people that suffer from autism. This systematic review will present an overview of the usage of AR technology, from a patients' perspective, where strengths and weaknesses of current approaches will be pointed out, to better understand good practices to follow when using AR for autism rehabilitation. The results from this systematic review will provide valuable information for further ideas and solutions regarding to enhance Autism Spectrum Disorder patients' wellbeing.

Keywords: Augmented Reality · Autism · Rehabilitation

1 Introduction

Augmented Reality (AR) technology, has been used for a wide range of areas, uncovering diverse benefits regarding to its usage, such as publicity [1], entertainment [2], education [3, 4], architecture [5], manufacturing [6], cultural heritage [7–9], or medicine [10–12]. Within each of these areas where we can find the use of AR and its features, several approaches are noticed with different purposes. Focusing on medicine, where AR is being used among the last few decades, gaining preponderance and with some visible results to enhance several different tasks – from surgery support [10] to treatment of phobias [13], this study intends to focus on the use of AR in one specific pathology: the rehabilitation of people that suffer from autism.

A simply literature review presents valuable systematic reviews of Virtual Reality (VR) effectiveness for particular health conditions. Regarding VR scenarios, a review on available studies that have used VR in the assessment and rehabilitation of given disabilities resulting from brain injury (such as executive dysfunction, memory impairments, spatial ability impairments, attention deficits, and unilateral visual neglect) was made, and the presented data validates VR for brain damage rehabilitation

© Springer Nature Switzerland AG 2019
J. M. R. S. Tavares and R. M. Natal Jorge (Eds.): VipIMAGE 2019, LNCVB 34, pp. 454–463, 2019.
https://doi.org/10.1007/978-3-030-32040-9_46

as an integral part of cognitive assessment and rehabilitation in the future [15]. Also covering the effect of VR on balance and gait ability in patients with Parkinson's disease was made and the presented data allows to state that the use of virtual reality enhances the balance of patients with Parkinson's disease [14]. Regarding to autism, a recent systematic review for autistic children and adolescents with VR-based treatments was performed [16]. Considering the use of VR technology, it was concluded that most of the studies are related to the improvement of activities in daily life and communication, especially social and emotional skills.

Following the VR approach for autism patients, the current review will present an overview of the usage of AR technology, from a patients' perspective, where strengths and weaknesses of current approaches will be pointed out, to better understand good practices to follow when using AR for autism rehabilitation.

Thus, based on the works identified for this systematic review, it is intended to identify the following issues: (1) which studies covered the usage of AR to improve autism patients' wellbeing; (2) how did they improve autism patients' lives – identifying the rehabilitation variables, such as social skills, learnability, or ability to complete tasks; (3) what AR technologies have been used; (4) what benefits were acquired; (5) and, what were the main limitations of the applied methodologies.

The results from this systematic review will provide valuable information for further ideas and solutions regarding to enhance ASD patients' wellbeing.

1.1 Autism Brief Description

According to the Autism Society[1] and Autism Speaks[2], Autism or Autism Spectrum Disorder (ASD), is a highly variable neurodevelopmental disorder that refers to a broad range of cognitive and physiological conditions. Autism is distinguished by a triad of symptoms, namely impairments in social interaction; impairments in communication; and restricted interests and repetitive behaviour. People with autism may be severely impaired in a specific physiological or physical aspect, but normal or even superior, in others when compared to people without autism.

Autism is a spectrum disorder, meaning that it affects people differently. In profound cases, young children may not interact with others, or treat people as objects. In milder cases it involves difficulty understanding and relating to others, and difficulty understanding other people's perspectives and emotions. A simpler definition, some children may have speech, whereas others may have little or no speech.

Without appropriate support, autistic children will not develop effective social skills and may speak or behave in ways that create severe challenges with everyone around them. The increase of research probing the possible causes of and potential treatments for autism has shed light and have improved patient health and therapeutic outcomes, and regarding this review, the usage of VR and AR on autistic patients.

---

[1] More information regarding to Autism Society available in the website http://www.autism-society.org/.

[2] More information regarding to Autism Speaks available in the website https://www.autismspeaks.org/.

1.2   AR Technology Characterization

To better understand the possibilities provided by AR, a brief characterization of this technology is necessary. While using VR, people can be totally immersed in a virtual environment, but with AR applications, the virtual information is mixed with the real scenario, where they perceive both real and virtual information in what was called by Milgram et al. [17], a *Virtuality Continuum*. AR technology is used to provide a given quantity of virtual elements that are being blended with reality. Consequently, AR is a collection of interactive technologies that merge these two elements – virtual and real – in real-time, providing accurate registration in all three dimensions [18]. The first technological device created was an optical see-through Head-Mounted Display developed by Sutherland in 1968. However, in addition to head-attached displays, other types of displays may be used to experience AR, such as hand-held displays or spatial displays [19, 20].

The definition of AR technology helps to better understand why it has been sparking the interest of its usage for autism patients. Evidence shows that children with autism suffer a deficit in sharing imaginative spontaneous pretend play, which is believed into reduced social interaction, known to be developmental links to key competences of their future lives [21]. Thus, an AR technology approach is found to promote spontaneous responses of children with autism in both solitary and socio-dramatic pretend play [22]. Kerdvibulvech et al. presented a three-dimensional human-computer interaction application based on AR technology for assisting children with special problems in communication for social innovation [23]. Lumbreras et al. is undergoing the evaluation of the learning process through a mobile augmented reality application, with the purpose of helping them in their relationships with the outside world [24]. Using AR systems as part of the teaching process, the ability to perform a chain task is studied on three elementary-age students with ASD [25]. Given the difficulty that children with ASD have to recognize facial expressions and to understand the associated emotions, a mobile AR application, implemented as a game, is evaluated to observe its impact on children interaction [26].

2   Methods

The systematic review method used for the current systematic review was based on the PRISMA methodology [27] to ensure a transparent and complete reporting of the surveyed topics.

2.1   Search Strategy and Study Selection

The literature was identified through online searches, by conducting an extensive search in the databases Science Citation Index on Web of Science (Clarivate), Elsevier Scopus, ACM Digital Library, IEEE Xplore, Wiley Online Library and ScienceDirect. The first search was run on 7 of January of 2019, being updated on 30 of July 2019. The search was performed to be equivalent to the following logical expression: Title/Keywords/Abstract contains ("Augmented Reality" OR "AR" OR "Mixed

Reality") AND ("Autism" or "ASD") AND ("Rehabilitation" OR "Rehab" OR "Assist" OR "Help" OR "Aid").

Following the aforementioned search method, whereby a total amount of 67 records were obtained, an eligibility assessment performed independently in a conventional unblinded standardised manner by two reviewers (A.M. and H.A.). Each paper was reviewed by these two reviewers to decide its eligibility, based on the title and abstract of each study, taking into consideration the exclusion criteria. When a record was rejected by one reviewer and accepted by the other, that record was kept for eligibility.

2.2   Data Collection Process and Quality Assessment

Previous literature, when analysing the effectiveness of VR systems for children and adolescents with autism [16], distinguished six areas: social skills, emotional skills, daily living skills, communication ability, attention, physical activity, and phobia or fear. Thus, the eligibility assessment performed across the 16 records included the referred areas for variables analysis. When analysing the records, a new area was established, namely, cognitive skills.

According to analysed records, emotional skills are stated to improve social skills [28], thus, these two areas previously proposed are collapsed into only one area named social skills. Also included in social skills are studies targeted for social communication, with non-verbal communication, such as eye contact. Communication ability area includes specific features related to verbal communication, such as speech-language [29].

Some examples of daily living skills are studies which aimed to improve social cues when meeting and greeting [30].

Due to the intention of the current study to present a systematic review and analysis to AR implementations for autism patients' wellbeing, a quality assessment was carried out. Following the guidelines to create a quality assessment checklist [31], and based on STROBE checklists [32], a set of questions was created to which pre-defined answers were settled – Yes (weight of 1), Partially (weight of 0.5), No (weight of 0). The questions established for this quality assessment are the following:

1. Clearly define all outcomes;
2. Clearly describe used technologies;
3. Report numbers of individuals at each stage of study;
4. Present variables to be handled in the study;
5. Describe all statistical methods;
6. Show and discuss results;
7. Report outcome events;
8. Discuss limitations of the study.

An average score of 0.8 was obtained, having as average deviation of 0.2. The high-quality papers having into account this quality assessment (with an average of 0.8 or higher) are marked with an asterisk in the table that summarises the 16 studies found applying AR for autistic patients (Table 1).

3   Results

The search in the identified databases, returned a total of 77 records, of which 2 were identified as being duplicate and were consequently removed, resulting in a total of 75 records. The title and abstracts of the unique 75 records were analysed, taking into account the eligibility criteria defined being that 28 records were excluded for not meeting such criteria. This resulted in 47 records eligible for full-text analysis. From those 47 records, 31 full-paper records were excluded based on the previously defined exclusion criteria, resulting in a total of 16 records that were included in the qualitative synthesis.

A total of 16 AR applications targeted for autistic patients were analysed, from where it was possible to extract valuable information regarding to its implementation, main findings obtained, and limitations found.

3.1   Qualitative and Quantitative Analysis

Among the 75 results screened, from where 16 were studies carefully analysed, Table 1 presents a summary of this review.

According to the presented data, we observe that, as noticed in VR applications [16], most part of AR applications were focused on improving patients' social skills. User tests included children, adolescents, and adults, having as the major focus, the children evaluations. None of the analysed studies were focused on physical activity neither on phobia and fear.

Table 1. A summary of the 16 studies found applying AR for autist patients.

Contribution	Patients	Variables	Technology	User sample	Positive findings
Escobedo et al. [33]*	Children	Social skills	Smartphone	3	Yes
Quintana et al. [34]	Children	Daily living skills	Smartphone	11	Yes
Escobedo et al. [35]*	Children	Attention	Smartphone	12	Yes
Lakshmiprabha et al. [36]	n.s.	Cognitive skills	Video projector	n.s.	Yes
Da Silva et al. [29]*	Children	Communication ability	Computer	4	Yes
Chen et al. [28]*	Adolescents	Social skills	Tablet	6	Yes
Chung et al. [37]	Children	Social skills	Computer	n.s.	Yes
Liu et al. [38]*	Children	Social skills	Smartglasses	2	Yes
Soares et al. [39]	Children	Social and cognitive skills	Google Cardboard	4	Yes
Zheng et al. [40]	Children	Social skills	Computer	10	Yes

(*continued*)

Table 1. (*continued*)

Contribution	Patients	Variables	Technology	User sample	Positive findings
Zheng et al. [41]	Children	Social skills	Video projector	10	Yes
Dragomir et al. [42]*	Children	Daily living skills	Computer	7	Yes
Keshav et al. [43]*	Adolescents	Social skills and attention	Smartglasses	1	Yes
Lee et al. [30]*	Children	Social skills	Tablet	3	Yes
Sahin et al. [44]*	Children and adults	Social skills	Smartglasses	18	Yes
Vahabzadeh, et al. [45]*	Children	Social skills	Smartglasses	4	Yes

n.s. *stands for "not specified"*.
* high-quality papers with an average of 0.8 or higher in que quality assessment score.

3.2 Discussion

Aiming to find AR applications used for enhancing the wellbeing of autism patients, 16 studies were found according to the eligibility criteria.

Most part of these studies, presented social skills as the variable to improve with the AR technology (56%). Daily living skills were identified when discriminating objects through object recognition and interaction gestures with it [34], as well as the ability to pretend play with toys [42]. Communication ability, for verbal communication, was evaluated by analysing benefits in speech-language [29]. Attention [35, 43], and Cognitive skills [36] which included psycho-pedagogical tasks [39], were also stated as getting improved when using AR systems.

The technologies used for the AR systems are diverse, having smartglasses (Empowered Brain [38, 43–45]) and Computer screens [29, 42] (specifying the use of a webcam [37, 40]) as the more common technologies used, where 4 cases of each technology were observed. Mobile devices were also common to find (3 smartphones [33–35] and 2 tablets [28, 30]), followed by 2 video projections [35] (specifying the use of a Kinect [41]) and one that implemented the AR using the Google cardboard [39].

The data collection was obtained through interviews (31%), video recording (22%), direct observation (17%), questionnaires (17%), and parental report (4%). About 9% of the reviewed studies did not specify the instrument used in their evaluation.

Most part of the studies were focused on children and adolescents with autism (88%), with ages between 3 and 14 years old.

All reviewed records presented positive main findings regarding to their presented goals when using the AR technology. Among these main findings, we found that the AR system facilitates practicing and learning social skills [30, 33, 37, 38, 43, 45], improves students' selective attention for more time [28, 29, 35, 42], helps children to learn new things [29, 36], and helps children to better understand the facial expressions and emotions [28, 40].

Limitations presented within these studies, are mainly focused on involving larger samples [30, 42, 43, 45], since these studies presented samples sizes between 1 and 12 participants, with an average sample size among all scanned records is 6.8, with an average deviation of 3.9. Other limitations point out the importance of considering individual differences in further researches [30, 43], and the need to explore longer-term effects of using AR smartglasses [44].

4 Conclusion

As mentioned before, AR is being used for the last few decades in several medical applications – from surgery support to treatment of phobias and other cognitive or physiological disorders. This study focused on the usage of AR in rehabilitation of people that suffer from autism. This systematic review presents an overview of the usage of AR technology, from a patients' perspective. Strengths and weaknesses of current approaches were identified, providing significant data to better understand how AR can be used for mitigating different distresses experienced by patients with autism.

A systematic review regarding to the usage of AR applications in patients with autism is presented, with a total of 16 implementations found in literature.

All analysed records presented positive main findings regarding to this technology with autistic patients, in particular, among children – the most common evaluated sample found.

According to these studies, several diverse technologies were used, namely smartglasses, mobile devices and video projection systems. The studies also state that the social skills, communication ability, attention and cognitive skills improved with the usage of AR technologies.

Some limitations are presented, namely the influence of the individual differences in the study along with the small sample sizes from 1 single participant, up to 12.

The results from this systematic review provide valuable information for further ideas and solutions regarding to enhance ASD patients' wellbeing with the usage of AR technologies.

References

1. Kim, Y., Kim, W.: Implementation of augmented reality system for smartphone advertisements. Int. J. Multimed. Ubiquit. Eng. 9(2), 385–392 (2014)
2. Hamasaki, M., Takeda, H., Nishimura, T.: Network analysis of massively collaborative creation of multimedia contents. In: UXTV 2008 Proceedings of the 1st International Conference on Designing Interactive User Experiences for TV and Video, pp. 165–168 (2008)
3. Yuen, S.C., Yaoyuneyong, G., Johnson, E.: Augmented reality: an overview and five directions for ar in education. J. Educ. Technol. Dev. Exch. 4(1), 119–140 (2011)
4. Bower, M., Howe, C., McCredie, N., Robinson, A., Grover, D.: Augmented reality in education – cases, places and potentials. EMI. Educ. Media Int. 51(1), 1–15 (2014)
5. Meža, S., Turk, Ž., Dolenc, M.: Measuring the potential of augmented reality in civil engineering. Adv. Eng. Softw. 90, 1–10 (2015)

6. Baratoff, G., Wilke, W., Artificial, H., Engineering, J.C.: Augmented reality projects in automotive and aerospace industry. IEEE Comput. Graph. Appl. **25**(6), 56–58 (2005)
7. Canciani, M., Conigliaro, E., Del Grasso, M., Papalini, P., Saccone, M.: 3D survey and augmented reality for cultural heritage. The case study of Aurelian Wall at Castra Praetoria in Rome. Int. Arch. Photogramm. Remote Sens. Spat. Inf. Sci. **41**, 931–937 (2016)
8. Fidas, C., Sintoris, C., Yiannoutsou, N., Avouris, N.: A survey on tools for end user authoring of mobile applications for cultural heritage. In: Proceedings of the 6th International Conference on Information, Intelligence, Systems and Applications (IISA), pp. 1–5 (2015)
9. Han, D., Jung, T., Gibson, A.: Dublin AR: implementing augmented reality (AR) in tourism. In: Proceedings of the Information and Communication Technologies in Tourism, vol. 523, pp. 511–523 (2014)
10. De Paolis, L.T.: Augmented visualization as surgical support in the treatment of tumors. In: Proceedings of the International Conference on Bioinformatics and Biomedical Engineering, pp. 432–443 (2017)
11. Quintana, E., Favela, J.: Augmented reality annotations to assist persons with Alzheimers and their caregivers. Pers. Ubiquitous Comput. **17**(6), 1105–1116 (2013)
12. García, A.: Interactive augmented reality as a support tool for Parkinson's disease rehabilitation programs, Paper 862, September 2012
13. Juan, M.C., Alcaniz, M., Monserrat, C., Botella, C., Baños, R.M., Guerrero, B.: Using augmented reality to treat phobias.pdf. IEEE Comput. Graph. Appl. **25**(6), 31–37 (2005)
14. Wang, B., Shen, M., Wang, Y., He, Z.: Effect of virtual reality on balance and gait ability in patients with Parkinson's disease: a systematic review and meta-analysis. Clin. Rehabil. **1**(9) (2019)
15. Rose, F.D., Brooks, B.M., Rizzo, A.A.: Virtual reality in brain damage rehabilitation: review. Cyberpsychol. Behav. **8**(3), 241–262 (2005)
16. Mesa-Gresa, P., et al.: Effectiveness of virtual reality for children and adolescents with autism spectrum disorder: an evidence-based systematic review. Sensors **18**(8), 2486 (2018)
17. Milgram, P., Takemura, H., Utsumi, A., Kishino, F.: Augmented reality: a class of displays on the reality-virtuality continuum. In: SPIE - The International Society for Optical Engineering, pp. 282–292 (1994)
18. Azuma, R.: A survey of augmented reality. Presence Teleoperators Virtual Environ. **6**(4), 355–385 (1997)
19. Azuma, R., Baillot, Y., Feiner, S., Julier, S., Behringer, R., Macintyre, B.: Recent advances in augmented reality. IEEE Comput. Graph. Appl. **21**(6), 34–47 (2001)
20. Bimber, O., Raskar, R.: Spatial Augmented Reality: Merging Real and Virtual Worlds. A K Peters, Wellesley (2005)
21. Bai, Z., Blackwell, A.F., Coulouris, G.: Making pretense visible and graspable: an augmented reality approach to promote pretend play. In: ISMAR 2012 - 11th IEEE International Symposium on Mixed and Augmented Reality 2012, Science and Technology Papers, pp. 267–268 (2012)
22. Bai, Z.: Augmenting imagination for children with autism. In: ACM International Conference Proceeding Series, pp. 327–330 (2012)
23. Kerdvibulvech, C., Wang, C.-C.: A new 3D augmented reality application for educational games to help children in communication interactively. In: Computational Science and Its Applications – ICCSA 2016, pp. 465–473 (2016)

24. Lumbreras, M.A.M., de Lourdes, M.T.M., Ariel, S.R.: Aura: augmented reality in mobile devices for the learning of children with ASD-augmented reality in the learning of children with autism. In: Augmented Reality for Enhanced Learning Environments, pp. 142–169 (2018)

25. McMahon, D.D., Moore, E.J., Wright, R.E., Cihak, D.F., Gibbons, M.M., Smith, C.: Evaluating augmented reality to complete a chain task for elementary students with autism. J. Spec. Educ. Technol. **31**(2), 99–108 (2016)

26. Cunha, P., Brandão, J., Vasconcelos, J., Soares, F., Carvalho, V.: Augmented reality for cognitive and social skills improvement in children with ASD. In: Proceedings of 2016 13th International Conference on Remote Engineering and Virtual Instrumentation, REV 2016, pp. 334–335 (2016)

27. Liberati, A., et al.: The PRISMA statement for reporting systematic reviews and meta-analyses of studies that evaluate health care interventions: explanation and elaboration. PLoS Med. **6**(7), 1–28 (2009)

28. Chen, C.-H., Lee, I.J., Lin, L.-Y.: Augmented reality-based video-modeling storybook of nonverbal facial cues for children with autism spectrum disorder to improve their perceptions and judgments of facial expressions and emotions. Comput. Human Behav. **55**, 477–485 (2016)

29. da Silva, C.A., Fernandes, A.R., Grohmann, A.P.: Star: speech therapy with augmented reality for children with autism spectrum disorders. In: Lecture Notes in Business Information Processing, vol. 227. pp. 379–396 (2015)

30. Lee, I.J., Chen, C.H., Wang, C.P., Chung, C.H.: Augmented reality plus concept map technique to teach children with ASD to use social cues when meeting and greeting. Asia-Pacific Educ. Res. **27**(3), 227–243 (2018)

31. Kitchenham, B., Charters, S.: Guidelines for performing systematic literature reviews in software engineering (2007)

32. University of Bern, STROBE Statement. https://www.strobe-statement.org/index.php?id=strobe-home. Accessed 19 Nov 2018

33. Escobedo, L., et al.: MOSOCO: a mobile assistive tool to support children with autism practicing social skills in real-life situations. In: Conference on Human Factors in Computing Systems - Proceedings, pp. 2589–2598 (2012)

34. Quintana, E., Ibarra, C., Escobedo, L., Tentori, M., Favela, J.: Object and gesture recognition to assist children with autism during the discrimination training. Lecture Notes in Computer Science (including subseries Lecture Notes in Artificial Intelligence and Lecture Notes in Bioinformatics), vol. 7441 LNCS. Department of Computer Science, CICESE, Mexico, pp. 877–884 (2012)

35. Escobedo, L., Tentori, M., Quintana, E., Favela, J., Garcia-Rosas, D.: Using augmented reality to help children with autism stay focused. IEEE Pervasive Comput. **13**(1), 38–46 (2014)

36. Lakshmiprabha, N.S., Santos, A., Mladenov, D., Beltramello, O.: An augmented and virtual reality system for training autistic children. In: Proceedings of ISMAR 2014 - IEEE International Symposium on Mixed and Augmented Reality - Science and Technology 2014, pp. 277–278 (2014)

37. Chung, C.H., Chen, C.H.: Augmented reality based social stories training system for promoting the social skills of children with autism. In: Advances in Intelligent Systems and Computing, vol. 486. pp. 495–505 (2017)

38. Liu, R.P., Salisbury, J.P., Vahabzadeh, A., Sahin, N.T.: Feasibility of an autism-focused augmented reality smartglasses system for social communication and behavioral coaching. Front. Pediatr. **5**, 145 (2017)

39. Soares, K.P., Burlamaqui, A.M.F., Gonçalves, L.M.G., Da Costa, V.F., Cunha, M.E., Da Silva Burlamaqui, A.A.R.S.: Preliminary studies with augmented reality tool to help in psycho-pedagogical tasks with children belonging to autism spectrum disorder. IEEE Lat. Am. Trans. **15**(10), 2017–2023 (2017)
40. Zheng, C., et al.: Toon-chat: a cartoon-masked chat system for children with autism (2017)
41. Zheng, C., et al.: KinToon: a kinect facial projector for communication enhancement for ASD children, pp. 201–203 (2017)
42. Dragomir, M., Manches, A., Fletcher-Watson, S., Pain, H.: Facilitating pretend play in autistic children: results from an augmented reality app evaluation. In: ASSETS 2018 - Proceedings of the 20th International ACM SIGACCESS Conference on Computers and Accessibility, pp. 407–409 (2018)
43. Keshav, N.U., et al.: Longitudinal socio-emotional learning intervention for autism via smartglasses: qualitative school teacher descriptions of practicality, usability, and efficacy in general and special education classroom settings. Educ. Sci. **8**(3), 107 (2018)
44. Sahin, N.T., Keshav, N.U., Salisbury, J.P., Vahabzadeh, A.: Second version of google glass as a wearable socio-affective aid: positive school desirability, high usability, and theoretical framework in a sample of children with autism. J. Med. Internet Res. **20**(1) (2018)
45. Vahabzadeh, A., Keshav, N., Abdus-Sabur, R., Huey, K., Liu, R., Sahin, N.: Improved socio-emotional and behavioral functioning in students with autism following school-based smartglasses intervention: multi-stage feasibility and controlled efficacy study. Behav. Sci. (Basel) **8**(10), 85 (2018)

Simulation and Modeling

Minimally Invasive Transforaminal Lumbar Interbody Fusion Surgery at Level L5-S1

D. S. Fidalgo[1], B. Areias[2], Luisa C. Sousa[2(✉)], M. Parente[2],
R. Natal Jorge[2], H. Sousa[3], and J. M. Goncalves[4]

[1] DEMec, FEUP, Universidade do Porto, Porto, Portugal
[2] INEGI/DEMec, FEUP, Universidade do Porto, Porto, Portugal
lcsousa@fe.up.pt
[3] Centro Hospitalar de Vila Nova de Gaia/Espinho, Vila Nova de Gaia, Portugal
[4] Hospital da Luz Arrábida, Porto, Portugal

Abstract. There are various techniques to obtain lumbar interbody fusion (LIF) at L5-S1 functional unit (FSU) level. This paper presents a finite element analysis to investigate the biomechanical changes caused by transforaminal lumbar interbody fusion (TLIF) at the L5-S1 FSU level. The information collected can be used to choose the best chirurgical treatment to solve several pathologies. This approach places the implant into the anterior portion of the intervertebral disc space, which stabilizes the flexion–extension movement across the operated level. Numerical results of segmental motion show that torsion and extension load cases yield the highest rotation angles during the early and long-term postoperative phases, respectively. Furthermore, a decrease in autogenous bone graft volume results in a reduction of displacements and rotation angles.

Keywords: Minimally invasive lumbar fusion · Transforaminal approach · T-PAL cage · Finite element

1 Introduction

Spine surgery today is different from the spine surgery of yesterday. Surgeons use minimally invasive techniques as an alternative to open spine surgery. Minimally invasive spine surgery results in a smaller scar, less time in the hospital, and quickly return to daily activities.

TLIF allows the stabilization and fusion of adult spinal deformity being a good treatment option for symptomatic spinal instability, spondylolisthesis, spinal stenosis, and degenerative scoliosis. TLIF approach presents some advantages as a decrease in potential neurological injury, improvement in lordotic alignment due to graft placement within the anterior column, and preservation of posterior column integrity [1]. Furthermore, grafts are placed near the center of rotation for a spinal motion segment, which promotes more stability. Through an incision above the pedicle it is possible to access the foramen and remove disc material [1]. The unilateral nature of TLIF approach results in less destruction of the posterior elements and less destabilization of the spine, which will maximize fusion stability. This technique also allows less

© Springer Nature Switzerland AG 2019
J. M. R. S. Tavares and R. M. Natal Jorge (Eds.): VipIMAGE 2019, LNCVB 34, pp. 467–473, 2019.
https://doi.org/10.1007/978-3-030-32040-9_47

perineural retraction, scarring, and preserves more bony surface for posterior spinal fusion. The annulus must be preserved to provide additional support for the implant and prevent migration of bone graft into the spinal canal.

The potential complications of this procedure include the risk of nerve root injury during retraction, and instability due to an extensive posterior decompression. Furthermore, there is a risk of posterior extrusion of the graft that may origin neural damage [2]. Implants have radiopaque marker pins to follow the exact position of the implant, during and after surgeries, which increases radiation exposure [3].

TLIF numerical simulations were carried out in order to obtain information about the biomechanical behaviour of the functional unity L5-S1.

2 Materials and Methods

Transforaminal lumbar interbody fusion can be performed via a minimally invasive technique. The chosen implant replaces lumbar interver-tebral disk after the unilateral removal of the pars interarticularis and facet complex disc. It is inserted via the posterior approach and usually is used in combination with supplemental posterior fixation. In order to simulate the mechanical behaviour of the spine segment a finite element model of the L5-S1 FSU was created using computed tomography images. The geometry of the two vertebrae surface was created with Mimics software (version 10.0, Materialise Inc., Belgium) and the smoothed surface was imported into Abaqus/CAE (2007) software to get volume meshing with four node tetrahedral elements C3D4 and finite element analysis. The geometry of the intervertebral disc was defined using the intervertebral space between the two vertebrae surfaces, the lower surface of L5 and the upper surface of S1. Both ligaments and intervertebral joint layers were modelled: tension-only spring connector elements, truss elements T3D2, were used to mesh ligaments and eight-noded hexahedral elements with a hybrid formulation C3D8H for joint layers.

Figure 1 shows the implant, T-PAL cage (DePuy Synthes GmbH) which geometry is characterized by 28 mm length, insertion depth 10 mm, total depth 14 mm, 11 mm thickness and a lordotic angulation equal to 0°. It is manufactured with a biocompatible polymer (PEEK) and Titanium alloy marker pins. 14217 nodes and 65775 four node tetrahedrical elements, C3D4 were used to mesh the implant.

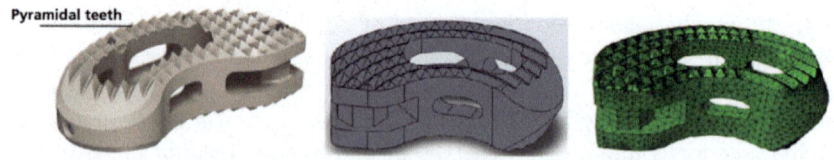

Fig. 1. T-PAL cage shape, SolidWorks model and finite element mesh.

A discectomy was performed at the L5-S1 FSU level: the left inferior facet joint of L5, the superior facet joint of S1 and the nucleous pulposes, were removed and only

1/3 of the annulus fibrosus was preserved. Consequently, left facet capsular ligament was not considered in this model. The implant central window and the anterior disk space were filled with autogenous bone graft to improve fusion rate and then the implant was placed in the anterior intervertebral place between L5 and S1. Figure 2 shows a section of the intervertebral space with the T-PAL implant and the autogenous bone graft volume represented in red.

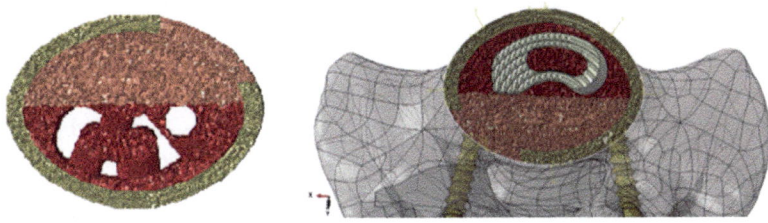

Fig. 2. L4-L5 T-PAL cage positioned between l5 and S1 vertebrae with greater autogenous bone graft volume.

Two TLIF models with different volume of autogenous bone graft were considered in this study in order to study the effect of bone graft volume on the FSU motion after surgery. The model M1, with greater autogenous bone graft volume is represented in Fig. 2, while the model M2, with less amount of bone graft is shown in Fig. 3.

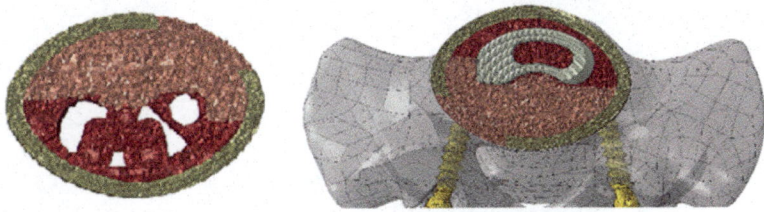

Fig. 3. L4-L5 T-PAL cage positioned between l5 and S1 vertebrae with less autogenous bone graft volume.

All components were modelled considering elastic behaviour [4, 5], except the preserved part of the annulus fibrosus where fibres were considered as an anisotropic hyperelastic material [6]. Boundary conditions were defined considering that all nodes of the inferior end plate of the S1 vertebra were constrained from moving in any direction. In order to apply the loads a reference point was defined in the centre of the superior surface of L5. The loads applied were based on the literature [7]. For compression a load of 10 N was considered and for the other load cases a moment of 10 Nm was applied. These values represent the maximum forces applied without causing spinal injuries.

Two postoperative stages were considered in this study, an early and a long-term postoperative stage; each stage was modelled considering different contact conditions type. Using Abaqus software, it is possible to apply a "surface to surface" contact or a "tie" constraint. A "surface to surface" contact is a condition that allows relative motion between the involved surfaces. On the other hand, a "tie" contact ties two separate surfaces together with no relative motion between them. For both postoperative stages, it was used a "tie" between the annulus layers and the vertebral endplates. In the early postoperative phase, a "surface to surface" contact interaction between the vertebral endplates and the autogenous bone graft/T-PAL cage was applied. To simulate the long-term postoperative phase, since the intervertebral fusion has already occurred, it was necessary to apply a "tie" contact condition between the vertebral endplates and the autogenous bone graft/T-PAL cage. In both stages the interaction between ligaments insertion points, and between pedicle screw and vertebra bone surface was also modelled with "tie" constraint; the same contact condition was used to simulate the interaction between pedicle screw and pedicle bar. The interaction between the intervertebral articulations was modelled with a "surface to surface" contact.

3 Results and Discussion

Figures 5, 6 and 7 present L5-S1 FSU axial displacements fields for compression, extension, flexion, lateral bending and torsion load cases considering the TLIF model M1 with greater autogenous bone graft volume.

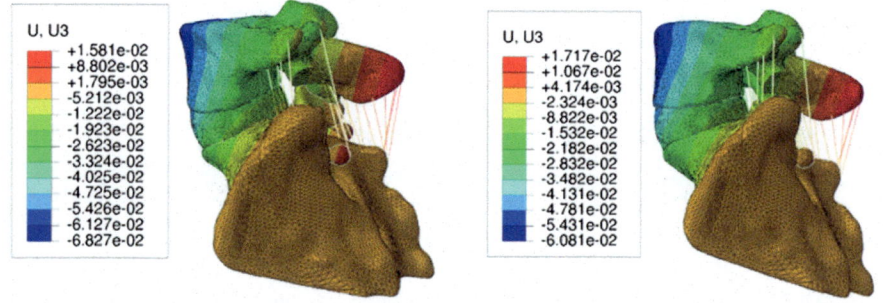

Fig. 5. Compression load case (model M1) - displacement field of the L5-S1 functional unit.

In all figures, the early and the long-term postoperative phase displacement fields are shown on the left and right side respectively.

As expected, it was found that the early postoperative axial displacements are greater than the ones found on the long-term postoperative stage. It can also be noticed that lateral bending load case presents the highest axial displacement for both postoperative phases, probably due to fact that the left posterolateral annulus region and the left facet capsular ligament were removed.

Figures 8 and 9 present the axial displacement fields of the TLIF model M2 with less autogenous bone graft volume for the extension and torsion load cases

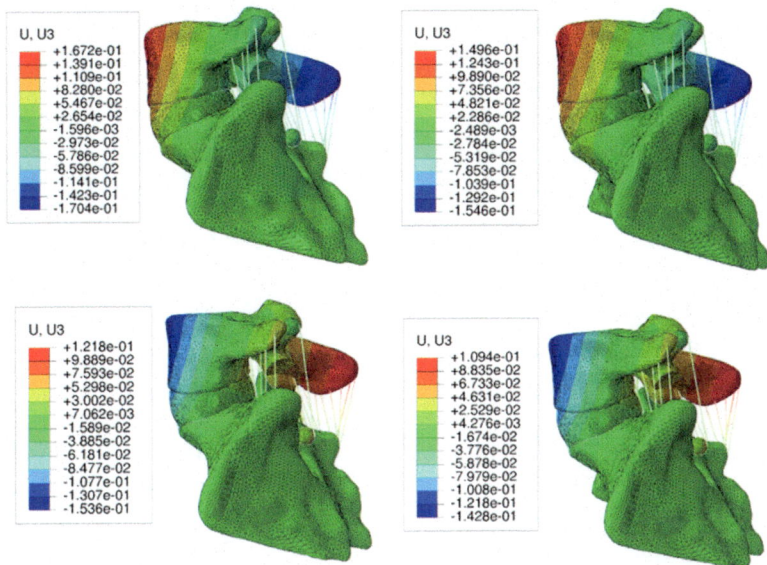

Fig. 6. Displacement field of the L5-S1 functional unit (model M1): extension load case (top) and flexion load case (bottom).

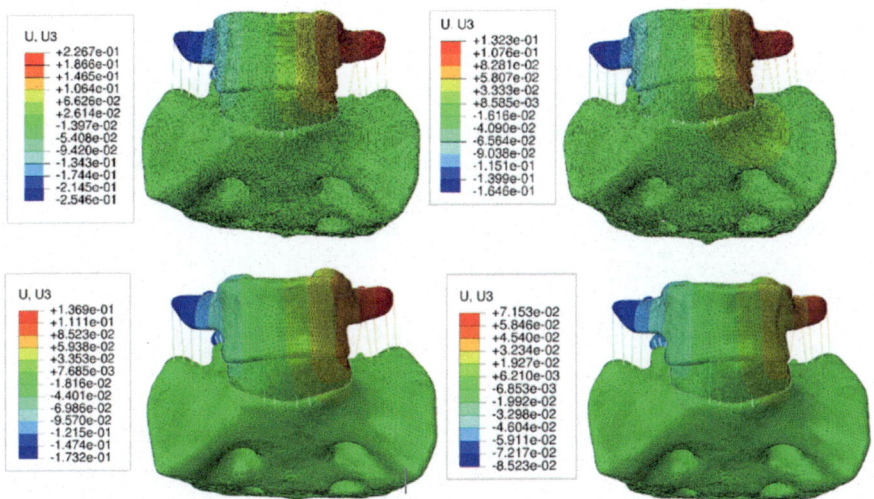

Fig. 7. Displacement field of the L5-S1 functional unit (model M1): lateral bending load case (top) and torsion load case (bottom).

respectively. As expected, the maximum axial displacement values are greater than the corresponding ones for the model M1, with greater bone graft volume.

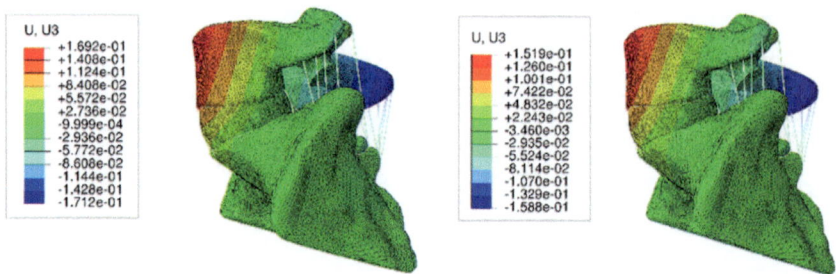

Fig. 8. Axial displacements for the extension load case (model M2): early postoperative stage (left) and long-term postoperative (right).

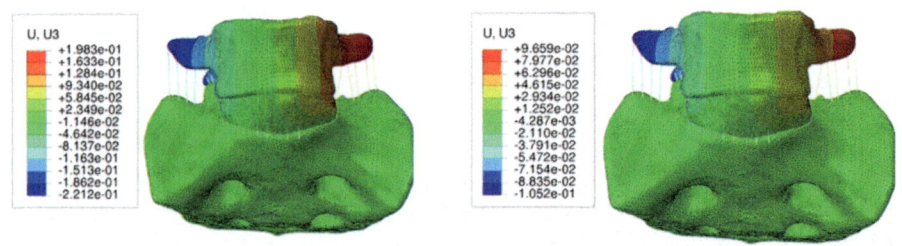

Fig. 9. Axial displacements for the torsion load case (model M2): early postoperative stage (left) and long-term postoperative stage (right).

The rotation angle of the reference node of the L5 vertebra surface is shown in Fig. 10. Regarding the two postoperative stages, rotation angles are greater in the early postoperative phase as fusion has not yet occurred. During the early recovered stage, torsion load case yields the highest rotation angle while, in long-term postoperative phase, extension presents the highest rotation angle.

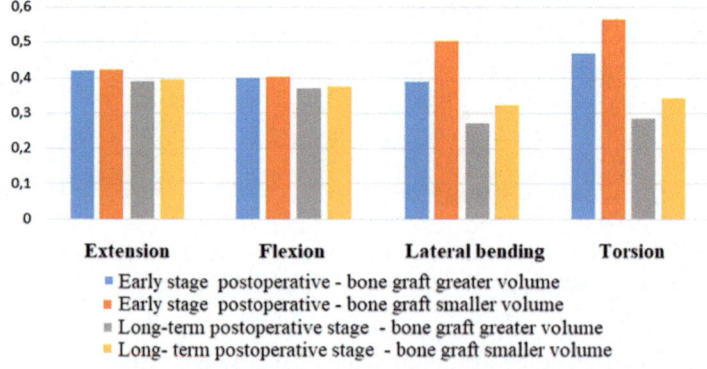

Fig. 10. Rotation angles for each TLIF model and postoperative stage.

Comparing the two TLIF models with different autogenous bone graft volume, the model M2, with smaller volume of bone graft, presents higher rotation angle values for all load cases and the two postoperative stages. As expected the difference between the two models is greater in the early recovered phase when fusion has not yet occurred. Lateral bending and torsion are the load cases that yield higher differences probably because these two loads are more influenced by the bone graft surface area variation.

4 Conclusion

A finite element analysis was performed to study spine stability after a minimally invasive TLIF using a T-PAL cage. Two models of the L5-S1 FSU, presenting different bone graft volume, were subjected to loads corresponding to daily activities. In this study, the TLIF model M1, with greater autogenous bone graft volume, yields lower axial displacements and rotation angles for all load cases, chiefly in the early postoperative phase, when fusion has not yet occurred. On the other hand, and for the two models, torsion load case yields the highest rotation angle during the early recovered stage, while in long-term postoperative phase the highest rotation angle occurs under extension load case.

Acknowledgments. The authors gratefully acknowledge the funding by FCT, Portugal, the Research Unit of LAETA-INEGI, Engineering Faculty of University of Porto, and to Dr. Maia Gonçalves of Hospital da Luz Arrábida.

References

1. Cole, C.D., McCall, T.D., Schmidt, M.H., Dailey, A.T.: Comparison of low back fusion techniques: transforaminal lumbar interbody fusion (TLIF) or posterior lumbar interbody fusion (PLIF) approaches. Curr. Rev. Musculoskelet. Med. **2**, 118–126 (2009)
2. Krishna, M., Pollock, R.D., Bhatia, C.: Incidence, etiology, classification, and management of neuralgia after posterior lumbar interbody fusion surgery in 226 patients. Spine J. **8**, 374–379 (2008)
3. Skovrlj, B.: Minimally invasive procedures on the lumbar spine. World J. Clin. Cases **3**(1), 1–9 (2015)
4. Zhang, Z., Li, H., Fogel, G.R., Liao, Z., Li, Y., Liu, W.: Biomechanical analysis of porous additive manufactured cages for lateral lumbar interbody fusion: a finite element analysis. World Neurosurg. **111**, e581–e591 (2017)
5. Cheung, J.T.-M., Zhang, M., Chow, D.H.-K.: Biomechanical responses of the intervertebral joints to static and vibrational loading: a finite element study. Clin. Biomech. **18**(9), 790–799 (2003)
6. Momeni Shahraki, N., Fatemi, A., Goel, V.K., Agarwal, A.: On the use of biaxial properties in modeling annulus as a Holzapfel–Gasser–Ogden material. Front. Bioeng. Biotechnol. **3**, 1–9 (2015)
7. Zhong, Z.-C., Feng, C.-K., Chen, C.-S., Wei, S.-H.: Finite element analysis of the lumbar spine with a new cage using a topology optimization method. Med. Eng. Phys. **28**(1), 90–98 (2005)

Osseous Fracture Patterns:
From CT to a Plane Representation

Adrián Luque-Luque[1]([✉]), Juan-Roberto Jiménez-Pérez[2],
and Juan José Jiménez-Delgado[3]

[1] Universidad de Jaén, Campus Las Lagunillas A3-103, Jaén, Spain
alluque@ujaen.es
[2] Universidad de Jaén, Campus Las Lagunillas A3-113, Jaén, Spain
rjimenez@ujaen.es
[3] Universidad de Jaén, Campus Las Lagunillas A3-142, Jaén, Spain
juanjo@ujaen.es

Abstract. The extraction of bone fracture patterns for a posterior
reproduction of the fracture on different bone models is almost an unex-
plored field of research. These templates can recreate a similar fracture
on other areas or bones, with applications on medical training of trauma-
tologist and automatic bone fracture reduction algorithms. This paper
is focused on the process of generating a fracture pattern taking computed tomography scans as a starting point. In addition, several representations of the templates, for different purposes, are presented and
discussed. Finally, the potential uses of the bone fracture patterns are
described.

Keywords: Bone fracture · Fracture generation · Fracture pattern ·
Fracture zone · Segmentation · Template

1 Introduction

A fracture pattern is an abstract model that represent the fractured area of a
bone. It acts as a template to enable a replication of a fracture with similar
characteristics in another area of the same bone, or in a different bone. The
ability to extract theses patterns opens the possibility to procedurally create
fractured models of bones for different purposes, e.g. medical training.

There is no much research on this issue in the literature. In this paper, a
method to build patterns from computed tomography (CT) scans is presented.
Starting from the segmentation of the CT 3D images, different representations
of the fracture pattern are generated. These representations come from the same
fracture but are used for different goals. A surface formed by the triangulation
of points of a fracture is better for visualization purposes, but a representation
based on spherical coordinates accelerates the application of the template on
other bones.

© Springer Nature Switzerland AG 2019
J. M. R. S. Tavares and R. M. Natal Jorge (Eds.): VipIMAGE 2019, LNCVB 34, pp. 474–481, 2019.
https://doi.org/10.1007/978-3-030-32040-9_48

The paper is structured as follows: Sect. 2 briefly introduces previous research, Sect. 3 explains our proposal, detailing the stages of the process and the main features of the generated fracture pattern models, Sect. 4 shows the results of our approach on a real fractured bone, finally, Sect. 5 highlights the conclusions of this study and the next challenges.

2   Previous Works

In the literature, different approaches have been proposed to compute the fracture area in order to achieve an automatic reduction of a fracture. Most of the time, the fractured area is not computed as such, but as an approximation for performing a fracture reduction. Usually this information of the fractured area is discarded once it has been used on the reduction procedure.

Some studies [1,2] delegate in the users the selection of points that will act as seeds to automatically expand a region of fracture. Also curvature detection algorithms have been applied to detect [3] or to discard [4] points of interest of the fracture zone.

Others approaches use the shape of the bones to detect the fractured area. Winkelbach et al. [5] proposed a method to identify vertices of the fractured area comparing its normals to the bone axis, Willis et al. [6] address the problem with a statistical solution that classifies the points whether as belonging to the fracture region or not.

Paulano et al. [7] discussed the potential advantages of the generation of fractures on geometric models that represent bone structures in computer-assisted methods that support specialist in fracture reduction interventions. Our goal in this study is to create a nexus between the detection of bone fracture zones and the use of these in the creation of realistic fractures in healthy bone models.

3   Steps for Obtaining a Fracture Pattern

In this section, the phases developed to obtain a fracture pattern are explained in detail. The segmentation of a CT scan and the detection of the fractured zone is induced by the work developed by Paulano-Godino et al. [4], so as that it is included at the first steps.

3.1   CT Scan Segmentation

As the data source of this process is a scan where only the density of the materials are registered, the trabecular tissue of the bone can not be discriminated from other soft tissue as muscle or tendons. In order to segment the cortical tissue the method proposed by Paulano-Godino et al. [4] is used. This segmentation uses manually placed seeds to separate near fragments which will be combined when different algorithms are used.

The output is a hollow model with a thick shell of cortical tissue of each fragment. The inclusion of the trabecular region would lead to an overgrowth of the model to soft-tissue areas. As a consequence, we restrict the final template to the cortical area.

3.2   Fractured Zone Identification

In their original form, the 3D models are processed to locate candidate points for fractured zone identification [4]. An oriented bounding box encloses the model and it is discretized using a grid, being the direction of the OBB defined by the vector that join the centroids of every pair of models, Fig. 1.

Afterwards, the grid is traversed through the parallel columns to the sweep vector and the points that belong to the first populated voxel in each column are marked as candidate points. Finally, the candidate points are filtered based on the distance between each point and the candidate points of the opposite fragment. They are discarded if the distance is greater than a predefined value. This threshold depends on the fracture and it is determined by experience.

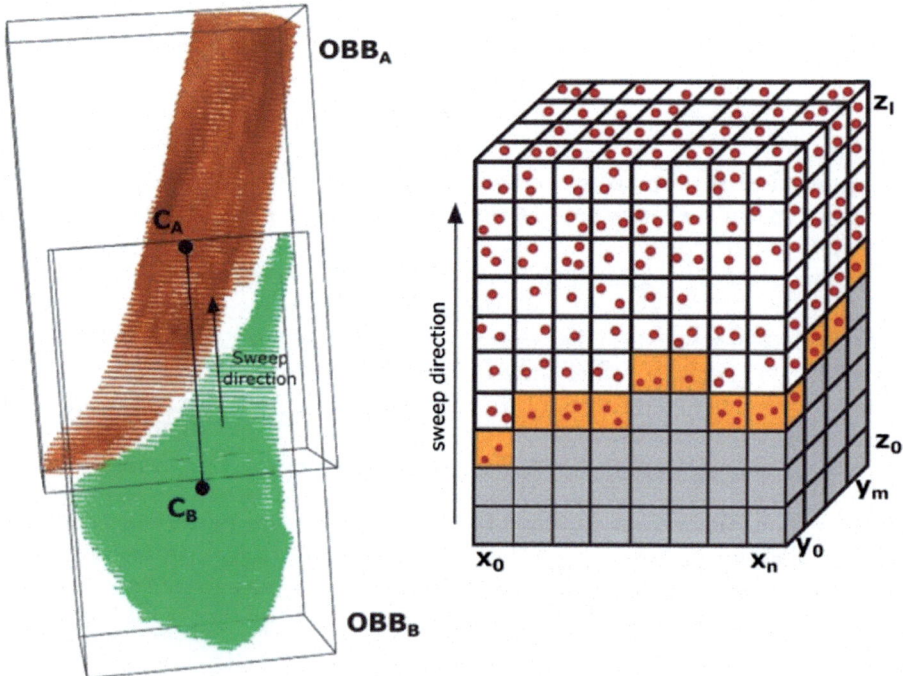

Fig. 1. Left, OBBs and centroids associated with two fragments. Sweep direction used to calculate the fractured zone of fragment A. Right, 3D representation of the sweeping procedure. Checked voxels that contain no points are displayed in grey. Checked voxels that contain points candidate to belong to the contact area are displayed in orange. White cells are not checked. Taken from [4]

In our work, as we only have a fragment, we have modified the original algorithm in order to work properly with a single fragment. To do that, the OBB is correctly aligned to the fracture. The used filter is based on the curvature:

points with a curvature greater than a threshold are discarded. As seen in Fig. 2, the bottom plane of the OBB is oriented to the fractured zone. This arrangement provides more points that helps to generate a smoother surface.

Fig. 2. Left, original OBB alignment. Right, modified alignment.

3.3   Point Cloud Cleaning

Some of the points identified in Sect. 3.2 are false positives that would distort the result template. In order to delete those points, two automatics cleaning methods are implemented. The first method uses an statistic algorithm included in the Point Cloud Library [8] to detect outliers, these outliers are small groups of points isolated from the rest of the cloud.

The second method takes advantage of the plane described in the Sect. 3.4, to remove points located further than a user-defined threshold. These algorithms are useful to find and delete the majority of the erroneous points. Unfortunately, it is difficult to automatically identify points near the surface of the fracture that are not part of it, so a manual filtering is needed as last task in this step.

3.4   Plane Fitting

Once the fractured area has been cleaned, the best fitting plane is calculated by the decomposition of a $3 \times N$ matrix where N is the number of points and each row represent the coordinates of a point, the normal vector of this plane is the left singular vector corresponding to the least singular value. The point of application of the plane is the center of mass of the point cloud, the mean of all the points.

This mathematical approximation returns the plane that minimizes the distance from the plane to all points of the cloud. The main purpose of the resulting plane is to define a reference system for the following phases.

3.5   Border Detection

The border detection is performed with a projected copy of the point cloud on the calculated plane. These points are processed by a 2D alpha shapes algorithm [9] that returns the cortical surface of the pattern. The alpha parameter is different for each point cloud. It needs to be manual-tuned to precisely delimit the area, so as not to get an over or undersized area.

3.6   Fracture Pattern Representation

The information obtained have been used to create different ways to represent the fracture. The first one is the surface itself, Fig. 4e, the purpose of this representation is to check visually for the correctness of the fracture zone obtained. For visualization purposes additional representations can be used, Fig. 3.

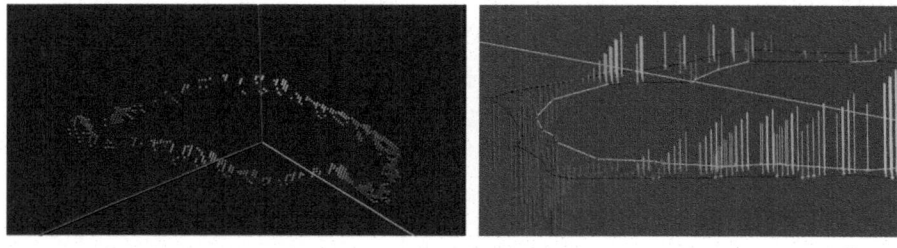

Fig. 3. Left, colored 3D point cloud of the fractured area, the lower the point in relation to the plane, the bluer the point, the higher the greener. Right, detail of the point cloud with the border of the fracture and lines that represent the projection of the points to the plane.

Additionally, with the goal of better storage and application in other bone models, two more representation are used. A 2D image of the exterior border of the fracture is saved as a raster image. In this case, we can maintain certain information, like the interior border. The advantage is the easy manner with which these textures can be applied to 3d models, with the possibility of doing changes just by modifying the texture coordinates, such as stretching or rotating.

Finally, the surface can be translated from cartesian to spherical coordinates. This transformation speed up the use of the template on another model. Once the application point is defined, the points of the interior border are projected to the interior border of the target model, doing the same process with the exterior border. These points allow recreating visually and geometrically a similar fracture on any bone model.

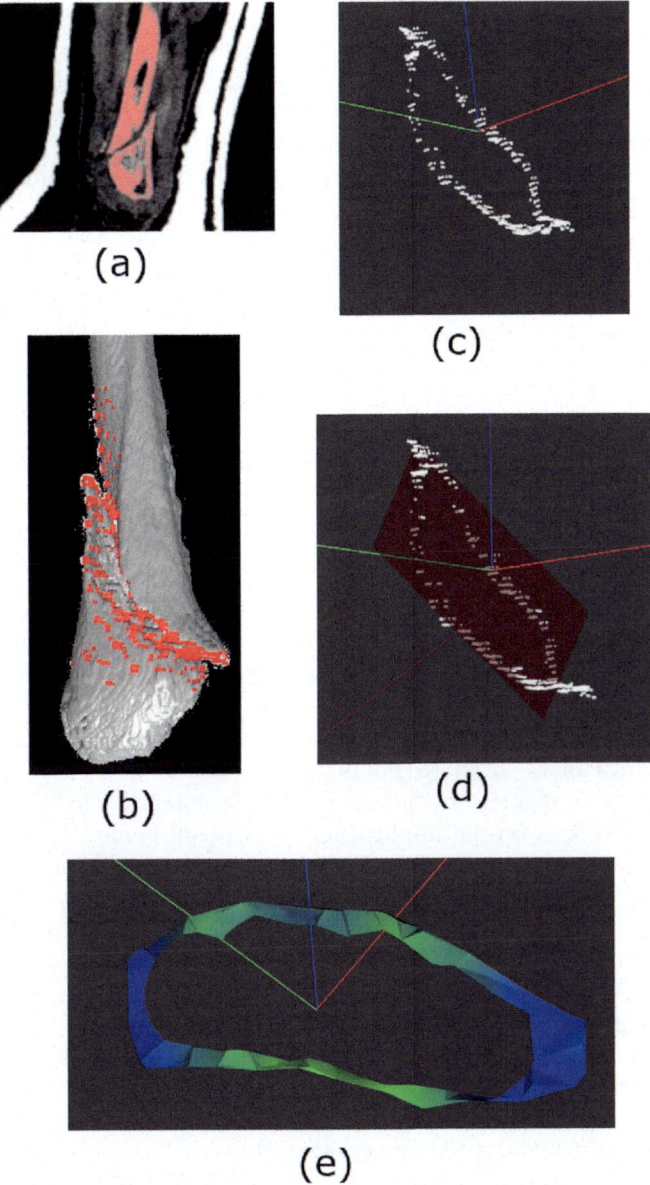

Fig. 4. Principal steps for obtaining a fracture pattern: (a) Segmentation of a CT scan, (b) points belonging to the fracture area, (c) cleaned point cloud, (d) best fitting plane and (e) surface representation of the fracture pattern.

4   Results

In our study, a broken fibula bone has been processed through all the steps, obtaining a clean three-dimensional surface of the fractured area. Figure 4 shows the most relevant phases and the final result. All tests were performed on a standard desktop computer, none of the steps of the process required more than a few seconds of time to run.

The initial scan has been segmented and the upper section of the broken fibula separated from the rest of the tissues, Fig. 4a. This generates a 3D model consisting of 28.358 points and 56.563 triangular faces, Fig. 4b shows the 1684 candidate points of the fracture. After the cleaning process there are 803 points that actually belongs to the fracture, Fig. 4c, around 60% of the points were cleaned automatically and the remaining outliers cleaned by hand. The best fitting plane, Fig. 4d, has been used to create the final representations. The surface fracture representation is illustrated in Fig. 4e.

5   Conclusions

In this study, we have proposed an approach to extract fracture patterns from CT scans to replicate them on different 3D bone models. Starting with a CT scan, a segmentation and a fracture identification of a bone was done. Then the resulting point cloud was cleaned and processed in order to define a fracture surface. Finally, different representations or templates have been obtained.

Our results can be used to populate a database of fractures and models in order to feed an algorithm that procedurally generate random, yet realistic, cracks on bones. A potential application of this tool is medical training, either for traumatologists or automatic bone fracture reduction algorithms.

As future work, an algorithm to reverse the process could be developed, these will serve as a validation to our approach and to improve the medical training, providing simulated CT scans from 3D models of broken bones.

This paper has been focused on bone fractures, however, it can be extended to a wide range of research areas such as identification and classification of vase fractures in archaeology or damage of structural elements on buildings and industrial structures.

Acknowledgments. This work has been funded by the Spanish Ministry of Economy and Competitiveness and the European Union (via ERDF funds) through the research project DPI2015-65123-R.

References

1. Chowdhury, A.S., Bhandarkar, S.M., Robinson, R.W., Jack, C.Y.: Virtual multi-fracture craniofacial reconstruction using computer vision and graph matching. Comput. Med. Imaging Graph. **33**(5), 333–342 (2009)
2. Chowdhury, A.S., Bhandarkar, S.M., Robinson, R.W., Yu, J.: Virtual craniofacial reconstruction from computed tomography image sequences exhibiting multiple fractures. In: International Conference on Image Processing 2006, pp. 1173–1176 (2006)
3. Okada, T., Iwasaki, Y., Koyama, T., Sugano, N., Chen, Y.-W., Yonenobu, K., Sato, Y.: Computer-assisted preoperative planning for reduction of proximal femoral fracture using 3-D-CT data. IEEE Trans. Biomed. Eng. **56**(3), 749–759 (2009)
4. Paulano-Godino, F., Jiménez-Delgado, J.J.: Identification of fracture zones and its application in automatic bone fracture reduction. Comput. Methods Programs Biomed. **141**, 93–104 (2017)
5. Winkelbach, S., Westphal, R., Goesling, T.: Pose estimation of cylindrical fragments for semi-automatic bone fracture reduction. In: Joint Pattern Recognition Symposium, pp. 566–573 (2003)
6. Willis, A., Anderson, D., Thomas, T., Brown, T., Marsh, J.L.: 3D reconstruction of highly fragmented bone fractures. In: Medical Imaging 2007: Image Processing, vol. 6512, p. 65121P (2007)
7. Paulano, F., Jiménez-Pérez, J.R., Jiménez, J.J.: Simulation of bone fractures via geometric techniques: an overview. Comput. Methods Biomech. Biomed. Eng.: Imaging Vis. 1–6 (2018). https://doi.org/10.1080/21681163.2018.1498393
8. Rusu, R.B., Cousins, S.: Removing outliers using a statistical outlier removal filter. http://pointclouds.org/documentation/tutorials/statistical_outlier.php. Accessed 25 Mar 2019
9. Akkiraju, N., Edelsbrunner, H., Facello, M., Fu, P., Mucke, E., Varela, C.: Alpha shapes: definition and software. In: Proceedings of the 1st International Computational Geometry Software Workshop, vol. 63, p. 66 (1995)

Perineal Pressure in Equestrians: Application to Pudendal Neuralgia

Sébastien Murer[1]([⊠]), Guillaume Polidori[1], Fabien Bogard[1], Fabien Beaumont[1], Élisa Polidori[2], and Marion Kinne[2]

[1] GRESPI, Research Group in Engineering Sciences,
University of Reims Champagne-Ardenne,
Campus Moulin de la Housse, 51100 Reims, France
sebastien.murer@univ-reims.fr
[2] ESO Paris SUPOSTEO, Higher School of Osteopathy,
Campus Descartes, 8 rue Alfred Nobel, 77420 Champs-sur-Marne, France

Abstract. The present study takes place in a broader work aimed at investigating the triggering and aggravating factors of pudendal neuralgia (PN) in female equestrians. PN occurs in the case of damage to the pudendal nerve, the branches of which innervate the anus, rectum, perineum, lower urinary tract and genitalia. Although uncommon, PN often results in chronic perineal pain, with drastic consequences in the everyday life of patients and sometimes leading to suicide. More specifically, the main purpose of this work is to extend current knowledge of the pressure distribution at the perineal level during the practice of horseback riding. The preliminary works described herein focus on the measurement of contact pressure over the whole saddle/rider interface, for a female horse rider in a static stance corresponding to sitting trot. The pressure mapping is compared to a reference case, highlighting major discrepancies in the pressure distribution. Three-dimensional scans performed in parallel will be used in an upcoming numerical simulation which will implement the experimental pressure distribution as boundary conditions. This model will hopefully help gain insight in the possible improvements that may be brought to the practice of horseback riding, so as to decrease the risk of pudendal neuralgia in equestrians.

1 Introduction

The pudendal nerve is a mixed nerve composed of sensory, motor and autonomic fibers originating from the medulla, more specifically from the second, third and fourth vertebrae (S2–S4) at the sacrum level. It spans laterally out of the pelvis through the greater sciatic notch and under the piriformis muscle. The gluteal path is short and goes dorsally around the sciatic spine, or generally around the distal insertion of the sacrospinous ligament [20]. The pudendal nerve gives rise to the rectal nerve which in particular innervates the external anal sphincter, to the dorsal nerve of the clitoris in women and of the penis in men, to the perineal nerve innervating the scrotum, labia majora, part of the muscles elevating the

© Springer Nature Switzerland AG 2019
J. M. R. S. Tavares and R. M. Natal Jorge (Eds.): VipIMAGE 2019, LNCVB 34, pp. 482–491, 2019.
https://doi.org/10.1007/978-3-030-32040-9_49

anus, the ischio- and bulbocavernosus muscles as well as the bulb of the penis (see Fig. 1). Therefore, it plays a major role in the erection process and in both urinal and anal continence.

Pudendal neuralgia (PN), also known as Alcock canal syndrome, is a rare pathology characterized by a damage in the pudendal nerve located in the Alcock canal. The first symptom drawing attention on pudendal distress is pain, mainly located in the perineal region and in various forms: burning, torsion or striction sensations, electrical shocks of varying intensities and/or sensitivity disorders [15]. According to the French Association for Pudendal Neuralgia (Association des Malades d'Algie Pudendale), the living conditions of patients suffering from this form of neuropathy is drastically altered, with severe socio-professional consequences, including suicide. The illness is recognized by the French Social Security.

PN was first described by Zuelzer in 1915 and then long forgotten, until Aramenco took over in 1987 [2]. The literature is sparse regarding the prevalence and the numbers reported (4% of all consultants in pain management structures) are probably unreliable. The explanation is twofold: first, diagnosis remains misunderstood by health professionals and second, patients often do not dare seek medical attention, due to the affected body area. For instance, the International Pudendal Neuropathy Foundation estimates that the rate may be around 1/100000 of the world population, 2/3 of which being female patients [12] and that patients are aged between 50 and 70. The predominance of female patients guided the choice of the test subject in the present study.

Although etiology remains to be scientifically proved, it seems that this neuralgia is connected to one or several of the following:

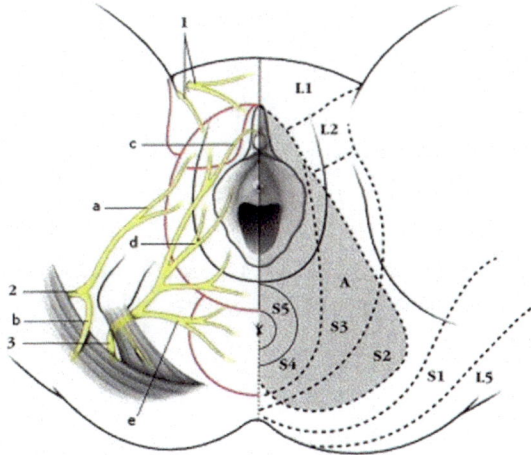

Fig. 1. Innervation of the perineum showing the pudendal nerve (3) and its ramifications: dorsal nerve of the clitoris (c), perineal nerve (d) and inferior rectal nerve (e). From [5].

- extensive cycling activities, or to a lesser extent horseback riding, leading to compression of the pudendal nerve
- pelvic or orthopedic surgery
- nerve stretching during childbirth
- anomalies in the sacroiliac joint
- some professions requiring prolonged sitting position

All these situations may lead to the Alcock syndrome [19].

In the field of sports, the practice of cycling is often cited as one of the main causes of PN: it has been reported that 6 to 8% of all cyclists would suffer from it consequently to practice over long distances [3]. As a matter of fact, cyclists undergo repeated impacts causing high levels of perineal pressure; an indirect consequence being compression of the pudendal nerve which in turn increases friction of the nerve in the Alcock canal [16]. This compression directly emanates from tissue compression between the bike seat and pubic symphisis.

The solicitations stemming from the practice of horseback riding are somehow similar, and the medical literature often mentions this practice as a triggering or aggravating factor of pudendal neuralgia and related pathologies [1,3,4,9,17,21]. Yet no studies has addressed the physiological and mechanical relations between horse riding and PN. From a physiological viewpoint, the perineal region is limited at the front by the edge of the pubic symphysis, at the back by the top of the coccyx and laterally by the ischial tuberosities. Its overall shape is convex. The geometric peculiarity of a horse saddle is that this seat system is concave, and matches the convex shape of the perineal zone as closely as possible. It is easy to understand in these conditions that the entire perineal zone is subjected to the pressures exerted by the saddle, especially since the opening of the rider's legs favours a wider exposure of this perineal zone. Therefore, although the pressure levels at the saddle/horse rider interaction are previously unknown, they may play a major role in the trapping or crushing of the pudendal nerve in the perineal area.

2   Materials and Methods

Test Subject

A 30 year-old national-level female rider (1.72 m, 69 kg) practicing show jumping has agreed to participate in the study and gave informed consent. The saddle used in the experiments is her own, and was tailor-made based on her pelvis shape.

Electromechanical Horse Simulator

Since our study involves the use of a 3D scanner, we opted for an electrome- chanical horse riding simulator (P. Klavins, Le Simulateur Équestre Français, Changé, France). Indeed, the acquisition requires still subjects, and a horse may have been scared by the blinking emanating from the scanner. Moreover, riding simulators have been found to be adequate for analyzing the rider's seat, position and movements [13].

3D Image Acquisition and CAD

The present study will be complemented by finite element simulations which will serve as a basis in the investigations regarding the stresses and strains undergone by the pudendal nerve. As seen in Fig. 2, the different branches of the pudendal nerve spread over a broad area and lie at various depths beneath the skin. Furthermore, each one of them is surrounded by organs with heterogeneous mechanical properties. As a consequence, complex mechanical phenomena occur when pressure is exerted by the saddle on the perineal area, and finite element simulations will be very helpful in this respect.

The first step in the creation of the numerical model is the acquisition of the geometry to be studied. To this end, a Creaform GO!SCAN 50® 3D scanner (resolution 0.5 mm) was employed to capture various geometries of interest (see Fig. 3). At first, a pelvis from a medical anatomical human skeleton model was scanned, as well as the saddle. Then, the rider's lower body and saddle were acquired while placed on the riding simulator (see Fig. 4).

The CAD geometry of the pudendal nerve, which will be the focus of the numerical model, was created using SolidWorks® software. An assembly of all these geometries is displayed in Fig. 5.

Pressure Mat

In the frame of rider/horse interaction, most studies focus on the pressure levels at the horse back level, in order to investigate the possible horse dorsal traumas

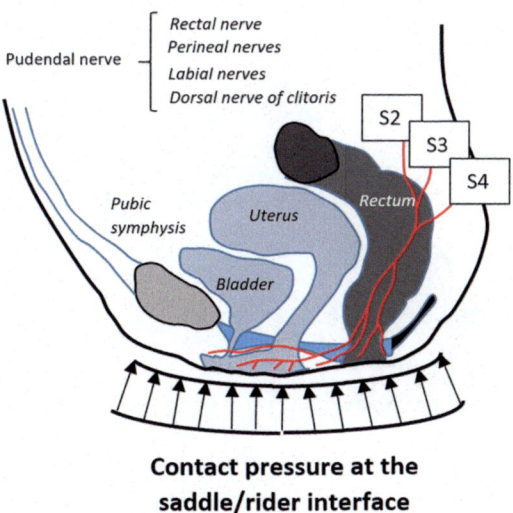

Fig. 2. Schematic drawing of the pudendal nerve and surrounding organs

486   S. Murer et al.

[8,11,14]. In this study, the main goal is to measure and analyze the pressure distribution in the perineal area during the practice of horseback riding.

The measurement device is a TexiSense Teximat® pressure mat, containing 1024 pressure sensors. The rider sat directly atop the electronic pressure mat, which was positioned on the saddle. Color-coded spatial pressure maps were generated by the acquisition software. It is worth noting that the sensors only detect the normal component of the force exerted on the mat. In such case, the vertical force representing the rider's weight is underestimated, especially in areas where saddle tilt is important. Nevertheless, this underestimation does not have significant influence on the pressure measured in the perineal area, since it is mostly horizontal therefore perpendicular to the body weight direction.

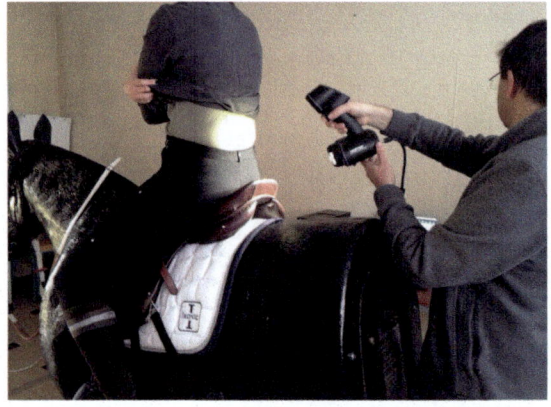

Fig. 3. Acquisition of 3D scans of the equestrian and saddle on the electromechanical horse simulator

Fig. 4. 3D scan of the rider and saddle on the horse simulator

3 Discussion

The pressure maps measured experimentally using the protocol described above are depicted in Fig. 6, alongside those corresponding to the reference case of the same rider seated on a wooden chair.

Sitting on a hard wooden chair (Fig. 6(a) and (b)) is associated with pressure concentration beneath the ischial tuberosities, while pressure is distributed far more homogeneously in the perineal area due to greater deformation of soft tissues. The imprint of the test subject is clearly observed.

Pressure map is quite dissimilar on a horse saddle (Fig. 6(c) and (d)): stress concentrations at the ischial tuberosities disappear and the distribution is homogeneous over an extended area. The explanation lies in the fact that the saddle material undergoes deformation, which increases the contact area between the rider and the saddle. Therefore body weight spreads over a broader area and contact pressure decreases. To a lesser extent, adaptation of the saddle to the rider's morphology also increases the contact area, as mentioned earlier.

More quantitatively, maximum contact pressures are shown to be 26.2 kPa for the horse saddle and 51.2 kPa for the chair, respectively. The results presented by Nicol et al. [18] highlight similar mean peak pressures, in the range 22–27 kPa. Interestingly, Fig. 6(c) reveals that pressure levels in the perineal area (12 to 16 kPa) are significantly higher than those observed on a hard wooden chair (4 kPa). The explanation is similar to that mentioned above: deformation of the saddle results in direct contact, thus contact forces, between the perineal

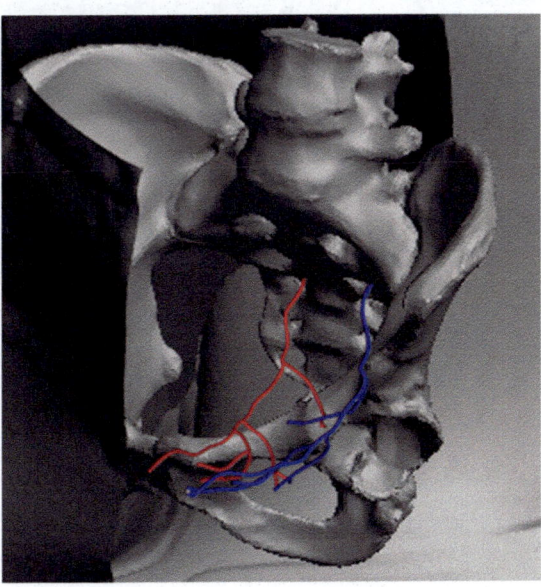

Fig. 5. 3D cut view of the rider's lower body, saddle, pelvis and CAD model of the left (blue) and right (red) pudendal nerves

area and the saddle. Conversely, the higher stiffness of a wooden chair decreases deformation and prevents direct contact in this location. Nonetheless, higher pressure levels in the pudendal nerve territory are at least partly responsible for nerve strain and damage. The upcoming numerical model built upon 3D scans and pressure measurements will also help test this hypothesis.

Regarding the contact pressure associated with horseback riding, Chang and Siegreg [7] showed that vascular occlusion may occur when the applied pressure exceeds the mean capillary pressure of 4.7 kPa in the case of prolonged sitting position. Bystrom et al. [6] report that mean pressure exceeding 11 kPa is likely to cause back pain injury in horse riding. There is no denying that even for a lightweight rider such as the one in this study, these thresholds are exceeded. The statement opens interesting perspectives in the understanding of pudendal neuralgia.

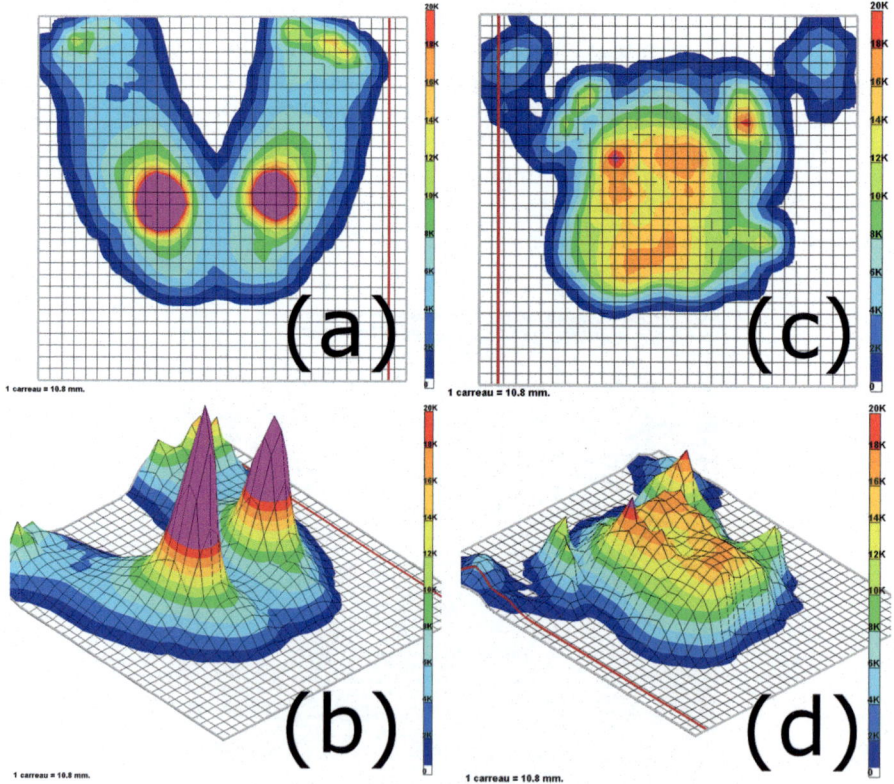

Fig. 6. Pressure maps: reference case of a wooden chair in 2D (a) and 3D (b); horse saddle in 2D (c) and 3D (d). Maximum pressure 20 kPa in all cases, grid size 10.8 mm.

4 Conclusions and Prospects

In this paper, we have presented preliminary studies aimed at improving knowledge in the mechanisms responsible for pudendal neuralgia, an uncommon yet serious disease which etiology remains unknown. The various branches of the pudendal nerve spread over the most intimate parts of the human body, constituting a major hurdle in the diagnosis which accounts for the prevalence uncertainty.

The pressure maps at the rider/saddle interface highlight significant discrepancies, both qualitatively and quantitatively, compared to those measured on a hard wooden chair. With the rider seated on the saddle in a sitting trot situation, the pressure distribution is more uniform over the measured area: pressure concentrations at the ischial tuberosities vanish due to saddle deformation. However, we found that the values of contact pressure in the vicinity of the perineum are noticeably higher than those recorded in the reference test case (wooden chair).

Our investigations into this area are still ongoing and we are currently in the process of building a numerical model based on the 3D scans described earlier and the aforementioned experimental pressure measurements. This model will certainly help gain insight with regard to the mechanical phenomena occurring inside the rider's lower body. To the best of our knowledge, the numerical studies reported in the scientific literature focus on stress urinary incontinence from 3D female pelvic floor modeling [10,22].

Acknowledgments. The authors would like to thank Arnaud Ravaux, head of the Centre Équestre de Rethel, and Céline Fouvry, head of the Centre de Formation d'Apprentis, without whose help this work would not have been possible.

References

1. Alanee, S., Heiner, J., Liu, N., Monga, M.: Horseback riding: impact on sexual dysfunction and lower urinary tract symptoms in men and women. Urology **73**(1), 109–114 (2009). https://doi.org/10.1016/j.urology.2008.07.058
2. Amarenco, G., Lanoe, Y., Perrigot, M., Goudal, H.: A new canal syndrome: compression of the pudendal nerve in Alcock's canal or perinal paralysis of cyclists. Presse Médicale **16**(8), 399 (1987)
3. Asplund, C., Barkdull, T., Weiss, B.D.: Genitourinary problems in bicyclists. Curr. Sports Med. Rep. **6**(5), 333–339 (2007)
4. Battaglia, C., Nappi, R.E., Mancini, F., Cianciosi, A., Persico, N., Busacchi, P.: Ultrasonographic and doppler findings of subclinical clitoral microtraumatisms in mountain bikers and horseback riders. J. Sex. Med. **6**(2), 464–468 (2009). https://doi.org/10.1111/j.1743-6109.2008.01124.x
5. Bolandard, F., Bonnin, M., Duband, P., Mission, J.P., Bazin, J.E.: Techniques d'anesthésie locorégionale du périnée: indications en gynécologie, en proctologie et en obstétrique. Annales Françaises d'Anesthésie et de Réanimation **25**(11–12), 1127–1133 (2006). https://doi.org/10.1016/j.annfar.2006.06.014

6. Byström, A., Stalfelt, A., Egenvall, A., von Peinen, K., Morgan, K., Roepstorff, L.: Influence of girth strap placement and panel flocking material on the saddle pressure pattern during riding of horses. Equine Vet. J. **42**, 502–509 (2010). https:// doi.org/10.1111/j.2042-3306.2010.00173.x

7. Chang, W., Seigreg, A.: Prediction of ulcer formation on the skin. Med. Hypotheses **53**(2), 141–144 (1999). https://doi.org/10.1054/MEHY.1998.0733

8. Clayton, H.M., Belock, B., Lavagnino, M., Kaiser, L.J.: Forces and pressures on the horse's back during bareback riding. Vet. J. **195**(1), 48–52 (2013). https://doi. org/10.1016/j.tvjl.2012.06.002

9. Devers, K.G., Heckman, S.R., Muller, C., Joste, N.E.: Perineal nodular induration: a trauma-induced mass in a female equestrian. Int. J. Gynecol. Pathol. **29**(4), 398–401 (2010). https://doi.org/10.1097/PGP.0b013e3181ce1341

10. Dias, N., Peng, Y., Timm, G.W., Zhang, Y., Sweet, R.M., Khavari, R., Erdman, A.G., Boone, T.B., Nakib, N.A.: Pelvic floor dynamics during high-impact athletic activities: a computational modeling study. Clin. Biomech. **41**, 20–27 (2017). https://doi.org/10.1016/j.clinbiomech.2016.11.003

11. Geutjens, C., Clayton, H., Kaiser, L.: Forces and pressures beneath the saddle during mounting from the ground and from a raised mounting platform. Vet. J. **175**(3), 332–337 (2008). https://doi.org/10.1016/J.TVJL.2007.03.025

12. Hibner, M., Desai, N., Robertson, L.J., Nour, M.: Pudendal neuralgia. J. Minim. Invasive Gynecol. **17**(2), 148–153 (2010). https://doi.org/10.1016/j.jmig.2009.11. 003

13. Ille, N., von Lewinski, M., Aurich, C., Erber, R., Wulf, M., Palme, R., Greenwood, B., Aurich, J.: Riding simulator training induces a lower sympathetic response in riders than training with horses. J. Equine Vet. Sci. **35**(8), 668–672 (2015). https:// doi.org/10.1016/j.jevs.2015.06.018

14. Kotschwar, A.B., Baltacis, A., Peham, C.: The effects of different saddle pads on forces and pressure distribution beneath a fitting saddle. Equine Vet. J. **42**(2), 114–118 (2010). https://doi.org/10.2746/042516409X475382

15. Labat, J.J., Delavierre, D., Sibert, L., Rigaud, J.: Approche symptomatique des douleurs pudendales chroniques. Progrès en Urologie **20**(12), 922–929 (2010). https://doi.org/10.1016/J.PUROL.2010.08.055

16. Leibovitch, I., Mor, Y.: The vicious cycling: bicycling related urogenital disorders. Eur. Urol. **47**(3), 277–287 (2005). https://doi.org/10.1016/j.eururo.2004.10.024

17. Mulhall, K.J., Khan, Y., Ahmed, A., O'Farrell, D., Burke, T.E., Moloney, M.: Diastasis of the pubic symphysis peculiar to horse riders: Modern aspects of pelvic pommel injuries. Br. J. Sports Med. **36**(1), 74–75 (2002). https://doi.org/10.1136/ bjsm.36.1.74

18. Nicol, G., Arnold, G.P., Wang, W., Abboud, R.J.: Dynamic pressure effect on horse and horse rider during riding. Sports Eng. **17**(3), 143–150 (2014). https://doi.org/ 10.1007/s12283-014-0149-z

19. Pérez-López, F.R., Hita-Contreras, F.: Management of pudendal neuralgia. Climacteric: J. Int. Menopause Soc. **17**(6), 654–656 (2014). https://doi.org/10.3109/ 13697137.2014.912263

20. Robert, R., Labat, J.J., Riant, T., Louppe, J.M., Hamel, O.: The pudendal nerve: clinical and therapeutic morphogenesis, anatomy, and physiopathology. Neurochirurgie **55**(4–5), 463–469 (2009). https://doi.org/10.1016/j.neuchi.2009.07.004

21. Turgut, A.T., Kosar, U., Kosar, P., Karabulut, A.: Scrotal sonographic findings in equestrians. J. Ultrasound Med. **24**(7), 911–917 (2005). https://doi.org/10.7863/jum.2005.24.7.911
22. Zhang, Y., Kim, S., Erdman, A.G., Roberts, K.P., Timm, G.W.: Feasibility of using a computer modeling approach to study SUI induced by landing a jump. Ann. Biomed. Eng. **37**(7), 1425–1433 (2009). https://doi.org/10.1007/s10439-009-9705-2

Macroescale Fracturation of Osseous Models

Francisco Daniel Pérez Cano[1][(✉)] and Juan José Jiménez Delgado[2]

[1] Universidad de Jaén, Campus Las Lagunillas A3-103, Jaén, Spain
fdperez@ujaen.es
[2] Universidad de Jaén, Campus Las Lagunillas A3-142, Jaén, Spain
juanjo@ujaen.es

Abstract. Fracturing of osseous models is a field that allows to obtain advances in computer assisted medical simulations, as well as to prototype fragile objects as is the case with bones. This field of study includes the use of physical features of bones. Not all fracturing methods are valid with osseous models, due to the importance of considering the hierarchical structure of the bones. In this work, we focus on the study of bone fractures at the macroscale level and on obtaining fractures with a realistic appearance. Therefore, we have developed a tool that allows the fracture of osseous models, obtained through a 3D segmentation or scanning process, by using patterns represented with a 2D image. The different approaches for the fracturing of osseous models are also showed, and discussed the need to obtain more information about the fracturing process.

Keywords: Bone model · Bone fragments · Fracture pattern · Fracture methods · Geometry · Hierarchical structure of bone · Interactive tool · Physical simulation · Segmentation

1 Introduction

The advantages provided by the fracture and deformation of geometric models, in fields such as architecture or medicine, has made it a topic of reference in computer graphics in recent years. In the field of traumatology, technological advances appoint new frontiers in the simulation and prototyping of fragile objects (as osseous tissue), as well as the evaluation of resistance and resilience studies by using geometric models [1].

A fracture can be defined as a phenomenon resulting in the breakage of a solid object as a result of a blow, force or traction exceeding its limit of elasticity. A fracture provides a great deal of information about the properties of a material and its relationship to other elements. The fracture of a osseous model also implies that the fragments have been obtained through a physical simulation applied to a heterogeneous material such as bone. This material is composed of a series of hierarchical structures that influence the entire fracturing process.

© Springer Nature Switzerland AG 2019
J. M. R. S. Tavares and R. M. Natal Jorge (Eds.): VipIMAGE 2019, LNCVB 34, pp. 492–500, 2019.
https://doi.org/10.1007/978-3-030-32040-9_50

We have developed a tool for fracturing a bone model at a macroscale level using patterns based on the fracture algorithm described by Müller et al. [6]. In addition to obtaining fractures with great detail it is possible to extend the algorithm so that new types of patterns can be included, thus increasing the detail and information about the fracture zone, enabling real time fracturation and performing fractures based on forces or physics. The purpose of this work is to obtain fractured models which can be used as input models for more detailed physical simulations, as prefractured bones, with a realistic appearance, using patterns obtained through the study of real fractures. These patterns will allow the creation of a large data set of realistic fractured bone models. In a future, we could use these fractured models obtained by physical simulation and compare them with real fractures. We could also consider the effect of different physical phenomena, as well as the training of intelligent systems for the automatic generation of fractures, so that advances could be achieved in the modelling and study of bone fractures.

2   Previous Works

There is a wide range of techniques to simulate the fracture of geometric models. Muguercia et al. [1] classifies methods for fracture simulation of geometric models into three approaches:

- Based on geometry: it focuses on the generation of suitable patterns, for the simulation of geometric fractures, obtaining a high degree of control over the fracture in aspects such as the size or shape of the different fragments. This approach does not focus on the physical description of the fracture phenomenon [1]. This type of fracture is faster to simulate than other approaches, although its main drawback is the loss of information and the difficulty of simulating the fracture in real time.
- Based on physics: this approach aims to obtain more realistic fractures through simulation. It is a very complex approach because the mechanical properties of the materials to be fractured need to be identified.
- Based on examples: it consists of a set of methods in order to obtain a fracture with real appearance, copying the behaviour of a natural fracture phenomenon [1]. The main drawback lies in obtaining real bone fractures.

In addition to the different ways of fracturing osseous models, it is also important to take into account the hierarchical structure of the bones, which is very complex, and the influence of the different microstructures that compose them in the fracture process. Furthermore to determining its mechanical behaviour, they also provide it with great hardness and rigidity. The most important levels are: the macroscale representing the entire bone, the mesoscale which distinguish between cortical and trabecular tissue and the microscale where several functional units are observed, highlighting the osteons and trabeculae.

In terms of fracturing at different scale levels, the macroscale simulations focus mainly on predicting the fracture risk at the whole bone level, whereas the microscale simulations evaluate the influence of microstructural features on crack propagation behaviour in the bone [2,10]. One of the most important studies at the macroscale level was realized by Hambli et al. [3] where a finite element model (FEM) is developed based on continuous damage to a model to simulate a bone fracture using force displacement curves (Fig. 1). The fracture begins in the femoral neck area and as the load increases, the damage grows rapidly following an oblique path to the top surface. The fracture pattern is calculated through the different levels of propagation of the cracks in relation to the force displacement curve. These studies predict bone strength to predict the risk of osteoporotic fractures. The problem with these approaches is that the results obtained can differ considerably with the real shape of a fracture.

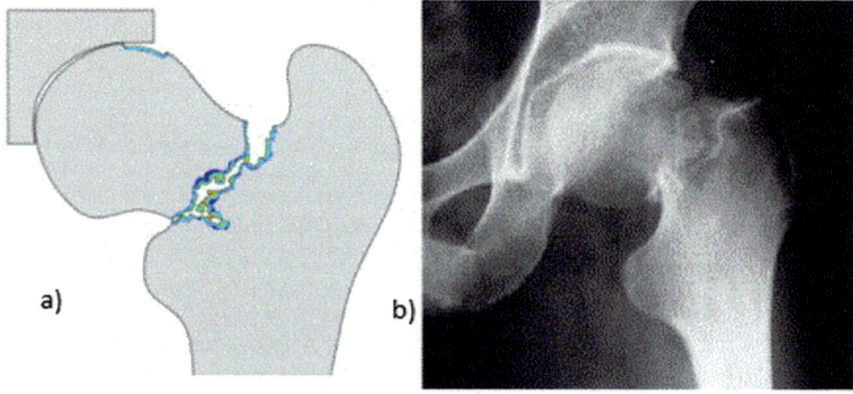

Fig. 1. Qualitative comparison between predicted fracture patterns and actual fracture examples in the Hambli et al. study [3]. (a) Expected fracture profile (full fracture). (b) Example of an X-ray of a 56-year-old man with a displaced type A fracture.

3   Materials and Methods

There are different libraries with algorithms for the simulation of fractures. They can be based on geometry such as Blast or Bullet or based on physics such as LibMesh or Vega. Blast is the reference library for geometric destruction developed by NVIDIA [4]. It replaces the APEX destruction module and has been designed from the ground up, focusing on resolving performance, scalability and flexibility deficiencies. It consists of three layers: a low-level layer, a high-level layer, and extensions. The extensions allow to complete the library's functionalities. This library is fully integrated with the Unreal Engine 4 (UE4) graphics engine as a separate plugin, so there is an option to work with the graphics engine without worrying about the graphical user interface (GUI) or extensions. Pérez et al. [5] analysed the use of this library for fracturing osseous models and concluded that Blast has been designed to assist in the destruction of geometric models but has no specific method for fracturing osseous models or for fracturing models in real time. Also concludes that it serves as a basis for the development of a bone fracture tool.

3.1 Methods for Fracturing Models with Blast

The library allows to fracture models with the following methods:

- Voronoi fracture: this method is based on generating Voronoi cells from cloud of internal points. It is composed of different variants to make a fracture. These variants range from specifying a number of pieces to be generated in the model to using a radial pattern.
- Uniform slicing fracture: this approach focuses on using a grid like a pattern. It is possible to adjust the number of cuts and to set a certain variation on the angle of slice. In addition, it can be activated a noise in the new faces that are generated when cutting the model.
- Cut fracture: the method consists in cutting the model with a specific plane.
- Cutout fracture: the user has to indicate a clipping pattern through a black and white image. This pattern is a texture where black regions assumed to be model projections and white lines are boundaries between projections. Noise can be applied to this fracture type as in the uniform slicing fracture method.

3.2 Algorithm for Fracturing

Although Blast provides a method to realistically perform fractures using patterns, this method is not adapted to osseous fracture and can be improved. The algorithm provided by Blast's Library is still under development so it has certain bugs that prevent the correct functioning of the tools. Therefore, we have developed our aim tool for fracturing osseous models based on the fracture algorithm described by Müller et al. [6] using Blast. This algorithm consists in the projection of a fracture pattern on a geometric model of bone to obtain a prefractured model. Then the use of forces in specific areas that affect the prefractured model cause the creation of different chunk depending on the damage in a certain area. In this article we focus on obtaining prefractured bone models while in the future we will use the forces to perform a deeper analysis of the fracture zone. The proposed method makes it possible to solve problems related to obtaining fractures with a realistic appearance.

For fracturing osseous models, we need two entries: the representation of the model and a pattern to cut out the model. For the representation we have studied two different approaches: 3D scanning of osseous models and segmentation of bones from computer tomography (CT) images. The biggest problems of the first approach are the time it takes to perform the scan and posterior segmentation of the model as well as the loss of information. The second approach solves most of the problems, and although obtaining medical images cannot be possible in all situations, it is a much faster and more precise process. Paulano [7] conducted a study focusing on segmentation of fractured bones from CT images and concludes that traditional methods of segmentation work well for segmentation of healthy bones, but are unable to identify fractured bones. As the input data contain healthy and fractured bones, we have used the method proposed by Paulano [7] based on 2D region growing [8,9].

3.3   Fracture Pattern

We use a 2D image that represents the fracture as the pattern to fracture the model. These patterns have been obtained through a process of manual analysis of the fracture zone of real bones after a process of mechanical experimentation. (Fig. 2a) represents just the moment of the fracture of a human femur while image (Fig. 2b) shows points that form the fracture limits view from the back of the bone. The pattern is composed of a series of regions indicating the fracture limits seen from one of the different axes of the bone. The white region indicate the limits of the fracture (Fig. 2c). The tool developed adjusts the texture to the osseous model and projects it through one of the axis to obtain the cutting area (Figs. 4a and c). Although we have focused on the projection of the pattern over a bone model and we use noise to obtain realistic-looking fractures, the model could be extended to accept different types of patterns obtained after analyzing the fracture area of a bone that could provide more precise patterns when fracturing osseous models. These patterns could be represented through spherical coordinates, through a 2D texture that includes the entire fractured area, or through a 3D representation using a map of heights.

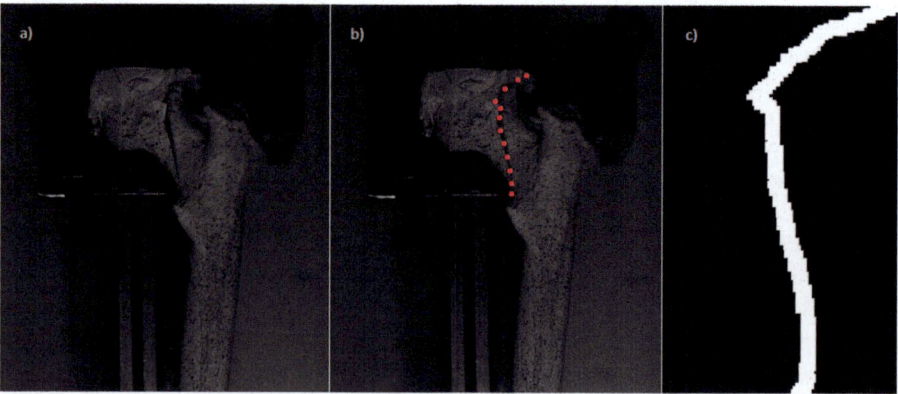

Fig. 2. Extraction of the fracture pattern through the fracture of a femur after a process of mechanical experimentation. Figure a represents the femur at the time the fracture occurs, figure b indicates the fracture zone with red c and figure c the fracture pattern obtained.

4   Results

The tool developed is an extension of the Blast fracture tool with integration
into Unreal Engine 4.18 (Fig. 3). This tool consists of a top menu with basic
options related to the fracture such as the fracture algorithm used, the model
reset, the optimization or the importation of a fracture. In this menu we have
added an option to apply basic transformations on the model to be fractured
that was not included in the default tool. On the left side there is a field with the
different fragments included in the editor. Normally only one entry appears until
it is fractured. In the menus on the right there are different options to customize
the fractures to be generated: from general options, such as the material to be
used for the new faces generated, to the properties needed for a specific billing
method. Our fracture method contains the following input fields:

- A dropdown to select the axis on which we want to project the fracture.
- The 2D image selector.
- A field to indicate the precision of the pattern to be used.
- A field to indicate the number of Voronoi regions to create and improve
  accuracy.

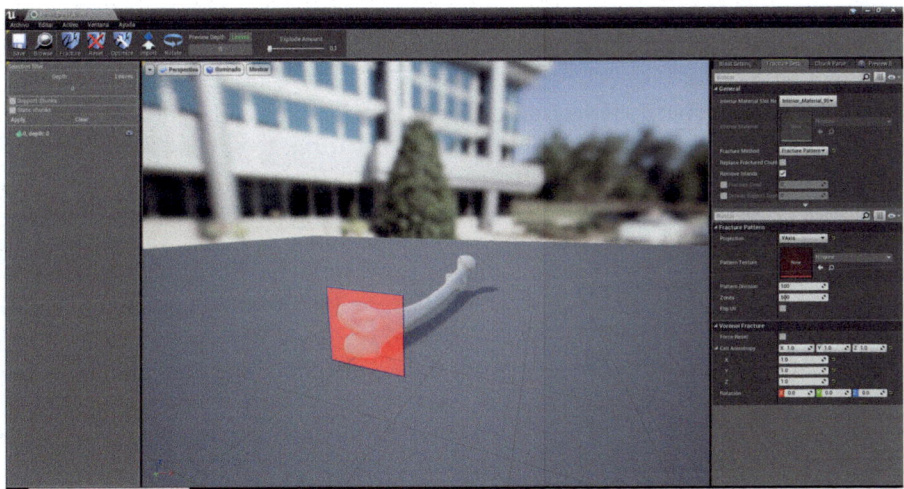

Fig. 3. Image of the editor with our method integrated in the Blast version for the
Unreal Engine.

In our study we have focused on fracturing geometric models representing
femurs. The examples below use two different patterns on the Y axis. In the
image (Figs. 4a and c) the fracture pattern can be seen. This pattern is already
projected on the Y axis of the model represented with a red mesh. As result we
have obtained two fragments in each case with some noise in the internal faces

since we have used the noise functionality that includes Blast so that it is not a flat fracture and has a more realistic aspect. As shown at the results (Figs. 4b and d) the method allows us to obtain realistic fractured bone models following the pattern provided as input.

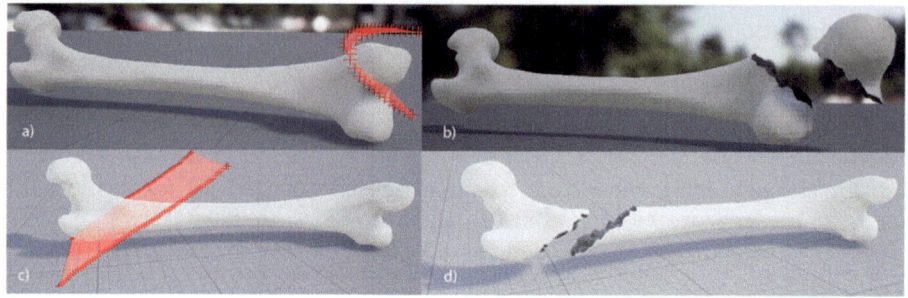

Fig. 4. Figure a and c represent the bone prepared to be fractured with the fracture patterns already projected onto the Y axis. Figure b and d represent the result obtained after being fractured.

5   Discussion

Blast has different methods to fracture geometric models. The problem with the approaches available with Blast library to fracturing geometric models is that it is not possible to perform fractures with a realistic appearance. The result obtained differs considerably from what is expected in a bone fracture in terms of appearance and morphology. In the last versions of Blast they added the cutout mode using a pattern that allows to obtain results similar to those of our tool. This mode is under development and there are several bugs currently preventing a smooth use. The method allows to obtaining prefractured models with realistic appearance, but not fractured models in real time.

Our tool focus on Müller et al. [6] fracture technique, which consists of the division of the geometric model into a large number of convex polygons. These polygons together with the fracture pattern allow to determine the different fracture zones of a model. This algorithm allows, in addition to obtaining the fractured model, to study the fracture process through the use of a force. Also allows to study the fracture effect it would produce on the model taking into account this prefractured model, because it is designed to produce fractures in real time. Although the result of fracturing a osseous model at the macroscale level results in a fracture with a realistic appearance, the main problem of fracturing it at this level or following a geometric approach is that it can be morphologically different from a fracture of a real bone.

Relative to the fracture pattern, we have talked about different types of patterns that can be used for fracturing osseous models. Our tool uses as a basis the representation of a fracture seen from one side of the bone. It is the simplest pattern to obtain although it is also the least precise although it is enough to obtain a realistic looking fracture. The representation of the fracture zone by spherical coordinates, 2D texture or a map of heights will provide a greater precision when fracturing a geometric model. The problem in obtaining these patterns is that an exhaustive and precise analysis of the fracture zone of a geometric model is required.

6 Conclusions and Future Work

The fracturing of osseous models is a very important field on traumatology with the aim of obtaining a varied database, with fragments of realistic aspect and relevant information for future studies. Therefore, there is a need to study the fracturing of osseous models at different scales. In our work we have focused on the development of a tool for fracturing osseous models, obtaining simulations of fractures with realistic aspects by using patterns.

The method used is designed to perform fractures in real time, so we could realize a physical simulation and study of the fracturing process in real time. Faced with the loss of information, it is necessary to include different levels of detail. Studies at the mesoscale and microscale levels will provide us with morphological information that we cannot obtain with studies at the macroscale level. Furthermore, it would be interesting not to be limited to the use of perfect shapes at this microscopics levels when representing structures such as osteons. In future studies, it will be recommendable to take into account the deformation of the bone contours in order to obtain more accuracy. In this way, the importance of these parameters has been demonstrated in the fracturing process.

The development of the proposed tool for the fracturing of models in a realistic looking is intended to lay the foundations and serve as a frame of reference for the fracturing of osseous models. Finally, the expert validation of the fragments obtained after this process by using the comparison with real cases is also pending for future work.

Acknowledgments. This work has been funded by the Spanish Ministry of Economy and Competitiveness and the European Union (via ERDF funds) through the research project DPI2015-65123-R.

References

1. Muguercia, L., Bosch, C., Patow, G.: Fracture modeling in computer graphics. Comput. Graph. **45**, 86–100 (2014). http://www.sciencedirect.com/science/article/pii/S0097849314000806
2. Ural, A.: Cohesive modeling of bone fracture at multiple scales. In: 11th International Conference on the Mechanical Behavior of Materials (ICM11), Procedia Engineering, vol. 10, pp. 2827–2832 (2011). http://www.sciencedirect.com/science/article/pii/S1877705811006588

3. Hambli, R., Bettamer, A., Allaoui, S.: Finite element prediction of proximal femur fracture pattern based on orthotropic behaviour law coupled to quasi-brittle damage. Med. Eng. Phys. **34**(2), 202–210 (2012). https://doi.org/10.1016/j.medengphy. 2011.07.011

4. Nvidia gameworks. https://developer.nvidia.com/what-is-gameworks. Accessed 20 Jan 2019

5. Pérez, F.D., Jiménez, J.J., Jiménez, J.R.: Blast features and requirements for fracturing osseous models. In: Spanish Computer Graphics Conference (CEIG) (2018). https://diglib.eg.org/handle/10.2312/ceig20181161

6. Müller, M., Chentanez, N., Kim, T.-Y.: Real time dynamic fracture with volumetric approximate convex decompositions. ACM Transact. Graph. **32**(4), 1 (2013). https://doi.org/10.1145/2461912.2461934

7. Paulano, F., Jiménez, J.J., Pulido, R.: 3D segmentation and labeling of fractured bone from CT images. Vis. Comput. **30**(6–8), 939–948 (2014). https://doi.org/10. 1007/s00371-014-0963-0

8. Justice, R., Stokely, E., Strobel, J., Ideker, R., Smith, W.: Medical image segmentation using 3D seeded region growing. Med. Imaging 1997 Image Process. **3034**, 900–910 (1997)

9. Fan, J., Zeng, G., Body, M., Hacid, M.-S.: Seeded region growing: an extensive and comparative study. Pattern Recogn. Lett. **26**(8), 1139–1156 (2005). https:// doi.org/10.1016/j.patrec.2004.10.010

10. Pérez Cano, F.D., Jiménez, J.J.: Towards the modelling of osseous tissue. In: Proceedings of the 14th International Joint Conference on Computer Vision, Imaging and Computer Graphics Theory and Applications - Volume 1: GRAPP, pp. 340–345 (2019)

Influence of Transversal Resolution on Reconstructing Atherosclerotic Plaque Components

Ondřej Lisický[1]([✉]), Aneta Malá[2], and Jiří Burša[1]

[1] Institute of Solid Mechanics, Mechatronics and Biomechanics,
Brno University of Technology, Brno, Czech Republic
`161238@vutbr.cz`, `bursa@fme.vutbr.cz`
[2] Institute of Scientific Instruments, The Czech Academy of Science,
Brno, Czech Republic
`malaa@ISIBrno.cz`

Abstract. Assessment of atherosclerotic plaque vulnerability is nowadays closely related to computational modelling. Mechanical loading is considered as trigger for plaque rupture and requires patient specific model geometry. Here segmentation of each cross section obtained from medical imaging is used for model reconstruction. However, the number of input images depends on the chosen method. The influence of transversal resolution on surface and volume deviations is quantified in this study. The influence of smoothing process was also considered. A specimen harvested during carotid endarterectomy was imaged by *ex vivo* high-resolution 9.4 T MRI. The slice thickness varied from 1.5 mm to 0.25 mm with fixed in-plane resolution 0.078 mm. Segmented cross sections were then combined along the arterial axis. Significant differences were found in both surface and volume analyses. Smoothing was found as insignificant parameter within the slice thickness. Findings of this study show that transversal resolution influences the model, although the influence on stresses should still be investigated. *Ex vivo* imaging with a significantly better resolution may help us with understanding a plaque rupture mechanism and serve as verification for *in vivo* based models.

Keywords: Atherosclerosis · Carotid artery ·
Computational modelling · Magnetic resonance · Segmentation

1 Introduction

Atherosclerosis is a severe cardiovascular disease with high mortality. Cumulation of lipids leads to plaque creation which results in artery narrowing also called stenosis [1,2]. The degree of stenosis is still the only validated diagnostic criterion for risk assessment [3].

Computational modelling shows its potential to become an appropriate indicator of plaque vulnerability. Biomechanics helps us to capture its mechanical

© Springer Nature Switzerland AG 2019
J. M. R. S. Tavares and R. M. Natal Jorge (Eds.): VipIMAGE 2019, LNCVB 34, pp. 501–508, 2019.
https://doi.org/10.1007/978-3-030-32040-9_51

behaviour, morphological and geometrical factors which may influence its stress-strains response [4–7]. Two dimensional (2D) analyses are gradually replaced with 3D idealized and patient-specific (PS) models [8,9]. Idealized models allow us to localize critical parameters such as size of the lipid core (LC), fibrous cap thickness, etc. Both Cilla et al. [8] and then Yuan et al. [9] confirmed the influence of those components on the stress-strain state. However, PS models are considered to be more suitable for plaque vulnerability assessment [10]. However, their geometry is expressly dependent on the imaging method. There is a large number of *in vivo* methods used for plaque visualization such as intravascular ultrasound, computed tomography (preferably angiography (CT-A)), optical coherent tomography, or magnetic resonance imaging (MRI) [10]. MRI is currently taken as only non-invasive method suitable to quantify plaque components with good correlation with histology [11,12]. However, the distance between input images (transversal resolution) is still far from in-plane resolution which leads to relatively inaccurate approximation between slices in reconstructing model geometry. Nieuwstadt et al. [13] found the importance of an anisotropic resolution with a relatively better in-plane resolution to improve carotid plaque quantification. However, Nieuwstadt et al. [6] already reported that a decreased number of axial images may lead to overestimation of some plaque components included in computational model. Decrease in slice thickness may not improve plaque quantification, although it is an important parameter for 3D PS models. Model used in those papers [6,13] were based on histology. Histology is taken as a gold standard for plaque composition although the sample preparation leads to considerable specimen deformation; it disables reconstruction of the 3D geometry of artery with significant curvature [4,6]. Groen et al. [4] demonstrated a methodology of adjusting histological sections with *ex vivo* imaging for appropriate representation of a curved shape. Model geometry is crucial input for computational modelling and if we want to better understand the mechanism of plaque vulnerability we have to focus on this issue although it is still based on *ex vivo* imaging and thus not yet feasible in decision making of surgeons. This study aims at quantification of deviations between model geometries differing in transversal sampling and ways of volume creation using high resolution MRI (9.4 T), and also at evaluation of their appropriateness for computational modelling; subsequently they could become etalons for comparison with *in vivo* models.

2   Materials and Methods

2.1   Magnetic Resonance Imaging

Carotid plaque sample from patient who was scheduled for endarterectomy (man, 61 years, provided by St. Ann University Hospital with his informed consent) underwent post-operative imaging within a day after the intervention. The sample was measured in a 9.4 T NMR system (Bruker-BioSpec 94/30 USR, Ettlingen, Germany). The images were acquired in two series: the first one with the

RARE [14] technique (repetition time TR = 2500 ms, effective echo time TEeff = 10.77 ms), the second one with IR-RARE [15] (TR = 3000 ms, TEeff = 6.66 ms, inversion time TI = 950 ms). In each series, the slice thickness and the number of slices were varied: for thickness 0.25, 0.5, 0.75, 1 and 1.5 mm we used 50, 25, 16, 12, and 10 slices, respectively. Matrix size of 256×256, field of view of 20×20 mm^2, in-plane (2D) resolution of 0.078×0.078 mm^2, and RARE-factor = 2 were used in all measurements. To achieve a sufficient image quality, we always set the interslice gaps equal to the slice thickness and obtained the missing information by a similar measurement with all slices shifted by the slice thickness.

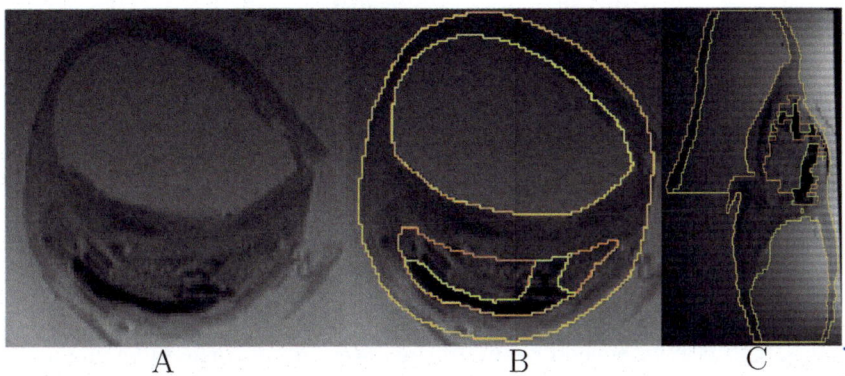

Fig. 1. *Ex vivo* MRI images of carotid plaque: (A) T1w image; (B) segmented transversal contour; (C) segmented sagittal contour

2.2 Image Segmentation and 3D Geometry Reconstruction

Each cross section was manually segmented using RETOMO software (BETA CAE Systems). Lipid core (LC), region with calcifications (CAL) and fibrous tissue (FT) were distinguished (Fig. 1). Images were segmented for each slice thickness to capture information modalities. A build-in algorithm was used to polygonize process often used to capture the object surface. High anisotropy of resolution leads to a stepped shape. The information from the image is stretched within specific slice thickness.

A smooth surface is usually presented within biological structures. As a large transversal resolution may lead to non-smooth structure, some approximation between slices is used such as NURBS [16]. In this study the Laplacian smoothing algorithm was used. Fixed parameters were used for each sample. Smooth model is also a more suitable basis for discretization.

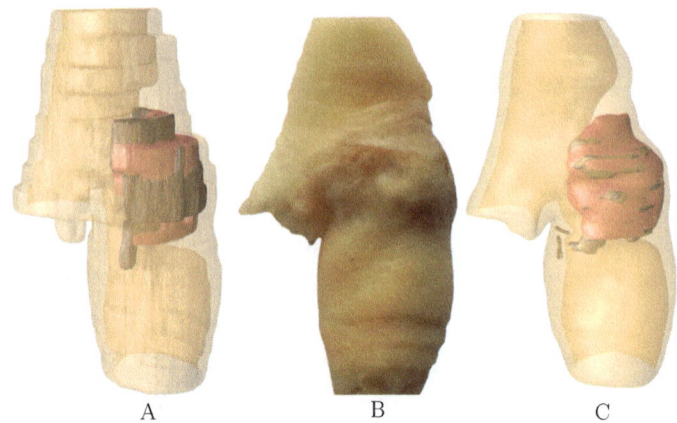

A   B   C

Fig. 2. 3D plaque model. Red: lipid core; grey: calcification region. (A) shows non-smoothed model based on 1.5 mm slice thickness; (B) shows the image of carotid plaque sample which was used for *ex vivo* imaging; (C) shows smoothed model based on 0.25 mm slice thickness.

2.3 Analysis

Smoothed and non-smoothed configurations were used in the following analyses. Surface deviations were defined as mean positive and mean negative differences between the reference model and the model in question. For this purpose, the smoothed model created from the 0.25 mm thick slices was chosen as reference. CAL were frequently localized as separated locations within the LC. LC and CAL were considered as connected components for surface deviation analysis. Volume of each component was also compared as another variance indicator. One way ANOVA and specifically Tukey comparison test were used to determine if there was a significant difference between the created models.

3 Results

The investigated greatly calcified atherosclerotic plaque was classified as type VII by scheme of the American Heart Association. Quantitative comparisons were performed with overall ten models, five raw models with visible stepped transitions and five after the Laplacian smoothing procedure. Figure 2 shows a view of the harvested plaque and its reference 0.25 mm smoothed model, as well as the non-smoothed model based on 1.5 mm slice thickness.

3.1 Surface Deviations

Surface deviations of the reference model compared with its non-smoothed configuration were found to be on the same order as the in-plane resolution for each plaque component. Figure 3 shows the comparison among the models.

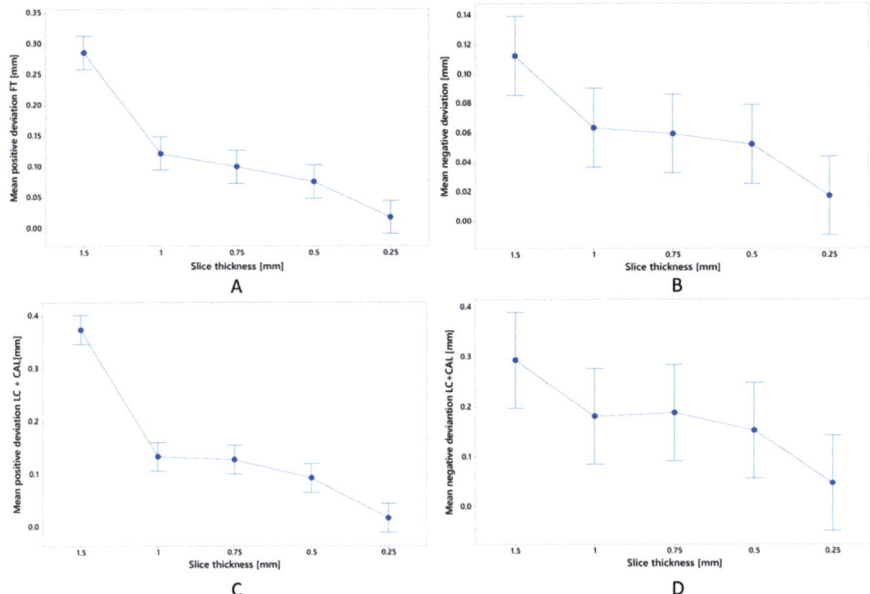

Fig. 3. Statistical analysis: (A–D) show surface deviations in dependence on slice thickness. A and B are for FT while C and D are for the joint region of CAL and LC. For all the models, both non-smoothed and smoothed models are compared with the reference model, and mean and variance are calculated from these two values. For slice thickness of 0.25 mm one of these values is zero (reference model itself).

Although differences between smoothed and non-smoothed models were insignificant (compared to vessel dimension), differences among slice thickness in both positive and negative surface deviations differed significantly ($p < 0.05$). From the graphs in Fig. 3 it is apparent that for slice thickness between 0.5 and 1 mm the differences are insignificant, while the models created on the basis of the largest (1.5 mm) and the lowest (0.25 mm) thickness differ significantly.

3.2   Volume Comparison

A volume analysis of each component confirmed significant differences among the models ($p < 0.05$). Generally, only 1 mm and 0.5 mm model may be considered as identical. Differences in volume are presented in Fig. 4.

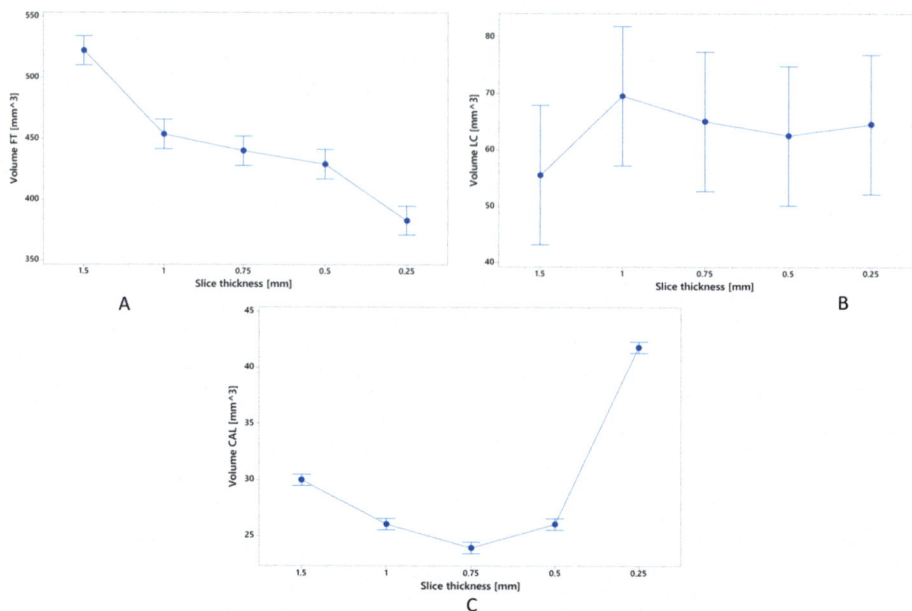

Fig. 4. Statistical analysis: (A–C) shows volume dependency on slice thickness for each component. (A) FT; (B) LC; (C) CAL.

4   Discussion

This study aimed to explore the influence of transversal resolution on the model geometry as a needed input for computational modelling in plaque vulnerability prediction. The results show that the smoothing process does not influence the model surface deviations and volume within the same slice thickness. Significant differences ($p < 0.05$) were found when comparing the models obtained with different slice thickness. However, for surface deviations these differences were on the order of 10^{-2} mm which is far below the model dimensions. Generally, the volume of FT was decreasing with decreasing slice thickness. This may be explained with the polygonization method used in this study which approximates the image content within the slice thickness linearly. LC and CAL volumes were nearly the same except for the 0.25 mm slice thickness. These results cannot be directly compared with the study of [6] since their analysis was focused on peak cap stress, although it was found that lower axial sampling has a profound influence on accuracy of the computed stresses. The sample used in this study contained a single LC. However, plaques may obtain smaller LC which may be visualized appropriately only by using lower slice thickness. Since differences between the corresponding non-smoothed and smoothed models were insignificant, it can be concluded that the used Laplacian smoothing algorithm could be used in creating the geometries for computational modelling. However, since significant differences were found among the models with different slice thickness, its influence on stress-strain analysis is to be expected and a deeper analysis of this dependence should be performed with more samples.

This study points to necessity of choice of an appropriate input for creation of model geometry of atheroma and its further computational modelling. Post-operative extracted samples may serve as a basis for those models. The use of high resolution MRI in connection with histological slices enables us to reconstruct the model geometries more accurately than in case of using only histological slices. With further development of imaging devices and their improved resolution, such models could be created on the basis of *in vivo* imaging only in future.

Acknowledgements. This research was funded by Czech Science Foundation project No. 18-13663S and by the MEYS project (LM2015062 Czech-BioImaging).

References

1. Ross, R.: The pathogenesis of atherosclerosis: a perspective for the 1990s. Nature (1993). https://doi.org/10.1038/362801a0
2. Guyton, J.R., Klemp, K.F.: Development of the lipid-rich core in human atherosclerosis. Arterioscler. Thromb. Vasc. Biol. **16**, 4–11 (1996)
3. Lusis, A.J.: Atherosclerosis. Nature (2010). https://doi.org/10.1038/35025203
4. Groen, H.C., et al.: Three-dimensional registration of histology of human atherosclerotic carotid plaques to in-vivo imaging. J. Biomech. **43**(11), 2087–2092 (2010). https://doi.org/10.1016/j.jbiomech.2010.04.005
5. Maldonado, N., et al.: A mechanistic analysis of the role of microcalcifications in atherosclerotic plaque stability: potential implications for plaque rupture. AJP: Heart Circ. Physiol. **303**(5), 619–628 (2012). https://doi.org/10.1152/ajpheart.00036.2012
6. Nieuwstadt, H.A., et al.: The influence of axial image resolution on atherosclerotic plaque stress computations. J. Biomech. (2012). https://doi.org/10.1016/j.jbiomech.2012.11.042
7. Huang, Y., et al.: The influence of computational strategy on prediction of mechanical stress in carotid atherosclerotic plaques: comparison of 2D structure-only, 3D structure-only, one-way and fully coupled fluid-structure interaction analyses. J. Biomech. (2014). https://doi.org/10.1016/j.jbiomech.2014.01.030
8. Cilla, M., Peña, E., Martínez, M.A.: 3D computational parametric analysis of eccentric atheroma plaque: influence of axial and circumferential residual stresses. Biomech. Model. Mechanobiol. (2012). https://doi.org/10.1007/s10237-011-0369-0
9. Yuan, J., et al.: Influence of material property variability on the mechanical behaviour of carotid atherosclerotic plaques: a 3D fluid-structure interaction analysis. Int. J. Numer. Methods Biomed. Eng. (2015). https://doi.org/10.1002/cnm.2722
10. Holzapfel, G.A., et al.: Computational approaches for analyzing the mechanics of atherosclerotic plaques: a review. J. Biomech. **47**(4), 859–869 (2014). https://doi.org/10.1016/j.jbiomech.2014.01.011
11. Hatsukami, T.S., et al.: Visualization of fibrous cap thickness and rupture in human atherosclerotic carotid plaque in vivo with high-resolution magnetic resonance imaging. Circulation **102**(9), 959–964 (2000). https://doi.org/10.1161/01.CIR.102.9.959

12. Saam, T., et al.: Quantitative evaluation of carotid plaque composition by in vivo MRI. Arterioscler. Thromb. Vasc. Biol. **25**(1), 234–239 (2005). https://doi.org/10.1161/01.ATV.0000149867.61851.31
13. Nieuwstadt, H.A., et al.: A computer-simulation study on the effects of MRI voxel dimensions on carotid plaque lipid-core and fibrous cap segmentation and stress modelling. PLoS ONE **10**(4), e0123031 (2015). https://doi.org/10.1371/journal.pone.0123031
14. Hennig, J., Nauerth, A., Friedburg, H.: RARE imaging: a fast imaging method for clinical MR. Magn. Reson. Med. **3**, 823–833 (1986)
15. Bydder, G.M., Young, I.R.: MR imaging: clinical use of the inversion recovery sequence (1985)
16. Holzapfel, G.A.: A layer-specific three-dimensional model for the simulation of balloon angioplasty using magnetic resonance imaging and mechanical testing. Ann. Biomed. Eng. (2002). https://doi.org/10.1114/1.1492812

3D-FEM Modeling of Iso-Concentration Maps in Single Trabecula from Human Femur Head

Fabiano Bini[1(✉)], Andrada Pica[1], Simone Novelli[1,2],
Andrea Marinozzi[3], and Franco Marinozzi[1]

[1] Department of Mechanical and Aerospace Engineering,
"Sapienza" University of Rome, Rome, Italy
fabiano.bini@uniroma1.it
[2] Institute for Liver and Digestive Health, University College London (UCL),
London, UK
[3] Orthopedy and Traumatology Area, "Campus Bio-Medico" University,
Rome, Italy

Abstract. In the present study, a 3D finite element model of a single trabecula from the human femur head is developed with the aim of evaluating the dynamics of water diffusion through bone tissue and the subsequent swelling of the specimen due to water uptake. Numerical results in terms of dimensional changes of the trabecula are in good agreement with experimental observations reported in literature. The results show iso-concentration maps at different instants of time. Furthermore, we determined the spatial distribution of the effective diffusion coefficients of water along the length (L), width (W) and thickness (T) direction of the trabecula. The findings suggest that the model could provide valuable information for bone microstructure characterization.

Keywords: Single trabecula · Swelling · Iso-concentration map · Iso-diffusivity map · 3D finite element method

1 Introduction

Bone is a composite material with a complex hierarchical structure that influences tissue properties. Investigation of bone tissue behaviour during physiological processes, as mass transport, could provide a thorough understanding of its characteristics.

Mass transport represents a fundamental phenomenon in bone physiology, since it allows the exchange of nutrients and signalling molecules that contribute to bone remodelling and mineralization process. Experimental analysis of diffusion within trabecular bone specimen has led to new insights for bone tissue characterization up to the length scale of its principal components, i.e. collagen and apatite mineral [1, 2]. The hygroexpansion of a single human trabecula consequent to water sorption was measured by Marinozzi and co-workers [1]. Namely, an air-dried single trabecula from human femur head was completely immersed in water and the displacements along the three main axes, Length (L), Width (W) and Thickness (T) of the trabecula were measured by means of a high accuracy dilatometer.

© Springer Nature Switzerland AG 2019
J. M. R. S. Tavares and R. M. Natal Jorge (Eds.): VipIMAGE 2019, LNCVB 34, pp. 509–518, 2019.
https://doi.org/10.1007/978-3-030-32040-9_52

The results of the experimental data were processed applying a genetic algorithm. This approach allowed to determine the effective diffusion coefficients of water (D_L, D_W, D_T) along the three main axes, taking into account the influence of the collagen apatite porosity [2].

In the present contribution, we reproduce by means of the finite element method (FEM) the experimental investigation of mass transport within the single trabecula described in the study of Marinozzi et al. [1]. Thus, we exploit the finite element analysis to assess the temporal evolution of water concentration within the trabecula. Moreover, an optimization approach allows to predict the effective diffusion coefficient in every point of the specimen giving further insight into the dynamics of water diffusion within the bone tissue.

2   Materials and Methods

The water diffusion and subsequent hygroexpansion of the single trabecula is modelled in a multiphysical framework. A 3D-FEM of the single trabecula is developed by approximating the single trabecula to a parallelepiped with W = 2 mm, T = 0.5 mm and L = 9 mm. We assumed that the origin of the Cartesian coordinate system is in the centre of the specimen (Fig. 1). The generated geometry of the trabecula is converted into a mapped mesh with 200 k elements. The elements are characterized by a size of 12.5 µm along the thickness direction, 40 µm along the width direction and 90 µm along the length direction. Consequently, the 3D model is characterized by 1.644.381 degrees of freedom.

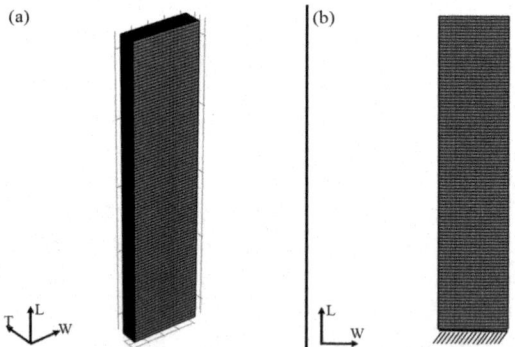

Fig. 1. (a) 3D model of the single trabecula: width W = 2 mm, thickness T = 0.5 mm and length L = 9 mm (b) 2D view of the fixed constraint at the bottom surface of the model.

In order to closely mimic the swelling experiment, displacement and rotations of the bottom surface of the specimen are constrained (Fig. 1b).

To model the water diffusion within the trabecula, Fick's law is applied (Eq. 1) [3]:

$$\frac{\partial c}{\partial t} = \nabla \cdot (D \nabla c) \tag{1}$$

where c is the water concentration $(mol \cdot m^{-3})$ at point (x, y, z) and at time t (s), D is the diffusivity matrix and it is composed as follows:

$$D = \begin{pmatrix} D_W & 0 & 0 \\ 0 & D_T & 0 \\ 0 & 0 & D_L \end{pmatrix} \tag{2}$$

where D_W, D_T, D_L are the effective diffusion coefficients $(m^2 \cdot s^{-1})$ of water along the width (W), thickness (T) and length (L) directions of the trabecula, respectively.

The initial and boundary conditions complete the mathematical description of the FEM. We assumed that initially, the trabecula specimen is at room temperature of $27 \pm 1 °C$ and relative humidity of $41 \pm 3\%$ RH. Subsequently, we make the hypothesis that the specimen is totally submerged in distilled water. The transient period represented by immersion and by the initial time able to measure the swelling of the trabecula, roughly of 10 s, is modelled by means of a smooth step function. The equilibrium water concentration (C_{eq}) is calculated as follows:

$$C_{eq} = \frac{m_{water}}{M_w \cdot V_{reservoir}} \tag{3}$$

where m_{water} (kg) is the mass of water in the reservoir, M_w $(kg \cdot mol^{-1})$ is the molar mass of water and $V_{reservoir}$ (m^3) is the volume of the reservoir.

The hygroexpansion of the trabecular volume due to water imbibition is modelled assuming linear elastic properties for the specimen. The governing equations for structural mechanics are reported in the following:

$$\rho \frac{\partial^2 d}{\partial t} = \nabla \cdot \sigma \tag{4}$$

$$\sigma = C : \varepsilon_{el} \tag{5}$$

$$C = C(E, v) \tag{6}$$

$$\varepsilon_{el} = \varepsilon - \varepsilon_{sw} \tag{7}$$

$$\varepsilon = \frac{1}{2} \left((\nabla d)^T + (\nabla d) \right) \tag{8}$$

where ρ is the density of the trabecular bone, d is the displacement field, σ is the Cauchy stress tensor, E and v are the Young's modulus and the Poisson's ratio of the trabecula, respectively. The elastic strain tensor (ε_{el}) is the difference between the total strain (ε) and the swelling strain (ε_{sw}).

Hygroscopic swelling (ε_{sw}) results in a strain proportional to the difference between the water concentration (c) and the strain-free reference concentration (c_{ref}) in the specimen:

$$\varepsilon_{sw} = \beta \cdot M_w (c - c_{ref}) \tag{9}$$

where β (m$^3 \cdot$kg^{-1}) is the coefficient of hygroscopic swelling defined as follows:

$$\beta = \begin{pmatrix} \beta_W & 0 & 0 \\ 0 & \beta_T & 0 \\ 0 & 0 & \beta_L \end{pmatrix} \tag{10}$$

where β_W, β_T and β_L are the coefficients of hygroscopic swelling in the width (W), thickness (T) and length (L) directions of the trabecula, respectively.

The input parameters necessary for implementing the FEM are listed in Table 1.

Subsequently, the concentration at point (x, y, z) and at time t (c_{FEM}) computed by FEM model is compared with the analytical solution for the diffusion problem in the case of a parallelepiped with constant surface concentration C_{eq} [3]. The water concentration c_{an} at point (x, y, z) and at time t is expressed as follows [3]:

$$\frac{c_{an}}{C_{eq}} = 1 - \frac{64}{\pi^3} \cdot \sum_{l=0}^{\infty} \sum_{m=0}^{\infty} \sum_{n=0}^{\infty} \frac{(-1)^{1+m+n}}{(2l+1)(2m+1)(2n+1)} \cdot \cos\frac{(2l+1)\pi x}{W} \cdot \cos\frac{(2m+1)\pi y}{T}$$
$$\cdot \cos\frac{(2n+1)\pi z}{L}$$
$$\cdot \exp\left(-t \cdot \frac{\pi^2}{4} \cdot \left(D_W\left(\frac{2l+1}{W/2}\right)^2 + D_T\left(\frac{2m+1}{T/2}\right)^2 + D_L\left(\frac{2n+1}{L/2}\right)^2\right)\right) \tag{11}$$

Table 1. Finite element model parameters.

Parameter	Value	Ref.
Trabecular bone density	1000 kg$\cdot$m^{-3}	[4, 5]
Trabecular bone Young's modulus	1 GPa	[4]
Trabecular bone Poisson's ratio	0.3	[4]
Trabecular bone porosity	0.8	[6]
Water diffusion coefficient D_W	1.26$\cdot$10^{-10} m$^2\cdot$s^{-1}	[2]
Water diffusion coefficient D_T	1.16$\cdot$10^{-11} m$^2\cdot$s^{-1}	[2]
Water diffusion coefficient D_L	1.03$\cdot$10^{-9} m$^2\cdot$s^{-1}	[2]
Coefficient of hygroscopic swelling β_W	7.02$\cdot$10^{-6} m$^3\cdot$kg^{-1}	[1]
Coefficient of hygroscopic swelling β_T	1.86$\cdot$10^{-5} m$^3\cdot$kg^{-1}	[1]
Coefficient of hygroscopic swelling β_L	2.61$\cdot$10^{-6} m$^3\cdot$kg^{-1}	[1]
Water density at 27 °C	996.5 kg$\cdot$m^{-3}	[7]
Molecular mass of water	0.018 kg$\cdot$mol^{-1}	[7]
Initial concentration c_{ref}	0.55 mol m^{-3}	[7]
Equilibrium concentration C_{eq}	55361 mol m^{-3}	[7]
Reservoir volume	10^{-5} m^{-3}	[1, 2]

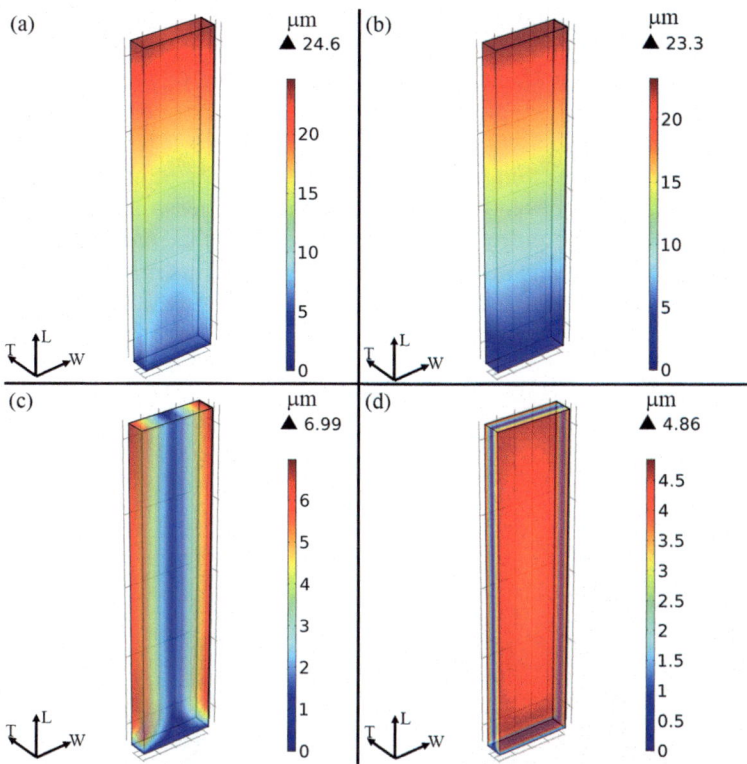

Fig. 2. Results of displacement due to water uptake: (a) total displacement, displacement component in the longitudinal (b), width (c) and thickness (d) direction at time t = 2000 s.

An optimization procedure is developed in order to obtain the effective diffusion coefficients D_W, D_T, D_L at point (x, y, z). An optimum value of the diffusion coefficient can be achieved by minimizing Eq. 12:

$$f_{min} = (c_{an}(t) - c_{FEM}(t))^2 \tag{12}$$

where $c_{an}(t)$ is the concentration at time t calculated through Eq. 11 and $c_{FEM}(t)$ is the concentration obtained from the FEM simulation. The optimization was performed applying the Nelder Mead algorithm implemented in Mathematica software v 11.3.

3   Results

The water concentration distribution and the subsequent swelling due to water imbibition are analysed for the time interval (0; 2000 s), in accordance with the previous experimental study [1]. The problem is discretized in time by means of Backward Difference Formulas (BDF) of order 1 and 2. Subsequently, the system of equations achieved from the BDF method is solved using the Generalized Minimal Residual (GMRES) method by employing an algebraic multigrid left preconditioner.

The FEM simulation is able to provide quantitative analysis of the swelling due to water diffusion within the single trabecula (Fig. 2). Furthermore, typical iso - concentration maps of water within the trabecula at four instants of time and for the central slice orthogonal to the direction of the main percent strain, i.e. thickness direction, are presented in Fig. 3. The concentration values calculated in each node of the model are utilized in the optimization procedure in order to determine the corresponding effective diffusion coefficients. Figures 4, 5 and 6 show the spatial distribution of D_W, D_T and D_L, respectively, on orthogonal planes to the direction relative to each component of the matrix diffusivity.

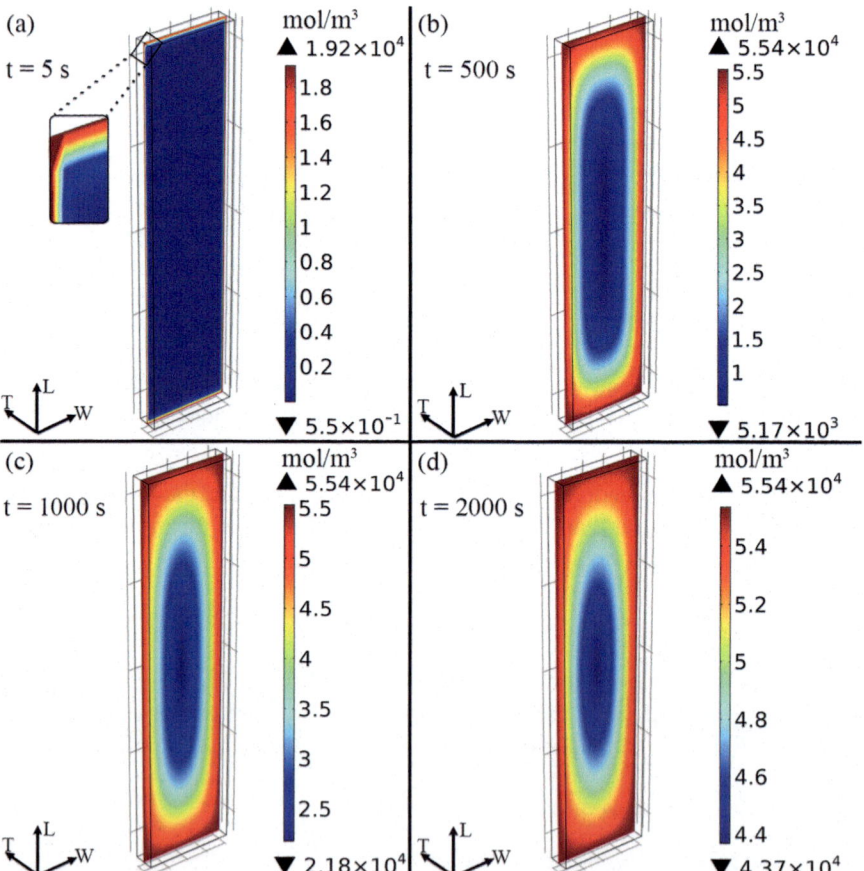

Fig. 3. Iso-concentration map of water in the central cross-section of the trabecula orthogonal to the direction of the main percent strain, i.e. thickness direction. The slice is sited at position $T = 0$ mm, at time (a) $t = 5$ s, (b) $t = 500$ s, (c) $t = 1000$ s and (d) $t = 2000$ s.

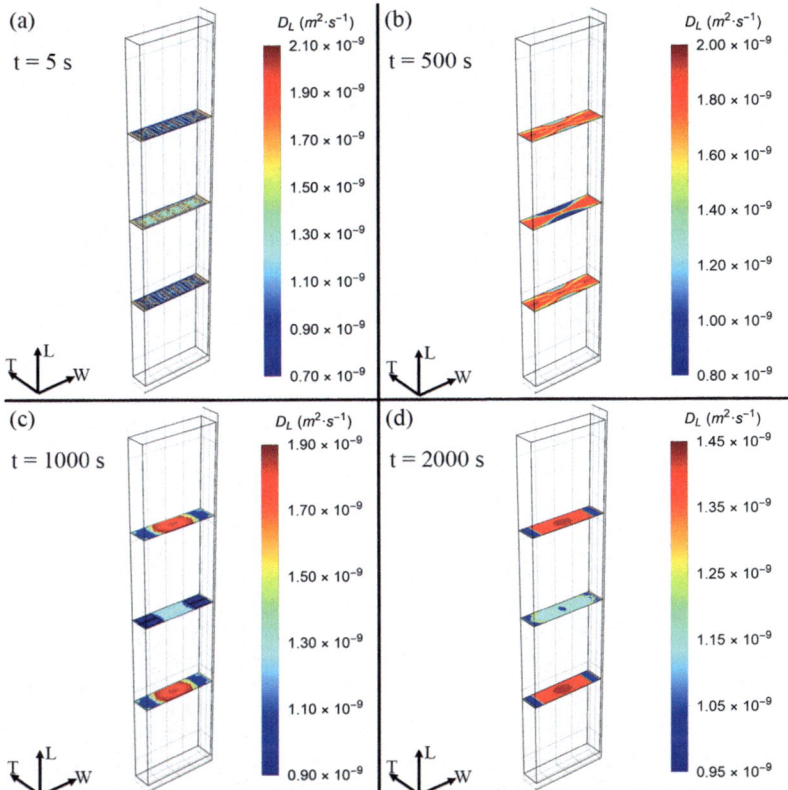

Fig. 4. Iso-diffusivity map corresponding to the water diffusion coefficient D_L on normal planes to the longitudinal direction at position L = −2.25; 0; 2.25 mm at time (a) t = 5 s, (b) t = 500 s, (c) t = 1000 s and (d) t = 2000 s.

4   Discussion

In the present study, a 3D computational model of a single trabecula has been developed in order to provide further insight into the mass transport phenomena within the single human trabecula.

It is worth pointing out that the selected input parameters allow to compute appropriate values of hygro-expansion due to water imbibition, in agreement with the study of Marinozzi et al. [1] (Fig. 2). The diffusion coefficients used in the FEM model were achieved from the study of Marinozzi et al. [2], while the hygroscopic swelling matrix was estimated from the mean dimensional changes measured by [1].

Therefore, the computational model could represent a valuable tool in order to highlight 3D features of the water diffusion phenomenon at any spatial coordinates within the trabecula specimen and at any time of the process. In accordance with the physical framework, the iso-concentration maps are characterized by a symmetric pattern with values close to the equilibrium concentration C_{eq} at the boundaries of the specimen and lower values in the centre of the trabecula (Fig. 3). However, at time t = 5 s (Fig. 3a), the concentration on the boundaries is significantly lower than the equilibrium concentration C_{eq}. This behaviour is due to the effects of the transient period, time interval 0–10 s, of the specimen immersion according to the experimental observations [1]. Moreover, the inner region of the specimen is characterized by the initial strain free concentration of water (c_{ref}), that corresponds to the water concentration in the room, at the conditions of $27 \pm 1°C$ and relative humidity of $41 \pm 3\%$ RH.

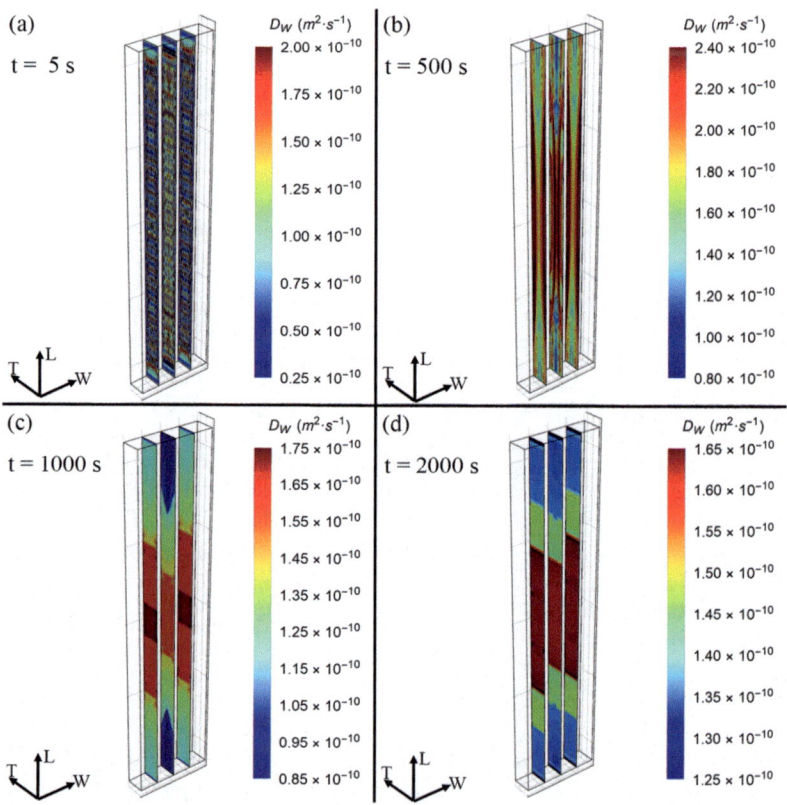

Fig. 5. Iso-diffusivity map corresponding to the water diffusion coefficient D_W on normal planes to the width direction at positions W = −0.5; 0; 0.5 mm at time (a) t = 5 s, (b) t = 500 s, (c) t = 500 s and (d) t = 2000 s.

An advantage of the local assessment of water concentration is the possibility to evaluate the diffusion coefficient at every spatial coordinate of the trabecula specimen through the algorithm described by Eqs. 10 and 11. Several numerical methods, e.g. genetic algorithm [2], Monte Carlo simulations [8, 9], were adopted in order to determine the components of the matrix diffusivity of water within the trabecula. However, these methods were not adequate to achieve spatial distribution maps of the diffusivity matrix components. The computational model represents a preliminary study that allows to determine the local variations of the effective diffusion coefficient (Figs. 4, 5 and 6). The diffusivity values obtained from the optimized approach show a good agreement with the outcomes of the previous studies [2, 8, 9]. It should be mentioned that in this 3D FEM model, the diffusion phenomenon takes into account only the porosity of the trabecula, whilst structural factors like tortuosity and constrictivity [8, 9] were not considered explicitly. Thus, a future improvement of the model should consist on the introduction of the influence of these factors on the diffusion phenomenon.

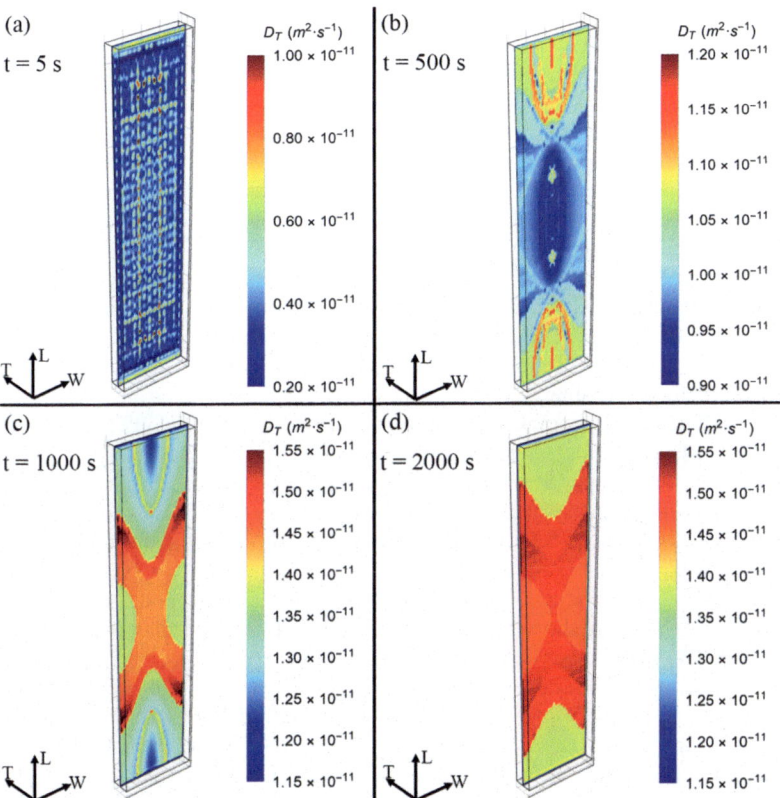

Fig. 6. Iso-diffusivity map corresponding to the water diffusion coefficient D_T on a normal plane to the thickness direction at position $T = 0$ mm, at time (a) $t = 5$ s, (b) $t = 500$ s, (c) $t = 500$ s and (d) $t = 2000$ s.

5   Conclusion

The FEM model provides the temporal evolution of the water concentration within the trabecula and the related swelling due to the water uptake. Future studies could investigate the dynamic behaviour of water diffusion within the trabecula in presence of loading conditions that simulate bone stress during both physiological and pathological gait [10, 11].

In addition to experimental [1, 2, 5] and numerical studies [8, 9] of assessment of bone matrix parameters, the FEM model could offer new insights for more accurate tissue characterization which could be of support in the improvement of mechanical properties of bone repair process, especially if combined with biosensors [12].

References

1. Marinozzi, F., Bini, F., Marinozzi, A.: Water uptake and swelling in single trabeculæ from human femur head. Biomatter **4**(1), e28237 (2014)
2. Marinozzi, F., Bini, F., Quintino, A., Corcione, M., Marinozzi, A.: Experimental study of diffusion coefficients of water through the collagen-apatite porosity in human trabecular bone tissue. Biomed. Res. Int. **2014**(796519), 8 (2014)
3. Crank, J.: The Mathematics of Diffusion, 2nd edn. Oxford University Press, Oxford (1957)
4. Marinozzi, F., Bini, F., Marinozzi, A.: Evidence of entropic elasticity of human bone trabeculae at low strains. J. Biomech. **44**(5), 988–991 (2011)
5. Marinozzi, F., Bini, F., Marinozzi, A., Zuppante, F., De Paolis, A., Pecci, R., Bedini, R.: Technique for bone volume measurement from human femur head samples by classification of micro-CT image histograms. Ann. I. Super. Sanità. **49**(3), 300–305 (2013)
6. Daish, C., Blanchard, R., Gulati, K., Losic, D., Findlay, D., Harvie, D.J.E., Pivonka, P.: Estimation of anisotropic permeability in trabecular bone based on microCT imaging and pore-scale fluid dynamics simulations. Bone Rep. **6**, 129–139 (2017)
7. Ebbing, D.D., Gammon, S.D.: General Chemistry, 9th edn. Houghton Mifflin Company, Boston (2009)
8. Bini, F., Pica, A., Marinozzi, A., Marinozzi, F.: 3D diffusion model within the collagen apatite porosity: an insight to the nanostructure of human trabecular bone. PLoS ONE **12**(12), e0189041 (2017)
9. Bini, F., Pica, A., Marinozzi, A., Marinozzi, F.: A 3D model of the effect of tortuosity and constrictivity on the diffusion in mineralized collagen fibril. Sci. Rep. **9**(1), 2658 (2019)
10. Iosa, M., Morone, G., Bini, F., Fusco, A., Paolucci, S., Marinozzi, F.: The connection between anthropometry and gait harmony unveiled through the lens of the golden ratio. Neurosci. Lett. **612**, 138–144 (2016)
11. Serrao, M., Chini, G., Iosa, M., Casali, C., Morone, G., Conte, C., Bini, F., Marinozzi, F., Coppola, G., Pierelli, F., Draicchio, F., Ranavolo, A.: Harmony as a convergence attractor that minimizes the energy expenditure and variability in physiological gait and the loss of harmony in cerebellar ataxia. Clin. Biomech. **48**, 15–23 (2017)
12. Araneo, R., Rinaldi, A., Notargiacomo, A., Bini, F., Marinozzi, F., Pea, M., Lovat, G., Celozzi, S.: Effect of the scaling of the mechanical properties on the performances of ZnO piezo-semiconductive nanowires. nanoforum 2013. AIP Conf. Proc. **1603**, 14–22 (2014)

Cardiovascular, Cerebrovascular and Orthopaedic diseases Imaging and Modelling

Analysis of Sequential Transverse B-Mode Ultrasound Images of the Carotid Artery Bifurcation

Ricardo Fitas[1], Catarina F. Castro[1,2(✉)], Luisa C. Sousa[1,2],
Carlos C. António[1,2], Rosa Santos[3,4], and Elsa Azevedo[3,4]

[1] Faculty of Engineering of the University of Porto (FEUP), Porto, Portugal
ccastro@fe.up.pt
[2] Institute of Science and Innovation in Mechanical and Industrial Engineering
(INEGI), Porto, Portugal
[3] Faculty of Medicine of the University of Porto (FMUP), Porto, Portugal
[4] Hospital de S. João, Porto, Portugal

Abstract. The study presented here addresses segmentation methodologies applied to Doppler ultrasound images of the carotid bifurcation acquired as a sequence of 2D ultrasound transverse images. Initial identification of image elements representing transversal cuts of the artery followed by parameter analysis enables a fast methodology for lumen and artery wall identification and segmentation for the complete image sequence. Different arterial geometries, along with the presence of plaque, artefacts and speckle noise, can modify the expected circular shapes of the arteries making identification extremely difficult. Initial identified parameters are crucial for the analysis of image behavior across the bifurcation suggesting methodologies for image correction and irregularity detection on the artery representation. Statistical analysis of the accuracy of the bifurcation modelling from transverse image segmentation validates the methodology results as compared to results from medical manual segmentation and results from segmentation of longitudinal Doppler ultrasound images.

Keywords: Doppler ultrasound · Carotid artery · B-mode images ·
Segmentation algorithms

1 Introduction

Image assessment of the arterial system plays an important role in the diagnosis of cardiovascular diseases. Segmentation and identification of the carotid bifurcation in B-mode ultrasound images is an important step in evaluating arterial disease severity [1, 2]. New segmentation methodologies work towards the user-independent ability to assign the correct number of regions of interest [3]. This ability indicates the possibility of implementing a fully automatic segmentation algorithm even in images corrupted by noise [4]. The difficulty remains on the correct identification of the artery centers for extremely irregular shapes. The aim of the proposed work is to segment the lumen and plaque contours of carotid arteries automatically, and reduce the physician's workload [5, 6].

© Springer Nature Switzerland AG 2019
J. M. R. S. Tavares and R. M. Natal Jorge (Eds.): VipIMAGE 2019, LNCVB 34, pp. 521–530, 2019.
https://doi.org/10.1007/978-3-030-32040-9_53

The proposed methodology is validated considering manual lumen segmentation by the medical team and analysis of the lumen area variation period to discard arterial wall movement due to cardiac cycle. Circularity index, MR, irregularity index, Ir, and centrality index, Id, are also included as parameter variation analysis along the full set of recorded images.

2 Materials and Methods

Specifically for this project, qualified medical team from Hospital de S. João, Porto, Portugal, selected data from patients with severe carotid disease acquired during medical examinations using a new ultrasound system Philips Affiniti 50G with a broadband linear array L12-4 transducer. Each data set contains a few hundred of B-mode ultrasound transverse images required to build the 3D bifurcation volume for a specific patient [5, 6]. Each set of transverse images are numbered according to the acquisition methodology, starting from the proximal common carotid artery (CCA), through the bifurcation and ending at the distal images of the extra-cranial internal (ICA) and external carotid (ECA) arteries. In order to model the carotid bifurcation, the implemented code (Matlab 2018a) follows three steps depending on the subset under consideration. First step enables the detection of the central axis for a single and almost circular lumen section of the proximal CCA. The second step considers bifurcation as an enlarged non-circular lumen suggesting splitting of the central axis into two-separated central axis as in an overlap of two ellipses. Finally independent lumen areas setting independent central axis for proximal ICA and ECA. It is very important that on the first subset of images viewing the proximal region of the CCA, the central axis is identified correctly. In order to guarantee that, mathematical indexes [3, 4], such as circularity index, MR, irregularity index, Ir, and centrality index, Id, are calculated. Indexes are due to take their best values for correct identification of the arterial lumen among all possible areas of minimum luminance. Then the central axis will be the center of the lumen area. The applied methodology becomes dependent on the correct assignment of CCA axis since for the subsequent frames, the assumption of continuity of central axis by minimizing the distance between consecutive points is considered. Further stretch correction is applied keeping the segmentation process in the correct track. Finally, the central axis from a set of B-mode images is modeled using B-spline methodology.

Sequential images are analyzed in order to reconstruct the bifurcation of the carotid artery, plaque detection or artefacts existence. Due to the complexity of interpreting transversal images, this work proposes to find patterns and similarities along one or more sequences of expansion and contraction of artery images. To overcome lumen segmentation problems, a methodology based on statistical analysis of artery wall parameters, as for instance circularity, observed on transverse images is considered.

It is important to explore regions of interest, some of them presenting extremely irregular shapes. For solving this, an elliptical approximation for each possible identification, turns easier its characterization. As non-static images are sequential, the final aim would be to draw continuous lines, in order to obtain the bifurcation geometry. However, it is necessary to overlook artifacts. A minimum of 6 points defines a conic, in this case, an ellipse, and arteries are necessarily circulars; so identifying parameters will allow knowing eccentricity values. A least square approximation to these points will define the ellipse. If the determinant of the matrix of the coefficients generated by this method is zero, this method can be insufficient. Image points must not form a straight line and to overcome the complexity of the algorithm an automatic change of Gaussian filters will identify new position for the points of interest.

The general conic equation is as follows:

$$Ax^2 + Bxy + Cy^2 + Dx + Ey + F = 0$$

This equation can take a parametric format:

$$X = X_0 + T(\alpha) * \begin{bmatrix} a\,\cos\theta \\ b\,\sin\theta \end{bmatrix} = \begin{bmatrix} x_0 \\ y_0 \end{bmatrix} + \begin{bmatrix} \cos\alpha & -\sin\alpha \\ \sin\alpha & \cos\alpha \end{bmatrix} * \begin{bmatrix} a\,\cos\theta \\ b\,\sin\theta \end{bmatrix}$$

where $T(\alpha)$ is the transformation matrix, being α the ellipse orientation, X_0 are the coordinates of the center and a and b the principal diagonals of the ellipse. This last equation characterizes more easily the segments with the disadvantage of using angle θ as a variable. Extending the initial form of conic equation, and writing coefficients A to F, as a function of orientation α, it is obtained

$$\begin{cases}
A = (cos\ \alpha)^2 b^2 + (\sin\ \alpha)^2 a^2 \\
B = (\sin 2\alpha)(b^2 - a^2) \\
C = (cos\ \alpha)^2 a^2 + (\sin\ \alpha)^2 b^2 \\
D = -2b^2 x_0(cos\ \alpha)^2 - y_0(\sin 2\alpha)b^2 + (\sin 2\alpha)y_0 a^2 - 2x_0 a^2(\sin\ \alpha)^2 \\
E = -2b^2 y_0(\sin\ \alpha)^2 - x_0(\sin 2\alpha)b^2 + (\sin 2\alpha)x_0 a^2 - 2y_0 a^2(cos\ \alpha)^2 \\
F = Ax_0^2 + Bx_0 y_0 + Cy_0^2 - a^2 b^2
\end{cases}$$

Through the implementation of this set of equations, it is possible to calculate the approximated area of the ellipse, Area $= ab\pi$, its eccentricity, $e = \sqrt{1 - \left(\frac{b}{a}\right)^2}$, and an approximated error.

In fact, it is important to validate the possibility of approximating segmented lines from regions of interest by ellipses. It was important to consider doing a statistical analysis to the nearest points and compare to manual medical segmentation.

Let be $S \in \mathbb{R}^2 : S = \{(x_0, y_0), (x_1, y_1), \ldots, (x_N, y_N)\}$ a set of segmented image points aimed to approximate an ellipse with an approximation error R formulated as follows:

$$R = 1 - \frac{1}{N} \sum_{i=1}^{N} \frac{d_{ri} - d_{ai}}{\max(\{d_{ri}, d_{ai}\})}$$

with

$$d_{ri} = \sqrt{(x_r - x_0)^2 + (y_r - y_0)^2}$$

and

$$d_{ai} = \sqrt{(x_i - x_0)^2 + (y_i - y_0)^2}$$

As there is a finite set of points, it is important to use another M set of points (X, Y) to define an ellipse with limits deduced by the following expression:

$$X_{1,M} = \frac{\pm 2\sqrt{CFB^2 - BDE + C^2D^2 + AE^2 - 4AFC^2} - BE + 2CD}{B^2 - 4AC}$$

and Y_j appears as a function of $X_j, j = 1, \cdots, M$:

$$Y_j = \frac{\pm \left[-BX_j - E + \sqrt{(BX_j + E)^2 - 4C\left(AX_j^2 + Di + F\right)} \right]}{2C}$$

being

$$D_r = \min_{j=1...M} \sqrt{(X_j - x_i)^2 + (Y_j - y_i)^2}$$

Even though, the standard deviation is another process to verify the precision of the elliptical approximation, this information only assigns a dispersion parameter to these points and cannot be used for comparing the real size of the regions.

In order to validate the methodology, a statistical study of a set of 52 manually segmented images was performed. Statistical data, error R, mean of the minimum distance between points, $\bar{x}$ and an approximation of a unique value of the mean diameter D_{max}, are considered for the analysis are resumed in Table 1.

Table 1. Statistical analysis of similar parameters for evaluation of precision of the elliptical approximation.

	μ	σ	Min	Max	Skewness	Kurtosis
R	0.994	0.003	0.983	0.99	1.094	1.183
$\bar{x}$	9.013	1.993	6.092	18.5	2.226	7.921
D_{max}	83.11	25.11	44.25	160.	0.596	0.178

This study also identifies the correlation between the error R and the mean of the minimum distance between points, $\bar{x}$. The R values are always closed to 1, however it is necessary to validate the value for maximum distance to the center of the ellipse, even because standard deviation value for parameter R is small. The value of D_{max} was used as an approximation of a unique value of the mean diameter. As standard deviation value for the biggest diameter of ellipses is high, it is just needed to verify the correlation of R with $\bar{x}$ to validate the significance of the diameter value when this value is high. The correlation values that are shown in Table 2 (Pearson and R^2) validates the methodology as results enhance the high existing correlation.

Table 2. Correlation values Pearson and R^2 between parameters

	Constant	β_1	R^2	Pearson
R-$\bar{x}$'	−0.032	1.002	0.907	0.952

The application of ellipse methodology to detect contour lines in regions of interest methodology is mainly usable when images have low values of luminance and considered, as it is important to reduce the segmented area. However, it can also be used to approach to the best values for circularity, increasing limit values of luminance. In fact, finding centers of contour lines requires the shrinkage of the segment what is not a great approximation. With this methodology, each data point is not able to find the values of origin, due to algorithm complexity, so the best procedure is to find the minimum distance between these points and the points of origin. Points, which are initially detected due to bad image quality, will no longer be considered, because they are the furthest of the center.

The remaining issue of this methodology deals with the knowledge of how far the segmentation will allow the contour line or lumen area is to be shrunk. Statistical results of manual medical segmentation for the maximum diameter of several ellipses will be considered within the area of interest, and the confident interval estimated. Based on that interval, and based on the sequence of the images, these values can be implemented as correct and secured.

3 Results

The approach presented in this work estimates the CCA lumen axis in a very reliable way. In fact, the methodology is quite robust to strong noise and detected vesel-like structures surrounding the carotid artery.

Figure 1 shows an example of stretch correction mode for selecting the central axis of the carotid arteries. Beginning with an original image, the transformation in a binary image allows artifact detection; the artifact is removed using a variation of filter parameters; then contraction of mesh created with white pixels and calculation of the minimum distance correspondent pixels detected using least square method; finally the overlap on the original image. For this particular example, a Gaussian filter [10 10] and $\sigma = 5$ was considered. This methodology is reliable on images of vessels with large echogenic plaques, supporting arteries with and without plaques (including highly translucent ones).

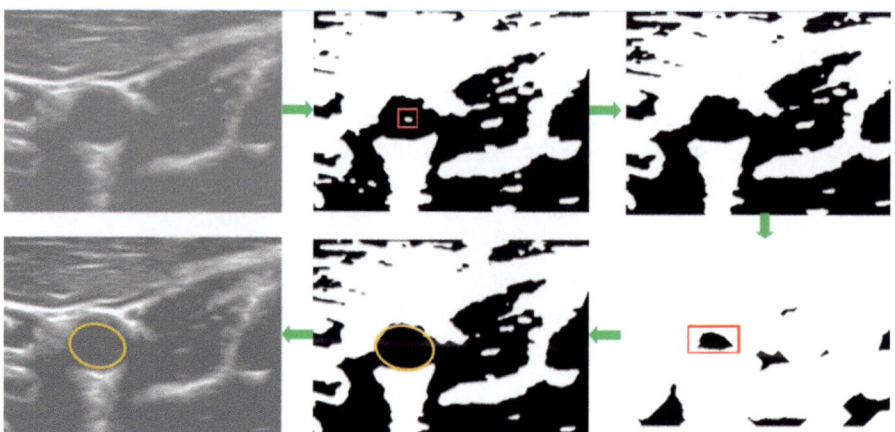

Fig. 1. Methodology of segmentation and elliptical approximation. Arrows indicate the step order along the procedure.

In order to select the regions of interest, a new formulation based on our previous report as well as following other author's publications has been implemented. The comparison of the manageable results based on the manually segmented some medical images (N = 52) uses a linear regression for both models.

Jodas et al. [4] following model:

$$E1 = MR + \frac{1}{Ir} + \frac{1}{Id}$$

and Castro et al. [3] presents another formulation:

$$E2 = MR + Ir2 + Id2 = MR + CP1 \ast CP2 + Id2$$

Please refer to previous publications in order to fully define the different parameters. Table 3 presents an analysis of the normal properties for some parameters allowing concluding for incoherence values used by Jodas et al. [4] expressions and validating the better performance of the second formulation by Castro et al. [3]. As noticeable, $CP2$ and $Id2$ parameters present no normal distribution and $\frac{1}{Ir}$ is close to normal. So, it becomes important to weight some coefficients in the formulation to validate the manual medical segmentation. If E1 and E2 were not pondered, selection will be not compatible with the set of medical segmentation. Table 4 presents β-values from linear regression for a new different formulation. The proposed method takes Castro et al. formulation [3] and substitutes only CP1 by eccentricity (e) value, making the best approximation to a normal distribution.

The standard deviation value is lower in the second formulation. It means that values are significantly close from each other, which validates the equally sum of the different parameters. It can be also calculated some confidence intervals for the lumen

Table 3. Distribution properties of selected parameters for artery identification.

Parameters	μ	σ	Skewness	Kurtosis
MR	0.916	0.052	−1.174	1.505
1/Ir	1.110	1.141	2.834	9.798
1/Id	0.017	0.026	3.366	12.039
CP1	0.997	0.004	−2.999	11.455
CP2	0.308	0.054	0.401	−0.944
Id2	0.438	0.254	0.568	−0.633
e	0.647	0.167	−0.441	0.131
NC2	0.198	0.057	−0.170	0.638

Table 4. β-values of regression of the different formulations and evaluation of dispersion through σ value

	Par. 1	Par. 2	Par. 3	σ (P1, P2, P3)
E1	0.044	0.967	0.022	0.5394
E2	0.201	0.210	0.984	0.4495
New	0.242	0.219	1.075	0.4877

artery area ([13437.7; 18775.8], with $\alpha = 95\%$) being the minimum value for R equal to 0.9114, and the maximum value of σ, equal to 4.6885.

Through these analyses, it is important to control variables, to apply the advised correction and to overpass artefacts issues. Figure 2 gives an example of the correct application of the new methodology using same region of interest as in Fig. 1, by changing filter parameters and turning elliptical form more circular.

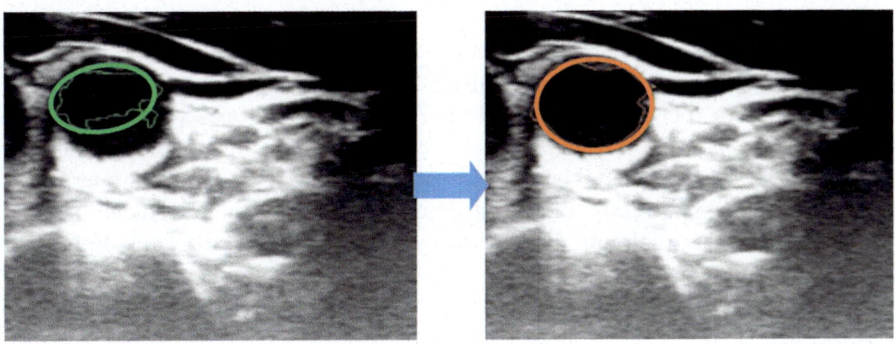

Fig. 2. Example of the application of different methodologies. Previous on left with limit value 0, [5 5] and σ = 10. MR = 0.814; new on right with limit value 14, [10 10] and σ = 20. MR = 0.913.

As a manner of parameter optimization, a study for best filter selection, dependent on image quality has been addressed. Gaussian filter were always applied and limit luminance value is the parameter with the largest effect, with an optimum value dependent on the presence of noise. Figure 3 (left) presents an analysis of different Gaussian filter parameters in order to optimize precision of image segmentation as compared to medical manual segmentation.

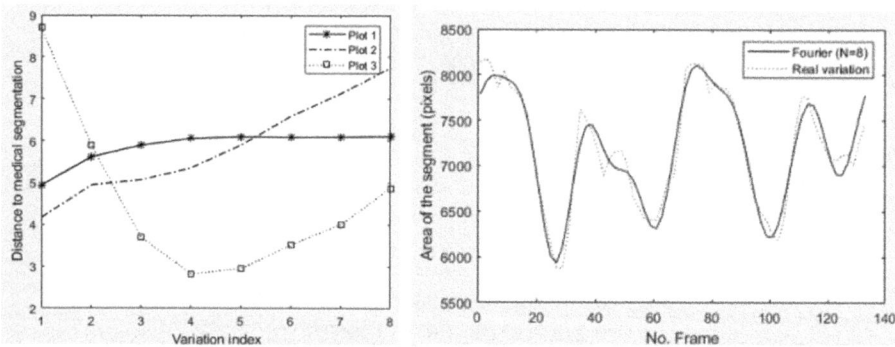

Fig. 3. On the left: analysis of filter parameters (Plot 1 – variation of σ with constant value of kernel; Plot 2 – kernel variation with limit luminance value of pixels; Plot 3 – variation of limit luminance value); On the right: area variation along sequential images of a CCA and Fourier series approximation.

Also, a better periodicity of artery enlargement caused by heartbeat, instead of persisting on noise influence. An example of 150 frames analyzed is presented in Fig. 3 (right), using an 8th order of Fourier series approximation.

Patient-specific 3D bifurcation models were reconstructed from segmented transversal B-mode images. Figure 4 presents an example of the results of central axis reconstruction using a set of B-mode transverse images from on specific patient.

The main difficulty is associated with bifurcation. All together, the transition from the central axis of the common carotid artery to the central axis of the internal and external carotid arteries is a challenge. However, spline methodology allows the optimization of a simple variable that is related with the interval of the interpolation. Two factors enable selection of the best plot: the sum of the gradient of centers interpolation and the minimum distance to the points. These two factors do not allow straight lines and, contrary to it, do not let noise accumulation happen.

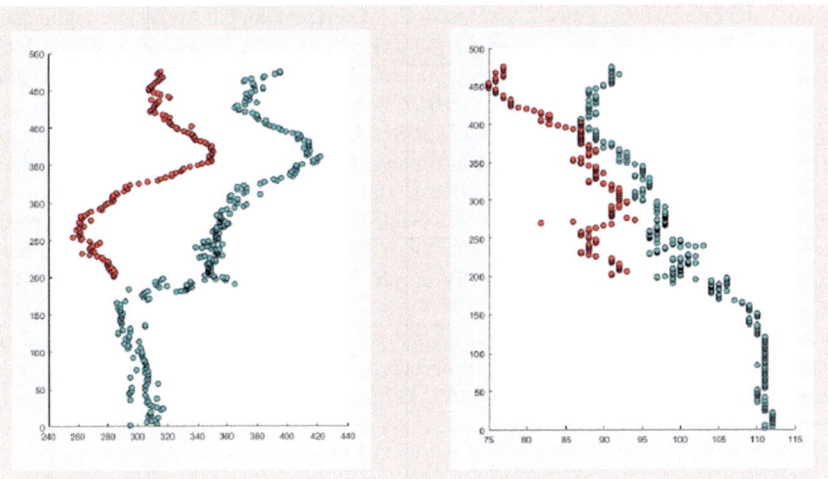

Fig. 4. Projections of the aligned central axis reconstruction of the bifurcation artery using B-spline methodology

4 Conclusion

The proposed method for the determination of the carotid axis in B-mode images supports arteries with or without plaques, heavy noise and the presence of other vessel-like structures. Despite of the good results further study is necessary to perform selection and segmentation with better accuracy. The proposed method is suited to real time processing, no user interaction is required, and the number of parameters is minimal and easy to determine. It was showed some statistical analysis to determinate some parameters with great approximation and in an easy way. In the future, it is important to apply some of these techniques in order to perform better results, mainly in relation to filter parameters. Continuous analysis of medical image segmentation improves modeling and accuracy allowing better results.

Acknowledgments. The authors gratefully acknowledge the funding by FCT, Portugal, of the Research Unit of LAETA-INEGI´.

References

1. Goddi, A., Bortolotto, C., Fiorina, I., Raciti, M.V., Fanizza, Turpini E., Boffelli, G., Calliada, F.: High-frame rate vector flow imaging of the carotid bifurcation. Insights Imaging **8**(3), 319–328 (2017)
2. Barratt, D.C., Ariff, B.B., Humphries, K.N., McG Thom, S.A., Hughes, A.D.: Reconstruction and quantification of the carotid artery bifurcation from 3-D ultrasound images. IEEE Trans. Med. Imaging **23**(5), 567–583 (2004)

3. Castro, C.F., Sousa, L.C., Fitas, R., António, C.A.C., Azevedo, E.: Automatic segmentation in transverse ultrasound B-mode images of the carotid artery. In: Proceedings of the 1st Iberic Conference on Theoretical and Experimental Mechanics and Materials/11th National Congress on Experimental Mechanics (2018), Porto/Portugal, 4–7 November 2018

4. Jodas, D.S., Pereira, A.S., Tavares, J.M.R.S.: Automatic segmentation of the lumen in magnetic resonance images of the carotid artery. Comput. Biol. Med. **79**, 233–242 (2018). https://doi.org/10.1016/j.compbiomed.2016.10.021

5. Sousa, L.C., Castro, C.F., António, C.C., Santos, A., Santos, R., Castro, P., Azevedo, E., Tavares, J.M.R.S.: Hemodynamic conditions of patient-specific carotid bifurcation based on ultrasound imaging. Comput. Methods Biomech. Biomed. Eng.: Imaging Visual. **2**(3), 157–166 (2014)

6. Castro, C.F., António, C.A.C., Sousa, L.C.: Vessel detection in carotid ultrasound images using artificial neural networks. In: Proceedings of the 6th International Conference on Integrity, Reliability and Failure, pp. 1169–1172 (2018)

Geometry Reconstruction of a Patient-Specific Right Coronary Artery with Atherosclerotic Plaque for CFD Study

I. S. Saraiva[1], Catarina F. Castro[1,2], Carlos C. António[1,2],
R. Ladeiras-Lopes[3,4], N. D. Ferreira[4], N. Bettencourt[3],
Luisa C. Sousa[1,2], and S. I. S. Pinto[1,2(✉)]

[1] Engineering Faculty, University of Porto, Porto, Portugal
spinto@fe.up.pt
[2] Institute of Science and Innovation in Mechanical and Industrial Engineering
(LAETA-INEGI), Porto, Portugal
[3] Cardiovascular R&D Unit, Faculty of Medicine, University of Porto, Porto,
Portugal
[4] Department of Cardiology, Gaia/Espinho Hospital Centre, Vila Nova de Gaia,
Portugal

Abstract. The geometry reconstruction of a patient-specific right coronary artery with atherosclerotic plaque, for hemodynamic study, is still a challenge. The reconstruction of the extremely irregular geometry of the RCA, of its side-branches and of the atherosclerotic plaque, as close as possible to the reality, is not an automatic method and requires several particular and rigorous steps. Then, the hemodynamic in that patient case was analysed using the relative residence time descriptor. This descriptor is the most important metric to evaluate the tendency of atherosusceptible regions. In this patient-specific case, there is an increase of 25% in the wall arterial area propitious to atherosclerotic plaque development, with a luminal stenosis increase from 20 to 70%.

Keywords: Atherosclerotic plaque · Image-based · Geometry reconstruction ·
Right coronary artery · Hemodynamic descriptors · Risk assessment

1 Introduction

The X-ray technique exists since the beginning of 20th century. However, medical images representing 3D dimension of the artery uniquely appeared in 1972 with the Computed Tomography (CT) [1]. So, in the 70's, a new modality for disease diagnosis, CT, was introduced in the clinical and hospital environment. Nowadays, the CT create images on digital format, Digital Imaging and Communication in Medicine (DICOM). Each DICOM image of the CT technique represents a "slice" in the body. The 3D reconstruction of the artery is obtained by stacking the slices of the all planes, interpolating the spaces between them, forming a volume [1]. The CT technique provides the possibility to visualize, the interior human body, in a non-invasive way and with a more accurate diagnosis. Moreover, the CT provides clear images of the vessels anatomy as well as the presence or absence of atherosclerotic disease [2].

© Springer Nature Switzerland AG 2019
J. M. R. S. Tavares and R. M. Natal Jorge (Eds.): VipIMAGE 2019, LNCVB 34, pp. 531–539, 2019.
https://doi.org/10.1007/978-3-030-32040-9_54

Atherosclerosis develops due to the accumulation of lipoproteins and other fat substances in the arterial wall, causing stenosis, which blocks the normal circulation of blood flow. In this way, cardiovascular disease has been the main cause of mortality and morbidity in developed countries [3]. CT scans give information about the geometry and location of the atherosclerotic disease; however, they do not explain the hemodynamic with detail. Therefore, numerical simulations have been an auxiliary tool for the prevention and treatment of such diseases.

Nevertheless, as far as we know, numerical studies of the hemodynamic in patient-specific right coronary arteries (RCA) with atherosclerotic plaque have not been well explored in the literature. Myers et al. 2001 [4] have suggested that the hemodynamic behaviour of the RCA is dominated by its complex geometric configuration. Therefore, the use of geometric models of the RCA as close as possible to the reality is essential to obtain an accurate hemodynamic through numerical simulations. However, the reconstruction of the extremely irregular geometry, of its side-branches and of the atherosclerotic plaque is not automatic and requires several particular and rigorous steps. The present work shows this difficulty before hemodynamic simulations and analyses.

2 Materials and Methods

From a population of symptomatic patients referred to Coronary CT Angiography at Gaia/Espinho Hospital Center, a male individual, good image quality, right coronary dominance, and 20% lumen stenosis was selected. This percentage was determined by using at the site of maximal narrowing the measurement of the luminal diameter and the estimated original width. The subject gave informed consent and the institutional ethical committee approved the present research.

The CT examination was performed with a third-generation 2×192-section dual-source CT system (SOMATOM Force; Siemens Healthcare Sector, Forchheim, Germany). Before the coronary CT angiography study, a nonenhanced prospectively ECG-triggered high-pitch spiral CT acquisition was performed in the patient to obtain the coronary calcium score. Optimal timing for acquisition start was determined by using a test-bolus protocol. A volume of 50–80 mL of Iopromide (Ultravist 370; Bayer Pharma AG, Berlin, Germany) was injected at a flow rate of 5–6 mL/s through an 18-gauge intravenous antecubital catheter, followed by a saline flush at the same flow rate. The coronary CT angiography data sets were acquired by using prospectively ECG-triggered high-pitch spiral acquisition during diastole.

The contrast-enhanced coronary CT angiography data were reconstructed with a section thickness of 0.6 mm in the axial plane, using a third-generation interactive reconstruction technique (advanced modeling iterative reconstruction, or ADMIRE; Siemens Healthcare Sector).

Then, the CT scans were transferred and imported to the *Mimics* software for segmentation and post-processing. The aorta (Fig. 1a), the starting point of the RCA (Fig. 1b) and the ending points of the side-branches, as conus (Fig. 1b), right-ventricular (Fig. 1d), acute-marginal (Fig. 1e), posterolateral and posterior descending (Fig. 1f) were manually selected. Since the right coronary artery has 20% lumen stenosis, the atherosclerotic plaque calcification was also selected (Fig. 1c) between the conus branch and the right-ventricular branch.

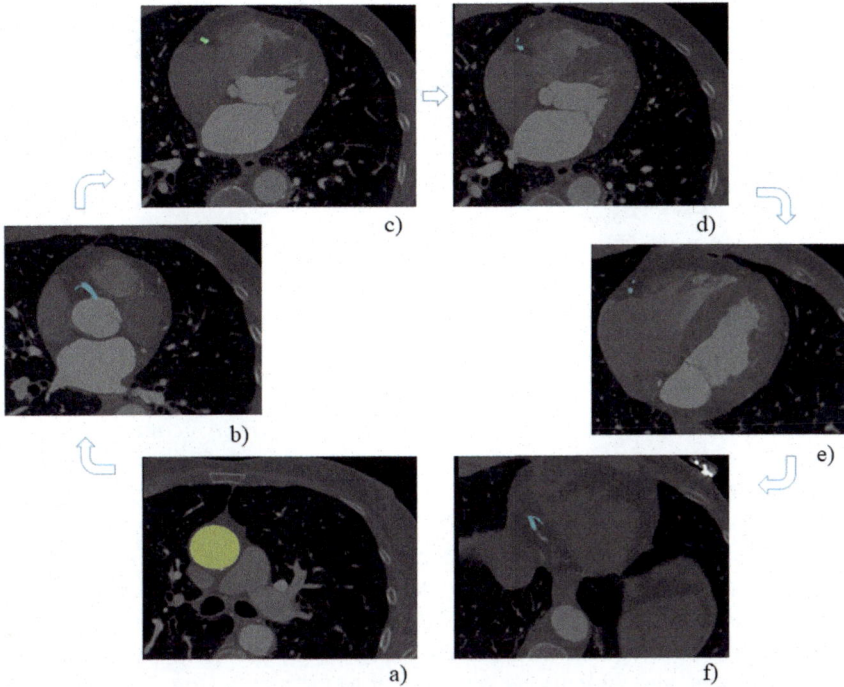

Fig. 1. Manual selection of (a) the aorta; (b) starting point of the RCA and conus; (c) atherosclerotic plaque calcification; (d) right-ventricular branch; (e) acute-marginal branch; (f) posterolateral and posterior descending branches of the RCA, through *Mimics* software.

The manual selection of all the RCA side-branches and the atherosclerotic plaque calcification results in a 3D mask (Fig. 2a) from the segmentation process. The segmentation involves the separation of an object of interest from other adjacent anatomic structures from different masks.

In order to obtain a geometry-free from irregularity forms and roughness, a method of improvement and general softening was applied in all the surface of the 3D mask previously created (Fig. 2a) – post-processing. This method is extremely important not only to acquire a more realistic geometry but also to provide easier handling in the following steps.

Thus, the 3D mask of the RCA, saved in STL (Standard Template Library) format, was imported to the *3-Matic* software, in order to globally decrease imperfections and abnormalities resulted from the segmentation process. Softening and refining the geometric elements will allow working in a more efficient way. Moreover, this method considers the non-modification of specific configurations of the real patient, namely shape and anatomical dimensions.

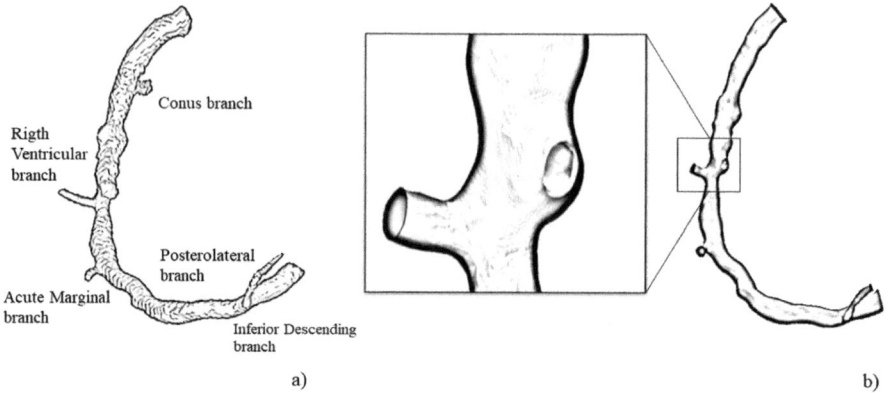

a)   b)

Fig. 2. (a) 3D mask of the RCA obtain through *Mimics* software; (b) 3D geometric model of the RCA after post-processing through *3-Matic* software.

The reconstruction of the atherosclerotic plaque of the RCA geometry is one of the most important steps to achieve the final geometric model. Since the segmentation process though *Mimics* software is a semi-automatic method, the atherosclerotic plaque is not obtained in its total shape. Normally, it looks like it is floating inside the lumen (Fig. 3a). The process consists in understanding which side of the atherosclerotic plaque can contact the arterial wall, in order to add material in the area between the plaque and the arterial wall (Fig. 3b). The last step is subtracting the atherosclerotic plaque from the coronary lumen in order to achieve a 3D geometric model with a stenosis with shape and dimensions very similar to the reality (Fig. 3c). A 20% lumen stenosis was obtained, in this specific case, which is in concordance with the medical team information.

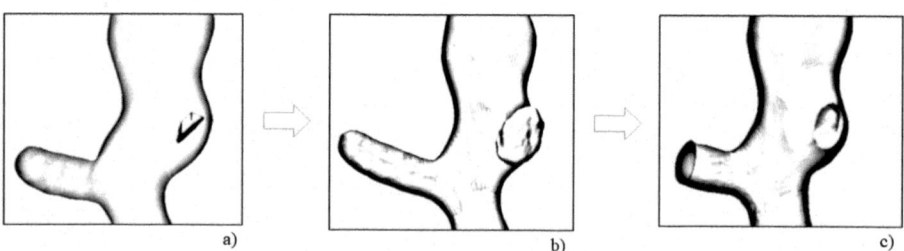

a)   b)   c)

Fig. 3. (a) Atherosclerotic plaque floating inside the lumen; (b) Addition of material in the area between the plaque and the arterial wall; (c) Subtraction of the atherosclerotic plaque from the coronary lumen.

In order to study the hemodynamic effects in this RCA if the stenosis is increased, other geometric models were reconstructed in order to acquire a higher lumen stenosis percentage: 40% lumen stenosis and 70% lumen stenosis. In these cases, the atherosclerotic plaque was manipulated. Material in plaque was added to increase the stenosis. However, the focus of maintaining the same shape of the 3D plaque is always present.

In the final treatment of the 3D geometry, perpendicular cuts relative to the blood flow direction must be done in each side-branch, starting point of the RCA and ending points of each side-branch, in order to facilitate the boundary condition definitions. This difference in the side-branch can be observed comparing Figs. 3b and 3c. Still, the geometry must be aligned with the axes to define the coordinate X perpendicular to the inlet boundary, and the coordinates Y and Z aligned with the inlet boundary plane.

The 3D geometric model of the RCA with the inlet and outlet boundaries, perpendicular to the flow, and the axis defined at the inlet is, from now, already reconstructed in order to import to *Ansys* software for computational fluid dynamic (CFD) simulations and hemodynamic analyses.

A tetrahedral mesh was reconstructed in *Meshing Ansys* software. A Path Independent method was used in order to uniform the elements and to obtain an accurate mesh (Fig. 4). Therefore, the statistical parameter Skewness was used to verify the accuracy of the mesh. A Maximum Skewness of 0 indicates an equilateral cell, the best case scenario. A Maximum Skewness equal of 1 indicates a completely degenerated cell, the worst case scenario. Following the tutorial of *Ansys*, the mesh is accurate when the Maximum Skewness is lower than 0.95 [5]. This patient-specific case has a Maximum Skewness equal to 0.59, which is considered accurate for numerical simulations.

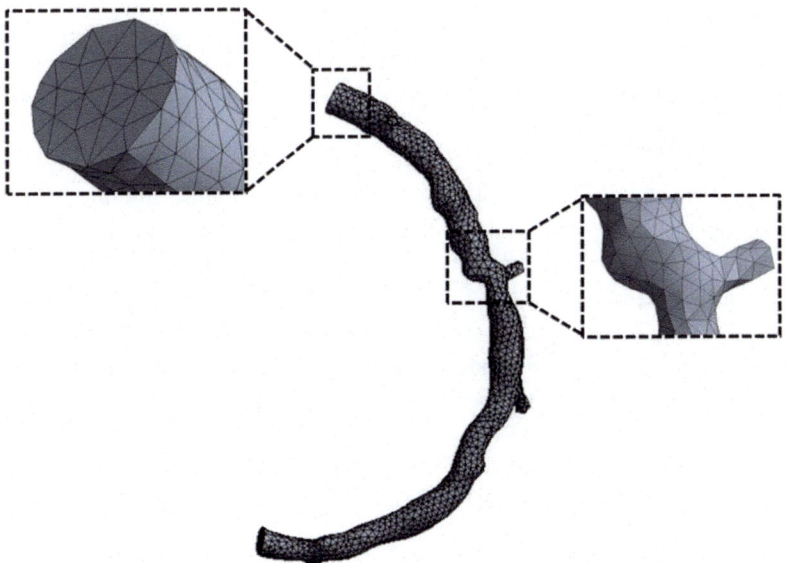

Fig. 4. 3D mesh of the patient-specific RCA obtained through *Meshing Ansys* software.

In this study, blood was considered isotropic, incompressible and homogeneous fluid with a constant density ($\rho = 1060$ kg/m^3). Carreau model was used to define the purely shear-thinning non-Newtonian property of blood at 37°C and parameters are well defined in the literature [6].

Boundary conditions must be imposed. At the inlet, Ostium of the RCA, a Womersley velocity profile was considered. This profile depends on the instant time of the cardiac cycle, the radial position at the inlet and the Womersley number:

$$\propto = R\sqrt{\frac{\rho\omega}{\mu}} \tag{1}$$

which depends on the radius of the patient-specific artery (R), the blood density (ρ), the viscosity of blood (μ) at infinite shear rate and the cardiac frequency (ω). At the outlet branches, pressure profiles were taken into account. This profile is dependent on the instant time of the cardiac cycle but radius-independent. The boundary conditions were implemented in User-Defined Functions (UDFs) in *Ansys* software [7]. Figure 5 shows the mean velocity profile imposed in the Ostium (inlet) and the pressure profile at the outlet branches.

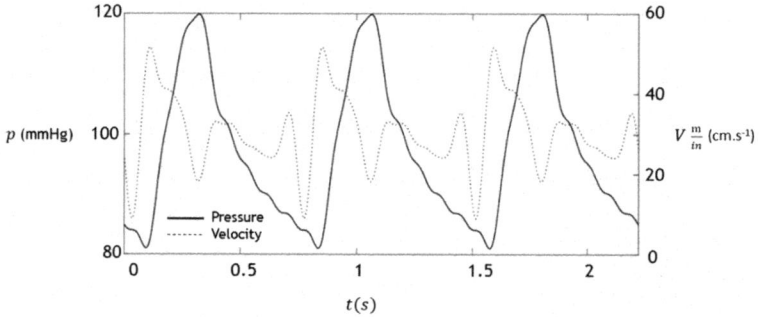

Fig. 5. Mean velocity profile imposed at the inlet (ostium) and pressure profile imposed at the outlet branches.

Then, *Ansys* software was used to perform computational fluid dynamic (CFD) simulations, blood flow simulations. Navier–Stokes equations were solved in a laminar regime, since Reynolds number in the systolic peak, maximum velocity, has a maximum of 1000. The velocity–pressure coupled equations were solved by SIMPLE algorithm. The momentum equations were discretized by the second-order upwind scheme. The time step size was considered equal to 0.005 s/time step, the time step number defined was 444 and the number of iterations for each time step was 20. Thus, the total time of the three cardiac cycles was 2.22 s (3 cardiac cycles × 0.74 s/cardiac cycle = 2.22 s). The simulation process was completed when the 444 time steps were achieved.

3 Results and Discussion

The wall shear-stress (WSS) hemodynamic descriptors are widely used to analyze prone regions of atherosclerosis appearance in a coronary artery. The relative residence time (RRT) is the strongest hemodynamic metric for assessing atherosclerotic plaque formation [8]. Regions higher than 8 Pa^{-1} are considered atherosusceptible regions [7]. Computational Fluid Dynamic (CFD) simulations and hemodynamic analyses were performed through *Ansys* software.

Figure 6a shows the RRT spatial distribution of the patient-specific case in study, with 20% lumen stenosis. Figures 6b and 6c represent the RRT for the manipulated stenosis, 40% and 70% lumen stenosis, respectively. Figure 6d shows the RRT with the hypothesis of no stenosis in the patient. The real 20% lumen stenosis was manipulated in order to verify the hemodynamic effects in this RCA when the stenosis is increased.

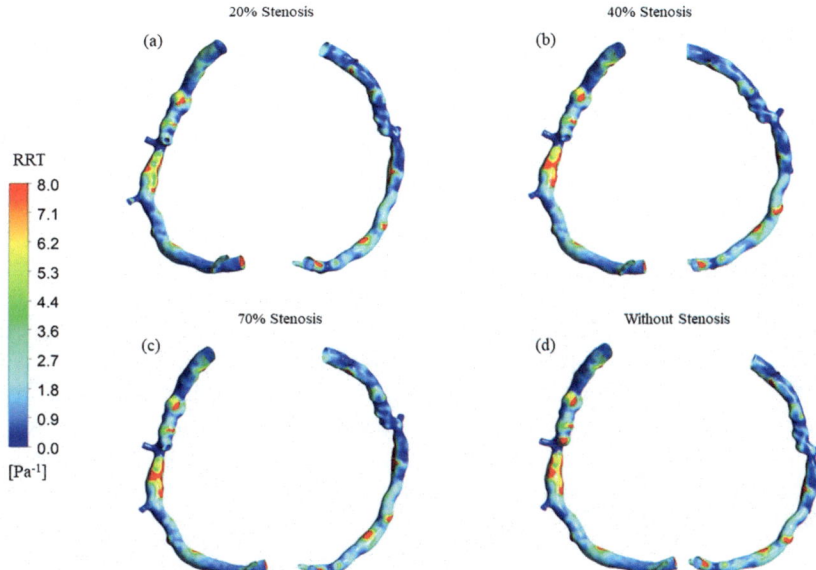

Fig. 6. RRT spatial distribution in a patient-specific RCA with (a) 20% lumen stenosis, (b) 40% lumen stenosis, (c) 70% lumen stenosis, (d) without stenosis.

In Figs. 6 and 7 is clearly evident the regions with the highest tendency for atherosclerosis appearance, regions with RRT higher than 8 Pa^{-1} represented at red, which are between the right ventricular branch and the acute marginal branch. However, in some other locations, propitious zones are also manifested. All these prone regions to atherosclerosis appearance, represented at red, are due to the tortuosity, curvature and roughness of the artery, meaning that the geometric configuration highly influence the development of atherosclerotic plaque.

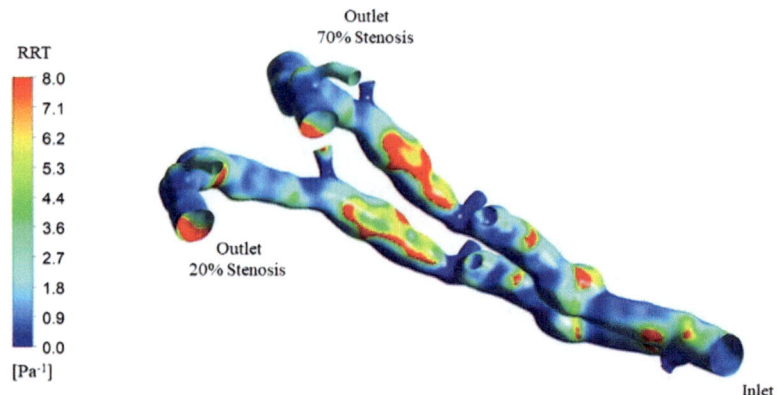

Fig. 7. RRT spatial distribution detail of the patient-specific RCA with 20% (real stenosis) and 70% lumen stenosis (manipulated stenosis).

Analyzing the RRT distribution without stenosis (Fig. 6d), there is a high tendency for atherosclerosis appearance immediately before the right-ventricular branch (red zone). That fact justifies the real location of the stenosis (Fig. 6a) just before the right-ventricular branch, in this patient case.

From 20 to 70% lumen stenosis, the tendency for atherosusceptible regions around the stenosis decreases (see Fig. 7). When the stenosis is small, there is a predisposition to increase it even more. However, from 20 to 70% stenosis, propitious regions to atherosclerosis increase significantly in other zones, mainly between the right ventricular branch and the acute marginal branch. This increase is much higher than the decrease around the stenosis. The percentage of RRT region, in the arterial wall, higher than 8 Pa^{-1} is 3.40% for a 20% lumen stenosis and 4.24% for a 70% lumen stenosis, which means a significant difference and increase, 25%, of atherosusceptible regions.

In future work, other patient-specific cases will be evaluated in order to obtain generalized conclusions on this matter. Authors would like to study if the tendency described previously is the same for any patient-specific case, i.e. independent on the location of the stenosis and on the geometry of the artery.

4   Conclusion

A 3D geometry of a patient-specific RCA with an atherosclerotic plaque was reconstructed for hemodynamic analysis. The use of a geometric model of the RCA as close as possible to the reality is essential to obtain an accurate hemodynamic through numerical simulations. However, the reconstruction of the extremely irregular geometry of the RCA, of its side-branches and of the atherosclerotic plaque is not automatic and requires several particular and rigorous steps. Then, the hemodynamic simulations and analyses were performed. The relative residence time descriptor, the strongest hemodynamic metric for assessing atherosclerotic plaque formation, was analysed for a patient-specific RCA with 20% luminal stenosis and for manipulated stenoses of 40 and 70% in the same patient. An increase of the stenosis percentage implies an increase in the tendency for atherosclerosis formation in many locations of the artery. In the present case, an increase of 25% was observed. Further studies in other patient-specific RCAs should be performed in order to generalize conclusion, i.e. if the tendency described previously is the same for any patient-specific case, independent on the location of the stenosis and on the geometry of the artery.

Acknowledgments. Authors gratefully acknowledge the financial support of the Foundation for Science and Technology (FCT), Portugal, the Engineering Faculty of University of Porto (FEUP), the Institute of Science and Innovation in Mechanical and Industrial Engineering (LAETA-INEGI), the Cardiovascular R&D Unit of the Medicine Faculty of University of Porto (FMUP) and the Cardiology Department of Gaia/Espinho Hospital Centre.

References

1. Yoo, T.S.: Insight into Images, 1st edn. A.K. Petters, Natick (2004)
2. Pianykh, O.S.: Digital Imaging and Communications in Medicine (DICOM), 1st edn. Springer Publishing, Heidelberg (2007)
3. Mozaffarian, D., Benjamin, E.J., Go, A.S., et al.: Heart disease and stroke statistics. Circulation **131**, e29–322 (2015)
4. Myers, J.G., Moore, J.A., Ojha, M., Johnston, K.W., Ethier, C.R.: Factors influencing blood flow patterns in the human right coronary artery. Ann. Biomed. Eng. **29**, 109–120 (2001)
5. Ansys_ Academic Research 15.0. ANSYS Fluent Tutorial Guide (2013)
6. Yilmaz, F., Gundogdu, M.Y.: A critical review on blood flow in large arteries; relevance to blood rheology, viscosity models and physiologic conditions. Korea-Aust. Rheol. J. **20**, 197–211 (2008)
7. Pinho, N., Castro, C.F., António, C.C., Bettencourt, N., Sousa, L.C., Pinto, S.I.S.: Correlation between geoemtric parameters of the left coronary artery and hemodynamic descriptors of atherosclerosis: FSI and statistical study. Med. Biol. Eng. Comput. **3**, 1–15 (2018)
8. Knight, J., Olgac, U., Saur, S.C., et al.: Choosing the optimal wall shear parameter for the prediction of plaque location – a patient-specific computational study in human right coronary arteries. Atherosclerosis **211**, 445–450 (2010)

A Statistical Shape Model Approach for Computing Left Ventricle Volume and Ejection Fraction Using Multi-plane Ultrasound Images

Dawei Liu[1], Isabelle Peck[2], Shusil Dangi[1], Karl Q. Schwarz[3], and Cristian A. Linte[1(✉)]

[1] Rochester Institute of Technology, 1 Lomb Memorial Drive, Rochester, NY 14623, USA
{dxl1169,calbme}@rit.edu, shusil.dangi@gmail.com
[2] Rensselaer Polytechnic Institute, 110 8th Street, Troy, NY 12180, USA
pecki@rpi.edu
[3] University of Rochester Medical Center, 601 Elmwood Ave, Rochester, NY 14642, USA
karl_schwarz@urmc.rochester.edu

Abstract. Assessing the left ventricular ejection fraction (LVEF) accurately requires 3D volumetric data of the LV. Cardiologists either have no access to 3D ultrasound (US) systems or prefer to visually estimate LVEF based on 2D US images. To facilitate the consistent estimation of the end-diastolic and end-systolic blood pool volume and LVEF based on 3D data without extensive complicated manual input, we propose a statistical shape model (SSM) based on 13 key anchor points—the LV apex (1), mitral valve hinges (6), and the midpoints of the endocardial contours (6)—identified from the LV endocardial contour of the tri-plane 2D US images. We use principal component analysis (PCA) to identify the principle modes of variation needed to represent the LV shapes, which enables us to estimate an incoming LV as a linear combination of the principle components (PC). For a new, incoming patient image, its 13 anchor points are projected onto the PC space; its shape is compared to each LV shape in the SSM based on Mahalanobis distance, which is normalized with respect to the LV size, as well as direct vector distance (i.e., PCA distance), without any size normalization. These distances are used to determine the weight each training shape in the SSM contributes to the description of the new patient LV shape. Finally, the new patient's LV systolic and diastolic volumes are estimated as the weighted average of the training volumes in the SSM. To assess our proposed method, we compared the SSM-based estimates of diastolic, systolic, stroke volumes, and LVEF with those computed directly from 16 tri-plane 2D US imaging datasets using the GE Echo-Pac PC clinical platform. The estimated LVEF based on Mahalanobis distance and PCA distance were within 6.8% and 1.7% of the reference LVEF computed using the GE Echo-Pac PC clinical platform.

© Springer Nature Switzerland AG 2019
J. M. R. S. Tavares and R. M. Natal Jorge (Eds.): VipIMAGE 2019, LNCVB 34, pp. 540–550, 2019.
https://doi.org/10.1007/978-3-030-32040-9_55

1   Introduction

Left ventricular ejection fraction (LVEF) is a vital measure of the ventricular contraction efficiency that is calculated as a ratio of LV stroke volume and its volume at end-diastole. Despite its humble definition, LVEF is one of the most critical variables not just in cardiology, but across all disciplines of medicine. LVEF is used for daily management of cardiac patients, as well as a criterion for subject group divisions in research studies. Ultrasound (US) imaging is a standard-of-care, non-invasive method for imaging the heart in real time and is used to assess the left ventricle (LV) filling and ejection capabilities. Accurate estimation of the LV blood-pool volume entails the segmentation of the endocardial border of the LV from 3D US images. This process is challenging, time consuming, and susceptible to high inter- and even intra-user variability, since different clinicians perform the task based on varying empirical knowledge or habits. Moreover, to not compromise frame rate, cardiologists prefer to use multi-view 2D images as opposed to 3D images, the latter of which raises the need to reconstruct a 3D model of the LV blood pool from the several multi-plane 2D US images. As a current practice in clinical settings, cardiologists often visually estimate LVEF based on the area changes of the blood pool as viewed in the real-time 2D US images of the heart. In our previous work [6,7], we demonstrated that the LVEF estimates based on area changes significantly underestimate the true LVEF computed based on volume changes, by as much as 13%. Therefore, an efficient method to calculate the true LVEF based on volume changes that minimize manual interaction to reduce inter- and intra-user variability is paramount.

Left ventricles can have innumerable shapes that pose challenges for developing a comprehensive model to encompass the variations in their geometries. A few geometric models such as cylindrical [4], truncated prolate spheroid [3], and paraboloid [15] were published to mathematically describe an LV shape, but they are not sufficiently accurate to represent patient-specific anatomy. Statistical shape models (SSMs) have been employed by several researchers in the field of cardiology to capture the characteristics of LV shapes and endocardial wall motion from a population of subjects. Previous works mainly focused on identifying characteristics of LV morphological changes associated with cardiac disease or treatment [1,5,8]. Generally, LV geometric coordinates are defined and assembled into a statistical model that comprises all of the coordinates in the population. A mean shape is then generated and the deviation from an individual LV geometry to the mean geometry is calculated. In multiple studies, researchers successfully distinguished LV shapes associated with cardiac disease or surgical intervention from healthy LV shapes. Moreover, SSM led to accuracies in the range of 83% to 98% when used to classify LV geometric morphology [12]. Due to the usually large number of modes of variation, multiple attempts [1,5,9,11,13,14] were made to reduce and analyze the modes of variation in SSM by using principle component analysis (PCA). Several researches discovered that the first three to five principle components calculated from PCA explain the majority of geometric variances such as size, sphericity, and concentricity [8,10].

Hence, SSMs constitute a promising method to characterize LV shapes even with a wide range of variability, this method shall enable us to efficiently estimate LV volumes with or without size effect.

Here we propose a method that relies on a SSM generated using a population of retrospective patient-specific cardiac US images. In compliance with Health Insurance Portability and Accountability Act (HIPPA) regulation, all patient information were de-identified. The SSM uses tri-plane 2D US images depicting the heart at both systole and diastole from 66 patients. In each tomographic view of the tri-plane images, five anchor points were identified by the user. Since one of the five anchor points is the apex of the LV, it co-exists at the same location in all three tomographic views, therefore leading to the LV endocardial shape representation consisting of 13 anchor points: the LV apex (1), mitral valve hinges (6), and the midpoints of the endocardial contours (6).

A ground truth segmentation of the blood pool, along with end-systolic and diastolic volumes, stroke volume, and LVEF computed using the GE Echo-Pac PC standard-of-care clinical platform were available for each dataset and served as gold standard metrics against which our methods were assessed. Once the SSM was generated from 50 of the 66 datasets, it is used to estimate the systolic volumes, diastolic volumes, and their corresponding LVEFs for the remaining 16 testing datasets. The estimated values were then compared to the ground truth values.

In this paper, we describe the construction of the SSM and demonstrate its feasibility to estimate blood pool volume and LVEF, assess their accuracy versus the ground truths, and further support the hypothesis that the estimated LVEF based on true volume measurement is more faithful than area-based LVEF estimate.

2   Methodology

2.1   Patient Data and Anchor Points

The first task of this project was to manually obtain the 13 anchor points for each tri-plane image dataset. In each tri-plane image dataset for each of the 66 patients, there were three images corresponding to three tomographic views of the LV acquired at roughly 60° apart: the 2-chamber view, 3-chamber view (or, sometimes, parasternal long-axis view (PLAX) view), and 4-chamber view. For each tomographic view in the tri-plane image set, five anchor points were selected by the user. The five anchor points are located at five landmarks of the LV endocardial border: apex, mitral valve hinges, and midpoints of the endocardial wall on each side. The LV apex remains stationary during a cardiac cycle, so only one unique set of coordinates exists for the apex in all three views, which resulted in a total of 13 anchor points describing each dataset. Of the 66 patient-specific image datasets in total, 50 were used as the training data for the SSM and the remaining 16 were used as testing data. The shapes characterized by the 13 anchor points, method of disc (MOD) volumes, and 3D reconstructed volumes of the 50 training data were used to predict the volumes and LVEFs of the 16 test datasets (Fig. 1).

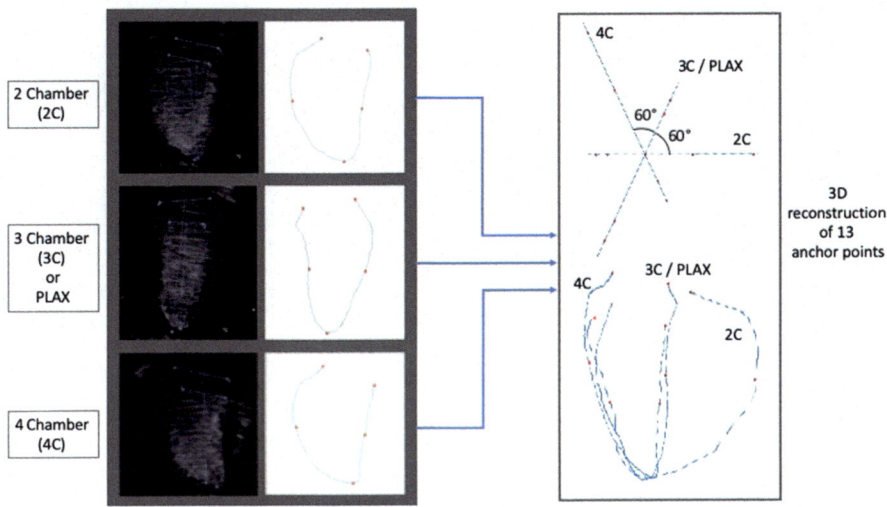

Fig. 1. Anchor point locations on an example diastolic LV endocardial border (the outer-most blue trace in the images with black background in the left column).

The 3D LV blood pool geometry of each patient consisted of a set of 13 anchor landmarks in systole and the same in diastole, while the SSM LV geometry consisted of 50×13 anchor landmarks in systole and same in diastole. Even though the anchor points were selected from 2D images, their transposition into the corresponding 3D volume was obtained under the clinically supported premise that the tri-plane 2D images were collected at 60° apart. Using a reconstruction algorithm previously developed and validated in our earlier work [2], all 2D anchor points were transformed into the 3D space, resulting in the SSM of dimension $50 \times 13 \times 3$ (Fig. 2).

2.2 LV Volume Estimation via Size Normalization and Mahalanobis Distance

Once the SSM was generated, it was used to estimate the systolic and diastolic volumes of an incoming test patient. We used the inverse co-variance matrix of the SSM to calculate the Mahalanobis distance between the test data and each of the training data in the SSM. The Mahalanobis distance can be calculated as defined below:

$$d_i = (test - train_i)^T C^{-1}(test - train_i), \tag{1}$$

where d_i is the Mahalanobis distance between a test data and each of the 50 training data with an index of i, C^{-1} is the inverse co-variance matrix of the SSM,

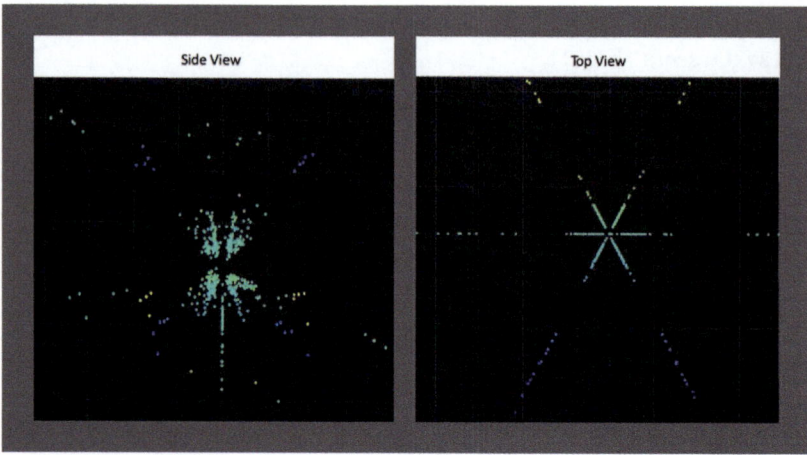

Fig. 2. The point cloud constructed based on 13 anchor points from each of the 50 training datasets.

and $()^T$ is the matrix transpose operation. The Euclidean distance $(test - train_i)$ between testing and training data in 3D space is calculated first, and its transpose is then multiplied by the co-variance matrix of the SSM, as well as the Euclidean distance.

The Mahalanobis distance between the test and each training data is used to calculate the weight according to which each training dataset contributes to the description of a new dataset. These weights were, in turn, used to estimate the systolic and diastolic volumes of the test data.

$$w_i = \frac{\sum_{i=0}^{50} d_i}{d_i} \tag{2}$$

Finally, either both systolic or diastolic volume of the test data can be estimated using the calculated weights multiplied by the corresponding ground truth volumes of each training dataset.

$$V_{estimated} = \sum_{i=0}^{50} w_i V_i \tag{3}$$

The ground truth LV volumes were calculated by method of disc (MOD) that were generated by the standard-of-care platform—GE EchoPac PC—used by our collaborator to acquire the patient-specific US tri-plane images. On each US image acquired and analyzed using EchoPac, there is a legend box that details the MOD volumes for the systolic and diastolic blood pools. The systolic or diastolic MOD volumes were automatically calculated by the EchoPac by averaging the volumes estimated from the three tomographic views of a patient's LV to obtain a single systolic or diastolic volume. Hence, the average MOD volume is defined by the following equation (Fig. 3).

$$V_{MOD/average} = \frac{V_{MOD/2C} + V_{MOD/3C(PLAX)} + V_{MOD/4C}}{3}. \tag{4}$$

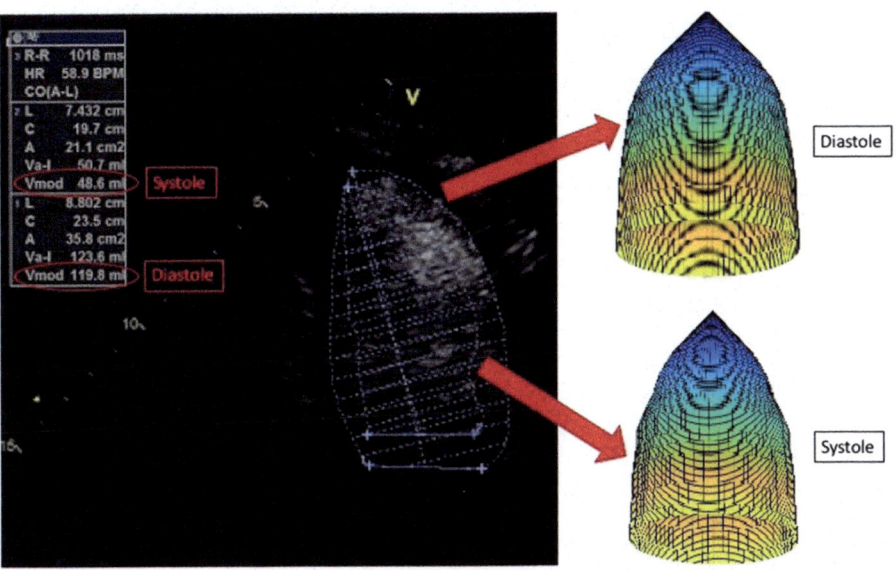

Fig. 3. An example US image captured on GE EchoPac PC with legend showing systolic and diastolic volumes using MOD algorithm, which assumes the LV volume is axisymmetric, as the simulated model shown in the right image panel.

2.3   LV Volume Estimation via Vector Distance Without Size Normalization

As opposed to using the Mahalanobis distance which removes the effect of size differences in the LV shape, we also employed an alternative, yet more traditional method to account for the size effect. The vector distance resulting from the dot product of the test patient's anchor points and each training patient's anchor points was used to determine the contributing weight of each training LV shape in the SSM for test patient LV volume estimations. To assist with PCA in the later steps, the matrix containing all 50 training patients' 3D anchor points was first reshaped from $50 \times 13 \times 3$ to 50×39.

$$A = 50 \times \begin{bmatrix} x_1 & y_1 & z_1 \\ x_2 & y_2 & z_2 \\ \vdots & \vdots & \vdots \\ x_13 & y13 & z_13 \end{bmatrix} \Rightarrow 50 \times \begin{bmatrix} x_1 \\ y_1 \\ z_1 \\ x_2 \\ y_2 \\ z_2 \\ \vdots \\ x_{13} \\ y_{13} \\ z_{13} \end{bmatrix} \tag{5}$$

Since obtaining real eigenvalues and eigenvectors requires the training patient anchor point matrix to be a symmetric matrix, we first computed the Gramian matrix as follows:

$$A' = [50 \times 39]^T [50 \times 39] = [39 \times 50] [50 \times 39] = [39 \times 39]. \tag{6}$$

By performing the eigen-decomposition of the training patient anchor point matrix, the eigenvalues and eigenvectors were calculated. The eigenvalues were sorted in descending order, and the eigenvectors were sorted correspondingly. The training and test patient anchor points were projected onto PCA space via multiplication by the eigenvector matrix, enabling the representation of each LV shape in terms of the 39 modes of variation identified via eigen-decomposition. Finally, the vector distance in PCA space between a test patient's anchor points and training patient's anchor points was calculated using the dot product operation. The vector distance was used as the weight according to which each training patient's diastolic and systolic volumes contribute to describing a test patient's corresponding volumes, similar to Eq. 3. Following this process, the systolic and diastolic volumes of all 16 test patients were estimated.

3   Results and Discussion

Table 1 summarizes the results for the estimated blood pool volumes and LVEFs based on Mahalanobis distance (Estimate 1) and vector distance (Estimate 2), versus their corresponding ground truth values. As shown, both SSM-based estimates were within 6.8% and 1.7% of the ground truth.

Examining the average systolic and diastolic volumes for the test data, the results are determined to be reasonable. The Estimate 1 (i.e., Mahalanobis distance following size normalization) results are lower than ground truth, overall, due to the contribution of some of the smaller volumes in the training data in the SSM. This trend can potentially be corrected by only using the several highest contributing modes of variation identified across all 50 training datasets to describe a new, incoming LV shape, instead of using all 39 modes of variation. Estimate 2 (i.e., vector distance without size normalization) results are much closer to the ground truth data, indicating that the size of LV shapes is a major mode of variation to be considered, and hence size normalization (i.e., Estimate 1) may fail to capture the proper size effect. The mean and standard deviations of the ground truth as well as estimated systolic and diastolic volumes and LVEF are summarized in Table 1.

Table 1. Comparison of LVEFs, diastolic volumes, and systolic volumes from the reference image data, estimation method 1, and estimation method 2.

Mean ± Std. Error[a]	Image data	Estimate 1	Estimate 2
LVEF [%]	55.7 ± 4.7	48.9 ± 4.3	57.4 ± 5.0
Diastolic volume [ml]	125.8 ± 17.9	117.3 ± 17.4	126.8 ± 18.1
Systolic volume [ml]	66.4 ± 17.7	68.7 ± 16.9	66.9 ± 17.9

[a] *Standard Error = Standard Deviation / $\sqrt{Number of Samples}$*

As an additional visual aide to analyze the results, the LVEF, diastolic volume, and systolic volume of all 16 test patient from reference image data, Estimate 1, and Estimate 2 are shown in Figs. 4 and 5 as box and whisker plots. The black 'X' symbols show the means of each data type. The gray boxes illustrate the range between the median and third quartile and the orange boxes illustrate the range between the median and the first quartile. The top whiskers illustrate the range between the third quartile to the maximums and the bottom whiskers illustrate the range between the first quartile to the minimums. The results from Estimate 1 and Estimate 2 are reasonably close to the ground truth image data.

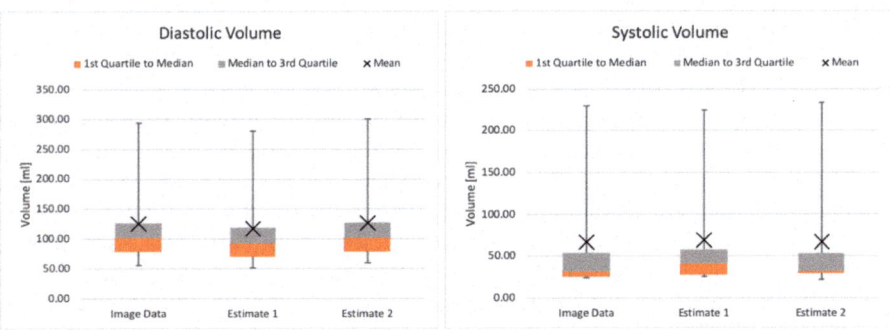

Fig. 4. Reference and estimated diastolic and systolic volumes based on MOD applied to image data, Mahalanobis distance (Estimate 1), and vector distance (Estimate 2).

As shown in Table 2, following an ANOVA comparison between the ground truth data, Estimate 1 and Estimate 2 data, the p-values are all higher than 0.05 and F values are all lower than the critical F values for diastolic volume, systolic volume, and LVEF, hence suggesting no statistically significant difference between the ground truth parameters and those estimated using the proposed SSM method.

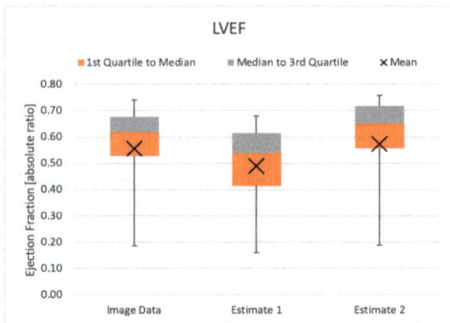

Fig. 5. Reference and estimated LVEFs based on MOD applied to image data, Mahalanobis distance (Estimate 1), and vector distance (Estimate 2).

Table 2. One-way ANOVA tests of LVEFs, diastolic volumes, and systolic volumes from the reference image data, estimation method 1, and estimation method 2.

Quantity	P-value	F value	F critical
LVEF	0.41	0.92	3.20
Diastolic volume	0.92	0.09	3.20
Systolic volume	0.99	0.01	3.20

4   Conclusion and Future Work

We described a method to estimate diastolic and systolic volumes, and LVEF using select key landmarks on the LV endocardial borders and a pre-trained SSM. This approach provides a viable means for quickly and accurately estimating the blood pool volume and LVEF with minimal manual interaction using volume rather than area-based measurements. This method enables clinicians to select only a few endocardial anchor points from the multi-plane 2D US images to estimate the 3D volume and LVEF and provides meaningful progress toward preventing crude visual estimations, often employed in current practice. Lastly, our study showed no statistically significant differences between the ground truth volume and LVEF parameters (obtained using the GE Echp-Pac PC clinical platform) and the corresponding parameters estimated using our proposed SSM-based approach. These estimates may be further improved with additional training data.

As part of our future work, we will also assess our SSM-based estimates against the same parameters computed using native 3D US data, which will also help assess whether the axisymmetric MOD estimates of the LV blood pool provided by the GE EchoPac PC platform is truly a sufficiently viable and accurate metric.

Acknowledgements. This work was supported by the National Institutes of Health under Award No. R35GM128877 and by the National Science Foundation under Award No. 1808530.

References

1. Bruse, J.L., Ntsinjana, H., Capelli, C., Biglino, G., McLeod, K., Sermesant, M., Pennec, X., Hsia, T.Y., Schievano, S., Taylor, A.: CMR-based 3D statistical shape modelling reveals left ventricular morphological differences between healthy controls and arterial switch operation survivors. J. Cardiovasc. Magn. Reson. **18** (2016)
2. Dangi, S., Ben-Zikri, Y.K., Cahill, N., Schwarz, K.Q., Linte, C.A.: Endocardial left ventricle feature tracking and reconstruction from tri-plane trans-esophageal echocardiography data. In: Medical Imaging 2015: Image-Guided Procedures, Robotic Interventions, and Modeling, vol. 9415, p 941505 (2015)
3. Domingues, J.S., Vale, M.D.P., Martinez, C.B.: New mathematical model for the surface area of the left ventricle by the truncated prolate spheroid. Sci. World J. **2017**, 6981515 (2017)
4. Dumesnil, J., Shoucri, R., Laurenceau, J., Turcot, J.: A mathematical model of the dynamic geometry of the intact left ventricle and its application to clinical data. Circulation **59**, 1024–1034 (1979). https://doi.org/10.1161/01.cir.59.5.1024
5. Farrar, G., Suinesiaputra, A., Gilbert, K., Perry, J.C., Hegde, S., Marsden, A., Young, A.A., Omens, J.H., McCulloch, A.D.: Atlas-based ventricular shape analysis for understanding congenital heart disease. Prog. Pediatr. Cardiol. **43**, 61–9 (2016)
6. Liu, D., Peck, I., Dangi, S., Schwarz, K., Linte, C.: Left ventricular ejection fraction: comparison between true volume-based measurements and area-based estimates. In: 2018 IEEE Western New York Image and Signal Processing Workshop (WNY-ISPW), pp. 1–5 (2018)
7. Liu, D., Peck, I., Dangi, S., Schwarz, K., Linte, C.: Left ventricular ejection fraction assessment: unraveling the bias between area- and volume-based estimates. In: Proceedings SPIE Medical Imaging - Ultrasonic Imaging and Tomography, vol. 10955, pp. 109550T–1–8 (2019)
8. Medrano-Gracia, P., Cowan, B.R., Ambale-Venkatesh, B., Bluemke, D.A., Eng, J., Finn, J.P., Fonseca, C.G., Lima, J.A., Suinesiaputra, A., Young, A.A.: Left-ventricular shape variation in asymptomatic populations: the multi-ethnic study of atherosclerosis. J. Cardiovasc. Magn. Reson. **16**, 56 (2014). https://doi.org/10.1186/s12968-014-0056-2
9. Medrano-Gracia, P., Cowan, B.R., Finn, J.P., Fonseca, C.G., Kadish, A.H., Lee, D.C., Tao, W., Young, A.A.: The cardiac atlas project: preliminary description of heart shape in patients with myocardial infarction. In: Camara, O., Pop, M., Rhode, K., Sermesant, M., Smith, N., Young, A. (eds.) Statistical Atlases and Computational Models of the Heart, pp. 46–53 (2010)
10. Piras, P., Teresi, L., Puddu, P., Concetta, T., Young, A., Suinesiaputra, A., Medrano-Gracia, P.: Morphologically normalized left ventricular motion indicators from MRI feature tracking characterize myocardial infarction. Sci. Rep. **7** (2017)
11. Remme, E.W., Young, A.A., Augenstein, K.F., Cowan, B., Hunter, P.J.: Extraction and quantification of left ventricular deformation modes. IEEE Trans. Biomed. Eng. **51**, 1923–31 (2004)

12. Suinesiaputra, A., Ablin, P., Albà, X., Alessandrini, M., Allen, J., Bai, W., Cimen, S., Claes, P., Cowan, B., D'hooge, J., Duchateau, N., Ehrhardt, J., Frangi, A., Gooya, A., Grau, V., Lekadir, K., Lu, A., Mukhopadhyay, A., Oksuz, I., Medrano-Gracia, P.: Statistical shape modeling of the left ventricle: myocardial infarct classification challenge. IEEE J. Biomed. Health Inform. **22**, 503–15 (2017)
13. Tejman-Yarden, S., Bratincsak, A., Bachner-Hinenzon, N., Khamis, H., Rzasa, C., Adam, D., Printz, B.F., Perry, J.C.: Left ventricular mechanical property changes during acute av synchronous right ventricular pacing in children. Pediatr. Cardiol. **37**, 106–111 (2016)
14. Zhang, X., Cowan, B.R., Bluemke, D.A., Finn, J.P., Fonseca, C.G., Kadish, A.H., Lee, D.C., Lima, J.A.C., Suinesiaputra, A., Young, A.A., Medrano-Gracia, P.: Atlas-based quantification of cardiac remodeling due to myocardial infarction. PLOS ONE **9**(10), 1–13 (2014)
15. Zhong, L., Su, Y., Yeo, S.Y., Tan, R.S., Ghista, D.N., Kassab, G.: Left ventricular regional wall curvedness and wall stress in patients with ischemic dilated cardiomyopathy. Am. J. Physiol.-Heart Circu. Physiol. **296**, H573–84 (2009)

Applications of Ontologies for Medical Image Analysis and Computer-Assisted Interventions

Towards a "Surgical GPS": Combining Surgical Ontologies with Physician-Designated Anatomical Landmarks

Austin Tapp[1](✉), Jason E. Blatt[2], H. Sheldon St-Clair[3],
and Michel A. Audette[4]

[1] Biomedical Engineering, Old Dominion University,
5115 Hampton Blvd, Norfolk, VA, USA
atapp001@odu.edu
[2] University of Florida, 1505 SW Archer Road, Gainesville, FL, USA
jason.blatt@neurosurgery.ufl.edu
[3] Children's Hospital of the King's Daughters, 601 Children's Ln,
Norfolk, VA, USA
harvey.st.clair@chkd.org
[4] Modeling and Simulation and Visualization Engineering,
Old Dominion University, 5115 Hampton Blvd, Norfolk, VA, USA
maudette@odu.edu

Abstract. In Computer Assisted Surgery (CAS) the application of generic surgical planning models on individual patient cases is often limited by the number of cases incorporated into the models, as well as the variation that exists between surgical departments in individual hospitals due to local practices. This study seeks to provide a software foundation for the integration of physician-designated anatomical landmarks with existing surgical ontologies for surgical process modeling, in conjunction with Resource Description Framework (RDF) models. Various ontologies were combined with physician-verified landmarks to form surgical process model components, encoded as RDF triples, using the software Karma. These generic, but approach-specific surgical procedure models have been augmented by the physician-submitted landmark information, which is unique for both the patient of interest and the workflows adopted by the physicians themselves. The landmark information is collected through a simple graphical user interface (GUI) using a basic form submission plugin from a website. The data are then integrated into existing ontologies as distinct values through Karma. These newly constructed RDFs are validated using the Apache Jena Fuseki software tool for interoperability, based on their compatibility with SPARQL (SPARQL Protocol and RDF Query Language). Anatomical landmark datasets were easily submitted through the Author's public website GUI, which also provides a baseline set of generic landmark information and notifies the user of changes made to this default dataset. These datasets and associated ontologies were then successfully joined in Karma to produce viable Terse RDF Triple Language (Turtle) entries. These Turtles specify surgeon-submitted anatomical landmarks, which are now represented by the object, the third part of the semantic triple. Apache Jena Fuseki accepted these models and successfully implemented SPARQL queries that allow users to obtain critical information based on content filtering. The information obtained

© Springer Nature Switzerland AG 2019
J. M. R. S. Tavares and R. M. Natal Jorge (Eds.): VipIMAGE 2019, LNCVB 34, pp. 553–567, 2019.
https://doi.org/10.1007/978-3-030-32040-9_56

through SPARQL may be utilized in conjunction with medical imaging-based multi-surface anatomical models to enhance patient care and descriptively guide surgeons through procedures by highlighting the anatomical landmarks along a surgical corridor. Augmented surgical ontologies may enhance the outcome of surgical interventions through their use within the field of computer-assisted surgery (CAS) and can provide a representation of the methods present in pre-operative, intra-operative and post-operative procedures. A combination of physician-determined, patient-specific, landmarks and verified ontologies, ultimately produce a Turtle that can be incorporated into any combination of surgical workflows. Upon development of the "Surgical GPS", which will utilize these patient-specific anatomical landmarks and personalized multi-surface anatomical models, surgeons will navigate a patient's surgical corridor through RDF-based highlighting of patient anatomy.

Keywords: Resource Description Framework (RDF) · Surgical process modeling · Robot-assisted surgery (RAS)/Computer assisted surgery (CAS) · Surgical ontologies · Surgical workflows

1   Introduction

1.1   Context

In recent years, concurrent developments have occurred in the fields of robot/computer-assisted surgery (RAS/CAS) and ontologies, which provide the explicit concepts that guide surgical process modeling (SPM) [1, 2]. Upper ontologies, which are ontologies that act as a foundation of general terms consistent across all domains, are used to create surgical ontologies, which represent the various elements of knowledge that are contained by a surgical process [3]. These elements of surgical ontologies may be utilized in CAS subsequent to the extracted representations found in upper ontologies such as GFO, BFO, YAMATO, PROTON, Cyc, and many others [4–9]. Following extraction, usually by OntoFox, and customization of the upper ontologies, surgical ontologies, like LapOntoSPM, BISON, Deep-Onto, etc. produced by different research groups around the world, are used in the emerging domain of surgical data science and guide SPM (Figs. 1 and 2) [10–12]. These models enhance patient care by optimizing the storage and querying of information related to preoperative or intraoperative data and images that influence the totality of the surgical process [13, 14]. Although the foundation of core ontologies for surgical process models is well established, their widespread application to many, unique patients and their broad utilization by physicians worldwide is challenging due to limitations produced by the nature of the surgical cases, the encoding of the local surgical practices these ontologies may be built upon, and disparities in vocabulary [15, 16]. A solution for the interconnectivity and incorporation of various ontologies in patient-specific interventions may lie in the use of a textual syntax for an RDF called Turtle, and its ability to enable data exchange within the semantic web [17–19].

This study does not suggest that previously used ontologies, upper or otherwise, are obsolete, but rather their integrity should be maintained and coupled with physician-determined anatomical landmarks that will also be integrated with labeling and segmentation in CAS/RAS [20–22]. These two modules, surgical ontologies and anatomical landmarks, should be merged to produce a Turtle that is manipulated to include a generic sense of shared surgical workflows as well as physician-designated information and patient-specific methodologies for surgical interventions [23]. The creation of Turtles in this manner provides an approach that yields more fluid SPMs and may bridge the gap between the institution or geographic variations in surgeons' canonical approaches.

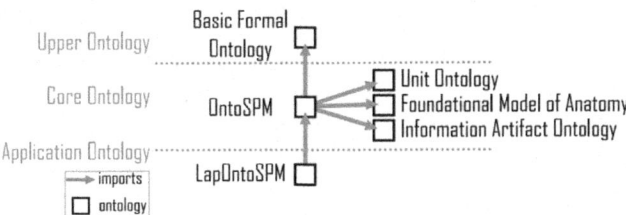

Fig. 1. This figure is a reproduction of Figure 1 from the publication by Katić et al. [10], and displays the relationships of a surgical ontology (LapOntoSPM), its core ontology (OntoSPM), and its upper ontology – Basic Formal Ontology (BFO).

1.2   Surgical Process Model Representation Using Turtles

RDFs, used for processing metadata, provide interoperability between applications that exchange machine-comprehensible information, while ontologies, in Web Ontology Language (OWL), act as a common framework for the semantic web, and may be understood to be a formal collection of terms. In turn, Turtles provide levels of compatibility with the N-Triples format and the triple pattern of SPARQL to encompass the benefits of both formalisms [24–27]. The utilization of a Turtle document frames an RDF graph in a compact textual form, which is physician-readable, and allows ontologies to be incorporated into the RDF graph while maintaining their strict conceptualizations [28]. Turtle entities have a distinct and important format that follows <subject> <predicate> <object> [24]. Recent work by Fetzer, et al., encourages the notion of individual clinical data integrated within a semantic data infrastructure using Turtle files, and this collaborative modeling supports the incorporation of additional existing ontologies (Fig. 3). However, the important distinction and focus of our study involves the declarations of diverse information, such as anatomical landmarks, to be designated by the physicians. Additionally, the subject, while initially assumed to be the surgeon, should instead be understood to be a physician's instrument. This designation would allow for the use of deep-learning algorithms to suggest the tools to be used based on the predicate and landmark designations of the physician, working backwards from the landmark being acted upon [12]. In the context of surgical intervention, this means that the surgeon may be assumed to always perform the action, so that a Uniform Resource Identifier (URI) could be designated to the instrument that the surgeon will use and each instrument would have an attributed action, i.e. "scalpel" "incises" <object> (some physician-designated landmark).

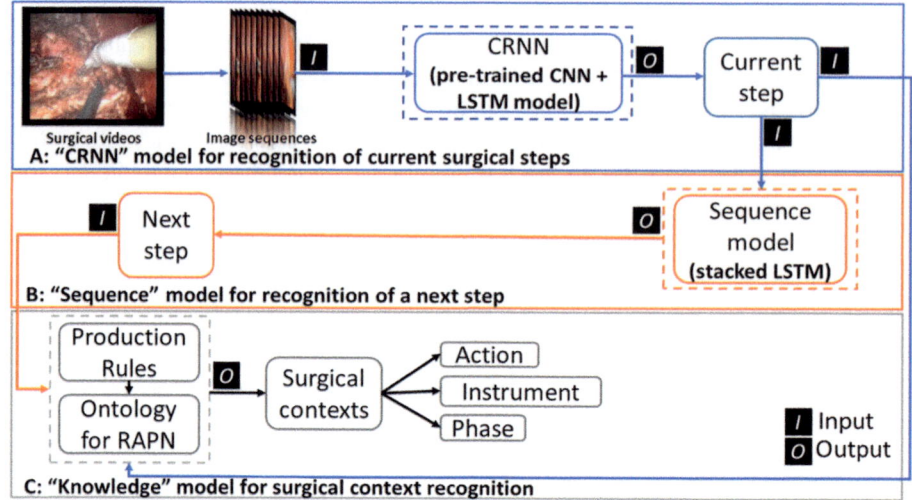

Fig. 2. This figure is a reproduction of Fig. 1 from the publication by Nakawala et al. [12], which displays the potential use for surgical ontologies within surgical workflows and their utilization in surgical process models. While this study covers these aforementioned aspects, "Deep-Onto" also incorporates context recognition through deep learning, which aims to be part of the future work that this study hopes to integrate at a later point in time.

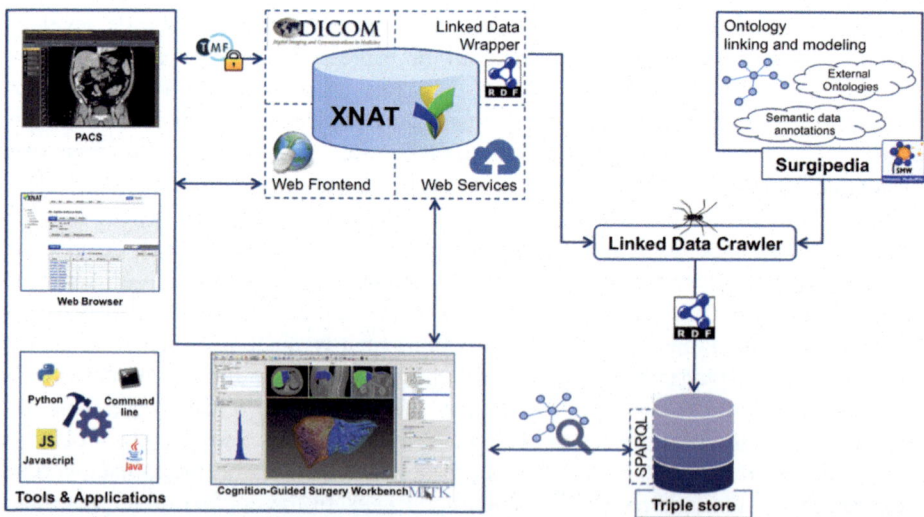

Fig. 3. This figure is a reproduction of Fig. 1 from the publication by Fetzer et al. [28], which displays the Extensible Neuroimaging Archive Toolkit (XNAT). XNAT provides flexible data storage with versatile interfaces and uses semantic information aggregated into a central triple store that is SPARQL accessible.

1.3   Toward a "Surgical GPS"

This study seeks to develop a method for community-based Turtle modeling, in which the surgeons themselves have the ability to update their actions and areas of anatomical interest within a coarsely-defined surgical ontology to be used with CAS (Fig. 4). While there are many canonical approaches used in surgical intervention, there are several schools of thought among physicians and their institutes, and thus variations in these approaches. The formation of these physician-augmented or institute-augmented ontologies will provide significant data on variations in patient therapies, which suggest the necessity of a generalized textbook-compatible coarse template that is readily

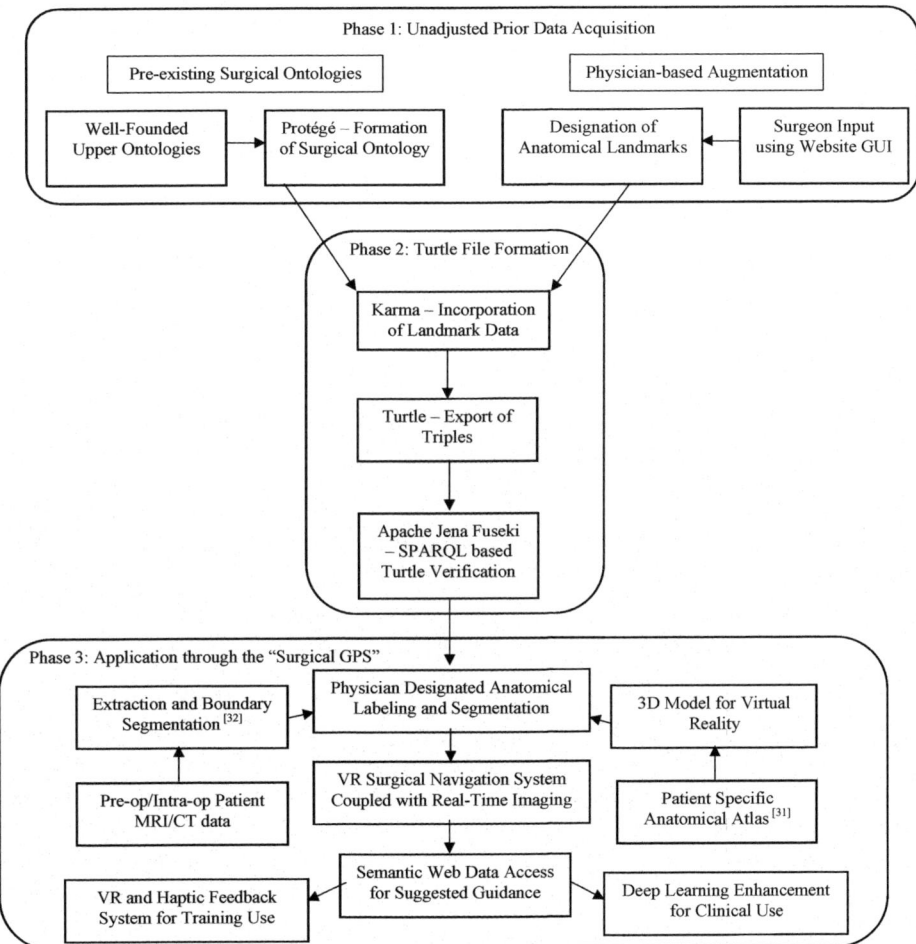

Fig. 4. This figure represents the entirety of the Turtles production and its use within the "Surgical GPS". Importing the data into Karma ultimately produces a SPARQL searchable file. The key differences occur between the types of anatomical landmarks: either (1) generalized by the surgical ontologies, or (2) patient-specific and determined by the physicians themselves.

enhanced through case specifics. Eventually, surgical process models, which make explicit anatomical landmarks (targeted structures or waypoints), from these community repositories will support the variations in canonical approaches that are so often tweaked for case-by-case use. Additionally, the conjunction of these landmarks with segmentation and labeling of patient image datasets, possibly informed by digital anatomical atlases, will lay the groundwork for the "Surgical GPS" [22, 28–31]. While the GPS is essentially understood in the mind of the surgeon dictating the landmarks, a fully descriptive map may be enhanced for use in training purposes and corroboration during surgical procedures, as a citation for malpractice lawsuits and as a reference for the wide-ranging workflows of physician communities. Furthermore, there is potential use for this knowledge contained by the public repository to be used comparatively, for similar patient profiles, which may enhance the collaborative abilities of surgeons for superior patient outcomes. Last, it can serve as a set of requirements for surgery simulation.

2 Methods

2.1 Instancing of Ontologies

As discussed previously, a number of verified ontologies were obtained from online repositories. These ontologies include LapOntoSPM, BISON, and BFO. While any OWL is compatible with Karma, these few were chosen because they are well documented, used by many ontology development groups and are founded on open biological and biomedical ontology principles [32]. Furthermore, a generically produced, non-unified and unverified ontology was created solely for the purposes of showing the versatility of this study's methodology. The ontology that was created is based on information obtained from a variety of surgeons and surgical websites describing neurosurgical procedures [33]. All of these ontologies have been created within Protege as previously described in "Ontology Development 101: A Guide to Creating Your First Ontology" and may be edited as needed [25].

2.2 Obtaining and Using Surgical Landmarks

There is limited documentation and few data repositories that describe physician-designated anatomical landmarks for surgical interventions. One example of an anatomical ontology includes The Foundation Model of Anatomy (FMA), but this ontology would require an extraction of the landmarks by the surgeons or assistant, based on the surgical workflow, and if the extraction is performed by a researcher, it may not encompass the needs of that surgeon's procedure at that particular time [34]. While some landmarks currently exist in a coarse, loosely defined manner on SurgicalWorkflows.com, the greater use of the site supports a positive feedback loop of surgical workflow description production. The landmarks followed by a previous physicians' general workflow may be reused and updated on the website and are then stored in a database (Figs. 5 and 6). These values are obtained as a comma-separated values spreadsheet, which can be seamlessly integrated into Karma [35]. Previously existing ontologies such as OntoSPM and BISON are then integrated into Karma along with the physician's verified landmark entries (Fig. 7). Users can then connect the terminology so that it can be mapped and formed into a complete model using links present from the verified ontologies, which reduces the margin for erroneous associ-ation. Additionally, the Turtle files may benefit from production by an automatic generation schema that is already built into Karma. This schema creates suggestions based on previous associations and stores mapping history. The Turtle files, since they exist in the RDF framework, can be checked using reasoners within Protege and Apache Jena Fuseki [36, 37]. The landmarks that are determined by the physician are now the RDFs third part of the triple, the object, upon which the surgical tool or instrument <subject> will perform an action <predicate>.

2.3   Turtles for Specific Surgical Interventions

When procedural anatomical landmarks are submitted into the repository, they are associated with particular procedural parameters. This updated dataset outlines the suggestions for the next physician who encounters a similar situation, while still pro-viding the ability of the physician to customize the procedure. Moreover, there is potential to produce a completely custom landmark dataset for a uniquely new pro-cedure, which can then be officially ratified and outlined to be used within the website by other physicians. For example, adaptations can occur as a result of topological variation of critical tissues such as the Circle of Willis or anastomoses in nerves. Ideally, this cycle continues for a significant number of surgical workflows and specific patient parameters to produce a wide range of repository data that can be integrated into existing ontologies, all of which would be verified by surgeons from their respective fields in a one-stop-shop-like manner.

Additionally, methodologies may be associated with the use of the action, also known as the predicate of the triple, that is performed by the instrument. While work needs to be done within the ontology to appropriately incorporate the predicate into the Turtle, unlike the seamless incorporation of the landmark, the potential for adjustments in these terms exists and can be submitted to a dataset using the same form found on a procedural page (Fig. 5). The Turtle file that is produced can be queried through SPARQL and may be introduced into the Apache Jena Fuseki server as a single RDF model or as multiple RDF models, which may be simultaneously uploaded, edited and manipulated [38]. The queries are achieved through a search that extracts the object values from the structured and semi-structured data. Data that exists between unknown relationships may also be joined using this server, and the object values may be interchanged for vocabulary purposes. Terminological adjustment that will better suit the situation necessary is a critical portion of anatomical landmark labeling that will come into play during surgical navigation. The ability of the Jena Fuseki server to annotate similar terms through vocabulary analysis is modulated through the use of NeuroLex and SNOMED CT, which provide variable substitutions for terminology [39, 40]. The terminology associated with the anatomical landmarks can be incorpo-rated into ontologies and published with URIs that are specific to the approach, the patient and/or the surgeon.

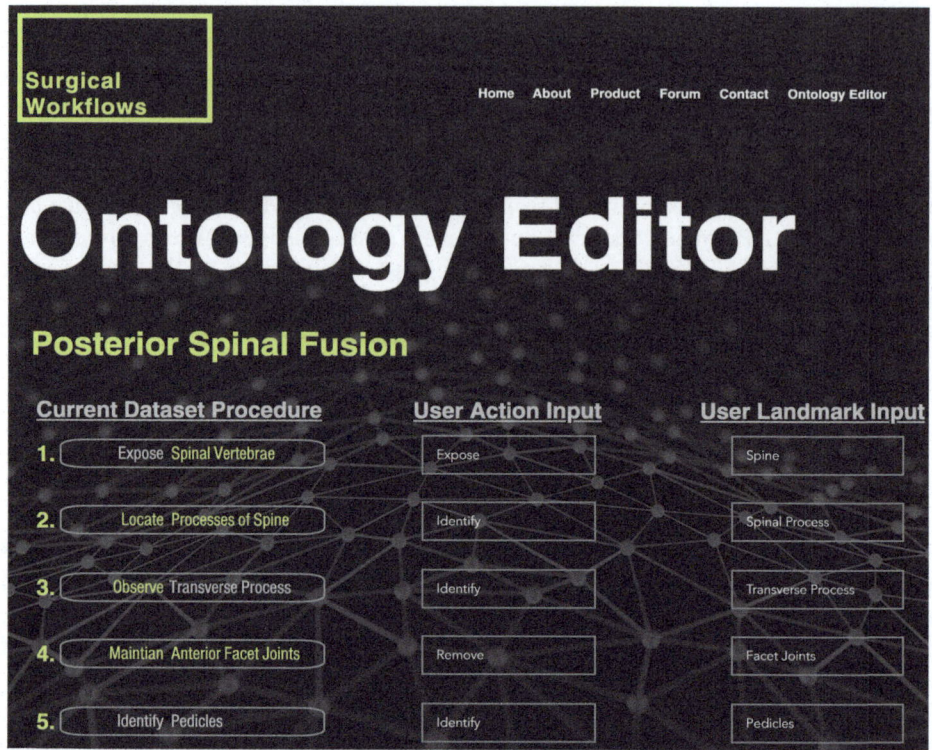

Fig. 5. This figure shows the website GUI from the view of a physician. Previously designated <predicate> and <object> values are present in the two rightmost columns. These values may be edited by the user and submitted into a new dataset unique to the physician and surgical intervention. The values returned in green display the adjustments made to the SPM, while gray values are representative of values that remain unchanged from the current default standard process.

		Title	Landmark1	Landmark2	Landmark3	Landmark4	Landmark5
1		Posterior Spinal Fusion	Spine	Spinal Process	Transverse Process	Facet Joints	Pedicles
2		Posterior Spinal Fusion	Spinal Vertebrae	Processes of Spine	Transverse Process	Anterior Facet Joints	Pedicles

Fig. 6. This figure displays the live data repository obtained from the anatomical landmark submission portion of the ontology editor. The first row is consistent with the anatomical landmarks that exist from a generalized outline of the procedure and are those provided in the website by default. The second row is consistent with the recent submission of input values displayed in Fig. 5 above.

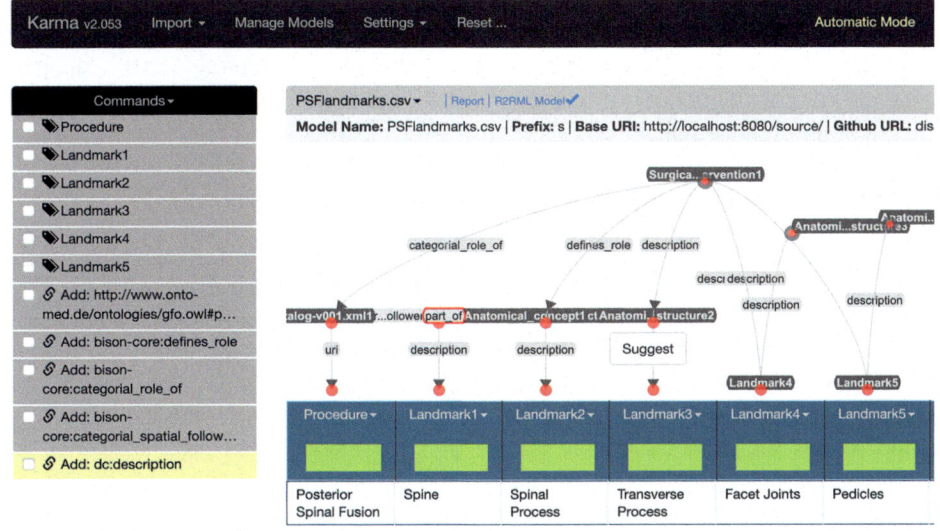

Fig. 7. This figure is a screenshot from Karma, which shows the landmarks extracted from the website have been incorporated with existing surgical or upper ontologies. The mapped features display the network of connections that ultimately form the <subject>, <predicate>, <object> triple set of a Turtle file.

3   Results

3.1   Turtle Model Verification

All of the ontologies that were downloaded could be used in Protege, where they were built initially. These ontologies can be adjusted as needed, and then were incorporated within Karma (Fig. 7). The physician-determined anatomical landmarks from the submission site are coupled within this software. Ultimately, multiple Turtle files were produced and successfully uploaded onto the Apache Jena Fuseki server for further use. In one case, the anatomical landmarks queried are those submitted by physicians who had identified these waypoints as key objects to be noted as they move through an ideal surgical corridor in various procedures (Fig. 8).

3.2   Surgical Process Model Comparison

The models that were formulated from the various surgical ontologies are compared in the table below (Table 1). All the Turtle files produced could be made from a suitable combination of ontologies (verified, unified or otherwise) and the comma-separated values spreadsheet introduced into Karma. In total, six complete, SPARQL compatible RDF models were produced: (1) OntoSPM with verified endoscopic endonasal (EE) surgical landmarks, (2) BISON with verified posterior spinal fusion (PSF) surgical landmarks, (3) BFO with verified surgical landmarks for both the PSF and EE datasets, (4) FMA ontology and all landmarks (anterior interhemispheric transcallosal and retrosigmoid added), (5) Non-unified endoscopic endonasal "ontology" with verified endoscopic endonasal landmarks and (6) Non-unified endoscopic endonasal "ontology" with verified posterior spinal fusion surgical landmarks. These models can all be integrated into a single queryable dataset and further searched simultaneously through the use of the Apache Jena Fuseki server using the methodology previously mentioned.

3.3   Surgical Process Model Expansions

The segmentation of these landmarks is applied for use in surgical planning, in the exact same way as previous studies that utilize the semantic web framework of an OWL [15, 40]. However, this use may be augmented using greater detail and adjusted by the physician and the physician community through their website submissions. Regardless of the ontology that was utilized or the landmarks that were determined, even if sham landmarks were incorporated, the surgical workflow production system proposed has the potential to form a set of RDF models that may be queried by CAS/RAS systems. Additional expansion may be incorporated into the models by uploading and editing them simultaneously with Apache Jena Fuseki Server. Following their subsequent conjunction within this SPARQL processor, they may be exported as a solitary file to be utilized as a single landmark dataset. This rapid integration would provide support for procedures performed by multiple surgeons, such as the endoscopic endonasal procedure. In this instance, the ENT surgeon and the Neurological surgeon would have two differing surgical processes, both of which may be conjoined to

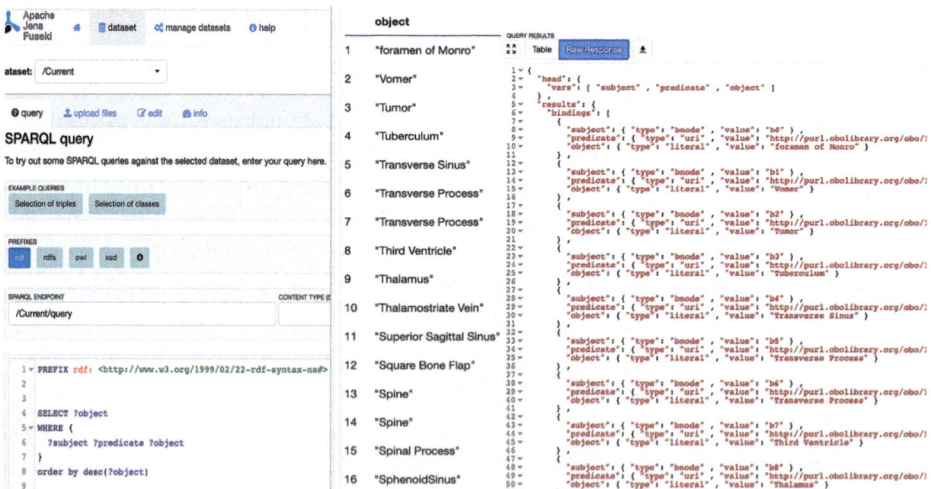

Fig. 8. This figure shows a representation of various anatomical landmarks displayed in both raw (right most image) and table (center image) views after executing a SPARQL search using some filtering options (left most image). This search verifies the ability of the landmarks to be inserted directly into Karma and then exported as a Turtle. These objects provide distinct examples that display the effectiveness of utilizing a form submission method for anatomical landmarks so they may be easily integrated into SPARQL queries and CAS software.

provide one, accentuated workflow based on their individual submissions of anatomical landmarks. This aggregation may occur after the two or more differing Turtle files are produced and then uploaded onto the server. Many other procedures, including ontologies for nurses who may be preparing the patient, can be conjoined in this way to produce a single solid workflow of queryable data that would not need or require multiple files after instantiation. This would enhance the flow of patient care from pre-operative to postoperative workflows of the healthcare providers throughout the institution.

Table 1. Turtle file formation overview

Ontology used	Landmark set	Turtle formed	SPARQL load	Landmarks queried
OntoSPM	Endoscopic Endonasal (EE)	Within Karma	Successful	16
BISON	Posterior Spinal Fusion (PSF)	Within Karma	Successful	12
BFO	EE and PSF	Within Karma	Successful	28
FMA	EE, PSF, AIT, Retrosigmoid	Within Karma	Successful	66
Non-unified	EE	Within Karma	Successful	16
Non-unified	PSF	Within Karma	Successful	12
All previous	All Previous	Previously (1–6)	All Successful	122

This table summarizes the various ontologies, their coupled landmark sets, their environmental formation into Turtle files, their acceptance into the Apache Jena Fuseki Server and the number of landmarks SPARQL successfully queried as an <object> file, the key aspect of the entity to be highlighted along the surgical corridor.

4   Conclusion

4.1   Summary

This study has presented a novel methodology for the incorporation of context-aware surgical interventions through the use of physician-designated anatomical landmarks. In healthcare, the reliance on standardized practices for surgical interventions results in an improved patient prognosis. However, rigidly canonical procedures are frequently adapted to match the specificities of the surgeon and the patient. While surgical ontologies can encapsulate a general surgical workflow for specific procedures, there is a difficulty matching small discrepancies between techniques of surgeons. In order to fit these adaptations in a rapid and easily reproducible way, existing surgical ontologies are manipulated using Karma to incorporate the anatomical landmarks of interest. Anatomical landmark information may be obtained using a GUI that originally rec-ommends some set of landmarks for physician selection and may accept new values that better suit the preferences of a particular case. The new landmarks reflect the patient's needs as well as the local workflows that may be present within specific surgical departments and between physicians. The use of community-verified surgical ontologies and textual anatomical landmarks provides surgeons with the ability to enhance their level of healthcare when coupling these models with guidance systems.

4.2 Future Work

Future work will focus on the enhancement of the SurgicalWorkflows.com website [41] and the development of the online surgical community. There will be a necessity to incorporate all types of specialty data into the website to provide a groundwork that will eventually build upon itself, improving every procedure available. This process begins with the elucidation of methodologies for neurological and orthopedic surgery. Additionally, it will be necessary to build a framework that will allow for the unstructured submission of anatomical landmarks for larger and more detailed procedures that have yet to be outlined by the Authors, as to not limit the surgical community by the pace of the website's development.

The ultimate purpose of this work is future incorporation of the SPMs into a "Surgical GPS". This "GPS" will include the culmination of anatomical information that is based on both physician-specific and community-based textual landmarks, in conjunction with digital atlas-based personalized geometric models of those landmarks, warped to patient preoperative and/or intraoperative MRI/CT data, all of which uses the RDF queries to suggest the next steps of any procedure.

Acknowledgements. I would like to express my appreciation to Professor Michel Audette for his support, supervision and criticisms regarding this study, and its many related and future prospects.

I am also grateful to Dr. Jason E. Blatt and Dr. H. Sheldon St. Clair, of University of Florida and The Children's Hospital of the Kings' Daughters, respectively, for their participation and validation of the procedures utilized within this study. Thank you for your time, and confidence in the future clinical applications of this project.

I wish to thank the Biomedical Engineering department at ODU and CHKD for the financial assistance that they provided to help fund this project.

Conflicts of Interest. The Authors have no conflicts of interest to declare at this time.

References

1. Lalys, F., Jannin, P.: Surgical process modelling: a review. Int. J. Comput. Assist. Radiol. Surg. (2014). https://doi.org/10.1007/s11548-013-0940-5
2. Gibaud, B., Forestier, G., Feldmann, C., Ferrigno, G., Gonçalves, P., Haidegger, T., Julliard, C., Katić, D., Kenngott, H., Maier-Hein, L., März, K.: Toward a standard ontology of surgical process models. Int. J. Comput. Assist. Radiol. Surg. **13**(9), 1397–1408 (2018). https://doi.org/10.1007/s11548-018-1824-5
3. Mudunuri, R., Burgert, O., Neumuth, T.: Ontological modelling of surgical knowledge. In: GI Jahrestagung (2009). http://subs.emis.de/LNI/Proceedings/Proceedings154/gi-proc-154-61.pdf

4. Herre, H., Heller, B., Burek, P., Hoehndorf, R., Loebe, F., Michalek, H.: General Formal Ontology (GFO) Part I: Basic Principles. Epidemiology (2006). www.onto-med.de
5. Grenon, P., Smith, B.: SNAP and SPAN: towards dynamic spatial ontology. Spat. Cognit. Compuat. **1**(4), 69–103 (2004)
6. Mizoguchi, R.: YAMATO: yet another more advanced top-level ontology (n.d.). http://www.ei.sanken.osaka-u
7. Casellas, N., Blázquez, M., Kiryakov, A., Casanovas, P., Poblet, M., Benjamins, R.: OPJK into PROTON: legal domain ontology integration into an upper-level ontology. In: Meersman, R., et al. (eds.) Proceedings of the 3rd International Workshop on Regulatory Ontologies (WORM 2005). LNCS, vol. 3762, pp. 846–855. Springer (2005)
8. Curtis, J., Baxter, D., Cabral, J.: On the application of the Cyc ontology to word sense disambiguation. In: Proceedings of the 19th International Florida Artificial Intelligence Research Society Conference, pp. 652– 657 (2006)
9. Mascardi, V., Cordì, V., Rosso, P., Mascardi, V., Valentina Cordì, P.R.: A comparison of upper ontologies. In: WOA 2007: Dagli Oggetti agli Agenti. 8th AI*IA/TABOO Joint Workshop "From Objects to Agents": Agents and Industry: Technological Applications of Software Agents, 24–25 September 2007. Seneca Edizioni Torino, Genova (2007). ISBN: 978-88-6122-061-4
10. Katić, D., Julliard, C., Wekerle, A.L., Kenngott, H., Müller-Stich, B.P., Dillmann, R., Speidel, S., Jannin, P., Gibaud, B.: LapOntoSPM: an ontology for laparoscopic surgeries and its application to surgical phase recognition. Int. J. Comput. Assist. Radiol. Surg. **10**(9), 1427–1434 (2015). https://doi.org/10.1007/s11548-015-1222-1
11. Siemoleit, S., Uciteli, A., Bieck, R., Herre, H.: Ontological modelling of situational awareness in surgical interventions. In: CEUR Workshop Proceedings, vol. 2050 (2017). http://ceur-ws.org/Vol-2050/ODLS_paper_2.pdf
12. Nakawala, H., Bianchi, R., Pescatori, L.E., De Cobelli, O., Ferrigno, G., De Momi, E.: Deep-Onto network for surgical workflow and context recognition. Int. J. Comput. Assist. Radiol. Surg. (2018). https://doi.org/10.1007/s11548-018-1882-8
13. Despinoy, F., Voros, S., Zemiti, N., Padoy, N., Jannin, P.: Towards unified dataset for modeling and monitoring of computer assisted medical interventions. In: Surgetica, Strasbourg, France, November 2017. lirmm-02105827
14. Kobayashi, S., Cho, B., Huaulmé, A., Tatsugami, K., Honda, H., Jannin, P., Hashizumea, M., Eto, M.: Assessment of surgical skills by using surgical navigation in robot-assisted partial nephrectomy. Int. J. Comput. Assist. Radiol. Surg. (2019). https://doi.org/10.1007/s11548-019-01980-8
15. Jannin, P., Morandi, X.: Surgical models for computer-assisted neurosurgery. NeuroImage **37**(3), 783–791 (2007). https://doi.org/10.1016/j.neuroimage.2007.05.034
16. Zaveri, A., Shah, J., Pradhan, S., Rodrigues, C., Barros, J., Ang, B.T., Pietrobon, R.: Center of excellence in research reporting in neurosurgery–diagnostic ontology. PLoS ONE **7**(5), e36759 (2012). https://doi.org/10.1371/journal.pone.0036759
17. Berners-Lee, T., Jaffe, J.: W3C, The World Wide Web Consortium (2018). https://www.w3.org/
18. Herman, I.: Semantic Web. Same Origin Policy - Web Security, Reuters Limited (2015). www.w3.org/standards/semanticweb/
19. Dunn, A.S., Markoff, B.: Physician-physician communication: what's the hang-up? J. Gen. Intern. Med. (2009). https://doi.org/10.1007/s11606-009-0913-0
20. Rubin, D.L., Talos, I.-F., Halle, M., Musen, M.A., Kikinis, R.: Computational neuroanatomy: ontology-based representation of neural components and connectivity. BMC Bioinform. **10**(Suppl. 2), S3 (2009). https://doi.org/10.1186/1471-2105-10-S2-S3

21. Carbonera, J.L., Abel, M.: Extended ontologies: a cognitively inspired approach. In: CEUR Workshop Proceedings, vol. 1442 (2015). http://ceur-ws.org/Vol-1442/paper_29.pdf

22. Raimbault, M., Jannin, P., Morandi, X., Riffaud, L., Gibaud, B.: Models of surgical procedures for multimodal image-guided neurosurgery. In: Studies in Health Technology and Informatics, vol. 95, pp. 50–55 (2003). https://doi.org/10.3233/978-1-60750-939-4-50

23. Tapp, A.R., Audette, M.A.: Practical applications of neurosurgical ontologies for various craniotomic approaches through computer assisted surgery. In: Modeling, Simulation and Visualization Student Capstone Conference (2018). https://sites.wp.odu.edu/capstone/wp-content/uploads/sites/11988/2018/11/CAPSTONE2018_Proceedings.pdf

24. Lassila, O., Swick, R.: W3C, The World Wide Web Consortium (2018). https://www.w3.org/TR/PR-rdf-syntax/#basicSyntax

25. Noy, N.F., McGuinness, D.L.: Ontology Development 101: A Guide to Creating Your First Ontology. Stanford Knowledge Systems Laboratory, 25 (2001). https://doi.org/10.1016/j.artmed.2004.01.014

26. Berners-Lee, T., Jaffe, J.: W3C, The World Wide Web Consortium (2018). https://www.w3.org/TR/turtle/

27. Guarino, N., Oberle, D., Staab, S.: Handbook on Ontologies. Handbook on Ontologies (2004). https://doi.org/10.1007/978-3-540-24750-0

28. Fetzer, A., Metzger, J., Katic, D., März, K., Wagner, M., Philipp, P., Engelhardt, S., Weller, T., Zelzer, S., Franz, A.M., Schoch, N.: Towards an open-source semantic data infrastructure for integrating clinical and scientific data in cognition-guided surgery. In: Medical Imaging 2016: PACS and Imaging Informatics: Next Generation and Innovations, vol. 9789, p. 97890O (2016). https://doi.org/10.1117/12.2217163

29. Jannin, P., Fleig, O.J., Seigneuret, E., Grova, C., Morandi, X., Scarabin, J.M.: A data fusion environment for multimodal and multi-informational neuronavigation. Comput. Aided Surg. 5(1), 1–10 (2000). https://doi.org/10.1002/(SICI)1097-0150(2000)5:1%3c1::AID-IGS1%3e3.0.CO;2-4

30. Viviani, R.: A digital atlas of middle to large brain vessels and their relation to cortical and subcortical structures. Front. Neuroanat. 10, 12 (2016). https://doi.org/10.3389/FNANA.2016.00012

31. Sultana, S., Blatt, J.E., Gilles, B., Rashid, T., Audette, M.A.: MRI-based medial axis extraction and boundary segmentation of cranial nerves through discrete deformable 3D contour and surface models. IEEE Trans. Med. Imaging 36(8), 1711–1721 (2017). https://doi.org/10.1109/TMI.2017.2693182

32. Grenon, P., Smith, B., Goldberg, L.: Biodynamic ontology: applying BFO in the biomedical domain. In: Pisanelli, D.M. (ed.) Ontologies in Medicine. Studies in Health Technology and Informatics, vol. 102, pp. 20–38 (2004)

33. Cornejo, V.F., Lopez, P.G., Olivas, J.A., Alshafai, N.S., Cardenas, E., Elbabaa, S.K., Kaen, A., Verdu, I.: 3D Neurosurgical Approaches (2014). http://3dneuroanatomy.com/category/3d-neurosurgical-approaches/page/2/. ISSN 2254-9595

34. Noy, N.F., Musen, M.A., Mejino, J.L.V, Rosse, C.: Pushing the Envelope : Challenges in a Frame-Based Representation of Human Anatomy 1 Are Frame Formalisms Expressive Enough?, pp. 1–31 (n.d.). http://citeseerx.ist.psu.edu/viewdoc/download?doi=10.1.1.18.3349&rep=rep1&type=pdf

35. Knoblock, C., Szekely, P., Burns, G.: "Karma." Karma: A Data Integration Tool, University of Southern California (2016). http://usc-isi-i2.github.io/karma/

36. Stanford Center for Biomedical Informatics Research. "Protégé." Protégé (2000). https://protege.stanford.edu/products.php

37. HP Labs. "Jena Ontology API." Apache Jena - Jena Ontology API (2010). https://jena.apache.org/documentation/ontology/

38. Apache Software Foundation. "Apache Jena." A free and open source Java framework for building Semantic Web and Linked Data applications (2011). https://jena.apache.org/index.html
39. Case, J.: Leading Healthcare Terminology, Worldwide. SNOMED International (2011). www.snomed.org/
40. Neurolex, Term Dashboard. University of California (2018). https://scicrunch.org/scicrunch/interlex/dashboard
41. https://tappaustin.wixsite.com/surgicalworkflows

Advances and Imaging Challenges in Micro and Nanofluidics

Blood Flow of Bubbles Moving
in Microchannels with Bifurcations

D. Bento[1], S. Lopes[2], I. Maia[3], A. I. Pereira[4,5], C. S. Fernandes[2],
J. M. Miranda[1(✉)], and R. Lima[6]

[1] CEFT, Faculdade de Engenharia da Universidade do Porto (FEUP),
Rua Dr. Roberto Frias, 4200-465 Porto, Portugal
jmiranda@fe.up.pt
[2] Polytechnic Institute of Bragança, ESTiG/IPB, C. Sta. Apolonia,
5301-857 Bragança, Portugal
[3] Instituto Superior Técnico, Universidade de Lisboa, Av. Rovisco Pais,
1049-001 Lisbon, Portugal
[4] Research Centre in Digitalization and Intelligent Robotics (CeDRI), Instituto
Politécnico de Bragança, C. Sta. Apolónia, 5301-857 Bragança, Portugal
[5] Algoritmi R & D Centre, University of Minho, Campus de Gualtar,
4710-057 Braga, Portugal
[6] MEtRiCS, Mechanical Engineering Department, University of Minho, Campus
de Azurém, 4800-058 Guimarães, Portugal

Abstract. The gas embolism is a well-known phenomenon. Previous studies
have been performed to understand the formation, the behavior and the influence
of air bubbles in microcirculation. This study aims to investigate the flow of
bubbles in a microchannel network with bifurcations. For that purpose, a
microchannel network was fabricated by soft lithography. The working fluids
used were composed by sheep red blood cells (RBCs) suspended in dextran 40
and two different hematocrits were studied, 5% and 10%. The in vitro blood
flow was analyzed for a flow rate of 10 µl/min, by using an inverted microscope
and a high-speed camera. It was possible to visualize the formation of the
bubbles and their behavior along the network. The results show that the passage
of air bubbles influences the cells local concentration, since a higher concentration of cells was seen upstream to the bubble and lower concentrations
downstream to the bubble.

Keywords: Blood flow · Gas embolism · Microcirculation · Bifurcations ·
Networks

1 Introduction

The presence of gas inside of a blood vessel is known as gas embolism. This phenomenon is most commonly due to medical errors and can lead to serious consequences to the patient, including death [1]. The origin of air inside of blood vessel can
be: (1) exogenous, when air is injected in the blood vessel, resulting from medical
procedures (e.g. surgeries [2, 3], hemodialysis [4] and embolotherapy [5]); (2) or
endogenous, when gas bubbles form inside the bloodstream (e.g. heart valves [6, 7],

© Springer Nature Switzerland AG 2019
J. M. R. S. Tavares and R. M. Natal Jorge (Eds.): VipIMAGE 2019, LNCVB 34, pp. 571–577, 2019.
https://doi.org/10.1007/978-3-030-32040-9_57

scuba diving [8], space operations [9] and high altitude flights [10] incidents). Understanding the formation of bubbles, their behavior in the microcirculation, as well as their influence on the blood flow are the objectives of several studies already performed [11–13]. Bubbles in small vessels flow as tubular bubbles [5, 13, 14]; they can be trapped on bifurcations or in small branches [5] or dissolve in blood, depending on their size, composition and shape [11].

It was observed in a previous work [13] that bubbles disturb the RBCs distribution leading to regions of high and of low concentrations of RBCs. In the present work, this phenomenon is further explored by analyzing the effect of bubbles in the RBCs distribution in the blood flow in a microchannel network.

2   Materials and Methods

2.1   Working Fluids and Network Geometry

For the preparation of the working fluids, blood was collected from a healthy sheep using a tube containing ethylenediaminetetraacetic acid (EDTA) to prevent coagulation. The RBCs were then separated from blood samples by centrifugation and aspiration and suspended in Dextran 40 (Dx40) to make the samples with hematocrits (Hct) of 5% and 10%. Detailed procedure can found elsewhere [13, 15]. The microfluidic device used in this study was produced in PDMS by a conventional soft-lithography technique from SU-8 molds (Microliquid, Spain). The geometry used to perform the microfluidic experiments is shown in Fig. 1. The geometry consists in two inlets, one for the fluid and the other for the air. In addition, there is a network of successive bifurcations and at the end there is an outlet.

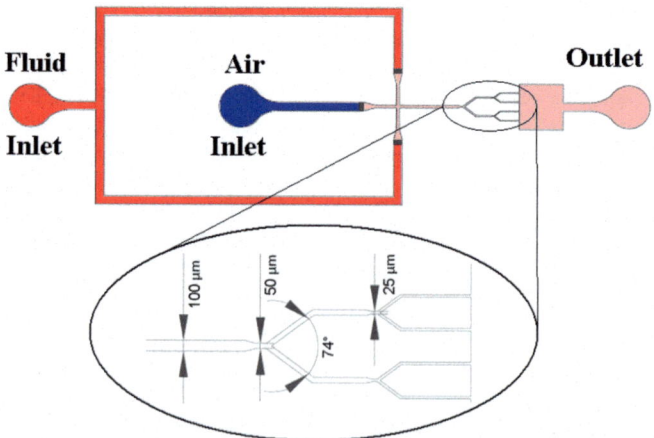

Fig. 1. Main dimensions of the geometry used in this study.

2.2   Experimental Set-Up

The experimental set up comprises an inverted microscope (IX71, Olympus, Japan) combined with a high-speed camera (Fastcam SA3, Photron, USA). The PDMS microchannel was placed on the stage of the inverted microscope and a syringe pump (PHD ULTRA, Harvard Apparatus) was used to control the flow rate of the working fluids. The liquid flow rate was set to 10 μL/min. Additionally, the air pressure was controlled by a pressure pump (Eleveflow PG1113). By adjusting the pressure, the pump has injected air droplets in the microchannel.

2.3   Image Processing

The images were recorded at the midplane of the microchannel using a frame rate of 2000 frames/second. All the frames were analyzed using the tool *Plot z-axis profile* of the ImageJ software. This function allows, after selecting a particular area of the video, to obtain a diagram with the tonality in the region of interest through time. By using this methodology, it was possible to evaluate the passage air bubbles in a region of interest (ROI) for a certain time.

3   Results and Discussion

In Fig. 2 we can see the formation of a bubble in the microchannel. Figure 2(a) shows the bubble before detachment whereas Fig. 2(b) shows the bubble after detachment.

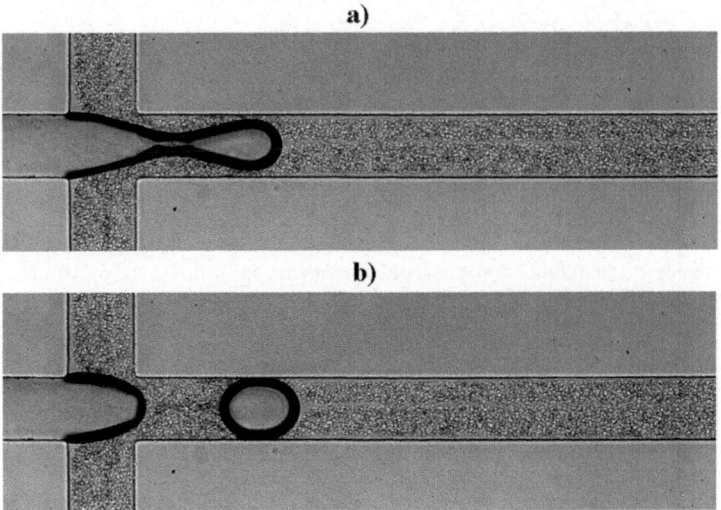

Fig. 2. Sequential images of the formation of a bubble air (a) bubble before detachment (b) bubble after detachment.

The obtained videos were, firstly, carefully examined. It was observed that circulating air bubbles influence the local haematocrit located downstream and upstream to the bubble. Figure 3 shows this phenomenon at two different instants.

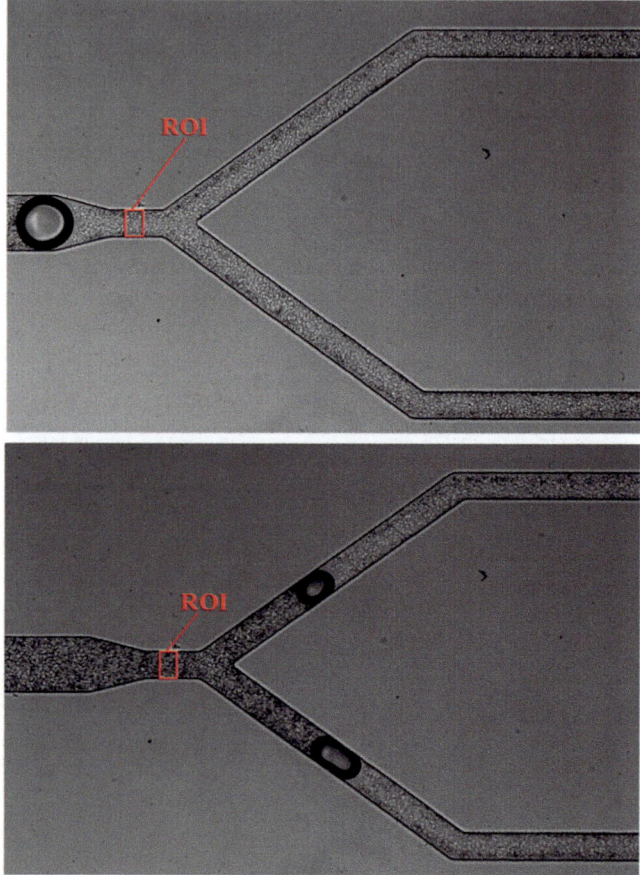

Fig. 3. Influence of air bubbles on the local hematocrit for a flow rate of 10 μl/min and 10% Hct.

By using the *Plot z-axis profile* tool from ImageJ, we were able to investigate the influence of the circulating bubbles on the local Hct. As shown in Fig. 4, it was possible to examine the variation of the Hct along a certain time, in a selected region (ROI) of the microchannel. The variation of the local Hct was analyzed by measuring the intensity of the pixels at the region where the *plot Z-axis profile* function was applied.

When the intensity of the pixels is constant, it means that the hematocrit that passes through this region is constant. If the intensity goes up, it indicates a decrease of the hematocrit at that region of interest (ROI). It is also possible to observe a drastic

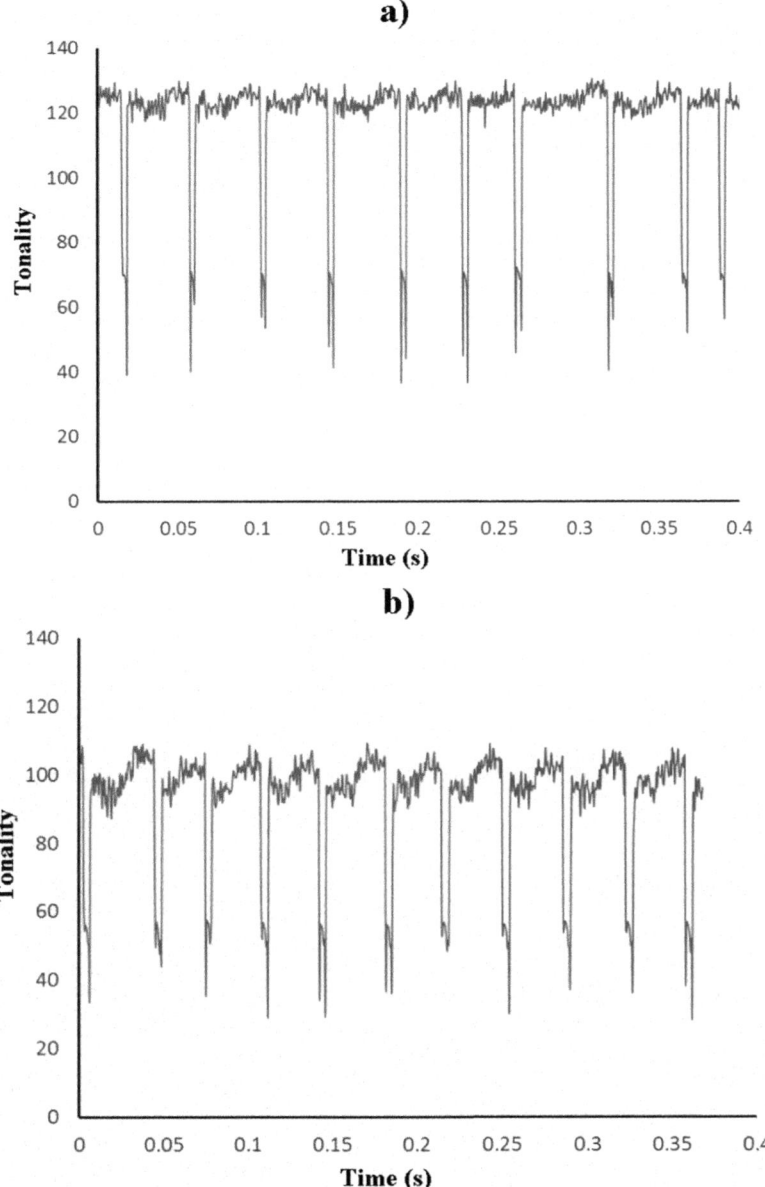

Fig. 4. Influence of air bubbles on the local Hct for a flow rate of 10 μl/min and two different feed Hcts: (a) 5% Hct and (b) 10% Hct.

reduction of the tonality that indicates the passage of the air bubbles at the ROI. After that, the tonality tends to increase until it stabilizes and returns to the values initially observed, i.e., just after the air bubble passes through the ROI there is an increase of the local Hct, which tends to decrease until it stabilizes and returns to its initial values.

From the results obtained in Fig. 4, it is possible to conclude that there is a strong influence of the air bubbles on the local Hct. These measurements show clearly that the local Hct is higher at the upstream of the bubble and lower at the downstream of the air bubble. In addition, for the 10% Hct sample, the difference between the local Hct located before and after the bubble was more evident than the difference presented in the samples with 5% Hct.

4 Conclusions and Future Directions

In this study, we have investigated the influence of air bubbles in the RBCs distribution when flowing through a microchannel network. The results indicate that the air bubbles strongly affect the local Hct at the regions ahead and behind them. When a bubble has approached a certain region of the microchannel, the local Hct has decreased at the downstream region. After the bubble has passed, the Hct tends to increase, and then it tends to decrease until it reaches the inlet Hct.

In the near future, we intend to verify if the difference of hematocrit before and after the bubble results only due to the method of injection of the bubble air. In addition, we also intend to study the behavior of bubbles in geometries with confluences.

Acknowledgments. This work was supported by Fundação para a Ciência e a Tecnologia (FCT), Portugal, under the strategic grants UID/EMS/04077/2019, UID/EEA/04436/2019 and UID/EMS/00532/2019. The authors are also grateful for the partial funding of FCT through the projects POCI-01-0145-FEDER-016861 (ref: PTDC/QEQ-FTT/4287/2014), NORTE-01-0145-FEDER-029394, NORTE-01-0145-FEDER-030171, funded by COMPETE2020, NORTE2020, PORTUGAL2020, and FEDER. D. Bento acknowledges the PhD scholarship SFRH/BD/91192/2012 granted by FCT.

References

1. Muth, C.M., Shank, E.S.: Gas embolism. New Engl. J. Med. **342**(7), 476–482 (2000)
2. Borger, M.A., et al.: Neuropsychologic impairment after coronary bypass surgery: effect of gaseous microemboli during perfusionist interventions. J. Thorac. cardiovasc. Surg. **121**(4), 743–749 (2001)
3. Abu-Omar, Y., et al.: Solid and gaseous cerebral microembolization during off-pump, on-pump, and open cardiac surgery procedures. J. Thorac. cardiovasc. Surg. **127**(6), 1759–1765 (2004)
4. Bischel, M.D., Scoles, B.G., Mohler, J.G.: Evidence for pulmonary microembolization during hemodialysis. Chest **67**(3), 335–337 (1975)
5. Samuel, S., et al.: In vivo microscopy of targeted vessel occlusion employing acoustic droplet vaporization. Microcirculation **19**(6), 501–509 (2012)
6. Deklunder, G., et al.: Microemboli in cerebral circulation and alteration of cognitive abilities in patients with mechanical prosthetic heart valves. Stroke **29**(9), 1821–1826 (1998)
7. Milo, S., et al.: Mitral mechanical heart valves: in vitro studies of their closure, vortex and microbubble formation with possible medical implications. Eur. J. Cardiothorac. Surg. **24**(3), 364–370 (2003)

8. Papadopoulou, V., et al.: A critical review of physiological bubble formation in hyperbaric decompression. Adv. Coll. Interface Sci. **191**, 22–30 (2013)

9. Karlsson, L.L., et al.: Venous gas emboli and exhaled nitric oxide with simulated and actual extravehicular activity. Respir. Physiol. Neurobiol. **169**, S59–S62 (2009)

10. Foster, P.P., Butler, B.D.: Decompression to altitude: assumptions, experimental evidence, and future directions. J. Appl. Physiol. **106**(2), 678–690 (2009)

11. Papadopoulou, V., et al.: Circulatory bubble dynamics: from physical to biological aspects. Adv. Coll. Interface Sci. **206**, 239–249 (2014)

12. Bento, D., Rodrigues, R., Faustino, V., Pinho, D., Fernandes, C., Pereira, A., Garcia, V., Miranda, J., Lima, R.: Deformation of red blood cells, air bubbles, and droplets in microfluidic devices: flow visualizations and measurements. Micromachines **9**, 151 (2018)

13. Bento, D., Sousa, L., Yaginuma, T., Garcia, V., Lima, R., Miranda, J.M.: Microbubble moving in blood flow in microchannels: effect on the cell free layer and cell local concentration. Biomed. Microdevices **19**(1), 6 (2017)

14. Branger, A.B., Eckmann, D.M.: Accelerated arteriolar gas embolism reabsorption by an exogenous surfactant. Anesthesiol.: J. Am. Soc. Anesthesiol. **96**(4), 971–979 (2002)

15. Pinho, D., Campo-Deano, L., Lima, R., Pinho, F.T.: In vitro particulate analogue fluids for experimental studies of rheological and hemorheological behavior of glucose-rich RBCs suspensions. Biomicrofluidics **11**(5), 054105 (2017)

Magnetic PDMS Microparticles for Biomedical and Energy Applications

Rui Lima[1,2(✉)], E. J. Vega[3], V. F. Cardoso[4], G. Minas[4],
and J. M. Montanero[3]

[1] Metrics, Mechanical Engineering Department, University of Minho,
Campus de Azurém, 4800-058 Guimarães, Portugal
rl@dem.uminho.pt
[2] CEFT, Faculdade de Engenharia da Universidade do Porto (FEUP),
R. Dr. Roberto Frias, 4200-465 Porto, Portugal
[3] Depto. de Ingeniería Mecánica, Energética y de los Materiales and Instituto de
Computación Científica Avanzada (ICCAEx), Universidad de Extremadura,
06006 Badajoz, Spain
[4] CMEMS-UMinho, Universidade do Minho, Campus de Azurém,
4800-058 Guimarães, Portugal

Abstract. Polydimethylsiloxane (PDMS) is one of the most widely used
polymers in microfluidics. Furthermore, magnetic nanoparticles (MNPs), due
their superior thermal properties, are also gaining a great interest among the
industry and microfluidic scientific community. In this work, a technique based
on a flow focusing principle was used to produce magnetic PDMS microparticles. A microvisualization system composed by digital video cameras and
optical lenses was used to control and measure the size of the obtained
microparticles. To the best of our knowledge, this is the first work that shows
magnetic PDMS microparticles able to be used for both biomedical and energy
applications.

Keywords: PDMS microparticles · Magnetic nanoparticles · Flow focusing ·
Nanofluids

1 Introduction

Polydimethysiloxane (PDMS) is a polymer widely used in many biomedical applications due to its remarkable properties, such as good optical transparency, flexibility,
biocompatibility and permeability to gases [1, 2]. This material is widely used to
fabricate *in vitro* arterial models [3, 4], microfluidic devices to perform blood cells
separation [5, 6], measure cells deformability [7, 8], to culture endothelial cells [9–11]
and to study blood flow phenomena happening in microvessels [12–14]. Recently, by
using a flow focusing technique [15] PDMS microparticles were successfully produced
with dimensions close to those of red blood cells [16, 17] and to be used to produce
particulate blood analogues.

In the past few years, advances in nanotechnology have led to a new generation of
thermofluids known as nanofluids (NFs). Generally, this fluid is a dispersion of

© Springer Nature Switzerland AG 2019
J. M. R. S. Tavares and R. M. Natal Jorge (Eds.): VipIMAGE 2019, LNCVB 34, pp. 578–584, 2019.
https://doi.org/10.1007/978-3-030-32040-9_58

nanoparticles (NPs) suspended in a base fluid [18, 19]. The introduction of NPs into the base fluid are claimed to increase the NF thermal conductivity and convective heat transfer and as a result has become the object of research for several industrial applications [18, 19]. However, only few NFs were successfully implemented in industry [18–20]. The present work shows a flow focusing technique able to produce magnetic PDMS microparticles. This new NF may solve the problems linked to the formation of clusters and improve the stability of the NFs.

2   Materials and Methods

This section presents the magnetic nanoparticles (MNPs) synthesis, along with the experimental setup and the methodology to generate the magnetic PDMS microparticles.

Magnetic Nanoparticles (MNPs) Synthesis

The magnetic iron oxide NPs were synthesized by using a co-precipitation method. This method is known to be cost effective and appropriate for mass production. Figure 1 shows the produced NPs and representative TEM image. A detailed description of this method can be found at Cardoso *et al.* [21].

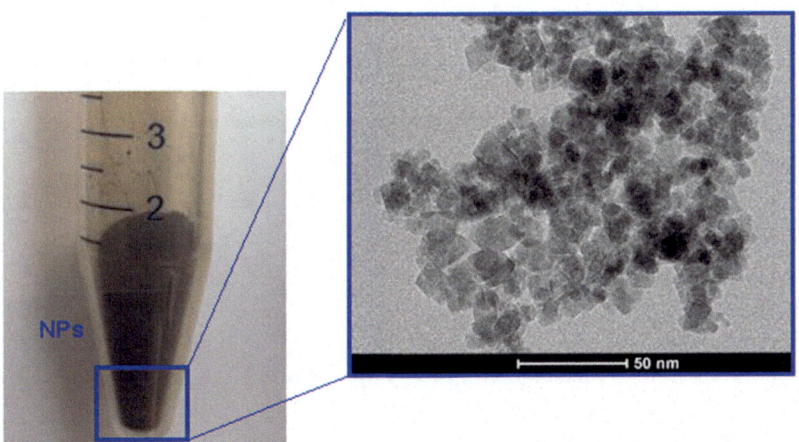

Fig. 1. Magnetic iron oxide NPs and representative TEM image, adapted from [21].

The MNPs feature an average size of 11 ± 2 nm and are consistent with the magnetite Fe_3O_4. Moreover, the nanoparticles show a normalized magnetization of ~ 69 emu g^{-1} at ~ 10 kOe, which also corresponds to the saturation magnetization, and a superparamagnetic behavior with an extremely low coercivity of 1.6 Oe [21].

Experimental Setup and Methodology to Generate Magnetic PDMS Microparticles

Muñoz-Sanchez *et al.* [16] proposed a flow-focusing technique to produce flexible PDMS microparticles to develop biomimetic fluids such as blood analogues. In this work, we adapt this technique to produce magnetic PDMS microparticles by mixing MNPs with the elastomer pre-polymer. Figure 2 shows the experimental setup to produce magnetic PDMS liquid droplets.

Fig. 2. Experimental setup to produce magnetic PDMS liquid droplets: detail of the needle located inside the capillary and jet of the PDMS mixture (left side); visualization of magnetic PDMS liquid droplets about 16 µm in diameter formed after the jet breakage.

Generally, the main steps to produce the magnetic PDMS microparticles are the following:

(1) A PDMS elastomer kit (*Dow Corning Sylgard 186*) formed by a base pre-polymer (Part A) and a curing agent (Part B) was used. The iron oxide MNPs were mixed with Part A, and Part B was then added to obtain at the end a PDMS precursor mixture. Note that, the ratio Part A:Part B was 6:4 and the amount of MNPs was 5% by weight.

(2) By using a syringe pump the PDMS mixture was injected through hypodermic needle (*Becton Dickinson Microlance 3 30G 1/2*) with a constant flow rate of 0.1 ml/h. The needle was introduced inside a glass cylindrical capillary (200 µm in diameter) where both were immersed in a bath with a mixture (9:1) of glycerol and surfactant Brij to avoid the coalescence of the droplets at the downstream region of the system. To produce the focusing effect, the outer bath was withdrawn across the capillary at a constant flow rate of 2 ml/h by means of another syringe pump.

(3) The obtained sample containing magnetic PDMS liquid droplets was placed into a glass optical cell. This container was heated at about 70 °C for about 14 h by means of a hot plate. Then, when the solution was cooled down to the room temperature, the cured PDMS microparticles were visualized and tested by a video microscopy system combined with a static magnetic field.

3   Results and Discussion

This section presents the comparison between the size of the liquid droplets and solid particles just after curing. During the production time, several images were recorded to measure the size distribution of the liquid droplets formed after the jet breakage (see Fig. 2). Regarding the cured PDMS microparticles the measurements were only performed to those having MNPs. For that, a needle immersed within the fluid sample was brought into contact with 20 magnets to produce a static magnetic field of about 285 mT (see Fig. 3). By using a video microscopy system, it was possible to not only measure the size of the magnetic PDMS microparticles but also to track the trajectories of the particles that were attracted to the magnetic needle (see Fig. 4). In addition, in Fig. 3 it is clear that the amount of PDMS particles around the tip of the needle tends to increase with time, supporting the evidence that the generated PDMS microparticles are magnetic.

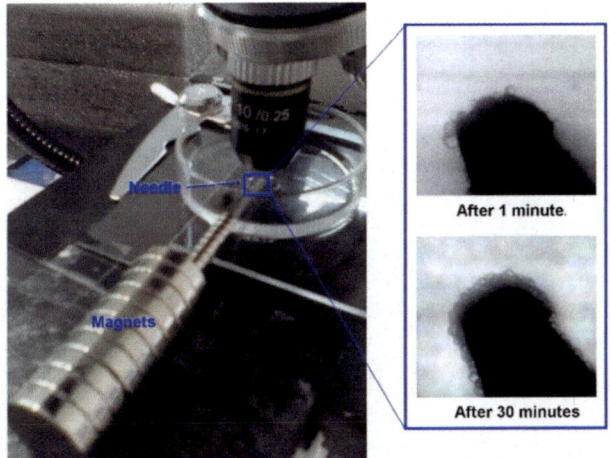

Fig. 3. Needle with multiple magnets (20), to create a static magnetic field of about 285 mT. Detailed images showing the amount of magnetic PDMS microparticles attached to the needle tip for two different time periods.

In Fig. 4, it is shown the trajectories of several magnetic PDMS microparticles located at different optical planes. From these visualizations it was possible to notice that the velocities tend to increase as the particles approach the needle and there is a relatively high degree of polydispersity. For instance, within the same sample it was possible to measure particles with diameters as big as 24 μm and down to 2 μm (see Fig. 5). The main reason for this degree of polydispersity might be due to some instabilities observed during the production process. In the future, we expect to obtain more insights into the generation of magnetic PDMS particles with a better degree of monodispersity.

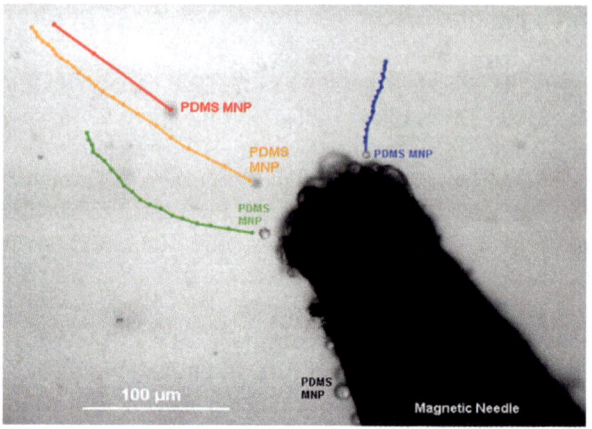

Fig. 4. Trajectories of several magnetic PDMS microparticles located at different optical planes.

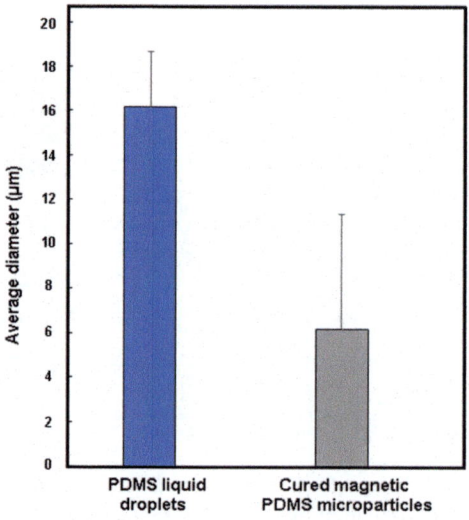

Fig. 5. Average size of the generated liquid droplets and cured magnetic PDMS microparticles.

Finally, Fig. 5, shows the average size of the generated liquid droplets and cured magnetic PDMS microparticles. The results clearly show that the liquid droplets tend to suffer a pronounced shrinkage while curing. This result corroborate previous works performed with both transparent [16] and colour [17] PDMS.

To the best of our knowledge, the present work shows for the first time magnetic PDMS microparticles. Although, the current results need further research, we believe that these microparticles have the potential to be used for both biomedical and energy applications.

Acknowledgment. This work was supported by *Fundação para a Ciência e a Tecnologia* (FCT) under the strategic grants UID/EMS/04077/2019, UID/EEA/04436/2019 and UID/EMS/00532/2019. The authors are also grateful for the funding of FCT through the projects POCI-01-0145-FEDER-016861, POCI-01-0145-FEDER-028159, NORTE-01-0145-FEDER-029394, NORTE-01-0145-FEDER-030171, funded by COMPETE2020, NORTE2020, PORTUGAL2020, and FEDER, and the PhD grant SFRH/BD/91192/2012. The authors also acknowledge to FCT for partially financing the research under the framework of the project UTAP-EXPL/CTE/0064/2017, *financiado no âmbito do Projeto 5665 - Parcerias Internacionais de Ciência e Tecnologia*, UT Austin Programme. Partial support from the Spanish Ministry of Science and Education (grant no. DPI2016-78887) and Junta de Extremadura (grants no. GR15014 and IB18005, partially financed by FEDER funds) are gratefully acknowledged too.

References

1. Mata, A., Fleischman, A., Roy, S.: Characterization of polydimethylsiloxane (PDMS) properties for biomedical micro/nanosystems. Biomed. Microdevice **7**(4), 281–293 (2005)
2. Lima, R., Wada, S., Tanaka, S., Takeda, M., Ishikawa, T., Tsubota, K., Imai, Y., Yamaguchi, T.: In vitro blood flow in a rectangular PDMS microchannel: experimental observations using a confocal micro-PIV system. Biomed. Microdevices **10**(2), 153–167 (2008)
3. Rodrigues, R.O., Pinho, D., Bento, D., Lima, R., Ribeiro, J.: Wall expansion assessment of an intracranial aneurysm model by a 3D digital image correlation system. Measurement **88**, 262–270 (2016)
4. Cardoso, C., Fernandes, C.S., Lima, R., Ribeiro, J.: Biomechanical analysis of PDMS channels using different hyperelastic numerical constitutive models. Mech. Res. Commun. **90**, 26–33 (2018)
5. Tanaka, T., Ishikawa, T., Numayama-Tsuruta, K., Imai, Y., Ueno, H., Matsuki, N., Yamaguchi, T.: Separation of cancer cells from a red blood cell suspension using inertial force. Lab Chip **12**, 4336–4343 (2012)
6. Faustino, V., Catarino, S.O., Lima, R., Minas, G.: Biomedical microfluidic devices by using low-cost fabrication techniques: a review. J. Biomech. **49**(11), 2280–2292 (2016)
7. Bento, D., Rodrigues, R., Faustino, V., Pinho, D., Fernandes, C., Pereira, A., Garcia, V., Miranda, J., Lima, R.: Deformation of red blood cells, air bubbles, and droplets in microfluidic devices: flow visualizations and measurements. Micromachines **9**, 151 (2018)
8. Faustino, V., Catarino, S.O., Pinho, D., Lima, R.A., Minas, G.: A passive microfluidic device based on crossflow filtration for cell separation measurements: a spectrophotometric characterization. Biosensors **8**, 125 (2018)
9. Shin, M., Matsuda, K., Ishii, O., Terai, H., Kaazempur-Mofrad, M., Borenstein, J., Detmar, M., Vacanti, J.P.: Endothelialized networks with a vascular geometry in microfabricated poly(dimethyl siloxane). Biomed. Microdevices **6**(4), 269–278 (2004)
10. Ohashi, T., Sato, M.: Endothelial cell responses to fluid shear stress: from methodology to applications. In: Single and Two-Phase Flows on Chemical and Biomedical Engineering, pp. 579–599. Bentham Science Publishers (2012)
11. Huh, D., Torisawa, Y.S., Hamilton, G.A., Kim, H.J., Ingber, D.E.: Microengineered physiological biomimicry: organs-on-chips. Lab Chip **12**(12), 2156–2164 (2012)
12. Abkarian, M., Faivre, M., Horton, R., Smistrup, K., Best-Popescu, C.A., Stone, H.A.: Cellular-scale hydrodynamics. Biomed. Mater. **3**, 034011 (2008)

13. Bento, D., Fernandes, C.S., Miranda, J.M., Lima, R.: In vitro blood flow visualizations and cell-free layer (CFL) measurements in a microchannel network. Exp. Thermal Fluid Sci. **109**, 109847 (2019)

14. Pinho, D., Rodrigues, R.O., Faustino, V., Yaginuma, T., Exposto, J., Lima, R.: Red blood cells radial dispersion in blood flowing through microchannels: the role of temperature. J. Biomech. **49**, 2293–2298 (2016)

15. Acero, A.J., Rebollo-Muñoz, N., Montanero, J.M., Gañán-Calvo, A.M., Vega, E.J.: A new flow focusing technique to produce very thin jets. J. Micromech. Microeng. **23**, 065009 (2013)

16. Muñoz-Sánchez, B.N., Silva, S.F., Pinho, D., Vega, E.J., Lima, R.: Generation of micro-sized PDMS particles by a flow focusing technique for biomicrofluidics applications. Biomicrofluidics **10**, 014122 (2016)

17. Anes, C.F., Pinho, D., Muñoz-Sánchez, B.N., Vega, E.J., Lima, R.: Shrinkage and colour in the production of micro-sized PDMS particles for microfluidic applications. J. Micromech. Microeng. **28**, 075002 (2018)

18. Pang, C., Lee, J.W., Tae, Y.T.: Review on combined heat and mass transfer characteristics in Nanofluids. Int. J. Therm. Sci. **87**, 49–67 (2015)

19. Sarkar, J., Ghosh, P., Adil, A.: A review on hybrid nanofluids: recent research, development and applications. Renew. Sustain. Energy Rev. **43**, 164–177 (2015)

20. Saidur, R., Leong, K.Y., Mohammad, H.A.: A review on applications and challenges of nanofluids. Renew. Sustain. Energy Rev. **15**(3), 1646–1668 (2011)

21. Cardoso, V.F., Irusta, S., Navascues, N., Lanceros-Mendez, S.: Comparative study of sol-gel methods for the facile synthesis of tailored magnetic silica spheres. Mater. Res. Express **3**, 7 (2016)

Red Blood Cells Separation in a Curved T-Shaped Microchannel Fabricated by a Micromilling Technique

Miguel Madureira[1], Vera Faustino[2,3,4], Helmut Schütte[1],
Diana Pinho[2,3,4,5], G. Minas[2(✉)], Stefan Gassmann[1], and Rui Lima[5]

[1] Jade University of Applied Science, Wilhelmshaven, Germany
[2] Microelectromechanical Systems Research Unit (CMEMS-UMinho),
University of Minho, 4800-058 Guimarães, Portugal
gminas@dei.uminho.pt
[3] MEtRiCS, Mechanical Engineering Department, University of Minho,
Campus de Azurém, 4800-058 Guimarães, Portugal
[4] CEFT, Faculty of Engineering, University of Porto, Rua Dr. Roberto Frias,
4200-465 Porto, Portugal
[5] Research Centre in Digitalization and Intelligent Robotics (CeDRI),
Instituto Politécnico de Bragança, Campus de Santa Apolónia, 5300-253
Bragança, Portugal

Abstract. Biomedical microfluidic devices are fabricated using different fabrication technologies. The most popular method is the soft-lithography manly due their main attraction, its high resolution capabilities and low material cost. However, usually, the fabrication of the moulds to produce microfluidic devices, is performed in a cleanroom environment and with specialized equipment that can be quite time consuming and costly. The micromilling is an alternative process that demonstrated potential to address some of the challenges in microdevices fabrication. It is a precise method, capable of creating complex channel geometries with specific measurements while ensuring a high level of resolution and a small error of tolerance, at the micron scale level. In fact, the non-necessity of a clean room, and being a fast and cheap manufacturing method makes it a great alternative to the traditional lithography process. Thus, in the present work, we show the ability of a micromilling machine to manufacture complex microchannels such as a curved T-shaped microchannel. By using a high-speed microscopic video system, flow visualizations and measurements were performed at four separation regions. Overall, the results show that the curved T-shaped microchannel is able to perform partial separation of blood cells from plasma.

Keywords: Micromilling process · Microfluidic devices · Cell-free layer ·
Cells separation · Low cost microfabrication

© Springer Nature Switzerland AG 2019
J. M. R. S. Tavares and R. M. Natal Jorge (Eds.): VipIMAGE 2019, LNCVB 34, pp. 585–593, 2019.
https://doi.org/10.1007/978-3-030-32040-9_59

1 Introduction

A critical step in the development of microfluidic systems is the selection of the proper microfabrication method. To select the most suitable fabrication process it is necessary taking into account not only the micro-structure parts to be fabricated but also the used material and their properties. Soft-photolithography techniques are the most popular way to produce polydimethylsiloxane (PDMS) microfluidic devices but this technique, despite in some cases that could be processed without a clean room facility, it involves multiple steps of fabrication, such as mask design, substrate pretreatment, photoresist development, UV exposure etc., [1–3]. Micromachining has several advantages among the manufacturing processes and the machine tools for micro-machining have achieved a high proficiency level in industry and research [4, 5], providing flexibility in design, saving experimental time and allowing the use of poly(methyl methacrylate) (PMMA) or acrylic, like materials as a substrate which is available at low cost. Also milling can be used to either directly machine microfluidic devices or to machine the moulds used to make these devices. One way to decide which of these options is the best for a particular situation is to consider the manufacturing volume [4]. Direct machining is probably more economical for small numbers of microfluidic devices, while milling a mould would make sense for higher-volume device production [4].

In this work, a direct machining is used to obtain a PMMA microfluidic device that combines to types of design, a curved inlet and a T-shaped bifurcation. The use of this two designs aimed increased the blood cells separation from plasma by increasing the cell-free layer (CFL) downstream the T-shaped bifurcation. In microcirculation at both *in vivo* and *in vitro* environments, the formation of a CFL is mainly caused by the tendency of red blood cells (RBCs) to migrate toward the center of the microchannel [6, 7]. The presence of a cell-depleted region adjacent to the wall has been used to extract plasma from blood samples by using microfluidic devices [6–11]. Recently, several researchers have demonstrated that is possible to perform cell separation by using nonlithographic techniques, such as xurography as Pinto *et al.* [12] described in their work and by micromilling [13, 14]. Lopes *et al.* [14] and Singhal *al.* [13] by using a micromilling fabrication technique, they have shown that the CFL could be enhanced by using a channel containing a constriction followed by a sudden expansion and thus able to separate plasma from a blood sample containing RBCs. However, they were only able to perform a partial separation and, as a result, alternative strategies and geometries need to be tested to have at the end a low cost microfluidic device able to perform a full separation of blood cells from plasma.

A previous work performed with a simple T-shaped geometry and with a single outlet has shown the existence of a considerable CFL due to the effect of the geometry [15]. In this work, by using a micromilling machine, a curved T-shaped microchannels were manufactured and investigated to explore the ability of this geometry to enhance the CFL around the walls and, consequently, to separate the plasma directly from an *in vitro* blood sample using multiple outlets through the microchannel.

2   Materials and Methods

2.1   PMMA Microchannel Design and Fabrication

The microfluidic devices were fabricated by using a micromilling machine, but first, their design was created using a CAD software. Figure 1 presents the geometry design and Fig. 2 shows the main dimensions of the studied regions of the tested microfluidic devices. The section A has exactly the same dimensions has the previous ones developed in the study [15]. Briefly, the inlet microchannel has 200 μm (W_5) and the outlet 1 microchannel located downstream the T-shaped contraction has 400 μm (W_3). The remaining microchannels outlets have widths of 120 μm (see Fig. 2). The microchannel depth is 50 μm.

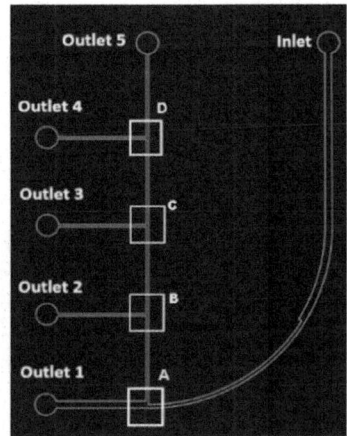

Fig. 1. Curved T-shaped microchannel design tested in this study.

The micromilling process was done by using the micromilling machine Minitech Mini-Mill/GX, as shown in Fig. 3. After drawing the channels, the software *Visual Mill* was used to create the numerical code, in order to be possible to define several process parameters such as the depth, velocity and angle of the tool.

For the curved T-shaped microchannel fabrication, two different milling tools were used, one with the diameter of 100 μm and another with 1 mm (see Fig. 3b). Additionally, a small microscope was used in order to observe the movements of the tools. More detailed information about the micromilling process can be found elsewhere [9, 11].

The inlets and outlets of the microfluidic devices were made by stainless steel tubes with an inner diameter of about 1 mm and with a length of about 1 cm. For assembling, epoxy glue was used to fix the tubes to the Plexiglass devices (acrylic glass, Plexiglas XT 0A000). Once it got done, the devices were clean by using an ultrasonic equipment, with water and detergent, for about 20 min. Finally, the microchannels were dried with pressurized air and sealed with an adhesive film for PCR plates Polyester 50 μm heat-resistant, ultra-clear, non-sterile.

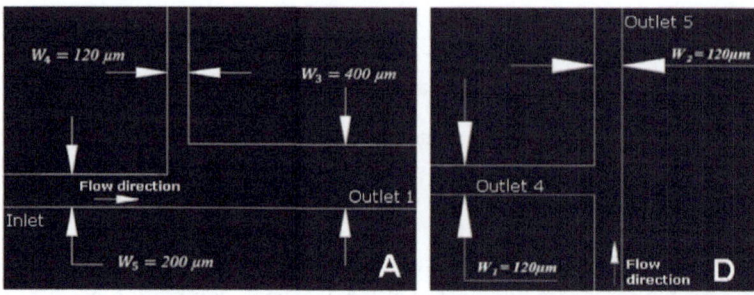

Fig. 2. Detail showing the dimensions of the regions A (left image) and D (right image).

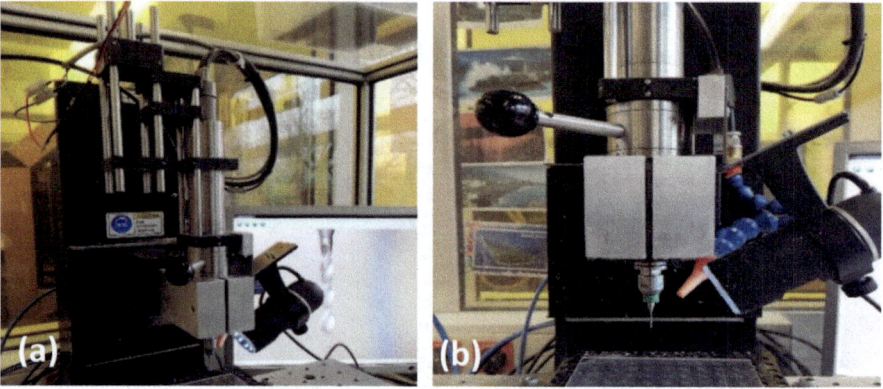

Fig. 3. (a) Micromilling machine Minitech Mini-Mill/GX, used to fabricate the microchannels; (b) close view of the milling tool.

2.2   Working Fluid and Experimental Set-Up

The working fluid used to test the microfluidic devices was dextran 40 (Dx40) with 5% hematocrit (Hct) of ovine RBCs. Briefly, blood was collected from a healthy ovine into a collection tube with heparin, in order to prevent coagulation. The RBCs were separated from the other blood constituents by centrifugation (2000 rpm for 15 min at 4 °C) and the plasma and buffy coat were removed by aspiration. The RBCs were then washed twice with a physiological saline solution and diluted with Dx40 to make up the required RBC concentration. All blood samples were stored hermetically at 4 °C until the experiments were performed at room temperature of approximately 20 °C.

The high-speed microscopic video system used in this study is demonstrated in Fig. 4. This system consisted of an inverted microscope (IX71, Olympus) equipped with a high-speed camera (FASTCAM SA3, Photron). The microfluidic device was placed on the stage of the inverted microscope and a syringe pump (Cetoni NEME-SYS) was used to inflow the working fluid at a constant flow rate within the microfluidic device.

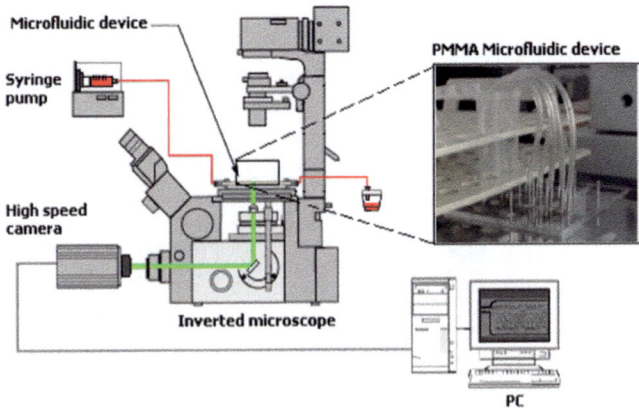

Fig. 4. Representation of the high-speed microscopic video system used to perform the experimental microfluidic studies.

2.3   Image Analysis

To better visualize the CFL and quantify the cells separation in the microchannel, the captured videos were converted to a sequence of static images (stack) and then, by using the plugin *Zproject* from the image processing program *ImageJ* [16], it was possible to obtain a single image having a sum of all static images. Then, by applying the *Plot profile* plugin to the result image, a two dimensional graph of image intensity along a defined region is obtained. Through this graph is possible to define the region of the CFL, and quantify their thickness.

3   Results and Discussion

The curved T-shaped microchannel was tested under the flow rate of 150 μL/min and with the working fluid with 5% of Hct suspended in Dx40. To evaluate the effect of the proposed microchannel design in the CFL, and consequently quantify the cells separation from plasma, a sequence of images was obtained at the bifurcation regions of interest, i.e., regions A, B, C and D (see Figs. 5, 6, 7 and 8). Thus, by using the *Plot profile* plugin, it was possible to measure the CFL thickness along the microchannel wall.

From the qualitative images captured in the region A and treated at the ImageJ, it is possible to observe that at upstream of the bifurcation (outlet 1) there is no CFL, while right after the bifurcation there is a clear CFL (see Fig. 5) formation. This is an indication that the T-shaped channel can increase the CFL downstream of the T-shaped bifurcation.

In the following sections an increase of the CFL thickness is observed, Figs. 6, 7 and 8, achieving at the region D (outlet 5), the higher CFL thickness and consequently the higher cells separation from plasma, although not full cell separation is obtained, see Fig. 8.

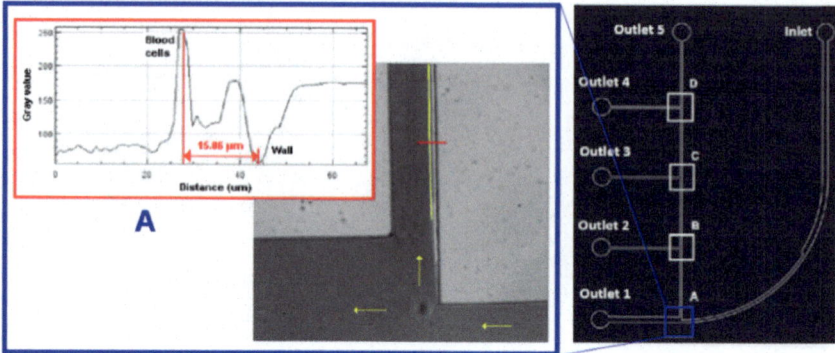

Fig. 5. Qualitative images obtained by a high-speed video microscopy system and treated by using the *Zproject* and *Plot profile* plugin, at the region A. The plot profile graph is obtained along the red line.

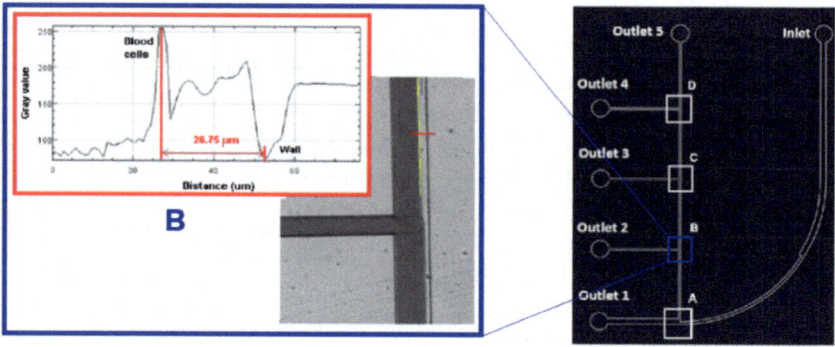

Fig. 6. Plot profile graph along the red line from the qualitative image obtained and treated by using the *Zproject* and *Plot profile* plugins at the region B.

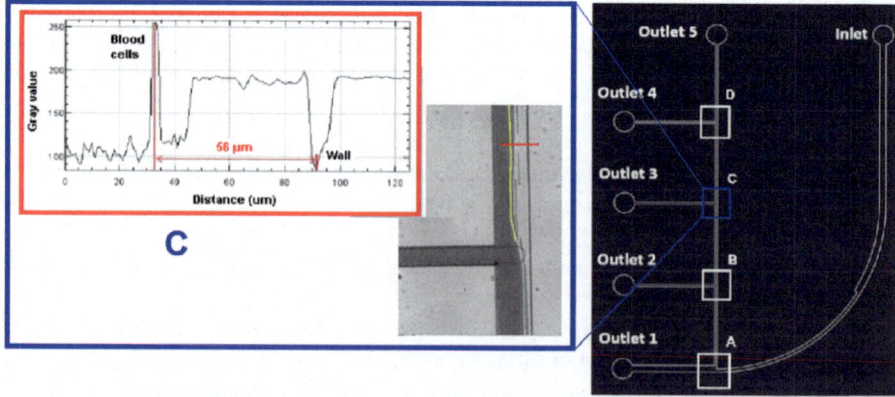

Fig. 7. Qualitative image obtained by using the *Zproject* plugin at the region C and respective plot profile graph along the red line.

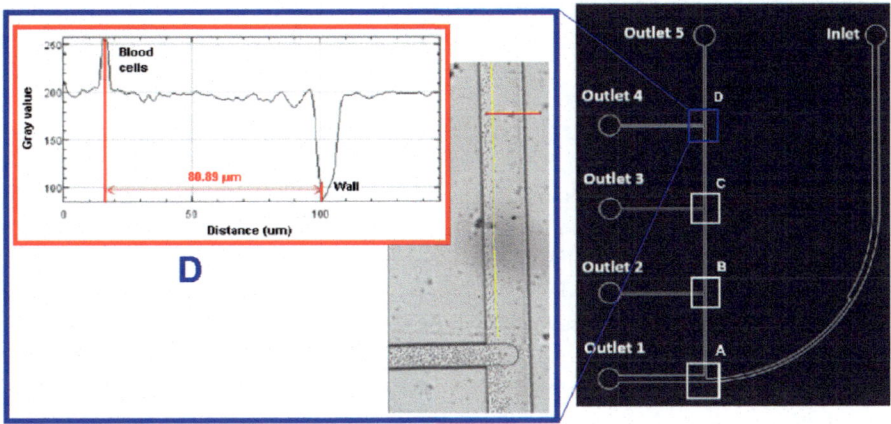

Fig. 8. Qualitative images obtained by the high-speed video microscopy system and by performing image analysis in the ImageJ, at the region D. Respective plot profile graph along the red line.

In addition, these results demonstrate that at the curved T-shape microchannel, the cells tend to move towards the right wall due to centrifugal forces and, as a result, the region of plasma is increased. These qualitative results show clearly that the curved channel enhances the CFL and, consequently, may result in an efficient separation method.

Figure 9 shows the average measurements of the CFL thickness at the different regions from A to D. These results demonstrate clearly that the T-shaped curved

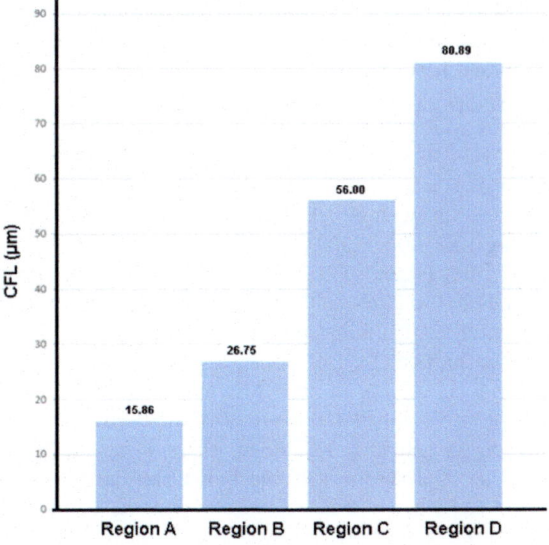

Fig. 9. Average measurements of the CFL thickness at the different regions from A to D.

geometry enhances the CFL thickness, increasing it along the regions. It is in the region D, last outlet, that we obtained the higher CFL thickness.

4 Conclusions and Future Directions

From this study we can conclude that by using a micromilling machine, it is possible to manufacture complex microchannels such as curved T-shape geometries with more than one outlet. Thus, micromilling is an alternative method with potential to address some of the challenges in microdevices fabrication. Additionally, by using a high-speed microscopic video system, we have shown the ability of the microfluidic device, having the five T-shape geometry outlets, to enhance the CFL downstream of the microchannel bifurcation. With the geometry presented in this work, we were able to produce good results regarding the enhancement of the CFL, allowing the RBC's separation from plasma in almost his full content. Ongoing experiments to achieve full RBC's separation from plasma are currently being performed and results will be published in due time.

Acknowledgments. This work was supported by *Fundação para a Ciência e a Tecnologia* (FCT) under the strategic grants UID/EEA/04436/2019, UID/EMS/04077/2019, and UID/EMS/00532/2019. The authors are also grateful for the funding of FCT through the projects POCI-01-0145-FEDER-016861, NORTE-01-0145-FEDER-029394, NORTE-01-0145-FEDER-030171, funded by COMPETE2020, NORTE2020, PORTUGAL2020, and FEDER, and the PhD grant SFRH/BD/91192/2012.

References

1. Faustino, V., Catarino, S.O., Lima, R., Minas, G.: Biomedical microfluidic devices by using low-cost fabrication techniques: a review. J. Biomech. **49**(11), 2280–2292 (2016). https://doi.org/10.1016/j.jbiomech.2015.11.031
2. Hossain, M.M., Rahman, T.: Low cost micro milling machine for prototyping plastic microfluidic devices. Proceedings **2**(13), 707 (2018). https://doi.org/10.3390/proceedings2130707
3. Pinto, V.P., Sousa, P.J., Cardoso, V.F., Minas, G.: Optimized SU-8 processing for low-cost microstructures fabrication without cleanroom facilities. Micromachines **5**(3), 738–755 (2014). https://doi.org/10.3390/mi5030738
4. Guckenberger, D.J., de Groot, T.E., Wan, A.M.D., Beebe, D.J., Young, E.W.K.: Micromilling: a method for ultra-rapid prototyping of plastic microfluidic devices. Lab Chip **15**(11), 2364–2378 (2015). https://doi.org/10.1039/c5lc00234f
5. Rincon Ardila, L., Wasnievski, L., Del Conte, E., Abackerli, A., Picarelli, T., Perroni, F., Schützer, K., Mewis, J., Uhlmann, E.: Micro-milling process for manufacturing of microfluidic moulds (2015). https://doi.org/10.20906/CPS/COB-2015-1250
6. Faivre, M., Abkarian, M., Bickraj, K., Stone, H.A.: Geometrical focusing of cells in a microfluidic device: an approach to separate blood plasma. Biorheology **43**(2), 147–159 (2006)
7. Lima, R.: Analysis of the blood flow behavior through microchannels by a confocal micro-PIV/PTV system. Ph.D. (Eng) (2007)

8. Pinho, D., Yaginuma, T., Lima, R.: A microfluidic device for partial cell separation and deformability assessment. BioChip J. **7**(4), 367–374 (2013). https://doi.org/10.1007/s13206-013-7408-0

9. Rodrigues, R.O., Lopes, R., Pinho, D., Pereira, A.I., Garcia, V., Gassmann, S., Sousa, P.C., Lima, R.: In vitro blood flow and cell-free layer in hyperbolic microchannels: visualizations and measurements. BioChip J. **10**(1), 9–15 (2016). https://doi.org/10.1007/s13206-016-0102-2

10. Saadatmand, M., Shimogonya, Y., Yamaguchi, T., Ishikawa, T.: Enhancing cell-free layer thickness by bypass channels in a wall. J. Biomech. **49**(11), 2299–2305 (2016). https://doi.org/10.1016/j.jbiomech.2015.11.032

11. Sollier, E., Cubizolles, M., Fouillet, Y., Achard, J.-L.: Fast and continuous plasma extraction from whole human blood based on expanding cell-free layer devices. Biomed. Microdevices **12**(3), 485–497 (2010). https://doi.org/10.1007/s10544-010-9405-6

12. Pinto, E., Faustino, V., Rodrigues, R.O., Pinho, D., Garcia, V., Miranda, J.M., Lima, R.: A rapid and low-cost nonlithographic method to fabricate biomedical microdevices for blood flow analysis. Micromachines **6**(1), 121–135 (2015)

13. Singhal, J., Pinho, D., Lopes, R., Sousa, P.C., Garcia, V., Schütte, H., Lima, R., Gassmann, S.: Blood flow visualization and measurements in microfluidic devices fabricated by a micromilling technique. Micro Nanosyst. **7**(3), 148–153 (2015). https://doi.org/10.2174/1876402908666160106000332

14. Lopes, R., Rodrigues, R.O., Pinho, D., Garcia, V., Schütte, H., Lima, R., Gassmann, S.: Low cost microfluidic device for partial cell separation: micromilling approach. In: 2015 IEEE International Conference on Industrial Technology (ICIT), 17–19 March 2015, pp. 3347–3350 (2015). https://doi.org/10.1109/ICIT.2015.7125594

15. Madureira, M., Faustino, V., Schütte, H., Gassmann, S., Lima, R.: Blood cells separation in a T-shaped microchannel fabricated by a micromilling technique **2** (2018). https://doi.org/10.24243/jmeb/2.5.171

16. Abràmoff, M.D., Magalhães, P.J., Ram, S.J.: Image processing with ImageJ. Biophotonics Int. **11**(7), 36–42 (2004)

Computational Vision and Image Processing Applied to Dentistry

Color Changes of a Polyolefin Thermoplastic Resin Used for Flexible Partial Dentures

Margarida Sampaio-Fernandes, João Galhardo[(✉)], Sandra Campos,
Susana Oliveira, José Carlos Reis-Campos,
and Maria Helena Figueiral

Faculty of Dental Medicine, University of Porto, Porto, Portugal
`joaogalhardo96@outlook.pt`

Abstract. This study aims to evaluate the color changes of samples of a flexible resin used as base of removable dentures in contact with coloring agents found in the diet. Twelve samples of iFlex™ by TCS® were placed in three solutions – coffee, curry and Coca-cola®. Color measurements were performed using the PCE-CSM 5 colorimeter programmed in the CIE system (L* a* b*), before the immersion, after 24 h and 10 days. Color difference (ΔE*) was evaluated and the NBS values determined. Statistical analysis was performed using the IBM® SPSS® Statistics program. After 1 day, there was statistically significant color differences in coffee and curry groups. On the 10th day, there was no clinical perception regarding the color change of the samples immerged in coffee and coca-cola but in the curry group the differences were marked. The flexible resin tested does not exhibit color stability when immersed in some coloring substances, which is a critical factor for the maintenance of the prostheses appearance and durability.

Keywords: Flexible prosthesis · Polypropylene · Polyolefins · Color stability · Thermoplastic materials · Removable prosthesis · Ethylene-polypropylene

1 Introduction

Removable dental prostheses are indicated for the restoration of masticatory, phonetic and esthetic functions, improving the quality of life of edentulous patients [1, 2]. As an alternative to Polymethylmethacrylate (PMMA), we can search for thermoplastic materials, such as flexible resins, which have competing properties: better esthetics (unlike conventional dentures using clasps and metal structures); absence of allergic reactions derived from metals or PMMA; reduced need for adjustments; improved comfort in comparison to conventional acrylic prosthesis [3].

Polyamides, polyolefins, polyesters and acrylic thermoplastic resins are used in the manufacturing of flexible prostheses [4, 5]. Polyolefins, such as polypropylene or polyethylene, are odor resistant materials characterized by great stability over time in terms of color, shape and mechanical strength [6].

Oral rehabilitation with flexible dentures is preferred in cases of previous prosthetic rehabilitations, patients in growth phase, in temporary prostheses after implant

© Springer Nature Switzerland AG 2019
J. M. R. S. Tavares and R. M. Natal Jorge (Eds.): VipIMAGE 2019, LNCVB 34, pp. 597–605, 2019.
https://doi.org/10.1007/978-3-030-32040-9_60

placement, in elderly or disabled patients, patients with high esthetic requirements, or individuals presenting PMMA allergy [7, 8].

The ideal properties of a denture base material are esthetic appearance, color stability, dimensional stability, low water absorption, ease of manufacturing/repair and biocompatibility [9]. Regarding the esthetics, color stability is a very important parameter, and there is often dissatisfaction of the patients and the need to perform a new rehabilitation, due to the pigmentation or discoloration of the prosthesis [10].

The oral environment causes degradation and aging of dental prostheses due to constant contact with saliva, food components and beverages. In recent years there is a growing number of studies attempting to explore the dental staining of conventional restorations and prostheses by different products and their consequences in relation to their preservation and longevity [4–6, 11–15]. However, there are few studies that refer to this influence on the surface of denture materials based on acrylic and flexible resins [4].

This study aims to evaluate the color changes of samples of a flexible resin used as denture base of removable prostheses (ethylene-polypropylene copolymer - iFlex™ by TCS®) in the presence of coloring agents found in the diet.

2 Materials and Methods

After a literature research about color change tests of dental materials, the following protocol was defined:

2.1 Preparation of Samples and Solutions

The flexible resin tested was iFlex™ by TCS® – light pink color, reference 5000-01/M2 25 mm. The samples were based on an ".stl" file, measuring 25 mm (length) 15 mm (width) × 2 mm (thickness). A 3D printer produced the models for the flasking procedure and the iFlex resin was injected according to the manufacturer's instructions. Twelve samples were produced. After removing excess material, the samples were polished with an Edenta 070 tungsten carbide cutter, ISO reference No. 658 104 273 534 070.

Prior to the preparation of the dye solutions, the samples were individually identified and immersed in distilled water for 48 h at 37 °C.

Three dye solutions were tested in comparison with a control solution of distilled water. Coffee (Nescafé® Classic) and curry (Margão® Caril) solutions were prepared by diluting 3 g of the corresponding powder per 100 mL of boiling water. Coca-cola (Coca-cola® Original Flavor) was not diluted. After their preparation, all solutions were placed in an incubator to reach a temperature of 37 °C.

Three samples were immersed in coffee (CF group), three in coca-Cola (CC group), and three in a curry solution (CU group). The remaining 3 samples were maintained in distilled water as the control group (group C).

In total, the samples were immersed in the respective solutions within a period of 10 days, only being withdrawn to replace the solutions (every 3 days) and to perform the color measurements [9, 13–15].

2.2   Color Stability Assessment

Prior to any measurement, and aiming to evaluate their homogeneity, the samples were visually inspected for irregularities or deformations - Subjective Visual Inspection.

Color evaluation was performed using the PCE-CSM 5 colorimeter (PCE instruments®) programmed in the CIE system (L* a* b*), as presented in Fig. 1. Prior to any measurement, the equipment was calibrated using a standard white part supplied by the manufacturer. The measurements were performed under a standardized D65 illumination (corresponding to the average daily light) and a blank base (corresponding to one sheet of paper) was always used.

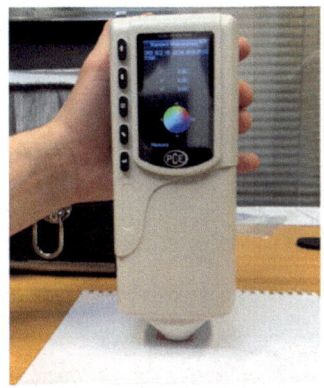

Fig. 1. Color evaluation using the PCE-CSM 5 colorimeter (PCE instruments®).

Color differences were evaluated objectively, subjectively and clinically.

Objectively, color differences were calculated according to the formula $\Delta E* = \left[(\Delta L*)^2 + (\Delta a*)^2 + (\Delta b*)^2\right]^{1/2}$. From the measurements on four different spots in each sample, an average value of L*, a*, and b* was calculated. $\Delta L*$, $\Delta a*$ and $\Delta b*$ were calculated from the initial measurements (after 48 h of immersion in distilled water – T0) and the measurements after 24 h (T1) and 10 days (T2), respectively.

Then, the National Bureau of Standard Units (NBS) values were determined by the equation NBS (U) = $\Delta E* \times 0.92$. These units represent clinically acceptable color change and the interpretation of these results is called Objective Visual Inspection. There are only significantly visual differences for NBS $\geq$ 3, as shown in Table 1 [9, 13–15].

2.3   Statistical Analysis

Statistical analysis was performed using the IBM® SPSS® Statistics program (Armonk NY: IBM Corp. 2017). The results were statistically analyzed using the multivariate repeated measures analysis of variance (ANOVA) model with post hoc analysis using Tukey's HSD test ($P = 0.05$).

Table 1. National Bureau of Standards (NBS) ratings [4].

NBS unit	Critical remarks of color differences	
0.0–0.5	Trace	Extremely slight change
0.5–1.5	Slight	Slight change
1.5–3.0	Noticeable	Perceivable
3.0–6.0	Appreciable	Marked change
6.0–12.0	Much	Extremely marked change
12.0 or more	Very much	Change to other color

3 Results

The results of the visual inspection are presented in Table 2. The existence of porosities, as well as the absence of color homogeneity, was observed in all samples. Only the CU group presented perceptible color differences; in the other groups color differences were not obvious.

Table 2. Photograph of one specimen of each group: (a) initial; (b) after 10 days.

The color of the samples was evaluated with the colorimeter at 3 time points. The mean was calculated per sample and then per group. In the scale L* a* b* (first value of each group refers to L*, second to a* and third to b*), the following mean values were calculated:

- CC group: 44.34 - 13.32 - 7.76 at T0; 43.52 - 13.18 - 8.23 at T1 and 43.70 - 13.50 - 8.22 for T2;
- CF group: 47.30 - 13.00 - 7.46 at T0; 46.78 - 11.71 - 7.51 at T1 and 47.50 - 11.78 - 7.61 at T2;
- CU group: 45.48 - 12.50 - 7.34 at T0; 44.59 - 12.34 - 10.71 at T1 and 44.36 - 12.52 - 14.13 at T2;
- Group C: 45.56 - 12.61 - 7.47 at T0; 45.19 - 12.14 - 7.53 at T1 and 45.68 - 11.35 - 7.73 at T2.

3.1   Color Difference (ΔE^*) and NBS Units

Table 3 shows the mean ΔE^* and NBS values calculated for each group for T0–T1, T1–T2 and T0–T2 time intervals. The most significant color difference was observed in the samples placed in the curry solution, presenting the highest ΔE^* value at all periods assessed. Moreover, in the curry solution the color difference was more pronounced after 1 day of immersion (3.60) compared to the remaining 9 days (3.49), reaching a ΔE^* of 6.90 at the end of 10 days. When converted to NBS Units, these values represent marked changes to the human eye (clinically sensitive, NBS higher than 3).

Table 3. Means of color difference (ΔE^*) and NBS calculated for each group.

Groups	T0–T1		T1–T2		T0–T2	
	ΔE^* (mean ± SD)	NBS (mean)	ΔE^* (mean ± SD)	NBS (mean)	ΔE^* (mean ± SD)	NBS (mean)
CC	1.06 ± 0.62	0.98	0.84 ± 0.53	0.77	0.89 ± 0.65	0.82
CF	2.06 ± 0.39	1.90	1.42 ± 0.46	1.31	1.86 ± 1.69	1.71
CU	3.60 ± 0.56	3.32	3.49 ± 1.68	3.21	6.90 ± 2.17	6.35
C	0.66 ± 0.11	0.61	1.69 ± 0.54	1.56	1.79 ± 0.54	1.65

SD – Standard deviation; CC – coca-cola group; CF – coffee group; CU – curry group; C – control group.

After 24 h of immersion in the different solutions, there were statistically significant color differences ($p < 0.05$) between the curry group and the control, coca-cola and coffee groups, and between the coffee and control group. So, on average, in the first period, the color change was slight for the control and coca-cola groups, noticeable for the coffee group, and appreciable for the curry group.

For the other periods, there are statistically significant color differences ($p < 0.05$) between CU group and CC group. Between the 1st and the 10th day, color changes were generally considered slight for the control group, coca cola, and coffee, but appreciable for the curry group. Finally, the color differences between initial (T0) and the final (T2) are aggravated, since they are considered slight for the coca-cola group, noticeable for the coffee group, and appreciable/marked for the curry group. Then, there was no clinical perception (NBS < 3) regarding the color change of the samples immerged in coffee and coca-cola on the 10th day.

4 Discussion

Comparing to conventional removable dentures, flexible removable prostheses have more favorable esthetics because they have no metal components, the translucency allows to visualize the underlying soft tissues and are almost imperceptible in the oral cavity. The improved flexibility properties enable a better adaptation to the tissues and more comfort for the patients, besides having a high biocompatibility due to the absence of residual monomer [6, 11, 12].

According to the literature, thermoplastic materials have a great color stability [4, 6] and this is a critical factor for prostheses to maintain their appearance and durability. However, this study suggests that the flexible resin tested (iFlex^{TM} by TCS$^{®}$) does not exhibit color stability upon contact with some coloring agents.

In this study, the color measurement were obtained using the CIE system L* a* b*. Besides being the most used system, it permits to evaluate the degree of color difference based on visual human perception of color difference. All samples were subjected to the same measurement conditions, following the standard D65, corresponding to natural light. This parameter was guaranteed by the colorimeter that emits a light beam (flash type) in all the measurements, canceling any light emitted around the samples that could bias the results. In addition, the measuring place marked on the sheet of paper (background) was always the same [12, 18, 19].

Ethylene-polypropylene is a flexible and transparent material, characterized by its durability and resistance to temperature and pressure [14]. The ethylene present in the formation of this material breaks the regular structure of the polypropylene, which results in a reduction of the crystalline uniformity of the polymer, causing areas of larger and smaller stains and rendering the samples non-uniform in terms of color [14]. To circumvent this heterogeneity, several measurements were taken in each sample, including lighter and darker areas to better characterized each sample. It should be underlined that measurements are always made in an area (in this case a sensor with 4 mm of diameter), that records the average color of the zone. Therefore, when different areas are measured in the same sample, the color values will always be different, which explains the discrepancy of some average measurement values. This constitutes a limitation in this type of *in vitro* studies, and may have biased the results. One way to overcome this problem would be to use sensors with thinner diameters, so that the values of the area to be measured were less variable, thus producing more reliable data. Another alternative would be to improve the homogeneity of the polymers by using a more thorough polishing procedure. Studies have shown that more polished surfaces are subject to less impregnation of water or other substances and are more difficult to color [12]. One could also choose to use polymers with a higher polypropylene purity in comparison to the flexible materials used today, so that the homogeneity of the material is maximized and the polymer chains are more compact and uniform.

The ethylene-polypropylene copolymer has very specific discoloration characteristics that could explain our results. Phenolic antioxidants (responsible for its great stability, physical and thermal resistance) react with free radicals, residual catalyst, humidity and heat, forming colored quinoid compounds (usually yellow) [15, 16]. Being a porous polymer, however, water absorption could not be neglected. In fact, solvents such as water, can diffuse into the polymeric matrix, separate the monomer chains, thus changing its color, usually making it whiter (lower L* value) [10]. This phenomenon might explain the color difference recorded in the control group.

Moreover, each solution tested has particular characteristics that lead to resin staining. Coca-Cola®, being an acidic agent, causes degradation of the material, because H^+ ions act on its surface (presence of phosphoric acid). In addition, the color of this beverage is due to the caramel present in its constitution, whose color ranges from light yellow to dark brown [15]. Despite the fact that this beverage has dyes in its constitution and is the most acidic solution of the three, there were no significant differences in the samples immersed in coca-cola in the period of time studied (10 days), with the changes observed being considered "slight" according to the NBS scale.

In the case of coffee, its coloring ability is associated with the presence of yellow dyes with different polarities. The tannic acid present in coffee plays an important role in the staining of many polymers. In parallel, it has been argued in several studies that coffee staining exists due to either adsorption or dye absorption, the compatibility of the polymers with the coffee coloring substances, or the absorption of water present in the solution of coffee, respectively [5, 16]. In our study, significant color differences were observed regarding the color change of the iFlex polymer (only between T0 and T1), probably due to the affinity of the coffee dyes by the polymer phase of the polypropylene, adhering and staining more during the first day of immersion.

Regarding the staining in the curry solution, some molecules naturally react with the polymer and create stain compounds. These compounds, namely curcumines (orange/yellow substances), have high affinity to the polymeric phase of the studied polymer, [17] which may explain the accentuated staining observed at the end of the first day (T1) of immersion, stabilizing thereafter until day-10 (T2).

In various studies of color stability in dental materials, it has been established that the value for which the color becomes clinically perceptible to the human eye and therefore unacceptable from the esthetic point of view is 3 (or more) NBS units [9, 13–15]. In this investigation, the color difference observed in the samples immersed in curry, after 10 days, was considered "marked".

Some studies referred color changes in PMMA and acrylic resins caused by coffee and red wine. Sepúlveda-Navarro and colleagues [12] verified that the PMMA used in the manufacturing of removable dentures is significantly stained when subjected to coffee and red wine solutions, in which the red wine presented higher ΔE* value at the end of the study. Another work that evaluated color stability of three resins, two acrylic resins and one polyamide, when subjected to coffee and green tea, also studied how water absorption modifies the samples color [19]. Color stability was proven (color differences not significant) despite slight changes (more significant in coffee), and the water absorption did not show significant influence in color differences [19]. As there is no study available in the literature on the specific material tested in this investigation, it is impossible to compare our results with other research made with ethylene-polypropylene copolymer.

5   Conclusion

The ethylene-polypropylene copolymer tested (iFlex™ by TCS®) showed no color stability when subjected to certain coloring substances, namely curry. Thus, additional studies are necessary to clarify the potential discoloration and behavior of this type of material when submitted to coloring agents present in the diet.

Acknowledgments. The authors would like to thank Prof. Jorge Lino (Faculty of Engineering of University of Porto), for providing the colorimeter used; to TCS Dental, for providing the polymeric material; and the dental laboratory Dentalmar (Póvoa de Varzim, Portugal) for manufacturing the 12 samples.

References

1. De Carvalho Dias, K., Da Fonte Porto Carreiro, A., Bastos Machado Resende, C.M., Soares Paiva Tôrres, A.C., Mestriner Júnior, W.: Does a mandibular RDP and new maxillary CD improve masticatory efficiency and quality of life in patients with a mandibular Kennedy class I arch? Clin. Oral Invest. **20**(5), 951–957 (2016)
2. Singh, K., Aeran, H., Kumar, N., Gupta, N.: Flexible thermoplastic denture base materials for aesthetical removable partial denture framework. J. Clin. Diagn. Res. **7**(10), 2372–2373 (2013)
3. Chuchulska, B., Yankov, S., Hristov, I., Aleksandrov, S.: Thermoplastic materials in the dental practice: a review. Int. J. Sci. Res. **6**(12), 1074–1076 (2018)
4. Nagakura, M., Tanimoto, Y., Nishiyama, N.: Color stability of glass-fiber-reinforced polypropylene for non-metal clasp dentures. J. Prosthodont. Res. **62**(1), 31–34 (2018)
5. Silva, A.L.N., Rocha, M.C.G., Guimarães, M.J.O.C., Lovisi, H., Coutinho, F.M.B., Santa Maria, L.C.: Desenvolvimento de Materiais à Base de Poliolefinas e Elastômeros Metalocênicos. Polímeros **11**(3), 135–141 (2001)
6. Patrocínio, B.M.G., Antenor, A.M., Haddad, M.F.: Prótese Parcial Removível Flexível – revisão de literatura. Archiv. Health Invest. **6**(6), 258–263 (2017)
7. Fueki, K., Ohkubo, C., Yatabe, M., Arakawa, I., Arita, M., Ino, S., et al.: Clinical application of removable partial dentures using thermoplastic resin—Part I: definition and indication of non-metal clasp dentures. J. Prosthodont. Res. **58**(1), 3–10 (2014)
8. Takabayashi, Y.: Characteristics of denture thermoplastic resins for non-metal clasp dentures. Dent. Mater. J. **29**(4), 353–361 (2010)
9. Kim, J.H., Choe, H.C., Son, M.K.: Evaluation of adhesion of reline resins to the thermoplastic denture base resin for non-metal clasp denture. Dent. Mater. J. **33**(1), 32–38 (2014)
10. Hersek, N., Canay, S., Uzun, G., Yildiz, F.: Color stability of denture base acrylic resins in three food colorants. J. Prosthet. Dentist. **81**(4), 375–379 (1999)
11. Guler, A.U., Yilmaz, F., Kulunk, T., Guler, E., Kurt, S.: Effects of different drinks on stainability of resin composite provisional restorative materials. J. Prosthet. Dentist. **94**(2), 118–124 (2005)
12. Sepúlveda-Navarro, W.F., Arana-Correa, B.E., Ferreira Borges, C.P., Habib Jorge, J., Urban, V.M., Campanha, N.H.: Color stability of resins and nylon as denture base material in beverages. J. Prosthodont. **20**(8), 632–638 (2011)

13. Akinyamoju, C.A., Ogunrinde, T.J., Taiwo, J.O., Dosumu, O.O.: Comparison of patient satisfaction with acrylic and flexible partial dentures. Niger. Postgrad. Med. J. **24**(3), 143–149 (2017)
14. Maddah, H.A.: Polypropylene as a promising plastic: a review. Am. J. Polymer Sci. **6**(1), 1–11 (2016)
15. Kikuchi, T., Ohtake, Y., Tanaka, K.: Discoloration phenomenon induced by the combination of phenolic antioxidants & hindered amine light stabilisers. Nippon Gomu Kyokaishi **9**, 283–288 (2012)
16. Hendrickson, L., Connole, K.B.: Review of stabilization of polyolefin insulated conductors. Part I: theory and factors affecting stability. Polymer Eng. Sci. **35**, 211–217 (1995)
17. Gupta, G., Gupta, T.: Evaluation of the effect of various beverages and food material on the color stability of provisional materials - an in vitro study. J. Conserv. Dentist. **14**(3), 287–292 (2011)
18. Sagsoz, N.P., Yanıkoglu, N., Ulu, H., Bayındır, F.: Color changes of polyamid and polymetyhl methacrylate denture base materials. Open J. Stomatol. **4**(10), 489–496 (2014)
19. Jang, D.E., Lee, J.Y., Jang, H.S., Lee, J.J., Son, M.K.: Color stability, water sorption and cytotoxicity of thermoplastic acrylic resin for non metal clasp denture. J. Adv. Prosthodont. **7**(4), 278–287 (2015)

Thermoplastic Materials for Infrastructures in Prosthodontic Rehabilitation: A Review

Margarida Sampaio-Fernandes, Sandra Campos, João Galhardo[(✉)],
Susana Oliveira, José Carlos Reis-Campos,
and Maria Helena Figueiral

Faculty of Dental Medicine, University of Porto, Porto, Portugal
joaogalhardo96@outlook.pt

Abstract. Conventionally, chromium-cobalt (Cr-Co) and titanium (Ti) alloys represent the most common metallic materials used in oral rehabilitation. However, due to esthetic concerns and adverse effects in oral tissues (eg. galvanism and allergic reactions), the development of metal-free infrastructures has been an expanding research area in Prosthodontics. In this context, a range of thermoplastic polymers such as polymethylmethacrylate (PMMA), polyamide, polycarbonate, acetal resin or polyoxymethylene (POM), polyether ether ketone (PEEK) and polyether ketone ketone (PEKK), has attracted attention for its versatility and mechanical properties. This article aims to present a literature review of the thermoplastic polymers most frequently used for prosthetic infrastructures, outlining their properties, advantages and disadvantages.

Article search was conducted in PubMed, Scopus and Google Scholar databases, including articles published from 2008 to 2018, written in English and Portuguese, selected according to their abstracts and relevance to the theme.

The properties and applications of each polymeric material were explored individually, establishing comparisons whenever appropriate.

Overall, each material has diverse clinical applications in the oral rehabilitation field. Thermoplastic material selection and design should be performed upon a complete study of the properties and limitations of each polymer, using a case-by-case approach. Although more studies are necessary to fully characterize their spectrum of applications, PEEK and PEKK appear as promising candidates for the production of prosthetic infrastructures.

Keywords: Framework · Infrastructure · Prosthodontics · CAD-CAM · PMMA · Polyamide · Polycarbonate · Acetal resin · PEEK · PEKK

1 Introduction

Prosthodontics, a field of Dental Medicine in continuous development, aims to replace missing teeth and oral supporting tissues, thus restoring health, function, aesthetics and self-esteem of patients. Multiple treatment options are available, ranging from complete or partial dentures, to tooth- or implant-supported removable or fixed prostheses. As a common feature, all these prosthetic devices include in their design a framework or infrastructure [1].

© Springer Nature Switzerland AG 2019
J. M. R. S. Tavares and R. M. Natal Jorge (Eds.): VipIMAGE 2019, LNCVB 34, pp. 606–614, 2019.
https://doi.org/10.1007/978-3-030-32040-9_61

Traditionally, metal-based infrastructures have been the gold-standard for both removable and fixed strategies. Despite their benefits, disadvantages associated with galvanism, hypersensitivity or allergic reactions and poor aesthetics, have expanded the development of alternative non-metallic materials. Along with ceramics, the thermoplastic polymers have attracted particular attention due to their mechanical properties, improved aesthetics, elastic behavior, cost-effectiveness, manufacturing flexibility and improvement potential [2, 3]. The advent of new polymers is also benefiting from the digital workflow in the design and fabrication of dental prostheses (intraoral scanning, CAD-CAM technology, high precision 3D printers) [4].

This review has focused upon the thermoplastic polymers used in the construction of infrastructures in Prosthodontics – polymethylmethacrylate (PMMA), polyamide, polycarbonate, acetal resin or polyoxymethylene (POM), polyether ether ketone (PEEK) and polyether ketone ketone (PEKK) – outlining their mechanical properties, advantages, drawbacks and clinical applications.

2 Materials and Methods

Article search was conducted in PubMed, Scopus and Google Scholar databases, using the keywords: "acetal resin AND removable prosthesis AND partial dentures", "peek AND cad cam AND removable prosthesis AND framework AND partial dentures", "pekk AND prosthodontics", "pmma resin AND cad cam AND prostheses AND removable AND resins digital dentures", "polyamide AND removable prosthesis", "polycarbonate AND prosthodontics".

Inclusion criteria were as follows: articles published from 2008 to 2018, written in English and Portuguese, with full-text available and containing the search terms in the title and/or abstract. The list of exclusion criteria comprised: articles beyond the scope of the current review, redundant content, books or book chapters and conferences. A total of 746 publications were identified, with only 64 included in the current work. The original keyword-based search was further complemented by references cited in the former articles.

3 Results

Based on the 64 publications selected for this review, the properties and applications of each polymeric material were explored individually, establishing comparisons whenever appropriate. For the purposes of the current review, thermoplastic materials were clustered into two groups: classic, including the commodity thermoplastics more widely used is dental medicine, and high performance polymers, encompassing the uppermost class of plastics. The former includes PMMA, polyamide, polycarbonate and POM, with the latter comprising PEEK and PEKK.

A list of some mechanical properties of each material is displayed in Table 1, while Table 2 includes a summarized description of their main applications, advantages and disadvantages. To illustrate the use of thermoplastic polymers in prosthodontic infrastructures, Fig. 1 depicts a bar and clip attachment system for overdentures made of POM resin.

Table 1. Some physical and mechanical properties of the classic and high-performance thermoplastic materials addressed in this review [5–17].

Materials	Properties	Flexural strength (MPa)	Flexural modulus (MPa)	Elastic modulus (GPa)	Roughness (μm)
Classic thermoplastic polymers	Conventional PMMA	62,38	1,55	3,9	0,12-0,92
	CAD-CAM PMMA	34,05	2,85	4,1	0,37-2,38
	Polyamide	13,7-83,6	612-1381	0,8-1,639	0,21-0,24
	Polycarbonate	29,6-100	2245	2,19-4,659	
	POM		2,30	2,9	
High-performance thermoplastic polymers	PEEK		4,0	3-4	0,018
	PEEK (*Bio-HPP*)			18	
	PEKK	200	4,5	5,1	
Human tissues	Enamel			40-83	
	Dentine	212,9		15	
	Cortical bone			14	

Table 2. Summary of the most relevant applications, advantages and disadvantages of the classic and high-performance polymers [3, 5–7, 9, 11–13, 16–24]. RP: Removable Prosthodontics; FP: Fixed Prosthodontics; RPD: Removable partial denture.

Material		Applications	Advantages	Disadvantages
Classic thermoplastic polymers	PMMA	**RP:** • Tooth-, muco- and implant-supported RPDs; • Maxilofacial prosthesis; • Denture relining. **FP:** • Coating of bridges and crowns.	• Low cost • Easily repaired • Simple processing techniques • Long-term color stability • Compatible with CAD-CAM production	• Residual MMA release (mechanical and biocompatibility problems) • Low dimensional stability • Surface porosity • Biofilm adhesion; *C. albicans* colonization • Low impact resistance • Difficult path of insertion in severe soft and hard undercuts
	Polyamide	**RP:** • Flexible RPDs; • Clasps of RPDs; • Temporary dental prosthesis. **Indications:** anterior teeth rehabilitation and mouth opening limitation. **Contraindications:** distal extension rehabilitation, high number of missing teeth, major connector and occlusal rests.	• Pleasant esthetics • High flexibility • Compatible with severe hard and soft undercuts • Good dimensional stability • Cost-effective • High resistance to fatigue, impact and abrasion • Requests less tooth preparation • Lighter-weight and thinner prostheses (increased comfort)	• Denture base deformation during function • Low vertical stability of dentures • Possible residual ridge resorption • Discoloration over-time • Polishing dificulties • Higher rugosity • *C. albicans* colonization • Difficult to repair and reline

Classic thermoplastic polymers	Polycarbonate	**RP:** • Flexible RPDs; • Clasps of RPDs; **FP:** • Preformed temporary dental crowns. **Indications:** anterior teeth rehabilitation, retentive areas, low occlusal forces. **Contraindications:** rehabilitation of supporting areas with high occlusal forces; high number of missing teeth.	• Easy polishing • Wear resistance • Low water absorption (dimensional stability) • Less biofilm adhesion • Pleasant esthetics • Less susceptible to discoloration than other flexible resins	• High stress transmission to abutment teeth • Possible bisphenol A release • High fracture risk • Difficult to repair and reline
	POM	**RP:** • Tooth-, muco- and implant-supported RPDs and complete dentures; • Clasps of RPDs. **FP:** • Implant abutments; • Infrastructures for implant-supported overdentures. **Indications:** esthetic or periodontally compromised areas, anterior clasps. **Contraindications:** major connector and long-term use (> 6 months)	• Resistance to organic solvents • Fracture resistance • Mechanical stability • Low thermal conductivity • Biocompatibility • Lightweight • High elasticity and flexibility • Acceptable esthetics • Compatible with self-cured repair resins	• Greater cross-sectional diameter of clasps • Higher biofilm accumulation • Unfavorable properties of supporting elements (alveolar ridge and abutment teeth stress) • Processing time and cost • Discoloration overtime

High-performance thermoplastic polymers	PEEK	**RP:** • Frameworks for RD; • Clasps of RPD; • Temporary dental prosthesis. **FP:** • Infrastructures for FP; • Crowns and endo-crowns; • 3-element bridges; • Implant infrastructures. **Indications:** teeth with reduced periodontal support and distal extension rehabilitations.	• Lightweight and low density • Elastic properties identical to human bone • Low stress transmission to abutment teeth • Compatible with heat and radiation sterilization • High resistance to fracture, wear, abrasion and corrosion • Radiolucent • Biocompatibility • Compatible with CAD-CAM	• Poor esthetics (veneering materials are necessary) • Limited repair capacity with self-cured resins • Low bioactivity
	PEKK	**RP:** • Frameworks for RPD; • Clasps of RPD; • Temporary dental prosthesis. **FP:** • Crowns; • Intraradicular post and core systems; • Tooth-supported bridges with one pontic; • Tooth- and implant-supported FP; • Denture attachments; • Implant abutments. **Contraindications:** bruxism patients, distal extension rehabilitations.	• High resistance to compression and fatigue • Lightweight • Compatible with diverse coating materials • High stress distribution/absorption capacity • Radiolucent • Lower risk of fracture of endodontically treated teeth.	• Poor esthetics (veneering materials are necessary) • Possible increase of stress transmission to implants and peri-implant bone under tensile stress conditions • Difficult to repair • Limited long-term stability

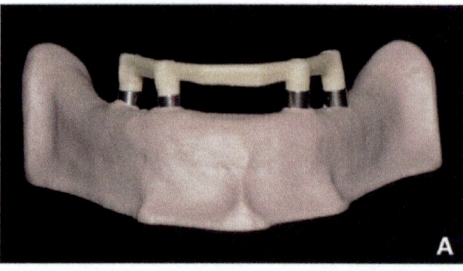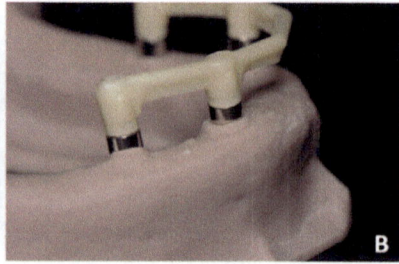

Fig. 1. Bar and clip attachment system for implant-supported overdentures made of POM resin. A, Frontal view. B, Lateral view (detail).

4 Discussion and Conclusions

The continuous development of polymeric materials for dental medicine applications, allied to the remarkable evolution of CAD-CAM systems, have raised the possibility of new treatment options and clinical workflows. The importance of thermoplastic polymers in the manufacturing of prosthodontic infrastructures has long been recognized, with the search for reliable metal-free restorations being fostered by the increasing esthetic demands of patients.

With the CAD-CAM technology, traditional materials such as PMMA can now be used with improved biomechanical properties and decreased residual monomer content [5, 6]. The introduction of thermoplastic polymers with lower elastic modulus in clinical practice (eg.: polyamide, polycarbonate and POM) enabled the rehabilitation of areas with severe undercuts, also obviating the need of metal clasps display. Despite these advantages, the failure of relining/repairing procedures, coupled with possible occlusal instability and residual ridge resorption, still constitute important limitations in the prosthodontic treatment of total edentulism and Kennedy class I and II cases [7, 9, 19–23].

More recently, the high-performance polymers PEEK and PEKK, emerged as attractive materials for prosthetic frameworks, due to their excellent biocompatibility, biomechanical properties, light weight, as well as manufacturing and application versatility [11, 12, 17, 18, 24]. Although promising, more studies are necessary to validate their clinical use in Prosthodontics.

Considering the diversity of thermoplastic polymers and their multiple clinical applications, selecting the proper material for prosthetic treatment depends on a thorough understanding of their properties, advantages and limitations. Such decision should be evidence-based, evaluating each case individually.

References

1. Campbell, S.D., Cooper, L., Craddock, H., Hyde, T.P., Nattress, B., Pavitt, S.H., et al.: Removable partial dentures: the clinical need for innovation. J. Prosthet. Dent. **118**(3), 273–280 (2017)
2. Rosca, B., Ramalho, S., Sampaio-Fernandes, J.C., Portugal, J.: Reparability of two different CAD/CAM polymer materials using a light-cured composite and universal adhesives. Revista Portuguesa de Estomatologia, Medicina Dentária e Cirurgia Maxilofacial **57**(4), 189–196 (2016)
3. Osada, H., Shimpo, H., Hayakawa, T., Ohkubo, C.: Influence of thickness and undercut of thermoplastic resin clasps on retentive force. Dent. Mater. J. **32**(3), 381–389 (2013)
4. Bilgin, M.S., Baytaroglu, E.N., Erdem, A., Dilber, E.: A review of computer-aided design/computer aided manufacture techniques for removable denture fabrication. Eur. J. Dent. **10**(2), 286–291 (2016)
5. Ayman, A.D.: The residual monomer content and mechanical properties of CAD\CAM resins used in the fabrication of complete dentures as compared to heat cured resins. Electron. Phys. **9**(7), 4766–4772 (2017)
6. Srinivasan, M., Gjengedal, H., Cattani-Lorente, M., Moussa, M., Durual, S., Schimmel, M., et al.: CAD/CAM milled complete removable dental prostheses: an in vitro evaluation of biocompatibility, mechanical properties, and surface roughness. Dent. Mater. J. **37**(4), 526–533 (2018)
7. Fueki, K., Ohkubo, C., Yatabe, M., Arakawa, I., Arita, M., Ino, S., et al.: Clinical application of removable partial dentures using thermoplastic resin. Part II: material properties and clinical features of non-metal clasp dentures. J. Prosthodont. Res. **58**(2), 71–84 (2014)
8. Lee, H.H., Lee, J.H., Yang, T.H., Kim, Y.J., Kim, S.C., Kim, G.R., et al.: Evaluation of the flexural mechanical properties of various thermoplastic denture base polymers. Dent. Mater. J. **37**(6), 950–956 (2018)
9. Fitton, J.S., Davies, E.H., Howlett, J.A., Pearson, G.J.: The physical properties of a polyacetal denture resin. Clin. Mater. **17**(3), 125–129 (1994)
10. Arda, T., Arikan, A.: An in vitro comparison of retentive force and deformation of acetal resin and cobalt-chromium clasps. J. Prosthet. Dent. **94**(3), 267–274 (2005)
11. Al-Rabab'ah, M., Hamadneh, W., Alsalem, I., Khraisat, A., AbuKaraky, A.: Use of high performance polymers as dental implant abutments and frameworks: a case series report. J. Prosthodont. **28**(4), 365–372 (2017)
12. Zoidis, P., Papathanasiou, I., Polyzois, G.: The use of a modified poly-ether-ether-ketone (PEEK) as an alternative framework material for removable dental prostheses: a clinical report. J. Prosthodont. **25**(7), 580–584 (2016)
13. Harb, I.E., Abdel-Khalek, E.A., Hegazy, S.A.: CAD/CAM constructed poly(etheretherketone) (PEEK) framework of Kennedy class I removable partial denture: a clinical report. J. Prosthodont. **28**(2), e595–e598 (2019)
14. Wiesli, M.G., Ozcan, M.: High-performance polymers and their potential application as medical and oral implant materials: a review. Implant Dent. **24**(4), 448–457 (2015)
15. Lee, K.S., Shin, J.H., Kim, J.E., Kim, J.H., Lee, W.C., Shin, S.W., et al.: Biomechanical evaluation of a tooth restored with high performance polymer PEKK post-core system: a 3D finite element analysis. Biomed. Res. Int. **2017**, 1373127 (2017)
16. Han, K.H., Lee, J.Y., Shin, S.W.: Implant- and tooth-supported fixed prostheses using a high-performance polymer (Pekkton) framework. Int. J. Prosthodont. **29**(5), 451–454 (2016)
17. Najeeb, S., Zafar, M.S., Khurshid, Z., Siddiqui, F.: Applications of polyetheretherketone (PEEK) in oral implantology and prosthodontics. J. Prosthodont. Res. **60**(1), 12–19 (2016)

18. Abuzar, M.A., Bellur, S., Duong, N., Kim, B.B., Lu, P., Palfreyman, N., et al.: Evaluating surface roughness of a polyamide denture base material in comparison with poly (methyl methacrylate). J. Oral Sci. **52**(4), 577–581 (2010)
19. Vojdani, M., Giti, R.: Polyamide as a denture base material: a literature review. J. Dent. (Shiraz) **16**(1 Suppl.), 1–9 (2015)
20. Takabayashi, Y.: Characteristics of denture thermoplastic resins for non-metal clasp dentures. Dent. Mater. J. **29**(4), 353–361 (2010)
21. Nasution, H., Kamonkhantikul, K., Arksornnukit, M., Takahashi, H.: Pressure transmission area and maximum pressure transmission of different thermoplastic resin denture base materials under impact load. J. Prosthodont. Res. **62**(1), 44–49 (2018)
22. Meenakshi, A., Gupta, R., Bharti, V., Sriramaprabu, G., Prabhakar, R.: An evaluation of retentive ability and deformation of acetal resin and cobalt-chromium clasps. J. Clin. Diagn. Res. **10**(1), ZC37–ZC41 (2016)
23. Jiao, T., Chang, T., Caputo, A.A.: Load transfer characteristics of unilateral distal extension removable partial dentures with polyacetal resin supporting components. Aust. Dent. J. **54**(1), 31–37 (2009)
24. Dawson, J.H., Hyde, B., Hurst, M., Harris, B.T., Lin, W.S.: Polyetherketoneketone (PEKK), a framework material for complete fixed and removable dental prostheses: a clinical report. J. Prosthet. Dent. **119**(6), 867–872 (2018)

Photogrammetry Technique for the 3D Digital Impression of Multiple Dental Implants

Luís Azevedo[1]([⊠]), Pedro Molinero-Mourelle[1],
José L. Antonaya-Martín[1], Jaime del Río-Highsmith[1],
André Correia[2], and Miguel Gómez-Polo[1]

[1] Department of Conservative Dentistry and Orofacial Prosthodontics,
Faculty of Dentistry, Complutense University of Madrid, Madrid, Spain
luisazevedo.2005@gmail.com, pedromol@ucm.es,
antonayam@hotmail.com, {jrh,mgpolo}@odon.ucm.es
[2] Institute of Health Sciences - Viseu.
Universidade Católica Portuguesa, Viseu, Portugal
correia.andre@gmail.com

Abstract. In the computer-aided design and manufacturing (CAD/CAM) dentistry, intra-oral scanners (IOS) are usually used to capture the teeth/implants position and adjacent soft tissues for fabrication of prostheses. However, scanning errors are associated with IOS for full arch impressions and teeth/implants with divergent alignments. While small degree of errors may be absorbed by the periodontal ligament of abutment teeth and the cement layer under the cement-retained prostheses, passive seating is particularly difficult to achieve for implant supported screw-retained prostheses in which implants are ankylosed and little leeway allowed for screw system. In this clinical report, five patients requested prosthodontic treatment for her compromised functional and aesthetic situation. Photogrammetry system and a conventional irreversible hydrocolloid impression were used to capture the three-dimensional implant position and the associated soft tissue topography, respectively. The two digital files were subsequently superimposed using a best-fit alignment function to generate the definitive digital model with information on teeth, soft tissues, and implants. An immediately digital set up was designed and a poly (methyl methacrylate) wax try-in was milled to check the esthetics parameters, the occlusal relations and to evaluate its passive fit. After this, the framework was manufactured using a milled-sintered cobalt-chromium and then coated with ceramic. A CAD/CAM implant supported screw-retained prosthesis with a good fit and esthetic was manufactured. The described technique suggests certain advantages over conventional techniques, however further studies and long-term clinical data are needed to evaluate the accuracy of photogrammetry in comparison with the other available techniques.

Keywords: Dental implants · Photogrammetry · Dental impression technique · Computer-aided design · Computer-aided manufacturing

© Springer Nature Switzerland AG 2019
J. M. R. S. Tavares and R. M. Natal Jorge (Eds.): VipIMAGE 2019, LNCVB 34, pp. 615–619, 2019.
https://doi.org/10.1007/978-3-030-32040-9_62

1 Introduction

One of the major problems of immediate implant-supported fixed dental prostheses is to obtain a correct adaptation between the prosthesis and the implants [1]. The prosthesis must have a good passive fit, in order to maintaining successful osseointegration and avoiding mechanical complications [1, 2].

To achieve a good passive fit, the accuracy of the employed clinical protocol from the impressions to framework manufacturing is crucial [3].

Photogrammetry is a technique that collects three-dimensional coordinate measurements to record the geometric properties of objects and their spatial position from photographic images by an extra-oral receiver [3–7].

It is a reliable technique used in many fields and it was introduced in dentistry by Lie and Jemt [8] in 1994 to study the distortion of implant frameworks.

This technique can be used as a reliable alternative to record the positions and angulations of multiple dental implants [3–9].

This study describes the use of photogrammetry as a reliable technique to record three-dimensional implant positions for fixed implant-supported rehabilitations.

2 Materials and Methods

Five patients attended the Complutense University Dental Clinic Faculty in Madrid and requested prosthodontic treatment for her compromised functional and aesthetic situation.

The implants positions were recorded using a photogrammetry technique (PIC Camera; PicDental) (Fig. 1).

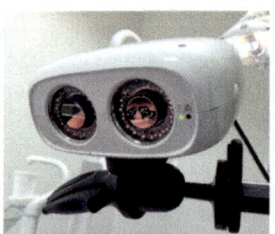

Fig. 1. PIC camera

PIC abutments were screwed into the transepithelial abutments and the PIC camera took the pictures (Fig. 2).

The information furnished by the photogrammetric abutments and gathered by the camera was processed by a software (Pic Cam Soft v1.1; PIC dental) that creates a standard tessellation language (STL) file (PIC file) showing the three-dimensional localization of the implants and their angulation.

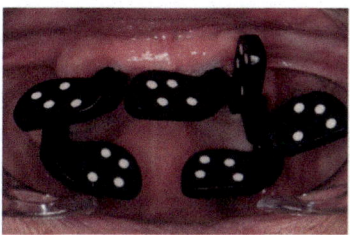

Fig. 2. PIC abutments

A second conventional irreversible hydrocolloid impression was made and digitally scanned, to record the patient's soft tissues and to create a second STL file to ensure best-fit alignment (Fig. 3).

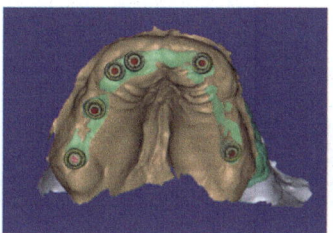

Fig. 3. Patient's soft tissues impression

These STL files (the PIC file and the digital impression of soft tissues) were subsequently aligned and merged using a dental CAD software (Exocad; exocad GmbH, Germany) to obtain a new digital archive integrating both.

A denture try-in was designed and milled from a polymethylmethacrylate (PMMA) resin to evaluate the aesthetics parameters, teeth positions and vertical dimension

The frameworks were designed with Exocad software in STL format (Fig. 4) and passive fit was checked in the patient's mouth. All tests suggested a correct fit between the implants and the frameworks.

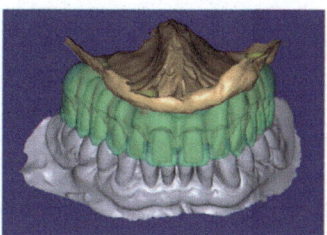

Fig. 4. Digital design of the framework.

The definitive prosthesis was screwed according with the manufacturer's recommended torque (Figs. 5, 6).

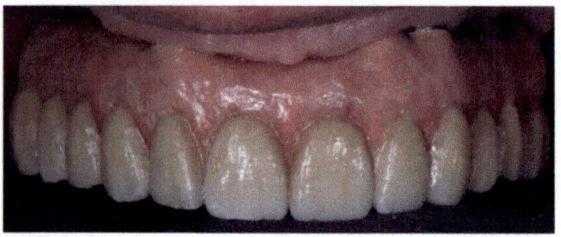

Fig. 5. Definitive implant-supported metal-ceramic maxillary prosthesis

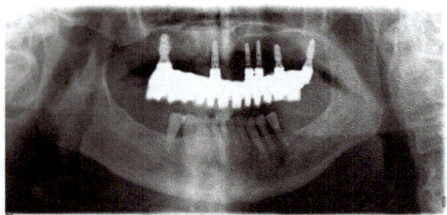

Fig. 6. Post-treatment panoramic radiograph

3   Results and Discussion

The 1-week, 1-months, 6-months and 1-year follow-up appointments revealed no mechanical or biological complications. Patients satisfaction was positive in all appointments.

Nowadays, the reliability of intraoral scanners remains questionable when they are used for the impression of multiple implants when they distributed along the whole arch [10, 11].

With photogrammetry, there is no need for making overlapping images (like with intraoral scanners), by using an extraoral receiver. Also, the presence of blood and saliva does not affect measurement precision of PIC camera [3, 7]. Therefore, this technique can be used as a reliable alternative to record the positions and angulations of multiple dental implants and even for producing immediate prostheses with a predictably proper passive fit. Nevertheless, one limitation of this technique is that it does not register the soft tissues, which requires a second STL file and specific equipment to provide this information.

In this case series, with photogrammetry technology we successfully created a CAD/CAM framework with a good fit. At 1-year of follow-up, any mechanical complications were observed. This study suggests that this technique provides accurate passive fit and minimize the possibility of posttreatment complications.

4   Conclusions

Photogrammetry technique suggests certain advantages in comparison with conventional techniques in the impression of multiple dental implants for immediate implant-supported fixed dental prosthesis. It provides accurate passive fit and minimizing the risks of posttreatment complications.

However, further studies with control groups are necessary to ensure rigorous scientific support and to determine the advantages of these new technologies.

Acknowledgments. The authors are grateful to Adrián Hernández, CEO of PiC dental and for his support in this work.

References

1. Wee, A.G., Aquilino, S.A., Schneider, R.L.: Strategies to achieve fit in implant prosthodontics: a review of the literature. Int. J. Prosthodont. **12**, 167–178 (1999)
2. Gómez-Polo, M., Gómez-Polo, C., del Río, J., Ortega, R.: Stereophotogrammetric impression making for polyoxymethylene, milled immediate partial fixed dental prostheses. J. Prosthet. Dent. **19**, 506–510 (2018)
3. Suarez, M.J., Paisal, I., Rodriguez-Alonso, V., Lopez-Suarez, C.: Combined stereophotogrammetry and laser-sintered, computer-aided milling framework for an implant-supported mandibular prosthesis: a case history report. Int. J. Prosthodont. **31**, 60–62 (2018)
4. Agustín-Panadero, R., Peñarrocha-Oltra, D., Gomar-Vercher, S., Peñarrocha-Diao, M.: Stereophotogrammetry for recording the position of multiple implants: technical description. Int. J. Prosthodont. **28**, 631–636 (2015)
5. Sánchez-Monescillo, A., Sánchez-Turrión, A., Vellon-Domarco, E., Salinas-Goodier, C., Prados-Frutos, J.C.: Photogrammetry impression technique: a case history report. Int. J. Prosthodont. **29**, 71–73 (2016)
6. Ey-Chmielewska, H., Chrusciel-Nogalska, M., Fraczak, B.: Photogrammetry and its potential application in medical science on the basis of selected literature. Adv. Clin. Exp. Med. **24**, 737–741 (2015)
7. Pradíes, G., Ferreiroa, A., Özcan, M., Giménez, B., Martínez-Rus, F.: Using stereophotogrammetric technology for obtaining intraoral digital impressions of implants. J. Am. Dent. Assoc. **145**, 338–344 (2014)
8. Jemt, T., Back, T., Petersson, A.: Photogrammetry – an alternative to conventional impressions in implant dentistry? A clinical pilot study. Int. J. Prosthodont. **12**, 363–368 (1999)
9. Peñarrocha-Oltra, D., Agustín-Panadero, R., Pradíes, G., Gomar-Verccher, S., Peñarrocha-Diago, M.: Maxillary full-arch inmmediately loaded implant-supported fixed prosthesis designed and produced by photogrammetry and digital printing: a clinical report. J. Prosthodont. **26**, 75–81 (2015)
10. Wenz, H.J., Hans-Ulrich, R., Hertrampf, K.: Accuracy of impressions and casts using different implant impression techniques in a multi-implant system with an internal hex connection. Int. J. Oral Maxillofac. Implants **23**, 39–47 (2008)
11. Eliasson, A., Örtorp, A.: The accuracy of an implant impression technique using digitally coded healing abutments. Clin. Implant Dent. Relat. Res. **14**, 30–38 (2012)

The Use of QR Code as an Identification Method in Complete Dentures

Bruno Valentim[1(✉)], Cristina Figueiredo[1,2], João Andrade[1],
Catarina Oliveira[1], and André Correia[1,2]

[1] Institute of Health Sciences (ICS), Viseu - Universidade Católica Portuguesa,
Viseu, Portugal
brunovalentim96@hotmail.com
[2] Center for Interdisciplinary Research in Health (CIIS),
Universidade Católica Portuguesa, Viseu, Portugal

Abstract. The identification process in cases of natural disasters and crimes is a necessity in every investigation process. With the technology evolving, new techniques have been developed in order to facilitate identification in an easier and efficient way. In edentulous individuals, dentures are one of the most valuable tools due to the incorporation of elements that can allow identification. In the last years, QR code have been studied for this purpose, since it presents more advantages than other methods.

In order to minimize the size of a QR code, titanium and cobalt-chrome pieces were milled in different sizes. Two different ways of reading the QR codes were used. Despite not being capable to read the three different sizes, two of them were perfectly readable.

The QR code use in dentures can play an important role, since the information included in it can save lives or help to identify human beings. However, more studies must be done in order to make this a common procedure.

Keywords: Forensic sciences · Forensic dentistry · Prosthodontics · QR code · Patient identification systems · Denture identification marking · Edentulism

1 Introduction

Forensics is the medicine's field that contributes the most to human identification in alive or deceased individuals. In cases where DNA and fingerprint databases are absent, dental identification continues to be crucial [1, 2]. Forensic Dentistry was defined in 1970 as the "proper handling and examination of the dental evidence, in the interests of justice, so that the dental finding may be properly presented and evaluated" [3]. However, forensic odontology has become more and more relevant in forensics sciences not only because it can help in crimes and violence, but also because it can help to identify people in natural disasters such as tsunamis and fires.

Dental identification has the principle of comparing the *post-mortem* remains with the *ante-mortem* records, which is called the comparative identification method. However, when there are no *ante-mortem* records, it's impossible to use the previous method. In this cases, it is necessary to use the reconstructive method, which consists in

© Springer Nature Switzerland AG 2019
J. M. R. S. Tavares and R. M. Natal Jorge (Eds.): VipIMAGE 2019, LNCVB 34, pp. 620–627, 2019.
https://doi.org/10.1007/978-3-030-32040-9_63

collecting all the data available, in order to establish an individual profile and then compare that information with an unknown individual [4, 5].

According to the American Board of Forensic Dentistry, to establish a positive identification, the data available must be accurate without discrepancies. For this evaluation, the experts must focus the analysis on the qualitative aspects of the material [6, 7].

Although most dental identifications are based on combinations of dental characteristics such as restorations, root canal treatments, tooth decay and tooth presence/absence, it is impossible to obtain concordant *ante-mortem* and *post-mortem* data in edentulous individuals, since there is a high rate of dental absence and high bone resorption. In addition, the current aging of the world population and the increase of hospitalized and institutionalized elderly have reinforced the need for new identification techniques through dental prostheses. These techniques are already mandatory by law in some countries [8, 9].

Denture marking has been a helpful tool in identification in natural disasters such as tsunamis and earthquakes, as well as in the identification of patients with dementia and other diseases that lead to memory loss. In a near future, with the introduction of new technologies, the denture marking will play an important role in identification of individuals in a daily basis [10, 11]. Denture marking systems have evolved over the years. However, there are two main categories: the surface methods and the inclusion methods.

Although the surface methods have some advantages such as being cheap, easy to prepare and to engrave, they are not permanent and need to be reapplied after some time [12, 13].

The inclusion methods can provide a predictable outcome without affecting the resistance of the denture, generally are not visible and they don't have the need to be reapplied. This system includes methods as barcode, lenticular card, microchip and memory card [14–16].

For the last years, new methods such as *Quick Response Code* (*QR Code*) have been studied. The *QR Code* is a two-dimensional code that can be easily read by any device with a photographic camera and can support up to 4926 alphanumeric characters [17, 18]. In denture identification, this is the method that presents more advantages, such as:

1. Can be impressed in any surface;
2. Extended availability;
3. High storage capacity;
4. Even if the code is destroyed up to 30%, it can be read [17, 19].

2 Materials and Methods

In order to make a comfortable denture with a *QR code*, it is fundamental to find a way of minimizing the size described in the literature as the minimum readable (10 × 10 mm) [20, 21].

For this purpose, we've impressed 9 *QR codes* in titanium and chromium-cobalt with different sizes, by the following protocol:

1° Step – Draw

The pieces where the *QR code* will later be impressed, were drawn in the software 3-Matic® (Fig. 1).

3 different Standard Tessellation Language (STL) files were generated - 5 × 5 mm, 7,5 × 7,5 mm and 10 × 10 mm. All the files had 0.5 mm thickness.

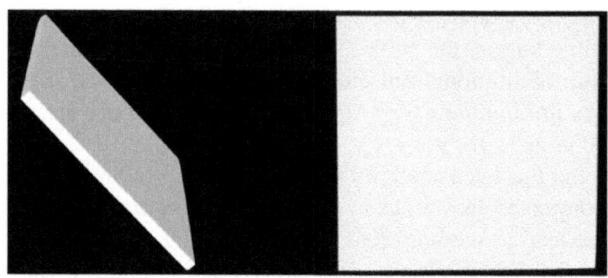

Fig. 1. Example of an *STL* file drawn in 3-Matic®

2° Step - Milling

The files were sent to a dental lab and the pieces were milled in the milling machine Coritec® 350i from Imes-Icore®. To do so, dental alloys discs of both materials were used (Fig. 2).

The titanium alloy TiAl6V4 disc Starbond® Ti5 (Scheftner®) is composed 89.4% by titanium, 6.25% with aluminum, 4% vanadium and 0,4% carbon, hydrogen and iron. The cobalt-chrome disc is from Zirlux® NP, and is composed by 61,6% cobalt, 27,8% chrome, 8,5% tungsten, 1,6% silicon, 0,3% manganese and 0,2% iron.

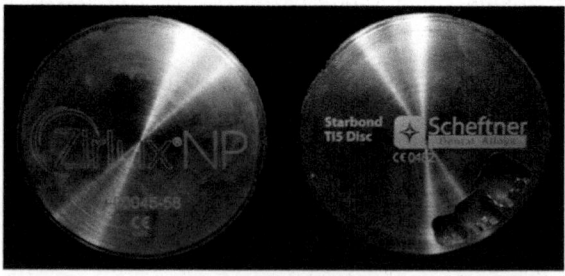

Fig. 2. Dental alloy discs: cobalt-chrome (in the left) and titanium (in the right)

The final result consisted in 6 titanium and 3 cobalt-chrome pieces, used as follow:

- 3 titanium pieces used as "test" to help us find the best laser definitions for each material;
- 3 titanium and 3 cobalt-chrome pieces to test the reading of the QR code (Fig. 3).

Fig. 3. Comparison between the three different sizes

Then the QR code was generated in a free website https://www.qr-code-generator. com (Fig. 4).

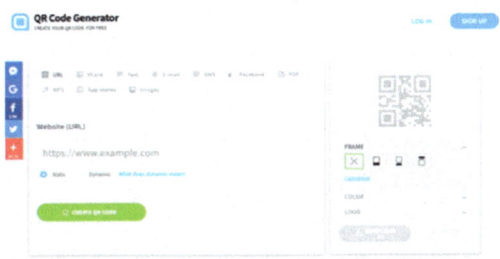

Fig. 4. Website used to generate the QR Code

3° Step – Engraving

The same *QR code* were laser engraved in the first 3 titanium pieces, in order to choose the best laser characteristics to do so (Fig. 5). The laser machine used was Big Smark® 200F, from Sisma®.

Fig. 5. Titanium pieces for laser engraving test

After the tests, three titanium and three cobalt-chrome pieces were engraved with a QR code containing data complied by a patient of an University Dental Clinic (Figs. 6 and 7).

Fig. 6. Titanium pieces with the engraved QR code

Fig. 7. Cobalt-chrome pieces with the engraved QR code

<u>4° Step – Reading Test</u>
The different pieces were read in 2 distinct ways:

1. Direct reading by the camera to test if the proposed sizes allow the information stored in the QR code to be available;
2. Indirect reading through photography, using the same camera and using the zoom tool, to test if it is possible to read the QR codes with the smallest enlarged sizes in the photograph.

A 12 megapixels (MP) camera with f 1,8, digital zoom up to 5× and optical image stabilization (OIS) from the Iphone® XR was used to test the two hypothesis.

3   Results

The laser engraving method was a difficult process because the machine used had to be capable to engrave the detail necessary in such small pieces as the ones used in this investigation. So, it was necessary the help of a jewelry company, which has this kind of machines. The engraving was successfully done in all the pieces and the definition of the engraving was very high quality (Table 1).

The reading of the QR codes was made, reaching to some results:

- 7,5 mm × 7,5 mm and 10 mm × 10 mm QR codes were readable directly in the both materials as well as in the indirect reading method;
- 5 mm × 5 mm QR codes were not readable by any of the methods used.

Table 1. QR codes reading;

	Titanium	**Cobalt-chrome**	
5mm x 5mm **(0,25cm^2)**	Non-readable	Non-readable	1
	Non-readable	Non-readable	2
7,5mm x 7,5mm **(56,25cm^2)**	Readable	Readable	1
	Readable	Readable	2
10mm x 10mm **(1cm^2)**	Readable	Readable	1
	Readable	Readable	2

1 – direct reading; 2 – indirect reading

4 Discussion

The Portuguese Statistics National Institute showed in 2018 that almost 8.2% of national population was toothless in 2018. This data is very worrying, since teeth are an important source of information in forensics [22]. However, the available data shows that the edentulous population without rehabilitation are less, which indicates that the identification through dentures could play an important role in human identification [23].

Fernandes et al. [20] found that the 10 mm x 10 mm (1 cm^2) QR code was the minimum readable and that the Cobalt-chrome was the best material to engrave a QR code. Although, in our study, the minimum directly readable size was 7,5 mm $\times$ 7,5 mm (56,25 cm^2), which is an upgrade in inferior dentures. It's easy to believe that in a near future it will be easy to reduce the QR code size even more, since the laser machines and the camera phone technology are evolving at the speed of light. However, the 7,5 mm $\times$ 7,5 mm QR code can be included in almost every inferior denture.

Cobalt-chrome is a non-noble metal with a melting point between 1300 °C and 1500 °C. However, titanium alloys such as TiAl6V4 are used in aerospace and medical components due to tensile modulus, mechanical resistance and melting point between 1878 °C and 1933 °C, as well as the lower weight compared to the cobalt-chrome [24]. In disasters such as Pedrogão Grande, in Portugal (2017), temperatures can reach up to 900 °C, which are inferior to the melting-points of both materials [25].

Furthermore, at the milling level of the materials, it is easier to machine titanium than cobalt chromium because of its hardness (2.94 GPa vs 2.80 GPa) [26, 27], since it allows to reduce the wear of the drills and the length of the process.

With the analysis made in this investigation, it is concluded that the lower mass of the TiAl6V4 titanium alloy relative to cobalt-chrome gives this material an extra advantage. However, both metals have properties that allow them to be used for the inclusion of a QR code in a prosthesis. The choice of the material to use will always depend on the availability of the prosthesis laboratory chosen by each dentist and the cost for the patient.

5 Conclusion

Beyond the limitations of this study, the following conclusions were made:

1. The process that guided this investigation was successfully implemented. The QR code use in dentures can play an important role, since the information included in it can save lives or help to identify human beings.
2. The success in decreasing the QR code in 44% of its area in relation to the literature represents a fundamental advance for its use also in mandibular prostheses;
3. Although both titanium and chromium-cobalt can be used to include the QR code, it has been found that titanium has advantages in density and ease of milling, which may be very relevant for its execution;

However, more studies must be done in order to make this a common procedure, especially in countries such as Portugal, where denture identification methods are not regulated by law.

References

1. Pretty, I.A., Sweet, D.: A look at forensic dentistry - part 1: the role of teeth in the determination of human identity. Br. Dent. J. **190**(7), 359–366 (2001)
2. Divakar, K.P.: Forensic odontology: the new dimension in dental analysis. Int. J. Biomed. Sci. **13**(1), 1–5 (2017)
3. Keiser-Neilsen, S.: Person identification by means of teeth. Am. J. Forensic Med. Pathol. **2**, 189 (1981)
4. Adams, C., Carabott, R., Evans, S. (eds.): Forensic Odontology - An Essential Guide, 1st edn, pp. 1–339. Wiley Blackwell, Hoboken (2014)
5. Kumar, S., Rathore, S., Pandey, A., Verma, A.: Role of dental expert in forensic odontology. Natl. J. Maxillofac. Surg. **5**(1), 2 (2014)
6. American Board of Forensic Odontology. Diplomates Reference Manual - Section IV : Standards & Guidelines [Internet] (2017). http://abfo.org/wp-content/uploads/2012/08/ABFO-DRM-Section-4-Standards-Guidelines-Sept-2017-New-page-numbers.pdf
7. Priyadharsini, C., Masthan, K.M.K., Balachander, N., Babu, N., Jimson, S.: Evolution of forensic odontology: an overview. J. Pharm. Bioallied Sci. **7**(5), 178 (2015)
8. Nogueira, T.E., Bandeira, A.C., Leles, C.R., Silva, R.F.: Use of QR code as personal identification of complete dentures – literature review. Rev. Bras Odontol. Leg. **5**(1), 61–67 (2018)
9. Gosavi, S., Gosavi, S.: Forensic odontology: a prosthodontic view. J. Forensic Dent. Sci. **4**, 38–41 (2012)
10. Mishra, S.K., Mahajan, H., Sakorikar, R., Jain, A.: Role of prosthodontist in forensic odontology. A literature review. J. Forensic. Dent. Sci. **6**(3), 154–159 (2014)
11. Kareker, N., Aras, M., Chitre, V.: A review on denture marking systems: a mark in forensic dentistry. J. Indian Prosthodont. Soc. **14**, 4–13 (2014)

12. Sudheendra, U.S., Sowmya, K., Vidhi, M., Shreenivas, K., Prathamesh, J.: 2D barcodes: a novel and simple method for denture identification. J. Forensic Sci. **58**(1), 170–172 (2013)
13. Datta, P., Sood, S.: The various methods and benefits of denture labeling. J. Forensic Dent. Sci. **2**, 53–58 (2010)
14. Bansal, P., Bhanot, R., Sharma, A.: Denture labeling: a new approach. Contemp. Clin. Dent. **2**(2), 76 (2011)
15. Chandran, C.R.: A new alternative technique for denture identification. World J. Dent. **1**, 188–192 (2015)
16. Rathee, M., Yadav, K.: Denture identification methods: a review. J. Dent. Med. Sci. **13**(10), 58–61 (2014)
17. Agülolu, S., Zortuk, M., Beydemir, K.: Denture barcoding: a new horizon. Br. Dent. J. **206** (11), 589–590 (2009)
18. Uzun, V., Bilgin, S.: Evaluation and implementation of QR Code Identity Tag system for Healthcare in Turkey. Springerplus **5**(1), 1454 (2016)
19. Kalyan, A., Clark, R.K.F., Radford, D.R.: Denture identification marking should be standard practice. Br. Dent. J. **216**(11), 615–617 (2014)
20. Fernandes, A., Correia, A., Silva, A.M., Figueiredo, C.: Forensic identification tool in dental removable prosthodontics. In: Silva Gomes, J.F., Meguid, S.A., (eds.) INEGI/FEUP, Lisbon, pp. 1441–1444 (2018)
21. Fernandes, A.M.A.: Identificação em desdentados totais - marcação codificada de próteses totais. Universidade Católica Portuguesa, Viseu (2017)
22. Instituto. Instituto Nacional de Estatística [Internet]. Cited 27 January 2018. https://www.ine.pt/xportal/xmain?xpid=INE&xpgid=ine_destaques&DESTAQUESdest_boui=134582847&DESTAQUESmodo=2
23. Ordem dos Médicos Dentistas. Barómetro de Saúde Oral (2018)
24. Trevisan, F., Calignano, F., Aversa, A., Marchese, G., Lombardi, M., Biamino, S., et al.: Additive manufacturing of titanium alloys in the biomedical fields: processes, properties and applications. J. Appl. Biomater. Funct. Mater. **16**(2), 57–67 (2018)
25. Observador. Temperaturas chegaram aos 900 graus na estrada da morte (2017)
26. Schein® H. Zirlux® NP Technical sheet [Internet]. Cited 17 June 2019. https://www.zirlux.com/assets/17ZX7572-Zirlux-NP.pdf
27. Scheftner®. Starbond Disc Ti5 Technical Sheet [Internet]. Cited 17 June 2019. https://www.scheftner.dental/starbond-ti5-disc-en.html

3D Printed Splint – Clinical Cases

Bruno Valentim[1]([⊠]), João Andrade[1], Catarina Oliveira[1],
Tiago Marques[1,2], and Patrícia Fonseca[1,2]

[1] Institute of Health Sciences (ICS), Viseu - Universidade Católica Portuguesa,
Viseu, Portugal
brunovalentim96@hotmail.com
[2] Center for Interdisciplinary Research in Health (CIIS),
Universidade Católica Portuguesa, Lisbon, Portugal

Keywords: Temporomandibular disorders · Centric relation splint · Digital
dentistry · Workflow

1 Introduction

Temporomandibular disorders (TMD) are a set of disorders related to the masticatory
muscles, a temporomandibular joint and adjacent structures (articular pathology) [1–4].

These pathologies may be related to systemic conditions, especially at somatic and
neurological level. For a correct diagnosis of TMD, it is important to have a deep
knowledge of the symptoms of temporomandibular disorders and orofacial pain, as
well as the systemic pathologies that may be associated [5].

The centric relation splint is a frequent treatment technique to muscular and/or
articular temporomandibular disorders capable of stabilizing the temporomandibular
joint (TMJ), protect the teeth, relax the muscles, reduce bruxism activity and allow the
balance of the bite forces. Beyond the dentist and the dental technician's consumption
of time, the process surrounding the fabrication of an occlusal splint is susceptible to
human errors and to the limitations of the materials used [6, 7].

The new technologies, as Intraoral Scanners and 3D Printers, have been showing
their utility in the different areas of Dentistry. Their application can also be used in
Occlusion, helping to minimize manufacturing errors and working time and to provide
consistency [7].

The aim of this work is to present the workflow in the fabrication of 3D printed
splints through TMD clinical cases.

2 Materials and Methods

Two patients with 20 and 22 years old were observed in the University Dental Clinic,
Universidade Católica Portuguesa - Viseu, Portugal, with pain complaints and temporomandibular disfunction. After the diagnosis of Temporomandibular disorder
established by the Research Diagnostic Criteria for Temporomandibular Disorders
(RDC/DTM), they both present indication for a centric relation splint.

© Springer Nature Switzerland AG 2019
J. M. R. S. Tavares and R. M. Natal Jorge (Eds.): VipIMAGE 2019, LNCVB 34, pp. 628–631, 2019.
https://doi.org/10.1007/978-3-030-32040-9_64

After arches digitalized (Straumann Dentalwings®), the technological procedures/ steps that give the splint a peculiar confection are its digital drawing (Meshmixer®) and 3D printing (Anycubic Photon®) in Harzlabs Dental® resin (Figs. 1 and 2). After placement of the splint (Fig. 3), occlusal centric contacts were evaluated as well as function movements (Figs. 4, 5 and 6). Clinical follow-ups were performed at 1 week, 1 and 3 months with significant signs and symptoms improvements.

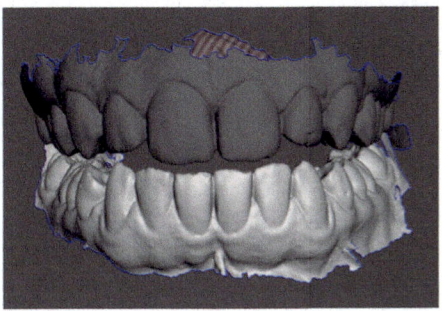

Fig. 1. Digital image of the arches digitalized

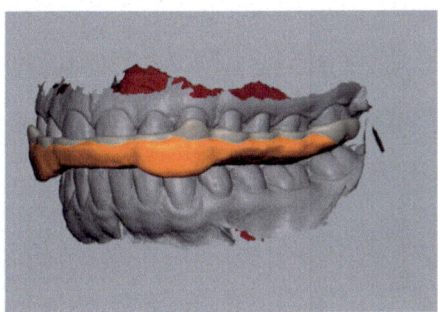

Fig. 2. Digital image of the arches digitalized with the splint waxing (grey) and the digital splint (orange)

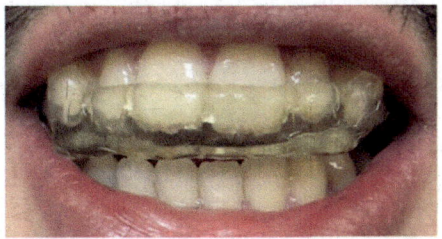

Fig. 3. Intercuspation - frontal view

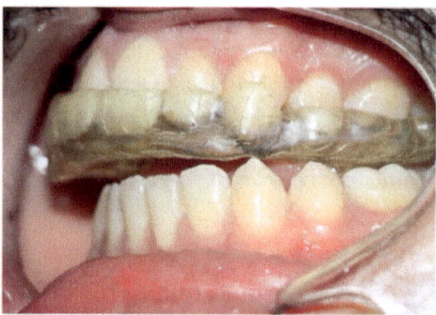

Fig. 4. Left lateral movement (left canine guidance)

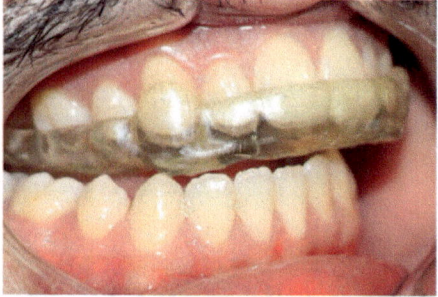

Fig. 5. Right lateral movement (right canine guidance)

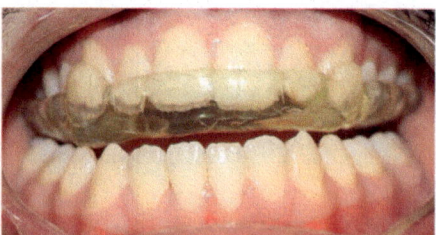

Fig. 6. Protrusive movement (Anterior guidance)

3   Results and Discussion

Although the two first printed splints of the University Dental Clinic of Viseu were able to be made, the constraints or limitations found prevented the process from being totally digital. In order to establish the desired disocclusion, intermaxilar centric relation and to create the eccentric movement guides, it was necessary to assemble semi-adjustable articulator models and to wax the splint. Thus, the desired decrease in time was not achieved, but the accuracy of the splint's fitting and occlusal adjustments were substantially improved.

It is a method that can bring many advantages and reduce human error during the manufacturing process of the splint, but it needs more clinical cases in order to make the digital processing more effective, cheaper and less time consuming.

4   Conclusion

We are in an early phase of new technologies application in the splints production, but by the results obtained in these clinical cases, we can affirm without doubt that we have the clinical and laboratory conditions and the digital skills necessary to improve this technique so that, in a near future, this procedure can be adopted as first choice.

Bibliography

1. Carrara, S., Conti, P., Barbosa, J.: Termo do 1º Consenso em Disfunção Temporomandibular e Dor Orofacial. Dent. Press J. Orthod. **15**, 114–120 (2010)
2. Gremillion, H.A.: Preface. Dent. Clin. North Am. **51**(1) (2007)
3. Schmitter, M., Rammelsberg, P., Hassel, A.: The prevalence of signs and symptoms of temporomandibular disorders in very old subjects. J. Oral Rehabil. **32**(7), 467–473 (2005)
4. Yadav, S., Yang, Y., Dutra, E.H., Robinson, J.L., Wadhwa, S.: Temporomandibular joint disorders in older adults. J. Am. Geriatr. Soc. **66**(6), 1213–1217 (2018)
5. Sharma, S, Pal, U., Gupta, D., Jurel, S.: Etiological factors of temporomandibular joint disorders. Natl. J. Maxillofac. Surg. **2**(2), 116 (2011). http://www.njms.in/text.asp?2011/2/2/116/94463
6. Lauren, M., McIntyre, F.: A new computer-assisted method for design and fabrication of occlusal splints. Am. J. Orthod. Dentofac. Orthop. **133**(4 Suppl.), 130–135 (2008)
7. Muric, A., Gokcen Röhlig, B., Ongul, D., Evlioglu, G.: Comparing the precision of reproducibility of computer-aided occlusal design to conventional methods. J. Prosthodont. Res. **63**(1), 110–114 (2019). https://doi.org/10.1016/j.jpor.2018.10.002

Intraoral Scan and 3D Analysis of Periodontal Plastic Surgery Outcomes

Sara Ramos[1](✉), Tiago Marques[2,4](✉), N. M. Santos[2,4], M. Sousa[4], G. Fernandes[2,4], and A. Correia[3,4]

[1] Catholic Portuguese University (UCP), Viseu, Portugal
spfframos@hotmail.com
[2] Periodontology Disciplinary Area, UCP, Viseu, Portugal
tiagomarques@gmail.com
[3] Oral Rehabilitation Disciplinary Area, UCP, Viseu, Portugal
[4] Interdisciplinary Investigation Health Center (CIIS) – UCP, Viseu, Portugal

Abstract. Intraoral scanning it's an emergent asset in Dentistry. 3D image capture has its guaranteed place in the future, replacing old traditional impression methods, consuming less time regarding to the planning process with exceptional precision. In this study, we present one of the many advantages that come with the use of this devices and their complement with digital programs, giving us a tridimensional full analysis on periodontal plastic surgery procedure outcomes, helping surgeons to understand the healing dynamics process.

Keywords: Periodontal plastic surgery · Tridimensional analysis · Intraoral scan · Wound healing · Healing dynamics

1 Introduction

In order to correct root exposures, different techniques have been recommended, among them, the vestibular incision subperiosteal tunnel access (VISTA) technique described by Zadeh and the modified tunnel technique proposed by Zuhr et al. The addition of a connective tissue graft (CTG) allows root coverage and gingival thickening and provides long-term stability to the gingival margin. For the CTG harvesting, some authors suggest a single incision technique, promoting primary intention healing at the donor site, inducing less postoperative morbidity to the patient. Others propose the de-epithelialized tissue graft, having a better view of the surgical site and advantages in gingival thickness. Our goal it's to do a full analysis of the healing tridimensional process in both surgical sites, comparing different techniques, using an intraoral scan and digital programs that allows us to overlap STL files and capture their intersections.

© Springer Nature Switzerland AG 2019
J. M. R. S. Tavares and R. M. Natal Jorge (Eds.): VipIMAGE 2019, LNCVB 34, pp. 632–638, 2019.
https://doi.org/10.1007/978-3-030-32040-9_65

2 Materials and Methods

This is a prospective cohort study, over a period of 6 months. Three evaluation times were made - surgery day (T0), 3 months (T1) and 6 months postoperative (T2).
Sample:

- Patient A – multiple type 1 recessions (Cairo classification), treated with modified tunnel technique by Zuhr *et al.* [1] with the addition of a de-epithelialized tissue graft harvested using the technique by Zucchelli *et al.* [2].
- Patient B – multiple type 1 recessions (Cairo classification), treated with vestibular incision subperiosteal tunnel access by Zadeh [3] with the addition of a de-epithelialized tissue graft harvested using the technique by Zucchelli *et al.* [2].

Digital evaluation protocol: patient's casts were recorded at T0, T1 and T2. The casts were digitalized by an intra-oral scanner (DentalWings®), obtaining an STL file for each situation. In the computer programs, Geomagic Control X® and Materialise Magics®, the three-dimensional digital analysis of the intervened areas was recorded.
For the recessions the following variables were evaluated:

- Gingival recession area and dimension
- Mean thickness gain
- Maximum thickness gain
- General volume gain

For the palate the following variables were evaluated:

- Mean thickness loss
- Maximum thickness loss
- Mean maximum thickness loss
- General volume loss

3 Results

3.1 Recession Evaluation

On Table 1 the initial (T0) measurements of GR dimensions and area of each recession treated in both patients are presented (Figs. 1 and 2).
On Tables 2 and 3 measurements of GR dimensions and area, maximum tissue gain (MáxTG), mean tissue gain (MTG) and volume gain (VG) of each recession treated in both patients are presented. An evaluation was made by comparing the pre-operative file with post-operative files obtain from the 3 (T1) and 6 (T2) months follow-ups, respectively. At this point, we are able to calculate a relative recession reduction (RRR).

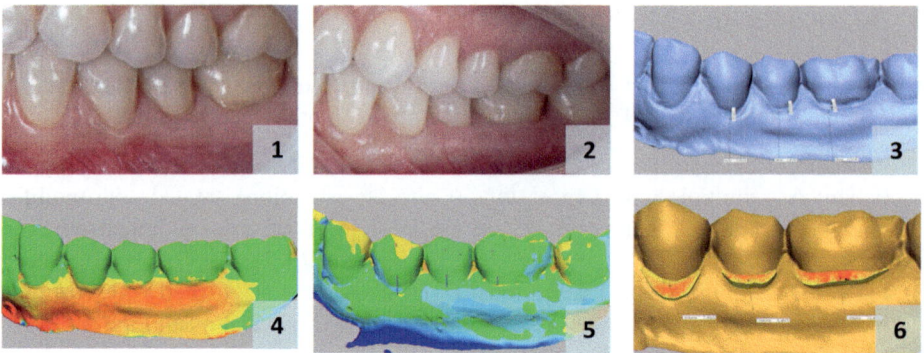

Fig. 1. Patient A recession site results: (1). Recession's appearance at T0; (2). Recession's appearance at T1; (3). Recession's measurements at T0; (4). Volume gain T0-T1; (5). Volume loss T1-T2.

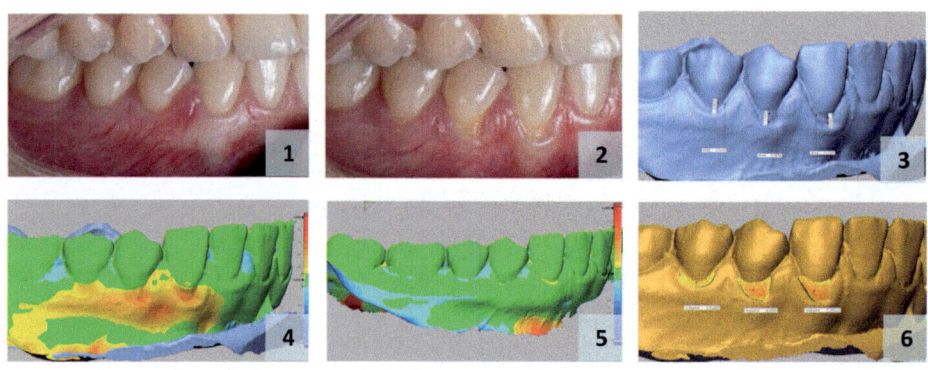

Fig. 2. Patient B recession site results: (1). Recession's appearance at T0; (2). Recession's appearance at T1; (3). Recession's measurements at T0; (4). Volume gain T0-T1; (5). Volume loss T1-T2.

Table 1. Recession's variable results on T0 evaluation.

	Tooth	GR[a] dimension (mm)	GR area (mm)	MáxTG[b] (mm)	MTG[c] (mm)	VG[d] (mm^3)
Patient A	34	0,99	3,91	–	–	–
	35	0,62	2,91	–	–	–
	36	1,1	6,5	–	–	–
Patient B	43	1,25	2,33	–	–	–
	44	1,54	3,38	–	–	–
	45	0,37	1,02	–	–	–

Table 2. Recession's variable results on T0-T1 evaluation.

	Tooth	GR dimension (mm)	GR area (mm)	RRR (%)	MáxTG (mm)	MTG (mm)	VG (mm³)
Patient A	34	0	0	100	1,5319	1,011	7,4926
	35	0	0	100	1,5272	0,9913	6,4476
	36	0	0	100	1,5344	0,9322	15,4015
Patient B	43	0	0	100	0,6224	1,2128	5,0512
	44	0	0	100	0,671	1,1517	4,858
	45	0	0	100	0,2957	0,814	1,5285

Table 3. Recession's variable results on T0-T2 evaluation.

	Tooth	GR dimension (mm)	GR area (mm)	RRR (%)	MáxTG (mm)	MTG (mm)	VG (mm³)
Patient A	34	0	0	100	1,3426	0,4674	6,9339
	35	0	0	100	1,2808	0,3779	5,217
	36	0	0	100	1,9881	0,5771	14,0439
Patient B	43	0	0	100	1,0863	0,4275	5,8461
	44	0	0	100	1,9532	0,4131	4,9959
	45	0	0	100	1,7857	0,1815	1,1452

3.2   Palatal Evaluation

On Table 4 are presented measurements regarding mean tissue loss (MTL), maximum thickness loss point (MáxTL), mean maximum thickness loss points (MMáxTL) and volume gain (VL) from both palatal donor sites. These results where obtained by comparing pre-operative file with post-operative files obtained from the 3 (T1) months follow-up.

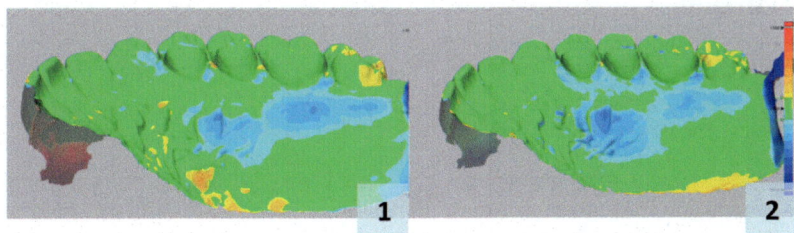

Fig. 3. Patient B STL files overlapping revealing volume changes through a colored map.

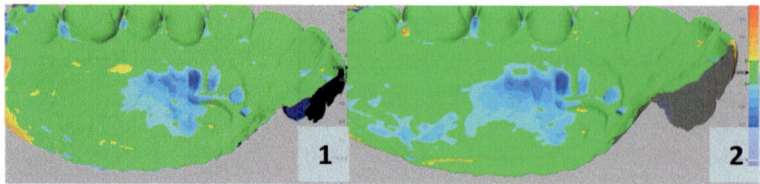

Fig. 4. Patient A STL files overlapping revealing volume changes through a colored map.

Table 4. Palatal variable results on T0-T1 evaluation

	MTL (mm)	MáxTL (mm)	MMáxTL (mm)	VL (mm³)
Patient A	−0,25 ± 0,13	−0,79	−0,41 ± 0,05	⁶41,50
Patient B	−0,21 ± 0,18	−1,31	−0,6175 ± 0,04	35,54

Table 5. Palatal variable results on T0-T2 evaluation

	MTL (mm)	MáxTLP (mm)	MMáxTL (mm)	VL (mm³)	RVD (%)
Patient A	−0,29 ± 0,14	−0,83	−0,50 ± 0,02	48,18	−16,08
Patient B	−0,17 ± 0,18	−1,37	−0,61 ± 0,04	35,75	−0,59

On Table 5 measurements regarding mean tissue loss (MTL), maximum thickness loss point (MáxTL), mean maximum thickness loss points (MMáxTL) and volume gain (VL) from both palatal donor sites are presented. These results where obtained by comparing pre-operative file with post-operative files obtain from the 3 (T1) and 6 (T2) months follow-ups, respectively. At this point, we are able to calculate a relative volume difference (RVD), giving us tissue changes information.

4   Discussion

4.1   Recession Evaluation

- Recession measurements

Recession clinical measurements were obtain through digitalized cast models for each patient at three evaluation times. Periodontal probes are still the gold standard measurement method, but they are associated to two potential errors: precision [4] and the LAC location (the upper reference for recession measurement) [5]. Periodontal probes have approximately a 25% error, due to millimeters approximation [5], Geomagic control X® measurements approximate to the hundredths of a millimeter (0,01 mm), reducing the error to 0,25% [6].

In all the evaluated cases, RRR% demonstrated a clinical success, since a 100% reduction was obtained, having 0 mm measurements at T1 and T2 evaluations.

- Healing dynamics

In all of the cases, MTG decreased from T1 to T2; VG suffered a reduction in all the treated recessions of patient A with tunnel technique, but on patient B, treated with VISTA technique, a volume increase was observed at 43 and 44 recessions, having also a MTG increase in all of them.

It has been speculated that que healing process seems to be completed after 6 months [7]. To understand if it actually stabilizes, a 12 month evaluation should have been made.

4.2   Palatal Evaluation

- Healing dynamics

The blue color spectrum observed reveals tissue loss (Figs. 3 and 4) at both evaluated timelines. Following these observations, the variables reveal negative values, with a total VL of 41,50 mm^3 and 35,54 mm^3 in patient A and B, respectively. At T2 post-operative evaluation, it was noticed a continuous volume loss, although minimal. As so, the continuity of the healing process and the importance of its long-term evaluation was demonstrated.

Soft tissue undergoes relevant three-dimensional alterations. Thickness decrease can be a detrimental factor when a CTG it's collected, therefore, bone exposure (in the case of the DE technique) [8] or epithelial tissue necrosis (in the case of the SE technique) may occur [2, 9].

Unfortunately, the scanner used on this study requires powder, which doesn't allow us to obtain the tissue graft volume, that can be captured intraoperatively.

5   Conclusion

Both tunnel and VISTA techniques allows us to obtain substantial increase of gingival thickness and CTG harvesting induces soft tissue alterations, whose appear not to stabilize at the first 3 months post-operative.

Introducing tridimensional digital evaluation, this study has shown to be innovative, allowing us to 3D measure and evaluate periodontal surgery outcomes, not only at the donor site as also at the intervened teeth.

Recent technological advances provide powder free intraoral scan access, allowing us to capture intraoperative precise images from the donor site, giving us the initial volume loss, eliminating limitations associated with the intraoral scan and cast models.

Acknowledgments. Greater thanks to the periodontology team from the Portuguese Catholic University, Viseu, Portugal.

References

1. Zuhr, O., Fickl, S., Wachtel, H., Bolz, W., Hurzeler, M.B.: Covering of gingival recessions with a modified microsurgical tunnel technique: case report. Int. J. Periodontics Restorative Dent. **27**(5), 457–463 (2007)
2. Zucchelli, G., Mele, M., Stefanini, M., Mazzotti, C., Marzadori, M., Montebugnoli, L., et al.: Patient morbidity and root coverage outcome after subepithelial connective tissue and de-epithelialized grafts: a comparative randomized-controlled clinical trial. J. Clin. Periodontol. **37**(8), 728–738 (2010)
3. Zadeh, H.H.: Minimally invasive treatment of maxillary anterior gingival recession defects by vestibular incision subperiosteal tunnel access and platelet-derived growth factor BB. Int. J. Periodontics Restorative Dent. **31**(6), 653–660 (2011)
4. Hefti, A.F.: Periodontal probing. Critical Rev. Oral Biol. Med. **8**(3), 336–356 (1997)
5. Kerner, S., Etienne, D., Malet, J., Mora, F., Monnet-Corti, V., Bouchard, P.: Root coverage assessment: validity and reproducibility of an image analysis system. J. Clin. Periodontol. **34** (11), 969–976 (2007)
6. Zuhr, O., Rebele, S.F., Schneider, D., Jung, R.E., Hurzeler, M.B.: Tunnel technique with connective tissue graft versus coronally advanced flap with enamel matrix derivative for root coverage: a RCT using 3D digital measuring methods. Part I. Clinical and patient-centred outcomes. J. Clin. Periodontol. **41**(6), 582–592 (2014)
7. Rebele, S.F., Zuhr, O., Schneider, D., Jung, R.E., Hurzeler, M.B.: Tunnel technique with connective tissue graft versus coronally advanced flap with enamel matrix derivative for root coverage: a RCT using 3D digital measuring methods. Part II. Volumetric studies on healing dynamics and gingival dimensions. J. Clin. Periodontol. **41**(6), 593–603 (2014)
8. Brasher, W.J., Rees, T.D., Boyce, W.A.: Complications of free grafts of masticatory mucosa. J. Periodontol. **46**(3), 133–138 (2013)
9. Carnio, J., Koutouzis, T.: Palatal augmentation technique: a predictable method to increase the palatal connective tissue at donor sites— a consecutive case series. Int. J. Periodontics Restorative Dent. **35**(5), 707–713 (2017)

Dimensions of the Dentogingival Unit
in the Anterior Maxilla – A Study with CBCT

D. Figueiredo[1(✉)], Tiago Marques[1,2], and A. Correia[1,2]

[1] Institute of Health Sciences, Universidade Católica Portuguesa, Viseu,
Portugal
daniela_figueiredo77@hotmail.com
[2] Center of Interdisciplinary Research in Health (CIIS), Institute of Health
Sciences, Universidade Católica Portuguesa, Viseu, Portugal

Keywords: Dentogingival unit · Gingival thickness · CBCT

1 Introduction

The dentogingival unit (DGU) can be defined as the anatomical complex formed by the gingival membrane, gingival sulcus, junctional epithelium and connective tissue insertion, also known as the "biological space" [1]. It is considered of great importance, and its invasion may cause the collapse and apical migration of the fixation apparatus. Biological space is considered essential for the maintenance of gingival health, and the success of various dental procedures like restorations [2] and consequent periodontal health. There are numerous methods to measure the dimensions of the DGU but studies using cone beam computed tomography (CBCT) are still very scarce.

The CBCT is often used in dentistry for its advantages over conventional radiography [3, 4] and it uses a conical beam of radiation with a centralized round or rectangular shape on a two-dimensional sensor to perform a 360-degree scan around the patient's head. During the scan, a series of 360 exposures or projections is acquired, one at each degree of rotation, which provides digital data for the reconstruction of the exposed volume by computational algorithm. In this way, a reproduction of the anatomical structure is obtained in any one of the planes of the space. Thus, with this 3D technique we were able to generate volumetric images with axial, coronal and sagittal sections that allow us to do a more detailed analysis than any type of two-dimensional radiography. The CBCT has the advantage that a single rotation is sufficient to acquire the desired image and the collimation of the X-ray beam in the area of interest minimizes the radiation dose; in addition to this, it has a good resolution and a high degree of accuracy [4].

In current CBCTs it is possible to control the field of view (FOV) that describes the scanning volume of a particular equipment and depends on the size and shape of the bulb and the collimation of the radiation beam [4, 5]. This collimation limits the exposure to ionizing radiation of the region under study. Thus, a 'large' FOV is used to analyze the entire skull, 'medium' to analyze both arches and 'small' to an area such as a quadrant or a sextant, which is the case used in this study. Thus, it is of interest to reduce the exposure area to minimize radiation and achieve a high-quality image.

© Springer Nature Switzerland AG 2019
J. M. R. S. Tavares and R. M. Natal Jorge (Eds.): VipIMAGE 2019, LNCVB 34, pp. 639–645, 2019.
https://doi.org/10.1007/978-3-030-32040-9_66

Regarding the voxel size in CBCT images, it is noted that this is isotropic, which means all sides have the same dimensions as a cube, which is an advantage over other equipment. It is therefore important to note that the larger the voxel size, the lower the image quality.

Nowadays, cone-beam computed tomography has been widely used in several areas within the field of dentistry, including periodontology, orthodontics and implantology, and has become an essential diagnostic tool. CBCT technology provides high quality diagnostic imaging and is currently the most powerful tool available for maxillofacial complex hard tissue evaluation. However, it has been described in the literature that CBCT is not suitable for evaluating soft tissues due to its low resolution of density and contrast, necessary in discriminating soft tissues such as muscle and salivary glands [6, 7]. Another disadvantage is that artifacts may exist due to amalgam restorations.

Thus, CBCT can be classified as a quantitative but not qualitative method. CBTC also has the advantage of using lower doses of radiation compared to conventional medical computerized axial tomography (CAT), so it has better risk-benefit ratio.

In addition to the above, this technique has the advantage that the measurements can be repeated many times without any discomfort for the patient and even without the patient's presence. Thus, it is left to the Dentist to choose the measurements from various angles and different cuts to better study the area in focus [8].

However, this technology has disadvantages such as being a relatively expensive method and subjecting the patient to radiation that, although low, always carries some degree of risk. The major concern in terms of medical radiology is stochastic effects, which are effects that may appear regardless of radiation and may lead to pathologies such as neoplasms. Although these effects are very rare, it is very important to make a good case selection and measure the risk-benefit as well as to make a good clinical history of the patient by asking for personal and family history [8].

Thus, proper treatment planning of a case with aesthetic implications requires careful assessment of the patient and factors such as gum thickness and underlying alveolar bone. Cone-beam computed tomography measurements were confirmed to be an accurate representation of them [9]. The objective of this research is to characterize the gingival phenotype by determining the dimensions of the dentogingival unit in the anterior maxilla with CBCT analysis.

2   Materials and Methods

The total study population consists of 50 elements, from which a preliminary random sample of 20 was collected. A CBCT of the anterior maxilla of each participant was done with Planmeca ProMax® 3D MID equipment with the same specifications/ conditions: 3D and HD capture and incidence field focused on the maxilla.

All participants used a "lip retractor" of the same genre to create a space between the soft tissues for their quantification. Two cotton rolls were placed between the arches to allow separation of the upper and lower teeth, so as to better isolate the area of interest, i.e. the anterior maxilla. All patients were also be positioned in the same way, with the chin supported and holding the lateral supports with the hands. The bipupillary plane and occlusal plane were positioned parallel to the floor using the "laser" orientation of the device.

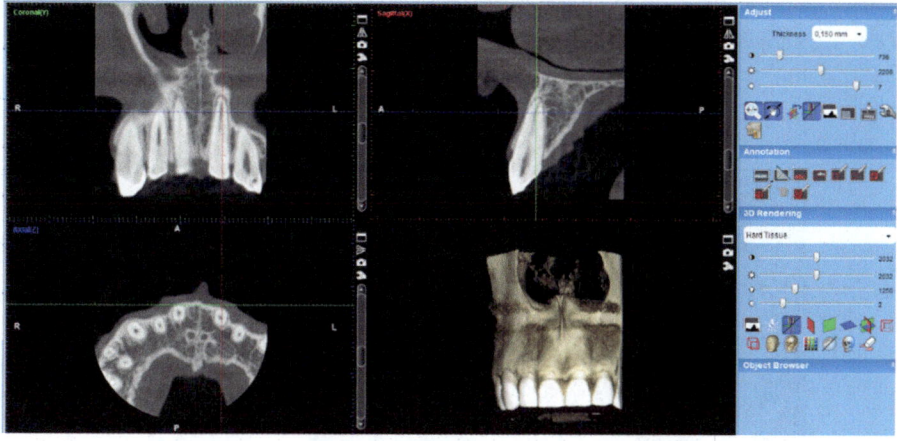

Fig. 1. Example of the type of image generated by the Planmeca software

The resulting images were collected by the Planmeca Romexis® software with dimensions 4.0 × 5.0 cm, voxel size of 150 µm, 90 kV, 8 mA, 15-second exposure and DAP (mGycm2) of 422 and the measurements were made using the software Planmeca Romexis Viewer version 5.1.0.R.

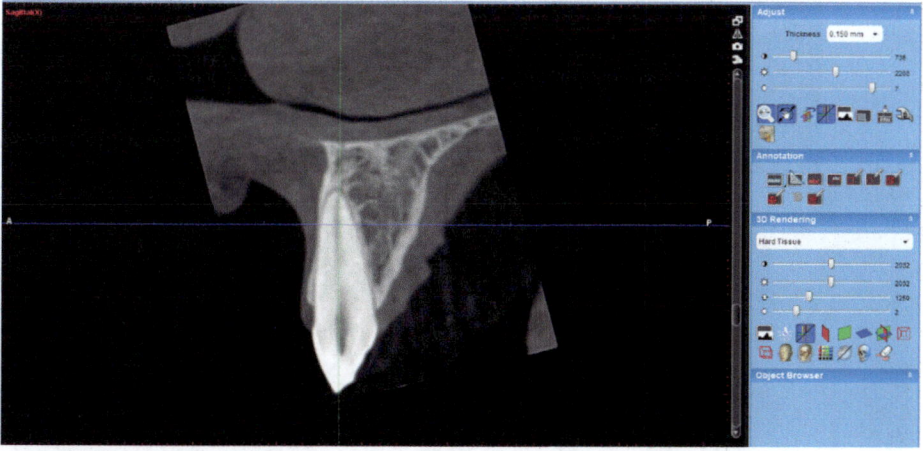

Fig. 2. Isolation of the focus area for the measurements

Before starting the measurements, the thickness of 0.150 mm and the same conditions of contrast and luminosity were selected. The *Show Annotation Overlay* option was selected for orientation, the sagittal cut was selected and the image was rotated until the long axis of the tooth overlapped the vertical orientation line. For all teeth, the most central sagittal cut of the teeth was sought. The entire process was repeated for teeth 21 and 11.

The characterization of the dentogingival unit was performed by measuring the following parameters: Thickness of Free Gingiva (TFG); Thickness of Attached Gingiva (TAG); Gingiva Thickness in Supracrestal Insertion (GTSI); Alveolar Bone Thickness (ABT); Distance from the Amelo-Cementum Line to the Alveolar Crest (DACAC);

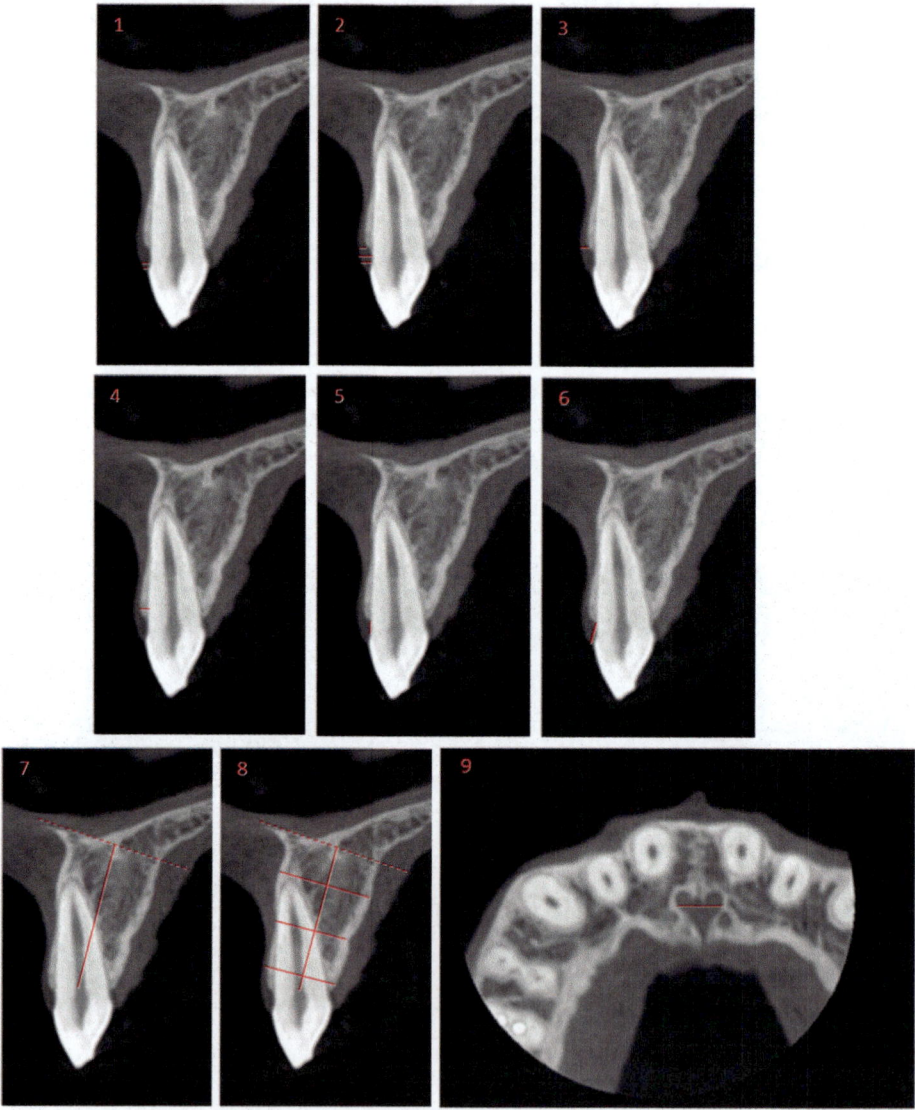

Fig. 3. Measurements of the study parameters: 1 - Thickness of Free Gingiva (TFG); 2 - Gingiva Thickness in Supracrestal Insertion (GTSI); 3 - Thickness of Attached Gingiva (TAG); 4 - Alveolar Bone Thickness (ABT); 5 - Distance from the Amelo-Cementum Line to the Alveolar Crest (DACAC); 6 - Distance from the Amelo-Cementum Line to the Free Gingival Margin (DACFGM); 7 - Alveolar Height (AH); 8 - Alveolar Width (AW); and 9 - Mesio-distal diameter of incisal foramen (DIF).

Distance from the Amelo-Cementum Line to the Free Gingival Margin (DACFGM); Alveolar Height (AH); Alveolar Width (AW); on the sagittal plane and Mesio-distal diameter of incisional foramen (DIF) on the coronal plane (Figs. 1, 2 and 3).

3   Results

The measurements collected on teeth 11 and 21 are presented on following Tables 1 and 2:

Table 1. Measurements of tooth 11

Measurement	Mean value	SD
Thickness of Free Gingiva	1,02 mm	0,16 mm
Gingiva Thickness in Supracrestal Insertion	1,36 mm	0,28 mm
Thickness of Attached Gingiva	0,86 mm	0,21 mm
Alveolar Bone Thickness	0,82 mm	0,18 mm
Distance Amelo-Cementum line to Alveolar Crest	1,57 mm	0,49 mm
Distance Amelo-Cementum line to Free Gingival Margin	2,81 mm	0,71 mm
Alveolar Height	18,96 mm	1,91 mm
Alveolar Width	8,86 mm	0,93 mm
Mesio-distal Diameter of Incisal Foramen	2,53 mm	0,74 mm

Table 2. Measurements of tooth 21

Measurement	Mean value	SD
Thickness of Free Gingiva	1,05 mm	0,25 mm
Gingiva Thickness in Supracrestal Insertion	1,32 mm	0,32 mm
Thickness of Attached Gingiva	0,90 mm	0,30 mm
Alveolar Bone Thickness	0,87 mm	0,30 mm
Distance Amelo-Cementum line to Alveolar Crest	1,67 mm	0,46 mm
Distance Amelo-Cementum line to Free Gingival Margin	2,80 mm	0,69 mm
Alveolar Height	19,12 mm	1,86 mm
Alveolar Width	8,81 mm	0,96 mm
Mesio-distal Diameter of Incisal Foramen	2,54 mm	0,74 mm

4   Discussion

The results correspond to the analysis of 20 cone-beam tomographies, corresponding to 20 participants of the present study. A total of 40 teeth, corresponding to teeth 11 and 21, were studied. The images were processed in the Planmeca Romexis® software and the mean values and standard deviation were calculated.

Due to the quality of the obtained images with the CBCT, it was very easy, feasible and accurate to measure all the parameters discussed in the previous sections.

In several clinical situations the knowledge of the dimensions of the masticatory mucosa is very relevant. A thin gingival biotype is prone to the development of recessions after some procedures like crown lengthening and orthodontic movement.

These preliminary results show that the value of the thickness of free gingiva are on average 1.02 mm for the tooth 11 and 1.05 mm for the tooth 21 and in general, the gingiva was found to be slightly thinner in females than men. The measurements of gingiva thickness in supracrestal insertion were of 1,36 on tooth 11 and 1,32 on 21, which is in accordance with Gargiulo et al. [1]. The thickness of attached gingiva was also determined, and it was between 0,86 mm and 0,90 mm, again the men showed thicker mucosa.

In regards to the distance of amelo-cementum line to alveolar crest and free gingival margin, the measurements were 1,57 mm and 2,81 mm and 1,67 mm and 2,80 mm to teeth 11 and 21 respectively, which makes sense since the participants were young and have a healthy periodontium.

The alveolar bone thickness was also evaluated and the mean values of this sample were 0,82 and 0,87 and interestingly, it was noted that the study participants who had orthodontic procedures had a thinner vestibular cortical, that is, the values measured were lower. This is probably related to the buccal movement of the teeth during the orthodontic alignment.

The alveolar height and width were 18,96 mm and 8,86 mm for tooth 11 and 19,12 mm and 8,81 mm for tooth 21. Currently, there a lack of information in regards to the alveolar dimensions and morphology in the maxillary region, a recent study by Zhang et al. [10], determined that these values ranged between 18.83–19.07 mm and 8.30–9.62 mm, for the selected population, which is in accordance with the results of the present study.

The mesio-distal diameter of incisal foramen was also measured and it is on average of 2,54 mm which is slightly lower than the mean value determined by Khojastepour et al. [11] that was of 3.17 mm; it's worth noting though that the mentioned study had a bigger population sample.

5   Conclusion

According to the evaluation of the preliminary sample, we concluded that the CBCT allows measurements with high precision, without bias of subjectivity or necessity of invasion. It is, therefore, a method with high efficacy and low intolerance. However, the evaluation of the total sample will give a more representative picture of the Portuguese population.

References

1. Gargiulo, A.W., Wentz, F.M., Orban, B.: Dimensions and relations of the dentogingival junction in humans. J. Periodontol. **32**, 12–35 (1961)
2. Ingber, J.S., Rose, L.F., Coslet, J.G.: The "biologic width"–a concept in periodontics and restorative dentistry. Alpha Omegan **70**(3), 62–65 (1977)
3. Pauwels, R., Araki, K., Siewerdsen, J.H., Thongvigitmanee, S.S.: Technical aspects of dental CBCT: state of the art. Dentomaxillofac Radiol. **44**(1), 20140224 (2015)
4. Acar, B., Kamburoğlu, K.: Use of cone beam computed tomography in periodontology. World J. Radiol. **6**(5), 139–147 (2014)
5. Galgali, S.R., Gontiya, G.: Evaluation of an innovative radiographic technique–parallel profile radiography–to determine the dimensions of dentogingival unit. Indian J. Dent. Res. **22**(2), 237–241 (2011)
6. Borges, G.J., Ruiz, L.F., De alencar, A.H., Porto, O.C., Estrela, C.: Cone-beam computed tomography as a diagnostic method for determination of gingival thickness and distance between gingival margin and bone crest. Sci. World J. **2015** (2015). Article ID 142108
7. Cao, J., Hu, W.J., Zhang, H., et al.: A novel technique for measurement of dentogingival tissue by cone beam computed tomography. Oral Surg. Oral Med. Oral Pathol. Oral Radiol. **119**(2), e82–e87 (2015)
8. Bhusari, B.M., Chelani, L.R., Suthar, N.J., Anjankar, J.P.: Gingival biotypes. J. Med. Dent. Sci. Res. **2**(11), 07–10 (2015)
9. Benavides, E., Rios, H.F., Ganz, S.D., et al.: Use of cone beam computed tomography in implant dentistry: the International Congress of Oral Implantologists consensus report. Implant Dent. **21**(2), 78–86 (2012)
10. Zhang, W., Skrypczak, A., Weltman, R.: Anterior maxilla alveolar ridge dimension and morphology measurement by cone beam computerized tomography (CBCT) for immediate implant treatment planning. BMC Oral Health **15**, 65 (2015)
11. Khojastepour, L., Haghnegahdar, A., Keshtkar, M.: Morphology and dimensions of nasopalatine canal: a radiographic analysis using cone beam computed tomography. J. Dent. (Shiraz). **18**(4), 244–250 (2017)

Parameterization of Reconstructed
Organ Models

A Semi-automatic Parameter Extraction Process for Modeling of Lumbar Vertebral Body Segments

Oğulcan Güldeniz[✉], Volkan Karadağ, and A. Fethi Okyar

Department of Mechanical Engineering, Faculty of Engineering,
Yeditepe University, Atasehir, Istanbul, Turkey
guldenizogulcan@gmail.com, guldenizogulcan@gmail.com

Abstract. For those studying the biomechanical response of the lumbar spine, anatomical meshes obtained from medical imaging data is quite important. However, such models are generally fixed and can only represent a single subject's geometry. The objective of this study was to improve our previous lumbar vertebral CAD model such that the parameters are now extracted from the CT scan using a semi-automatic procedure. To illustrate the procedure, first, the transverse cross-sections of vertebral bodies were obtained from an individual at three levels, superior, middle and inferior. Parametric contour curves were fitted onto the vertebral body boundaries using an optimization procedure and the fitting errors are reported here. Five lumbar vertebral bodies were then formed using the lofting operation. Hausdorff distance between the CAD and segmentation models was used to assess the accuracy of the resulting models. The means and standard deviations of Hausdorff distances are also reported here. The adopted optimization process was observed to be resulting coefficient of determination values as high as 0.978. Our new model is expected to lead to dramatic reductions in both time and effort required to build a patient-specific biomechanical model.

Keywords: Parametric model · Segmentation · Patient specific ·
Lumbar spine · Vertebral parameters · Spinal curve · Medical image processing

1 Introduction

For those studying the biomechanical response of the spine, processing medical imaging data into CAD (computer-aided design) models is a valuable tool. Segmentation processes are widely used to generate such models. Generally, the output of segmentation is fixed in nature and can only represent a single subject's geometry (Ayturk and Puttlitz 2011), while the process is time-consuming and tedious. Alternatively, feature-based parametric modeling yields CAD models driven by parameters, which can also be turned into a finite element analysis (FEA) mesh almost instantaneously. Moreover, feature-based modeling also provides model reshaping and resizing capabilities which allows for independent tuning of feature shapes and sizes (Laville et al. 2009). It is important to set up a scheme that will allow determination of CAD model parameters either by optimization or machine learning algorithms.

© Springer Nature Switzerland AG 2019
J. M. R. S. Tavares and R. M. Natal Jorge (Eds.): VipIMAGE 2019, LNCVB 34, pp. 649–658, 2019.
https://doi.org/10.1007/978-3-030-32040-9_67

Recently, a novel attempt towards the parameterization of the lumbar portion and its FEA analysis was presented in (Lavecchia et al. 2018). Their models captured the spinal parameters quite well but their design philosophy differed from ours. In (Okyar et al. 2019), we described a model using full parameterization and semi-automatic CT-based visual parameter extraction. It was expected to become a platform that can easily be adopted by other machine learning algorithms.

Here, we explain our following work on parametric lumbar vertebra modeling approach and introduce changes to it. Instead of visually inspecting four parameters we adopted a new conceptual contour design and utilized machine learning algorithms to determine a best fit with imaging data. The new model is called type-II while the previous one is coined as type-I. In the next section, the construction process using CT-scan data is outlined where only the vertebral bodies are considered. In Results, correlations based on the Hausdorff measure between types I and II as well as the segmentation model is assessed. Effects of the recent update is explained in Discussion along with further elaborations.

2   Materials and Methods

Here, the three steps to convert the CT data of lumbar vertebrae into the CAD model of the vertebral bodies (Fig. 1) will be described under subsections.

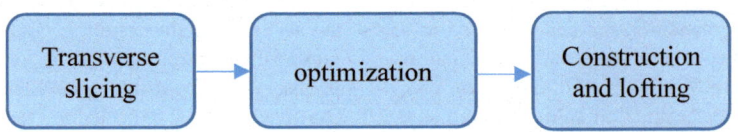

Fig. 1. The three step methodology to create a CAD model from CT data.

2.1   Transverse Slicing

The CT scans can be manipulated in several ways by powerful biomedical imaging software such as 3d-slicer (Fedorov et al. 2012). Slicing is an operation in which the section view along a plane with an arbitrary orientation is obtained. Our design approach requires the acquisition of section from transverse planes of each vertebral body at three levels, superior (S), middle (M), and inferior (I). During acquisition three reference points, called the reference triad, are marked in each section and recorded for later use.

Transverse sections from the vertebral CT scan set of L1–L5 have been obtained from a subject. Three sample sections (S), (M) and (I) were obtained from the L1 level as shown in Fig. 2. Image manipulations such as cropping, painting and thresholding are applied to the determine the section area centroid, x_μ, and its' principal axes (P.A.1 and P.A.2). Note that, P.A.1 coincides with the sagittal plane of symmetry and it is oriented at an angle θ_B from the horizontal (x) axis.

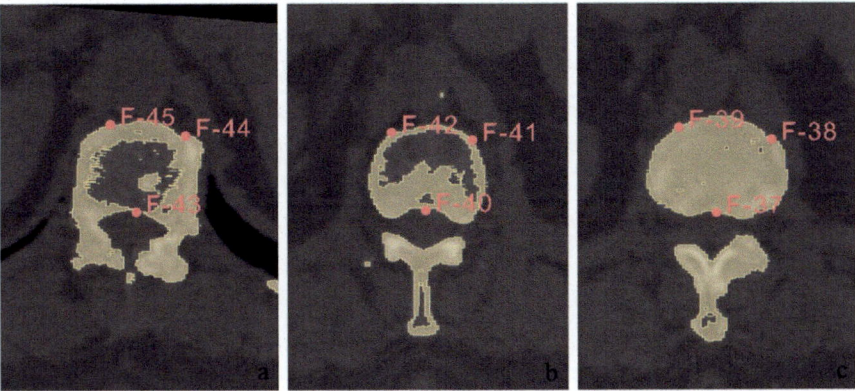

Fig. 2. Transverse slices are taken from each vertebra at three levels superior (a), middle (b), and inferior (c).

Finally, a secondary thresholding was held to determine the N pixels defining the section's external contour, Γ, called the scanned point set $\{X_i \in \Gamma : i = 1, 2, \ldots, N\}$.

2.2   Optimization

The transverse cross section of the vertebral body has a distinct oval shape that resembles a kidney. A group of 31 subjects' CT scans were visually inspected. Observations resulted with a conceptual contour design composed of four distinct tangent circles, two of which are laterally symmetric with respect to the sagittal plane (coincident with the first principal axis of the area, P.A.1), and the remaining anterior and posterior circles centered at x_A and x_P are coincident with P.A.1 (see Fig. 3). The left and right symmetric circles centered at x_L and x_R, respectively, were called lateral (L). The circles centered on the sagittal plane were called anterior (A) and posterior (P).

A total of 18 variables consisting of nine angles (θ_B, θ_{P1}, θ_{P2}, $\ldots$, θ_{L1}, θ_{L2}) and nine coordinates (corresponding to the four circle centers x_P, x_R, x_A, x_L and the vertical coordinate x_{B2}) were required to sketch the contour $\bar{\Gamma}$. However, this number reduced to a set Λ: (x_{P1}, r_P, r_R, θ_{P2}, θ_{R2}, θ_{A2}) of only six independently assigned parameters by taking advantage of existing geometric relations between the sketch entities (tangency, centroid, etc.). The sketch construction began from x_P moving counter clock-wise visiting the rest of the circle centers. Solutions to a linear and a non-linear algebraic set of equations were also used in the sketch construction process.

Evaluating the contour function at M discrete points yielded the point set $\{\bar{X}_j = \bar{X}(\Lambda, u_j) \in \bar{\Gamma}, u \subset [0, 1], j = (1, 2, \ldots, M)\}$. The radial distances from the centroid to the contour points $\bar{X}_j - x_\mu$ with respect to their angular position θ_j from the horizontal are shown in Fig. 4. Note that this mapping was found to be more advantageous in terms of measuring the mean square fit error, $E(X_i - \bar{X}_j, \Delta\theta)$. Here, $\Delta\theta$ represents the size of error calculation window. Error calculations were based on window averaging because of the angular mismatch between the point sets X_i and $\bar{X}_j$.

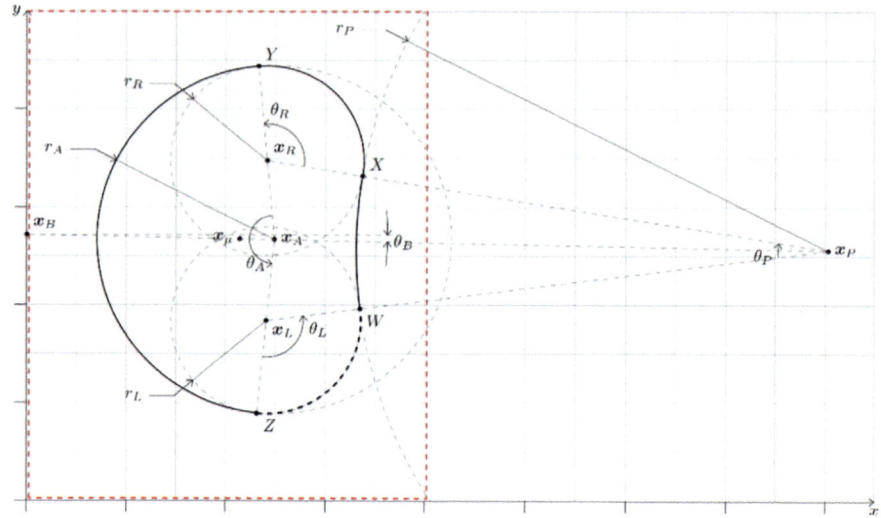

Fig. 3. Geometric construction of the vertebral body cross-section using four tangent circles.

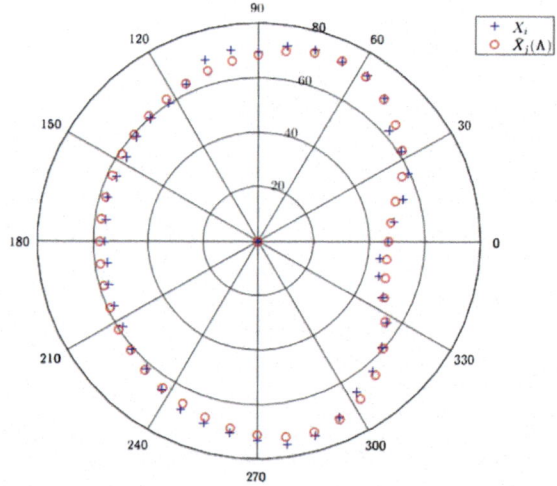

Fig. 4. Circumferential mapping of the evaluated point set $\bar{X}_j$ and the scanned point set $\bar{X}_i$, (Coefficient of determination, $R^2 = 0.978$)

A Matlab (*MATLAB* 2015) function to calculate the mean square fit error E corresponding to a given set of parameters Λ was written. This was supplied to an unconstrained optimization problem as the cost function to be minimized. The parameters Λ^* that yielded $\min_\Lambda E$ were searched for by the fminsearch function of Matlab. The illustration of the algorithm is shown in Fig. 5, whereas the resulting footprints of one such parameter search is shown in Fig. 6. In the latter, inferior section of the L1 vertebra of a subject was used.

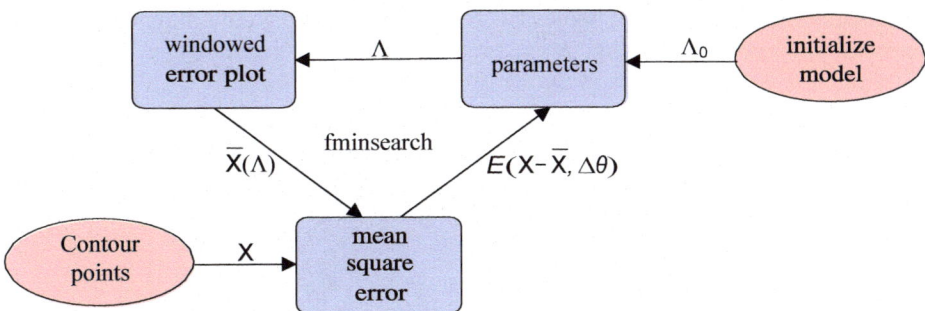

Fig. 5. The optimization loop to find the parameter set S that best fits the point set X_i

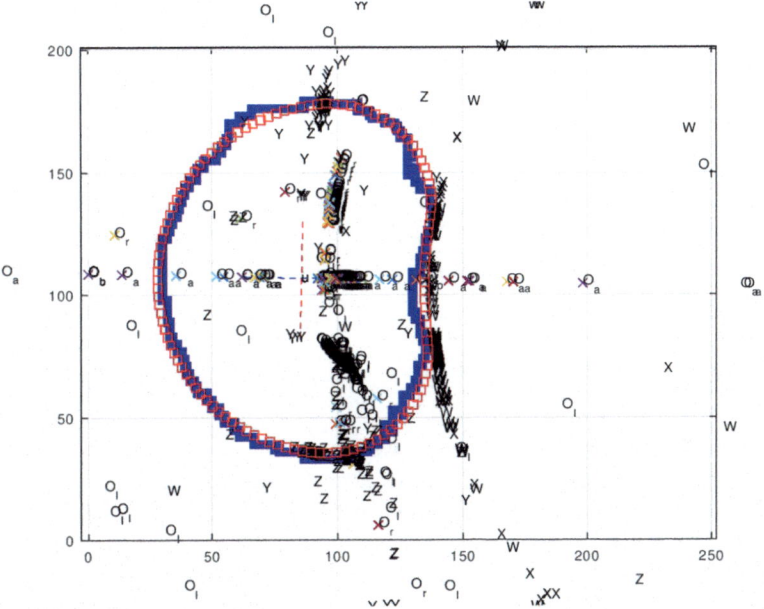

Fig. 6. The footprints marked during the fminsearch run. Note the cloud of trial points around the circle centers as well as the tangency points

2.3   Construction and Lofting

Upon completion of the parameter optimization in the above step the contour construction parameters Λ^* are recorded for the 15 distinct planes of the L1–L5 segment. Each plane's reference triad coordinates recorded in Sect. 2.1 are used to create their corresponding construction plane in the CAD system. The section contour is then constructed based on the sagittal axis P.A.1, the parameters Λ^*, and geometric relations between the sketch entities.

Lofting is an advanced CAD operation that combines cross-sectional sketches into a single swept volume. There are three section constructions (S), (M) and (I) per vertebral body. Five vertebral bodies L1–L5 are formed using the lofting operation. L2 vertebral body is shown in Fig. 7.

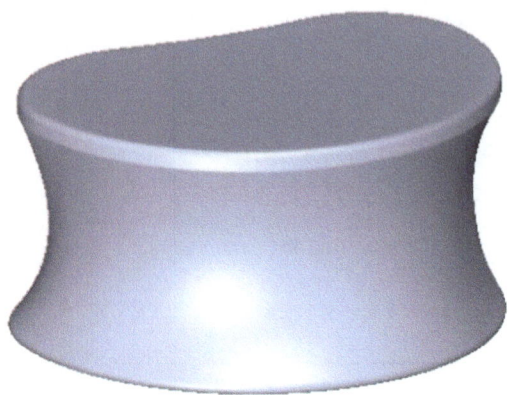

Fig. 7. L1 vertebra of a subject obtained using the procedure described

Comparison of model types I and II with original segmentation model using Hausdorff distance measurements taken from Meshlab (Cignoni et al. 2008) and in house calculations from Matlab are presented in Results. Recorded improvements will be given in the Discussion.

3   Results

An unconstrained optimization search was conducted via fminsearch function of Matlab. 15 pairs of parametric contours were matched with respective edges from CT data. The converged parameters are shown partly in Tables 1 and 2.

A pair of linear and non-linear systems of equations resulting from the problem of geometric closing of the contour did not pose a convergence difficulty in the optimization.

For the demonstration and validation of our new methodology, 49 years old healthy male individual NA's one CT scan was used. Reference slices for superior endplates, inferior endplates and mid planes were determined and recorded from the CT scans. Afterwards, three reference points for each reference slice were also obtained from reference slices and recorded.

In order to calculate Hausdorff distances between the CAD models and the STL model obtained from the segmentation process, both were imported into Matlab. The closest distances between the vertices of the CAD model and the STL surface were recorded and then imported into Gnuplot (Williams and Kelley 2013) for further processing. The kernel density function of Gnuplot was used to depict the distribution of Hausdorff distance shown in Table 3.

Color contours of Hausdorff distances between the CAD mesh and the STL model from the segmentation were obtained from Meshlab and shown in Fig. 8. Statistical

Table 1. Model type-II's construction tangent circle's point data (relative to principal axes), and principal axis transformation angle data for superior (S), medium (M), and inferior (I) vertebral body cross-sections. All units in mm.

		$x_{\mu 1}$	$x_{\mu 2}$	θ_B	x_{A1}	x_{B1}	x_{R1}	x_{R2}	x_{L1}	x_{L2}
L1	S	90.67	105.89	0.12	3.91	77.67	2.29	9.96	5.43	−12.7
	M	89.84	106.2	0.05	3.86	76.23	1.99	7.63	1.92	−7.55
	I	85.7	106.84	0.02	4.62	78.98	3.42	10.82	3.73	−11.11
L2	S	77.15	102.45	−0.11	3.57	82.2	4.86	10.25	1.05	−6.47
	M	81.16	103.94	0.07	2.76	78.79	1.98	7.12	2.78	−7.88
	I	78.25	105.27	−0.01	5.57	81.69	4.45	12.48	2.96	−10.97
L3	S	70.76	103.29	−0.1	2.63	79.32	4.25	8.99	1.71	−6.77
	M	69.83	103.94	−0.04	3.34	80.24	3.15	9.48	4.06	−10.26
	I	66.88	107.42	−0.05	5.84	83.07	4.91	13.38	1.39	−9.48
L4	S	63.09	103.72	0.03	10.22	95.55	6.96	13.09	6.44	−12.5
	M	59.11	106.09	0.06	4.39	81.78	2.26	8.64	3.48	−9.92
	I	58.66	107.11	0.04	9	84.14	5.99	14.13	5.95	−14.09
L5	S	53.66	105.81	0.05	11.75	91.27	10.25	11.53	7.5	−8.26
	M	52.58	109.57	0.06	12.76	86.54	9.36	11.04	9.08	−10.71
	I	55.65	105.46	−0.06	11.1	87.53	11.56	15.7	9.73	−14.14
Mean		70.2	105.53	0.01	6.35	83	5.18	10.92	4.48	−10.19
Standard deviation		12.63	1.79	0.06	3.45	5.12	2.99	2.37	2.67	2.4

Table 2. Model type-II's construction tangent circle's radius data, and their mean and standard deviation values for superior (S), medium (M), and inferior (I) vertebral body cross-sections. All units in mm.

		Radii (mm)			
		r_A	r_P	r_R	r_L
L1	I	23.62	62.51	13.52	10.83
	M	20.89	61.60	13.03	13.09
	S	23.91	63.30	13.03	12.76
L2	I	23.51	64.84	13.18	16.57
	M	20.69	63.61	13.52	12.80
	S	25.53	65.24	13.00	14.25
L3	I	22.65	62.08	13.52	15.82
	M	21.97	65.18	12.49	11.68
	S	26.49	66.22	13.08	16.01
L4	I	27.44	75.60	13.95	14.38
	M	22.56	66.32	13.66	12.60
	S	27.35	66.51	12.90	12.93
L5	I	25.40	68.06	13.77	16.11
	M	24.29	65.22	12.74	12.97
	S	29.11	64.18	13.40	14.90
Mean		24.36	65.36	13.25	13.85
Standard deviation		2.41	3.24	0.39	1.68

Table 3. Comparison between the CT and CAD mesh files for types I and II, and Kernel densities of Hausdorff distances between types I and II and the segmentation model, respectively.

		Type-I	Type-II	Hausdorff distance - kernel density
Number of vertices	CT mesh	37892	37892	
	CAD mesh	24902	25936	
Mean distance	Normal	1.49	1.33*	
	Meshlab	1.28	1.04*	
	Lognormal	0.18	0.08*	
Standard deviation	Normal	1.16	0.99	
	Meshlab	1.76	1.41*	
	Lognormal	0.65	0.63*	
%95 limit	Lognormal	0.19	0.08*	
	Histogram	0.18	0.17*	
Max distance	Matlab	8.36	8.02	
	Meshlab	8.32	8.15	

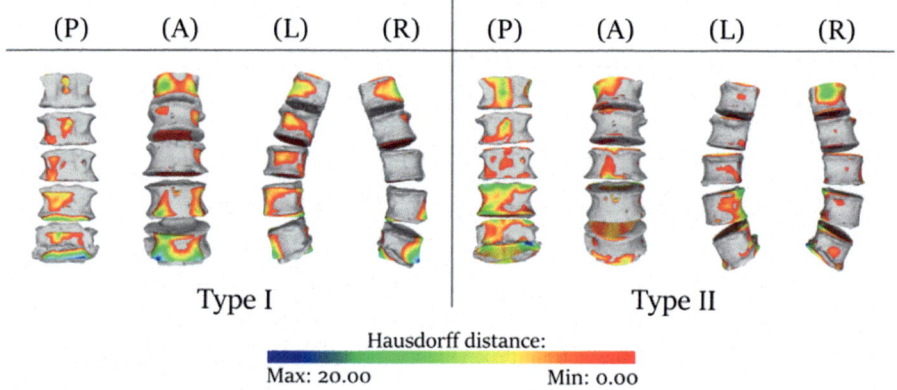

Fig. 8. Sequential images obtained from the Meshlab software depicting the posterior, anterior, left and right views of the vertebral bodies.

data related with meshes, number or vertices, mean and standard deviation of Hausdorff distances (according to lognormal and normal distributions), and maximum Hausdorff distances are also given in Table 3.

4 Discussion

The patient-specific measurement of vertebral body model parameters requires processing of the subjects' CT scans. Manual processing of the CT scans for model parameter extraction is a time consuming and tedious task. Alternatively, optimization of model parameters that yield the lowest error between the contours obtained from CT sections and the model is a promising method in order to automate the parameter extraction process.

Type-I was based on classical morphometrical dimensions which are generally implicit functions of the geometric entities used in the construction sketches. Thus, it was not convenient in measuring the error between the actual data and the model. Instead, the parameters directly related with the construction geometry were used in type-II, observed to be much more convenient to measure the error of a given fit. The parameters that yielded the minimum fit error were solved for by a derivative-free unconstrained multivariable optimization routine.

The CAD model of the lumbar vertebral body is based on three transverse cross-sectional sketches obtained from the superior, middle and inferior levels. The cross-sectional geometric constructions are designed to capture the actual cross-sectional contours obtained from the CT scans. Although, increasing the complexity of the construction sketch potentially allows it to better fit onto the contour, this would also lead to an increase in the dimension of the parameter space. Obviously an increase in the parameter space in an optimization study is a challenging task, making it harder to converge to a global minimum. Striking just the necessary number of free parameters that yield the required accuracy without compromising the convergence behavior is thus of crucial importance.

Depending on the comparison data presented on Table 3, and the coefficient of determination value of our optimization process, the design was found to be robust and accurate in representing the geometry of vertebral bodies.

In conclusion, a semi-automatic procedure to obtain a patient-specific vertebral body model with dramatic reductions in build time and effort has successfully been implemented.

References

Ayturk, U.M., Puttlitz, C.M.: Parametric convergence sensitivity and validation of a finite element model of the human lumbar spine. Comput. Methods Biomech. Biomed. Eng. **14**(8), 695–705 (2011). https://doi.org/10.1080/10255842.2010.493517

Cignoni, P., Callieri, M., Corsini, M., Dellepiane, M., Ganovelli, F., Ranzuglia, G.: MeshLab: an open-source mesh processing tool. In: Eurographics Italian Chapter Conference (2008). https://s3.amazonaws.com/academia.edu.documents/41716786/MeshLab_an_Open-Source_Mesh_Processing_T20160129-30045-1kk7i64.pdf?AWSAccessKeyId=AKIAIWOWYYGZ2Y53UL3A&Expires=1553071509&Signature=wdSMd1uw70o8rnKVEInZRqyZAF4%3D&response-content-disposition=inline

Fedorov, A., Beichel, R., Kalpathy-Cramer, J., Finet, J., Fillion-Robin, J.-C., Pujol, S., Bauer, C., Jennings, D., Fennessy, F., Sonka, M., Buatti, J., Aylward, S., Miller, J.V., Pieper, S., Kikinis, R.: 3D slicer as an image computing platform for the quantitative imaging network. Magn. Reson. Imaging **30**, 1323–1341 (2012). https://doi.org/10.1016/j.mri.2012.05.001

Lavecchia, C.E., Espino, D.M., Moerman, K.M., Tse, K.M., Robinson, D., Lee, P.V.S., Shepherd, D.E.T.: Lumbar model generator: a tool for the automated generation of a parametric scalable model of the lumbar spine. J. R. Soc. Interface **15**(138), 20170829 (2018). https://doi.org/10.1098/rsif.2017.0829

Laville, A., Laporte, S., Skalli, W.: Parametric and subject-specific finite element modelling of the lower cervical spine. Influence of geometrical parameters on the motion patterns. J. Biomech. **42**(10), 1409–1415 (2009). https://doi.org/10.1016/j.jbiomech.2009.04.007

MATLAB. MathWorks Inc., Natick (2015)

Okyar, F., Guldeniz, O., Atalay, B.: A holistic parametric design attempt towards geometric modeling of the lumbar spine. Comput. Methods Biomech. Biomed. Eng. Imaging Vis. 1–11 (2019). https://doi.org/10.1080/21681163.2019.1574606

Williams, T., Kelley, C.: Gnuplot: an interactive plotting program (2013). gnuplot.sourceforge. net

Computer Simulations and Visualization Applied to Tissue Engineering

Numerical Characterization of a Hyperelastic Material to Shear Stress

Andrews V. Souza[1], João E. Ribeiro[2(✉)], and Fernando Araújo[3]

[1] Instituto Politécnico de Bragança, Bragança, Portugal
andrews.va.souza@alunos.ipb.pt
[2] CIMO, Instituto Politécnico de Bragança, Bragança, Portugal
jribeiro@ipb.pt
[3] Federal Center of Technological Education of Rio de Janeiro, CEFET/RJ
Campus Angra dos Reis, Rio de Janeiro, Brazil
Fernando.araujo@cefet-rj.br

Abstract. In last years, some studies have been proved that there is an association between the wall shear stress with intracranial aneurysm rupture, however, is very difficult to understand the mechanical tissue behaviour when subjected to shear stresses. In this work, it is implemented numerical simulations to characterise the polydimethylsiloxane (PDMS) material when it is subjected to a shear solicitation. For this, it was initially necessary to perform some experimental tests to characterize the mechanical behaviour of the material. Based on these results, several numerical simulations were performed with the most common constitutive models in the simulation of hyperelastic materials by varying numerical factors and parameters of the numerical models.

Keywords: Numerical simulation · Shear stress · Hyperelastic material · PDMS · Ansys®

1 Introduction

Some biological tissues, like soft tissues, are composed by several layers with different compositions and they are known to support large reversible deformations, also called hyperelastic behaviour. It is considered that exist four typical soft tissues: muscular tissue, neuronal tissue, epithelial tissue and neuronal tissue [1]. The mechanical behaviour can be described by hyperelastic constitutive equations or models. The hyperelastic constitutive models can be anisotropic or isotropic and it is, generally, expressed in terms of strain components or strain invariants [2]. In last decades have been developed materials with hyperelastic behaviour applied in biomedicine [3] some of them are used as prosthesis [4] other are used as *in vitro* models to study and analyse some pathologies [5]. So, one the most popular hyperelastic materials used in the biomedical industry is the elastomer polydimethylsiloxane (PDMS). The most important reasons of the PDMS popularity is related with its biocompatibility and biomechanical behaviour, similar to biological tissues, with applications in the study of aneurysm behaviour [6] and devices such as: micro pumps, optical systems, microfluidic devices [7], among others.

© Springer Nature Switzerland AG 2019
J. M. R. S. Tavares and R. M. Natal Jorge (Eds.): VipIMAGE 2019, LNCVB 34, pp. 661–670, 2019.
https://doi.org/10.1007/978-3-030-32040-9_68

PDMS belongs to the group of siloxanes, however, in its advent, it was called silicocetones or silicones, but since as there was no double bond of Si = S, its name was later replaced by a specific nomenclature and its basic unit has become known as siloxanes. The most known material of this group is PDMS, a synthetic polymer whose main axis is made from the repetition of silicon and oxygen bonds and methyl groups [8]. PDMS is a material that has good microstructural characteristics, good manufacturing ability and a low cost. In addition, PDMS is thermally stable, optically transparent [9], works as a thermal and electrical insulation [10], has good chemical stability and degrades quickly in the natural environment when compared to other polymers, and it presents no environmental problem. However, the main disadvantage from the biomedical point of view is the difficulty of wetting its surface with aqueous solvents.

Many researchers have been observed that there is an association between the wall shear stress with intracranial aneurysm rupture [11]. Hemodynamics plays a central role throughout intracranial aneurysm natural history, and shear stress has emerged as an important determinant of arterial physiological characteristics [12]. However, the analysis of wall shear stress *in vivo* is very difficult, being *in vitro* solution a valid and interesting approach. In this sense, the use of PDMS to create an aneurysm model have been developed in last years by the scientific community [13, 14]. For this reason, it is very important to understand the mechanical behaviour of this material when subjected to shear stresses. So, in the present work we have carried out numerical simulations to analyse the shear stress field in the PDMS. The numerical analysis was based on a finite element method, a computational technique that, due to the development of robust and optimized algorithms, allows simulations with high accuracy and precision results. The numerical method used in the present study, allows to test the characteristics of previously known hyperelastic materials by using mathematical models suitable for these kind of materials.

2   Numerical Simulation

Hyperelastic models have been widely used to model the nonlinear and anisotropic behaviour of materials, since these under large deformations often recover their elasticity. The constitutive behaviour of hyperelastic materials is defined in terms of energy potential. Among all constitutive models for hyperelastic materials, in this work was used only the most common to simulate the mechanical behaviour of PDMS, which are the Mooney-Rivlin, Ogden and Yeoh [15, 16]. However, its formulations need constants and coefficients that can be determined by fitting a suitable experimental stress-strain curve. This curve was obtained from tensile test of a PDMS specimen. In Fig. 1 is possible to observe the stress-strain curve implemented with a specimen of Sylgard 184, which the geometry and dimensions were in agreement with the BS 2782 standard.

One the most commons tests to analyse and evaluate the shear stress is implemented by using a single lap joint. In this case, the single lap joint was used to transfer the loading from the substrate to the PDMS. In this joint, the material of adherent was the 6061 aluminium alloy and for the adhesive was the PDMS. The geometry and

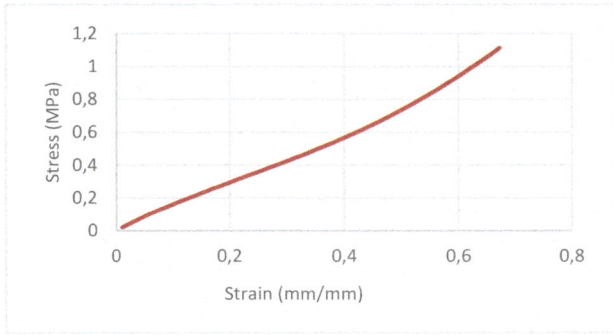

Fig. 1. Stress-strain curve of PDMS (Sylgard 184) specimen.

dimensions of the joint have been chosen and determined based on ASTM D1002-10 standard and can be seen in Fig. 2.

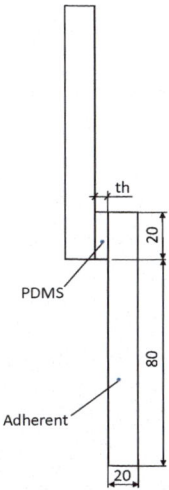

Fig. 2. Dimensions and geometry of the single lap joint.

The most important goals of this study are to analyse the influence of PDMS thickness and the applied displacement in the PDMS shear stress distribution. For this reason, was implemented nine different numerical simulations in agreement with the Table 1.

The numerical simulation was implemented using a commercial finite element method (FEM) software ANSYS®.

To perform the numerical simulation, it was necessary to create a model with a geometry similar to that of the specimen and boundary conditions matching the experimental testing and to discretize the domain finite element mesh. For the material properties, a nonlinear hyperelastic behaviour, based on the constitutive models of

Table 1. Simulation variables.

Simulation	PDMS thickness [mm]	Applied displacement [mm]
1	2	2
2	2	3
3	2	4
4	4	2
5	4	3
6	4	4
7	6	2
8	6	3
9	6	4

Mooney-Rivlin, Yeoh and Ogden, was considered. The application of these models required the determination of several constants, which were identified from the experimental curve of the tensile test. Nevertheless, the adherents were considered isotropic with a linear elastic behaviour which the mechanical properties are: E = 70 GPa and a Poisson's ratio of 0.3. The common nodes (adherent and PDMS) are the same, so, the properties are the average of both materials.

A bi-dimensional finite-element mesh, with 2790 parametric structural solid elements (PLANE183) was used and is shown in Fig. 3. In relation to the boundary conditions of the numerical model, a uniform displacement was applied to the upper lips, stretching the single lap joint sample (Fig. 3). The simulations were carried out for different values of displacement, according to Table 1.

3 Results

The numerical simulations implemented in this work allow to analyse the shear stress variation on the PDMS for different conditions (Table 1) and constitutive models. In Fig. 4 is represented, as an example, the results for the simulation number nine and Ogden constitutive model.

To compare the evolution of shear stresses for the different simulations it is more suitable and intuitive to define a path on the centre of the PDMS plate. In Fig. 5 is presented the chosen path to compare the obtained results. This region was chosen because is expected that the shear stress reaches the highest values.

Despite all the 27 simulations had been implemented, the authors will analyse in this paper the simulations 1, 3, 4, 6, 7 and 9 with three constitutive models. These chosen simulations constitute the situations with the highest amplitude between the variables, for example, in the simulations 1 and 3, the PDMS thickness is the same (2 mm), but the applied displacement are the minimum (2 mm) and the maximum (4 mm), the same principle has been used for the other analysed simulations.

In Figs. 6 and 7, is possible to observe the shear stress variation along to the chosen path, defined in Fig. 5 by the d direction for the three constitutive models.

Fig. 3. Finite element mesh and boundary conditions.

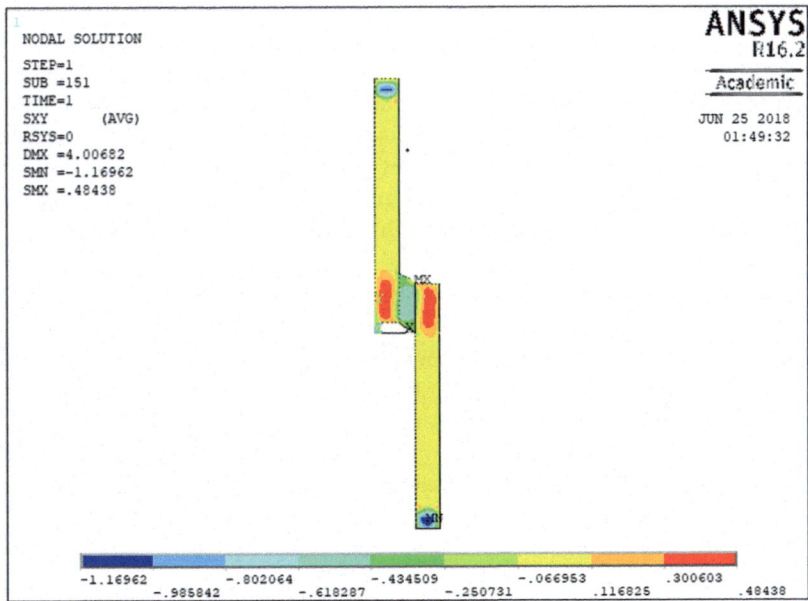

Fig. 4. Shear stress map obtained numerically for the condition 9: 6 mm of PDMS, 4 mm of applied displacement; and the Ogden constitutive model.

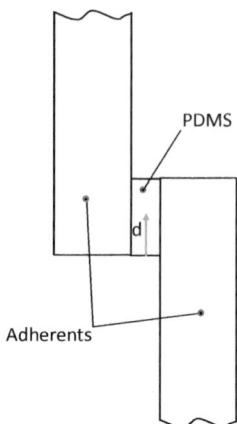

Fig. 5. The chosen path or distance (d) to analyse the evolution of shear stress on the centre of PDMD plate.

Observing both Figs. 6 and 7 it is verified that the constitutive models have a strong influence in the obtained results. There isn't any case in which the shear stress is the same for different constitutive models. This difference is higher when the applied displacement is higher, particularly, when is associated with the lower values of thickness. In Fig. 7(a) it is verified that the value of shear stress for the Yeoh is, approximately, the double of the results obtained by the simulation implemented with Ogden, i.e., −4.5 MPa and −9 MPa, respectively. A similar event happens in the Fig. 7 (b). Is, also, observed in Fig. 6, that the values of shear stress obtained with Yeoh simulation are always lower than those obtained by the constitutive model of Mooney-Rivlen and the Ogden is between the two. However, in Fig. 7, it does not happen the same rule as in the previous case, so, in Fig. 7(a) the maximum values of shear stress happed with the Yeoh and the minimum with the Ogden simulation, but, for the Fig. 7 (b) the maximum value occurs for Mooney-Rivlen constitutive model and the minimum values comes up for the simulation implemented with the Yeoh model, although, in Fig. 7(c) is possible to observe that behaviour is also different from the Fig. 7(a).

Analysing the influence of PDMS thickness and applied displacement in the shear stress, is possible to verify that for the same displacement the maximum shear stress decreases when the thickness increases. In Fig. 6, considering the Mooney-Rivlen numerical simulation, the shear stress goes from −1.2 MPa, for 2 mm of PDMD thickness, to −0.24 MPa when the thickness rises to 6 mm. The same phenomenon is verified in Fig. 7, the maximum shear stress decrease, proximally, 15 times when the PDMS thickness grow 3 times.

Naturally, with the increase of applied displacement the shear stress also grows. Comparing the values show in Figs. 6(a) and 7(a), where the PDMS thickness is same (2 mm) and the displacement is 2 mm and 6 mm, respectively, the maximum shear stress increases from −1.2 MPa to −9 MPa. This behaviour is repeated for the other values of thickness.

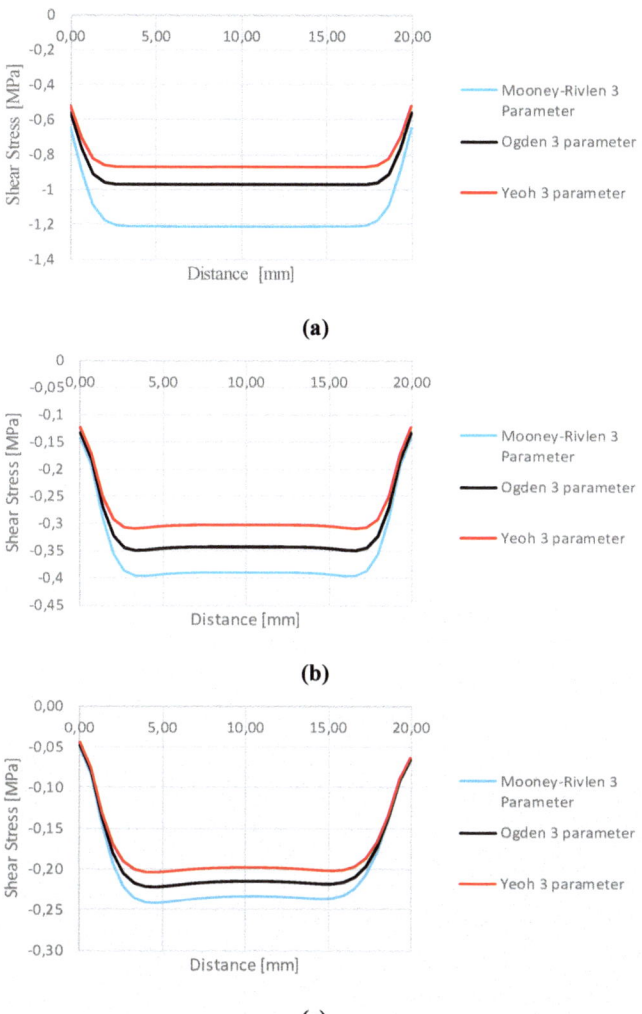

Fig. 6. Shear stress variation on PDMS along *d* direction (distance) for simulations: (a) 1, (b) 4 and (c) 7.

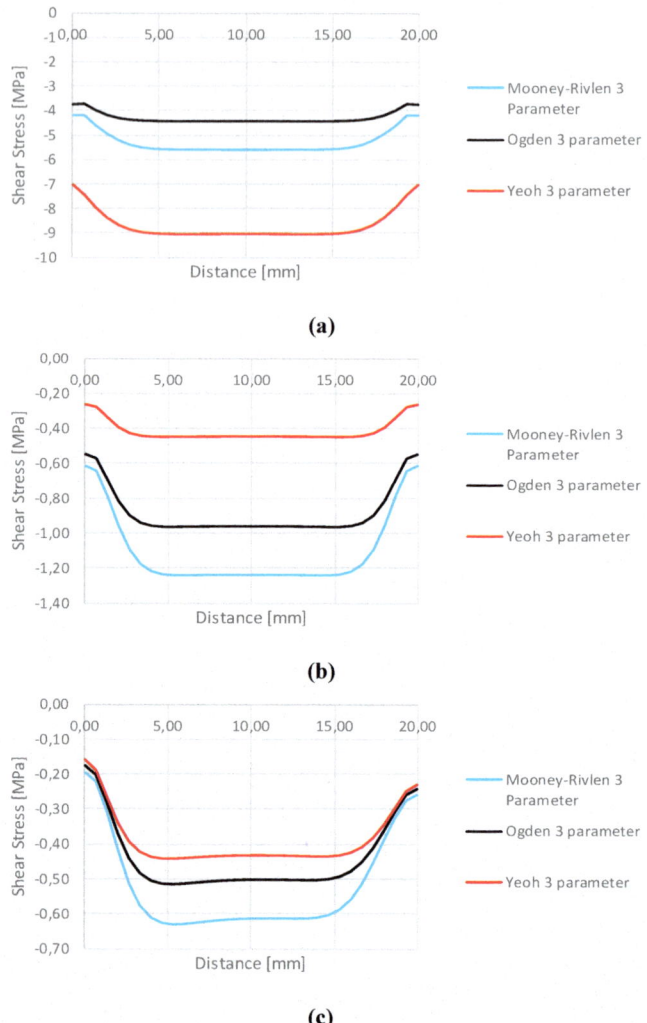

Fig. 7. Shear stress variation on PDMS along d direction (distance) for simulations: (a) 3, (b) 6 and (c) 9.

4   Conclusions

The numerical simulations presented in this work were implemented with the commercial element finite code ANSYS®. The main goals of this work were achieved, i.e., the simulations allow to analyse the influence of thickness and applied displacement in the shear stress for hyperelastic material (PDMS). To implemented these numerical simulations were used three constitutive models: Mooney-Rivlen, Ogden and Yeoh.

For each constitutive model was obtained a different result of shear stress. In the majority of simulations results the constitute model of Mooney-Rivlen is the most conservative, as such the highest values of shear stresses were obtained with it and the maximum values of shear stresses are among −0.25 MPa and −1.25 MPa, depending of simulation variables. On the other hand, the last conservative is the Yeoh, the numerical simulations with this constitutive models result in smallest values of shear stress, from −0.2 MPa to − 0.9 MPa. The only exception to this rule happened with the simulation 3, where the maximum value of −9 MPa was obtained with Yeoh model and the minimum (−4.2 MPa) was computed with the constitutive model of Ogden.

The influence of the PDMS thickness and applied displacement in the shear stress, was observed that for the same displacement, the maximum shear stress decreases when the thickness increases. However, for the same PDMS thickness when the applied displacement increases the shear stress also raises.

Acknowledgments. The authors gratefully acknowledge the funding by PORTUGAL2020: SI I&DT Individual – Call 16/SI/2015 (Proj. nº 9703): "Automatização de Processos de Soldadura de Estruturas Hiperestáticas em Ligas de Alumínio (APSEHAL)".

References

1. Marieb, E.G., Hoehn, K.: Human Anatomy & Physiology. Pearson Education, Upper Saddle River (2010)
2. Boyce, M.C., Arruda, E.M.: Constitutive models of rubber elasticity: a review. Rubber Chem. Technol. **73**, 504–523 (2000)
3. Kanyanta, V., Ivankovi, A.: Mechanical characterization of polyurethane elastomer for biomedical applications. J. Mech. Behav. Biomed. Mater. **3**(1), 51–62 (2010)
4. Garcia-Gonzalez, D., Garzon-Hernandez, S., Arias, A.: A new constitutive model for polymeric matrices: application to biomedical materials. Compos. B Eng. **139**, 117–129 (2018)
5. Rodrigues, R.O., Pinho, D., Bento, D., Lima, R., Ribeiro, J.: Wall expansion assessment of an intracranial aneurysm model by a 3D digital image correlation system. Measurement **88**, 262–270 (2016)
6. Pinho, D., Bento, D., Ribeiro, J., Lima, R., Vaz, M.: An in vitro experimental evaluation of the displacement field in an intracranial aneurysm model. In: Flores, P., Viadero, F. (eds.) New Trends in Mechanism and Machine Science, Mechanisms and Machine Science, vol. 24, pp. 261–268. Springer, Heidelberg (2015)
7. Faustino, V., Catarino, S.O., Lima, R., Minas, G.: Biomedical microfluidic devices by using low-cost fabrication techniques: a review. J. Biomech. **49**(11), 2280–2292 (2016)
8. Kuncová-Kallio, J., Kallio, P.J.: PDMS and its suitability for analytical microfluidic devices. In: Proceedings of Annual International Conference on IEEE Engineering in Medicine and Biology, pp. 2486–2489 (2006)
9. Martin, S., Bhushan, B.: Transparent, wear-resistant, superhydrophobic and superoleophobic poly(dimethylsiloxane) (PDMS) surfaces. J. Coll. Inter. Sci. **488**, 118–126 (2017)
10. Cherney, E.A.: Silicone rubber dielectrics modified by inorganic fillers for outdoor high voltage insulation applications. IEEE Trans. Dielectr. Electr. Insul. **12**, 1108–1115 (2005)
11. Zhou, G., Zhu, Y., Yin, Y., Su, M., Li, M.: Association of wall shear stress with intracranial aneurysm rupture: systematic review and meta-analysis. Sci. Rep. **7**(1), 5331 (2017)

12. Xiang, J., Tutino, V.M., Snyder, K.V., Meng, H.: CFD: computational fluid dynamics or confounding factor dissemination? The role of hemodynamics in intracranial aneurysm rupture risk assessment. AJNR Am. J. Neuroradiol. **35**, 1849–1857 (2014)
13. Sugiu, K., Martin, J., Jean, B., Gailloud, P., Mandai, S., Rüfenacht, D.: Artificial cerebral aneurysm model for medical testing, training, and research. Neurol. Med. Chir. **43**, 69–72 (2003)
14. Cardoso, C., Fernandes, C., Lima, R., Ribeiro, J.: Biomechanical analysis of PDMS channels using different hyperelastic numerical constitutive models. Mech. Res. Commun. **90**, 26–33 (2018)
15. Nunes, L.: Mechanical characterization of hyperelastic polydimethylsiloxane by simple shear test. Mater. Sci. Eng. **528**, 1799–1804 (2011)
16. Ribeiro, J., Lopes, H., Martins, P., César, M.B.: Mechanical analysis of PDMS material using biaxial test. AIMS Mater. Sci. **6**(1), 97–110 (2019)

Manufacturing Process of a Brain Aneurysm Biomodel in PDMS Using Rapid Prototyping

Andrews V. Souza[1], João E. Ribeiro[2], and Rui Lima[3,4(✉)]

[1] Polytechnic Institute of Bragança, Bragança, Portugal
andrews.va.souza@alunos.ipb.pt
[2] CIMO, Polytechnic Institute of Bragança, Bragança, Portugal
jribeiro@ipb.pt
[3] MEtRiCS, Mechanical Engineering Department, University of Minho,
Guimaraes, Portugal
rl@dem.uminho.pt
[4] CEFT, Faculdade de Engenharia da Universidade do Porto (FEUP),
Porto, Portugal

Abstract. Cerebral aneurysm is an abnormal dilatation of the blood vessel into a saccular form. They can originate in congenital defects, weakening of the arterial wall with increasing age, atherosclerotic changes, trauma and infectious emboli.

The in vivo experiments are an effective way of investigating the appearance, validating new practices and techniques, but beyond ethical issues, these types of experiments are expensive and have low reproducibility. Thus, to better understand the pathophysiological and geometric aspects of an aneurysm, it is important to fabricate in vitro models capable of improving existing endovascular treatments, developing and validating theoretical and computational models. Another difficulty is in the preoperative period of the non-ruptured cerebral aneurysm, known for the success of the skilled acts because there is an anatomical structure of the aneurysm as its current position. Although there are technologies that facilitate three-dimensional video visualization in the case of aneurysms with complex geometries the operative planning is still complicated, so the development of the real three-dimensional physical model becomes advantageous. In this work, the entire process of manufacturing an aneurysm biomodel using polydimethylsiloxane (PDMS) is disassembled by rapid prototyping technology. The manufactured biomodels are able to perform different hemodynamic studies, validate theoretical data, numerical simulations and assist in the preoperative planning.

Keywords: PDMS biomodels · Cerebral aneurysm · In vitro models

1 Introduction

Cerebral aneurysm or intracranial aneurysm is the abnormal dilatation of the blood vessel in a saccular form [1]. It may originate in congenital defects, weakening of the arterial wall with increasing age [2], atherosclerotic changes, trauma and infectious emboli. Other factors may be related to the onset of the aneurysm, such as

© Springer Nature Switzerland AG 2019
J. M. R. S. Tavares and R. M. Natal Jorge (Eds.): VipIMAGE 2019, LNCVB 34, pp. 671–676, 2019.
https://doi.org/10.1007/978-3-030-32040-9_69

hypertension, smoking, and excessive alcohol use. It is estimated that intracranial aneurysm occurs in 2.6% of the population and the cause of 85% of subarachnoid hemorrhage is the rupture of a saccular intracranial aneurysm [3].

In vivo experimental methods are an effective way to validate new practices and techniques for diseases such as the aneurysm, but in addition to ethical issues, these kinds of experiments are expensive and have low reproductively [4]. Thus, in order to better understand the pathophysiological and geometric aspects of an aneurysm, it is important to manufacture *in vitro* models to improve existing endovascular treatments and to develop and validate theoretical and computational models [5, 6].

Recently Pinho et al. developed a model with geometry and dimensions based on clinical data of a common saccular intracranial aneurism. The model was fabricated with the help of a 3D printer and produced with the polydimethylsiloxane (PDMS) which is a biocompatible material [7]. The PDMS has an elastic behavior similar to arteries [8] and with numerous applications in biomedical [9–16]. The model was subsequently subjected to different experimental tests and numerical simulations [2, 17].

In order to improve the existing model, in this work we propose a novel process of manufacturing a cerebral aneurysm biomodel obtained through an angiography. The image was processed by means of the Scanlp software and the mould was obtained by a three-dimensional printing process (TDP) combined with a PDMS gravity casting process. The biomodel fabricated in this work allows the performance of different hemodynamic studies, test different aneurysm repair techniques and validate numerical approaches.

2   Manufacturing Process

The first process used to obtain the biomodel of the cerebral aneurysm was the angiography of the anatomical zone where the aneurysm is located. For the image treatment, it was performed the segmentation of the image with the objective to isolate the internal carotid artery from the others structures presented in the imaging examination.

The software used was Scanlp, which read and import the angiography images, translating to DICOM (Digital Imaging and Communications in Medicine) format, allowing the visualization of the 3D image. In order to isolate the structure of interest, we used the technique of binarization (Thresholding) and the thickness of the layer (Thickness layer) was 0.0889. Note that, it was the smallest available thickness and as a result it was obtained the best possible precision. In Fig. 1 we can visualize the image after this treatment step.

Finally, a mask was applied to the aneurysm and arteries. After this procedure the images were converted to stl format. The 3D model has been converted to sli. (slice) to be laminated into thin slices parallel to each other and perpendicular to the Z axis. The file was generated and sent to the three-dimensional printer (Fig. 2).

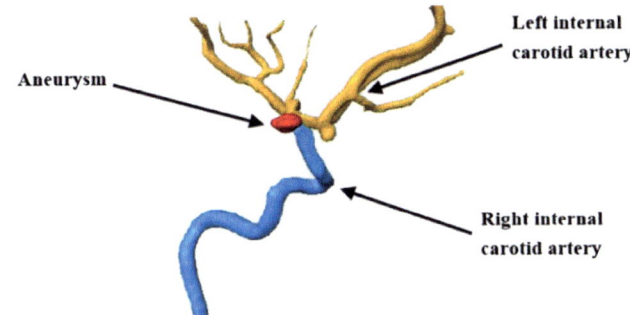

Fig. 1. Mask image processing step.

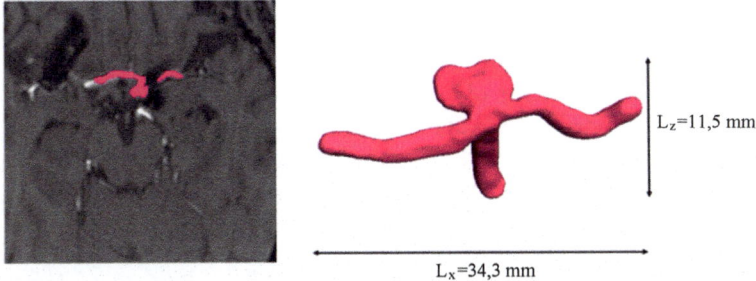

$L_z=11,5$ mm

$L_x=34,3$ mm

Fig. 2. Final model of the aneurysm.

3 Rapid Prototyping

Rapid Prototyping (RP) is a technology that builds physical structures layer by layer from a virtual model, these structures can be made of various materials and for a range of applications. This technology was initially developed for the manufacturing industry, designing components for automobiles, aircraft and computers. Rapid prototyping was primarily used in the field of medicine in 1990 when a model of cranial bone anatomy was produced demonstrating complete internal details with computed tomography image data [18]. At the present moment, industrial and academic interest in three-dimensional (3D) prints has grown and 3D-printed products have been gaining space in pre-surgical planning and operative use [19].

In this study, this technology was used to generate the physical model of the aneurysm, manufactured by using the 3DSystems ProJet 1200 printer (3DSystems, USA). This 3D printer prints with a Visijet FTX Green cartridge, making it possible to obtain a anerurism model with a good surface roughness (Fig. 3).

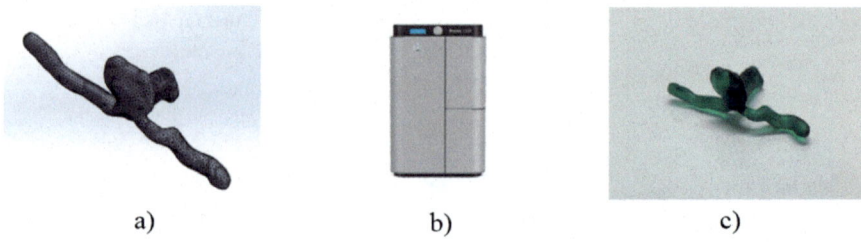

a)   b)   c)

Fig. 3. (a) stl format, (b) 3DSystems ProJet 1200 printer, (c) Mold matrix Visijet FTX Green.

4   In Vitro PDMS Biomodel

After obtaining the wax physical model, we have prepared 10:1 PDMS, i.e., 10 g of the base polymer correspond to 1 g of the curing agent. After, performing the mixture, it was used a vacuum pump for remove the bubbles and subsequently the liquid was poured into the mold by means of a gravity casting process. The curing process was performed at room temperature for about 24 h. After this period the mold was placed in pure acetone and the matrix was removed, obtaining at the end the final biomodel as shown in Fig. 4.

a)   b)   c)

Fig. 4. (a) Biomodel in PDMS, (b) Biomodel immersed in pure acetone, (c) Final PDMS biomodel with a cerebral aneurism to perform fluid and pre-clinical experiments.

5   Conclusions

One of the major difficulties affecting the long-term success of aneurysm repair is the migration of the graft (stent), which may generate internal leakage and even rupture. To avoid this phenomenon and other problems related to the aneurysm treatment, there is the need to perform different kinds of tests, in order to improve techniques and materials, which are very difficult to be carried out in vivo due to the difficulty related to the organs authorization, high costs and reproducibility. The low cost of the *in vitro* model manufacturing process proposed in this study aims to help and simplify the tests in aneurysms. This process is a simple method able to replicate different configurations of aneurysms, which can be used not only for medical and preoperative diagnosis but also to perform in vitro flow experiments to validate and improve existent numerical models.

Acknowledgments. The authors gratefully acknowledge the funding by PORTUGAL2020: SI I&DT Individual – Call 16/SI/2015 (Proj. nº 9703): "Automatização de Processos de Soldadura de Estruturas Hiperestáticas em Ligas de Alumínio (APSEHAL)".

This work was partially supported by *Fundação para a Ciência e a Tecnologia* (FCT) under the strategic grants UID/EMS/04077/2019 and UID/EMS/00532/2019. The authors are also grateful for the funding of FCT through the projects POCI-01-0145-FEDER-016861, NORTE-01-0145-FEDER-029394, NORTE-01-0145-FEDER-030171, funded by COMPETE2020, NORTE2020, PORTUGAL2020, and FEDER.

References

1. Rodriguez-Régent, C.: Non-invasive diagnosis of intracranial aneurysms. Diagn. Interv. Imaging **95**, 1163–1174 (2014)
2. Cardoso, C., Fernandes, C.S., Lima, R., Ribeiro, J.: Biomechanical analysis of PDMS channels using different hyperelastic numerical constitutive models. Mech. Res. Commun. **90**, 26–33 (2018)
3. Sathyan, S.: Association of Versican (VCAN) gene polymorphisms rs251124 and rs2287926 (G428D), with intracranial aneurysm. Meta Gene **2**, 651–660 (2014)
4. Bosman, W.M.P.F.: Aortic customize: an in vivo feasibility study of a percutaneous technique for the repair of aortic aneurysms using injectable elastomer. Eur. J. Vasc. Endovasc. Surg. **40**, 65–70 (2010)
5. Bosman, W.M.P.F.: The effect of injectable biocompatible elastomer (PDMS) on the strength of the proximal fixation of endovascular aneurysm repair grafts: an in vitro study. J. Vasc. Surg. **52**, 152–158 (2010)
6. Nam, S.W., Choi, S., Cheong, Y., Kim, Y.H., Park, H.K.: Evaluation of aneurysm-associated wall shear stress related to morphological variations of circle of Willis using a microfluidic device. J. Biomech. **48**, 348–353 (2015)
7. Hassler, C., Boretius, T., Stieglitz, T.: Polymers for neural implants. J. Polym. Sci. Part B Polym. Phys. **49**, 18–33 (2011)
8. Lima, R.: In vitro blood flow in a rectangular PDMS microchannel: experimental observations using a confocal micro-PIV system. Biomed. Microdevices **10**, 153–167 (2008)
9. Maitz, M.F.: Applications of synthetic polymers in clinical medicine. Biosurf. Biotribol. **1**, 161–176 (2015)
10. Unger, M.A., Chou, H.P., Thorsen, T., Scherer, A., Quake, S.R.: Monolithic microfabricated valves and pumps by multilayer soft lithography. Science **288**, 113–116 (2000)
11. Bozukova, D., Pagnoulle, C., Jérôme, R., Jérôme, C.: Polymers in modern ophthalmic implants - historical background and recent advances. Mater. Sci. Eng. R Rep. **69**, 63–83 (2010)
12. Wang, W., Fang, J.: Variable focusing microlens chip for potential sensing applications. IEEE Sens. J. **7**, 11–17 (2007)
13. Faustino, V., Catarino, S.O., Lima, R., Minas, G.: Biomedical microfluidic devices by using low-cost fabrication techniques: a review. J. Biomech. **49**(11), 2280–2292 (2016)
14. Muñoz-Sánchez, B.N., Silva, S.F., Pinho, D., Vega, E.J., Lima, R.: Generation of micro-sized PDMS particles by a flow focusing technique for biomicrofluidics applications. Biomicrofluidics **10**, 014122 (2016)
15. Pinho, D., Campo-Deano, L., Lima, R., Pinho, F.T.: In vitro particulate analogue fluids for experimental studies of rheological and hemorheological behavior of glucose-rich RBCs suspensions. Biomicrofluidics **11**(5), 054105 (2017)

16. Bento, D., Rodrigues, R., Faustino, V., Pinho, D., Fernandes, C., Pereira, A., Garcia, V., Miranda, J., Lima, R.: Deformation of red blood cells, air bubbles, and droplets in microfluidic devices: flow visualizations and measurements. Micromachines **9**, 151 (2018)

17. Rodrigues, R.O., Pinho, D., Bento, D., Lima, R., Ribeiro, J.: Wall expansion assessment of an intracranial aneurysm model by a 3D digital image correlation system. Measurement **88**, 262–270 (2016)

18. Kim, M.S., Hansgen, A.R., Wink, O., Quaife, R.A., Carroll, J.D.: Rapid prototyping: a new tool in understanding and treating structural heart disease. Circulation **117**, 2388–2394 (2008)

19. Jakus, A.E., Rutz, A.L., Shah, R.N.: Advancing the field of 3D biomaterial printing. Biomed. Mater. **11**, 14102 (2016)

Direct Digital Fabrication in Medicine from Digital Data to Physical Models

Choosing the Best Direction of Printing for Additive Manufacturing Process in Medical Applications Using a New Geometric Complexity Model Based on Part CAD Data

Sabrine Ben Amor[1(✉)], Salah Abdellaoui[2], Antoine Tahan[3(✉)],
Borhen Louhichi[1(✉)], and João Manuel R. S. Tavares[4(✉)]

[1] Laboratoire de Mécanique de Sousse, Ecole Nationale d'Ingénieurs de Sousse,
Université de Sousse, Sousse, Tunisie
Benamor.Sabrine@eniso.u-sousse.tn,
borhen.louhichi@etsmtl.ca
[2] Institut supérieur des sciences appliquées et de technologie de Sousse,
Université de Sousse, Sousse, Tunisie
abdellaouisalah95@gmail.com
[3] École de Technologie Supérieure, Montréal, QC, Canada
antoine.tahan@etsmtl.ca
[4] Faculdade de Engenharia da Universidade do Porto (FEUP), Porto, Portugal
tavares@fe.up.pt

Abstract. Additive manufacturing processes is now experiencing significant growth and is at the origin of intense research activity (optimization of topology, biomedical applications, etc.). One of the characteristics of this method is that the geometric complexity is free. The complexity of a CAD model is also a field of research. The basic idea is that the complexity of a component has implications in design and especially in manufacturing. Indeed, industrial competitiveness in the mechanical field generated the need to produce increasingly complex systems and parts (in terms of geometry, topology …). Part deposition orientation is also very important factor of additive manufacturing as it effects build time, support structure, dimensional accuracy, surface finish and cost of the part. A number of layered manufacturing process specific parameters and constraints have to be considered while deciding the part deposition orientation. Determination of an optimal part deposition orientation is a difficult and time consuming task as one has to trade-off among various contradicting objectives like part surface finish and build time. This paper describes and compares various attempts made to determine part deposition orientation of orthoses using geometric complexity model and part CAD information.

Keywords: 3D printing · Orthoses · Computer-Aided Design · Deposition orientation · Complexity metrics

© Springer Nature Switzerland AG 2019
J. M. R. S. Tavares and R. M. Natal Jorge (Eds.): VipIMAGE 2019, LNCVB 34, pp. 679–692, 2019.
https://doi.org/10.1007/978-3-030-32040-9_70

1   Introduction, Background and Motivations

Additive Manufacturing (AM) is defined as the process of adding materials to fabricate parts from 3D model data, usually layer-by-layer [1]. Contrary to material-removal manufacturing methods, such as traditional machining, AM can produce very high complexity parts for different types of materials (plastic, metal, ceramic, etc.) and without the loss of material. The techniques and methods of AM are developing, on the one hand, in terms of applications and, on the other hand, in terms of areas of use (automotive, medical, aeronautics …). The AM is growing strongly for the next few years. The most used materials in AM are plastics. While in recent years, there is a significant trend towards metallic materials.

The part orientation problem can be defined as changing the orientation of the part to maximize or minimize one or more manufacturing considerations. This can be done on either the STL file or on the CAD model itself [2]. Depending on the application or purpose of the part, certain features may be more important than others. This makes the orientation of the part a design challenge as much as an optimization challenge [3]. Orientation can affect build time, quality of the part, and mechanical properties of anisotropic parts. Depending on the process, there may even be manufacturing constrains that must be considered such as supports or deposition properties [4].

The determination of the optimal part orientation is essential for all AM processes. However, the way in which a part's orientation affects the manufactured part is process dependent. Different factors that can be considered while choosing the build direction for AM [2].

Figure 1 shows how different orientations affect support material. The amount of part area touching the base of the build platform can also affect the quality of the part surface, so should be minimized [2].

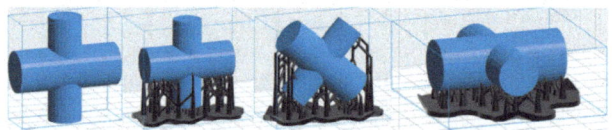

Fig. 1. Original model and different orientations requiring different support structures [5].

Many different methods have been used to find an optimal part orientation in the literature.

Frank and Fadel [6] suggested that many factors get affected by the part deposition orientation. Some of these factors are surface quality, build time, amount of support structures, shrinkage, curling, distortion, roundness/flatness, resin flow, material cost and trapped volume etc. Frank and Fadel [6] considered the first three factors namely surface quality, build time and amount of support structures to obtain a better part deposition orientation.

Cheng et al. [7] presented an approach for determining suitable part deposition orientation for SL parts considering dimensional accuracy as well as build time as objectives. Part accuracy was treated as primary objective and was estimated using the different weight factors for different types of surface geometries.

Lan et al. [8] determined part deposition orientation for parts based on the considerations of surface quality, build time or complexity of the support structures. Surface quality was evaluated either by maximizing the area of non-stepped surfaces or by minimizing the area of worst quality surfaces. The maximization of area of non-stepped surfaces was achieved by selecting a part deposition orientation that maximizes the total area of perpendicular and horizontal faces, which do not offer stair stepping.

Alexander et al. [9] determined suitable part deposition orientation for better part accuracy and lower cost.

McClurkin and Rosen [10] developed statistical models to predict build time, accuracy and surface finish of SL parts using response surface methodology.

In sum, Different published works for part deposition orientation determination are presented. Most of the researchers have considered part accuracy, build time, support structure and cost as a criterion to determine part deposition orientation.

In recent years, 3D printing technology was introduced to the medical field, specifically for medical orthoses production. The goal was to bring new revolutionary medical devices to the market. As known, the traditional orthoses fabrication processes have always been labeled slow, expensive, skill depending and time-consuming, which generally results in a poor orthoses quality.

In this paper, we propose to choose the best direction of printing medical orthoses using the complexity metric model (C), the cost and the mass of material of the part.

Finally, this paper is structured as follows: Sect. 2 presents the complexity metric model. Section 3 then proceeds to describe the methodology used to choose the best direction of 3D printing and case studies.

2   Complexity Model Presentation

In order to develop our product complexity metric model (C), we have inspired from the principle of AM. He treats the part layer by layer as build. This model is calculated after a slicing of the part following the direction of printing. It is composed of five parameters (Eq. (1)).

$$C = \sum_{j=1}^{5} C_j \tag{1}$$

The first metric C_1 represent the mean of C_{1i}, n represent the total number of layers in part after slicing. The coefficient C_{1i} is designed to describe the complexity of the part between two consecutive layers along the z-axis:

$$C_1 = \frac{1}{n-1} \sum_{i=2}^{n} C_{1i} \tag{2}$$

C_{1i} is calculated using the difference between tow surfaces.

$$C_{1i} = \ln(|S_{net} - S_{net-1}| + 1) \tag{3}$$

Where S_{net} represent the net area of a given section and S_{net-1} the area of the previous section.

The second metric C_2 represent the mean of C_{2i}.

$$C_2 = \frac{1}{n-1} \sum_{i=2}^{n} C_{2i} \tag{4}$$

The coefficient C_{2i} is designed to describe the complexity of the part between two consecutive layers along the z-axis. It consists in studying the difference between their centers of gravities. (X_g, Y_g) are the center of gravity coordinates of a given section and (X_{g-1}, Y_{g-1}) are the center of gravity coordinates of the previous one.

$$C_{2i} = 25 \; ln\sqrt{(X_g - X_{g-1})^2 + (Y_g - Y_{g-1})^2 + 1} \tag{5}$$

The metric C_3 represents the mean of C_{3i}:

$$C_3 = \frac{1}{n} \sum_{i=1}^{n} C_{3i} \tag{6}$$

The coefficient C_{3i} is designed to describe the amount of cavities in the part. It consists of dividing S_{max} by S_{net}. Where S_{max} is the area of a section including the cavities and S_{net} is the net area of a given section. This coefficient C_{3i} is similar to the method of [11] who described the complexity in extrusion in the same way.

$$C_{3i} = \frac{S_{max}}{S_{net}} \tag{7}$$

The fourth metric C_4 represent the arithmetic mean of C_{4i}:

$$C_4 = \frac{1}{n} \sum_{i=1}^{n} C_{4i} \tag{8}$$

The complexity factor C_{4i} is inspired by the work of Joshi and Ravi [12], who described the complexity as being on the surface of the piece with respect to the surface of a sphere of the same volume. Indeed, in our case C_{4i} describes the two-dimensional complexity. Therefore, we did the projection of the work. This projection gave the perimeter instead of surface (P_{tot}[mm]) and the net surface (S_{net}) of a given layer [mm^2] at a given height.

$$C_{4i} = \frac{P_{tot}}{2\sqrt{\pi S_{net}}} \tag{9}$$

The fifth metric C_5 represent the arithmetic mean of C_{5i}:

$$C_5 = \frac{1}{n}\sum_{i=1}^{n} C_{5i} \tag{10}$$

The coefficient C_{5i} consists in studying the sum of the numbers of the independent surfaces and the cavities in a given section.

$$C_{5i} = \ln(N_p + N_h) \tag{11}$$

Where N_p and N_h are respectively the number of independent surfaces and cavities in a given section.

3 Case Studies

In order to choose the best direction of 3D printing for medical orthoses, we start by calculating the complexity metric model using the methodology described in (Fig. 2) then we move to ideamaker ® to calculate the cost and the mass of the material of the part.

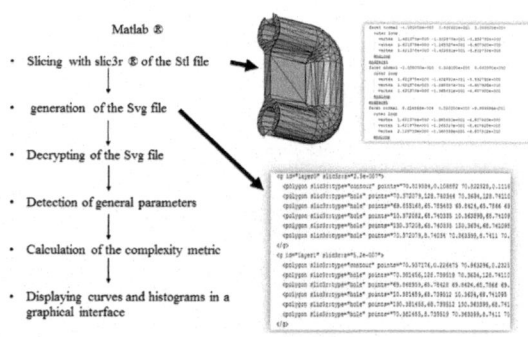

- Slicing with slic3r ® of the Stl file
- generation of the Svg file
- Decrypting of the Svg file
- Detection of general parameters
- Calculation of the complexity metric
- Displaying curves and histograms in a graphical interface

Fig. 2. The methodology to calculate the complexity metric

Five medical orthoses were treated (Table 1) in this part in different direction of slicing to choose the best orientation deposition of each part.

3.1 The Result of Part 1

Part 1 is a Foot Orthosis (Fig. 3) manufactured by 3DP, three directions are treated and a choice of the better direction for 3D printing has been done.

3.1.1 Result in the First Direction

In this part we will see the general parameters and the metric coefficients in the first direction. In this direction, after slicing, we had 901 layers.

Table 1. The studied parts

	Part 1
	Part 2
	Part 3
	Part 4
	Part 5

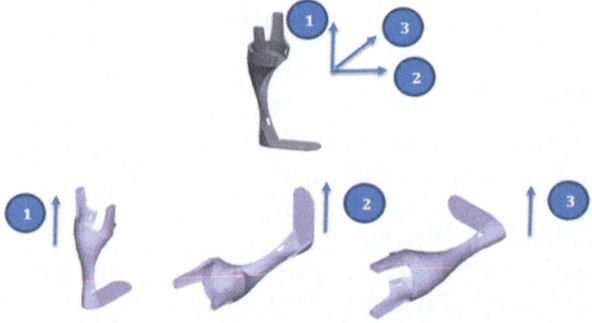

Fig. 3. Part 1 in three direction

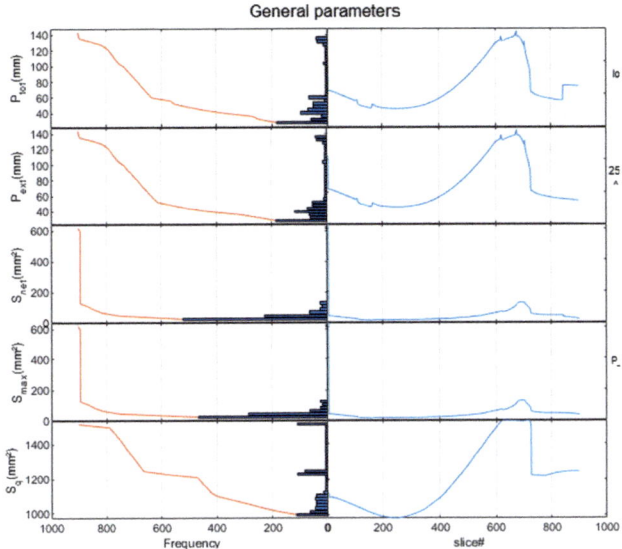

Fig. 4. The general parameters of part 1 in the first direction

Figure 4 represents the different general parameters of part 1 in the first direction. The right part of the figure shows its parameters according to the number of the layer. The section on the left illustrates the frequency of these parameters with the cumulative curve.

Figure 5 shows the different candidate metrics for part 1 in the first direction. The right part of the figure shows its metrics based on the number of the layer. The left side shows the frequency of these parameters with the cumulative curve.

3.1.2   Result in the Second Direction

In this part we will see the general parameters and the metric coefficients in the second direction. In this direction, after slicing, we had 487 layers.

Figure 6 represents the different general parameters of part 1 in the second direction. The right part of the figure shows its parameters according to the number of the layer. The section on the left illustrates the frequency of these parameters with the cumulative curve.

Figure 7 shows the different candidate metrics for part 1 in the second direction. The right part of the figure shows its metrics based on the number of the layer. The left side shows the frequency of these parameters with the cumulative curve.

3.1.3   Result in the Third Direction

In this part we will see the general parameters and the metric coefficients in the third direction. In this direction, after slicing, we had 1732 layers.

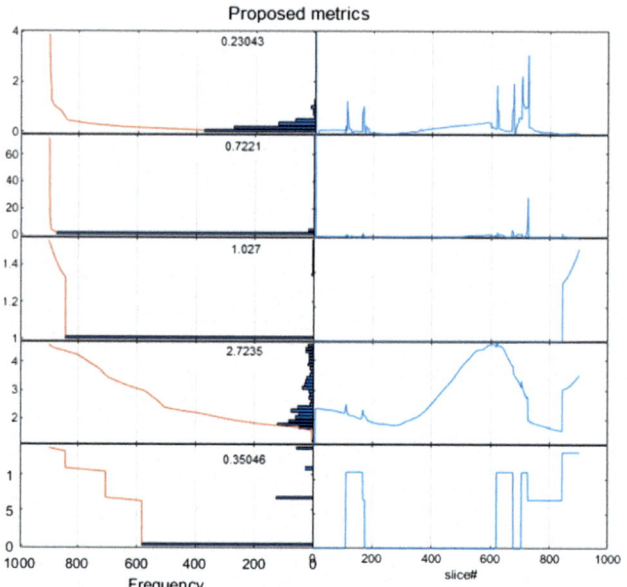

Fig. 5. The proposed metrics of part 1 in the first direction

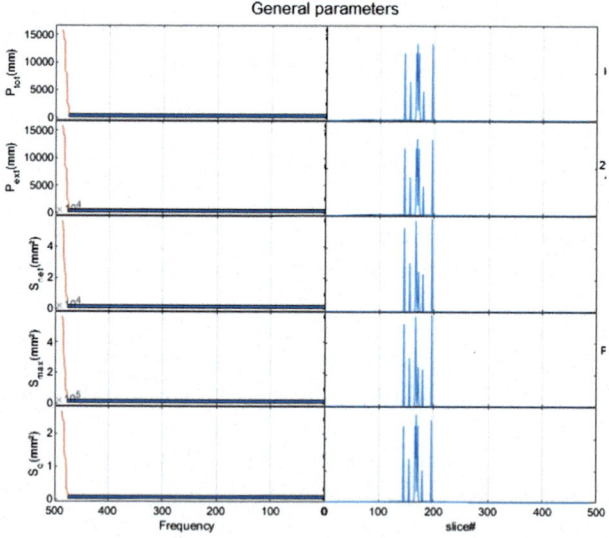

Fig. 6. The general parameters of part 1 in the second direction

Figure 8 represents the different general parameters of part 1 in the third direction. The right part of the figure shows its parameters according to the number of the layer. The section on the left illustrates the frequency of these parameters with the cumulative curve.

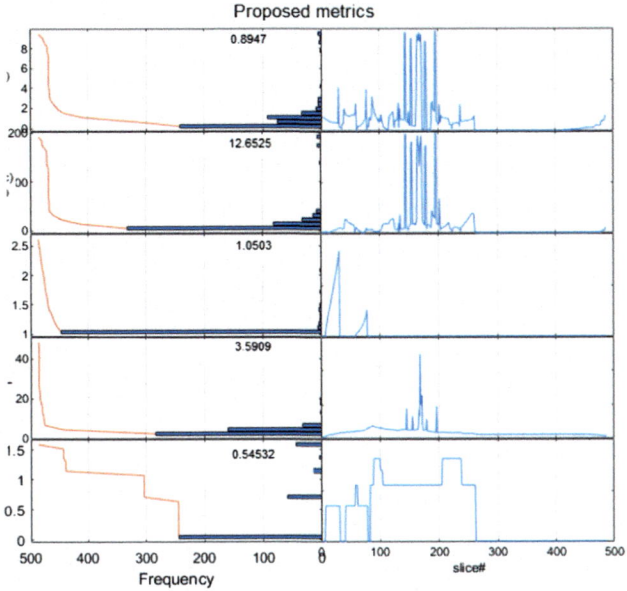

Fig. 7. The proposed metrics of part 1 in the second direction

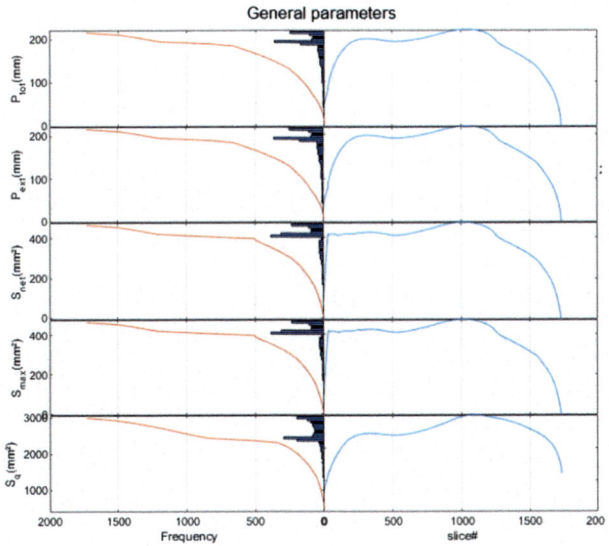

Fig. 8. The general parameters of part 1 in the third direction

Figure 9 shows the different candidate metrics for part 1 in the third direction. The right part of the figure shows its metrics based on the number of the layer. The left side shows the frequency of these parameters with the cumulative curve.

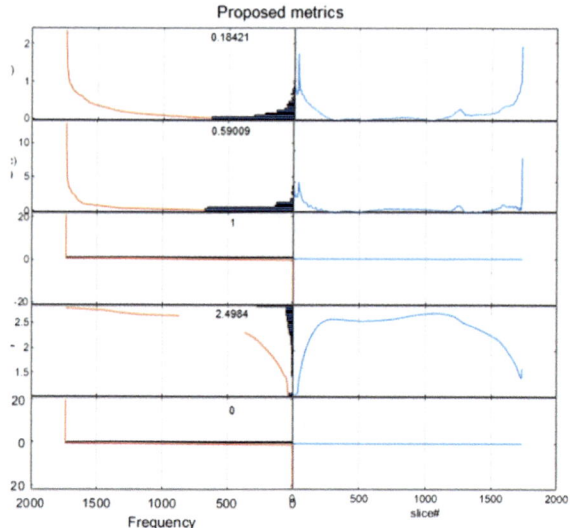

Fig. 9. The proposed metrics of part 1 in the third direction

Table 2. The results of part 1

	C_1	C_2	C_3	C_4	C_5	C	Cost ($)	Mass of material (g)
Direction 1	0.230	0.722	1.027	2.723	0.350	5.05	0.2	6.7
Direction 2	0.894	12.652	1,050	3.590	0.545	18.7	0.16	5.2
Direction 3	0.184	0.590	1	2.498	0	4.2	0.16	5.2

Table 2 shows the average values of these metrics in the three directions. These values reflect that the direction 3 is the best in the direction of surface variation.

This method was used to get an idea about the geometry of the part and to geometrically comparison in different directions. But you have to consider the amount of fastening material and the cost to choose the best print direction. For this part, we can choose the third direction to print the part. This direction has the lowest geometric complexity, the lowest cost and the lowest mass of material.

3.2   The Result of Part 2

Part 2 is a Foot Orthosis (Fig. 10) manufactured by 3DP, Three directions are treated and a choice of the better direction for 3D printing has been done.

Table 3 shows the average values of the complexity metrics, the cost and mass of material in the three directions. For this part, we can choose the first direction to print the part. This direction has the lowest geometric complexity, the lowest cost and the lowest mass of material.

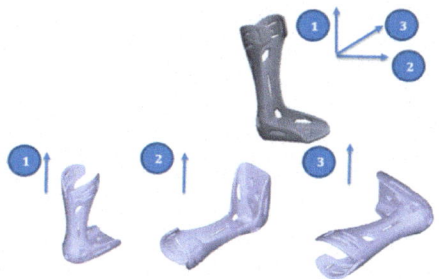

Fig. 10. Part 2 in three direction

Table 3. The results of part 2

	C_1	C_2	C_3	C_4	C_5	C	Cost (\$)	Mass of material (g)
Direction 1	0.468	1.434	1.002	2.707	0.870	6.48	0.95	31.7
Direction 2	0.771	3.752	1,012	3.461	1.263	10.2	0.97	32.5
Direction 3	1.173	8.008	1.002	3.293	1.119	14.5	1.03	34.4

3.3   The Result of Part 3

Part 3 (Fig. 11) manufactured by 3DP, Three directions are treated and a choice of the better direction for 3D printing has been done.

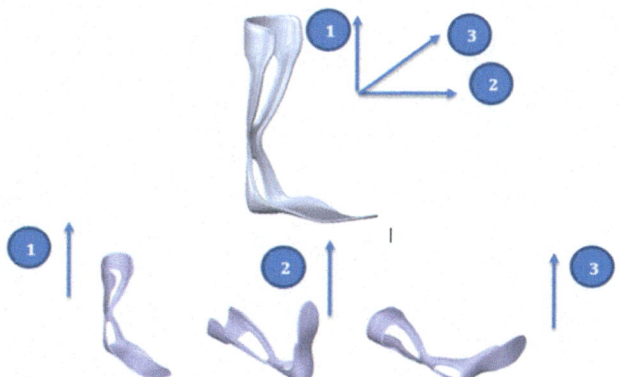

Fig. 11. Part 3 in three direction

Table 4 shows the average values of the complexity metrics, the cost and the mass of material in the three directions. For this part, we can choose the first direction to print the part. This direction has the lowest geometric complexity, the lowest cost and the lowest mass of material.

Table 4. The results of part 3

	C_1	C_2	C_3	C_4	C_5	C	Cost ($)	Mass of material (g)
Direction 1	0.289	1.949	1	2.483	0.458	6.18	0.67	22.3
Direction 2	0.375	2.752	1	2.990	0.370	7.48	0.67	22.3
Direction 3	0.850	9.055	1	4.205	1.034	16.1	0.7	23.2

3.4   The Result of Part 4

Part 4 (Fig. 12) manufactured by 3DP, Three directions are treated and a choice of the better direction for 3D printing has been done.

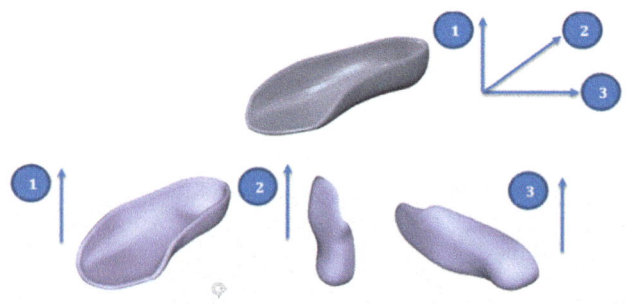

Fig. 12. Part 4 in three direction

Table 5. The results of part 4

	C_1	C_2	C_3	C_4	C_5	C	Cost ($)	Mass of material (g)
Direction 1	0.936	5.790	1.020	3.701	0.139	11.50	1.3	43.3
Direction 2	0.184	0.590	1	2.498	0	4.27	1.16	38.6
Direction 3	0.574	3.559	1	2.796	0	7.92	1.27	42.4

Table 5 shows the average values of the complexity metrics, the cost and the mass of material in the three directions. For this part, we can choose the second direction to print the part. This direction has the lowest geometric complexity, the lowest cost and the lowest mass of material.

3.5   The Result of Part 5

Part 5 (Fig. 13) manufactured by 3DP, Three directions are treated and a choice of the better direction for 3D printing has been done.

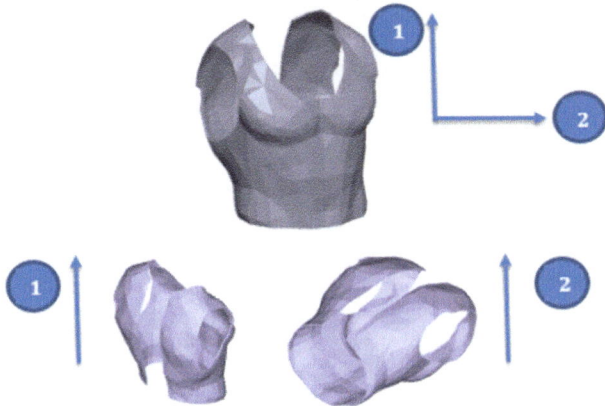

Fig. 13. Part 5 in two direction

Table 6. The results of part 5

	C_1	C_2	C_3	C_4	C_5	C	Cost ($)	Mass of material (g)
Direction 1	0,465	1,683	1	6,135	0,672	9,95	0,47	15,8
Direction 2	0,717	4,790	1,043	5,610	1,2	13,3	0,49	16,2

Table 6 shows the average values of the complexity metrics, the cost and the mass of material in the three directions. For this part, we can choose the first direction to print the part. This direction has the lowest geometric complexity, the lowest cost and the lowest mass of material.

4   Conclusion

Additive manufacturing process has shown great manufacturing performance in the medical applications. They are able to produce very high complexity parts. The part orientation problem can be defined as changing the orientation of the part to maximize or minimize one or more manufacturing considerations. This work was done to choose the best direction of printing medical orthoses using the geometric complexity model in AM, the cost and the mass of material.

References

1. Obaton, A.F., Bernard, A., Taillandier, G., Moschetta, J.M.: Fabrication additive: état de l'art et besoins métrologiques engendrés Additive manufacturing: state of the art and generated metrological needs. Revue française de métrologie **37**, 133 (2015)
2. Kulkarni, P., Marsan, A., Dutta, D.: A review of process planning techniques in layered manufacturing. Rapid Prototyp. J. **6**(1), 18–35 (2000)

3. Gibson, I., Rosen, D.W., Stucker, B.: Design for additive manufacturing. In: Additive Manufacturing Technologies, pp. 299–332. Springer, Boston (2010)
4. Pandey, P.M., Reddy, N.V., Dhande, S.G.: Part deposition orientation studies in layered manufacturing. J. Mater. Process. Technol. **185**(1–3), 125–131 (2007)
5. Oropallo, W., Piegl, L.A.: Ten challenges in 3D printing. Eng. Comput. **32**(1), 135–148 (2016)
6. Frank, D., Fadel, G.: Expert system-based selection of the preferred direction of build for rapid prototyping processes. J. Intell. Manuf. **6**(5), 339–345 (1995)
7. Cheng, W., Fuh, J.Y.H., Nee, A.Y.C., Wong, Y.S., Loh, H.T., Miyazawa, T.: Multi-objective optimization of part-building orientation in stereolithography. Rapid Prototyp. J. **1**(4), 12–23 (1995)
8. Lan, P.T., Chou, S.Y., Chen, L.L., Gemmill, D.: Determining fabrication orientations for rapid prototyping with stereolithography apparatus. Comput. Aided Des. **29**(1), 53–62 (1997)
9. Alexander, P., Allen, S., Dutta, D.: Part orientation and build cost determination in layered manufacturing. Comput. Aided Des. **30**(5), 343–356 (1998)
10. McClurkin, J.E., Rosen, D.W.: Computer-aided build style decision support for stereolithography. Rapid Prototyp. J. **4**(1), 4–13 (1998)
11. Qamar, S.Z., Arif, A.F.M., Sheikh, A.K.: A new definition of shape complexity for metal extrusion. J. Mater. Process. Technol. **155**, 1734–1739 (2004)
12. Joshi, D., Ravi, B.: Quantifying the shape complexity of cast parts. Comput. Aided Des. Appl. **7**(5), 685–700 (2010)

Layer Thickness Evaluation Between Medical Imaging and Additive Manufacturing

Henrique A. Almeida[1,2(✉)], Joel Vasco[1,3], Mário Correia[1,4],
and Rui Ruben[1,5]

[1] School of Technology and Management,
Polytechnic Institute of Leiria, Leiria, Portugal
henrique.almeida@ipleiria.pt
[2] Computer Science and Communication Research Centre,
Polytechnic Institute of Leiria, Leiria, Portugal
[3] Institute for Polymers and Composites, University of Minho, Guimarães,
Portugal
[4] Centre for Mechanical Engineering, Materials and Processes,
University of Coimbra, Coimbra, Portugal
[5] CSRsp, Polytechnic Institute of Leiria, Marinha Grande, Portugal

Abstract. Additive manufacturing (AM) applied to the orthopaedic and surgical domains provided access to newer solutions for customised implants, customised scaffolds or even organ printing. These solutions are based on imagological data, gathered from CT-scans and/or MRI-scans. The compromise between patient's radiation exposition and resolution along the focus direction plays an important role for the successful production of biological products. Scan detail can be increased with a thinner scanning thickness to obtain the required biological features for replication. On the other hand, a thinner scan thickness requires more scanning slices and therefore, higher exposition time to radiation. Literature shows that a maximum amount of radiation is admissible for humans, requiring an optimized approach concerning the acquisition of internal details of the human body tissues. State-of-the-art scans offer focus resolution in the range of 100 to 200 μm, although such resolution is not used for the patient's protection. In fact, in clinical practice distance between slices are in the range of 1 to 6 mm. Concerning the AM processes that are able to use biocompatible materials, different layer thicknesses are available upon the final application in spite of the digital layer thickness that was applied during the medical imaging exam. The layer thickness during production also details the mechanical properties of the implant, with an additional aspect that the geometric data is based on information of a different layer thickness. This study intends to discuss the layer thickness used in medical imaging, the layer thickness used in AM systems providing public awareness of the operating gap between medical imaging systems and AM systems. A brief description of the accumulated errors is also presented.

Keywords: Additive manufacturing · Digital fabrication · Medical imaging · CT-scan · MRI-scan

© Springer Nature Switzerland AG 2019
J. M. R. S. Tavares and R. M. Natal Jorge (Eds.): VipIMAGE 2019, LNCVB 34, pp. 693–701, 2019.
https://doi.org/10.1007/978-3-030-32040-9_71

1 Introduction

Over the past few years have emerged various manufacturing and processing technologies that have revolutionized the production of customized products. These technologies have exceeded the limits and geometric design limitations that existed previously in conventional manufacturing technologies. However, these technologies have emerged into areas where the requirements are constantly more demanding such as in the medical sector (Almeida et al. 2018).

The use of Additive Manufacturing (AM) in the medical field is expanding very fast due to the ability to produce complex, low weight, anatomically shaped and personalized medical devices in a wide range of biocompatible, degradable and non-degradable biomaterials, biological materials (e.g. cells, pharmaceuticals and growth factors), functionally graded, etc. (Tibbitt et al. 2015). Additive manufacturing technologies are capable of producing medical devices and implants with several types of specifications (Fig. 1). In the case of medical devices, it is possible to produce surgical training devices and surgical devices and other medical devices that interact directly with patients such as orthosis and exoskeletons (Almeida et al. 2018). In the case of medical implants, they may be of two types, namely externally to the body and internally to the body. Examples of external implants are prosthesis for the leg, foot, hand, eye, etc. (Almeida et al. 2018). Regarding internal implants, they may also be divided in two additional categories, namely permanent and temporary. Examples of internal permanent implants are metallic implants such as hip and knee prostheses etc. (Almeida et al. 2018). Examples of internal temporary implants are scaffolds which are used for repairing, restoring, and regenerating tissue or organ functions in either hard or soft tissue engineering and regenerative medicine applications where they are absorbed by the body during their degradation and tissue regeneration (Ligon et al. 2017).

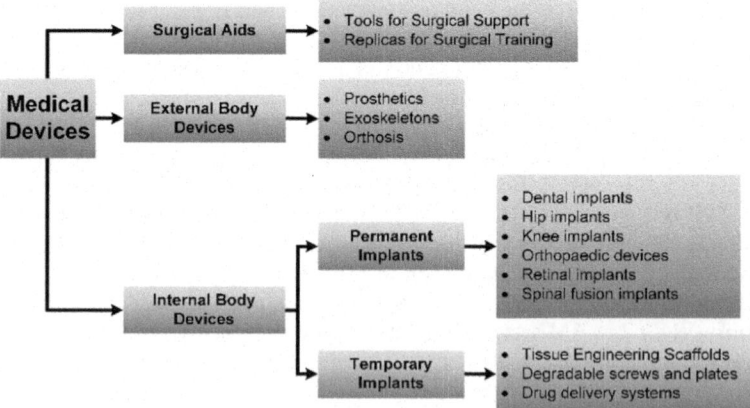

Fig. 1. Process flowchart for the production of passive/non-biological and active/biological structures for medical applications using AM.

These solutions are based on imagological data, gathered from CT-scans and/or MRI-scans. The compromise between patient's radiation exposition and resolution along the focus direction plays an important role for the successful production of biological products. Scan detail can be increased with a thinner scanning thickness to obtain the required biological features for replication. On the other hand, a thinner scan thickness requires more scanning slices and therefore, higher exposition time to radiation. Literature shows that a maximum amount of radiation is admissible for humans, requiring an optimized approach concerning the acquisition of internal details of the human body tissues (Almeida et al. 2016). State-of-the-art scans offer focus resolution in the range of 100 to 200 µm, although such resolution is not used for the patient's protection (Almeida et al. 2017). In fact, in clinical practice, distance between slices are in the range of 1 to 6 mm (Almeida et al. 2016). Concerning the AM processes that are able to use biocompatible materials, different layer thicknesses are available upon the final application in spite of the digital layer thickness that was applied during the medical imaging exam. The biomaterial type, geometric form and designed functionality influences the mechanical performance of the medical implant. Furthermore, the layer thickness during production also details the mechanical properties of the implant, with an additional aspect that the geometric data is based on information of a different layer thickness (Almeida et al. 2018).

This study intends to discuss the layer thickness used in medical imaging, the layer thickness used in AM systems providing public awareness of the operating gap between medical imaging systems and AM systems. A brief description of the accumulated errors is also presented.

2 Overview of Medical Imaging Systems

State-of-the-art medical scans (computer tomography or magnetic resonance imaging) offer focus resolution in the range of 100 to 200 µm, although such resolution is not used for the patient's protection. In clinical practice, pixel spacing used is between 0.5 and 0.8 mm and slice thickness between 0.6 and 7 mm (Almeida et al. 2016). Slice thickness is a major issue in this kind of exams. In fact, a smaller value increases radiation exposure of patients. However, nowadays some CT-scan machines are able to follow low dose protocols with a natural radiation exposure (Almeida et al. 2017). But, radiation is not the major issue. The major issue is the cost of these exams. An increase on slice thickness from 1 to 6 mm has a big impact on the exam's cost and time. So, normally exams have a pixel size of 0.7 mm and a slice thickness around 6 mm, producing volumetric data (voxel-based data). When this kind of parameters is used, it is difficult to avoid the staircase effect in the medical image segmentation procedure.

3 Overview of Medical Imaging Processing Software

Medical imaging processing is a normal step in order to have better information for diagnosis. For instance, segmentation of organs gives a better anatomical view of the patient. After segmentation, a virtual surgery plan can be done and also some specific long-term prediction of a surgery can be performed (Almeida et al. 2017). This is an area of strong research, but there are at least three commercial softwares that can be used to segment medical images, namely: Mimics (Materialise), InVesalius (CTI-Brazil) and ITK-Snap (itksnap.org). Mimics is a powerfull software with virtual surgery capabilities. InVesalius and ITK-Snap are free softwares and their source code is available for new developers.

4   Overview of Additive Manufacturing Systems

By definition, AM is the process of joining materials in order to make objects from 3D digital models in a layer-by-layer fashion, as opposed to conventional manufacturing technologies (subtractive or shape forming) (Chua and Leong 2014). The main features of AM are its ability to produce parts of virtually any shape and high complexity in one process step, with less material and energy, reducing assembly requirements by consolidating parts into a single component. It disrupts the traditional supply chain, allowing for goods to be produced closer to the point of use at the time of need and dramatically shrinking the time between design creation and prototyping (Gibson et al. 2015). It is also the ideal technology to create lightweight structures without requiring expensive tooling. Multi-material objects, geometric or material functionally gradient structures can also be produced at multiple scales.

According to the ISO17296-2:2015, AM processes comprise the following techniques (Almeida and Correia 2016; Gibson et al. 2015):

1. Vat photopolymerization (e.g., SLA) – an AM process in which a liquid photopolymer in a vat is selectively cured by light-activated polymerization.
2. Material jetting (e.g., Polyjet) – an AM process in which droplets of build material are selectively deposited.
3. Binder jetting (e.g., 3D printers using powder and binder) – an AM process in which a liquid bonding agent is selectively deposited to join powder materials.
4. Material extrusion (e.g., FDM) – an AM process in which material is selectively dispensed through a nozzle.
5. Powder bed fusion (e.g., SLS) – an AM process in which thermal energy selectively fuses regions of a powder bed.
6. Sheet lamination (e.g., Sheet Forming) – an AM process in which laminated sheets of material are consecutively cut and added to the previous layer.
7. Directed energy deposition (e.g., LENS) – an AM process in which material is fused and added during the production of the part.

The precision accuracies and the layer thicknesses for the most representative AM systems in the medical field is presented in Table 1.

For a better understanding of AM, Fig. 2 provides a general overview of the necessary steps to produce physical components, in this particular case, medical devices such as passive/non-biological and active/biological structures. The first step is the generation of the corresponding digital model. In this step, one may use medical imaging techniques (for the definition of internal medical devices), 3D scanning systems (for the definition of external body geometries), conventional CAD modelling

Table 1. Generic dimensional accuracies and layer thicknesses of AM systems (Redwood et al. 2017).

Technology	SLA/DLP	Material jetting	Binder jetting	FDM	SLS	SLM
Dimensional accuracy	±0.5% (lower limit ± 0.15 mm)	±0.1 mm	±0.1 mm (polymer) ±0.2 mm (metal)	±0.5% (lower limit ± 0.5 mm)	±0.3% (lower limit ± 0.3 mm)	±0.1 mm
Common layer thickness [μm]	25–100	16–30	100	50–400	100	50

softwares, or combinations between them (Almeida and Bártolo 2013). Depending on the type of product, two different routes can be considered. The fabrication of passive products usually requires the use of non-degradable materials and the fabrication process follows the steps commonly used to produce any kind of products using AM. Active products usually require the use of cells. They can be directly printed embedded in hydrogels or seeded on biodegradable 3D structures. In most cases constructs are pre-cultured before implantation (Almeida et al. 2018).

5 Digital Process Steps and Accumulated Errors

Figure 2 presents the general overview of AM, but considering a summarised version regarding the digital information, Fig. 3 presents the information chain of the overall process.

It is possible to observe that the digital information undergoes several transformations, being the most critical when the source is based on medical imaging data. According to Fig. 3, the upper row summarizes a process that begins with an accurate 3D model from a CAD system. The lower row includes the acquisition of medical data, that both by approximation and resolution failures, resulting in the first error source. In this case, the digital model is defined on voxel data according to the pixel size and slice thickness. The higher the slice thickness, higher the error associated with the medical imagining exam, creating the initial error.

After processing the medical data, the following step requires a conversion into STL file format (triangular based facets). According to the medical imaging slice thickness, the second error arises (Tessellation error) as the software creates the STL triangular facets on information with several millimetres apart. With STL processing software, it is possible to correct STL errors and smoothing the digital model, but this may either reduce or increase the geometric error of either approximating or distancing even more from the geometry, increasing the chordal error. Additionally, the smoothing process requires the creation of more data points to obtain smaller triangular facets which requires extra computing resources due to the resulting file size.

The next and final step regards the production of the physical model. The digital model is imported into the AM software that creates the information necessary for the

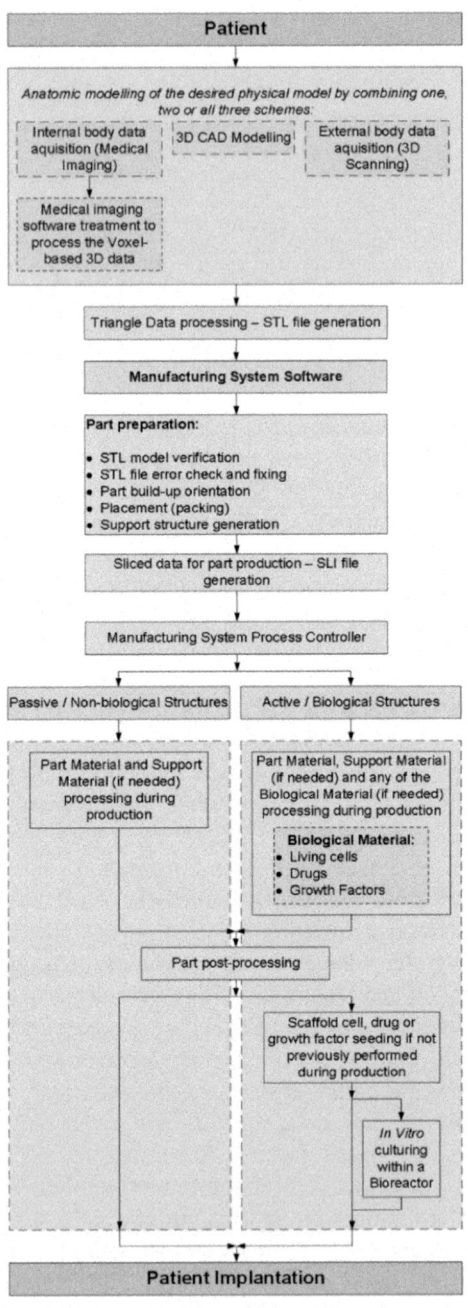

Fig. 2. Process flowchart for the production of passive/non-biological and active/biological structures for medical applications using AM (Almeida et al. 2018).

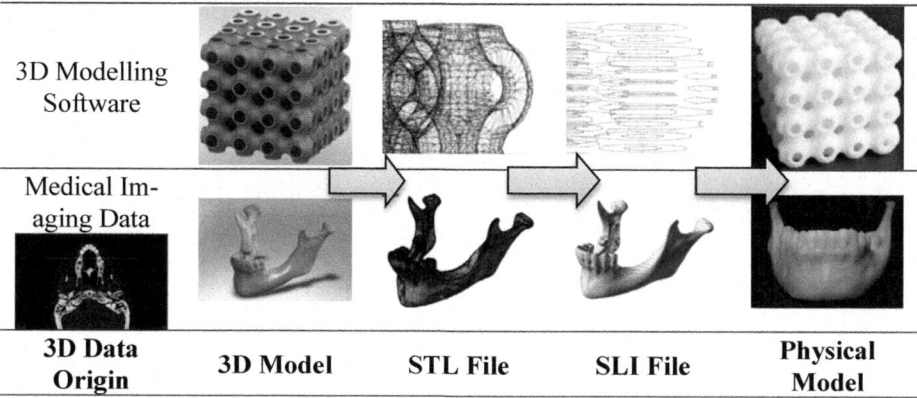

3D Modelling Software				
Medical Imaging Data				
3D Data Origin	**3D Model**	**STL File**	**SLI File**	**Physical Model**

Fig. 3. Flowchart of digital information to produce physical models (Almeida et al. 2018).

machine to produce the physical model. According to the type of AM system, different layer thicknesses is required, therefore, different SLI files (sliced data) are obtained, creating again a staircase effect according to the layer thickness. The error associated to this staircase effect is less significant when compared to the initial staircase effect in the medical imaging exam due to the differences of layer thickness. While the medical imaging slice thickness is in millimetres, the AM layer thickness is in micrometres. In spite of this error being less significant, when associated with other production errors that way occur, the third error is created.

The accumulation of these errors are illustrated in Fig. 4.

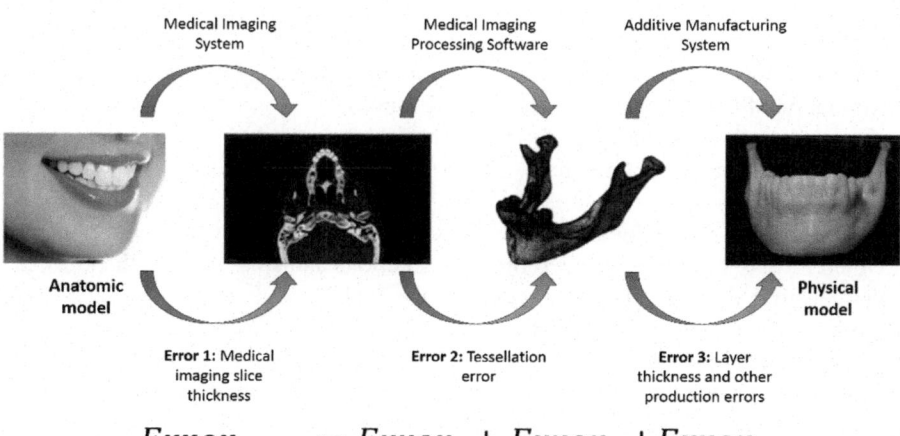

$$Error_{Total} = Error_1 + Error_2 + Error_3$$

Fig. 4. Error classification from anatomic model to physical model.

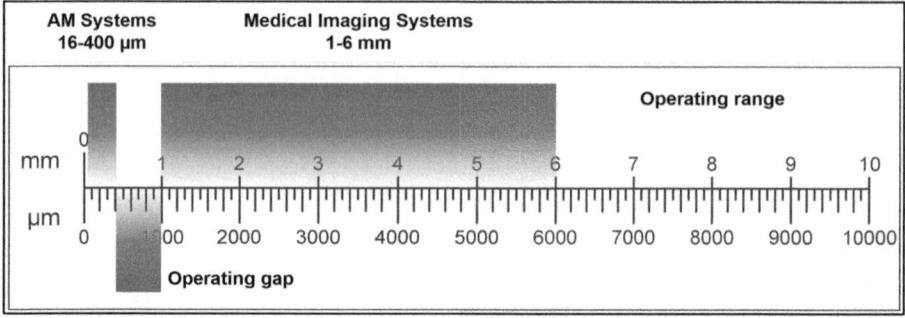

Fig. 5. Operating range for AM systems and medical imaging systems.

6 Conclusions

The use of AM in the medical field is expanding very fast due to the ability to produce complex, low weight, anatomically shaped and personalized medical devices in a single development process. In fact, the compromise between the increased technological capacities and human patients' capacity to be exposed to radiation is the main issue. Clinical practice is focused on quick results which requires less exposure, less details, faster and less expensive imaging exams. However, such approach may not be quite suitable for reconstruction since it includes several error sources.

The development process of a medical device is composed by several development stages to proceed from the original model to the new physical model. As mentioned before, the first error originates from the semi-automatic transformation of medical imaging data into a STL file. In this case, the user is unable to control any of the STL generation parameters. On the other hand, a medical device originated from a CAD system, includes the chordal error when converted to STL. Such error may be decreased by the CAD user by increasing the number of triangles, however, increasing the limitations for file handling in further operations.

Concerning the manufacture of medical devices through AM systems, the operating range for AM processes compared with medical imaging systems, shows an existing operating gap (Fig. 5).

The operating range for AM systems is not only driven by medical applications and, therefore, is strongly pushed to higher resolutions and smaller layer thicknesses by other industrial domains. However, despite the much lower operating range of AM systems, it is not enough to overcome the low resolution of medical imaging data due to the high slice thickness. Another fact to consider is that imaging data for 3D model reconstruction should be substantially more detailed than common imaging data for clinical diagnosis. The need to cope with the accumulated errors between all development stages poses the question of how detailed should be the final model.

This study intended to discuss the layer thickness used in medical imaging, the layer thickness used in AM systems and providing public awareness of the operating gap between both medical imaging and AM systems.

References

Almeida, D., Ruben, R.B., Folgado, J., Fernandes, P., Gamelas, J., Verhegghe, B., et al.: Automated femoral landmark extraction for optimal prosthesis placement in total hip arthroplasty. Int. J. Numer. Methods Biomed. Eng. **33**, e2844 (2017)

Almeida, D.F., Ruben, R.B., Folgado, J., Fernandes, P.R., De Beule, M., Verhegghe, B.: Fully automated segmentation of femurs with medullary canal definition for low resolution CT scans. Med. Eng. Phys. **38**, 1474–1480 (2016)

Almeida, H.A., Bártolo, P.J.: Computational technologies in tissue engineering. In: Kiss, R., Brebbia, C.A. (eds.) Modelling in Medicine and Biology X, pp. 117–129. Wit Press (2013)

Almeida, H.A., Correia, M.S.: Sustainability impact evaluation of support structures in the production of extrusion based parts. In: Muthu, S.S., Savalani, M.M. (eds.) Handbook of Sustainability in Additive Manufacturing, vol. I, pp. 7–30. Springer (2016). https://doi.org/10.1007/978-981-10-0549-7_2. ISBN 978-981-10-0549-7

Almeida, H.A., Costa, A.F.D., Ramos, C.A.R., Torres, C., Minondo, M., Bartolo, P.J.S., Nunes, A.A., Takanori, D.K., da Silva, J.V.L.: Additive manufacturing systems for medical applications: case studies. In: Pei, E., Verona, M.M., Bernard, A. (eds.) Additive Manufacturing - Developments in Training and Education, pp. 187–209. Springer (2018). (ISBN 978-3-319-76083-4 (Print) 978-3-319-76084 1 (Online))

Chua, C.K., Leong, K.F.: 3D Printing and Additive Manufacturing – Principles and Applications. 4th edn. World Scientific Publishing (2014)

Gibson, I., Rosen, D., Stucker, B.: Additive Manufacturing Technologies – 3D Printing, Rapid Prototyping and Direct Digital Manufacturing, 2nd edn. Springer, New York (2015)

Ligon, S.C., Liska, R., Stampfl, J., Gurr, M., Mülhaupt, R.: Polymers for 3D printing and customized additive manufacturing. Chem. Rev. **117**(15), 10212–10290 (2017)

Redwood, B., Schoffer, F., Garret, B.: The 3D Printing Handbook – Technologies, Design and Applications. 3D Hubs B.V. Amsterdam, The Netherlands (2017). ISBN 978-90-827485-0-5

Tibbitt, M.W., Rodell, C.B., Burdick, J.A., Anseth, K.S.: Progress in material design for biomedical applications. Proc. Natl. Acad. Sci. U.S.A. **112**(47), 14444–14451 (2015)

Author Index

© Springer Nature Switzerland AG 2019
J. M. R. S. Tavares and R. M. Natal Jorge (Eds.): VipIMAGE 2019, LNCVB 34, pp. 703–706, 2019.
https://doi.org/10.1007/978-3-030-32040-9

Printed by Printforce, the Netherlands